Life

The Science of Biology

NINTH EDITION

Sinauer Associates, Inc.

W. H. Freeman and Company

NINTH EDITION Life The Science of Biology

DAVID SADAVA
The Claremont Colleges
Claremont, California

DAVID M. HILLIS
University of Texas
Austin, Texas

H. CRAIG HELLER
Stanford University
Stanford, California

MAY R. BERENBAUM
University of Illinois
Urbana-Champaign, Illinois

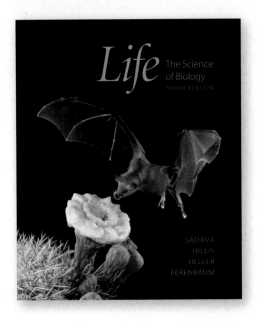

About the Cover

The cover shows a southern long-nosed bat (*Leptonycteris curasoae*) drinking the nectar of a saguaro cactus flower (*Carnegiea gigantea*). The bat, an endangered species, depends on the copious nectar produced by the cactus for its survival. The cactus, in turn, depends on the bat for pollination, and thus for its reproduction; these two species have coevolved during a long history of mutualistic association. The bat's ability to hover as it drinks from the flower is an example of adaptation of body form and physiology. All of these themes—nutrition, reproduction, integrated form and function, species interactions, adaptation, and evolution—echo throughout the chapters of *Life*. Photograph copyright © Dr. Merlin D. Tuttle/Photo Researchers, Inc.

The Frontispiece

Blue wildebeest (*Connochaetes taurinus*) and Burchell's zebra (*Equus quagga burchellii*) migrate together through Serengeti National Park, Tanzania. Copyright © Art Wolfe, www.artwolfe.com.

LIFE: The Science of Biology, Ninth Edition

Address editorial correspondence to:
Sinauer Associates Inc., 23 Plumtree Road, Sunderland, MA 01375 U.S.A.
www.sinauer.com
publish@sinauer.com

Address orders to:
MPS / W. H. Freeman & Co., Order Dept., 16365 James Madison Highway,
U.S. Route 15, Gordonsville, VA 22942 U.S.A.
Examination copy information: 1-800-446-8923
Orders: 1-888-330-8477

Planet Friendly Publishing
✔ Made in the United States
✔ Printed on Recycled Paper
Text: 10%   Cover: 10%
Learn more: www.greenedition.org

SUSTAINABLE FORESTRY INITIATIVE
Certified Chain of Custody
Promoting Sustainable Forest Management
www.sfiprogram.org

This SFI label applies to the text and cover papers.

Library of Congress Cataloging-in-Publication Data
Life, the science of biology / David Sadava .. [et al.]. — 9th ed.
    p. cm.
    Includes index.
    ISBN 978-1-4292-1962-4 (hardcover) — ISBN 978-1-4292-4645-3 (pbk. : v. 1) —
ISBN 978-1-4292-4644-6 (pbk. : v. 2) — ISBN 978-1-4292-4647-7 (pbk. : v. 3)
1. Biology. I. Sadava, David E.
    QH308.2.L565 2011
    570—dc22   2009036693

Printed in U.S.A.
Third Printing June 2011
The Courier Companies, Inc.

To Bill Purves and Gordon Orians,
extraordinary colleagues, biologists, and teachers,
and the original authors of LIFE

The Authors

CRAIG HELLER   DAVID HILLIS   MAY BERENBAUM   DAVID SADAVA

DAVID SADAVA is the Pritzker Family Foundation Professor of Biology, Emeritus, at the Keck Science Center of Claremont McKenna, Pitzer, and Scripps, three of The Claremont Colleges. In addition, he is Adjunct Professor of Cancer Cell Biology at the City of Hope Medical Center. Twice winner of the Huntoon Award for superior teaching, Dr. Sadava taught courses on introductory biology, biotechnology, biochemistry, cell biology, molecular biology, plant biology, and cancer biology. In addition to *Life: The Science of Biology*, he is the author or coauthor of books on cell biology and on plants, genes, and crop biotechnology. His research has resulted in many papers coauthored with his students, on topics ranging from plant biochemistry to pharmacology of narcotic analgesics to human genetic diseases. For the past 15 years, he has investigated multi-drug resistance in human small-cell lung carcinoma cells with a view to understanding and overcoming this clinical challenge. At the City of Hope, his current work focuses on new anti-cancer agents from plants.

DAVID HILLIS is the Alfred W. Roark Centennial Professor in Integrative Biology and the Director of the Center for Computational Biology and Bioinformatics at the University of Texas at Austin, where he also has directed the School of Biological Sciences. Dr. Hillis has taught courses in introductory biology, genetics, evolution, systematics, and biodiversity. He has been elected into the membership of the National Academy of Sciences and the American Academy of Arts and Sciences, awarded a John D. and Catherine T. MacArthur Fellowship, and has served as President of the Society for the Study of Evolution and of the Society of Systematic Biologists. His research interests span much of evolutionary biology, including experimental studies of evolving viruses, empirical studies of natural molecular evolution, applications of phylogenetics, analyses of biodiversity, and evolutionary modeling. He is particularly interested in teaching and research about the practical applications of evolutionary biology.

CRAIG HELLER is the Lorry I. Lokey/BusinessWire Professor in Biological Sciences and Human Biology at Stanford University. He earned his Ph.D. from the Department of Biology at Yale University in 1970. Dr. Heller has taught in the core biology courses at Stanford since 1972 and served as Director of the Program in Human Biology, Chairman of the Biological Sciences Department, and Associate Dean of Research. Dr. Heller is a fellow of the American Association for the Advancement of Science and a recipient of the Walter J. Gores Award for excellence in teaching. His research is on the neurobiology of sleep and circadian rhythms, mammalian hibernation, the regulation of body temperature, the physiology of human performance, and the neurobiology of learning. Dr. Heller has done research on a huge variety of animals and physiological problems ranging from sleeping kangaroo rats, diving seals, hibernating bears, photoperiodic hamsters, and exercising athletes. Some of his recent work on the effects of temperature on human performance is featured in the opener to Chapter 40, "Physiology, Homeostasis, and Temperature Regulation."

MAY BERENBAUM is the Swanlund Professor and Head of the Department of Entomology at the University of Illinois at Urbana-Champaign. She has taught courses in introductory animal biology, entomology, insect ecology, and chemical ecology, and has received awards at the regional and national level for distinguished teaching from the Entomological Society of America. A fellow of the National Academy of Sciences, the American Academy of Arts and Sciences, and the American Philosophical Society, she served as President of the American Institute for Biological Sciences in 2009. Her research addresses insect–plant coevolution, from molecular mechanisms of detoxification to impacts of herbivory on community structure. Concerned with the practical application of ecological and evolutionary principles, she has examined impacts of genetic engineering, global climate change, and invasive species on natural and agricultural ecosystems. Devoted to fostering science literacy, she has published numerous articles and five books on insects for the general public.

Contents in Brief

Investigating Life/Tools for Investigating Life

INVESTIGATING LIFE

TOOLS FOR INVESTIGATING LIFE

Preface

Biology is a dynamic, exciting, and important subject. It is dynamic because it is constantly changing, with new discoveries about the living world being made every day. (Although it is impossible to pinpoint an exact number, approximately 1 million new research articles in biology are published each year.) The subject is exciting because life in all of its forms has always fascinated people. As active scientists who have spent our careers teaching and doing research in a wide variety of fields, we know this first hand.

Biology has always been important in peoples' daily lives, if only through the effects of achievements in medicine and agriculture. Today more than ever the science of biology is at the forefront of human concerns as we face challenges raised both by recent advances in genome science and by the rapidly changing environment.

Life's new edition brings a fresh approach to the study of biology while retaining the features that have made the book successful in the past. A new coauthor, the distinguished entomologist May R. Berenbaum (University of Illinois at Urbana-Champaign) has joined our team, and the role of evolutionary biologist David Hillis (University of Texas at Austin) is greatly expanded in this edition. The authors hail from large, medium-sized, and small institutions. Our multiple perspectives and areas of expertise, as well as input from many colleagues and students who used previous editions, have informed our approach to this new edition.

Enduring Features

We remain committed to blending the presentation of core ideas with an emphasis on introducing students to the *process of scientific inquiry*. Having pioneered the idea of depicting seminal experiments in specially designed figures, we continue to develop this here, with 79 INVESTIGATING LIFE figures. Each of these figures sets the experiment in perspective and relates it to the accompanying text. As in previous editions, these figures employ a structure: Hypothesis, Method, Results, and Conclusion. They often include questions for further research that ask students to conceive an experiment that would explore a related question. Each *Investigating Life* figure has a reference to BioPortal (*yourBioPortal.com*), where citations to the original work as well as additional discussion and references to follow-up research can be found.

A related feature is the TOOLS FOR INVESTIGATING LIFE figures, which depict laboratory and field methods used in biology. These, too, have been expanded to provide more useful context for their importance.

Over a decade ago—in *Life's* Fifth Edition—the authors and publishers pioneered the much-praised use of BALLOON CAPTIONS in our figures. We recognized then, and it is even truer today, that many students are visual learners. The balloon captions bring explanations of intricate, complex processes directly into the illustration, allowing students to integrate information without repeatedly going back and forth between the figure, its legend, and the text.

Life is the only introductory textbook for biology majors to begin each chapter with a story. These OPENING STORIES provide historical, medical, or social context and are intended to intrigue students while helping them see how the chapter's biological subject relates to the world around them. In the new edition, all of the opening stories (some 70 percent of which are new) are revisited in the body of the chapter to drive home their relevance.

We continue to refine our well-received *chapter organization*. The chapter-opening story ends with a brief IN THIS CHAPTER preview of the major subjects to follow. A CHAPTER OUTLINE asks questions to emphasize scientific inquiry, each of which is answered in a major section of the chapter. A RECAP at the end of each section asks the student to pause and answer questions to review and test their mastery of the previous material. The end-of-chapter summary continues this inquiry framework and highlights key figures, bolded terms, and activities and animated tutorials available in BioPortal.

New Features

Probably the most important new feature of this edition is *new authorship*. Like the biological world, the authorship team of *Life* continues to evolve. While two of us (Craig Heller and David Sadava) continue as coauthors, David Hillis has a greatly expanded role, with full responsibility for the units on evolution and diversity. New coauthor May Berenbaum has rewritten the chapters on ecology. The perspectives of these two acclaimed experts have invigorated the entire book (as well as their coauthors).

Even with the enduring features (see above), this edition has a different look and feel from its predecessor. A fresh *new design* is more open and, we hope, more accessible to students. The extensively *revised art program* has a contemporary style and color palette. The information flow of the figures is easier to follow, with numbered balloons as a guide for students. There are new conceptual figures, including a striking visual timeline for the evolution of life on Earth (Figure 25.12) and a single overview figure that summarizes the information in the genome (Figure 17.4).

In response to instructors who asked for more real-world data, we have incorporated a feature introduced online in the Eighth Edition, WORKING WITH DATA. There are now 36 of these exercises, most of which relate to an *Investigating Life* figure. Each is referenced at the end of the relevant chapter and is available online via BioPortal (*yourBioPortal.com*). In these exercises, we describe in detail the context and approach of the research paper that forms the basis of the figure. We then ask the student to examine the data, to make calculations, and to draw conclusions.

We are proud that this edition is a *greener Life,* with the goal of reducing our environmental impact. This is the first introductory biology text to be printed on paper earning the Forest Stewardship Council label, the "gold standard" in green paper products, and it is manufactured from wood harvested from sustainable forests. And, of course, we also offer *Life* as an eBook.

The Ten Parts

We have reorganized the book into ten parts. **Part One, The Science of Life and Its Chemical Basis,** sets the stage for the book: the opening chapter focuses on biology as an exciting science. We begin with a startling observation: the recent, dramatic decline of amphibian species throughout the world. We then show how biologists have formed hypotheses for the causes of this environmental problem and are testing them by carefully designed experiments, with a view not only to understanding the decline, but reversing it. This leads to an outline of the basic principles of biology that are the foundation for the rest of the book: the unity of life at the cellular level and how evolution unites the living world. This is followed by chapters on the basic chemical building blocks that underlie life. We have added a new chapter on nucleic acids and the origin of life, introducing the concepts of genes and gene expression early and expanding our coverage of the major ideas on how life began and evolved at its earliest stages.

In **Part Two, Cells,** we describe the view of life as seen through cells, its structural units. In response to comments by users of our previous edition, we have moved the chapter on cell signaling and communication from the genetics section to this part of the book, with a change in emphasis from genes to cells. There is an updated discussion of ideas on the origin of cells and organelles, as well as expanded treatment of water transport across membranes.

Part Three, Cells and Energy, presents an integrated view of biochemistry. For this edition, we have worked to clarify such challenging concepts as energy transfer, allosteric enzymes, and biochemical pathways. There is extensive revision of the discussions of alternate pathways of photosynthetic carbon fixation, as well as a greater emphasis on applications throughout these chapters.

Part Four, Genes and Heredity, is extensively revised and reorganized to improve clarity, link related concepts, and provide updates from recent research results. Separate chapters on prokaryotic genetics and molecular medicine have been removed and their material woven into relevant chapters. For example, our chapter on cell reproduction now includes a discussion of how the basic mechanisms of cell division are altered in cancer cells. The chapter on transmission genetics now includes coverage of this phenomenon in prokaryotes. New chapters on gene expression and gene regulation compare prokaryotic and eukaryotic mechanisms and include a discussion of

epigenetics. A new chapter on mutation describes updated applications of medical genetics.

In **Part Five, Genomes,** we reinforce the concepts of the previous part, beginning with a new chapter on genomes—how they are analyzed and what they tell us about the biology of prokaryotes and eukaryotes, including humans. This leads to a chapter describing how our knowledge of molecular biology and genetics underpins biotechnology (the application of this knowledge to practical problems). We discuss some of the latest uses of biotechnology, including environmental cleanup. Part Five finishes with two chapters on development that explore the themes of molecular biology and evolution, linking these two parts of the book.

Part Six, The Patterns and Processes of Evolution, emphasizes the importance of evolutionary biology as a basis for comparing and understanding all aspects of biology. These chapters have been extensively reorganized and revised, as well as updated with the latest thinking of biologists in this rapidly changing field. This part now begins with the evidence and mechanisms of evolution, moves into a discussion of phylogenetic trees, then covers speciation and molecular evolution, and concludes with the evolutionary history of life on Earth. An integrated timeline of evolutionary history shows the timing of major events of biological evolution, the movements of the continents, floral and faunal reconstructions of major time periods, and depicts some of the fossils that form the basis of the reconstructions.

In **Part Seven, The Evolution of Diversity,** we describe the latest views on biodiversity and evolutionary relationships. Each chapter has been revised to make it easier for the reader to appreciate the major changes that have evolved within the various groups of organisms. We emphasize understanding the big picture of organismal diversity, as opposed to memorizing a taxonomic hierarchy and names (although these are certainly important). Throughout the book, the tree of life is emphasized as a way of understanding and organizing biological information. A *Tree of Life Appendix* allows students to place any group of organisms mentioned in the text of our book into the context of the rest of life. The web-based version of this appendix provides links to photos, keys, species lists, distribution maps, and other information to help students explore biodiversity of specific groups in greater detail.

After modest revisions in the past two editions, **Part Eight, Flowering Plants: Form and Function,** has been extensively reorganized and updated with the help of Sue Wessler, to include both classical and more recent approaches to plant physiology. Our emphasis is not only on the basic findings that led to the elucidation of mechanisms for plant growth and reproduction, but also on the use of genetics of model organisms. There is expanded coverage of the cell signaling events that regulate gene expression in plants, integrating concepts introduced earlier in the book. New material on how plants respond to their environment is included, along with links to both the book's earlier descriptions of plant diversity and later discussions of ecology.

Part Nine, Animals: Form and Function, continues to provide a solid foundation in physiology through comprehensive coverage of basic principles of function of each organ system and then emphasis on mechanisms of control and integration. An important reorganization has been moving the chapter on immunology from earlier in the book, where its emphasis was on molecular genetics, to this part, where it is more closely allied to the information systems of the body. In addition, we have added a number of new experiments and made considerable effort to clarify the sometimes complex phenomena shown in the illustrations.

Part Ten, Ecology, has been significantly revised by our new coauthor, May Berenbaum. A new chapter of biological interactions has been added (a topic formerly covered in the community ecology chapter). Full of interesting anecdotes and discussions of field studies not previously described in biology texts, this new ecology unit offers practical insights into how ecologists acquire, interpret, and apply real data. This brings the book full circle, drawing upon and reinforcing prior topics of energy, evolution, phylogenetics, Earth history, and animal and plant physiology.

Exceptional Value Formats

We again provide *Life* both as the full book and as a cluster of *paperbacks*. Thus, instructors who want to use less than the whole book can choose from these split volumes, each with the book's front matter, appendices, glossary, and index.

Volume I, The Cell and Heredity, includes: Part One, The Science of Life and Its Chemical Basis (Chapters 1–4); Part Two, Cells (Chapters 5–7); Part Three, Cells and Energy (Chapters 8–10); Part Four, Genes and Heredity (Chapters 11–16); and Part Five, Genomes (Chapters 17–20).

Volume II, Evolution, Diversity, and Ecology, includes: Chapter 1, Studying Life; Part Six, The Patterns and Processes of Evolution (Chapters 21–25); Part Seven, The Evolution of Diversity (Chapters 26–33); and Part Ten, Ecology (Chapters 54–59).

Volume III, Plants and Animals, includes: Chapter 1, Studying Life; Part Eight, Flowering Plants: Form and Function (Chapters 34–39); and Part Nine, Animals: Form and Function (Chapters 40–53).

Responding to student concerns, we offer two options of the entire book at a *significantly reduced cost.* After it was so well received in the previous edition, we again provide *Life* as a *loose-leaf version.* This shrink-wrapped, unbound, 3-hole punched version fits into a 3-ring binder. Students take only what they need to class and can easily integrate any instructor handouts or other resources.

Life was the first comprehensive biology text to offer the entire book as a truly robust *eBook.* For this edition, we continue to offer a flexible, interactive ebook that gives students a new way to read the text and learn the material. The ebook integrates the student media resources (animations, quizzes, activities, etc.) and offers instructors a powerful way to customize the textbook with their own text, images, Web links, documents, and more.

Media and Supplements for the Ninth Edition

The wide range of media and supplements that accompany *Life*, Ninth Edition have all been created with the dual goal of helping students learn the material presented in the textbook more efficiently and helping instructors teach their courses more effectively. Students in majors introductory biology are faced with learning a tremendous number of new concepts, facts, and terms, and the more different ways they can study this material, the more efficiently they can master it.

All of the *Life* media and supplemental resources have been developed specifically for this textbook. This provides strong consistency between text and media, which in turn helps students learn more efficiently. For example, the animated tutorials and activities found in BioPortal were built using textbook art, so that the manner in which structures are illustrated, the colors used to identify objects, and the terms and abbreviations used are all consistent.

For the Ninth Edition, a new set of Interactive Tutorials gives students a new way to explore many key topics across the textbook. These new modules allow the student to learn by doing, including solving problem scenarios, working with experimental techniques, and exploring model systems. All new copies of the Ninth Edition include access to the robust new version of BioPortal, which brings together all of *Life's* student and instructor resources, powerful assessment tools, and new integration with Prep-U adaptive quizzing.

The rich collection of visual resources in the Instructor's Media Library provides instructors with a wide range of options for enhancing lectures, course websites, and assignments. Highlights include: layered art PowerPoint® presentations that break down complex figures into detailed, step-by-step presentations; a collection of approximately 200 video segments that can help capture the attention and imagination of students; and PowerPoint slides of textbook art with editable labels and leaders that allow easy customization of the figures.

For a detailed description of all the media and supplements available for the Ninth Edition, please turn to "*Life's* Media and Supplements," on page xvii.

Many People to Thank

"If I have seen farther, it is by standing on the shoulders of giants." The great scientist Isaac Newton wrote these words over 330 years ago and, while we certainly don't put ourselves in his lofty place in science, the words apply to us as coauthors of this text. This is the first edition that does not bear the names of Bill Purves and Gordon Orians. As they enjoy their "retirements," we are humbled by their examples as biologists, educators, and writers.

One of the wisest pieces of advice ever given to a textbook author is to "be passionate about your subject, but don't put your ego on the page." Considering all the people who looked over our shoulders throughout the process of creating this book, this advice could not be more apt. We are indebted to many people who gave invaluable help to make this book what it is. First and foremost are our colleagues, biologists from over 100 institutions. Some were users of the previous edition, who suggested many improvements. Others reviewed our chapter drafts in detail, including advice on how to improve the illustrations. Still others acted as accuracy reviewers when the book was almost completed. All of these biologists are listed in the Reviewer credits.

Of special note is Sue Wessler, a distinguished plant biologist and textbook author from the University of Georgia. Sue looked critically at Part Eight, Flowering Plants: Form and Function, wrote three of the chapters (34–36), and was important in the revision of the other three (37–39). The new approach to plant biology in this edition owes a lot to her.

The pace of change in biology and the complexities of preparing a book as broad as this one necessitated having two developmental editors. James Funston coordinated Parts 1–5, and Carol Pritchard-Martinez coordinated Parts 6–10. We benefitted from the wide experience, knowledge, and wisdom of both of them. As the chapter drafts progressed, we were fortunate to have experienced biologist Laura Green lending her critical eye as in-house editor. Elizabeth Morales, our artist, was on her third edition with us. As we have noted, she extensively revised almost all of the prior art and translated our crude sketches into beautiful new art. We hope you agree that our art program remains superbly clear and elegant. Our copy editors, Norma Roche, Liz Pierson, and Jane Murfett, went far beyond what such people usually do. Their knowledge and encyclopedic recall of our book's chapters made our prose sharper and more accurate. Diane Kelly, Susan McGlew, and Shannon Howard effectively coordinated the hundreds of reviews that we described above. David McIntyre was a terrific photo editor, finding over 550 new photographs, including many new ones of his own, that enrich the book's content and visual statement. Jefferson Johnson is responsible for the design elements that make this edition of *Life* not just clear and easy to learn from, but beautiful as well. Christopher Small headed the production department—Joanne Delphia, Joan Gemme, Janice Holabird, and Jefferson Johnson—who contributed in innumerable ways to bringing *Life* to its final form. Jason Dirks once again coordinated the creation of our array of media and supplements, including our superb new Web resources. Carol Wigg, for the ninth time in nine editions, oversaw the editorial process; her influence pervades the entire book.

W. H. Freeman continues to bring *Life* to a wider audience. Associate Director of Marketing Debbie Clare, the Regional Specialists, Regional Managers, and experienced sales force are effective ambassadors and skillful transmitters of the features and unique strengths of our book. We depend on their expertise and energy to keep us in touch with how *Life* is perceived by its users. And thanks also to the Freeman media group for eBook and BioPortal production.

Finally, we are indebted to Andy Sinauer. Like ours, his name is on the cover of the book, and he truly cares deeply about what goes into it. Combining decades of professionalism, high standards, and kindness to all who work with him, he is truly our mentor and friend.

DAVID SADAVA

DAVID HILLIS

CRAIG HELLER

MAY BERENBAUM

Reviewers for the Ninth Edition

Between-Edition Reviewers

David D. Ackerly, University of California, Berkeley

Amy Bickham Baird, University of Leiden

Jeremy Brown, University of California, Berkeley

John M. Burke, University of Georgia

Ruth E. Buskirk, University of Texas, Austin

Richard E. Duhrkopf, Baylor University

Casey W. Dunn, Brown University

Erika J. Edwards, Brown University

Kevin Folta, University of Florida

Lynda J. Goff, University of California, Santa Cruz

Tracy A. Heath, University of Kansas

Shannon Hedtke, University of Texas, Austin

Richard H. Heineman, University of Texas, Austin

Albert Herrera, University of Southern California

David S. Hibbett, Clark University

Norman A. Johnson, University of Massachusetts

Walter S. Judd, University of Florida

Laura A. Katz, Smith College

Emily Moriarty Lemmon, Florida State University

Sheila McCormick, University of California, Berkeley

Robert McCurdy, Independence Creek Nature Preserve

Jacalyn Newman, University of Pittsburgh

Juliet F. Noor, Duke University

Theresa O'Halloran, University of Texas, Austin

K. Sata Sathasivan, University of Texas, Austin

H. Bradley Shaffer, University California, Davis

Rebecca Symula, Yale University

Christopher D. Todd, University of Saskatchewan

Elizabeth Willott, University of Arizona

Kenneth Wilson, University of Saskatchewan

Manuscript Reviewers

Tamarah Adair, Baylor University

William Adams, University of Colorado, Boulder

Gladys Alexandre, University of Tennessee, Knoxville

Shivanthi Anandan, Drexel University

Brian Bagatto, University of Akron

Lisa Baird, University of San Diego

Stewart H. Berlocher, University of Illinois, Urbana-Champaign

William Bischoff, University of Toledo

Meredith M. Blackwell, Louisiana State University

David Bos, Purdue University

Jonathan Bossenbroek, University of Toledo

Nicole Bournias-Vardiabasis, California State University, San Bernardino

Nancy Boury, Iowa State University

Sunny K. Boyd, University of Notre Dame

Judith L. Bronstein, University of Arizona

W. Randy Brooks, Florida Atlantic University

James J. Bull, University of Texas, Austin

Darlene Campbell, Cornell University

Domenic Castignetti, Loyola University, Chicago

David T. Champlin, University of Southern Maine

Shu-Mei Chang, University of Georgia

Samantha K. Chapman, Villanova University

Patricia Christie, MIT

Wes Colgan, Pikes Peak Community College

John Cooper, Washington University

Ronald Cooper, University of California, Los Angeles

Elizabeth Cowles, Eastern Connecticut State University

Jerry Coyne, University of Chicago

William Crampton, University of Central Florida

Michael Dalbey, University of California, Santa Cruz

Anne Danielson-Francois, University of Michigan, Dearborn

Grayson S. Davis, Valparaiso University

Kevin Dixon, Florida State University

Zaldy Doyungan, Texas A&M University, Corpus Christi

Ernest F. Dubrul, University of Toledo

Roland Dute, Auburn University

Scott Edwards, Harvard University

William Eldred, Boston University

David Eldridge, Baylor University

Joanne Ellzey, University of Texas, El Paso

Susan H. Erster, State University of New York, Stony Book

Brent E. Ewers, University of Wyoming

Kevin Folta, University of Florida

Brandon Foster, Wake Technical Community College

Richard B. Gardiner, University of Western Ontario

Douglas Gayou, University of Missouri, Columbia

John R. Geiser, Western Michigan University

Arundhati Ghosh, University of Pittsburgh

Alice Gibb, Northern Arizona University

Scott Gilbert, Swarthmore College

Matthew R. Gilg, University of North Florida

Elizabeth Godrick, Boston University

Lynda J. Goff, University of California, Santa Cruz

Elizabeth Blinstrup Good, University of Illinois, Urbana-Champaign

John Nicholas Griffis, University of Southern Mississippi

Cameron Gundersen, University of California, Los Angeles

Kenneth Halanych, Auburn University

E. William Hamilton, Washington and Lee University

Monika Havelka, University of Toronto at Mississauga

Tyson Hedrick, University of North Carolina, Chapel Hill

Susan Hengeveld, Indiana University, Bloomington

Albert Herrera, University of Southern California

Kendra Hill, South Dakota State University

Richard W. Hill, Michigan State University

Erec B. Hillis, University of California, Berkeley

Jonathan D. Hillis, Carleton College

William Huddleston, University of Calgary

Dianne B. Jennings, Virginia Commonwealth University

Norman A. Johnson, University of Massachusetts, Amherst

William H. Karasov, University of Wisconsin, Madison

Susan Keen, University of California, Davis

Cornelis Klok, Arizona State University, Tempe

Olga Ruiz Kopp, Utah Valley University

William Kroll, Loyola University, Chicago

Allen Kurta, Eastern Michigan University

Rebecca Lamb, Ohio State University

Brenda Leady, University of Toledo

Hugh Lefcort, Gonzaga University

Sean C. Lema, University of North Carolina, Wilmington

Nathan Lents, John Jay College, City University of New York

Rachel A. Levin, Amherst College

Donald Levin, University of Texas, Austin

Bernard Lohr, University of Maryland, Baltimore County

Barbara Lom, Davidson College

David J. Longstreth, Louisiana State University

Catherine Loudon, University of California, Irvine

Francois Lutzoni, Duke University

Charles H. Mallery, University of Miami

Kathi Malueg, University of Colorado, Colorado Springs

Richard McCarty, Johns Hopkins University

Sheila McCormick, University of California, Berkeley

Francis Monette, Boston University

Leonie Moyle, Indiana University, Bloomington

Jennifer C. Nauen, University of Delaware

Jacalyn Newman, University of Pittsburgh

Alexey Nikitin, Grand Valley State University

Shawn E. Nordell, Saint Louis University

Tricia Paramore, Hutchinson Community College

Nancy J. Pelaez, Purdue University

Robert T. Pennock, Michigan State University

Roger Persell, Hunter College

Debra Pires, University of California, Los Angeles

Crima Pogge, City College of San Francisco

Jaimie S. Powell, Portland State University

Susan Richardson, Florida Atlantic University

David M. Rizzo, University of California, Davis

Benjamin Rowley, University of Central Arkansas

Brian Rude, Mississippi State University

Ann Rushing, Baylor University

Christina Russin, Northwestern University

Udo Savalli, Arizona State University, West

Frieder Schoeck, McGill University

Paul J. Schulte, University of Nevada, Las Vegas

Stephen Secor, University of Alabama

Vijayasaradhi Setaluri, University of Wisconsin, Madison

H. Bradley Shaffer, University of California, Davis

Robin Sherman, Nova Southeastern University

Richard Shingles, Johns Hopkins University

James Shinkle, Trinity University

Richard M. Showman, University of South Carolina

Felisa A. Smith, University of New Mexico

Ann Berry Somers, University of North Carolina, Greensboro

Ursula Stochaj, McGill University

Ken Sweat, Arizona State University, West

Robin Taylor, Ohio State University

William Taylor, University of Toledo

Mark Thogerson, Grand Valley State University

Sharon Thoma, University of Wisconsin, Madison

Lars Tomanek, California Polytechnic State University

James Traniello, Boston University

Jeffrey Travis, State University of New York, Albany

Terry Trier, Grand Valley State University

John True, State University of New York, Stony Brook

Elizabeth Van Volkenburgh, University of Washington

John Vaughan, St. Petersburg College

Sara Via, University of Maryland

Suzanne Wakim, Butte College (Glenn Community College District)

Randall Walikonis, University of Connecticut

Cindy White, University of Northern Colorado

Elizabeth Willott, University of Arizona

Mark Wilson, Humboldt State University

Stuart Wooley, California State University, Stanislaus

Lan Xu, South Dakota State University

Heping Zhou, Seton Hall University

Accuracy Reviewers

John Alcock, Arizona State University

Gladys Alexandre, University of Tennessee, Knoxville

Lawrence A. Alice, Western Kentucky University

David R. Angelini, American University

Fabia U. Battistuzzi, Arizona State University

Arlene Billock, University of Louisiana, Lafayette

Mary A. Bisson, State University of New York, Buffalo

Meredith M. Blackwell, Louisiana State University

Nancy Boury, Iowa State University

Eldon J. Braun, University of Arizona

Daniel R. Brooks, University of Toronto

Jennifer L. Campbell, North Carolina State University

Peter C. Chabora, Queens College, CUNY

Patricia Christie, MIT

Ethan Clotfelter, Amherst College

Robert Connour, Owens Community College

Peter C. Daniel, Hofstra University

D. Michael Denbow, Virginia Polytechnic Institute

Laura DiCaprio, Ohio University

Zaldy Doyungan, Texas A&M University, Corpus Christi

Moon Draper, University of Texas, Austin

Richard E. Duhrkopf, Baylor University

Susan A. Dunford, University of Cincinnati

Brent E. Ewers, University of Wyoming

James S. Ferraro, Southern Illinois University

Rachel D. Fink, Mount Holyoke College

John R. Geiser, Western Michigan University

Elizabeth Blinstrup Good, University of Illinois, Urbana-Champaign

Melina E. Hale, University of Chicago

Patricia M. Halpin, University of California, Los Angeles

Jean C. Hardwick, Ithaca College

Monika Havelka, University of Toronto at Mississauga

Frank Healy, Trinity University

Marshal Hedin, San Diego State University

Albert Herrera, University of Southern California

David S. Hibbett, Clark University

James F. Holden, University of Massachusetts, Amherst

Margaret L. Horton, University of North Carolina, Greensboro

Helen Hull-Sanders, Canisius College

C. Darrin Hulsey, University of Tennessee, Knoxville

Timothy Y. James, University of Michigan

Dianne B. Jennings, Virginia Commonwealth University

Norman A. Johnson, University of Massachusetts, Amherst

Susan Jorstad, University of Arizona

Ellen S. Lamb, University of North Carolina, Greensboro

Dennis V. Lavrov, Iowa State University

Hugh Lefort, Gonzaga University

Rachel A. Levin, Amherst College

Bernard Lohr, University of Maryland, Baltimore County

Sharon E. Lynn, College of Wooster

Sarah Mathews, Harvard University

Susan L. Meacham, University of Nevada, Las Vegas

Mona C. Mehdy, University of Texas, Austin

Bradley G. Mehrtens, University of Illinois, Urbana-Champaign

James D. Metzger, Ohio State University

Thomas W. Moon, University of Ottowa

Thomas M. Niesen, San Francisco State University

Theresa O'Halloran, University of Texas, Austin

Thomas L. Pannabecker, University of Arizona

Nancy J. Pelaez, Purdue University

Nicola J. R. Plowes, Arizona State University

Gregory S. Pryor, Francis Marion University

Laurel B. Roberts, University of Pittsburgh

Anjana Sharma, Western Carolina University

Richard M. Showman, University of South Carolina

John B. Skillman, California State University, San Bernadino

John J. Stachowicz, University of California, Davis

Brook O. Swanson, Gonzaga University

Robin A. J. Taylor, Ohio State University

William Taylor, University of Toledo

Steven M. Theg, University of California, Davis

Mark Thogerson, Grand Valley State University

Christopher D. Todd, University of Saskatchewan

Jeffrey Travis, State University of New York, Albany

Joseph S. Walsh, Northwestern University

Andrea Ward, Adelphi University

Barry Williams, Michigan State University

Kenneth Wilson, University of Saskatchewan

Carol L. Wymer, Morehead State University

LIFE's Media and Supplements

BIO P(3)RTAL featuring Prep-U

yourBioPortal.com

BioPortal is the new gateway to all of *Life's* state-of-the-art on-line resources for students and instructors. BioPortal includes the breakthrough quizzing engine, Prep-U; a fully interactive eBook; and additional premium learning media. The textbook is tightly integrated with BioPortal via in-text references that connect the printed text and media resources. The result is a powerful, easily-managed online course environment. Bio-Portal includes the following features and resources:

Life, Ninth Edition eBook

- Integration of all activities, animated tutorials, and other media resources.
- Quick, intuitive navigation to any section or subsection, as well as any printed book page number.
- In-text links to all glossary entries.
- Easy text highlighting.
- A bookmarking feature that allows for quick reference to any page.
- A powerful Notes feature that allows students to add notes to any page.
- A full glossary and index.
- Full-text search, including an additional option to search the glossary and index.
- Automatic saving of all notes, highlighting, and bookmarks.

Additional eBook features for instructors:

- Content Customization: Instructors can easily add pages of their own content and/or hide chapters or sections that they do not cover in their course.
- Instructor Notes: Instructors can choose to create an annotated version of the eBook with their own notes on any page. When students in the course log in, they see the instructor's personalized version of the eBook. Instructor notes can include text, Web links, images, links to all Bio-Portal content, and more.

Smarter than the average quiz

Built by educators, Prep-U focuses student study time exactly where it should be, through the use of personalized, adaptive quizzes that move students toward a better grasp of the material—and better grades. For *Life,* Ninth Edition, Prep-U is fully integrated into BioPortal, making it easy for instructors to take advantage of this powerful quizzing engine in their course. Features include:

- Adaptive quizzing
- Automatic results reporting into the BioPortal gradebook

- Misconception index
- Comparison to national data

Student Resources

Diagnostic Quizzing. The diagnostic quiz for each chapter of *Life* assesses student understanding of that chapter, and generates a Personalized Study Plan to effectively focus student study time. The plan includes links to specific textbook sections, animated tutorials, and activities.

Interactive Summaries. For each chapter, these dynamic summaries combine a review of important concepts with links to all of the key figures from the chapter as well as all of the relevant animated tutorials, activities, and key terms.

Animated Tutorials. Over 100 in-depth animated tutorials, in a new format for the Ninth Edition, present complex topics in a clear, easy-to-follow format that combines a detailed animation with an introduction, conclusion, and quiz.

Activities. Over 120 interactive activities help students learn important facts and concepts through a wide range of exercises, such as labeling steps in processes or parts of structures, building diagrams, and identifying different types of organisms.

NEW! Interactive Tutorials. New for the Ninth Edition, these tutorial modules help students master key concepts through hands-on activities that allow them to learn through action. With these tutorials, students can solve problem scenarios by applying concepts from the text, by working with experimental techniques, and by using interactive models to discover how biological mechanisms work. Each tutorial includes a self-assessment quiz that can be assigned.

Interactive Quizzes. Each question includes an image from the textbook, thorough feedback on both correct and incorrect answer choices, references to textbook pages, and links to eBook pages, for quick review.

***BioNews from* Scientific American.** BioNews makes it easy for instructors to bring the dynamic nature of the biological sciences and up-to-the-minute currency into their course. Accessible from within BioPortal, BioNews is a continuously updated feed of current news, podcasts, magazine articles, science blog entries, "strange but true" stories, and more.

NEW! BioNavigator. This unique visual resource is an innovative way to access the wide variety of *Life* media resources. Starting from the whole-Earth view, instructors and students can zoom to any level of biological inquiry, encountering links to a wealth of animations, activities, and tutorials on the full range of topics along the way.

Working with Data. Built around some of the original experiments depicted in the Investigating Life figures, these exercises help build quantitative skills and encourage student interest in how scientists do research, by looking at real experimental data and answering questions based on those data.

Flashcards. For each chapter of the book, there is a set of flashcards that allows the student to review all the key terminology from the chapter. Students can review the terms in study mode, and then quiz themselves on a list of terms.

Experiment Links. For each Investigating Life figure in the textbook, BioPortal includes an overview of the experiment featured in the figure and related research or applications that followed, a link to the original paper, and links to additional information related to the experiment.

Key Terms. The key terminology introduced in each chapter is listed, with definitions and audio pronunciations from the glossary.

Suggested Readings. For each chapter of the book, a list of suggested readings is provided as a resource for further study.

Glossary. The language of biology is often difficult for students taking introductory biology to master, so BioPortal includes a full glossary that features audio pronunciations of all terms.

Statistics Primer. This brief introduction to the use of statistics in biological research explains why statistics are integral to biology, and how some of the most common statistical methods and techniques are used by biologists in their work.

Math for Life. A collection of mathematical shortcuts and references to help students with the quantitative skills they need in the laboratory.

Survival Skills. A guide to more effective study habits. Topics include time management, note-taking, effective highlighting, and exam preparation.

Instructor Resources

Assessment

- Diagnostic Quizzing provides instant class comprehension feedback to instructors, along with targeted lecture resources for those areas requiring the most attention.

- Question banks include questions ranked according to Bloom's taxonomy.

- Question filtering: Allows instructors to select questions based on Bloom's category and/or textbook section.

- Easy-to-use customized assessment tools allow instructors to quickly create quizzes and many other types of assignments using any combination of the questions and resources provided along with their own materials.

- Comprehensive question banks include questions from the test bank, study guide, textbook self-quizzes, and diagnostic quizzes.

Media Resources *(see Instructor's Media Library below for details)*

- Videos

- PowerPoint® Presentations (Textbook Figures, Lectures, Layered Art)

- Supplemental Photos
- Clicker Questions
- Instructor's Manual
- Lecture Notes

Course Management

- Complete course customization capabilities
- Custom resources/document posting
- Robust Gradebook
- Communication Tools: Announcements, Calendar, Course Email, Discussion Boards

Note: The printed textbook, the eBook, BioPortal, and Prep-U can all be purchased individually as stand-alone items, in addition to being available in a package with the printed textbook.

Student Supplements

Study Guide (ISBN 978-1-4292-3569-3)

Jacalyn Newman, *University of Pittsburgh*; Edward M. Dzialowski, *University of North Texas*; Betty McGuire, *Cornell University*; Lindsay Goodloe, *Cornell University*; and Nancy Guild, *University of Colorado*

For each chapter of the textbook, the *Life* Study Guide offers a variety of study and review tools. The contents of each chapter are broken down into both a detailed review of the Important Concepts covered and a boiled-down Big Picture snapshot. New for the Ninth Edition, Diagram Exercises help students synthesize what they have learned in the chapter through exercises such as ordering concepts, drawing graphs, linking steps in processes, and labeling diagrams. In addition, Common Problem Areas and Study Strategies are highlighted. A set of study questions (both multiple-choice and short-answer) allows students to test their comprehension. All questions include answers and explanations.

Lecture Notebook (ISBN 978-1-4292-3583-9)

This invaluable printed resource consists of all the artwork from the textbook (more than 1,000 images with labels) presented in the order in which they appear in the text, with ample space for note-taking. Because the Notebook has already done the drawing, students can focus more of their attention on the concepts. They will absorb the material more efficiently during class, and their notes will be clearer, more accurate, and more useful when they study from them later.

Companion Website www.thelifewire.com

(Also available as a CD, which can be optionally packaged with the textbook.)

For those students who do not have access to BioPortal, the *Life*, Ninth Edition Companion Website is available free of charge (no access code required). The site features a variety of resources, including animations, flashcards, activities, study ideas, help with math and statistics, and more.

CatchUp Math & Stats

Michael Harris, Gordon Taylor, and Jacquelyn Taylor (ISBN 978-1-4292-0557-3)

This primer will help your students quickly brush up on the quantitative skills they need to succeed in biology. Presented in brief, accessible units, the book covers topics such as working with powers, logarithms, using and understanding graphs, calculating standard deviation, preparing a dilution series, choosing the right statistical test, analyzing enzyme kinetics, and many more.

Student Handbook for Writing in Biology, Third Edition

Karen Knisely, *Bucknell University* (ISBN 978-1-4292-3491-7)

This book provides practical advice to students who are learning to write according to the conventions in biology. Using the standards of journal publication as a model, the author provides, in a user-friendly format, specific instructions on: using biology databases to locate references; paraphrasing for improved comprehension; preparing lab reports, scientific papers, posters; preparing oral presentations in PowerPoint®, and more.

Bioethics and the New Embryology: Springboards for Debate

Scott F. Gilbert, Anna Tyler, and Emily Zackin (ISBN 978-0-7167-7345-0)

Our ability to alter the course of human development ranks among the most significant changes in modern science and has brought embryology into the public domain. The question that must be asked is: Even if we can do such things, should we?

BioStats Basics: A Student Handbook

James L. Gould and Grant F. Gould (ISBN 978-0-7167-3416-1)

BioStats Basics provides introductory-level biology students with a practical, accessible introduction to statistical research. Engaging and informal, the book avoids excessive theoretical and mathematical detail, and instead focuses on how core statistical methods are put to work in biology.

Instructor Media & Supplements

Instructor's Media Library

The *Life,* Ninth Edition Instructor's Media Library (available both online via BioPortal and on disc) includes a wide range of electronic resources to help instructors plan their course, present engaging lectures, and effectively assess student comprehension. The Media Library includes the following resources:

Textbook Figures and Tables. Every image and table from the textbook is provided in both JPEG (high- and low-resolution) and PDF formats. Each figure is provided both with and without balloon captions, and large, complex figures are provided in both a whole and split version.

Unlabeled Figures. Every figure is provided in an unlabeled format, useful for student quizzing and custom presentation development.

Supplemental Photos. The supplemental photograph collection contains over 1,500 photographs (in addition to those in the text), giving instructors a wealth of additional imagery to draw upon.

Animations. Over 100 detailed animations, revised and enlarged for the Ninth Edition, all created from the textbook's art program, and viewable in either narrated or step-through mode.

Videos. A collection of over 200 video segments that covers topics across the entire textbook and helps demonstrate the complexity and beauty of life. Includes the Cell Visualization Videos.

PowerPoint® Resources. For each chapter of the textbook, several different PowerPoint presentations are available. These give instructors the flexibility to build presentations in the manner that best suits their needs. Included are:

- Textbook Figures and Tables
- Lecture Presentation
- Figures with Editable Labels
- Layered Art Figures
- Supplemental Photos
- Videos
- Animations

Clicker Questions. A set of questions written specifically to be used with classroom personal response systems, such as the iClicker system, is provided for each chapter. These questions are designed to reinforce concepts, gauge student comprehension, and engage students in active participation.

Chapter Outlines, Lecture Notes, and the complete ***Test File*** are all available in Microsoft Word® format for easy use in lecture and exam preparation.

Intuitive Browser Interface provides a quick and easy way to preview and access all of the content on the Instructor's Media Library.

Instructor's Resource Kit

The *Life,* Ninth Edition Instructor's Resource Kit includes a wealth of information to help instructors in the planning and teaching of their course. The Kit includes:

Instructor's Manual, featuring (by chapter):

- A "What's New" guide to the Ninth Edition
- Brief chapter overview
- Chapter outline
- Key terms section with all of the boldface terms from the text

Lecture Notes. Detailed notes for each chapter, which can serve as the basis for lectures, including references to figures and media resources.

Media Guide. A visual guide to the extensive media resources available with the Ninth Edition of *Life.* The guide includes thumbnails and descriptions of every video, animation, lecture PowerPoint®, and supplemental photo in the Media Library, all organized by chapter.

Overhead Transparencies

This set includes over 1,000 transparencies—including all of the four-color line art and all of the tables from the text—along with convenient binders. All figures have been formatted and color-enhanced for clear projection in a wide range of conditions. Labels and images have been resized for improved readability.

Test File

Catherine Ueckert, *Northern Arizona University;* Norman Johnson, *University of Massachusetts;* Paul Nolan, *The Citadel;* Nicola Plowes, *Arizona State University*

The Test File offers more than 5,000 questions, covering the full range of topics presented in the textbook. All questions are referenced to textbook sections and page numbers, and are ranked according to Bloom's taxonomy. Each chapter includes a wide range of multiple choice and fill-in-the-blank questions. In addition, each chapter features a set of diagram questions that involve the student in working with illustrations of structures, graphs, steps in processes, and more. The electronic versions of the Test File (within BioPortal, the Instructor's Media Library, and the Computerized Test Bank CD) also include all of the textbook end-of-chapter Self-Quiz questions, all of the BioPortal Diagnostic Quiz questions, and all of the Study Guide multiple-choice questions.

Computerized Test Bank

The entire printed Test File, plus the textbook end-of-chapter Self-Quizzes, the BioPortal Diagnostic Quizzes, and the Study Guide multiple-choice questions are all included in Wimba's easy-to-use Diploma® software. Designed for both novice and advanced users, Diploma enables instructors to quickly and easily create or edit questions, create quizzes or exams with a "drag-and-drop" feature, publish to online courses, and print paper-based assignments.

Course Management System Support

As a service for *Life* adopters using WebCT, Blackboard, or ANGEL for their courses, full electronic course packs are available.

www.whfreeman.com/facultylounge/majorsbio
NEW! The new Faculty Lounge for Majors Biology is the first publisher-provided website for the majors biology community that lets instructors freely communicate and share peer-reviewed lecture and teaching resources. It is continually updated and vetted by majors biology instructors—there is always something new to see. The Faculty Lounge offers convenient access to peer-recommended and vetted resources, including the following categories: Images, News, Videos, Labs, Lecture Resources, and Educational Research.

In addition, the site includes special areas for resources for lab coordinators, resources and updates from the *Scientific Teaching* series of books, and information on biology teaching workshops.

Developed for educators by educators, iclicker is a hassle-free radio-frequency classroom response system that makes it easy for instructors to ask questions, record responses, take attendance, and direct students through lectures as active participants. For more information, visit www.iclicker.com.

www.whfreeman.com/labpartner

NEW! LabPartner is a site designed to facilitate the creation of customized lab manuals. Its database contains a wide selection of experiments published by W. H. Freeman and Hayden-McNeil Publishing. Instructors can preview, choose, and re-order labs, interleave their original experiments, add carbonless graph paper and a pocket folder, and customize the cover both inside and out. LabPartner offers a variety of binding types: paperback, spiral, or loose-leaf. Manuals are printed on-demand once W. H. Freeman receives an order from a campus bookstore or school.

The Scientific Teaching Book Series is a collection of practical guides, intended for all science, technology, engineering and mathematics (STEM) faculty who teach undergraduate and graduate students in these disciplines. The purpose of these books is to help faculty become more successful in all aspects of teaching and learning science, including classroom instruction, mentoring students, and professional development. Authored by well-known science educators, the Series provides concise descriptions of best practices and how to implement them in the classroom, the laboratory, or the department. For readers interested in the research results on which these best practices are based, the books also provide a gateway to the key educational literature.

Scientific Teaching

Jo Handelsman, Sarah Miller, and Christine Pfund, *University of Wisconsin-Madison* (ISBN 978-1-4292-0188-9)

NEW! Transformations: Approaches to College Science Teaching

A Collection of Articles from CBE Life Sciences Education
Deborah Allen, *University of Delaware;* Kimberly Tanner, *San Francisco State University* (ISBN 978-1-4292-5335-2)

Contents

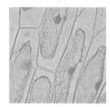

PART TWO

CELLS

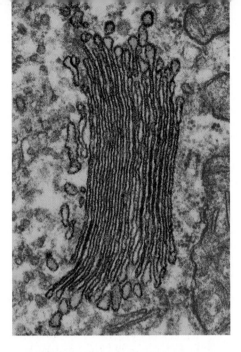

PART THREE

CELLS AND ENERGY

8 Energy, Enzymes, and Metabolism 148

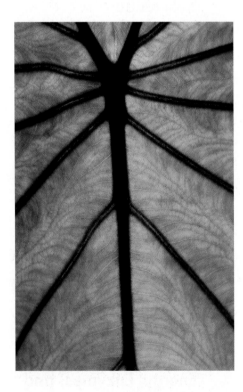

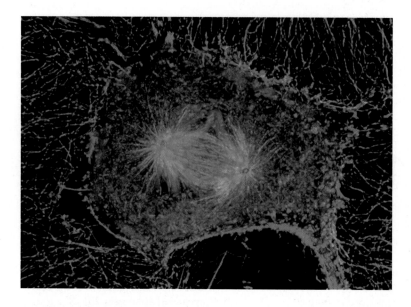

PART FIVE
GENOMES

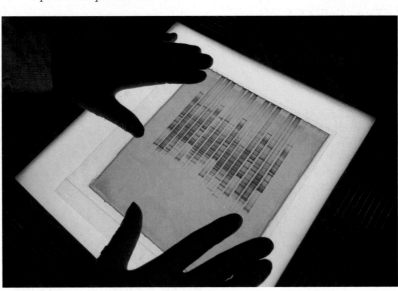

18 Recombinant DNA and Biotechnology 386

19 Differential Gene Expression in Development 405

20 Development and Evolutionary Change 426

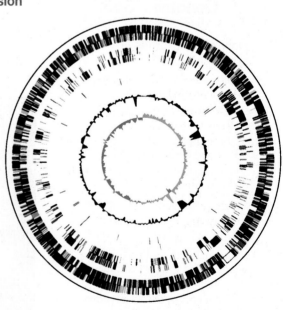

PART SIX
THE PATTERNS AND PROCESSES OF EVOLUTION

25 The History of Life on Earth 518

PART SEVEN
THE EVOLUTION OF DIVERSITY

26 Bacteria and Archaea: The Prokaryotic Domains 536

PART EIGHT
FLOWERING PLANTS: FORM AND FUNCTION

36 Plant Nutrition 755

37 Regulation of Plant Growth 771

PART NINE
ANIMALS: FORM AND FUNCTION

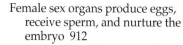

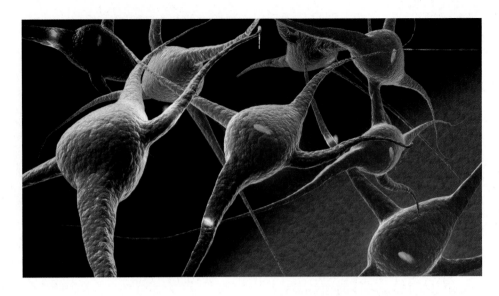

53 Animal Behavior 1113

PART TEN
ECOLOGY

Life

The Science of Biology

NINTH EDITION

1 Studying Life

Why are frogs croaking?

Amphibians—frogs, toads, and salamanders—have been around for a long time. They watched the dinosaurs come and go. But today amphibian populations around the world are in dramatic decline, with more than a third of the world's amphibian species threatened with extinction. Why?

Biologists work to answer this question by making observations and doing experiments. A number of factors may be involved, and one possible cause may be the effects of agricultural pesticides and herbicides. Several studies have shown that many of these chemicals tested at realistic concentrations do not kill amphibians. But Tyrone Hayes, a biologist at the University of California at Berkeley, probed deeper.

Hayes focused on atrazine, the most widely used herbicide in the world and a common contaminant in fresh water. More than 70 million pounds of atrazine are applied to farmland in the United States every year, and it is used in at least 20 countries. Atrazine is usually applied in the spring, when many amphibians are breeding and thousands of tadpoles swim in the ditches, ponds, and streams that receive runoff from farms.

In his laboratory, Hayes and his associates raised frog tadpoles in water containing no atrazine and in water with concentrations ranging from 0.01 parts per billion (ppb) up to 25 ppb. The U.S. Environmental Protection Agency considers environmental levels of atrazine of 10 to 20 ppb of no concern; the level it considers safe in drinking water is 3 ppb. Rainwater in Iowa has been measured to contain 40 ppb. In Switzerland, where the use of atrazine is illegal, the chemical has been measured at approximately 1 ppb in rainwater.

In the Hayes laboratory, concentrations as low as 0.1 ppb had a dramatic effect on tadpole development: it feminized the males. In some of the adult males that developed from these larvae, the vocal structures used in mating calls were smaller than normal, female sex organs developed, and eggs were found growing in the testes. In other studies, normal adult male frogs exposed to 25 ppb had a tenfold reduction in testosterone levels and did not produce sperm. You can imagine the disastrous effects these developmental and hormonal changes could have on the capacity of frogs to breed and reproduce.

But Hayes's experiments were performed in the laboratory, with a species of frog bred for laboratory use. Would his results be the same in nature? To find out, he and his students traveled from Utah to Iowa, sampling water and collecting frogs. They analyzed the water

Frogs Are Having Serious Problems An alarming number of species of frogs, such as this tiny leaf frog (*Agalychnis calcarifer*) from Ecuador, are in danger of becoming extinct. The numerous possible reasons for the decline in global amphibian populations have been a subject of widespread scientific investigation.

A Biologist at Work   Tyrone Hayes grew up near the great Congaree Swamp in South Carolina collecting turtles, snakes, frogs, and toads. Now a professor of biology at the University of California at Berkeley, he has more than 3,000 frogs in his laboratory and studies hormonal control of their development.

for atrazine and examined the frogs. In the only site where atrazine was undetectable in the water, the frogs were normal; in all the other sites, male frogs had abnormalities of the sex organs.

Like other biologists, Hayes made observations. He then made predictions based on those observations, and designed and carried out experiments to test his predictions. Some of the conclusions from his experiments, described at the end of this chapter, could have profound implications not only for amphibians but also for other animals, including humans.

IN THIS CHAPTER we identify and examine the most common features of living organisms and put those features into the context of the major principles that underlie all biology. Next we offer a brief outline of how life evolved and how the different organisms on Earth are related. We then turn to the subjects of biological inquiry and the scientific method. Finally we consider how knowledge discovered by biologists influences public policy.

1.1 What Is Biology?

Biology is the scientific study of living things. Biologists define "living things" as all the diverse organisms descended from a single-celled ancestor that evolved almost 4 billion years ago. Because of their common ancestry, living organisms share many characteristics that are not found in the nonliving world. Living organisms:

- consist of one or more cells
- contain genetic information
- use genetic information to reproduce themselves
- are genetically related and have evolved
- can convert molecules obtained from their environment into new biological molecules
- can extract energy from the environment and use it to do biological work
- can regulate their internal environment

This simple list, however, belies the incredible complexity and diversity of life. Some forms of life may not display all of these characteristics all of the time. For example, the seed of a desert plant may go for many years without extracting energy from the environment, converting molecules, regulating its internal environment, or reproducing; yet the seed is alive.

And what about viruses? Viruses do not consist of cells, and they cannot carry out physiological functions on their own; they must parasitize host cells to do those jobs for them. Yet viruses contain genetic information, and they certainly mutate and evolve (as we know, because evolving flu viruses require constant changes in the vaccines we create to combat them). The existence of viruses depends on cells, and it is highly probable that viruses evolved from cellular life forms. So, are viruses alive? What do you think?

This book explores the characteristics of life, how these characteristics vary among organisms, how they evolved, and how they work together to enable organisms to survive and reproduce. *Evolution* is a central theme of biology and therefore of this book. Through differential survival and reproduction, living systems evolve and become adapted to Earth's many environments. The processes of evolution have generated the enormous diversity that we see today as life on Earth.

Cells are the basic unit of life

We lay the chemical foundation for our study of life in the next three chapters, after which we will turn to cells and the processes by which they live, reproduce, age, and die. Some organisms are *unicellular*, consisting of a single cell that carries out

(A) *Sulfolobus*

(B) *Escherichia coli*

(C) Coccolithophore

4 μm

0.5 μm

0.6 μm

(D) Scarlet banksia

(E) Stinkhorn mushrooms

(F) Milkweed grasshopper

(G) Giant tortoise   Galápagos hawk

1.1 The Many Faces of Life The processes of evolution have led to the millions of diverse organisms living on Earth today. Archaea (A) and bacteria (B) are all single-celled, prokaryotic organisms, as described in Chapter 26. (C) Many protists are unicellular but, as discussed in Chapter 27, their cell structures are more complex than those of the prokaryotes. This protist has manufactured "plates" of calcium carbonate that surround and protect its single cell. (D–G) Most of the visible life on Earth is multicellular. Chapters 28 and 29 cover the green plants (D). The other broad groups of multicellular organisms are the fungi (E), discussed in Chapter 30, and the animals (F, G), covered in Chapters 31–33.

all the functions of life (**Figure 1.1A–C**). Others are *multicellular*, made up of many cells that are specialized for different functions (**Figure 1.1D–G**). Viruses are *acellular*, although they depend on cellular organisms.

The discovery of cells was made possible by the invention of the microscope in the 1590s by the Dutch spectacle makers Hans and Zaccharias Janssen (father and son). In the mid- to late 1600s, Antony van Leeuwenhoek of Holland and Robert Hooke of England both made improvements on the Janssens' technology and used it to study living organisms. Van Leeuwenhoek discovered that drops of pond water teemed with single-celled organisms, and he made many other discoveries as he progressively improved his microscopes over a long lifetime of research. Hooke put pieces of plants under his microscope and observed that they were made up of repeated units he called *cells* (**Figure 1.2**). In 1676, Hooke wrote that van Leeuwenhoek had observed "a vast number of small animals in his Excrements which were most abounding when he was troubled with a Loosenesse and very few or none when he was well." This simple observation

represents the discovery of bacteria—and makes one wonder why scientists do some of the things they do.

More than a hundred years passed before studies of cells advanced significantly. As they were dining together one evening in 1838, Matthias Schleiden, a German biologist, and Theodor Schwann, from Belgium, discussed their work on plant and animal tissues, respectively. They were struck by the similarities in their observations and came to the conclusion that the basic structural elements of plants and animals were essentially the same. They formulated their conclusion as the **cell theory**, which states that:

- Cells are the basic structural and physiological units of all living organisms.
- Cells are both distinct entities and building blocks of more complex organisms.

But Schleiden and Schwann also believed (wrongly) that cells emerged by the self-assembly of nonliving materials, much as crystals form in a solution of salt. This conclusion was in ac-

1.2 Cells Are the Building Blocks of Life The development of microscopes revealed the microbial world to seventeenth-century scientists such as Robert Hooke, who proposed the concept of cells based on his observations. (A) Hooke drew the cells of a slice of plant tissue (cork) as he saw them under his optical microscope. (B) A modern optical, or "light," microscope reveals the intricacies of cells in a leaf. (C) Transmission electron microscopes (TEMs) allow scientists to see even smaller objects. TEMs do not visualize color; here color has been added to a black-and-white micrograph of cells in a duckweed stem.

(A)
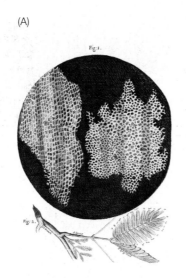

(B)
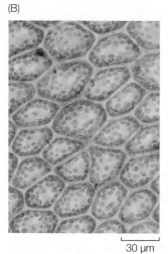

30 µm

(C)

5 µm

cordance with the prevailing view of the day, which was that life can arise from non-life by spontaneous generation—mice from dirty clothes, maggots from dead meat, or insects from pond water.

The debate continued until 1859, when the French Academy of Sciences sponsored a contest for the best experiment to prove or disprove spontaneous generation. The prize was won by the great French scientist Louis Pasteur, who demonstrated that sterile broth directly exposed to the dirt and dust in air developed a culture of microorganisms, but a similar container of broth not directly exposed to air remained sterile (see Figure 4.7). Pasteur's experiment did not prove that it was microorganisms in the air that caused the broth to become infected, but it did uphold the conclusion that life must be present in order for new life to be generated.

Today scientists accept the fact that all cells come from preexisting cells and that the functional properties of organisms derive from the properties of their cells. Since cells of all kinds share both essential mechanisms and a common ancestry that goes back billions of years, modern cell theory has additional elements:

- All cells come from preexisting cells.

- All cells are similar in chemical composition.

- Most of the chemical reactions of life occur in aqueous solution within cells.

- Complete sets of genetic information are replicated and passed on during cell division.

- Viruses lack cellular structure but remain dependent on cellular organisms.

At the same time Schleiden and Schwann were building the foundation for the cell theory, Charles Darwin was beginning to understand how organisms undergo evolutionary change.

All of life shares a common evolutionary history

Evolution—change in the genetic makeup of biological populations through time—is the major unifying principle of biol-

ogy. Charles Darwin compiled factual evidence for evolution in his 1859 book *On the Origin of Species*. Since then, biologists have gathered massive amounts of data supporting Darwin's theory that all living organisms are descended from a common ancestor. Darwin also proposed one of the most important processes that produce evolutionary change. He argued that differential survival and reproduction among individuals in a population, which he termed **natural selection**, could account for much of the evolution of life.

Although Darwin proposed that living organisms are descended from common ancestors and are therefore related to one another, he did not have the advantage of understanding the mechanisms of genetic inheritance. Even so, he observed that offspring resembled their parents; therefore, he surmised, such mechanisms had to exist. That simple fact is the basis for the concept of a **species**. Although the precise definition of a species is complicated, in its most widespread usage it refers to a group of organisms that can produce viable and fertile offspring with one another.

But offspring do differ from their parents. Any population of a plant or animal species displays variation, and if you select breeding pairs on the basis of some particular trait, that trait is more likely to be present in their offspring than in the general population. Darwin himself bred pigeons, and was well aware of how pigeon fanciers selected breeding pairs to produce offspring with unusual feather patterns, beak shapes, or body sizes (see Figure 21.2). He realized that if humans could select for specific traits in domesticated animals, the same process could operate in nature; hence the term *natural selection* as opposed to artificial (human-imposed) selection.

How would natural selection function? Darwin postulated that different probabilities of survival and reproductive success would do the job. He reasoned that the reproductive capacity of plants and animals, if unchecked, would result in unlimited growth of populations, but we do not observe such growth in nature; in most species, only a small percentage of offspring survive to reproduce. Thus any trait that confers even a small increase in the probability that its possessor will survive and reproduce would be spread in the population.

Many leaves are wide and flat, a configuration that presents a maximum of photosynthetic surface to the sun. Some trees, such as this Japanese maple, lose their leaves in response to cold or dry weather.

The leaves of many evergreen conifers, such as spruce trees, are waxy-coated needles that resist water loss and are not shed on a yearly basis.

These water lilies are rooted in the pond bottom; their large leaves are flat "pads" that float on the surface.

The leaves of pitcher plants form a vessel that holds water. The plant receives extra nutrients from the decomposing bodies of insects that drown in the pitcher.

The ability to climb can be advantageous to a plant, enabling it to reach above other plants to obtain more sunlight. Some of the leaves of this climbing cucumber are tightly furled tendrils that wrap around a stake.

1.3 Adaptations to the Environment The leaves of all plants are specialized for photosynthesis—the sunlight-powered transformation of water and carbon dioxide into larger structural molecules called carbohydrates. The leaves of different plants, however, display many different adaptations to their individual environments.

Because organisms with certain traits survive and reproduce best under specific sets of conditions, natural selection leads to **adaptations**: structural, physiological, or behavioral traits that enhance an organism's chances of survival and reproduction in its environment (**Figure 1.3**). In addition to natural selection, evolutionary processes such as sexual selection (selection due to mate choice) and genetic drift (the random fluctuation of gene frequencies in a population due to chance events) contribute to the rise of diverse adaptations. These processes operating over evolutionary history have led to the remarkable array of life on Earth.

If all cells come from preexisting cells, and if all the diverse species of organisms on Earth are related by descent with modification from a common ancestor, then what is the source of information that is passed from parent to daughter cells and from parental organisms to their offspring?

Biological information is contained in a genetic language common to all organisms

Cells are the basic building blocks of organisms, but even a single cell is complex, with many internal structures and many functions that depend on information. The information required for a cell to function and interact with other cells—the "blueprint" for existence—is contained in the cell's **genome**, the sum total of all the DNA molecules it contains. **DNA** (deoxyribonucleic acid) molecules are long sequences of four different subunits called **nucleotides**. The sequence of the nucleotides contains genetic information. **Genes** are specific segments of DNA encoding the information the cell uses to make **proteins** (**Figure 1.4**). Protein molecules govern the chemical reactions within cells and form much of an organism's structure.

By analogy with a book, the nucleotides of DNA are like the letters of an alphabet. Protein molecules are the sentences. Combinations of proteins that form structures and control biochemical processes are the paragraphs. The structures and processes that are organized into different systems with specific tasks (such as digestion or transport) are the chapters of the book, and the complete book is the organism. If you were to write out your own genome using four letters to represent the four nucleotides, you would write more than 3 billion letters. Using the size type you are reading now, your genome would fill about a thousand books the size of this one. The mechanisms of evolution, including natural selection, are the authors and editors of all the books in the library of life.

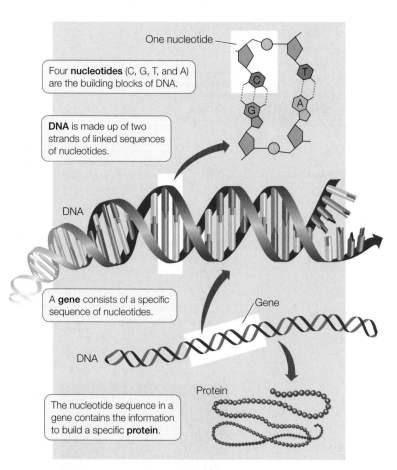

One nucleotide

Four **nucleotides** (C, G, T, and A) are the building blocks of DNA.

DNA is made up of two strands of linked sequences of nucleotides.

DNA

A **gene** consists of a specific sequence of nucleotides.

Gene

DNA

Protein

The nucleotide sequence in a gene contains the information to build a specific **protein**.

1.4 DNA Is Life's Blueprint The instructions for life are contained in the sequences of nucleotides in DNA molecules. Specific DNA nucleotide sequences comprise genes. The average length of a single human gene is 16,000 nucleotides. The information in each gene provides the cell with the information it needs to manufacture molecules of a specific protein.

of biochemical reactions that occur inside cells. Some of these reactions break down nutrient molecules into smaller chemical units, and in the process some of the energy contained in the chemical bonds of the nutrients is captured by high-energy molecules that can be used to do different kinds of cellular work.

One obvious kind of work cells do is mechanical—moving molecules from one cellular location to another, moving whole cells or tissues, or even moving the organism itself, as muscles do (**Figure 1.5A**). The most basic cellular work is the building, or *synthesis*, of new complex molecules and structures from smaller chemical units. For example, we are all familiar with the fact that carbohydrates eaten today may be deposited in the body as fat tomorrow (**Figure 1.5B**). Still another kind of work is the electrical work that is the essence of information processing in nervous systems. The sum total of all the chemical transformations and other work done in all the cells of an organism is its **metabolism**, or **metabolic rate**.

The myriad of biochemical reactions that go on in cells are integrally linked in that the products of one are the raw materials of the next. These complex networks of reactions must be integrated and precisely controlled; when they are not, the result is disease.

Living organisms regulate their internal environment

Multicellular organisms have an *internal environment* that is not cellular. That is, their individual cells are bathed in extracellular fluids, from which they receive nutrients and into which they excrete waste products of metabolism. The cells of multicellu-

All the cells of a multicellular organism contain the same genome, yet different cells have different functions and form different structures—contractile proteins form in muscle cells, hemoglobin in red blood cells, digestive enzymes in gut cells, and so on. Therefore, different types of cells in an organism must express different parts of the genome. How cells control gene expression in ways that enable a complex organism to develop and function is a major focus of current biological research.

The genome of an organism consists of thousands of genes. If the nucleotide sequence of a gene is altered, it is likely that the protein that gene encodes will be altered. Alterations of the genome are called *mutations*. Mutations occur spontaneously; they can also be induced by outside factors, including chemicals and radiation. Most mutations are either harmful or have no effect, but occasionally a mutation improves the functioning of the organism under the environmental conditions it encounters. Such beneficial mutations are the raw material of evolution and lead to adaptations.

Cells use nutrients to supply energy and to build new structures

Living organisms acquire *nutrients* from the environment. Nutrients supply the organism with energy and raw materials for carrying out biochemical reactions. Life depends on thousands

(A)

(B)

1.5 Energy Can Be Used Immediately or Stored (A) Animal cells break down and release the energy contained in the chemical bonds of food molecules to do mechanical work—in this kangaroo's case, to jump. (B) The cells of this Arctic ground squirrel have broken down the complex carbohydrates in plants and converted their molecules into fats, which are stored in the animal's body to provide an energy supply for the cold months.

lar organisms are specialized, or *differentiated*, to contribute in some way to the maintenance of the internal environment. With the evolution of specialization, differentiated cells lost many of the functions carried out by single-celled organisms, and must depend on the internal environment for essential services.

To accomplish their specialized tasks, assemblages of differentiated cells are organized into *tissues.* For example, a single muscle cell cannot generate much force, but when many cells combine to form the tissue of a working muscle, considerable force and movement can be generated (see Figure 1.5B). Different tissue types are organized to form *organs* that accomplish specific functions. For example, the heart, brain, and stomach are each constructed of several types of tissues. Organs whose functions are interrelated can be grouped into *organ systems*; the stomach, intestine, and esophagus, for example, are parts of the digestive system. The functions of cells, tissues, organs, and organ systems are all integral to the multicellular *organism.* We cover the biology of organisms in Parts Eight and Nine of this book.

Living organisms interact with one another

The internal hierarchy of the individual organism is matched by the external hierarchy of the biological world (**Figure 1.6**). Organisms do not live in isolation. A group of individuals of the same species that interact with one another is a *population*, and populations of all the species that live and interact in the same area are called a *community*. Communities together with their abiotic environment constitute an *ecosystem*.

Individuals in a population interact in many different ways. Animals eat plants and other animals (usually members of another species) and compete with other species for food and other resources. Some animals will prevent other individuals of their own species from exploiting a resource, whether it be food, nesting sites, or mates. Animals may also *cooperate* with members of their species, forming social units such as a termite colony or a flock of birds. Such interactions have resulted in the evolution of social behaviors such as communication.

Plants also interact with their external environment, which includes other plants, animals, and microorganisms. All terrestrial plants depend on complex partnerships with fungi, bacteria, and animals. Some of these partnerships are necessary to obtain nutrients, some to produce fertile seeds, and still others to disperse seeds. Plants compete with each other

1.6 Biology Is Studied at Many Levels of Organization
Life's properties emerge when DNA and other molecules are organized in cells. Energy flows through all the biological levels shown here.

yourBioPortal.com
GO TO **Web Activity 1.1 • The Hierarchy of Life**

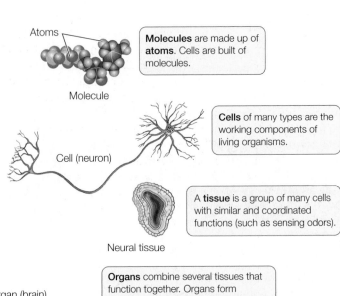

Atoms

Molecule

Molecules are made up of **atoms**. Cells are built of molecules.

Cell (neuron)

Cells of many types are the working components of living organisms.

Neural tissue

A **tissue** is a group of many cells with similar and coordinated functions (such as sensing odors).

Organ (brain)

Organs combine several tissues that function together. Organs form **systems**, such as the nervous system.

Organism (fish)

An **organism** is a recognizable, self-contained individual. Complex multicellular organisms are made up of organs and organ systems.

Population (school of fish)

A **population** is a group of many organisms of the same species.

Communities consist of populations of many different species.

Community (coral reef)

Biological communities in the same geographical location form **ecosystems**. Ecosystems exchange energy and create Earth's **biosphere**.

Biosphere

for light and water, and they have ongoing evolutionary interactions with the animals that eat them, evolving anti-predation adaptations or ways to attract the animals that assist in their reproduction. The interactions of populations of different plant and animal species in a community are major evolutionary forces that produce specialized adaptations.

Communities interacting over a broad geographic area with distinguishing physical features form ecosystems; examples might include an Arctic tundra, a coral reef, or a tropical rainforest. The ways in which species interact with one another and with their environment in communities and in ecosystems is the subject of *ecology* and of Part Ten of this book.

Discoveries in biology can be generalized

Because all life is related by descent from a common ancestor, shares a genetic code, and consists of similar building blocks—cells—knowledge gained from investigations of one type of organism can, with care, be generalized to other organisms. Biologists use **model systems** for research, knowing that they can extend their findings to other organisms, including humans. For example, our basic understanding of the chemical reactions in cells came from research on bacteria but is applicable to all cells, including those of humans. Similarly, the biochemistry of photosynthesis—the process by which plants use sunlight to produce biological molecules—was largely worked out from experiments on *Chlorella*, a unicellular green alga (see Figure 10.13). Much of what we know about the genes that control plant development is the result of work on *Arabidopsis thaliana*, a relative of the mustard plant. Knowledge about how animals develop has come from work on sea urchins, frogs, chickens, roundworms, and fruit flies. And recently, the discovery of a major gene controlling human skin color came from work on zebrafish. Being able to generalize from model systems is a powerful tool in biology.

1.1 RECAP

Living organisms are made of (or depend on) cells, are related by common descent and evolve, contain genetic information and use it to reproduce, extract energy from their environment and use it to do biological work, synthesize complex molecules to construct biological structures, regulate their internal environment, and interact with one another.

- Describe the relationship between evolution by natural selection and the genetic code. See pp. 6–7

- Why can the results of biological research on one species often be generalized to very different species? See p. 9

Now that you have an overview of the major features of life that you will explore in depth in this book, you can ask how and when life first emerged. In the next section we will summarize briefly the history of life from the earliest simple life forms to the complex and diverse organisms that inhabit our planet today.

1.2 How Is All Life on Earth Related?

What do biologists mean when they say that all organisms are *genetically related*? They mean that species on Earth share a *common ancestor*. If two species are similar, as dogs and wolves are, then they probably have a common ancestor in the fairly recent past. The common ancestor of two species that are more different—say, a dog and a deer—probably lived in the more distant past. And if two organisms are very different—such as a dog and a clam—then we must go back to the *very* distant past to find their common ancestor. How can we tell how far back in time the common ancestor of any two organisms lived? In other words, how do we discover the evolutionary relationships among organisms?

For many years, biologists have investigated the history of life by studying the *fossil record*—the preserved remains of organisms that lived in the distant past (**Figure 1.7**). Geologists supplied knowledge about the ages of fossils and the nature of the environments in which they lived. Biologists then inferred the evolutionary relationships among living and fossil organisms by comparing their anatomical similarities and differences. Frequently big gaps existed in the fossil record, forcing biologists to predict the nature of the "missing links" between two lineages of organisms. As the fossil record became more complete, those missing links were filled in.

Molecular methods for comparing genomes, described in Chapter 24, are enabling biologists to more accurately establish the degrees of relationship between living organisms and to use that information to interpret the fossil record. Molecular information can occasionally be gleaned from fossil specimens, such as recently deciphered genetic material from fossil bones of Ne-

1.7 Fossils Give Us a View of Past Life This fossil, formed some 150 million years ago, is that of an *Archaeopteryx*, the earliest known representative of the birds. Birds evolved from the same group of reptiles as the modern crocodiles.

anderthals that led to the conclusion that even though Neanderthals and modern humans coexisted, they did not interbreed.

In general, the greater the differences between the genomes of two species, the more distant their common ancestor. Using molecular techniques, biologists are exploring fundamental questions about life. What were the earliest forms of life? How did simple organisms give rise to the great diversity of organisms alive today? Can we reconstruct a family tree of life?

Life arose from non-life via chemical evolution

Geologists estimate that Earth formed between 4.6 and 4.5 billion years ago. At first, the planet was not a very hospitable place. It was some 600 million years or more before the earliest life evolved. If we picture the history of Earth as a 30-day month, life first appeared somewhere toward the end of the first week (**Figure 1.8**).

When we consider how life might have arisen from nonliving matter, we must take into account the properties of the young Earth's atmosphere, oceans, and climate, all of which were very different than they are today. Biologists postulate that complex biological molecules first arose through the random physical association of chemicals in that environment. Experiments simulating the conditions on early Earth have confirmed that the generation of complex molecules under such conditions is possible, even probable. The critical step for the evolution of life, however, had to be the appearance of molecules that could reproduce themselves and also serve as templates for the synthesis of large molecules with complex but stable shapes. The variation of the shapes of these large, stable molecules (described in Chapters 3 and 4) enabled them to participate in increasing numbers and kinds of chemical reactions with other molecules.

Cellular structure evolved in the common ancestor of life

The second critical step in the origin of life was the enclosure of complex biological molecules by *membranes* that contained them in a compact internal environment separate from the surrounding external environment. Fatlike molecules played a critical role because they are not soluble in water and they form membranous films. When agitated, these films can form spherical *vesicles*, which could have enveloped assemblages of biological molecules. The creation of an internal environment that concentrated the reactants and products of chemical reactions opened up the possibility that those reactions could be integrated and controlled. As described in Section 4.4, scientists postulate that this natural process of membrane formation resulted in the first cells with the ability to replicate themselves—the evolution of the first cellular organisms.

For more than 2 billion years after cells originated, all organisms consisted of only one cell. These first unicellular organisms were (and are, as multitudes of their descendants exist in similar form today) **prokaryotes**. Prokaryotic cells consist of DNA and other biochemicals enclosed in a membrane.

These early prokaryotes were confined to the oceans, where there was an abundance of complex molecules they could use as raw materials and sources of energy. The ocean shielded them from the damaging effects of ultraviolet light, which was intense at that time because there was little or no oxygen (O_2) in the atmosphere, and hence no protective ozone (O_3) layer.

Photosynthesis changed the course of evolution

To fuel their cellular metabolism, the earliest prokaryotes took in molecules directly from their environment and broke these small molecules down to release and use the energy contained in their chemical bonds. Many modern species of prokaryotes still function this way, and very successfully. During the early eons of life on Earth, there was no oxygen in the atmosphere. In fact, oxygen was toxic to the life forms that existed then.

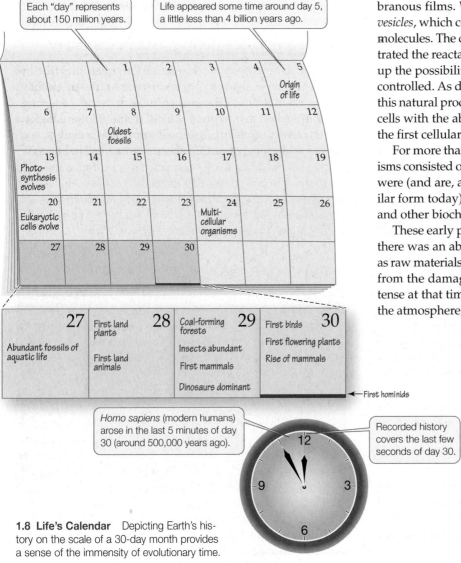

1.8 Life's Calendar Depicting Earth's history on the scale of a 30-day month provides a sense of the immensity of evolutionary time.

1.9 Photosynthetic Organisms Changed Earth's Atmosphere
These strands are composed of many cells of cyanobacteria. This modern species (*Oscillatoria tenuis*) may be very similar to the early photosynthetic prokaryotes responsible for the buildup of oxygen in Earth's atmosphere.

About 2.7 billion years ago, the evolution of **photosynthesis** changed the nature of life on Earth. The chemical reactions of photosynthesis transform the energy of sunlight into a form of biological energy that can power the synthesis of large molecules (see Chapter 10). These large molecules are the building blocks of cells, and they can be broken down to provide metabolic energy. Photosynthesis is the basis of much of life on Earth today because its energy-capturing processes provide food for other organisms.

Early photosynthetic cells were probably similar to present-day prokaryotes called *cyanobacteria* (**Figure 1.9**). Over time, photosynthetic prokaryotes became so abundant that vast quantities of O_2, which is a by-product of photosynthesis, slowly began to accumulate in the atmosphere. Oxygen was poisonous to many of the prokaryotes that lived at that time. Those organisms that did tolerate oxygen, however, were able to proliferate as the presence of oxygen opened up vast new avenues of evolution. *Aerobic metabolism* (energy production based on the conversion of O_2) is more efficient than *anaerobic* (non-O_2-using) *metabolism*, and today it is used by the majority of Earth's organisms. Aerobic metabolism allowed cells to grow larger.

Oxygen in the atmosphere also made it possible for life to move onto land. For most of life's history, ultraviolet (UV) radiation falling on Earth's surface was too intense to allow life to exist outside the shielding water. But the accumulation of photosynthetically generated oxygen in the atmosphere for more than 2 billion years gradually produced a layer of ozone in the upper atmosphere. By about 500 million years ago, the ozone layer was sufficiently dense and absorbed enough UV radiation to make it possible for organisms to leave the protection of the water and live on land.

Eukaryotic cells evolved from prokaryotes

Another important step in the history of life was the evolution of cells with discrete intracellular compartments, called **organelles**, which were capable of taking on specialized cellular functions. This event happened about 3 weeks into our calendar of Earth's history (see Figure 1.8). One of these organelles, the dense-appearing *nucleus* (Latin *nux,* "nut" or "core"), came to contain the cell's genetic information and gives these cells their name: **eukaryotes** (Greek *eu,* "true"; *karyon,* "kernel" or "core"). The eukaryotic cell is completely distinct from the cells of prokaryotes (*pro,* "before"), which lack nuclei and other internal compartments.

Some organelles are hypothesized to have originated by **endosymbiosis** when cells ingested smaller cells. The *mitochondria* that generate a cell's energy probably evolved from engulfed prokaryotic organisms. And *chloroplasts*—organelles specialized to conduct photosynthesis—could have originated when photosynthetic prokaryotes were ingested by larger eukaryotes. If the larger cell failed to break down this intended food object, a partnership could have evolved in which the ingested prokaryote provided the products of photosynthesis and the host cell provided a good environment for its smaller partner.

Multicellularity arose and cells became specialized

Until just over a billion years ago, all the organisms that existed—whether prokaryotic or eukaryotic—were unicellular. An important evolutionary step occurred when some eukaryotes failed to separate after cell division, remaining attached to each other. The permanent association of cells made it possible for some cells to specialize in certain functions, such as reproduction, while other cells specialized in other functions, such as absorbing nutrients and distributing them to neighboring cells. This **cellular specialization** enabled multicellular eukaryotes to increase in size and become more efficient at gathering resources and adapting to specific environments.

Biologists can trace the evolutionary tree of life

If all the species of organisms on Earth today are the descendants of a single kind of unicellular organism that lived almost 4 billion years ago, how have they become so different? A simplified answer is that as long as individuals within a population mate with one another, structural and functional changes can evolve within that population, but the population will remain one species. However, if something happens to isolate some members of a population from the others, the structural and functional differences between the two groups may accumulate over time. The two groups may diverge to the point where their members can no longer reproduce with each other and are thus distinct species. We discuss this evolutionary process, called *speciation*, in Chapter 23.

Biologists give each species a distinctive scientific name formed from two Latinized names (a **binomial**). The first name identifies the species' *genus*—a group of species that share a recent common ancestor. The second is the name of the species. For example, the scientific name for the human species is *Homo sapiens*: *Homo* is our genus and *sapiens* our species. *Homo* is Latin for "man"; *sapiens* is from the Latin for word for "wise" or "rational."

Tens of millions of species exist on Earth today. Many times that number lived in the past but are now extinct. Many millions of speciation events created this vast diversity, and the unfolding of these events can be diagrammed as an evolutionary "tree" whose branches describe the order in which populations split and eventually evolved into new species, as described in Chapter 22. Much of biology is based on comparisons among species, and these comparisons are useful precisely because we can place species in an evolutionary context relative to one another. Our ability to do this has been greatly enhanced in recent decades by our ability to sequence and compare the genomes of different species.

Genome sequencing and other molecular techniques have allowed *systematists*—scientists who study the evolution and classification of life's diverse organisms—to augment evolutionary knowledge based on the fossil record with a vast array of molecular evidence. The result is the ongoing compilation of *phylogenetic trees* that document and diagram evolutionary relationships as part of an overarching tree of life, the broadest categories of which are shown in **Figure 1.10**. (The tree is expanded in this book's Appendix; you can also explore the tree interactively at http://tolweb.org/tree.)

Although many details remain to be clarified, the broad outlines of the tree of life have been determined. Its branching patterns are based on a rich array of evidence from fossils, struc-

tures, metabolic processes, behavior, and molecular analyses of genomes. Molecular data in particular have been used to separate the tree into three major **domains**: Archaea, Bacteria, and Eukarya. The organisms of each domain have been evolving separately from those in the other domains for more than a billion years.

Organisms in the domains **Archaea** and **Bacteria** are single-celled prokaryotes. However, members of these two groups differ so fundamentally in their metabolic processes that they are believed to have separated into distinct evolutionary lineages very early. Species belonging to the third domain—**Eukarya**—have eukaryotic cells whose mitochondria and chloroplasts may have originated from the ingestion of prokaryotic cells, as described on page 11.

The three major groups of multicellular eukaryotes—plants, fungi, and animals—each evolved from a different group of the eukaryotes generally referred to as *protists*. The chloroplast-containing, photosynthetic protist that gave rise to plants was completely distinct from the protist that was ancestral to both animals and fungi, as can be seen from the branching pattern of Figure 1.10. Although most protists are unicellular (and thus sometimes called *microbial eukaryotes*), multicellularity has evolved in several protist lineages.

The tree of life is predictive

There are far more species alive on Earth than biologists have discovered and described to date. In fact, most species on Earth

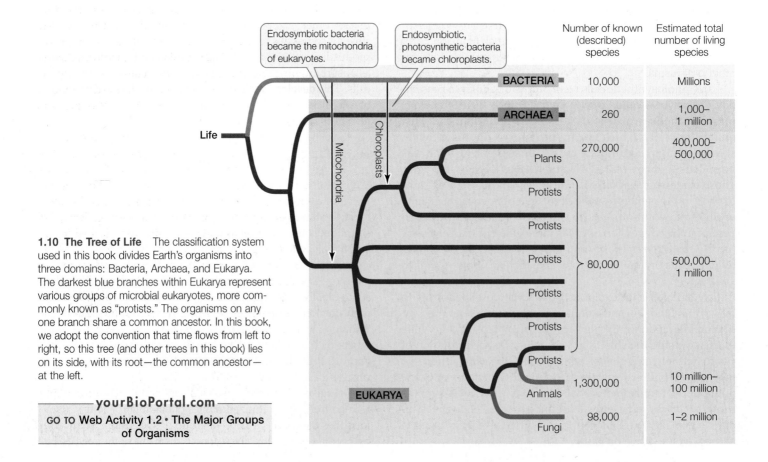

1.10 The Tree of Life   The classification system used in this book divides Earth's organisms into three domains: Bacteria, Archaea, and Eukarya. The darkest blue branches within Eukarya represent various groups of microbial eukaryotes, more commonly known as "protists." The organisms on any one branch share a common ancestor. In this book, we adopt the convention that time flows from left to right, so this tree (and other trees in this book) lies on its side, with its root—the common ancestor—at the left.

yourBioPortal.com
GO TO **Web Activity 1.2** • **The Major Groups of Organisms**

	Number of known (described) species	Estimated total number of living species
BACTERIA	10,000	Millions
ARCHAEA	260	1,000– 1 million
Plants	270,000	400,000– 500,000
Protists		
Protists		
Protists	80,000	500,000– 1 million
Protists		
Protists		
Protists		
Animals	1,300,000	10 million– 100 million
Fungi	98,000	1–2 million

Endosymbiotic bacteria became the mitochondria of eukaryotes.

Endosymbiotic, photosynthetic bacteria became chloroplasts.

Life

Mitochondria

Chloroplasts

EUKARYA

have yet to be discovered by humans (see Section 32.4 for a discussion of how we know this). When we encounter a new species, its placement on the tree of life immediately tells us a great deal about its biology. In addition, understanding relationships among species allows biologists to make predictions about species that have not yet been studied, based on our knowledge of those that have.

For example, until phylogenetic methods were developed, it took years of investigation to isolate and identify most newly encountered human pathogens, and even longer to discover how these pathogens moved into human populations. Today, pathogens that cause diseases such as the flu are identified quickly on the basis of their evolutionary relationships. Placement in an evolutionary tree also gives us clues about the disease's biology, possible effective treatments, and the origin of the pathogen (see Chapters 21 and 22).

1.2 RECAP

The first cellular life on Earth was prokaryotic and arose about 4 billion years ago. The complexity of the organisms that exist today is the result of several important evolutionary events, including the evolution of photosynthesis, eukaryotic cells, and multicellularity. The genetic relationships of all organisms can be shown as a branching tree of life.

- Discuss the evolutionary significance of photosynthesis. See pp. 10–11

- What do the domains of life represent? What are the major groups of eukaryotes? See p. 12 and Figure 1.10

In February of 1676, Robert Hooke received a letter from the physicist Sir Isaac Newton in which Newton famously re-

marked, "If I have seen a little further, it is by standing on the shoulders of giants." We all stand on the shoulders of giants, building on the research of earlier scientists. By the end of this course, you will know more about evolution than Darwin ever could have, and you will know infinitely more about cells than Schleiden and Schwann did. Let's look at the methods biologists use to expand our knowledge of life.

1.3 How Do Biologists Investigate Life?

Regardless of the many different tools and methods used in research, all scientific investigations are based on *observation* and *experimentation*. In both, scientists are guided by the *scientific method,* one of the most powerful tools of modern science.

Observation is an important skill

Biologists have always observed the world around them, but today our ability to observe is greatly enhanced by technologies such as electron microscopes, DNA chips, magnetic resonance imaging, and global positioning satellites. These technologies have improved our ability to observe at all levels, from the distribution of molecules in the body to the distribution of fish in the oceans. For example, not too long ago marine biologists were only able to observe the movement of fish in the ocean by putting physical tags on the fish, releasing them, and hoping that a fisherman would catch that fish and send back the tag—and even that would reveal only where the fish ended up. Today we can attach electronic recording devices to fish that continuously record not only where the fish is, but also how deep it swims and the temperature and salinity of the water around it (**Figure 1.11**). The tags download this information to a satellite, which relays it back to researchers. Suddenly we are acquiring a great deal of knowledge about the distribution of life in the oceans—information that is relevant to studies of climate change.

Technologies that enable us to *quantify* observations are very important in science. For example, for hundreds of years species were classified by generally qualitative descriptions of the physical differences between them. There was no way of objectively calculating evolutionary distances between organisms, and biologists had to depend on the fossil record for insight. Today our ability to rapidly analyze DNA sequences enables quantitative estimates of evolutionary distances, as described in Parts Five and Six of this book. The ability to gather quantitative observations adds greatly to the biologist's ability to make strong conclusions.

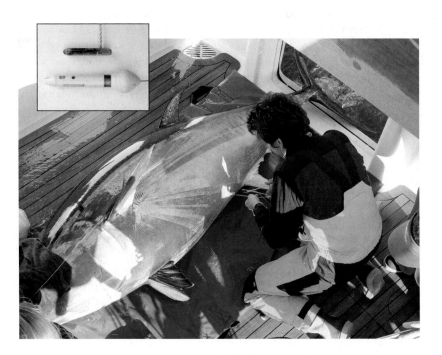

1.11 Tuna Tracking Marine biologist Barbara Block attaches computerized data recording tags (inset) to a live bluefin tuna before returning it to the ocean. Such tags make it possible to track an individual tuna wherever it travels in the world's oceans.

The scientific method combines observation and logic

Observations lead to questions, and scientists make additional observations and do experiments to answer those questions. The conceptual approach that underlies most modern scientific investigations is the **scientific method**. This powerful tool, also called the *hypothesis–prediction (H–P) method*, has five steps: (1) making *observations*; (2) asking *questions*; (3) forming *hypotheses*, or tentative answers to the questions; (4) making *predictions* based on the hypotheses; and (5) *testing* the predictions by making additional observations or conducting experiments (**Figure 1.12**).

After posing a question, a scientist uses *inductive logic* to propose a tentative answer. Inductive logic involves taking observations or facts and creating a new proposition that is compatible with those observations or facts. Such a tentative proposition is called a **hypothesis**. In formulating a hypothesis, scientists put together the facts they already know to formulate one or more possible answers to the question. For example, at the opening of

this chapter you learned that scientists have observed the rapid decline of amphibian populations worldwide and are asking why. Some scientists have hypothesized that a fungal disease is a cause; other scientists have hypothesized that increased exposure to ultraviolet radiation is a cause. Tyrone Hayes hypothesized that exposure to agricultural chemicals could be a cause. He knew that the most widely used chemical herbicide is atrazine; that it is mostly applied in the spring, when amphibians are breeding; and that atrazine is a common contaminant in the waters in which amphibians live as they develop into adults.

The next step in the scientific method is to apply a different form of logic—*deductive logic*—to make predictions based on the hypothesis. Deductive logic starts with a statement believed to be true and then goes on to predict what facts would also have to be true to be compatible with that statement. Based on his hypothesis, Tyrone Hayes predicted that frog tadpoles exposed to atrazine would show adverse effects of the chemical once they reached adulthood.

Good experiments have the potential to falsify hypotheses

Once predictions are made from a hypothesis, experiments can be designed to test those predictions. The most informative experiments are those that have the ability to show that the prediction is wrong. If the prediction is wrong, the hypothesis must be questioned, modified, or rejected.

There are two general types of experiments, both of which compare data from different groups or samples. A *controlled* experiment manipulates one or more of the factors being tested; *comparative* experiments compare unmanipulated data gathered from different sources. As described at the opening of this chapter, Tyrone Hayes and his colleagues conducted both types of experiment to test the prediction that the herbicide atrazine, a contaminant in freshwater ponds and streams throughout the world, affects the development of frogs.

In a **controlled experiment**, we start with groups or samples that are as similar as possible. We predict on the basis of our hypothesis that some critical factor, or **variable**, has an effect on the phenomenon we are investigating. We devise some method to manipulate *only that variable* in an "experimental" group and compare the resulting data with data from an unmanipulated "control" group. If the predicted difference occurs, we then apply statistical tests to ascertain the probability that the manipulation created the difference (as opposed to the difference being the result of random chance). **Figure 1.13** describes one of the many controlled experiments performed by the Hayes laboratory to quantify the effects of atrazine on male frogs.

The basis of controlled experiments is that one variable is manipulated while all others are held constant. The variable that is manipulated is called the *independent variable*, and the response that is measured is the *dependent variable*. A good controlled experiment is not easy to design because biological variables are so interrelated that it is difficult to alter just one.

A **comparative experiment** starts with the prediction that there will be a difference between samples or groups based on the hypothesis. In comparative experiments, however, we can-

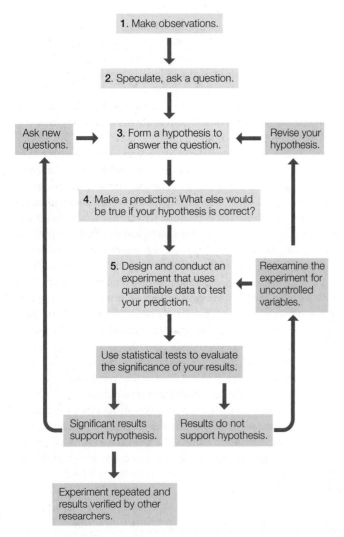

1.12 The Scientific Method The process of observation, speculation, hypothesis, prediction, and experimentation is the cornerstone of modern science. Answers gleaned through experimentation lead to new questions, more hypotheses, further experiments, and expanding knowledge.

INVESTIGATING LIFE

1.13 Controlled Experiments Manipulate a Variable

The Hayes laboratory created controlled environments that differed only in the concentrations of atrazine in the water. Eggs from leopard frogs (*Rana pipiens*) raised specifically for laboratory use were allowed to hatch and the tadpoles were separated into experimental tanks containing water with different concentrations of atrazine.

HYPOTHESIS Exposure to atrazine during larval development causes abnormalities in the reproductive system of male frogs.

METHOD

1. Establish 9 tanks in which all attributes are held constant except the water's atrazine concentrations. Establish 3 atrazine conditions (3 replicate tanks per condition): 0 ppb (control condition), 0.1 ppb, and 25 ppb.
2. Place *Rana pipiens* tadpoles from laboratory-reared eggs in the 9 tanks (30 tadpoles per replicate).
3. When tadpoles have transitioned into adults, sacrifice the animals and evaluate their reproductive tissues.
4. Test for correlation of degree of atrazine exposure with the presence of abnormalities in the reproductive systems of male frogs.

RESULTS

Abnormal testes development

Oocytes (eggs) in normal-size testis (sex reversal)

- Gonadal dysgenesis
- Testicular oogenesis

In the control condition, only one male had abnormalities.

Male frogs with gonadal abnormalities (%)

Atrazine (ppb)

0.1   25   0.0 Control

CONCLUSION Exposure to atrazine at concentrations as low as 0.1 ppb induces abnormalities in the male reproductive systems of frogs. The effect is not proportional to the level of exposure.

Go to **yourBioPortal.com** for original citations, discussions, and relevant links for all INVESTIGATING LIFE figures.

not control the variables; often we cannot even identify all the variables that are present. We are simply gathering and comparing data from different sample groups.

When his controlled experiments indicated that atrazine indeed affects reproductive development in frogs, Hayes and his colleagues performed a comparative experiment. They collected frogs and water samples from eight widely separated sites across the United States and compared the incidence of abnormal frogs from environments with very different levels of atrazine (**Figure 1.14**). Of course, the sample sites differed in many ways besides the level of atrazine present.

The results of experiments frequently reveal that the situation is more complex than the hypothesis anticipated, thus raising new questions. In the Hayes experiments, for example, there was no clear direct relationship between the *amount* of atrazine present and the percentage of abnormal frogs: there were fewer abnormal frogs at the highest concentrations of atrazine than at lower concentrations. There are no "final answers" in science. Investigations consistently reveal more complexity than we expect. The scientific method is a tool to identify, assess, and understand that complexity.

— **yourBioPortal.com** —
GO TO Animated Tutorial 1.1 • The Scientific Method

Statistical methods are essential scientific tools

Whether we do comparative or controlled experiments, at the end we have to decide whether there is a difference between the samples, individuals, groups, or populations in the study. How do we decide whether a measured difference is enough to support or falsify a hypothesis? In other words, how do we decide in an unbiased, objective way that the measured difference is significant?

Significance can be measured with statistical methods. Scientists use statistics because they recognize that variation is always present in any set of measurements. Statistical tests calculate the probability that the differences observed in an experiment could be due to random variation. The results of statistical tests are therefore probabilities. A statistical test starts with a **null hypothesis**—the premise that no difference exists. When quantified observations, or **data**, are collected, statistical methods are applied to those data to calculate the likelihood that the null hypothesis is correct.

More specifically, statistical methods tell us the probability of obtaining the same results by chance even if the null hypothesis were true. We need to eliminate, insofar as possible, the chance that any differences showing up in the data are merely the result of random variation in the samples tested. Scientists generally conclude that the differences they measure are significant if statistical tests show that the *probability of error* (that is, the probability that the same results can be obtained by mere chance) is 5 percent or lower.

Not all forms of inquiry are scientific

Science is a unique human endeavor that is bounded by certain standards of practice. Other areas of scholarship share with science the practice of making observations and asking ques-

INVESTIGATING LIFE

1.14 Comparative Experiments Look for Differences among Groups

To see whether the presence of atrazine correlates with reproductive system abnormalities in male frogs, the Hayes lab collected frogs and water samples from different locations around the U.S. The analysis that followed was "blind," meaning that the frogs and water samples were coded so that experimenters working with each specimen did not know which site the specimen came from.

HYPOTHESIS   Presence of the herbicide atrazine in environmental water correlates with reproductive system abnormalities in frog populations.

METHOD
1. Based on commercial sales of atrazine, select 4 sites (sites 1–4) less likely and 4 sites (sites 5–8) more likely to be contaminated with atrazine.
2. Visit all sites in the spring (i.e., when frogs have transitioned from tadpoles into adults); collect frogs and water samples.
3. In the laboratory, sacrifice frogs and examine their reproductive tissues, documenting abnormalities.
4. Analyze the water samples for atrazine concentration (the sample for site 7 was not tested).
5. Quantify and correlate the incidence of reproductive abnormalities with environmental atrazine concentrations.

RESULTS

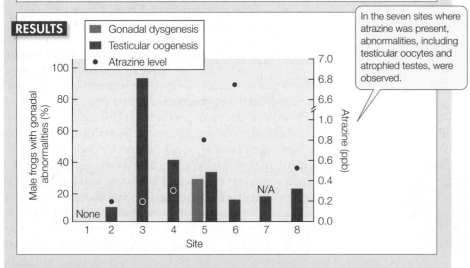

In the seven sites where atrazine was present, abnormalities, including testicular oocytes and atrophied testes, were observed.

CONCLUSION   Reproductive abnormalities exist in frogs from environments in which aqueous atrazine concentration is 0.2 ppb or above. The incidence of abnormalities does not appear to be proportional to atrazine concentration at the time of transition to adulthood.

FURTHER INVESTIGATION: The highest proportion of abnormal frogs was found at site 3, located on a wildlife reserve in Wyoming. What kind of data and observations would you need to suggest possible explanations for this extremely high incidence?

Go to **yourBioPortal.com** for original citations, discussions, and relevant links for all INVESTIGATING LIFE figures.

Scientific explanations for natural processes are objective and reliable because the hypotheses proposed *must be testable* and *must have the potential of being rejected* by direct observations and experiments. Scientists must clearly describe the methods they use to test hypotheses so that other scientists can repeat their results. Not all experiments are repeated, but surprising or controversial results are always subjected to independent verification. Scientists worldwide share this process of testing and rejecting hypotheses, contributing to a common body of scientific knowledge.

If you understand the methods of science, you can distinguish science from non-science. Art, music, and literature all contribute to the quality of human life, but they are not science. They do not use the scientific method to establish what is fact. Religion is not science, although religions have historically purported to explain natural events ranging from unusual weather patterns to crop failures to human diseases. Most such phenomena that at one time were mysterious can now be explained in terms of scientific principles.

The power of science derives from the uncompromising objectivity and absolute dependence on evidence that comes from *reproducible and quantifiable observations*. A religious or spiritual explanation of a natural phenomenon may be coherent and satisfying for the person holding that view, but it is not testable, and therefore it is not science. To invoke a supernatural explanation (such as a "creator" or "intelligent designer" with no known bounds) is to depart from the world of science.

Science describes the facts about how the world works, not how it "ought to be." Many scientific advances that have contributed to human welfare have also raised major ethical issues. Recent developments in genetics and developmental biology, for example, enable us to select the sex of our children, to use stem cells to repair our bodies, and to modify the

---

tions, but scientists are distinguished by what they do with their observations and how they answer their questions. Data, subjected to appropriate statistical analysis, are critical in the testing of hypotheses. The scientific method is the most powerful way humans have devised for learning about the world and how it works.

human genome. Although scientific knowledge allows us to do these things, science cannot tell us whether or not we should do them, or, if we choose to do so, how we should regulate them.

To make wise decisions about public policy, we need to employ the best possible ethical reasoning in deciding which outcomes we should strive for.

1.3 RECAP

The scientific method of inquiry starts with the formulation of hypotheses based on observations and data. Comparative and controlled experiments are carried out to test hypotheses.

- Explain the relationship between a hypothesis and an experiment. **See p. 14 and Figure 1.12**

- What is controlled in a controlled experiment? **See p. 14 and Figure 1.13**

- What features characterize questions that can be answered only by using a comparative approach? **See pp. 14–15 and Figure 1.14**

- Do you understand why arguments must be supported by quantifiable and reproducible data in order to be considered scientific? **See pp. 15–16**

The vast scientific knowledge accumulated over centuries of human civilization allows us to understand and manipulate aspects of the natural world in ways that no other species can. These abilities present us with challenges, opportunities, and above all, responsibilities.

1.4 How Does Biology Influence Public Policy?

Agriculture and medicine are two important human activities that depend on biological knowledge. Our ancestors unknowingly applied the principles of evolutionary biology when they domesticated plants and animals, and people have speculated about the causes of diseases and searched for methods to combat them since ancient times. Long before the microbial causes of diseases were known, people recognized that infections could be passed from one person to another, and the isolation of infected persons has been practiced as long as written records have been available.

Today, thanks to the deciphering of genomes and our newfound ability to manipulate them, vast new possibilities exist for controlling human diseases and increasing agricultural productivity, but these capabilities raise ethical and policy issues. How much and in what ways should we tinker with the genes of humans and other species? Does it matter whether the genomes of our crop plants and domesticated animals are changed by traditional methods of controlled breeding and crossbreeding or by the biotechnology of gene transfer? What rules should govern the release of genetically modified organisms into the environment? Science alone cannot provide all the answers, but wise policy decisions must be based on accurate scientific information.

Biologists are increasingly called on to advise government agencies concerning the laws, rules, and regulations by which society deals with the increasing number of challenges that have at least a partial biological basis. As an example of the value of scientific knowledge for the assessment and formulation of public policy, let's return to the tracking study of bluefin tuna introduced in Section 1.3. Prior to this study, both scientists and fishermen knew that bluefins had a western breeding ground in the Gulf of Mexico and an eastern breeding ground in the Mediterranean Sea (**Figure 1.15**). Overfishing had led to declining numbers of fish in the western-breeding populations, to the point of these populations being endangered.

1.15 Bluefin Tuna Do Not Recognize Boundaries It was assumed that tuna from western-breeding populations and those from eastern-breeding populations also fed on their respective sides of the Atlantic, so separate fishing quotas were established on either side of 45° W longitude (dashed line) to allow the endangered western population to recover. However, tracking data shows that the two populations *do not* remain separate after spawning, so in fact the established policy does not protect the western population.

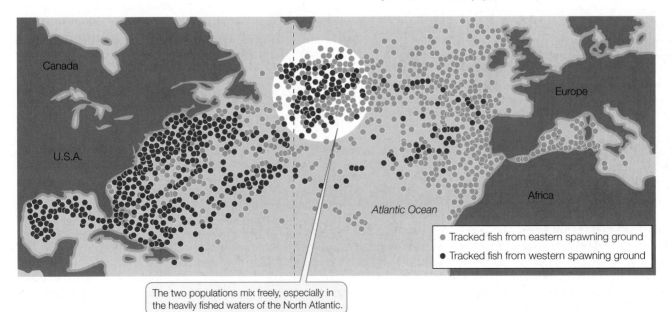

The two populations mix freely, especially in the heavily fished waters of the North Atlantic.

- Tracked fish from eastern spawning ground
- Tracked fish from western spawning ground

Initially it was assumed by scientists, fishermen, and policy makers alike that the eastern and western populations had geographically separate feeding grounds as well as separate breeding grounds. Acting on this assumption, an international commission drew a line down the middle of the Atlantic Ocean and established strict fishing quotas on the western side of the line, with the intent of allowing the western population to recover. New tracking data, however, revealed that in fact the eastern and western bluefin populations mix freely on their feeding grounds across the entire North Atlantic—a swath of ocean that includes the most heavily fished waters in the world. Tuna caught on the eastern side of the line could just as likely be from the western breeding population as the eastern; thus the established policy was not achieving its intended goal.

Policy makers take more things into consideration than scientific knowledge and recommendations. For example, studies on the effects of atrazine on amphibians have led one U.S. group, the Natural Resources Defense Council, to take legal action to have atrazine banned on the basis of the Endangered Species Act. The U.S. Environmental Protection Agency, however, must also consider the potential loss to agriculture that such a ban would create and has continued to approve atrazine's use as long as environmental levels do not exceed 30 to 40 ppb—which is 300 to 400 times the levels shown to induce abnormalities in the Hayes studies. Scientific conclusions do not always prevail in the political world.

Another reason for studying biology is to understand the effects of the vastly increased human population on its environment. Our use of natural resources is putting stress on the ability of Earth's ecosystems to continue to produce the goods and services on which our society depends. Human activities are changing global climates, causing the extinctions of a large number of species like the amphibians featured in this chapter, and spreading new diseases while facilitating the resurgence of old ones. The rapid spread of flu viruses has been facilitated by modern modes of transportation, and the recent resurgence of tuberculosis is the result of the evolution of bacteria that are resistant to antibiotics. Biological knowledge is vital for determining the causes of these changes and for devising wise policies to deal with them.

Beyond issues of policy and pragmatism lies the human "need to know." Human beings are fascinated by the richness and diversity of life, and most people want to know more about organisms and how they interact. Human curiosity might even be seen as an adaptive trait—it is possible that such a trait could have been selected for if individuals who were motivated to learn about their surroundings were likely to have survived and reproduced better, on average, than their less curious relatives. Far from ending the process, new discoveries and greater knowledge typically engender questions no one thought to ask before. There are vast numbers of questions for which we do not yet have answers, and the most important motivator of most scientists is curiosity.

CHAPTER SUMMARY

1.1 What Is Biology?

- **Biology** is the scientific study of living organisms, including their characteristics, functions, and interactions. Cells are the basic structural and physiological units of life. The **cell theory** states that all life consists of cells and that all cells come from preexisting cells.

- All living organisms are related to one another through descent with modification. **Evolution** by **natural selection** is responsible for the diversity of **adaptations** found in living organisms.

- The instructions for a cell are contained in its **genome**, which consists of **DNA** molecules made up of sequences of **nucleotides**. Specific segments of DNA called **genes** contain the information the cell uses to make **proteins**. Review Figure 1.4

- Living organisms regulate their internal environment. They also interact with other organisms of the same and different species. Biologists study life at all these levels of organization. Review Figure 1.6, **WEB ACTIVITY 1.1**

- Biological knowledge obtained from a **model system** may be generalized to other species.

1.2 How Is All Life on Earth Related?

- Biologists use fossils, anatomical similarities and differences, and molecular comparisons of genomes to reconstruct the history of life. Review Figure 1.8

- Life first arose by chemical evolution. Cells arose early in the evolution of life.

- **Photosynthesis** was an important evolutionary step because it changed Earth's atmosphere and provided a means of capturing energy from sunlight.

- The earliest organisms were **prokaryotes**. Organisms called **eukaryotes**, with more complex cells, arose later. Eukaryotic cells have discrete intracellular compartments, called **organelles**, including a nucleus that contains the cell's genetic material.

- The genetic relationships of **species** can be represented as an evolutionary tree. Species are grouped into three **domains**: **Archaea**, **Bacteria**, and **Eukarya**. Archaea and Bacteria are domains of unicellular prokaryotes. Eukarya contains diverse groups of protists (most but not all of which are unicellular) and the multicellular plants, fungi, and animals. Review Figure 1.10, **WEB ACTIVITY 1.2**

1.3 How Do Biologists Investigate Life?

- The **scientific method** used in most biological investigations involves five steps: making observations, asking questions, forming hypotheses, making predictions, and testing those predictions. Review Figure 1.12

- **Hypotheses** are tentative answers to questions. Predictions made on the basis of a hypothesis are tested with additional observations and two kinds of **experiments**: **comparative** and **controlled experiments**. Review Figures 1.13 and 1.14, **ANIMATED TUTORIAL 1.1**

- Statistical methods are applied to **data** to establish whether or not the differences observed are significant or whether they could be the result of chance. These methods start with the **null hypothesis** that there are no differences.

- Science can tell us how the world works, but it cannot tell us what we should or should not do.

1.4 How Does Biology Influence Public Policy?

- Biologists are often called on to advise government agencies on the solution of important problems that have a biological component.

FOR DISCUSSION

1. Even if we knew the sequences of all of the genes of a single-celled organism and could cause those genes to be expressed in a test tube, it would still be incredibly difficult to create a functioning organism. Why do you think this is so? In light of this fact, what do you think of the statement that the genome contains all of the information for a species?

2. Why is it so important in science that we design and perform tests capable of falsifying a hypothesis?

3. What features characterize questions that can be answered only by using a comparative approach?

4. Cite an example of how you apply aspects of the scientific method to solve problems in your daily life.

ADDITIONAL INVESTIGATION

1. The abnormalities of frogs in Tyrone Hayes's studies were associated with the presence of a herbicide in the environment. That herbicide did not kill the frogs, but it feminized the males. How would you investigate whether this effect could lead to decreased reproductive capacity for the frog populations in nature?

2. Just as all cells come from preexisting cells, all mitochondria—the cell organelles that convert energy in food to a form of energy that can do biological work—come from preexisting mitochondria. Cells do not synthesize mitochondria from the genetic information in their nuclei. What investigations would you carry out to understand the nature of mitochondria?

WORKING WITH DATA (GO TO yourBioPortal.com)

Feminization of Frogs   Analogous to the experiment shown in Figure 1.13, this exercise asks you to graph data about the size of the laryngeal (throat) muscles required to produce male mating calls in the frog *Xenopus laevis*. After plotting data from frogs exposed to different levels of the herbicide atrazine during their development, you will formulate conclusions about the effects of the herbicide on this physical attribute and speculate about what these effects might mean.

21 Evidence and Mechanisms of Evolution

Evolutionary theory leads to better flu vaccines

On November 11, 1918, an armistice agreement signed in France signaled the end of the First World War. But the death toll from four years of war was soon surpassed by the casualties of a massive influenza epidemic that began in the spring of 1918 among soldiers in a U.S. Army barracks. Over the next year and a half, this particular strain of flu virus spread across the globe in a true pandemic that killed more than 50 million people worldwide—more than twice the number of WWI-related combat deaths.

The 1918–1919 pandemic was also noteworthy because the death rate among young adults—who are usually less likely to die from influenza than are the elderly or the very young—was 20 times higher than in influenza epidemics before or since. Why was that particular flu virus so deadly, especially to typically hardy individuals?

Influenza viruses evolve constantly and rapidly. The 1918 strain triggered an especially intense reaction in the human immune system. This overreaction (called a "cytokine storm") meant that people with strong immune systems were likely to be more severely affected. Usually, however, our immune system helps fight flu viruses, and this immune response is the basis of *vaccination* (see Chapter 42). The first flu vaccines were offered in 1945, and since then immunization programs have helped keep the number and severity of outbreaks in check. But flu viruses evolve so rapidly that last year's vaccine is not effective against this year's virus.

Although many of the influenza strains circulating in a given season are closely related, they are not identical. *Genetic variation* ensures that there are always different strains, and these different strains compete with one another. The strains that are best able to escape detection by the immune systems of their hosts are most likely to spread, and thus have an advantage over other strains. But the immune system responds to counteract the virus, and last year's virus loses its advantage. If flu viruses did not evolve, we would become resistant and annual vaccination would be unnecessary.

Our immune system recognizes an influenza infection by detecting a protein called hemagglutinin on the viral surface. New mutations arise rapidly, and genetic sequence changes in the viral genome sometimes result in variation in the structure of hemagglutinin. Variants with altered hemagglutinin structure

Deadly Epidemic So many people were incapacitated during the flu epidemic of 1918–1919 that temporary hospitals had to be set up. Here the beds of flu-stricken patients cover the floor of the Dartmouth College gymnasium during the peak of the epidemic in the United States.

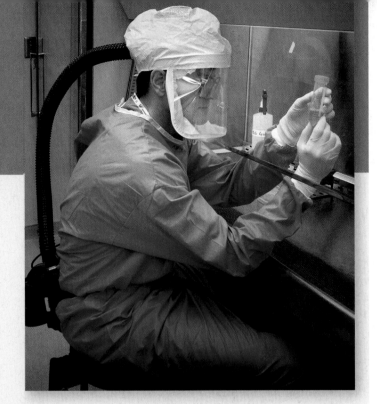

Reconstructing a Deadly Virus   Terrence Tumpey, of the Centers for Disease Control and Prevention, reconstructed the 1918 influenza virus in his lab to identify the characteristics that made it so deadly.

are more likely to escape detection by our immune system, and thus more likely to survive and replicate. The term *positive selection* describes the evolution of favored changes like those in influenza hemagglutinin protein.

In developing flu vaccines, biologists examine the hemagglutinin gene, particularly at certain sites known to be under positive selection for change. Viruses with the greatest number of changes at these sites are the ones most likely to cause next year's flu epidemic, and therefore are the best targets for new vaccines. Understanding evolutionary theory thus helps us determine the causes and find solutions for a potentially deadly disease.

IN THIS CHAPTER we will examine the factual basis of evolution and consider some of the mechanisms that result in evolutionary change. We will see how Charles Darwin developed his ideas on one such mechanism (natural selection). We will discuss the genetic basis of evolution and show how genetic variation in populations is measured. Throughout the chapter, we will discuss ways that evolution can be applied to practical problems and how it helps us understand the diversity of life.

21.1 What Facts Form the Basis of Our Understanding of Evolution?

The living world is constantly changing. Biologists observe many of these changes directly, both in laboratory experiments and in natural populations. Many other changes are recorded in the fossil record of life. We can measure the rate at which new mutations arise, observe the spread of new genetic variants through a population, and see the effects of genetic change on the form and function of organisms. In other words, evolution is a fact that we can observe directly. Biologists also have accumulated a large body of evidence about *how* evolutionary changes occur, and about *what* evolutionary changes have occurred in the past. The understanding and application of the mechanisms of evolutionary change to biological problems is known as evolutionary theory.

Evolutionary theory has many useful applications, such as the development of influenza vaccines described in the opening of this chapter. We use evolution to study, understand, and treat diseases; to develop better agricultural crops and practices; and to develop industrial processes that produce new molecules with useful properties. Knowledge of evolutionary principles has helped biologists understand how life diversified and how species interact. It also allows us to make predictions about the biological world.

In everyday speech, people tend to use the word "theory" to mean an untested hypothesis, or even a guess. But the term "evolutionary theory" does not refer to any single hypothesis, and it certainly is not guesswork. As used in science, "theory" refers to the entire body of work on the understanding and application of a field of knowledge. When we refer to "gravitational theory," we are not implying that gravity is an untested idea. No one doubts that gravity exists—we can see its effects all around us. Instead, we are referring to our understanding of the mechanisms that result in gravitational pull, and the use of that understanding to make predictions about the interactions of physical objects. Drop this book, and it will fall at a predicable rate, according to gravitational theory.

In a similar manner, when we refer to evolutionary theory, we are referring to our understanding of the mechanisms that result in biological changes in populations over time, and the use of that understanding to interpret changes and interactions of biological organisms. That biological populations evolve through time is not disputed by biologists. We can, and do, observe evolutionary change on a regular basis. We can directly observe the evolution of influenza viruses, but it is evolutionary theory that allows us to apply that information to the problem of developing more effective vaccines.

Several mechanisms of evolutionary change are recognized, and the scientific community is continually expanding its understanding of how and when these mechanisms apply to particular biological problems. Studying the mechanisms of evolution and their innumerable applications constitutes the active and exciting field of evolutionary theory.

Charles Darwin articulated the principle of natural selection

Today a rich array of geological, morphological, and molecular data support and enhance the factual basis of evolution. In the 1820s, however, it was not yet evident to the young Charles Darwin (or almost anyone else) that life had evolved. Darwin was passionately interested in both geology and natural history—the scientific study of how different organisms function and carry out their lives in nature. Despite these interests, he planned (at his father's behest) to become a doctor. But surgery conducted

without anesthesia nauseated Darwin, and he gave up medicine to study at Cambridge University for a career as a clergyman in the Church of England. Always more interested in science than in theology, he gravitated toward scientists on the faculty, especially the botanist John Henslow. In 1831, Henslow recommended Darwin for a position on the H.M.S. *Beagle*, which was preparing for a survey voyage around the world (**Figure 21.1**).

Whenever possible during the 5-year voyage, Darwin (who was often seasick) went ashore to study rocks and to observe and collect plants and animals. He noticed striking differences between the species he saw in South America and those from Europe. He observed that the species of the temperate regions of South America (Argentina and Chile) were more similar to those of tropical South America (Brazil) than they were to temperate European species. When he explored the islands of the

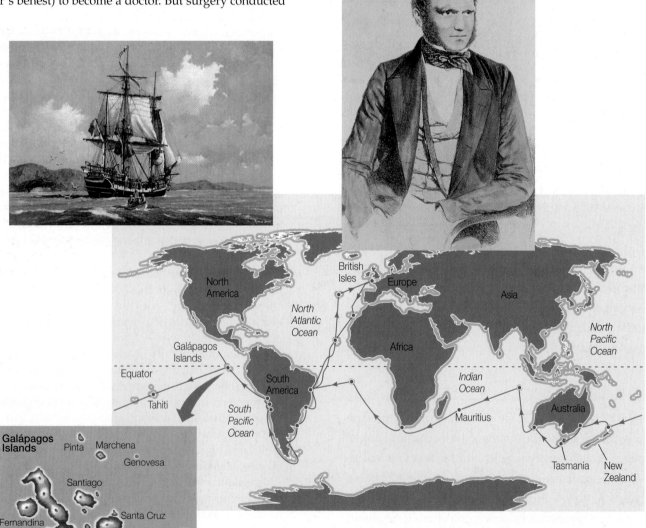

21.1 Darwin and the Voyage of the Beagle The mission of H.M.S. *Beagle* was to chart the oceans and collect oceanographic and biological information from around the world. The world map indicates the ship's path; the inset map shows the Galápagos Islands, whose organisms were an important source of Darwin's ideas on natural selection. The portrait is of Charles Darwin at age 24, shortly after the *Beagle* returned to England.

Galápagos Archipelago west of Ecuador, he noted that most of its animal species were found nowhere else, although they were similar to animals found on the mainland of South America. Darwin also observed that the animals of the Galápagos differed from island to island. He postulated that some animals had come to the archipelago from mainland South America and had subsequently undergone different changes on each of the islands. He then wondered what might account for these changes.

When he returned to England in 1836, Darwin continued to ponder his observations. Within a decade he had developed the major elements of an explanatory theory for evolutionary change based on three major propositions:

- Species are not immutable; they change over time.

- Divergent species share a common ancestor.

- The mechanism that produces changes in species is **natural selection**, or the differential survival and reproduction of individuals in a population based on variation in their traits.

The revolutionary assertions in Darwin's first two propositions were that evolution is a historical fact that can be demonstrated to have taken place, and that species are related to one another through common descent. In 1844, he wrote a long essay on natural selection, which he described as the mechanism of evolution (his third proposition), but despite urging from colleagues, he was reluctant to publish it, preferring to assemble more evidence first.

Darwin's hand was forced in 1858 when he received a letter and manuscript from another traveling naturalist, Alfred Russel Wallace, who was studying the biota of the Malay Archipelago. Wallace asked Darwin to evaluate his manuscript, which included an explanation of natural selection almost identical to Darwin's. Darwin was at first dismayed, believing Wallace had preempted his idea. But parts of Darwin's 1844 essay, together with Wallace's manuscript, were presented to the Linnaean Society of London on July 1, 1858, thereby crediting both men for the idea of natural selection. Darwin then worked quickly to finish his own book, *The Origin of Species*, which was published the next year.

Although Darwin and Wallace independently articulated the concept of natural selection, Darwin developed his ideas first.

Furthermore, *The Origin of Species* provided exhaustive evidence from many fields to support both natural selection and evolution itself. Thus both concepts are more closely associated with Darwin than with Wallace.

The facts that Darwin used to conceive and develop his explanation of evolution by natural selection were familiar to most contemporary biologists. His insight was to perceive the significance of relationships among these facts. Both Darwin and Wallace were influenced by the ideas of the economist Thomas Malthus, who in 1838 published *An Essay on the Principle of Population*. Malthus argued that because the rate of human population growth is greater than the rate of increase in food production, unchecked growth inevitably leads to famine. Darwin saw parallels throughout nature. He recognized that populations of all species have the potential to rapidly increase in number. To illustrate this point, he used the following example:

Suppose…there are eight pairs of birds, and that only four pairs of them annually…rear only four young, and that these go on rearing their young at the same rate, then at the end of seven years…there will be 2048 birds instead of the original sixteen.

Such increases are rarely seen in nature, though. Darwin therefore reasoned that death rates in nature must also be high. If they weren't, even the most slowly reproducing species would quickly reach enormous population sizes.

Darwin also observed that although offspring tend to resemble their parents, the offspring of most organisms are not identical to one another or to their parents. He suggested that slight variations among individuals affect the chance that a given individual will survive and reproduce. Darwin called this differential survival and reproduction of individuals natural selection.

Darwin may have used the words "natural selection" because he was familiar with the **artificial selection** of strains with certain desirable traits by animal and plant breeders. Many of Darwin's observations on the nature of variation came from domesticated plants and animals. Darwin was a pigeon breeder, and he knew firsthand the astonishing diversity in color, size, form, and behavior that breeders could achieve (**Figure 21.2**). He recognized

21.2 Artificial Selection Charles Darwin raised pigeons as a hobby, and he noted similar forces at work in artificial and natural selection. The "fancy" pigeons shown here represent three of the more than 300 varieties derived from the wild rock dove *Columba livia* (at left) by artificial selection for character traits such as color and feather distribution.

close parallels between selection by breeders and selection in nature. As he argued in *The Origin of Species*,

How can it be doubted, from the struggle each individual has to obtain subsistence, that any minute variation in structure, habits or instincts, adapting that individual better to the new conditions, would tell upon its vigour and health? In the struggle it would have a better chance of surviving; and those of its offspring which inherited the variation, be it ever so slight, would have a better chance.

That statement, written more than 150 years ago, still stands as a good expression of the process of evolution by natural selection.

It is important to remember that, as Darwin clearly understood, *individuals do not evolve; populations do*. A **population** is a group of individuals of a single species that live and interbreed in a particular geographic area at the same time. A major consequence of the evolution of populations is that their members become adapted to the environments in which they live. But what do biologists mean when they say that an organism is adapted to its environment?

—————— **yourBioPortal.com** ——————
GO TO Animated Tutorial 21.1 • Natural Selection

Adaptation has two meanings

In evolutionary biology, **adaptation** refers both to the *processes* by which characteristics that appear to be useful to their bearers evolve—that is, the evolutionary mechanisms that produce them—and to the *characteristics* themselves. With respect to the latter, an adaptation is a phenotypic characteristic that has made it more likely for an organism to survive and reproduce.

Biologists regard an organism as being adapted to a particular environment when they can demonstrate that a slightly different organism reproduces and survives less well in that environment. To understand adaptation, biologists compare the performance of individuals that differ in their traits. For example, biologists can assess the adaptive role of changes to the hemagglutinin protein of influenza viruses, as described in the opening of this chapter. By comparing the survival and proliferation rates of influenza viruses that have different hemagglutinin gene sequences, biologists can study adaptation of the viruses through time.

When Darwin proposed his ideas on evolution by natural selection, he could point to many examples of evolutionary mechanisms operating in nature, but none were supported by experiments. Since then, biologists have conducted thousands of observational and experimental studies that have confirmed the important role of natural selection as a mechanism of evolution. Biologists have also documented changes over time in the genetic composition and morphology of many populations, and our understanding of the mechanisms of inheritance has improved enormously since Darwin's time.

Population genetics provides an underpinning for Darwin's theory

Darwin had no knowledge of the mechanisms of genetic transmission, and his speculations on the topic proved to be incor-

rect. Biologists did not have a good understanding of the genetic details of how natural selection works until the field of transmission genetics was established in the early 1900s. At that time, the rediscovery of Gregor Mendel's publications (see Section 12.1) paved the way for the development in the 1930s and 1940s of the field of *population genetics*. As the principles of evolution were integrated with the principles of modern genetics during this period, a new understanding of evolutionary biology—known as the *Modern Synthesis*—emerged. This was when biologists began to study mechanistic aspects of evolution as well as the broad evolutionary patterns that were so evident in nature.

For a population to evolve, its members must possess heritable genetic variation, which is the raw material on which mechanisms of evolution act. In everyday life, we do not directly observe the genetic compositions of organisms. What we see are *phenotypes*, the physical expressions of organisms' genes (including interactions among genes). The features of a phenotype are its *characters*—eye color, for example. The specific form of a character, such as brown eyes, is a *trait*. A **heritable trait** is a characteristic that is at least partly determined by the organism's genes. The genetic constitution that governs a character is called its *genotype*. *A population evolves when individuals with different genotypes survive or reproduce at different rates.*

The field of population genetics has three main goals:

- To explain the patterns and organization of genetic variation
- To explain the origin and maintenance of genetic variation
- To understand the mechanisms that cause changes in allele frequencies in populations

The perspective of population genetics complements the insights into evolutionary processes provided by developmental biology, as described in Chapter 20.

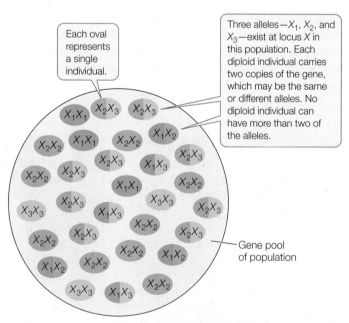

Each oval represents a single individual.

Three alleles—X_1, X_2, and X_3—exist at locus X in this population. Each diploid individual carries two copies of the gene, which may be the same or different alleles. No diploid individual can have more than two of the alleles.

Gene pool of population

21.3 A Gene Pool A gene pool is the sum of all the alleles found in a population, or for a particular locus. This figure shows the gene pool for one locus, X. The allele frequencies are 0.20 for X_1, 0.50 for X_2, and 0.30 for X_3.

21.4 Many Vegetables from One Species All of the crop plants shown here derive from a single wild mustard species. European agriculturalists produced these crop species by selecting and breeding plants with unusually large buds, stems, leaves, or flowers. The results substantiate the vast amount of variation present in a gene pool.

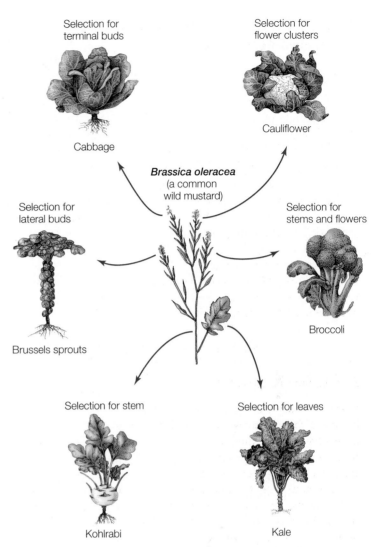

Different forms of a gene, known as **alleles**, may exist at a particular locus. At any particular locus, a single individual has only some of the alleles found in the population to which it belongs (**Figure 21.3**). The sum of all copies of all alleles at all loci found in a population constitutes its **gene pool**. (We can also refer to the "gene pool" for a particular locus or loci.) The gene pool is the source of the genetic variation that produces the phenotypic traits on which natural selection acts. To understand evolution and the role of natural selection, we need to know how much genetic variation populations have, what the sources of that genetic variation are, and how genetic variation changes in populations over space and time.

Most populations are genetically variable

Nearly all populations have genetic variation for many characters. Artificial selection on different characters in a single European species of wild mustard produced many important crop plants (**Figure 21.4**). Agriculturalists could achieve these results because the original mustard population had genetic variation for the characters of interest.

Laboratory experiments also demonstrate the existence of considerable genetic variation in populations. In one such experiment, investigators attempted to breed populations of the fruit fly *Drosophila melanogaster* with high or low numbers of bristles on their abdomens from an initial population with intermediate numbers of bristles. After 35 generations, all flies in both the high- and low-bristle lineages had bristle numbers that fell well outside the range found in the original population (**Figure 21.5**). Thus there must have been considerable genetic variation in the original fruit fly population on which selection could act.

Studying the genetic basis of natural selection is difficult because genotypes alone do not determine all phenotypes. With dominance, for example, a particular phenotype can be produced by more than one genotype (e.g., *AA* and *Aa* individuals may be phenotypically identical). Also, as we describe in Section 20.5, a given genotype can produce different phenotypes, depending on the environment encountered during development. For example, the cells of all the leaves on a tree or shrub are usually genetically identical, yet leaves of the same plant often differ in shape and size depending, for example, on the amount of ambient light they receive.

21.5 Artificial Selection Reveals Genetic Variation In experiments subjecting *Drosophila melanogaster* to artificial selection for bristle number, this trait evolved rapidly. The graphs show the number of flies with different numbers of bristles after 35 generations of artificial selection, which clearly diverged from the bristle numbers present in the original population (the blue bars in the center).

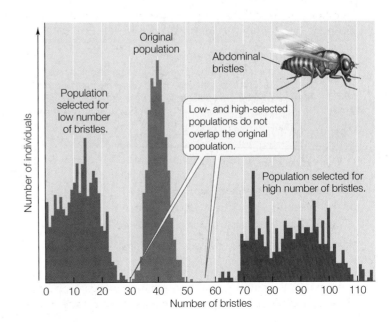

Evolutionary change can be measured by allele and genotype frequencies

Allele frequencies are usually estimated in locally interbreeding groups, called Mendelian populations, within a geographic population of a species. To measure allele frequencies in a Mendelian population precisely, we would need to count every allele at every locus in every individual in the population. By doing so, we could determine the frequencies of all alleles in the population. The word *frequency* in this case refers to an allele's proportion in the gene pool at a particular locus.

Fortunately, we do not need to make complete measurements, because we can reliably estimate allele frequencies for a given locus by counting alleles in a sample of individuals from the population. The sum of all allele frequencies at a locus is equal to 1, so measures of allele frequency range from 0 to 1.

An allele's frequency is calculated using the following formula:

$$p = \frac{\text{number of copies of the allele in the population}}{\text{sum of alleles in the population}}$$

If only two alleles (we'll call them A and a) for a given locus are found among the members of a diploid population, they may combine to form three different genotypes: AA, Aa, and aa. Such a population is said to be *polymorphic* at that locus, since there is more than one allele. Using the formula above, we can calculate the relative frequencies of alleles A and a in a population of N individuals as follows:

- Let N_{AA} be the number of individuals that are homozygous for the A allele (AA).
- Let N_{Aa} be the number that are heterozygous (Aa).
- Let N_{aa} be the number that are homozygous for the a allele (aa).

Note that $N_{AA} + N_{Aa} + N_{aa} = N$, the total number of individuals in the population, and that the total number of copies of both alleles present in the population is $2N$, because each individual is diploid. Each AA individual has two copies of the A allele, and each Aa individual has one copy of the A allele. Therefore, the total number of A alleles in the population is $2N_{AA} + N_{Aa}$. Similarly, the total number of a alleles in the population is $2N_{aa} + N_{Aa}$.

If p represents the frequency of A, and q represents the frequency of a, then

$$p = \frac{2N_{AA} + N_{Aa}}{2N}$$

and

$$q = \frac{2N_{aa} + N_{Aa}}{2N}$$

Figure 21.6 shows how these formulas can be used to calculate allele frequencies in two hypothetical populations, each containing 200 diploid individuals. Population 1 has mostly homozygotes (90 AA, 40 Aa, and 70 aa), whereas population 2 has mostly heterozygotes (45 AA, 130 Aa, and 25 aa).

The calculations in Figure 21.6 demonstrate two important points. First, notice that for each population, $p + q = 1$, which means that $q = 1 - p$. So when there are only two alleles at a given locus in a population, we can calculate the frequency of one allele and then obtain the second allele's frequency by subtraction. If there is only one allele at a given locus in a population, its frequency is 1: the population is then *monomorphic* at that locus, and the allele is said to be *fixed*.

The second thing to notice is that population 1 (consisting mostly of homozygotes) and population 2 (consisting mostly of heterozygotes) have the same allele frequencies for A and a. Thus they have the same gene pool for this locus. Because the alleles in the gene pool are distributed differently among individuals, however, the *genotype frequencies* of the two populations differ. Genotype frequencies are calculated as the number of individuals that have a given genotype divided by the total number of individuals in the population. Using the numbers in Figure 21.6, the genotype frequencies in population 1 would be 0.45 AA, 0.20 Aa, and 0.35 aa.

The frequencies of different alleles at each locus and the frequencies of different genotypes in a Mendelian population de-

TOOLS FOR INVESTIGATING LIFE

21.6   Calculating Allele Frequencies

Allele frequencies for any gene pool can be calculated using the equations in panel 1. When these equations are applied to the populations in panel 2, we find that the *frequencies* of alleles A and a in the two populations are the same, but the alleles are distributed differently between heterozygous and homozygous genotypes.

1 Determine the allele frequencies in the population.

In any population:

$$\text{Frequency of allele } A = p = \frac{2N_{AA} + N_{Aa}}{2N} \qquad \text{Frequency of allele } a = q = \frac{2N_{aa} + N_{Aa}}{2N}$$

where N is the total number of individuals in the population.

2 Compute allele frequencies for different populations.

For population 1 (mostly homozygotes):

$N_{AA} = 90$, $N_{Aa} = 40$, and $N_{aa} = 70$

so

$$p = \frac{180 + 40}{400} = 0.55$$

$$q = \frac{140 + 40}{400} = 0.45$$

For population 2 (mostly heterozygotes):

$N_{AA} = 45$, $N_{Aa} = 130$, and $N_{aa} = 25$

so

$$p = \frac{90 + 130}{400} = 0.55$$

$$q = \frac{50 + 130}{400} = 0.45$$

scribe that population's **genetic structure**. Allele frequencies measure the amount of genetic variation in a population; genotype frequencies show how a population's genetic variation is distributed among its members. Other measures, such as the proportion of polymorphic loci, are also used to measure variation in populations. With these measurements, it becomes possible to consider how the genetic structure of a population changes or remains the same over generations—that is, to measure evolutionary change.

The genetic structure of a population changes over time, unless certain restrictive conditions exist

In 1908, the British mathematician Godfrey Hardy and the German physician Wilhelm Weinberg independently deduced the conditions that must prevail if the genetic structure of a population is to remain the same over time. If the conditions they identified do not exist, then evolution will occur. The resulting principle, known as **Hardy–Weinberg equilibrium**, is a cornerstone of population genetics. Hardy–Weinberg equilibrium describes a model in which allele frequencies do not change across generations and genotype frequencies can be predicted from allele frequencies (**Figure 21.7**). The principles of Hardy–Weinberg equilibrium apply only to sexually reproducing organisms. Several conditions must be met for a population to be at Hardy–Weinberg equilibrium:

- *Mating is random.* Individuals do not preferentially choose mates with certain genotypes.

- *Population size is infinite.* The larger a population, the smaller will be the effect of **genetic drift**—random (chance) fluctuations in allele frequencies from one generation to another.

- *There is no* **gene flow** (movement of individuals into or out of the population, or reproductive contact with other populations).

- *There is no mutation.* There is no change to alleles in the population, and no new alleles are added to change the gene pool.

- *Selection does not affect the survival of particular genotypes.* There is no differential survival of individuals with different genotypes.

If these "ideal" conditions hold, two major consequences follow. First, the frequencies of alleles at a locus remain constant from generation to generation. Second, following one generation of random mating, the genotype frequencies occur in the following proportions:

Genotype	AA	Aa	aa
Frequency	p^2	$2pq$	q^2

Consider a population that is *not* in Hardy–Weinberg equilibrium, such as generation I of Figure 21.7. This could occur, for example, if the initial population is founded by migrants from several other populations, thus violating the Hardy–Weinberg assumption of no gene flow. In this example, "generation I" has more homozygous individuals and fewer heterozygous individuals than would be expected under Hardy–Weinberg equilibrium (a condition known as *heterozygote deficiency*).

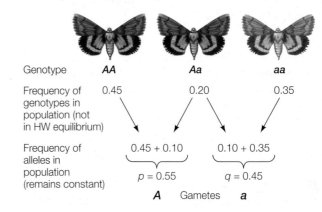

Generation I (Founder population)

Genotype	AA	Aa	aa
Frequency of genotypes in population (not in HW equilibrium)	0.45	0.20	0.35

Frequency of alleles in population (remains constant)

0.45 + 0.10   0.10 + 0.35
$p = 0.55$ $q = 0.45$

A Gametes a

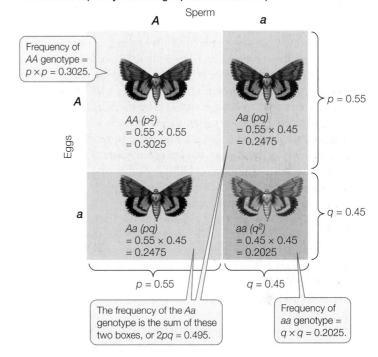

Generation II (Hardy–Weinberg equilibrium restored)

Sperm
A a

Eggs

Frequency of AA genotype = $p \times p = 0.3025$.

A

$AA\ (p^2)$
$= 0.55 \times 0.55$
$= 0.3025$

$Aa\ (pq)$
$= 0.55 \times 0.45$
$= 0.2475$

$p = 0.55$

a

$Aa\ (pq)$
$= 0.55 \times 0.45$
$= 0.2475$

$aa\ (q^2)$
$= 0.45 \times 0.45$
$= 0.2025$

$q = 0.45$

$p = 0.55$ $q = 0.45$

The frequency of the Aa genotype is the sum of these two boxes, or $2pq = 0.495$.

Frequency of aa genotype = $q \times q = 0.2025$.

21.7 One Generation of Random Mating Restores Hardy–Weinberg Equilibrium Generation I of this example population is founded by migrants from several source populations, and so is not initially in Hardy–Weinberg equilibrium. After one generation of random mating, the allele frequencies are unchanged and the genotype frequencies return to Hardy–Weinberg expectations. The length of the sides of each rectangle are proportional to the allele frequencies in the population; the areas of the rectangles are proportional to the genotype frequencies.

Even with a starting population that is not in Hardy–Weinberg equilibrium, we would predict that after a single generation of random mating, and without violating the other Hardy–Weinberg assumptions, the *allele frequencies* will remain unchanged but the *genotype frequencies* will return to Hardy–Weinberg expectations. Let's explore why this is true.

In generation I of Figure 21.7, the frequency of the A allele (p) is 0.55. Because we assume that individuals select mates at random, without regard to their genotype, gametes carrying A

or *a* combine at random—that is, as predicted by the allele frequencies of *p* and *q*. Thus in this example, the probability that a particular sperm or egg will bear an *A* allele is 0.55. In other words, 55 out of 100 randomly sampled sperm or eggs will bear an *A* allele. Because $q = 1 - p$, the probability that a sperm or egg will bear an *a* allele is $1 - 0.55 = 0.45$. (You may wish to review the discussion of probability in Section 12.1.)

To obtain the probability of two *A*-bearing gametes coming together at fertilization, we multiply the two independent probabilities of their occurring separately:

$$p \times p = p^2 = (0.55)^2 = 0.3025$$

Therefore, 0.3025, or 30.25 percent, of the offspring in generation II will have homozygous genotype *AA*. Similarly, the probability of bringing together two *a*-bearing gametes is

$$q \times q = q^2 = (0.45)^2 = 0.2025$$

Thus 20.25 percent of generation II will have the *aa* genotype.

There are two ways of producing a heterozygote: an *A* sperm may combine with an *a* egg, the probability of which is $p \times q$; or an *a* sperm may combine with an *A* egg, the probability of which is $q \times p$. Consequently, the overall probability of obtaining a heterozygote is 2*pq*, or 0.495. The frequencies of the *AA*, *Aa*, and *aa* genotypes in generation II of Figure 21.7 are now at Hardy–Weinberg expectations, and the frequencies of the two alleles (*p* and *q*) have not changed from generation I.

Under the assumptions of Hardy–Weinberg equilibrium, allele frequencies *p* and *q* remain constant for each generation. If Hardy–Weinberg assumptions are violated and the genotype frequencies in the parental generation are altered (say, by the loss of a large number of *AA* individuals from the population), then the allele frequencies in the next generation would be altered. However, based on the new allele frequencies, another generation of random mating is sufficient to restore the genotype frequencies to Hardy–Weinberg equilibrium.

---

yourBioPortal.com

GO TO **Animated Tutorial 21.2 • Hardy–Weinberg Equilibrium**

Deviations from Hardy–Weinberg equilibrium show that evolution is occurring

You probably have realized that populations in nature never meet the stringent conditions necessary to be at Hardy–Weinberg equilibrium. Why, then, is this model considered so important for the study of evolution? There are two reasons. First, the equation is often useful for predicting the approximate genotype frequencies of a population from its allele frequencies. Second—and crucially—the model describes the conditions required for there to be *no* evolution in a population.

Few if any of the Hardy–Weinberg model's conditions are ever met completely in real populations, and allele frequencies in all populations do in fact change through time—that is, populations *do* evolve. The specific patterns of deviation from Hardy–Weinberg equilibrium can help us identify the various mechanisms of evolutionary change.

We have briefly outlined Charles Darwin's vision of natural selection and adaptation and explained the mathematical basis of Hardy–Weinberg equilibrium and its importance for studying evolution. We'll now examine some of the forces that cause populations to deviate from equilibrium—the mechanisms of evolutionary change.

21.2 What Are the Mechanisms of Evolutionary Change?

Evolutionary mechanisms are forces that change the genetic structure of a population. Hardy–Weinberg equilibrium is a null hypothesis that assumes those forces are absent. The known evolutionary mechanisms include mutation, gene flow, genetic drift, nonrandom mating, and selection—each of which contradicts one of the five basic assumptions of Hardy–Weinberg equilibrium. We have already discussed Darwin's principal explanation for evolution, namely natural selection. Although natural selection is in many cases an important component of evolution, even Darwin recognized that it was not the only mechanism of evolution, and many additional evolutionary forces have been discovered since Darwin's time. Here we discuss some of the other mechanisms that result in evolution.

Mutations generate genetic variation

The origin of genetic variation is mutation. A **mutation**, as Section 14.6 describes, is any change in the nucleotide sequences of an organism's DNA. The process of DNA replication is not perfect, and changes appear almost every time a genome is replicated. Mutations occur randomly with respect to an organism's adaptive needs; it is selection acting on this random variation that results in adaptation. Most mutations are either harmful to their bearers or neutral. A few are beneficial, however, and previously harmful or neutral alleles may become advantageous if conditions change. In addition, mutations can restore to a population genetic variation that other evolutionary processes have

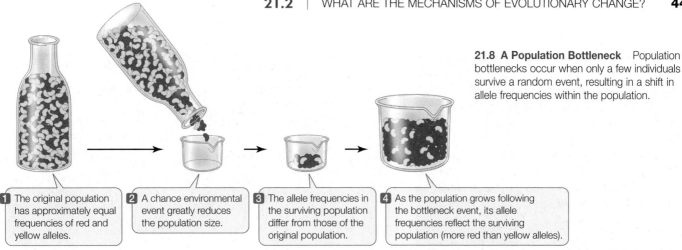

21.8 A Population Bottleneck Population bottlenecks occur when only a few individuals survive a random event, resulting in a shift in allele frequencies within the population.

1 The original population has approximately equal frequencies of red and yellow alleles.

2 A chance environmental event greatly reduces the population size.

3 The allele frequencies in the surviving population differ from those of the original population.

4 As the population grows following the bottleneck event, its allele frequencies reflect the surviving population (more red than yellow alleles).

removed. Thus mutations both create and help maintain genetic variation in populations.

Mutation rates can be high, as we saw with the influenza viruses described in the opening of this chapter, but in many organisms the mutation rate is very low (on the order of 10^{-8} to 10^{-9} changes per base pair of DNA per generation). Even low overall mutation rates, however, are sufficient to create considerable genetic variation, because each of a large number of genes may change, and populations often contain large numbers of individuals. For example, if the probability of a point mutation (an addition, deletion, or substitution of a single base) were 10^{-9} per base pair per generation, then each human gamete, the DNA of which contains 3×10^9 base pairs, would average three new point mutations ($3 \times 10^9 \times 10^{-9} = 3$)—and each zygote would carry an average of six new mutations. The current human population of about 7 billion people would be expected to carry about 42 billion new mutations that were not present one generation earlier. So even though the mutation rate in humans is quite low, human populations still contain enormous genetic variation on which selection can act.

One of the conditions for Hardy–Weinberg equilibrium is that there be no mutation. Although this condition is never strictly met, the rate at which mutations arise at a single locus is usually so low that mutations by themselves result in only small deviations from Hardy–Weinberg equilibrium. If large deviations are found, it is usually appropriate to dismiss mutation as the cause and to look for evidence of other evolutionary mechanisms acting on the population.

Gene flow may change allele frequencies

Few populations are completely isolated from other populations of the same species. Migration of individuals and movements of gametes between populations—a phenomenon called gene flow—can change allele frequencies in a population. If the arriving individuals survive and reproduce in their new location, they may add new alleles to the population's gene pool, or they may change the frequencies of alleles already present if they come from a population with different allele frequencies. For a population to be at Hardy–Weinberg equilibrium, there must be no gene flow from populations with different allele frequencies.

Genetic drift may cause large changes in small populations

In small populations, genetic drift—random changes in allele frequencies from one generation to the next—may produce large changes in allele frequencies over time. Harmful alleles may increase in frequency, and rare advantageous alleles may be lost. Even in large populations, genetic drift can influence the frequencies of alleles that do not affect the survival and reproductive rates of their bearers.

As an example, suppose we cross $Aa \times Aa$ fruit flies to produce an F_1 population in which $p = q = 0.5$ and in which the genotype frequencies are 0.25 AA, 0.50 Aa, and 0.25 aa. If we randomly select 4 individuals (= 8 copies of the gene) from the F_1 population to produce the F_2 generation, the allele frequencies in this small sample population may differ markedly from $p = q = 0.5$. If, for example, we happen by chance to draw 2 AA homozygotes and 2 heterozygotes (Aa), the allele frequencies in the sample will be $p = 0.75$ (6 out of 8) and $q = 0.25$ (2 out of 8). If we replicate this experiment 1,000 times, one of the two alleles will be missing entirely from about 8 of the 1,000 sample populations.

The same principles operate when a population is reduced dramatically in size. Populations that are normally large may occasionally pass through a period in which only a small number of individuals survive, a situation known as a **population bottleneck**. During population bottlenecks, genetic variation can be reduced by genetic drift. This is illustrated in **Figure 21.8**, in which red and yellow beans represent two different alleles of a gene. Most of the "surviving" beans in the small sample taken from the original population are, just by chance, red, so the new population has a much higher frequency of red beans than the previous generation had. In a real population, the allele frequencies would be described as having "drifted."

A population forced through a bottleneck is likely to lose much of its genetic variation. For example, when Europeans first arrived in North America, millions of greater prairie-chickens (*Tympanuchus cupido*) inhabited the prairies. As a result of hunting and habitat destruction by the new settlers, the Illinois population of this species plummeted from about 100 million birds in 1900 to fewer than 50 in the 1990s (**Figure 21.9A**). A comparison of DNA from birds collected in Illinois during the middle of the twentieth century with DNA from the surviving pop-

(A) *Tympanuchus cupido*

(B) *Washingtonia filifera*

21.9 Species with Low Genetic Variation (A) Greater prairie-chickens in Illinois lost most of their genetic variation when the population crashed from millions to fewer than 50 individuals. (B) The California fan palm, whose range has been reduced to a small area of southern California and neighboring Mexico, has little genetic variation.

ulation in the 1990s showed that Illinois prairie-chickens have lost most of their genetic diversity. The remaining population is experiencing low reproductive success. Similarly, the California fan palm (*Washingtonia filifera*) was once widespread in California and Mexico; today it is restricted to a few oases in extreme southern California and adjacent Mexico (**Figure 21.9B**). The species has little genetic variation: an average individual is heterozygous at fewer than 1 percent of its loci.

Genetic drift can have similar effects when a few pioneering individuals colonize a new region. Because of its small size, the colonizing population is unlikely to have all the alleles found among members of its source population. The resulting change in genetic variation, called a **founder effect**, is equivalent to that in a large population reduced by a bottleneck. For example, the current population of the pitcher plant *Sarracenia purpurea* on a small island in central Ohio arose from a single individual that was planted there in 1912. Today the population has only one detectable polymorphic locus in its entire genome.

Scientists were given an opportunity to study the genetic composition of founding populations when *Drosophila subobscura*, a well-studied species of fruit fly native to Europe, was discovered near Puerto Montt, Chile (in 1978), and at Port Townsend, Washington (in 1982). The *D. subobscura* founders probably reached Chile from Europe on a ship, and a few flies carried north from Chile on another ship founded the North American population. In both South and North America, populations of the flies grew rapidly and expanded their ranges. Today in North America, *D. subobscura* ranges from British Columbia to central California. In Chile it has spread across 15 degrees of latitude (**Figure 21.10**).

European populations of *D. subobscura* have 80 different *chromosomal inversions*, but the North and South American populations have only a subset of 20 of these inversions—and they are the same 20 on both continents. North and South American populations also have lower allele diversity at certain enzyme-producing genes compared with European populations. Only those alleles that have a frequency higher than 0.10 in European populations are present in the Americas. Thus, as expected for a small founding population, only a small part of the total genetic variation found in Europe reached the Americas. Geneticists estimate that somewhere between 4 and 100 flies founded the North and South American populations.

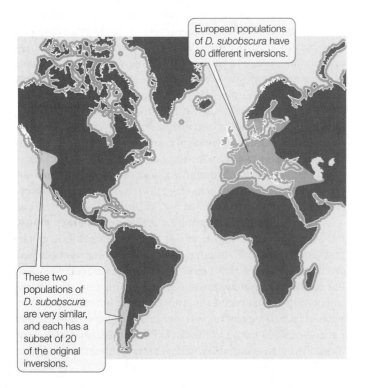

European populations of *D. subobscura* have 80 different inversions.

These two populations of *D. subobscura* are very similar, and each has a subset of 20 of the original inversions.

21.10 A Founder Effect Populations of the fruit fly *Drosophila subobscura* in North and South America contain less genetic variation than the European populations from which they came, as measured by the number of chromosome inversions in each population. Within two decades of arriving in the Americas, *D. subobscura* populations had increased dramatically and spread widely in spite of their reduced genetic variation.

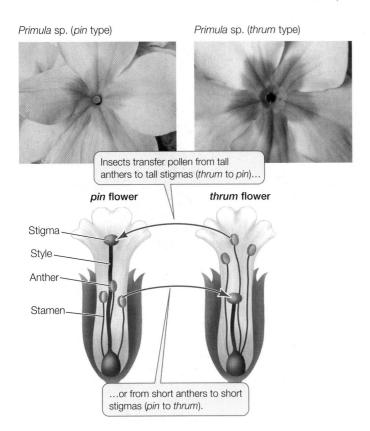

Primula sp. (*pin* type)   *Primula* sp. (*thrum* type)

Insects transfer pollen from tall anthers to tall stigmas (*thrum* to *pin*)…

pin flower   **thrum** flower

Stigma
Style
Anther
Stamen

…or from short anthers to short stigmas (*pin* to *thrum*).

21.11 Flower Structure Fosters Nonrandom Mating Differing floral structure within the same plant species, as illustrated by this primrose, ensures that pollination usually occurs between individuals of different genotypes.

Nonrandom mating can change genotype frequencies

Mating patterns may alter genotype frequencies if individuals in a population do not choose mates at random. For example, if they mate preferentially with individuals of the same genotype, then homozygous genotypes will be overrepresented and heterozygous genotypes underrepresented relative to Hardy–Weinberg expectations. Alternatively, individuals may mate primarily or exclusively with individuals of different genotypes.

Nonrandom mating is seen in some plant species, such as primroses (genus *Primula*), in which individual plants bear flowers of only one of two different types. One type, known as *pin*, has a long style (the stalk that supports the stigma, where pollen is received) and short stamens (the stalks ending in anthers, where pollen is produced). The other type, known as *thrum*, has a short style and long stamens (**Figure 21.11**). In many species with this reciprocal arrangement, pollen from one flower type can fertilize only flowers of the other type. Pollen grains from *pin* and *thrum* flowers are deposited on different parts of the bodies of insects that visit the flowers. When the insects visit other flowers, pollen grains from *pin* flowers are most likely to come into contact with stigmas of *thrum* flowers, and vice versa.

Self-fertilization (*selfing*), another form of nonrandom mating, is common in many groups of organisms, especially plants. Selfing reduces the frequencies of heterozygous individuals

from Hardy–Weinberg equilibrium and increases the frequencies of homozygotes, but it does not change allele frequencies.

Sexual selection is a particularly important form of nonrandom mating that *does* change allele frequencies and also often results in the evolution of significant differences between males and females of a species. We will discuss this important evolutionary mechanism in detail in the next section.

21.2 RECAP

Evolutionary mechanisms are processes that change the genetic structure of a population. Known evolutionary mechanisms include mutation, gene flow, genetic drift, nonrandom mating, and selection.

- Why do mutations, by themselves, result in only small deviations from Hardy–Weinberg equilibrium? See pp. 448–449

- Explain how genetic drift can cause large changes in small populations. See pp. 449–450 and Figure 21.8

- Why is it that some types of nonrandom mating alter genotype frequencies but not allele frequencies? See p. 451

The evolutionary mechanisms discussed so far influence the frequencies of alleles and genotypes in populations. Although all of these processes influence the course of biological evolution, only natural selection results in adaptation. For adaptation to occur, individuals that differ in heritable traits must survive and reproduce with different degrees of success.

21.3 How Does Natural Selection Result in Evolution?

Although evolution is defined as changes in the gene frequencies of a population from one generation to the next, natural selection acts on the *phenotype*—the physical features expressed by an organism with a given genotype—rather than directly on the genotype. The reproductive contribution of a phenotype to subsequent generations relative to the contributions of other phenotypes is called its **fitness**.

Changes in reproductive rate do not necessarily change the genetic structure of a population. For example, if all individuals in a population experience the same increase in reproductive rate (during an environmentally favorable year, for instance), the genetic structure of the population will not change. Changes in numbers of offspring are responsible for increases and decreases in the *size* of a population, but only changes in the *relative* success of different phenotypes in a population lead to changes in allele frequencies from one generation to the next. The fitness of individuals of a particular phenotype is a function of the probability of those individuals surviving multiplied by the average number of offspring they produce over their lifetimes. In other words, the *fitness of a phenotype is determined by the relative rates of survival and reproduction of individuals with that phenotype.*

Natural selection can change or stabilize populations

To simplify our discussion until now, we have considered only characters influenced by alleles at a single locus. As we describe in Section 12.3, however, most characters are influenced by alleles at more than one locus. Such characters are likely to show quantitative rather than qualitative variation. For example, the distribution of body sizes of individuals in a population, a character that is influenced by genes at many loci as well as by the environment, is likely to resemble the bell-shaped curves shown in the right-hand column of Figure 21.12.

Natural selection can act on characters with quantitative variation in any one of several different ways, producing quite different results:

● **Stabilizing selection** preserves the average characteristics of a population by favoring average individuals.

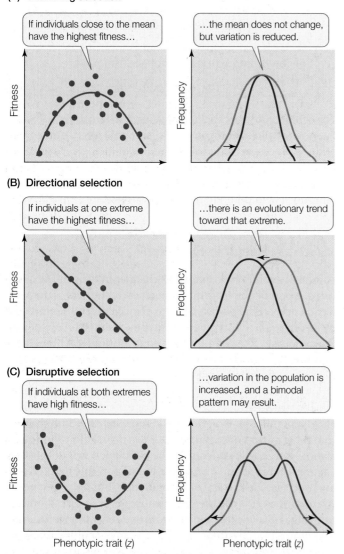

(A) **Stabilizing selection**

If individuals close to the mean have the highest fitness...

...the mean does not change, but variation is reduced.

Fitness

Frequency

(B) **Directional selection**

If individuals at one extreme have the highest fitness...

...there is an evolutionary trend toward that extreme.

Fitness

Frequency

(C) **Disruptive selection**

If individuals at both extremes have high fitness...

...variation in the population is increased, and a bimodal pattern may result.

Fitness

Frequency

Phenotypic trait (z)   Phenotypic trait (z)

21.12 Natural Selection Can Operate in Several Ways The graphs in the left-hand column show the fitness of individuals with different phenotypes of the same trait. The graphs on the right show the distribution of the phenotypes in the population before (light green) and after (dark green) the influence of selection.

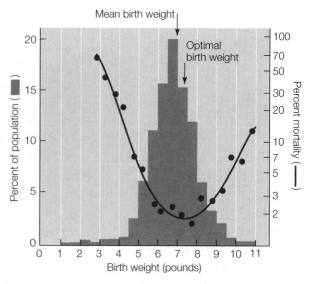

21.13 Human Birth Weight Is Influenced by Stabilizing Selection Babies that weigh more or less than average are more likely to die soon after birth than babies with weights close to the population mean.

● **Directional selection** changes the characteristics of a population by favoring individuals that vary in one direction from the mean of the population.

● **Disruptive selection** changes the characteristics of a population by favoring individuals that vary in both directions from the mean of the population.

STABILIZING SELECTION If the smallest and largest individuals in a population contribute fewer offspring to the next generation than do individuals closer to the average size, then stabilizing selection is operating on size (**Figure 21.12A**). Stabilizing selection reduces variation in populations, but it does not change the mean. Natural selection frequently acts in this way, countering increases in variation brought about by sexual recombination, mutation, or migration. Rates of evolution in many species are slow because natural selection is often stabilizing. Stabilizing selection operates, for example, on human birth weight. Babies born lighter or heavier than the population mean die at higher rates than babies whose weights are close to the mean (**Figure 21.13**). In discussions of specific genes, stabilizing selection is often called *purifying selection*, because there is selection against any deleterious mutations to the usual gene sequence.

DIRECTIONAL SELECTION Directional selection is operating when individuals at one extreme of a character distribution contribute more offspring to the next generation than other individuals do, shifting the average value of that character in the population toward that extreme. In the case of a single gene locus, directional selection may result in favoring a particular genetic variant (known as *positive selection* for that variant). By favoring one phenotype over another, directional selection results in an increase of the frequencies of alleles that produce the favored phenotype (as with the hemagglutinin gene of influenza in the opening of this chapter).

If directional selection operates over many generations, an *evolutionary trend* is seen in the population (**Figure 21.12B**). Directional evolutionary trends often continue for many genera-

21.14 Texas Longhorns Are the Result of Directional Selection Longer horns were advantageous for defending young calves from attacks by predators, so feral herds of Spanish cattle developed much longer horns between the early 1500s and the 1860s. The trend has been maintained in modern times by ranchers practicing artificial selection.

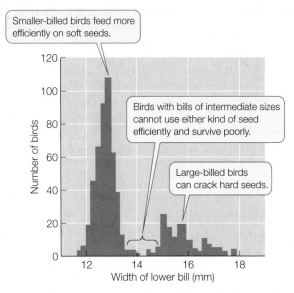

21.15 Disruptive Selection Results in a Bimodal Distribution The bimodal distribution of bill sizes in the black-bellied seedcracker of West Africa is a result of disruptive selection, which favors individuals with larger and smaller bill sizes over individuals with intermediate-sized bills.

tions, but they can be reversed if the environment changes and different phenotypes are favored, or halted when an optimal phenotype is reached or when trade-offs oppose further change. The character then falls under stabilizing selection.

Many cases of directional selection have been observed directly, and long-term examples abound in the fossil record. The long horns of Texas Longhorn cattle (**Figure 21.14**) are an example of a trait that has evolved through directional selection. Texas Longhorns are descendants of cattle that Christopher Columbus brought to the New World. Columbus picked up a few cattle in the Canary Islands and brought them to the island of Hispaniola in 1493. The cattle quickly multiplied, and their descendants were taken to the mainland of Mexico. As the Spanish explored what would later become Texas and the southwestern United States, they brought some of these cattle with them, some of which escaped and formed feral herds. Populations of these feral cattle increased greatly over the next few hundred years, but there was heavy predation from bears, mountain lions, and wolves, especially on the young calves. Cows with longer horns were more successful in protecting their calves against attacks, and over the next few hundred years the average horn length of cattle in the feral herds increased considerably. In addition, the cattle evolved resistance to endemic diseases of the Southwest, as well as higher fecundity and longevity. Texas Longhorn cows often live and produce calves well into their twenties, or about twice as long as many breeds of cattle that have been artificially selected by humans for traits such as high fat content or high milk production (which are examples of artificial directional selection).

DISRUPTIVE SELECTION When disruptive selection operates, individuals at opposite extremes of a character distribution contribute more offspring to the next generation than do individuals close to the mean, which increases variation in the population (**Figure 21.12C**).

The strikingly bimodal (two-peaked) distribution of bill sizes in the black-bellied seedcracker (*Pyrenestes ostrinus*), a West African finch (**Figure 21.15**), illustrates how disruptive selection can influence populations in nature. The seeds of two types of sedges (marsh plants) are the most abundant food source for these finches during part of the year. Birds with large bills can readily crack the hard seeds of the sedge *Scleria verrucosa*. Birds with small bills can crack *S. verrucosa* seeds only with difficulty; however, they feed more efficiently on the soft seeds of *S. goossensii* than do birds with larger bills.

Young finches whose bills deviate markedly from the two predominant bill sizes do not survive as well as finches whose bills are close to one of the two sizes represented by the distribution peaks. Because there are few abundant food sources in the environment, and because the seeds of the two sedges do not overlap in hardness, birds with intermediate-sized bills are less efficient in using either one of the principal food sources. Disruptive selection therefore maintains a bimodal bill size distribution.

Sexual selection influences reproductive success

Sexual selection acts on characteristics that determine reproductive success. In *The Origin of Species*, Darwin devoted only a few pages to sexual selection, but in 1871 he wrote an entire book about it: *The Descent of Man, and Selection in Relation to Sex*. Sexual selection was Darwin's explanation for the evolution of conspicuous characters that would appear to inhibit survival, such as bright colors, long tails, and elaborate courtship displays in males of many species. He hypothesized that these features either improved the ability of their bearers to compete for access to mates (*intrasexual selection*) or made their bearers more attractive to members of the opposite sex (*intersexual selection*). The concept of sexual selection was either ignored or questioned for many decades, but recent investigations have demonstrated its importance.

INVESTIGATING LIFE

21.16 Sexual Selection in Male Widowbirds

The extensive tail of the territorial male African long-tailed widowbird (*Euplectes progne*) actually inhibits its ability to fly. Darwin attributed the evolution of this trait to sexual selection. Behavioral ecologist Malte Andersson tested this hypothesis.

HYPOTHESIS Female widowbirds prefer to mate with the male that displays the longest tail; longer-tailed males thus are favored by sexual selection because they will father more offspring.

METHOD

1. Capture males and artificially lengthen or shorten tails by cutting or gluing on feathers. In a control group, cut and replace tails to their normal length (to control for the effects of tail-cutting).
2. Release the males to establish their territories and mate.
3. Count the nests with eggs or young on each male's territory.

RESULTS Male widowbirds with artificially shortened tails established and defended display sites sucessfully but fathered fewer offspring than did control or unmanipulated males. Males with artificially lengthened tales fathered the most offspring.

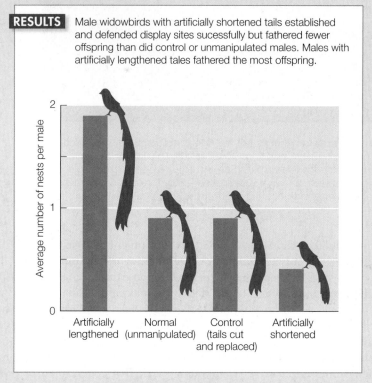

CONCLUSION Sexual selection in *Euplectes progne* has favored the evolution of long tails in the male.

Go to **yourBioPortal.com** for original citations, discussions, and relevant links for all INVESTIGATING LIFE figures.

Darwin devoted an entire book to sexual selection because he recognized that, whereas natural selection typically favors traits that enhance the survival of their bearers or their descendants, sexual selection is primarily about success in reproduction. Of course, an animal must survive to reproduce, but if it survives and fails to reproduce, it makes no contribution to the next generation. Thus sexual selection may favor traits that enhance an individual's chances of reproduction but reduce its chances of survival. For example, females may be more likely to see or hear males with a given trait (and thus be more likely to mate with those males), even though the favored trait may also increase the chances that the male will be seen or heard by a predator. In other cases, the sexual signal may indicate a successful genotype in the male. In many species of frogs, for example, females prefer males with low-frequency calls. Males' calls vary with body size, and a low-frequency call is indicative of a large-bodied frog. Frogs exhibit indeterminate growth— that is, they continue to grow indefinitely—so a large frog is a long-lived frog, which indicates high survivorship. In this case, the sexual signal represents what is known as an *honest signal* of the male's ability to survive in the local environment.

One example of a trait that Darwin attributed to sexual selection is the remarkable tail of the male African long-tailed widowbird (*Euplectes progne*), which is longer than the bird's head and body combined. Male widowbirds normally select, and defend from other males, a territory where they perform courtship displays to attract females. To investigate whether sexual selection drove the evolution of widowbird tails, Malte Andersson, a behavioral ecologist at Gothenburg University, Sweden, clipped the tails of some captured male widowbirds and lengthened the tails of others by gluing on additional feathers. He then cut and reglued the tail feathers of still other males, which served as controls. Both short- and long-tailed males successfully defended their display territories, indicating that a long tail does not confer an advantage in male–male competition. However, males with artificially elongated tails attracted about four times more females than did males with shortened tails (**Figure 21.16**).

Why do female widowbirds prefer males with long tails? One possibility is that ability to grow and maintain a costly feature such as a long tail may indicate that the male bearing it is vigorous and healthy, even though the tail impairs the bird's ability to fly. If so, then females that are attracted to long tails are in-

INVESTIGATING LIFE

21.17 Do Bright Bills Signal Good Health?

Female zebra finches (*Taeniopygia guttata*) preferentially choose mates with the brightest bill color. Does this preference increase their chances of mating with the healthiest males? This experiment made use of carotenoids (antioxidant pigment molecules believed to boost immune response) to test the hypothesis.

HYPOTHESIS The brightness of a male zebra finch's red bill is correlated with the strength of the bird's immune response and a corresponding likelihood of good health.

Taeniopygia guttata

METHOD
1. Provide carotenoids in the drinking water of experimental, but not control, males.
2. Challenge all males immunologically and measure responses.

RESULTS Experimental males responded more strongly to the immunological challenge. They also developed brighter bills than control males.

CONCLUSION Bill color is an indication of immunological strength and general health.

FURTHER INVESTIGATION: How would you test this same hypothesis in the field? What would constitute experimental and control birds?

Go to **yourBioPortal.com** for original citations, discussions, and relevant links for all INVESTIGATING LIFE figures.

directly attracted to vigorous, healthy males, which likely carry beneficial genes that would lead to higher survivorship of offspring. Although the manipulated males in Andersson's investigation did not have to pay the price of growing and supporting (except briefly) artificially long tails, the hypothesis that having well-developed ornamental traits signals vigor and health has been tested experimentally with captive zebra finches.

The bright red bills of male zebra finches (*Taeniopygia guttata*) are the result of red and yellow carotenoid pigments. Zebra finches (and most other animals) cannot synthesize carotenoids and must obtain them from their food. In addition to influencing bill color, carotenoids are antioxidants and components of the immune system. Males in good health may need to allocate fewer carotenoids to immune function than males in poorer health. If so, then females can use the brightness of a male's bill to assess his health.

Tim Birkhead and his colleagues at Sheffield University manipulated blood levels of carotenoids in genetically similar male zebra finches by giving experimental males drinking water with added carotenoids; they gave control males only distilled water. All the males had access to the same food. After one month, the experimental males had higher levels of carotenoids in their blood, had much brighter bills than the control males, and were preferred by female zebra finches.

Next, the investigators challenged both groups of males immunologically by injecting phytohemagglutinin (PHA) into their wings. PHA induces a response by T lymphocytes, a type of white blood cell that functions in the immune system to recognize and deactivate foreign substances (see Chapter 42). The injection results in an accumulation of white blood cells and thus a thickening of the skin at the injection site. Experimental males with enhanced carotenoid levels developed thicker skins because they responded more strongly to PHA than control males did, indicating that higher carotenoid levels are associated with stronger immune systems (**Figure 21.17**).

This experiment showed that when a female chooses a male with a bright red bill, she probably gets a mate with a healthy immune system. Such males are less likely to become infected with parasites and diseases, and are better able to assist with parental care.

21.3 RECAP

Variation in genotype can lead to variation in fitness. Fitness refers to the relative reproductive contribution of a phenotype to subsequent generations. Natural and sexual selection can both change and stabilize phenotypes within populations.

- Explain why natural selection that acts on a phenotype results in changes in genotype frequencies. See p. 451

- Describe the differences between stabilizing, directional, and disruptive selection, giving examples of each. See pp. 452–453 and Figure 21.12

- Why did Darwin devote an entire book to sexual selection? See pp. 453–455

Genetic drift, stabilizing selection, and directional selection all tend to reduce genetic variation within populations. Nevertheless, as we have seen, most populations harbor considerable genetic variation. What processes produce and maintain genetic variation within populations?

21.4 How Is Genetic Variation Maintained within Populations?

Genetic variation is the raw material on which mechanisms of evolution act. In this section we will discuss several processes—neutral mutations, sexual recombination, frequency-dependent selection, and heterozygote advantage—that operate to maintain genetic variation in populations, despite the action of other forces (such as genetic drift and many types of selection) that reduce variation. We will also show how genetic variation may be maintained over geographic space.

Neutral mutations accumulate in populations

As we discuss in Section 14.6, some mutations do not affect the function of the proteins encoded by the mutated genes. An allele that does not affect the fitness of an organism—that is, an allele that is no better or worse than alternative alleles at the same locus—is called a **neutral allele**. Neutral alleles are unaffected by natural selection. Even in large populations, neutral alleles may be lost or may increase in frequency, purely by genetic drift. Neutral alleles are added to a population over time through mutation, providing the population with considerable genetic variation.

Much of the phenotypic variation we are able to observe is not neutral. However, modern techniques enable us to measure neutral variation at the molecular level and provide the means to distinguish it from adaptive variation. Section 24.2 discusses how variation in neutral molecular traits can be used to study divergence among genes, populations, and species.

Sexual recombination amplifies the number of possible genotypes

In asexually reproducing organisms, each new individual is genetically identical to its parent unless there has been a mutation. When organisms reproduce sexually, however, offspring differ from their parents because of crossing over and independent assortment of chromosomes during meiosis as well as the combination of genetic material from two different gametes, as described in Chapter 11. Sexual recombination generates an endless variety of genotypic combinations that increase the evolutionary potential of populations—a long-term advantage of sex. Although many species may reproduce asexually most of the time, few are strictly asexual over long periods of evolutionary time. Almost all have some means of achieving genetic recombination.

The evolution of the mechanisms of meiosis and sexual recombination were crucial events in the history of life. Exactly how these attributes arose is puzzling, however, because sex has at least three striking disadvantages in the short term:

- Recombination breaks up adaptive combinations of genes.
- Sex reduces the rate at which females pass genes on to their offspring.
- Dividing offspring into separate genders greatly reduces the overall reproductive rate.

To see why this last disadvantage exists, consider an asexual female that produces the same number of offspring as a sexual female. Let's assume that both females produce two offspring, but that the sexual female produces 50 percent males. In the next generation, both asexual F_1 females will produce two more offspring, but there is only one sexual F_1 female to produce offspring. Thus, the effective reproductive rate of the asexual lineage is twice that of the sexual lineage. The evolutionary problem is to identify the advantages of sex that can overcome such short-term disadvantages.

A number of hypotheses have been proposed for the existence of sex, none of which are mutually exclusive. One is that sexual recombination facilitates repair of damaged DNA, because breaks and other errors in DNA on one chromosome can be repaired by copying the intact sequence from the homologous chromosome.

Another advantage of sexual reproduction is that it permits the elimination of deleterious mutations. As Section 13.4 describes, DNA replication is not perfect. Errors are introduced in every generation, and many or most of these errors result in lower fitness. Asexual organisms have no mechanism to eliminate deleterious mutations. Hermann J. Muller noted that the accumulation of deleterious mutations in a non-recombining genome is like a genetic ratchet. The mutations accumulate—"ratchet up"—at each replication: that is, a mutation occurs and is passed on when the genome replicates, then two new mutations occur in the next replication, so three mutations are passed on, and so on. Deleterious mutations cannot be eliminated except by the death of the lineage or a rare back mutation. This accumulation of deleterious mutations in lineages that lack genetic recombination is known as *Muller's ratchet*.

In sexual species, on the other hand, genetic recombination produces some individuals with more of these deleterious mutations and some with fewer. The individuals with fewer deleterious mutations are more likely to survive. Therefore, sexual reproduction allows natural selection to eliminate particular deleterious mutations from the population over time.

Another explanation for the existence of sex is that the great variety of genetic combinations created in each generation may be advantageous. For example, genetic variation can be a defense against pathogens and parasites. Most pathogens and parasites have much shorter life cycles than their hosts and can rapidly evolve counteradaptations to host defenses. Sexual recombination might give the host's defenses a chance to keep up.

Sexual recombination does not directly influence the frequencies of alleles; rather, *it generates new combinations of alleles on which natural selection can act*. It expands variation in a character influenced by alleles at many loci by creating new genotypes. That is why artificial selection for bristle number in *Drosophila* (see Figure 21.5) resulted in flies that had either more or fewer bristles than the flies in the initial population had.

Frequency-dependent selection maintains genetic variation within populations

Natural selection often preserves variation as a polymorphism (two or more variants of a trait present in the same population). A polymorphism may be maintained when the fitness of a given phenotype depends on its frequency in a population, a phenomenon known as **frequency-dependent selection**.

A small fish that lives in Lake Tanganyika in East Africa provides an example of frequency-dependent selection. Because of an asymmetrical jaw joint, the mouth of this scale-eating fish, *Perissodus microlepis*, opens either to the right or to the left; the direction is genetically determined (**Figure 21.18**). The scale-eater approaches its prey (another fish) from behind and dashes in to bite off several scales from its flank. "Right-mouthed" individuals always attack from the victim's left, and "left-mouthed" individuals always attack from the victim's right. The distorted mouth enlarges the area of teeth in contact with the prey's flank, but only if the scale-eater attacks from the appropriate side.

Prey fish are alert to approaching scale-eaters, so attacks are more likely to be successful if the prey must watch both flanks. Vigilance by prey thus favors equal numbers of right- and left-mouthed scale-eaters, because if attacks from one side were more common than the other, prey fish would pay more attention to potential attacks from that side. Over an 11-year study of this fish in Lake Tanganyika, the polymorphism was found to be stable: the two forms of *P. microlepis* remained at about equal frequencies.

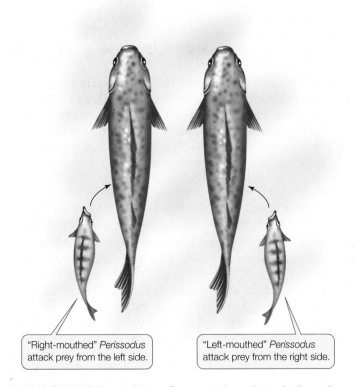

"Right-mouthed" *Perissodus* attack prey from the left side.

"Left-mouthed" *Perissodus* attack prey from the right side.

21.18 A Stable Polymorphism Frequency-dependent selection maintains equal proportions of left- and right-mouthed individuals of the scale-eating fish *Perissodus microlepis*.

Heterozygote advantage maintains polymorphic loci

In many cases, different alleles for a particular gene have advantages under different environmental conditions. Most organisms, however, experience a wide diversity of environments over time. A night is dramatically different from the preceding day. A cold, cloudy day differs from a clear, hot one. Day length and temperature change seasonally. For many genes, a single allele is unlikely to perform well under all these conditions. In such situations, a heterozygous individual (with two different alleles) is likely to outperform individuals that are homozygous for either one of the alleles.

Colias butterflies of the Rocky Mountains live in environments where dawn temperatures often are too cold, and afternoon temperatures too hot, for the butterflies to fly. Populations of this butterfly are polymorphic for the enzyme phosphoglucose isomerase (PGI), which influences how well the butterfly flies at different temperatures. Butterflies with certain PGI genotypes can fly better during the cold hours of early morning; others perform better during midday heat. The optimal body temperature for flight is 35°C to 39°C, but some butterflies can fly with body temperatures as low as 29°C or as high as 40°C. During spells of unusually hot weather, heat-tolerant genotypes are favored; during spells of unusually cool weather, cold-tolerant genotypes are favored.

Heterozygous *Colias* butterflies can fly over a greater temperature range than homozygous individuals, which should give them an advantage in foraging and finding mates. A test of this prediction did find a mating advantage in heterozygous males, and further, that this advantage maintains the polymorphism in the population (**Figure 21.19**). Of course, the heterozygotes can never become fixed in the population, because the offspring of two heterozygotes will include both classes of homozygotes in addition to heterozygotes.

Much genetic variation in species is maintained in geographically distinct populations

Much of the genetic variation in species is preserved as differences among members living in different places (populations). Populations often vary genetically because they are subjected to different selective pressures in different environments. Environments may vary significantly over short distances. For example, in the Northern Hemisphere, temperature and soil moisture differ dramatically between north- and south-facing mountain slopes. In the Rocky Mountains of Colorado, the proportion of ponderosa pines (*Pinus ponderosa*) that are heterozygous for a particular peroxidase enzyme is particularly high on south-facing slopes, where temperatures fluctuate dramatically, often on a daily basis. This heterozygous genotype performs well over a broad range of temperatures. On north-facing slopes and at higher elevations, where temperatures are cooler and fluctuate less strikingly, a peroxidase homozygote, which has a lower optimal temperature, is much more frequent.

Plant species may also vary geographically in the chemicals they synthesize to defend themselves against herbivores. Some individuals of the white clover (*Trifolium repens*) pro-

INVESTIGATING LIFE

21.19 A Heterozygote Mating Advantage

Among butterflies of the genus *Colias*, males that are heterozygous for two alleles of the PGI enzyme can fly farther under a broader range of temperatures than males that are homozygous for either allele. Does this ability give heterozygous males a mating advantage?

HYPOTHESIS   Heterozygous male *Colias* will have proportionally greater mating success than homozygous males.

METHOD
1. For each of two *Colias* species, capture butterflies in the field. In the laboratory, determine their genotypes and allow them to mate.
2. Determine the genotypes of the offspring, thus revealing paternity and mating success of the males.

RESULTS   For both species, the proportion of heterozygous males that mated successfully was higher than the proportion of all males seeking females ("flying").

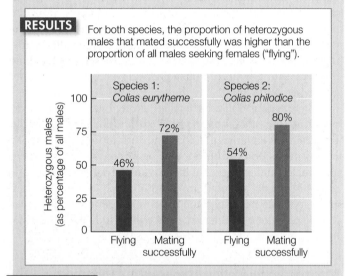

CONCLUSION   Heterozygous *Colias* males have a mating advantage over homozygous males.

Go to **yourBioPortal.com** for original citations, discussions, and relevant links for all INVESTIGATING LIFE figures.

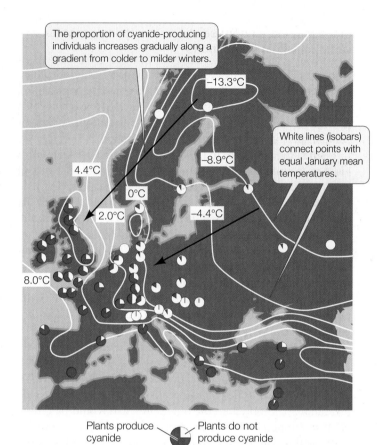

The proportion of cyanide-producing individuals increases gradually along a gradient from colder to milder winters.

White lines (isobars) connect points with equal January mean temperatures.

Plants produce cyanide   Plants do not produce cyanide

21.20 Geographic Variation in a Defensive Chemical   The frequency of cyanide-producing individuals in European populations of white clover (*Trifolium repens*) depends on winter temperatures.

duce the poisonous chemical cyanide. Poisonous individuals are less appealing to herbivores—particularly mice and slugs—than are nonpoisonous individuals. However, clover plants that produce cyanide are more likely to be killed by frost, because freezing damages cell membranes and releases cyanide into the plant's own tissues.

In European populations of *Trifolium repens*, the frequency of cyanide-producing individuals increases gradually from north to south and from east to west (**Figure 21.20**). This gradual change in phenotype across a geographic gradient is known as **clinal variation**. In this cline, poisonous plants make up a large proportion of clover populations only in areas where winters are mild. Cyanide-producing individuals are rare where winters are cold, even though herbivores graze clovers heavily in those areas.

21.4 RECAP

Neutral mutations, sexual recombination, frequency-dependent selection, and heterozygote advantage all act to maintain considerable genetic variation in most populations. Variation within species is also maintained among geographically distinct, genetically variable populations.

- Do you understand why sexual reproduction is so prevalent in nature, despite its having at least three short-term evolutionary disadvantages? See p. 456

- How does frequency-dependent selection act to maintain genetic variation in a population? See p. 457

The mechanisms of evolution have produced a remarkable variety of organisms, some of which are adapted to most environments on Earth. This natural variation, and the success of breeders attempting to produce desired traits in domesticated plants and animals, suggests that evolution can produce a wide variety of adaptive traits. But are there limits to the adaptations evolution can produce?

21.5 What Are the Constraints on Evolution?

We would be mistaken to assume that evolutionary mechanisms can produce any trait we might imagine. Evolution is constrained in many ways. Lack of appropriate genetic variation, for example, prevents the development of many potentially favorable traits. If the allele for a given trait does not exist in a population, that trait cannot evolve even if it would be highly favored by natural selection. Most possible combinations of genes and genotypes have never existed in any population, and so have never been tested under natural selection.

Constraints are imposed on organisms by the dictates of physics and chemistry. The size of cells, for example, is constrained by the stringencies of surface area-to-volume ratios (see Section 2.1). The ways in which proteins can fold are limited by the bonding capacities of their constituent molecules (see Section 3.2). And the energy transfers that fuel life must operate within the laws of thermodynamics (see Section 8.1). Keep in mind that evolution works within the boundaries of these universal constraints, as well as the constraints described here.

Developmental processes constrain evolution

As Section 20.5 explained, developmental constraints on evolution are paramount because *all evolutionary innovations are modifications of previously existing structures.* Human engineers seeking to power an airplane can start "from scratch" to design a completely new type of engine (powered by jet propulsion), to replace the previous type (powered by propellers). Evolutionary changes, however, cannot happen in this way. Current phenotypes of organisms are constrained by historical conditions and past selective pressures.

A striking example of such developmental constraints is provided by the evolution of fish that spend most of their time on the sea bottom. One lineage, the bottom-dwelling skates and rays, share a common ancestor with sharks, whose bodies were already somewhat ventrally flattened and whose skeletal frame is made of flexible cartilage. Skates and rays evolved a body type that further flattened their bellies, allowing them to swim along the ocean floor (**Figure 21.21A**).

By contrast, plaice, sole, and flounder are bottom-dwelling descendants of deep-bellied, laterally flattened ancestors with bony skeletons. The only way these fishes can lie flat is to flop over on their sides. Their ability to swim is thus curtailed, but their bodies can lie still and are well camouflaged. During development, one eye of these flatfishes moves so that both eyes are positioned on the same side of the body (**Figure 21.21B**). Such shifts in eye position have evolved several times, and shifts have happened in both directions (that is, both left- and right-eyed flatfishes have evolved independently). Small shifts in the position of one eye probably helped ancestral flatfishes see better, resulting in the flat body forms found today. This path to producing a flattened body may not be optimal, but the fishes' developmental capabilities constrain the pathways that evolution can take.

Trade-offs constrain evolution

Adaptations frequently impose both fitness costs and benefits. For an adaptation to evolve, the fitness benefits it confers must exceed the fitness costs it imposes—in other words, the **trade-off** must be worthwhile. For example, there are metabolic costs associated with developing and maintaining certain conspicuous features (such as antlers or horns) that males use to compete with other males for access to females. The fact that these features are common in many species suggests that the benefits derived from possessing them must outweigh the costs.

As a result of trade-offs, many traits that are adaptive in one context may be maladaptive in another. Consider the rough-skinned newt, *Taricha granulosa*, and one of its predators, the common garter snake, *Thamnophis sirtalis* (**Figure 21.22A**). The newt sequesters in its skin a potent neurotoxin called tetrodotoxin (TTX). TTX paralyzes nerves and muscles by blocking

(A) *Taeniura lymma*

(B) *Bothus lunatus*

21.21 Two Solutions to a Single Problem (A) This stingray, whose ancestors were dorsoventrally flattened, lies on its belly. Stingrays' bodies are symmetrical around the dorsal backbone. (B) This flounder, whose ancestors were laterally flattened, lies on its side. (The backbone of this individual is at the right.) Flounders' eyes migrate during development so that both are on the same side of the body.

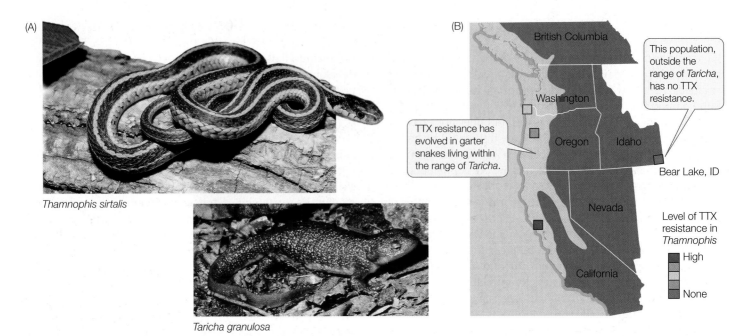

21.22 Resistance to a Toxin Comes at a Cost (A) Garter snakes (above) prey on newts (below). Rough-skinned newts counter with the ability to sequester a neurotoxin, TTX, in their skin. In turn, TTX-resistant sodium channels have evolved in some snake populations, allowing the snakes to eat toxic prey but resulting in slower movement by the snakes. (B) High resistance to TTX in garter snakes is only found in regions where snake and newt populations overlap (tan area).

───────── **yourBioPortal.com** ─────────
GO TO **Animated Tutorial 21.3** •
Assessing the Costs of Adaptation

sodium channels (see Section 6.3). Most vertebrates—including many garter snakes—will die if they eat a rough-skinned newt. But some snakes can eat rough-skinned newts and survive. In some populations of garter snakes, TTX-resistant sodium channels have evolved in the nerves and muscles (see Chapter 24 for another example of the evolution of sodium channels). However, the snakes pay a price for this attribute. For several hours after eating a newt, TTX-resistant snakes can move only slowly, and they never move as fast as nonresistant snakes. Thus resistant snakes are more vulnerable to their own predators than are TTX-sensitive snakes that simply don't encounter poisonous newts. Therefore, there is selection against TTX-resistant sodium channels in populations of garter snakes that occur outside the range of rough-skinned newts, but selection for TTX-resistance in many areas where newts are present (**Figure 21.22B**).

Short-term and long-term evolutionary outcomes sometimes differ

The short-term changes in allele frequencies within populations that we have emphasized in this chapter are an important focus of study for evolutionary biologists. These changes can be observed directly, they can be manipulated experimentally, and they demonstrate the actual processes by which evolution occurs. By themselves, however, they do not enable us to predict long-term evolutionary changes.

Long-term patterns of evolutionary change can be strongly influenced by events that occur so infrequently (a meteorite impact, for example) or so slowly (continental drift) that they are unlikely to be observed during short-term studies. The ways in which evolutionary processes act may change over time with changing environmental conditions. Even among the descendants of a single ancestral species, different lineages may evolve in different directions. Therefore, additional types of evidence, demonstrating the effects of rare and unusual events on trends in the fossil record, must be gathered if we wish to understand the course of evolution over billions of years.

21.5 RECAP

Developmental processes constrain evolution because all evolutionary innovations are modifications of previously existing structures. An adaptation can evolve only if the fitness benefits it confers exceed the fitness costs it imposes.

● Describe an example of an evolutionary trade-off in which the advantages of an adaptation outweigh its costs in the long run. See pp. 459–460

● Do you see why the presence of a great deal of genetic variation within a population could increase the chances that some members of the population would survive an unprecedented environmental change? Do you also understand why there is no guarantee that this would be the case?

CHAPTER SUMMARY

21.1 What Facts Form the Base of Our Understanding of Evolution?

- Charles Darwin attributed changes in species over time to the possession of advantageous traits by some individuals. He understood that it is not individuals that evolve but **populations**. A population evolves when individuals with favorable **heritable traits** survive and reproduce at higher rates than other members of the population.

- **Adaptation** refers both to characteristics of organisms and the way those characteristics are acquired via **natural selection**. ANIMATED TUTORIAL 21.1

- The sum of all copies of all alleles at all loci found in a population constitutes its **gene pool** and represents the genetic variation that results in different phenotypic traits on which natural selection can act. Review Figure 21.3

- **Artificial selection** and laboratory experiments demonstrate the existence of considerable genetic variation in most populations. Review Figure 21.5

- Allele frequencies measure the amount of genetic variation in a population; genotype frequencies show how a population's genetic variation is distributed among its members. Together, allele and genotype frequencies describe a population's **genetic structure**. Review Figure 21.6

- **Hardy–Weinberg equilibrium** predicts the allele frequencies in populations in the absence of evolution. Deviation from these frequencies indicates the work of evolutionary mechanisms. Review Figure 21.7, ANIMATED TUTORIAL 21.2

21.2 What Are the Mechanisms of Evolutionary Change?

- **Mutation** provides new genetic variants; favored variants increase in populations through natural selection.

- Migration or mating of individuals between populations results in **gene flow**.

- In small populations, **genetic drift**—the random loss of individuals and the alleles they possess—may produce large changes in allele frequencies from one generation to the next and greatly reduce genetic variation. Review Figure 21.8

- **Population bottlenecks** occur when only a few individuals survive a random event, resulting in a drastic shift in allele frequencies within the population and the loss of variation. Similarly, a population established by a small number of individuals colonizing a new region may lose variation via a **founder effect**.

- Nonrandom mating may result in genotype frequencies that deviate from Hardy–Weinberg equilibrium.

21.3 How Does Natural Selection Result in Evolution?

- **Fitness** is the reproductive contribution of a phenotype to subsequent generations relative to the contributions of other phenotypes.

- Changes in numbers of offspring are responsible for changes in the absolute size of a population, but only changes in the relative success of different phenotypes within a population lead to changes in allele frequencies.

- Natural selection can act on variable traits in several different ways, resulting in **stabilizing**, **directional**, or **disruptive selection**. Review Figure 21.12

- **Sexual selection** primarily affects success in reproduction, rather than success in survival. Review Figures 21.16 and 21.17

21.4 How Is Genetic Variation Maintained within Populations?

- Neutral mutations, sexual recombination, frequency-dependent selection, and heterozygote advantage can all maintain genetic variation within populations.

- **Neutral alleles** do not affect the fitness of an organism, are not affected by natural selection, and may accumulate or be lost by genetic drift.

- Despite short-term disadvantages, sexual reproduction generates countless genotypic combinations that increase the evolutionary potential and survivorship of populations.

- A polymorphism may be maintained by **frequency-dependent selection** when the fitness of a genotype depends on its frequency in a population.

- Genetic variation within species may be maintained by the existence of genetically distinct populations over geographic space. A gradual change in phenotype across a geographic gradient is known as **clinal variation**. Review Figure 21.20

21.5 What Are the Constraints on Evolution?

- Developmental processes constrain evolution because all evolutionary innovations are modifications of previously existing structures.

- Most adaptations impose costs. An adaptation can evolve only if the benefits it confers exceed the costs it imposes, a situation that leads to **trade-offs**. Review Figure 21.22, ANIMATED TUTORIAL 21.3

SELF-QUIZ

1. Long-horned cattle have greater difficulty moving through heavily forested areas compared with cattle that have short or no horns, but long-horned cattle are better able to defend their young against predators. This contrast is an example of
   a. an adaptation.
   b. genetic drift.
   c. natural selection.
   d. a trade-off.
   e. none of the above

2. Which of the following is true?
   a. Darwin and Wallace were both influenced by Malthus.
   b. Wallace proposed a theory of evolution by natural selection that was similar to Darwin's.
   c. Malthus claimed that because human population growth would outstrip any increases in food production, famine was a likely result.
   d. Darwin realized that all populations had the capacity to rapidly increase in numbers.
   e. All of the above

3. The phenotype of an organism is
   a. the type specimen of its species in a museum.
   b. its genetic constitution, which governs its traits.
   c. the chronological expression of its genes.
   d. the physical expression of its genotype.
   e. its adult form.

4. The appropriate unit for defining and measuring genetic variation is the
   a. cell.
   b. individual.
   c. population.
   d. community.
   e. ecosystem.

5. Which statement about allele frequencies is *not* true?
   a. The sum of all allele frequencies at a locus is always 1.
   b. If there are two alleles at a locus and we know the frequency of one of them, we can obtain the frequency of the other by subtraction.
   c. If an allele is missing from a population, its frequency in that population is 0.
   d. If two populations have the same allele frequencies at a locus, they must have the same proportion of homozygotes at that locus.
   e. If there is only one allele at a locus, its frequency is 1.

6. Which of the following is *not* required for a population at Hardy–Weinberg equilibrium?
   a. There is no migration between populations.
   b. Natural selection is not acting on the alleles in the population.
   c. Mating is random.
   d. Multiple alleles must be present at every locus.
   e. All of the above.

7. The fitness of a genotype is a function of the
   a. average rates of survival and reproduction of individuals with that genotype.
   b. individuals that have the highest rates of both survival and reproduction.

c. individuals that have the highest rates of survival.
d. individuals that have the highest rates of reproduction.
e. average reproductive rate of individuals with that genotype.

8. Laboratory selection experiments with fruit flies have demonstrated that
   a. bristle number is not genetically controlled.
   b. bristle number is not genetically controlled, but changes in bristle number are caused by the environment in which the fly is raised.
   c. bristle number is genetically controlled, but there is little variation on which natural selection can act.
   d. bristle number is genetically controlled, but selection cannot result in flies having more bristles than any individual in the original population had.
   e. bristle number is genetically controlled, and selection can result in flies having more, or fewer, bristles than any individual in the original population had.

9. Disruptive selection maintains a bimodal distribution of bill size in the West African seedcracker because
   a. bills of intermediate shapes are difficult to form.
   b. the birds' two major food sources differ markedly in size and hardness.
   c. males use their large bills in displays.
   d. migrants introduce different bill sizes into the population each year.
   e. older birds need larger bills than younger birds.

10. Which of the following is *not* a reason why trade-offs constrain evolution?
    a. Most adaptations impose both fitness costs and benefits.
    b. Structures such a horns and antlers are metabolically costly to produce, but result in more reproduction by the males that possess them.
    c. Changes in allele frequencies may be influenced by chance events.
    d. Ability to consume toxic prey may reduce mobility.
    e. Adaptations can evolve only if the fitness benefits they confer exceed the costs they impose.

FOR DISCUSSION

1. In what ways does artificial selection by humans differ from natural selection? Was Darwin wise to base so much of his argument for natural selection on the results of artificial selection?

2. In nature, mating among individuals in a population is never truly random, immigration and emigration are common, and natural selection is continuous. Why, then, is Hardy–Weinberg equilibrium, which is based on assumptions known generally to be false, so useful in our study of evolution? Can you think of other models in science that are based on false assumptions? How are such models used?

3. As far as we know, natural selection cannot adapt organisms to future events. Yet many organisms appear to respond to natural events before they happen. For example, many mammals go into hibernation while it is still quite warm. Similarly, many birds leave the temperate zone for their southern wintering grounds long before winter has arrived. How can such "anticipatory" behaviors evolve?

4. Populations of most of the thousands of species that have been introduced to areas where they were previously not found, including those that have become pests, began with a few individuals. Founding populations therefore begin with much less genetic variation than their parental populations have. If genetic variation is generally advantageous, why have so many of these species been successful in their new environments?

5. Why is it important that the ways in which males advertise their health and vigor to females reliably indicate their status?

6. As more humans live longer, many people face degenerative conditions such as Alzheimer's disease that (in most cases) are linked to advancing age. Assuming that some individuals may be genetically predisposed to successfully combat these conditions, is it likely that natural selection alone would act to favor such a predisposition in human populations? Why or why not?

ADDITIONAL INVESTIGATION

During the past 50 years, more than 200 species of insects that attack crop plants have become highly resistant to DDT and other pesticides. Using your recently acquired knowledge of evolutionary processes, explain the rapid and widespread evolution of resistance. What proposals concerning pesticide use would you make in order to slow down the rate of evolution of resistance? Explain why you think your proposals could work and how you might test them.

WORKING WITH DATA (GO TO yourBioPortal.com)

Testing for Significant Differences   In this hands-on exercise based on Figure 21.16, you will use a simple method for randomizing Malte Andersson's data to test for significant differences among the various experimental groups. You will also explore how sample size affects the power to make significant conclusions in experiments.

Female Mating Preference in Zebra Finches   In this exercise based on Figure 21.17, you will evaluate the data Jonathan Blount and his colleagues used to demonstrate female preference for males with bright bills among zebra finches. You will also consider the limitations to the experiment and explore alternative study designs.

Determining the Paternity of Butterfly Larvae   In working with a sample of the data collected by Ward et al. for the experiment described in Figure 21.19, you will consider how many larvae from a clutch of butterfly eggs must be examined in order to determine (with a high level of confidence) whether the clutch was fathered by a heterozygous or a homozygous male. You will also consider alternative hypotheses to the authors' conclusions and suggest how these alternative ideas could be tested.

Phylogenetic trees in the courtroom

Transmitting HIV, while irresponsible, is not usually prosecuted as a crime. But in one true-crime case, a woman we'll call "April" went to the police immediately upon learning she was HIV-positive. April believed she was the victim of an attempted murder by "Victor," a physician and her former boyfriend, who had repeatedly threatened violence when she tried to break up with him. April contended that Victor, under the pretense of administering vitamin therapy, had injected her with blood from one of his HIV-infected patients.

Police investigators discovered that Victor had drawn blood from one of his HIV-positive patients just before giving April the injection. The blood draw had no clinical purpose, and Victor had tried to hide the records of it. The police were convinced he might indeed have committed the alleged crime.

The district attorney, however, had to show that April's HIV infection had come from Victor's patient, and from no other source. To reconstruct the history of the infection, the district attorney turned to *phylogenetic analysis*—the study of the evolutionary relationships among a group of organisms.

The district attorney's task was complicated by the nature of HIV. HIV is a retrovirus, in which poor repair of replication errors leads to a high rate of evolution. Once a person is infected with HIV, the virus not only replicates quickly but evolves quickly, so that the infected individual is soon host to a genetically diverse population of viruses. Thus when one person transmits HIV to another, typically very few viral particles (often only one) initiate the infectious event. But the person who is the source of the infection may be host to a large, genetically diverse population of viruses—not just the variant he or she transmits to the recipient.

Enter molecular phylogeny. Samples of HIV from an infected individual can be sequenced to trace their evolutionary lineages back to the originally transmitted virus. The virus that is passed to the recipient will be very closely related to some of the viruses in the source individual and more distantly related to others. A reconstruction of the evolutionary history of the viruses in both individuals is needed to reveal not only whether the two individuals' viruses are closely related, but also who infected whom.

Human Immunodeficiency Virus A computer image of the human immunodeficiency virus (HIV), the cause of acquired immunodeficiency syndrome, or AIDS. To combat AIDS, it is also essential to understand the phylogeny of HIV.

A Source of the Virus AIDS is a zoonotic disease, meaning that the virus was transferred to humans from another animal. Phylogenetic analyses of immunodeficiency viruses show that humans acquired HIV-1 from chimpanzees (see Figure 22.9). Other forms of the virus have been passed to humans by different simians.

To prove attempted murder, the district attorney needed to demonstrate that April's HIV was more closely related to that of Victor's patient than to other HIV variants in her community. Samples of HIV were isolated from the blood of the patient, from April, and from other HIV-positive individuals in the community. Phylogenetic analysis revealed that April's HIV was indeed closely related to a subset of the patient's HIV, and more distantly related to the other HIV sources in the community. Given this fact, along with other evidence in the case, Victor was convicted of attempted murder.

IN THIS CHAPTER we will examine the field of systematics, the scientific study of the diversity of life. We will see how phylogenetic methods are used to reconstruct evolutionary history and to study diversity across genes, populations, species, and larger groups of organisms. We will see how systematists reconstruct the past and use phylogenies to make predictions in biology. We will end the chapter with a look at taxonomy, the theory and practice of classifying organisms.

22.1 What Is Phylogeny?

Phylogeny is the evolutionary history of relationships among organisms or their genes. A **phylogenetic tree** is a diagram that portrays a reconstruction of that history. Phylogenetic trees are commonly used to depict the evolutionary history of species, populations, and genes. Each split (or *node*) in a phylogenetic tree represents a point at which lineages diverged in the past. In the case of species, these splits represent past speciation events, when one lineage divided into two. Thus a phylogenetic tree can be used to trace the evolutionary relationships from the ancient common ancestor of a group of species, through the various speciation events (when lineages split), up to the present populations of the organisms (**Figure 22.1**). Over the past several decades, phylogenetic trees have become important tools for studying and describing evolutionary patterns, and for applying evolutionary theory throughout biology. You will need to understand phylogenetic trees to comprehend many articles and books about biology, including this one.

A phylogenetic tree may portray the evolutionary history of all life forms; of a major evolutionary group (such as the insects); of a small group of closely related species; or in some cases, even the history of individuals, populations, or genes within a species. The common ancestor of all the organisms in the tree forms the *root* of the tree. The phylogenetic trees in this book depict time flowing from left (earliest) to right (most recent) (**Figure 22.2A**). It is also common practice to draw trees with the earliest times at the bottom.

The timing of splitting events in lineages is shown by the position of nodes on a time axis, sometimes called a *divergence axis*. These splits represent events where one lineage diverged into two, such as a speciation event (for a tree of species), a gene duplication event (for a tree of genes), or a transmission event (for a tree of viral lineages transmitted through a host population). The divergence axis may have an explicit scale or simply show the relative timing of splitting events. In this book, the order of nodes along the horizontal (time) axis have meaning, but the vertical distance between the branches does not. Vertical distances are adjusted for legibility and clarity of presentation; they do not correlate with the degree of similarity or difference between groups. Note too that lineages can be rotated around nodes in the tree, so the vertical order of lineages is also largely arbitrary (**Figure 22.2B**). The important information in the tree is the branching order along the horizontal axis, as this indicates when the various lineages last shared a common ancestor.

Any group of species that we designate or name is called a **taxon** (plural *taxa*). Some examples of familiar taxa include humans, primates, mammals, and vertebrates (note that in this

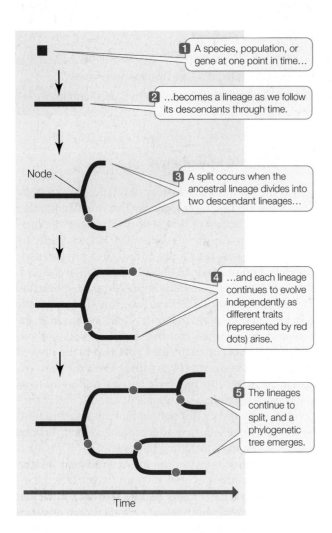

1 A species, population, or gene at one point in time...

2 ...becomes a lineage as we follow its descendants through time.

Node

3 A split occurs when the ancestral lineage divides into two descendant lineages...

4 ...and each lineage continues to evolve independently as different traits (represented by red dots) arise.

5 The lineages continue to split, and a phylogenetic tree emerges.

Time

22.1 A Phylogenetic Tree Evolutionary relationships among lineages, as well as the evolution of new traits, can be represented in a treelike diagram.

series, each taxon in the list is also a member of the next, more inclusive taxon). Any taxon that consists of all the evolutionary descendants of a common ancestor is called a **clade**. Clades can be identified by picking any point on a phylogenetic tree and then tracing all the descendant lineages to the tips of the terminal branches (**Figure 22.3**). Two species that are each other's closest relatives are called **sister species**; similarly, any two clades that are each other's closest relatives are called **sister clades**.

Before the 1980s, phylogenetic trees tended to be seen only in the literature on evolutionary biology, especially in the area of **systematics**: the study and classification of biodiversity. But almost every journal in the life sciences published during the last few years con-

22.2 How to Read a Phylogenetic Tree (A) A phylogenetic tree displays the evolutionary relationships among organisms. Such trees can be produced with time scales, as shown here, or with no indication of time. If no time scale is shown, then the branch lengths show relative rather than absolute times of divergence. (B) Lineages can be rotated around a given node, so the vertical order of taxa is largely arbitrary.

tains phylogenetic trees. Trees are widely used in molecular biology, biomedicine, physiology, behavior, ecology, and virtually all other fields of biology. Why have phylogenetic studies become so important?

All of life is connected through evolutionary history

In biology, we study life at all levels of organization—from genes, cells, organisms, populations, and species to the major divisions of life. In most cases, however, no individual gene or organism (or other unit of study) is exactly like any other gene or organism that we investigate.

Consider the individuals in your biology class. We recognize each person as an individual human, but we know that no two are exactly alike. If we knew everyone's family tree in detail, the genetic similarity of any pair of students would be more predictable. We would find that more closely related students have many more traits in common (from the color of their hair to their susceptibility or resistance to diseases). Likewise, biologists use phylogenies to make comparisons and predictions about shared traits across genes, populations, and species.

(A)

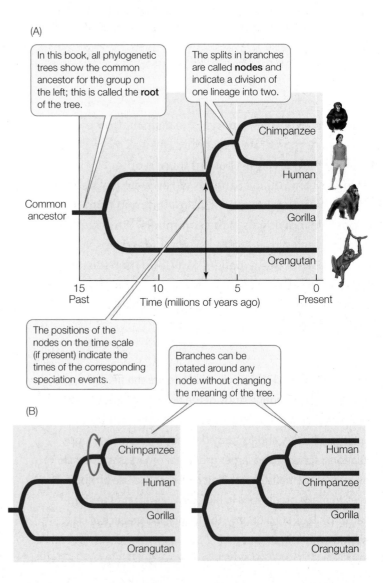

In this book, all phylogenetic trees show the common ancestor for the group on the left; this is called the **root** of the tree.

The splits in branches are called **nodes** and indicate a division of one lineage into two.

Common ancestor

The positions of the nodes on the time scale (if present) indicate the times of the corresponding speciation events.

Branches can be rotated around any node without changing the meaning of the tree.

Chimpanzee

Human

Gorilla

Orangutan

| 15 | 10 | 5 | 0 |

Past   Time (millions of years ago)   Present

(B)

Chimpanzee
Human
Gorilla
Orangutan

Human
Chimpanzee
Gorilla
Orangutan

22.3 Clades Represent All the Descendants of a Common Ancestor All clades are subsets of larger clades, with all of life as the most inclusive taxon. In this example, the groups called mammals, amniotes, tetrapods, and vertebrates represent successively larger clades. Only a few species within each clade are represented on the tree.

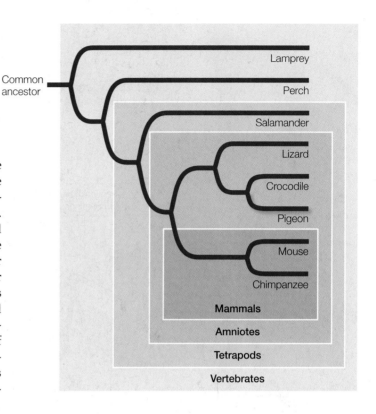

One of the great unifying concepts in biology is that all life is connected through its evolutionary history. The complete evolutionary history of life is known as the **tree of life**. Biologists estimate that there are tens of millions of species on Earth. Only about 1.8 million have been formally described and named. New species are being discovered and named all the time, and phylogenetic analyses reviewed and revised, but our knowledge of the tree of life is far from complete, even for known species. Yet knowledge of evolutionary relationships is essential for making comparisons in biology, so biologists build phylogenies for groups of interest as the need arises. The evolutionary relationships among species, as shown in the tree of life, form the basis for biological classification. This evolutionary framework allows biologists to make many predictions about the behavior, ecology, physiology, genetics, and morphology of species that have not yet been studied in detail.

Comparisons among species require an evolutionary perspective

When biologists make comparisons among species, they observe traits that differ within the group of interest and try to ascertain when these traits evolved. In many cases, investigators are interested in how the evolution of a trait depends on environmental conditions or selective pressures. For instance, scientists have used phylogenetic analyses to discover changes in the genome of HIV that confer resistance to particular drug treatments. The association of a particular genetic change in HIV with a particular treatment provides a hypothesis about the evolution of resistance that can be tested experimentally.

Any features shared by two or more species that have been inherited from a common ancestor are said to be **homologous**. Homologous features may be any heritable traits, including DNA sequences, protein structures, anatomical structures, and even some behavior patterns. Traits that are shared across a group of interest are likely to have been inherited from a common ancestor. For example, all living vertebrates have a vertebral column, and all known fossil vertebrates had a vertebral column. Therefore, the vertebral column is judged to be homologous in all vertebrates.

In tracing the evolution of a trait, biologists distinguish between *ancestral* and *derived* traits. A trait that was already present in the ancestor of a group is known as an **ancestral trait** for that group. A trait found in a descendent that differs from its ancestral form is called a **derived trait**. Derived traits that are shared among a group of organisms, and are viewed as evidence of the common ancestry of the group, are called **synapomorphies** (*syn*, "shared"; *apo*, "derived"; *morph*, "form," referring to the "form" of a trait). Thus the vertebral column is

considered a synapomorphy—a shared, derived trait—of the vertebrates.

Not all similar traits are evidence of relatedness, however. Similar traits in unrelated groups of organisms can develop for either of the following reasons:

- Independently evolved traits subjected to similar selection pressures may become superficially similar, a phenomenon called **convergent evolution**. For example, although the wing bones of bats and birds are homologous, having been inherited from a common ancestor, the wings of bats and the wings of birds are not homologous because they evolved independently from the forelimbs of different nonflying ancestors (**Figure 22.4**).

- A character may revert from a derived state back to an ancestral state in an event called an **evolutionary reversal**. For example, most frogs lack teeth in the lower jaw, but the ancestor of frogs did have such teeth. Teeth have been regained in the lower jaw of one South American species, and thus represent an evolutionary reversal in that species.

Similar traits generated by convergent evolution and evolutionary reversals are called *homoplastic traits* or **homoplasies**.

A particular trait may be ancestral or derived, depending on our point of reference in a phylogeny. For example, all birds have feathers, which are highly modified scales. We infer from this that feathers were present in the common ancestor of modern birds. Therefore, we consider the presence of feathers to be an *ancestral* trait for any particular group of modern birds, such as the songbirds. However, feathers are not present in any other living animals. If we were reconstructing a phylogeny of all

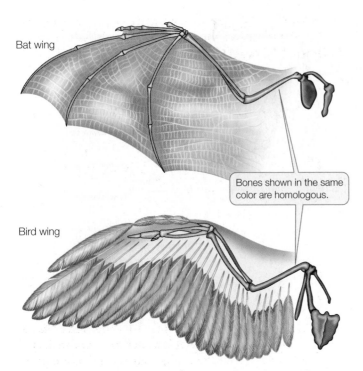

Bat wing

Bird wing

Bones shown in the same color are homologous.

22.4 The Bones Are Homologous, the Wings Are Not   The supporting bone structures of both bat wings and bird wings are derived from a common four-limbed ancestor and are thus homologous. However, the wings themselves—an adaptation for flight—evolved independently in the two groups.

living vertebrates, the presence of feathers would be a *derived* trait that is found only among birds (and thus a synapomorphy of the birds).

22.1 RECAP

A phylogenetic tree is a description of evolutionary relationships—how a group of genes, populations, or species have evolved from a common ancestor. All living organisms share a common ancestor and are related through the phylogenetic tree of life.

- Do you understand the different elements of a phylogenetic tree? See pp. 465–466 and Figure 22.2

- Explain the difference between an ancestral and a derived trait. See p. 467

- Do you see how similar traits might arise independently in species that are only distantly related? See p. 467 and Figure 22.4

Phylogenetic analyses have become increasingly important to many types of biological research in recent years, and they are the basis for the comparative nature of biology. For the most part, however, evolutionary history cannot be observed directly. How, then, do biologists reconstruct the past? One way is by using phylogenetic analyses to construct a tree.

22.2 How Are Phylogenetic Trees Constructed?

To illustrate how a phylogenetic tree is constructed, let's consider the eight vertebrate animals listed in **Table 22.1**: lamprey, perch, salamander, lizard, crocodile, pigeon, mouse, and chimpanzee. We will assume initially that a given derived trait evolved only once during the evolution of these animals (that is, there has been no convergent evolution), and that no derived traits were lost from any of the descendant groups (there has been no evolutionary reversal). For simplicity, we have selected traits that are either present (+) or absent (−).

In a phylogenetic study, the group of organisms of primary interest is called the **ingroup**. As a point of reference, an ingroup is compared with an **outgroup**. a closely related species or group known to be phylogenetically outside the group of interest. If the outgroup is known to have diverged before the ingroup, the outgroup can be used to determine which traits of the ingroup are derived (evolved within the ingroup) and which are ancestral (evolved before the origin of the ingroup). As we will see in Chapter 33, a group of jawless fishes called the lampreys is thought to have separated from the lineage leading to the other vertebrates before the jaw arose. Therefore, we have included the lamprey as the outgroup for our analysis. Because derived traits are traits acquired by other members of the vertebrate lineage *after* they diverged from the outgroup, any trait that is present in both the lamprey and the other vertebrates is judged to be ancestral.

We begin by noting that the chimpanzee and mouse share two derived traits—mammary glands and fur—that are absent in both the outgroup and in the other species of the ingroup. Therefore, we infer that mammary glands and fur are derived traits that evolved in a common ancestor of chimpanzees and mice after that lineage separated from the lineages leading to the other vertebrates. In other words, we provisionally assume that mammary glands and fur evolved only once among the animals in our ingroup. These characters are synapomorphies that unite chimpanzees and mice (as well as all other mammals, although we have not included other mammalian species in this example). By the same reasoning, we can infer that the other shared derived traits are synapomorphies for the various groups in which they are expressed. For instance, keratinous scales are a synapomorphy of the lizard, crocodile, and pigeon.

Table 22.1 also tells us that, among the animals in our ingroup, the pigeon has a unique trait: the presence of feathers. Feathers are a synapomorphy of birds, but since we only have one bird in this example, the presence of feathers provides no clues concerning relationships among the eight species of vertebrates we have sampled. However, gizzards are found in birds and crocodiles, so this trait is evidence of a close relationship between birds and crocodilians.

By combining information about the various synapomorphies, we can construct a phylogenetic tree. We infer, for example, that mice and chimpanzees, the only two animals that share fur and mammary glands in our example, share a more recent common ancestor with each other than they do with

TABLE 22.1
Eight Vertebrates Ordered According to Unique Shared Derived Traits

TAXON	DERIVED TRAIT[a]							
	JAWS	LUNGS	CLAWS OR NAILS	GIZZARD	FEATHERS	FUR	MAMMARY GLANDS	KERATINOUS SCALES
Lamprey (outgroup)	–	–	–	–	–	–	–	–
Perch	+	–	–	–	–	–	–	–
Salamander	+	+	–	–	–	–	–	–
Lizard	+	+	+	–	–	–	–	+
Crocodile	+	+	+	+	–	–	–	+
Pigeon	+	+	+	+	+	–	–	+
Mouse	+	+	+	–	–	+	+	–
Chimpanzee	+	+	+	–	–	+	+	–

[a]A plus sign indicates the trait is present, a minus sign that it is absent.

pigeons and crocodiles. Otherwise, we would need to assume that the ancestors of pigeons and crocodiles also had fur and mammary glands but subsequently lost them—unnecessary additional assumptions.

Figure 22.5 shows a phylogenetic tree for the vertebrates in Table 22.1, based on the shared derived traits we examined and the assumption that each derived trait evolved only once. This particular tree was easy to construct because the animals and characters we chose met the assumptions that derived traits appeared only once and were never lost after they appeared. Had we included a snake in the group, our second assumption would have been violated, because we know that the lizard ancestors of snakes had limbs that were subsequently lost. We would need to examine additional characters to determine that the lineage leading to snakes separated from the one leading to lizards long after the lineage leading to lizards separated from

22.5 Inferring a Phylogenetic Tree This phylogenetic tree was constructed from the information given in Table 22.1 using the parsimony principle. Each clade in the tree is supported by at least one shared derived trait, or synapomorphy.

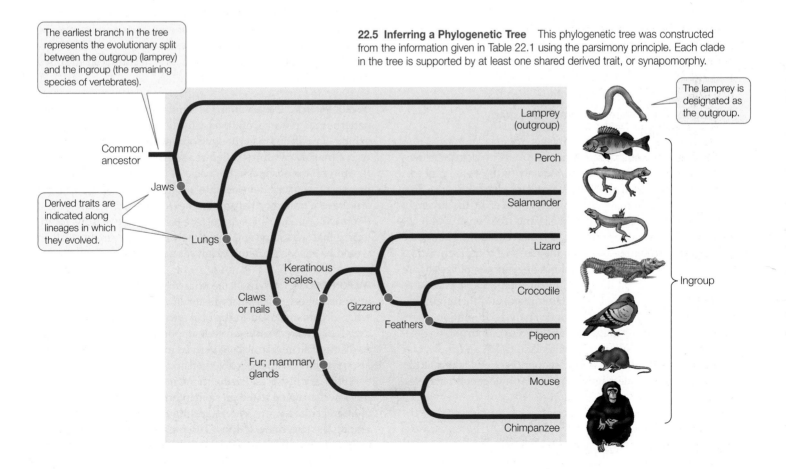

The earliest branch in the tree represents the evolutionary split between the outgroup (lamprey) and the ingroup (the remaining species of vertebrates).

Derived traits are indicated along lineages in which they evolved.

The lamprey is designated as the outgroup.

the others. In fact, the analysis of several characters shows that snakes evolved from burrowing lizards that became adapted to a subterranean existence.

Parsimony provides the simplest explanation for phylogenetic data

The phylogenetic tree shown in Figure 22.5 is based on only a very small sample of traits. Typically, biologists construct phylogenetic trees using hundreds or thousands of traits. With larger data sets, we would expect to observe some traits that have changed more than once, and thus we would expect to see some convergence and evolutionary reversal. How do we determine which traits are synapomorphies and which are homoplasies? One way is to invoke the principle of *parsimony*.

In its most general form, the **parsimony principle** states that the preferred explanation of observed data is the simplest explanation. Applying the principle of parsimony to the reconstruction of phylogenies entails minimizing the number of evolutionary changes that need to be assumed over all characters in all groups in the tree. In other words, the best hypothesis under the parsimony principle is one that requires the fewest homoplasies. This application of parsimony is a specific case of a general principle of logic called *Occam's razor*: the best explanation is the one that fits the data best while making the fewest assumptions.

We apply the parsimony principle in constructing phylogenetic trees not because all evolutionary changes always occurred parsimoniously, but because it is logical to adopt the simplest explanation that can account for the observed data. More complicated explanations are accepted only when the evidence requires them. Phylogenetic trees represent our best estimates about evolutionary relationships. They are continually modified as additional evidence becomes available.

Phylogenies are reconstructed from many sources of data

Naturalists have constructed various forms of phylogenetic trees for more than 150 years. In fact, the only figure in the first edition of Darwin's *Origin of Species* was a phylogenetic tree. Tree construction has been revolutionized, however, by the advent of computer software for trait analysis and tree construction, allowing us to consider far more data than could ever before be processed. Combining this with the massive comparative data sets being generated through studies of genomes, biologists are learning details about the tree of life at a remarkable pace.

Any trait that is genetically determined, and therefore heritable, can be used in a phylogenetic analysis. Evolutionary relationships can be revealed through studies of morphology, development, the fossil record, behavioral traits, and molecular traits such as DNA and protein sequences. Let's take a closer look at the types of data used in modern phylogenetic analyses.

---
yourBioPortal.com

GO TO **Web Activity 22.1** • Constructing a
Phylogenetic Tree

MORPHOLOGY An important source of phylogenetic information is *morphology*: the presence, size, shape, and other attributes of body parts. Since living organisms have been observed, depicted, and studied for millenia, we have a wealth of recorded morphological data as well as extensive museum and herbarium collections of organisms whose traits can be measured. New technological tools, such as the electron microscope and computed tomography (CT) scans, enable systematists to examine and analyze the structures of organisms at much finer scales than was formerly possible.

Most species are described and known primarily by their morphology, and morphology provides the most comprehensive data set available for many taxa. The features of morphology that are important for phylogenetic analysis are often specific to a particular group of organisms. For example, the presence, development, shape, and size of various features of the skeletal system are important for the study of vertebrate phylogeny, whereas floral structures are important for studying the relationships among flowering plants (*angiosperms*).

Although often useful, morphological approaches to phylogenetic analysis have some limitations. Some taxa exhibit little morphological diversity, despite great species diversity. For example, the phylogeny of the leopard frogs of North and Central America would be difficult to infer from morphological differences alone, because the many species look very similar, despite important differences in their behavior and physiology. At the other extreme, few morphological traits can be compared across distantly related species (consider earthworms and mammals, for instance). Some morphological variation has an environmental (rather than a genetic) basis and so must be excluded from phylogenetic analyses. An accurate phylogenetic analysis often requires information beyond that supplied by morphology.

DEVELOPMENT Observations of similarities in developmental patterns may reveal evolutionary relationships. Some organisms exhibit similarities in early developmental stages only. The larvae of marine creatures called sea squirts, for example, have a flexible gelatinous rod in the back—the *notochord*—that disappears as the larvae develop into adults. All vertebrate animals also have a notochord at some time during their development (**Figure 22.6**). This shared structure is one of the reasons for inferring that sea squirts are more closely related to vertebrates than would be suspected if only adult sea squirts were examined.

PALEONTOLOGY The fossil record is another important source of information on evolutionary history. Fossils show us where and when organisms lived in the past and give us an idea of what they looked like. Fossils provide important evidence that helps us distinguish ancestral from derived traits. The fossil record can also reveal when lineages diverged and began their independent evolutionary histories. Furthermore, in groups with few species that have survived to the present, information on extinct species is often critical to an understanding of the large divergences among the surviving species. The fossil record does have limitations, however. Few or no fossils have been found for some groups, and the fossil record for many groups is fragmentary.

Sea squirt larva

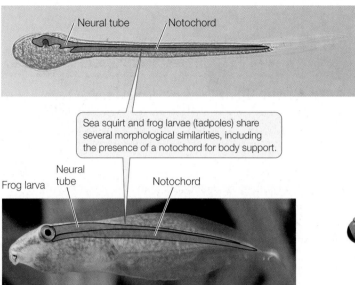

Neural tube   Notochord

Sea squirt and frog larvae (tadpoles) share several morphological similarities, including the presence of a notochord for body support.

Frog larva

Neural tube   Notochord

Adult

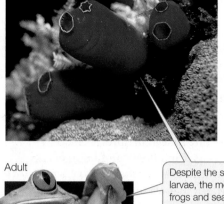

Adult

Despite the similarity of their larvae, the morphology of adult frogs and sea squirts provides little evidence of the common ancestry of these two groups.

22.6 The Evolutionary Relationship Between Sea Squirts and Vertebrates All chordates—a taxonomic group that includes sea squirts and frogs—have notochords at some stage of their development. The larvae share similarities that are not apparent in the adults. Such similarities in development can provide useful evidence of evolutionary relationships. The notochord is lost in adult sea squirts. In adult frogs, as in all vertebrates, the vertebral column replaces the notochord as the support structure.

BEHAVIOR Some behavioral traits are culturally transmitted and some are inherited. If a particular behavior is culturally transmitted, it may not accurately reflect evolutionary relationships (but may nonetheless reflect cultural connections). Bird songs, for instance, are often learned and may be inappropriate traits for phylogenetic analysis. Frog calls, however, are genetically determined and appear to be acceptable sources of information for reconstructing phylogenies.

MOLECULAR DATA All heritable variation is encoded in DNA, and so the complete genome of an organism contains an enormous set of traits (the individual nucleotide bases of DNA) that can be used in phylogenetic analyses. In recent years, DNA sequences have become among the most widely used sources of data for constructing phylogenetic trees. Comparisons of nucleotide sequences are not limited to the DNA in the cell nucleus. Eukaryotes have genes in their mitochondria as well as in their nuclei; plant cells also have genes in their chloroplasts. The chloroplast genome (cpDNA), which is used extensively in phylogenetic studies of plants, has changed slowly over evolutionary time, so it is often used to study relatively ancient phylogenetic relationships. Most animal mitochondrial DNA (mtDNA) has changed more rapidly, so mitochondrial genes have been used extensively to study evolutionary relationships among closely related animal species (the mitochondrial genes of plants evolve more slowly). Many nuclear gene sequences are also commonly analyzed, and now that several entire genomes have been sequenced, they too are used to construct phylogenetic trees. Information on gene products (such as the amino acid sequences of proteins) is also widely used for phylogenetic analyses, as we discuss in Chapter 24.

Mathematical models expand the power of phylogenetic reconstruction

As biologists began to use DNA sequences to infer phylogenies in the 1970s and 1980s, they developed explicit mathematical models describing how DNA sequences change over time. These models account for multiple changes at a given position in a DNA sequence. They also take into account different rates of change at different positions in a gene, at different positions in a codon, and among different nucleotides (see Section 24.1). For example, *transitions* (changes between two purines or between two pyrimidines) are usually more likely than are *transversions* (changes between a purine and pyrimidine).

Mathematical models can be used to compute how a tree might evolve given the observed data. A **maximum likelihood** method will identify the tree that most likely produced the observed data, given the assumed model of evolutionary change. Maximum likelihood methods can be used for any kind of characters, but they are most often used with molecular data, for which explicit mathematical models of evolutionary change are easier to develop. The principal advantages to maximum likelihood analyses are that they incorporate more information about evolutionary change than do parsimony methods, and they are easier to treat in a statistical framework. The principal disadvantages are that they are computationally intensive and require explicit models of evolutionary change (which may not be available for some kinds of character change).

The accuracy of phylogenetic methods can be tested

If phylogenetic trees represent reconstructions of past events, and if many of these events occurred before any humans were around to witness them, how can we test the accuracy of phylogenetic methods? Biologists have conducted experiments both in living organisms and with computer simulations that have demonstrated the effectiveness and accuracy of phylogenetic methods.

In one experiment designed to test the accuracy of phylogenetic analysis, a single viral culture of bacteriophage T7 was used as a starting point, and lineages were allowed to evolve from this ancestral virus in the laboratory (**Figure 22.7**). The initial culture was split into two separate lineages, one of which became the ingroup for analysis and the other of which became the outgroup for rooting the tree. The lineages in the ingroup were split in two after every 400 generations, and samples of the virus were saved for analysis at each branching point. The lineages were allowed to evolve until there were eight lineages in the ingroup. Mutagens were added to the viral cultures to increase the mutation rate so that the amount of change and the degree of homoplasy would be typical of the organisms analyzed in average phylogenetic analyses. The investigators then sequenced samples from the end points of the eight lineages, as well as from the ancestors at the branching points. They then gave the sequences from the end points of the lineages to other investigators to analyze, without revealing the known history of the lineages or the sequences of the ancestral viruses.

After the phylogenetic analysis was completed, the investigators asked two questions: Did phylogenetic methods reconstruct the known history correctly, and were the sequences of the ancestral viruses reconstructed accurately? The answer in both cases was yes: the branching order of the lineages was reconstructed exactly as it had occurred, more than 98 percent of the nucleotide positions of the ancestral viruses were reconstructed correctly, and 100 percent of the amino acid changes in the viral proteins were reconstructed correctly.

---
yourBioPortal.com
GO TO Animated Tutorial 22.1 • Using Phylogenetic Analysis to Reconstruct Evolutionary History
---

The experiment shown in Figure 22.7 demonstrated that phylogenetic analysis was accurate under the conditions tested, but it did not examine all possible conditions. Other experimental studies have taken other factors into account, such as the sensitivity of phylogenetic analysis to convergent environments and highly variable rates of evolutionary change. In addition, computer simulations based on evolutionary models have been used extensively to study the effectiveness of phylogenetic analysis. These studies have also confirmed the accuracy of phylogenetic methods and have been used to refine those methods and extend them to new applications.

INVESTIGATING LIFE

22.7 The Accuracy of Phylogenetic Analysis

To test whether analysis of gene sequences can accurately reconstruct evolutionary phylogeny, we must have an unambiguously known phylogeny to compare against the reconstruction. Will the observed phylogeny match the reconstruction?

HYPOTHESIS A phylogeny reconstructed from analysis of the DNA sequences of living organisms can accurately match the known evolutionary history of the organisms.

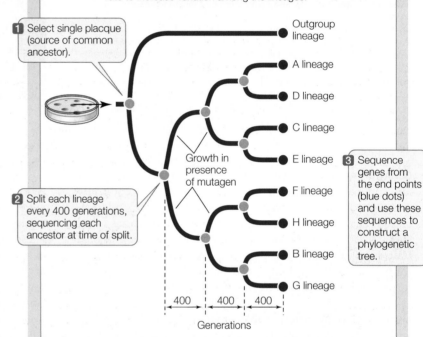

METHOD In the laboratory, researchers produced an unambiguous phylogeny of nine viral lineages, enhancing the mutation rate to increase variation among the lineages.

1 Select single placque (source of common ancestor).

Growth in presence of mutagen

2 Split each lineage every 400 generations, sequencing each ancestor at time of split.

Outgroup lineage
A lineage
D lineage
C lineage
E lineage
F lineage
H lineage
B lineage
G lineage

3 Sequence genes from the end points (blue dots) and use these sequences to construct a phylogenetic tree.

400 | 400 | 400
Generations

Viral sequences from the end points of each lineage (blue dots) were subjected to phylogenetic analysis by investigators who were unaware of the history of the lineages or the gene sequences of the ancestral viruses. These investigators reconstructed the phylogeny based solely on their analyses of the descendants' genomes.

RESULTS The true phylogeny and ancestral DNA sequences were accurately reconstructed solely from the DNA sequences of the viruses at the tips of the tree.

CONCLUSION Phylogenetic analysis of DNA sequences can accurately reconstruct evolutionary history.

FURTHER INVESTIGATION: The lineages in this experiment evolved under similar conditions. How might changing environmental conditions for some of the lineages affect the result?

Go to **yourBioPortal.com** for original citations, discussions, and relevant links for all INVESTIGATING LIFE figures.

22.2 RECAP

Phylogenetic trees can be constructed by using the parsimony principle to find the simplest explanation for the evolution of traits. Maximum likelihood methods incorporate more explicit models of evolutionary change to reconstruct evolutionary history.

- Do you understand how a phylogenetic tree is constructed? See pp. 468–471 and Figure 22.5

- Is there a way to test whether phylogenetic trees provide accurate reconstructions of evolutionary history? See p. 474 and Figure 22.7

Biologists in many fields now routinely reconstruct phylogenetic relationships. Let's examine some of the many uses of these phylogenetic trees.

22.3 How Do Biologists Use Phylogenetic Trees?

Information about the evolutionary relationships among organisms is useful to scientists investigating a wide variety of biological questions. In this section we will illustrate how phylogenetic trees can be used to ask questions about the past, and to compare aspects of the biology of organisms in the present.

Phylogenies help us reconstruct the past

Most flowering plants reproduce by mating with another individual—a process called *outcrossing*. Many outcrossing species have mechanisms to prevent self-fertilization, and so are referred to as *self-incompatible*. Individuals of some species, however, regularly fertilize themselves with their own pollen; they are termed *selfing* species, which of course requires that they be *self-compatible*. How can we tell how often self-compatibility has evolved in a group of plants? We can do so by conducting a phylogenetic analysis of outcrossing and selfing species and testing the species for self-compatibility.

The evolution of fertilization mechanisms was examined in *Linanthus*, a genus in the phlox family that exhibits a diversity of breeding systems and pollination mechanisms. The outcrossing species of *Linanthus* have long petals, are pollinated by long-tongued flies, and are self-incompatible. The self-pollinating species of *Linanthus*, in contrast, all have short petals and do not require insect pollinators to reproduce successfully. Investigators reconstructed a phylogeny for 12 species in the genus using nuclear ribosomal DNA sequences (**Figure 22.8**). They determined whether each species was self-compatible by artificially pollinating flowers with the plant's own pollen or with pollen from other individuals and observing whether viable seeds formed.

Several lines of evidence suggest that self-incompatibility is the ancestral state in *Linanthus*. Multiple origins of self-incompatibility have not been found in any flowering plant family to date. Self-incompatibility depends on physiological mechanisms in both the pollen and the stigma (the female organ on which pollen lands) and is under the control of least three different alleles. Therefore, a change from self-incompatibility to

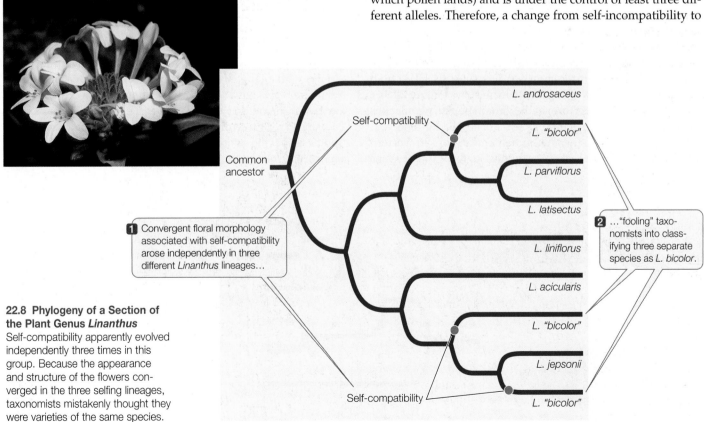

22.8 Phylogeny of a Section of the Plant Genus *Linanthus*
Self-compatibility apparently evolved independently three times in this group. Because the appearance and structure of the flowers converged in the three selfing lineages, taxonomists mistakenly thought they were varieties of the same species.

Common ancestor

Self-compatibility

1 Convergent floral morphology associated with self-compatibility arose independently in three different *Linanthus* lineages…

2 …"fooling" taxonomists into classifying three separate species as *L. bicolor*.

L. androsaceus
L. "bicolor"
L. parviflorus
L. latisectus
L. liniflorus
L. acicularis
L. "bicolor"
L. jepsonii
L. "bicolor"

Self-compatibility

22.9 Phylogenetic Tree of Immunodeficiency Viruses
Immunodeficiency viruses have been transmitted to humans from two different simian hosts: HIV-1 from chimpanzees and HIV-2 from sooty mangabeys. (SIV stands for simian immunodeficiency virus.)

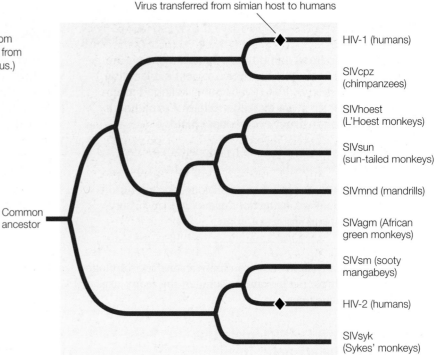

Virus transferred from simian host to humans

- HIV-1 (humans)
- SIVcpz (chimpanzees)
- SIVhoest (L'Hoest monkeys)
- SIVsun (sun-tailed monkeys)
- SIVmnd (mandrills)
- SIVagm (African green monkeys)
- SIVsm (sooty mangabeys)
- HIV-2 (humans)
- SIVsyk (Sykes' monkeys)

Common ancestor

self-compatibility would be easier than the reverse change. In addition, in all self-incompatible species of *Linanthus*, the site of pollen rejection is the stigma, even though sites of pollen rejection vary greatly among other plant families.

Assuming that self-incompatibility is the ancestral state, the reconstructed phylogeny suggests that self-compatibility evolved three times within this group of *Linanthus* (see Figure 22.8). The change to self-compatibility has been accompanied by the evolution of reduced petal size. Interestingly, the striking similarity of the flowers in the self-compatible groups once led to their being classified as members of a single species. The phylogenetic analysis using ribosomal DNA showed them to be members of three distinct lineages, however.

Reconstructing the past is important for understanding many biological processes. In the case of *zoonotic* diseases (diseases caused by infectious organisms transmitted to humans from another animal host), it is important to understand when, where, and how the disease first entered a human population. Human immunodeficiency virus (HIV) is the cause of such a zoonotic disease, acquired immunodeficiency syndrome, or AIDS. As we described in the opening to this chapter, phylogenetic analyses have become important for studying the transmission of viruses such as HIV. Phylogenies are also important for understanding the present global diversity of HIV and for determining the virus's origins in human populations. A broader phylogenetic analysis of immunodeficiency viruses shows that humans acquired these viruses from two different hosts: HIV-1 from chimpanzees, and HIV-2 from sooty mangabeys (**Figure 22.9**).

HIV-1 is the common form of the virus in human populations in central Africa, where chimpanzees are hunted for food, and HIV-2 is the common form in human populations in western Africa, where sooty mangabeys are hunted for food. Thus it seems likely that these viruses entered human populations through hunters who cut themselves while skinning chimpanzees and sooty mangabeys. The relatively recent global pandemic of AIDS occurred when these infections in local African populations rapidly spread through human populations around the world.

Phylogenies allow us to compare and contrast living organisms

Male swordtails—a group of fishes in the genus *Xiphophorus*—have a long, colorful tail extension (**Figure 22.10A**), and their reproductive success is closely associated with this appendage. Males with a long sword are more likely to mate successfully than are males with a short sword (an example of *sexual selection*; see Chapters 21 and 23). Several explanations have been advanced for the evolution of this structure, including the hypothesis that the sword simply exploits a preexisting bias in the sensory system of the females. This *sensory exploitation hypoth-*

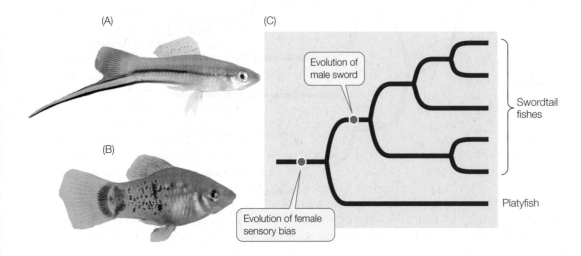

(A)

(B)

(C)

Evolution of male sword

Evolution of female sensory bias

Swordtail fishes

Platyfish

22.10 The Origin of a Sexually Selected Trait (A) The large tail of male swordtail fishes (genus *Xiphophorus*) apparently evolved through sexual selection, with females mating preferentially with males with a longer "sword." (B) A male platyfish, member of a related species. (C) Phylogenetic analysis reveals that the platyfishes split from the swordtails before the evolution of the sword. The independent finding that female platyfishes prefer males with an artificial sword further supports the idea that this appendage evolved as a result of a preexisting preference in the females.

esis suggests that female swordtails had a preference for males with long tails even before the tails evolved (perhaps because females assess the size of males by their total body length—including the tail—and prefer larger males).

To test the sensory exploitation hypothesis, a phylogeny was used to identify the swordtail relatives that had split most recently from their lineage before the evolution of sword extensions. These closest relatives turned out to be the platyfishes, another group of *Xiphophorus* (**Figure 22.10B**). Even though male platyfishes do not normally have swords, when researchers attached artificial swordlike structures to the tails of some male platyfishes, female platyfishes preferred the males with an artificial sword, thus providing support for the hypothesis that female *Xiphophorus* had a preexisting sensory bias favoring tail extensions even before the trait evolved (**Figure 22.10C**). Thus, a long tail became a sexually selected trait because of the preexisting preference of the females.

Ancestral states can be reconstructed

In addition to using phylogenetic methods to infer evolutionary relationships among lineages, biologists can use them to reconstruct the morphology, behavior, or nucleotide and amino acid sequences of ancestral species (as was demonstrated for the ancestral sequences of bacteriophage T7 in the experiment shown in Figure 22.7). For instance, a phylogenetic analysis was used to reconstruct an opsin protein in the ancestral archosaur (the most recent common ancestor of birds, dinosaurs, and crocodiles). Opsins are pigment proteins involved in vision; different opsins (with different amino acid sequences) are excited by different wavelengths of light. Knowledge of the opsin sequence in the ancestral archosaur would provide clues about the animal's visual capabilities and therefore about some of its probable behaviors. Investigators used phylogenetic analysis of opsin from living vertebrates to estimate the amino acid sequence of the pigment that existed in the ancestral archosaur. A protein with this same sequence was then constructed in the laboratory. The investigators tested the reconstructed opsin and found a significant shift toward the red end of the spectrum in the light sensitivity of this protein compared with that of most modern opsins. Modern species that exhibit similar sensitivity are adapted for nocturnal vision, so the investigators inferred that the ancestral archosaur might have been active at night. Thus, reminiscent of the movie *Jurassic Park*, phylogenetic analyses are being used to reconstruct extinct species, one protein at a time.

Molecular clocks help date evolutionary events

For many applications, biologists want to know not only the order in which evolutionary lineages split but also the timing of those splits. In 1965, Emile Zuckerkandl and Linus Pauling hypothesized that rates of molecular change were constant enough that they could be used to predict evolutionary divergence times—an idea that has become known as the *molecular clock hypothesis*.

Of course, different genes evolve at different rates, and there are also differences in evolutionary rates among species related

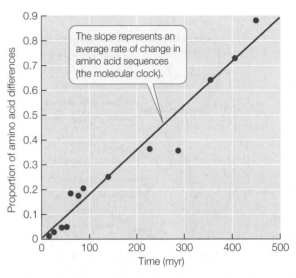

22.11 A Molecular Clock of the Protein Hemoglobin Amino acid replacements in hemoglobin have occurred at a relatively constant rate over nearly 500 million years of evolution. The graph shows the relationship between time of divergence and proportion of amino acid change for 13 pairs of vertebrate hemoglobin proteins. The average rate of change represents the molecular clock for hemoglobin in vertebrates.

to differing generation times, environments, efficiencies of DNA repair systems, and other biological factors. Nonetheless, among closely related species, a given gene usually evolves at a reasonably constant rate. Therefore, the protein encoded by the gene also accumulates amino acid substitutions at a relatively constant rate (**Figure 22.11**). A **molecular clock** uses the average rate at which a given gene or protein accumulates changes to gauge the time of divergence for a particular split in the phylogeny. Molecular clocks must be calibrated using independent data, such as the fossil record, known times of divergence, or biogeographic dates (such as the dates for separations of continents). Using such calibrations, times of divergence have been estimated for many groups of species that have diverged over millions of years.

Molecular clocks are not only used to date ancient events; they are also used to study the timing of comparatively recent events. Most samples of HIV-1 have been collected from humans only since the early 1980s, although a few isolates from medical biopsies are available from as early as the 1950s. But biologists can use the observed changes in HIV-1 over the past several decades to project back to the common ancestor of all HIV-1 isolates, and estimate when HIV-1 first entered human populations from chimpanzees. The clock can be calibrated using the samples from the 1980s and 1990s, and then tested using the samples from the 1950s. As shown in Figure 22.12C, a sample from a 1959 biopsy is dated by molecular clock analysis at 1957 ± 10 years. The molecular clock was also used to project back to the common ancestor of this group of HIV-1 samples. Extrapolation suggests a date of origin for this group of viruses of about 1930. Although AIDS was unknown to Western medicine until the 1980s, this analysis shows that HIV-1 was present (probably at very low frequency) in human populations in Africa for at least a half-century before its emergence as a

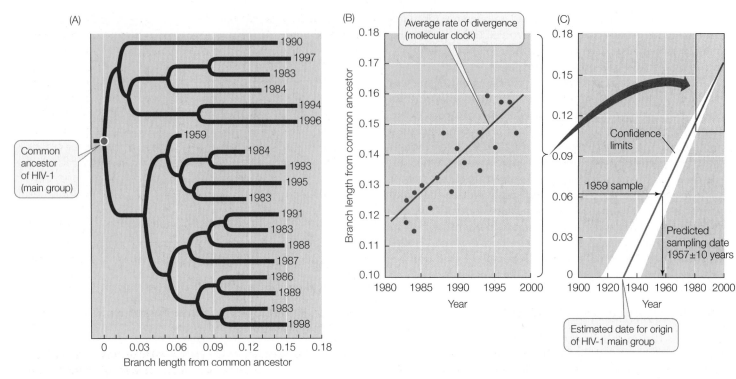

22.12 Dating the Origin of HIV-1 in Human Populations
(A) A phylogenetic analysis of the main group of HIV-1 viruses. The dates indicate the years in which samples were taken. (For clarity, only a small fraction of the samples that were examined in the original study are shown.) (B) A plot of year of isolation versus genetic divergence from the common ancestor provides an average rate of divergence, or a molecular clock. (C) The molecular clock is used to date a sample taken in 1959 (as a test of the clock) and the unknown date of origin of the HIV-1 main group (about 1930).

global pandemic (**Figure 22.12**). Biologists have used similar analyses to conclude that immunodeficiency viruses have been transmitted repeatedly into human populations from multiple primates for more than a century (see also Figure 22.9).

22.3 RECAP

Phylogenetic trees are used to reconstruct the past history of lineages, to determine when and where traits arose, and to make relevant biological comparisons among genes, populations, and species. They can also be used to reconstruct ancestral traits and to estimate the timing of evolutionary events.

- Explain how phylogenetic trees can help determine the number of times a particular trait evolved. See pp. 473–474 and Figure 22.8

- How does the reconstruction of ancestral traits help biologists explain the evolution of visual pigment proteins? See p. 475

- How do molecular clocks add a time dimension to phylogenetic trees? See p. 475 and Figure 22.12

All of life is connected through evolutionary history, and the relationships among organisms provide a natural basis for making biological comparisons. For these reasons, biologists use phylogenetic relationships as the basis for organizing life into a coherent classification system, described in the next section.

22.4 How Does Phylogeny Relate to Classification?

The biological classification system in widespread use today is derived from a system developed by the Swedish biologist Carolus Linnaeus in the mid-1700s. Linnaeus developed a naming system called **binomial nomenclature** that has allowed scientists throughout the world to refer unambiguously to the same organisms by the same names (**Figure 22.13**).

Linnaeus gave each species two names, one identifying the species itself and the other the genus to which it belongs. A **genus** (plural, *genera*) is a group of closely related species. Optionally, the name of the taxonomist who first proposed the species name may be added at the end. Thus *Homo sapiens* Linnaeus is the name of the modern human species. *Homo* is the genus to which the species belongs, and *sapiens* identifies the particular species in the genus *Homo*; Linnaeus proposed the species name *Homo sapiens*. You can think of the generic name *Homo* as equivalent to your surname and the specific name *sapiens* as equivalent to your first name. The name of the genus is always capitalized, and the name identifying the species is always lowercased. Both names are italicized, whereas common names of organisms are not. Rather than repeating the name of a genus when it is used several times in the same discussion, biologists often spell it out only once and abbreviate it to the initial letter thereafter (*D. melanogaster* rather than *Drosophila melanogaster*, for example).

(A) *Campanula rotundifolia*

(B) *Endymion non-scriptus*

(C) *Mertensia virginica*

22.13 Many Different Plants Are Called Bluebells   All three of these distantly related plant species are called "bluebells." Binomial nomenclature allows us to communicate exactly what is being described. (A) *Campanula rotundifolia*, found on the North American Great Plains, belongs to a larger group of bellflowers. (B) *Endymion non-scriptus*, English bluebell, is related to hyacinths. (C) *Mertensia virginica*, Virginia bluebell, belongs in a very different group of plants known as borages.

As we noted earlier, any group of organisms that is treated as a unit in a biological classification system, such as the genus *Drosophila*, or all insects, is called a *taxon*. In the Linnaean system, species and genera are further grouped into a hierarchical system of higher taxonomic categories. The taxon above the genus in the Linnaean system is the **family**. The names of animal families end in the suffix "-idae." Thus Formicidae is the family that contains all ant species, and the family Hominidae contains humans and our recent fossil relatives, as well as our closest living relatives, the chimpanzees and gorillas. Family names are based on the name of a member genus; Formicidae is based on the genus *Formica*, and Hominidae is based on *Homo*. The same rules are used in classifying plants, except that the suffix "-aceae" is used for plant family names instead of "-idae." Thus Rosaceae is the family that includes the genus of roses (*Rosa*) and its close relatives. In the Linnaean system, families

are grouped into **orders**, orders into **classes**, and classes into **phyla** (singular *phylum*), and phyla into **kingdoms**. However, Linnaean classification is often subjective; whether a particular taxon is considered, say, an order or a class is often a subjective decision. Today, Linnaean terms are used largely for convenience. Although families are always grouped within orders, orders within classes, and so forth, there is nothing that makes a "family" in one group equivalent (in number of genera or in evolutionary age, for instance) to a "family" in another group.

Linnaeus recognized the overarching hierarchy of life, but he developed his system before evolutionary thought had become widespread. Biologists today recognize the tree of life as the basis for biological classification and often name clades without placing them into any Linnaean rank. But regardless of whether they rank organisms into Linnaean categories or use unranked clade names, modern biologists use evolutionary relationships as the basis for distinguishing biological taxa.

Evolutionary history is the basis for modern biological classification

Biological classification systems are used to express relationships among organisms. The kind of relationship we wish to express influences which features we use to classify organisms. If, for instance, we were interested in a system that would help us decide what plants and animals were desirable as food, we might devise a classification based on tastiness, ease of capture, and the number of edible parts each organism possessed. Early Hindu classifications of organisms were designed according to these criteria. Such systems served the needs of the people who developed them, but are not adequate for formal scientific classification.

Taxonomists today use biological classifications to express the evolutionary relationships of organisms. Taxa are expected to be **monophyletic**, meaning that the taxon contains an ancestor and all descendants of that ancestor, and no other organisms (**Figure 22.14**). In other words, the taxon is an historical group of related species, or a complete branch on the tree of life (a *clade*). Although biologists seek to describe and name only monophyletic taxa, the detailed phylogenetic information needed to do so is not always available. A group that does not include its common ancestor is called a **polyphyletic** group. A group that does not include all the descendants of a common ancestor is called a **paraphyletic** group.

A true monophyletic group (i.e., a clade) can be removed from a phylogenetic tree by a single "cut" in the tree, as shown in Figure 22.14. Note that there are many monophyletic groups on any phylogenetic tree, and that these groups are successively smaller subsets of larger monophyletic groups. This hierarchy of biological taxa, with all of life as the most inclusive taxon and many smaller taxa within larger taxa, down to the individual species, is the modern basis for biological classification.

Virtually all taxonomists now agree that polyphyletic and paraphyletic groups are inappropriate as taxonomic units, because they do not correctly reflect evolutionary history. The classifications used today still contain such groups because some organisms have not been evaluated phylogenetically. As mistakes in prior classifications are detected, taxonomic names

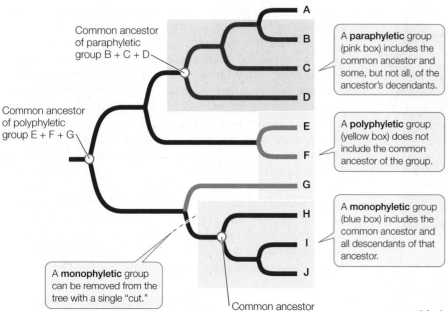

Common ancestor
of paraphyletic
group B + C + D

A **paraphyletic** group
(pink box) includes the
common ancestor and
some, but not all, of the
ancestor's decendants.

Common ancestor
of polyphyletic
group E + F + G

A **polyphyletic** group
(yellow box) does not
include the common
ancestor of the group.

A **monophyletic** group
(blue box) includes the
common ancestor and
all descendants of that
ancestor.

A **monophyletic** group
can be removed from the
tree with a single "cut."

Common ancestor
of monophyletic
group H + I + J

22.14 Monophyletic, Polyphyletic, and Paraphyletic Groups
Monophyletic groups are the basis of biological taxa in modern classifications. Polyphyletic and paraphyletic groups do not accurately reflect
evolutionary history.

---

yourBioPortal.com

GO TO **Web Activity 22.2 • Types of Taxa**

---

are revised and polyphyletic and paraphyletic groups are eliminated from the classifications.

Several codes of biological nomenclature govern the use of scientific names

Several sets of explicit rules govern the use of scientific names.
Biologists around the world follow these rules voluntarily to facilitate communication and dialogue. Although there may be
dozens of common names for an organism in many different lan-
guages, the rules of biological nomenclature are designed so that
there is only one correct scientific name for any single recognized
taxon and (ideally) a given scientific name applies only to a
single taxon (that is, each scientific name is unique). Sometimes
the same species is named more than once (when more than one
taxonomist has taken up the task); the rules specify that the valid

name is the first name that was proposed. If
the same name is inadvertently given to two
different species, then a replacement name
must be given to the species that was named
second.

Because of the historical separation of the
fields of zoology, botany (including, originally, the study of fungi), and microbiology,
different sets of taxonomic rules were developed for each of these groups. Yet another set
of rules for classifying viruses emerged later.
This has resulted in many duplicated names
in groups that are governed by different sets
of rules: *Drosophila*, for instance, is both a
genus of fruit flies and a genus of fungi, and
there are species in both groups that have
identical names. Until recently these duplicated names caused little confusion, since traditionally biologists who studied fruit flies
were unlikely to read the literature on fungi (and vice versa).
Today, however, given the use of large, universal biological databases (such as GenBank, which includes DNA sequences from
across all life), it is increasingly important that each taxon have
a unique name. Taxonomists are now working to develop common sets of rules that can be applied across all living organisms.

22.4 RECAP

Biologists organize and classify life by identifying
and naming monophyletic groups. Several sets of
rules govern the use of scientific names so that each
species and higher taxon can be identified and
named unambiguously.

- Explain the difference between monophyletic, paraphyletic, and polyphyletic groups. See p. 477 and
  Figure 22.14

- Do you understand why biologists prefer monophyletic groups in formal classifications? See p. 477

Now that we have seen how evolution occurs and how phylogenies can be used to study evolutionary relationships, we are
ready to consider the process of speciation. Speciation is what
leads to the splitting events on the tree of life, and is the process
that results in the millions of species that constitute biodiversity.

CHAPTER SUMMARY

22.1 What Is Phylogeny?

- **Phylogeny** is the history of descent of organisms from their
  common ancestor. Groups of evolutionarily related species are
  represented as related branches in a **phylogenetic tree**.
  Review Figure 22.2

- A group of species that consists of all the evolutionary descendants of a common ancestor is called a **clade**. Named clades
  and species are called **taxa**.

- **Homologies** are similar traits that have been inherited from a
  common ancestor. Review Figure 22.4

- A trait that is shared by two or more taxa and is derived through evolution from a common ancestral form is called a **synapomorphy**.
- Similar traits may occur among species that do not result from common ancestry. **Convergent evolution** and **evolutionary reversals** can give rise to such traits, which are called **homoplasies**.

22.2 How Are Phylogenetic Trees Constructed?
SEE WEB ACTIVITY 22.1

- Phylogenetic trees can be inferred from synapomorphies using the principle of **parsimony**. Review Figure 22.5
- Sources of phylogenetic information include morphology, patterns of development, the fossil record, behavioral traits, and molecular traits such as DNA and protein sequences.
- Phylogenetic trees can also be inferred with **maximum likelihood** methods, which calculate the probability that a particular tree will have generated the observed data.

22.3 How Do Biologists Use Phylogenetic Trees?

- Phylogenetic trees are used to reconstruct the past and understand the origin of traits. Review Figure 22.8
- Phylogenetic trees are used to make appropriate evolutionary comparisons among living organisms.
- Biologists can use phylogenetic trees to reconstruct ancestral states. **SEE ANIMATED TUTORIAL 22.1**
- Phylogenetic trees may include estimates of times of divergence of lineages determined by **molecular clock** analysis. Review Figure 22.12

22.4 How Does Phylogeny Relate to Classification?

- Taxonomists organize biological diversity on the basis of evolutionary history.
- Taxa in modern classifications are expected to be **monophyletic** groups. **Paraphyletic** and **polyphyletic** groups are not considered appropriate taxonomic units. Review Figure 22.14, **WEB ACTIVITY 22.2**
- Several sets of rules govern the use of scientific names, with the goal of providing unique and universal names for biological taxa.

SELF-QUIZ

1. A *clade* is
   a. a type of phylogenetic tree.
   b. a group of evolutionarily related species that share a common ancestor.
   c. a tool for constructing phylogenetic trees.
   d. an extinct species.
   e. an ancestral species.

2. Phylogenetic trees may be constructed for
   a. genes.
   b. species.
   c. major evolutionary groups.
   d. viruses.
   e. All of the above.

3. A shared derived trait, used as the basis for inferring a monophyletic group, is called
   a. a synapomorphy.
   b. a homoplasy.
   c. a parallel trait.
   d. a convergent trait.
   e. a phylogeny.

4. The parsimony principle can be used to infer phylogenetic trees because
   a. evolution is nearly always parsimonious.
   b. it is logical to adopt the simplest hypothesis capable of explaining the known facts.
   c. once a trait changes, it never reverses condition.
   d. all species have an equal probability of evolving.
   e. closely related species are always very similar to one another.

5. Convergent evolution and evolutionary reversal are two sources of
   a. homology.
   b. parsimony.
   c. synapomorphy.
   d. monophyly.
   e. homoplasy.

6. Which of the following are commonly used to infer phylogenetic relationships among plants but not among animals?
   a. Nuclear genes
   b. Chloroplast genes
   c. Mitochondrial genes
   d. Ribosomal RNA genes
   e. Protein-coding genes

7. Which of the following is *not* true of maximum likelihood or parsimony methods for inferring phylogeny?
   a. The maximum likelihood method requires an explicit model of evolutionary character change.
   b. The parsimony method is computationally easier than the maximum likelihood method.
   c. The maximum likelihood method is easier to treat in a statistical framework.
   d. The maximum likelihood method is most often used with molecular data.
   e. Parsimony is usually used to infer time on a phylogenetic tree.

8. Taxonomists strive to include taxa in biological classifications that are
   a. monophyletic.
   b. paraphyletic.
   c. polyphyletic.
   d. homoplastic.
   e. monomorphic.

9. Which of the following groups have separate sets of rules for nomenclature?
   a. Animals
   b. Plants and fungi
   c. Bacteria
   d. Viruses
   e. All of the above

10. If two scientific names are proposed for the same species, how do taxonomists decide which name should be used?
    a. The name that provides the most accurate description of the organism is used.
    b. The name that was proposed most recently is used.
    c. The name that was used in the most recent taxonomic revision is used.
    d. The first name to be proposed is used, unless that name was previously used for another species.
    e. Taxonomists use whichever name they prefer.

FOR DISCUSSION

1. Why are taxonomists concerned with identifying species that share a particular common ancestor?

2. How are fossils used to identify ancestral and derived traits of organisms? How can fossils be integrated into phylogenetic analyses?

3. The parsimony principle is often used to construct phylogenetic trees. What are the limitations of parsimony, and why do some biologists prefer model-based approaches, such as maximum likelihood methods?

4. A student of the evolution of frogs has proposed a strikingly new classification of frogs based on an analysis of a few mitochondrial genes from about 10 percent of frog species.

Should frog taxonomists immediately accept the new classification? Why or why not?

5. What are some of the assumptions that go into a molecular clock analysis? How could these assumptions be violated? How could molecular clock analyses be modified to consider these additional sources of variation?

6. Classification systems summarize much information about organisms and enable us to remember the traits of many organisms. From your general knowledge, how many traits can you associate with the following names: conifer, fern, bird, mammal?

ADDITIONAL INVESTIGATION

West Nile virus kills birds of many species and can cause fatal encephalitis (inflammation of the brain) in humans and horses. The virus was first isolated in Africa (where it is thought to be endemic) in the 1930s, and by the 1990s it had been found throughout much of Eurasia. West Nile virus was not found in North America until 1999, but since that time it has spread rapidly across most of the United States. The genome of West Nile virus evolves quickly. How could you use phylogenetic analysis to investigate the geographic origin of the West Nile virus that was introduced into North America in 1999?

WORKING WITH DATA (GO TO yourBioPortal.com)

Constructing a Phylogenetic Tree In this exercise based on Figure 22.7, you will use a subset of the DNA sequences from the experimental lineages to reconstruct the evolutionary relationships among the viruses. You will also use these data to reconstruct the DNA sequences of the viral ancestors.

Species and Their Formation

Catching speciation in the act

When biologists first explored the Cuatro Ciénegas basin of northern Mexico, they found many organisms that are not found anywhere else in the world. So far researchers have described about 150 *species* of plants and animals that are restricted to this small region. Even though Cuatro Ciénegas is in a desert, about 30 of these unique species are aquatic, living in the isolated springs and marshes of the basin. An unusual aquatic box turtle, beautiful cichlid fishes, and tiny crustaceans are among the many aquatic species that are confined to Cuatro Ciénegas. Why are so many different species found here and nowhere else?

Biologists and geologists found that, over the past several million years, this desert oasis has repeatedly been isolated by a succession of geological events that cut it off from the river systems and mountain ranges of northern Mexico. Many different *speciation events* associated with these geological events make Cuatro Ciénegas a natural laboratory for studying speciation by geographic isolation.

Each time gene flow between organisms in the basin and the surrounding areas ceased, populations living inside and outside the basin began to diverge from one another. Over thousands of generations of such isolation, new species developed. These new species no longer share the same gene pool, are adapted to different environments, and look different from one another. And—extremely important—the organisms have diverged to the point that they are no longer capable of reproducing with one another—one of the hallmarks of distinct species.

Although speciation is often studied in natural settings such as Cuatro Ciénegas, some aspects of speciation can be studied in controlled laboratory experiments, using organisms with short generation times. For example, William Rice and George Salt conducted an experiment in which fruit flies were allowed to choose food sources in different habitats, where mating also took place. The habitats were vials in different parts of an experimental cage. The habitats differed in three parameters: (1) light; (2) the direction in which the fruit flies could move (up or down); and (3) concentrations of two aromatic chemicals, ethanol and acetaldehyde. In just 35 generations, two groups of flies were genetically and reproductively isolated from one another because they had evolved distinct preferences for different habitats. In controlled experiments like these, biologists are beginning to study and understand the genetic details of speciation.

A Natural Laboratory A swimmer surveys several of the fish species that are isolated in the desert oasis of the Cuatro Ciénegas basin in northern Mexico.

Experimental Subjects Fruit flies of the genus *Drosophila* are easily reared in the laboratory. Their short generation time (7–10 days from newly laid egg to reproductive adult) makes them ideal subjects for controlled experiments on speciation.

The *origin of species*—the splitting and diverging of a single lineage into two or more distinct and evolutionarily independent lineages—is one of the most important phenomena in biological science. Charles Darwin recognized its preeminence when he chose the title of *The Origin of Species.* But without the underlying knowledge supplied by the modern science of genetics, Darwin was primarily viewing the consequences of speciation, not its underlying causes. Today biologists are actively searching for and finding answers to the many questions about the process of speciation, something biologists have been known to call "the mystery of mysteries."

IN THIS CHAPTER we will describe what species are and discuss how Earth's millions of species came into being. We will examine the mechanisms by which a lineage splits into new species and how such separations are maintained. Finally, we will look at different factors that can make speciation a rapid or a very slow process.

23.1 What Are Species?

Although "species" is a useful and commonly used term in biology, the concept of "species" sometimes varies among different biologists. Biologists are interested in several different aspects of the divergence of biological lineages. Different biologists think about species differently because they ask different questions about species: How can we recognize and identify species? How do new species arise? How do different species remain separate? Why do rates of speciation differ among groups? In answering these questions, biologists focus on different attributes of species, leading to several different ways of thinking about what species are and how they form. Most of the various *species concepts* proposed by biologists are not mutually exclusive—they are just different ways of approaching the question "What are species?"

We can recognize many species by their appearance

Biological diversity does not always vary in a smooth, incremental way; groups of organisms often differ in distinct, obvious ways. People have long recognized groups of similar organisms that mate with one another, and there are usually distinct morphological breaks between these groups. Groups of organisms that mate with one another are commonly called *species* (note that this is both the plural and singular form of the word). Someone who is knowledgeable about a group of organisms, such as birds or flowering plants, usually can distinguish the different species found in a particular area simply by looking at them. Standard field guides to birds, mammals, insects, and wildflowers are possible only because many species change little in appearance over large geographic distances. A casual birdwatcher can easily recognize male red-winged blackbirds (*Agelaius phoeniceus*) from the east and west coasts of North America as members of the same species (**Figure 23.1A**).

More than 250 years ago, the Swedish biologist Carolus Linnaeus developed the binomial system of biological nomenclature by which species are named today (see Section 22.4). Linnaeus described thousands of species, and because he knew nothing about genetics or the mating behavior of the organisms he was naming, he classified them only on the basis of their appearance. Linnaeus differentiated species using a **morphological species concept**, a construct that assumes a species consists of individuals that "look alike," and that individuals that don't look alike belong to different species. Although Linnaeus did not know it, members of many of the groups that he classified as species by their appearance look alike because they share many of the alleles that code for their body structures.

23.7 Allopatric Speciation among Darwin's Finches The descendants of the ancestral finch that colonized the Galápagos archipelago several million years ago evolved into 14 different species whose members are variously adapted to feed on seeds, buds, and insects. (The fourteenth species, not pictured here, lives in Cocos Island, farther north in the Pacific Ocean.)

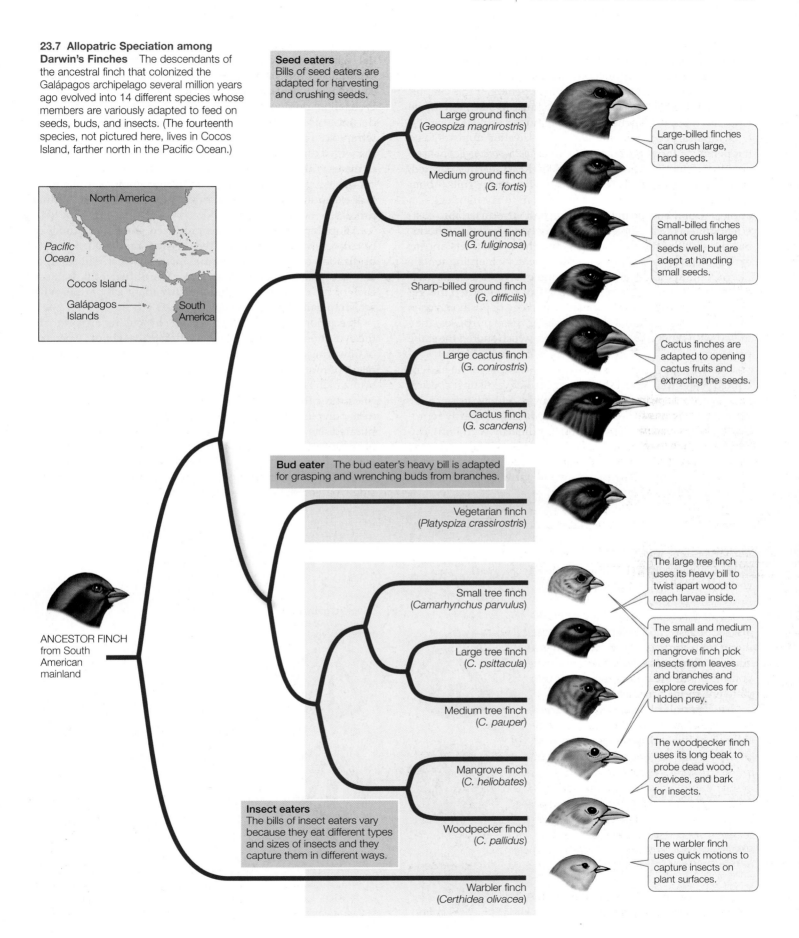

Seed eaters
Bills of seed eaters are adapted for harvesting and crushing seeds.

Large ground finch
(*Geospiza magnirostris*)

Medium ground finch
(*G. fortis*)

Small ground finch
(*G. fuliginosa*)

Sharp-billed ground finch
(*G. difficilis*)

Large cactus finch
(*G. conirostris*)

Cactus finch
(*G. scandens*)

Large-billed finches can crush large, hard seeds.

Small-billed finches cannot crush large seeds well, but are adept at handling small seeds.

Cactus finches are adapted to opening cactus fruits and extracting the seeds.

Bud eater The bud eater's heavy bill is adapted for grasping and wrenching buds from branches.

Vegetarian finch
(*Platyspiza crassirostris*)

Small tree finch
(*Camarhynchus parvulus*)

Large tree finch
(*C. psittacula*)

Medium tree finch
(*C. pauper*)

Mangrove finch
(*C. heliobates*)

Woodpecker finch
(*C. pallidus*)

Warbler finch
(*Certhidea olivacea*)

The large tree finch uses its heavy bill to twist apart wood to reach larvae inside.

The small and medium tree finches and mangrove finch pick insects from leaves and branches and explore crevices for hidden prey.

The woodpecker finch uses its long beak to probe dead wood, crevices, and bark for insects.

The warbler finch uses quick motions to capture insects on plant surfaces.

Insect eaters
The bills of insect eaters vary because they eat different types and sizes of insects and they capture them in different ways.

North America

Pacific Ocean

Cocos Island

Galápagos Islands

South America

ANCESTOR FINCH from South American mainland

to mate with one another? Sympatric speciation may occur with some form of disruptive selection in which certain genotypes have a preference for distinct microhabitats where mating takes place. The experiment described in the opening of this chapter shows that this kind of disruptive selection can take place in the laboratory, but does it also occur in nature?

Sympatric speciation via disruptive selection appears to be happening in the apple maggot fly (*Rhagoletis pomonella*) in eastern North America. Until the mid-1800s, *Rhagoletis* flies courted, mated, and deposited their eggs only on hawthorn fruits. About 150 years ago, some *Rhagoletis* flies began to lay their eggs on apples, which European immigrants had introduced into eastern North America. Apple trees are closely related to hawthorns, but the smell of the fruits differs, and the apple fruits appear earlier than those of hawthorns. Some early-emerging female *Rhagoletis* laid their eggs on apples and evolved a genetic preference for the smell of apples. Their offspring inherited this genetic preference for apples for mating and egg deposition. When the offspring sought out apple trees for these purposes, they mated with other flies reared on apples, which shared the same preferences.

Today the two groups of *Rhagoletis pomonella* in the eastern United States may be on the way to becoming distinct species. One group mates and lays eggs primarily on hawthorn fruits, the other on apples. The two incipient species are partly reproductively isolated because they mate primarily with individuals raised on the same fruit and because they emerge from their pupae at different times of the year. In addition, the apple-feeding flies have evolved so that they now grow more rapidly on apples than they originally did.

Sympatric speciation via ecological isolation, as appears to be happening in *Rhagoletis pomonella*, may be widespread among insects, many of which feed on only a single plant species. The most common means of sympatric speciation, however, is **polyploidy**, or the duplication of sets of chromosomes within individuals (see Section 11.5). Polyploidy can arise either from chromosome duplication in a single species (**autopolyploidy**) or from the combining of the chromosomes of two different species (**allopolyploidy**).

An autopolyploid individual originates when (for example) two accidentally unreduced diploid gametes (with two sets of chromosomes) combine to form a tetraploid individual (with four sets of chromosomes). Tetraploid and diploid individuals of the same species are reproductively isolated because their hybrid offspring are triploid and are usually sterile; they cannot produce viable gametes because their chromosomes do not segregate evenly during meiosis (**Figure 23.8**). So a tetraploid individual cannot produce viable offspring by mating with a diploid individual—but it *can* do so if it self-fertilizes or mates with another tetraploid. Thus polyploidy can result in complete reproductive isolation in two generations—an important exception to the general rule that speciation is a gradual process.

Allopolyploids may also be produced when individuals of two different (but closely related) species interbreed. Such hybridization often disrupts normal meiosis, which can result in chromosomal doubling. Allopolyploids are often fertile because each of the chromosomes has a nearly identical partner with which to pair during meiosis.

Speciation by polyploidy has been particularly important in the evolution of plants. Botanists estimate that about 70 percent of flowering plant species and 95 percent of fern species are the result of recent polyploidization. Some of these arose from hybridization between two species, followed by chromosomal duplication and self-fertilization. Many other species diverged from polyploid ancestors, so the new species also shared the duplicated sets of chromosomes. New species may arise by means of polyploidy more easily among plants than among animals because plants of many species can reproduce by self-fertilization. In addition, if polyploidy arises in several offspring of a single parent, the siblings can fertilize one another.

23.8 Tetraploids Are Reproductively Isolated from Their Diploid Ancestors
Even if the triploid offspring of a diploid and a tetraploid parent survives and reaches sexual maturity, most of the gametes it produces have aneuploid (unbalanced) numbers of chromosomes. Such triploid individuals are effectively sterile. (For simplicity, the diagram shows only three chromosomes; most species have many more than that.)

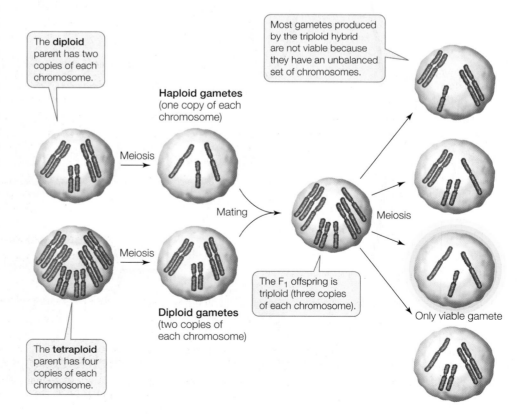

The **diploid** parent has two copies of each chromosome.

Haploid gametes (one copy of each chromosome)

Meiosis

Diploid gametes (two copies of each chromosome)

The **tetraploid** parent has four copies of each chromosome.

Mating

Meiosis

The F$_1$ offspring is triploid (three copies of each chromosome).

Meiosis

Most gametes produced by the triploid hybrid are not viable because they have an unbalanced set of chromosomes.

Only viable gamete

23.2 RECAP

Allopatric speciation results from the separation of populations by geographic barriers; it is the dominant mode of speciation among most groups of organisms. Sympatric speciation may result from ecological isolation, but among plants and some animals, polyploidy is the most common cause of sympatric speciation.

- How can speciation via polyploidy happen in two generations? See p. 488

- Explain why an effective barrier to gene flow for one species may not effectively isolate another species.

---

yourBioPortal.com

GO TO **Animated Tutorial 23.2 • Speciation Mechanisms**

Polyploidy, as we have just seen, can result in a new species that is completely reproductively isolated from its parent species in two generations, but most populations separated by a physical barrier become reproductively isolated only very slowly. Let's see how reproductive isolation may become established once two populations have separated from each other.

23.3 What Happens When Newly Formed Species Come Together?

As discussed in the previous section, once a barrier to gene flow is established, reproductive isolation can develop through genetic divergence. Over many generations, genetic differences accumulate that reduce the probability that members of the two populations can mate and produce viable offspring. In this way, reproductive isolation evolves as a by-product of the genetic changes in the two populations. What types of mechanisms prevent or reduce gene flow between populations, leading to reproductive isolation? Reproductive isolating mechanisms fall into two major categories: **prezygotic reproductive barriers** act *before* fertilization to prevent individuals of different species or populations from mating, whereas **postzygotic reproductive barriers** act *after* fertilization to prevent the development of viable offspring, or to reduce the offsprings' fertility.

Prezygotic barriers prevent fertilization

Prezygotic mechanisms come into play before fertilization and can involve several kinds of reproductive isolation.

HABITAT ISOLATION When individuals of different species evolve genetic preferences for different habitats in which they live or mate, they may never come into contact during their respective mating periods. The *Rhagoletis* flies in the eastern United States (discussed in Section 23.2) experienced such habitat isolation, as did the *Drosophila* in the experiment described in the opening of this chapter.

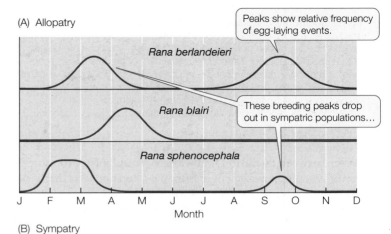

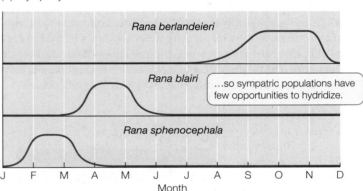

23.9 Temporal Isolation in the Breeding Seasons of Three Species of Frogs (A) The peak breeding seasons of three species of *Rana* overlap when the species are physically separated (allopatry). (B) When two or more species of *Rana* occupy the same territory (sympatry), overlap between peak breeding seasons of each species is greatly reduced or eliminated. In areas where only one species is found, the breeding seasons are broader. Selection against hybridization in areas of overlap helps reinforce the prezygotic isolating mechanism.

TEMPORAL ISOLATION Many organisms have distinct mating seasons. If two closely related species breed at different times of the year (or different times of day), the two may never have an opportunity to hybridize. For example, in sympatric populations of three closely related leopard frogs, each species breeds at a different time of year (**Figure 23.9**). Although there is some overlap in the breeding seasons, the opportunities for hybridization are minimized.

MECHANICAL ISOLATION Differences in the sizes and shapes of reproductive organs may prevent the union of gametes from different species. With animals, this may involve a match in the shape of reproductive organs between males and females, so that reproduction between species with mismatching structures is not physically possible. In plants, the mechanical isolation between species may involve a pollinator. For example, orchids of the genus *Cryptostylis* produce flowers that look and smell like the females of particular species of wasps (**Figure 23.10**). When a male wasp visits and attempts to mate with the flower (thinking it is a female wasp), the mating action results in transfer of

23.10 Mechanical Isolation through Mimicry Many orchid species maintain reproductive isolation because their flowers look and smell like a specific species of bee or wasp, inducing copulatory actions on the part of that specific pollinator insect. The placement of the anthers and stigmas on the flower results in transfer of pollen from the flower to the insect that "mates" with it, and then from the insect to the next flower with which it attempts to mate. Shown here are an Australian orchid (*Cryptostylis* sp.) and its pollinator, a male wasp of the genus *Lissopimpla*.

pollen between the flower and wasp as a result of appropriately configured anthers and stigmas on the flower. Insects that visit the flower but do not attempt to mate with it do not trigger the transfer of pollen between the insect and flower.

BEHAVIORAL ISOLATION   Individuals of a species may reject, or fail to recognize, individuals of other species as potential mating partners. For example, the breeding calls of male frogs quickly diverge between related species (**Figure 23.11**). Female frogs respond to and approach calls from males of their own species, but ignore the calls of even closely related species.

Sometimes the mate choice of one species is mediated by the behavior of individuals of other species. For example, whether two plant species hybridize may depend on the food preferences of their pollinators. The floral traits of plants, including their color and shape, can enhance reproductive isolation either by influencing which pollinators are attracted to the flowers or by altering where pollen is deposited on the bodies of pollinators. A plant whose flowers are pendant (**Figure 23.12A**) will be pollinated by an animal with different physical characteristics than a plant in which the flowers grow upright (**Figure 23.12B**). Because each pollinator prefers (and is adapted to) a different type of flower, the pollinators rarely transfer pollen from one plant species to the other.

Such isolation by pollinator behavior is seen in the case of two sympatric species of columbines (*Aquilegia*) in the mountains of California that have diverged in flower color, structure, and orientation. *Aquilegia formosa* has pendant flowers with short spurs (spikelike, nectar-containing structures) and is pollinated by hummingbirds (**Figure 23.12C**). *A. pubescens* has upright, lighter-colored flowers with long spurs and is pollinated by hawkmoths (**Figure 23.12D**). The difference in pollinators means that these two species are effectively reproductively isolated even though they populate the same geographic range.

GAMETIC ISOLATION   Sperm of one species may not attach to the eggs of another species because the eggs do not release the appropriate attractive chemicals, or the sperm may be unable to penetrate the egg because the two gametes are chemically in-

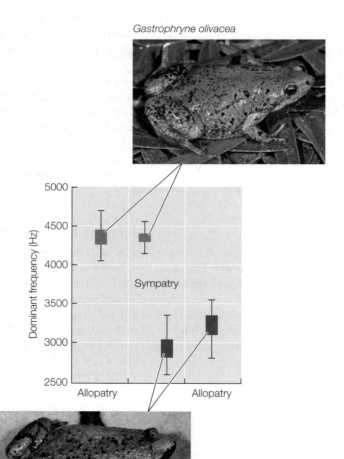

Gastrophryne olivacea

Gastrophryne carolinensis

23.11 Behavioral Isolation in the Mating Calls of Male Frogs   The males of most species of frogs produce species-specific calls. The calls of the two closely related frog species in this figure differ in their dominant frequency (a high-frequency sound wave results in a high-pitched sound; a low frequency results in a low-pitched sound). Female frogs are attracted to the calls of males of their own species. Note that the calls of the two species are more distinct in areas of sympatry than in areas of allopatry (an example of reinforcement).

(A)

(B)

(C) *Aquilegia formosa*

(D) *Aquilegia pubescens*

23.12 Floral Morphology is Associated with Pollinator Morphology (A) This hummingbird's morphology and behavior are adapted to approach plants whose flowers are pendant (hanging downward). (B) The nectar-extracting proboscis of this hawkmoth is adapted to flowers that grow upright. (C) *Aquilegia formosa* flowers are normally pendant and are pollinated by hummingbirds. (D) Flowers of *A. pubescens* are normally upright, which facilitates pollination by hawkmoths. In addition, the long floral spurs appear to restrict access by some other potential pollinators.

compatible. Thus, even though individuals of the two species may attempt to mate, the gametes never fuse into a zygote. For example, gametic isolation has arisen between species of sea urchins. A protein known as bindin occurs in sea urchin sperm and functions in attaching ("binding") the sperm to eggs. All sea urchins produce this egg-recognition protein, but the gene sequence diverges rapidly between species. The sperm protein evolves so that it will only bind to eggs of the same species, thus preventing interspecific hybridization.

Postzygotic barriers can isolate species after fertilization

If individuals of two different populations lack complete prezygotic reproductive barriers, postzygotic reproductive barriers may prevent the species from merging. Genetic differences that accumulate while the populations are isolated from each other may reduce the survival and reproduction of hybrid offspring in any of several ways:

- *Low hybrid zygote viability.* Hybrid zygotes may fail to mature normally, either dying during development or developing such severe abnormalities that they cannot mate as adults.

- *Low hybrid adult viability.* Hybrid offspring may simply have lower survivorship than offspring resulting from within-population matings.

- *Hybrid infertility.* Hybrids may mature normally but be infertile. For example, the offspring of matings between horses and donkeys—mules—are healthy but sterile; they produce no descendants.

Although natural selection does not directly favor the evolution of postzygotic reproductive barriers, if hybrid offspring survive poorly, natural selection may favor the evolution of prezygotic barriers. This happens because individuals that mate with individuals of the other population will leave fewer surviving descendants than individuals that mate only within their own population. In this case, individuals that can avoid mating with members of the other population have a selective advantage, and any trait that favors such avoidance will be favored by natural selection. Such strengthening of prezygotic barriers is known as **reinforcement**.

Donald Levin of the University of Texas noticed that individuals of *Phlox drummondii* in most of the range of the species in Texas have pink flowers. Where *P. drummondii* is sympatric with the pink-flowered *P. cuspidata*, however, *P. drummondii* has red flowers. No other *Phlox* species has red flowers. Levin performed an experiment whose results showed that reinforcement might explain the evolution of red flowers where the two species are sympatric (**Figure 23.13**).

Reinforcement can also be detected by comparing sympatric and allopatric populations of potentially hybridizing species. If reinforcement is occurring, then sympatric pairs of closely related species should evolve more effective prezygotic reproductive barriers than do allopatric populations of the same species. The examples of temporal isolation shown in Figure 23.9 and of behavioral isolation shown in Figure 23.11 illustrate reinforcement of prezygotic barriers. The breeding seasons of the sympatric populations of frogs (Figure 23.9) overlap much less than do those of the corresponding allopatric populations. Similarly, the frequencies of the frog mating calls illustrated in Figure 23.11 are more divergent in sympatric populations than in allopatric populations. In both cases, there appears to have been selection against hybrids in areas of sympatry, so individuals that do not produce hybrids are more likely to leave more genes to future generations.

INVESTIGATING LIFE

23.13 Flower Color Reinforces a Reproductive Barrier in *Phlox*

Most *Phlox drummondii* flowers are pink, but in regions where they are sympatric with *P. cuspidata*—which is always pink—most *P. drummondii* individuals are red. Most pollinators preferentially visit flowers of one color or the other. In this experiment, Donald Levin explored whether flower color reinforces a prezygotic reproductive barrier, lessening the chances of interspecific hybridization.

HYPOTHESIS Red-flowered *P. drummondii* are less likely to hybridize with *P. cuspidata* than are pink-flowered *P. drummondii*.

METHOD

1. Introduce equal numbers of red- and pink-flowered *P. drummondii* individuals into an area with many pink-flowered *P. cuspidata*.

P. cuspidata

P. drummondii

2. After the flowering season ends, measure hybridization by assessing the genetic composition of the seeds produced by *P. drummondii* plants of both colors.

RESULTS Of the seeds produced by pink-flowered *P. drummondii*, 38% were hybrids with *P. cuspidata*. Only 13% of the seeds produced by red-flowered individuals were genetic hybrids.

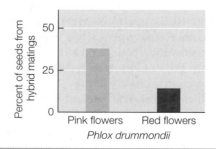

Phlox drummondii

CONCLUSION *P. drummondii* and *P. cuspidata* are less likely to hybridize if the flowers of the two species differ in color.

FURTHER INVESTIGATION: This experiment did not address the probable reproduction advantages for individual *Phlox* plants of donating and receiving primarily intraspecific pollen. Can you design an experiment to measure such an advantage?

Go to **yourBioPortal.com** for original citations, discussions, and relevant links for all INVESTIGATING LIFE figures.

During the process of reinforcement, closely related species may form hybrids in areas where their ranges overlap, and they may continue to do so for many years. Let's examine what happens when reproductive barriers do not completely prevent individuals from different populations from mating and producing offspring.

Hybrid zones may form if reproductive isolation is incomplete

If contact is reestablished between formerly isolated populations before complete reproductive isolation has developed, members of the two populations may interbreed. Three outcomes of such interbreeding are possible:

- If hybrid offspring are as fit as those resulting from matings within each population, hybrids will mate with individuals of both parental species. The gene pools will gradually become completely mixed, resulting in one species.

- If hybrid offspring are less fit, complete reproductive isolation may evolve as reinforcement strengthens prezygotic reproductive barriers.

- Even if hybrid offspring are at some disadvantage, a narrow **hybrid zone**—a region in which genetically distinct populations come together and produce offspring of mixed ancestry—may develop in the absence of reinforcement, or before reinforcement is complete.

When a hybrid zone first forms, most hybrids are offspring of crosses between purebred individuals of the two populations. However, subsequent generations include a variety of individuals with different proportions of their genes derived from the original two populations. Thus hybrid zones contain recombinant individuals resulting from many generations of hybridization. Detailed genetic studies can tell us much about why hybrid zones may be narrow and stable for long periods of time.

The hybrid zone between two species of European toads of the genus *Bombina* has been studied intensively. The fire-bellied toad (*B. bombina*) lives in eastern Europe; the closely related yellow-bellied toad (*B. variegata*) lives in western and southern Europe. The ranges of the two species overlap in a long but very narrow zone stretching 4,800 kilometers from eastern Germany to the Black Sea (**Figure 23.14**). Hybrids between the two species suffer from a range of defects, many of which are lethal. Those hybrids that survive often have skeletal abnormalities, such as misshapen mouths, ribs that are fused to vertebrae, and a reduced number of vertebrae. By following the fates of thousands of toads from the hybrid zone, investigators found that a hybrid toad is on average only half as fit as a purebred individual. The

23.14 Hybrid Zones May Be Long and Narrow The narrow zone in Europe where fire-bellied toads meet and hybridize with yellow-bellied toads stretches across Europe. This hybrid zone has been stable for hundreds of years and has never expanded, and no reinforcement has evolved.

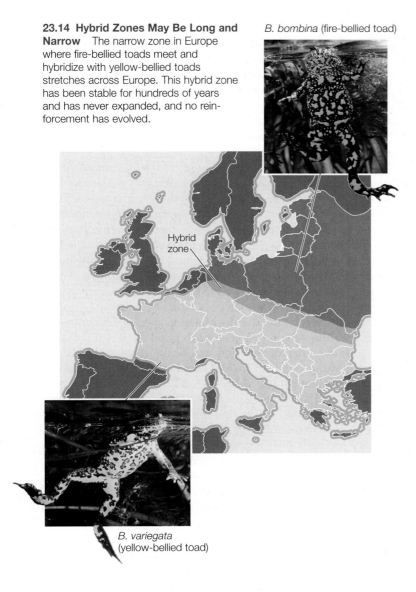

B. bombina (fire-bellied toad)

Hybrid zone

B. variegata (yellow-bellied toad)

hybrid zone thus remains narrow because there is strong selection against hybrids, and because adult toads do not move over long distances. The zone has persisted for hundreds of years, however, because many individuals of both species continue to move short distances into it, constantly replenishing the hybrid population.

23.3 RECAP

Reproductive isolation may result from prezygotic or postzygotic reproductive barriers. Lower fitness of hybrids in contact zones can lead to the reinforcement of prezygotic reproductive barriers.

- Describe various kinds of prezygotic and postzygotic reproductive barriers. See pp. 489–491

- Why is reinforcement of prezygotic barriers likely if hybrid offspring survive more poorly than offspring produced by within-population matings? See p. 491

Some groups of organisms have many species, others only a few. Hundreds of species of *Drosophila* evolved in the small area of the Hawaiian Islands over about 20 million years. In contrast, there are only a few species of horseshoe crabs in the world, and only one species of ginkgo tree, even though these latter groups have persisted for hundreds of millions of years. Why do different groups of organisms have such different rates of speciation?

23.4 Why Do Rates of Speciation Vary?

Rates of speciation (the proportion of existing species that split to form new species over a given period of time) vary greatly because many factors influence the likelihood that a lineage will split to form two or more species. What are some of the factors that influence the probability of a given lineage splitting into two?

Populations of species that have specialized diets may be more likely to diverge than are populations that have generalized diets. To investigate the effects of diet on rates of speciation, Charles Mitter and colleagues compared species richness in some closely related groups of true bugs (hemipterans). The common ancestor of these groups was a predator that fed on other insects, but a dietary shift to herbivory (eating plants) evolved at least twice in the groups under study. The herbivorous groups have many more species than those that are predatory (**Figure 23.15**). Herbivorous bugs typically specialize on one or a few closely related species of plants, whereas predatory bugs tend to feed on many different species of insects. High diversity of host plants can thus lead to a correspondingly high diversity in the herbivorous specialists.

Speciation rates in plants are faster in animal-pollinated than in wind-pollinated plants. Animal-pollinated groups have, on average, 2.4 times as many species as related groups pollinated by wind. Among animal-pollinated plants, speciation rates are correlated with pollinator specialization. In columbines (*Aquilegia*), the rate of evolution of new species has been about three times faster in lineages that have long nectar spurs than in lineages that lack spurs. Why do nectar spurs increase the spe-

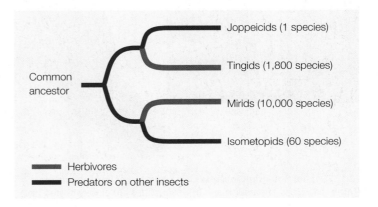

Joppeicids (1 species)

Common ancestor

Tingids (1,800 species)

Mirids (10,000 species)

Isometopids (60 species)

— Herbivores
— Predators on other insects

23.15 Dietary Shifts Can Promote Speciation Herbivorous groups of hemipteran insects have speciated several times faster than closely related predatory groups.

ciation rate? Apparently it is because having longer spurs restricts the number of pollinator species that visit the flowers, thus increasing opportunities for reproductive isolation (see Figure 23.12).

The mechanisms of sexual selection (see Section 21.2) appear to result in increased rates of speciation. Some of the most striking examples of sexual selection are found in birds with promiscuous mating systems. Bird-watchers travel thousands of miles to Papua New Guinea to witness the mating displays of male birds of paradise, some of which have long, brightly colored tail feathers and look distinctly different than the females (*sexual dimorphism*). In many of these 33 species, males assemble at display grounds, called *leks*, and females come there to choose a male with whom to copulate. After mating, the females leave the display grounds, build their nests, lay their eggs, and feed their offspring with no help from the males. The males remain to court more females (**Figure 23.16A**).

The closest relatives of the birds of paradise are the manucodes. Male and female manucodes differ only slightly in size and plumage (so they are *sexually monomorphic*). They form monogamous pair bonds, and both sexes contribute to raising the young. There are only 5 species of manucodes (**Figure 23.16B**), compared with 33 species of birds of paradise. This

is just one comparison, and by itself would not be convincing evidence that sexually dimorphic clades of birds have higher rates of speciation than do monomorphic clades. However, when biologists compare all the examples of birds in which one clade is sexually dimorphic, and the most closely related clade is sexually monomorphic, the sexually dimorphic clades are significantly more likely to contain more species. But why would sexual dimorphism be associated with a higher rate of speciation?

Animals with complex sexually selected behaviors are likely to form new species at a high rate because they make sophisticated discriminations among potential mating partners. They distinguish members of their own species from members of other species, and they make subtle discriminations among members of their own species on the basis of size, shape, appearance, and behavior. Such discriminations can greatly influence which individuals are most successful in mating and producing offspring, and may lead to rapid evolution of prezygotic reproductive barriers among populations.

Speciation rates are usually higher in species with poor dispersal abilities than in species with good dispersal abilities, because even narrow barriers can be effective in dividing a species whose members are highly sedentary. The Hawaiian Islands have about 1,000 species of land snails, many of which are restricted to a single valley. Because snails move only short distances, the high ridges that separate the valleys are effective barriers to their dispersal.

The proliferation of a large number of daughter species from a single ancestor is called an **evolutionary radiation**. If the rapid proliferation of species results in an array of species that live in a variety of environments and differ in the characteristics they use to exploit those environments, the radiation is said to be **adaptive**. Several remarkable adaptive radiations have occurred in the Hawaiian Islands. In addition to its 1,000 species of land snails, the native biota of the Hawaiian Islands includes 1,000 species of flowering plants, 10,000 species of insects, and more than 100 bird species. However, there were no amphibians, no terrestrial reptiles, and only one native terrestrial mammal—a bat—on the islands until humans introduced additional species. The 10,000 known native species of insects on Hawaii are believed to have evolved from only about 400 immigrant species; only 7 immigrant species are believed to account for all the native Hawaiian land birds. Similarly, as we saw earlier in this chapter, an adaptive radiation in the Galápagos archipelago resulted in the 14 species of Darwin's finches, which differ strikingly in the size and shape of their bills and, accordingly, in the food resources they use (see Figure 23.7).

The 28 species of Hawaiian sunflowers called silverswords are an impressive example of an adaptive radiation in plants. DNA sequences show that these species share a relatively recent common ancestor with a species of tarweed from the Pacific coast of North America (**Figure 23.17**). Whereas all mainland tarweeds are small, upright herbs (nonwoody plants), the silverswords include prostrate and upright herbs, shrubs, trees, and vines. Silversword species occupy nearly all the habitats of the Hawaiian Islands, from sea level to above timberline

(A) *Paradisaea minor*

(B) *Manucodia comrii*

23.16 Sexual Selection in Birds Can Lead to Higher Speciation Rates (A) Birds of paradise and (B) manucodes are closely related bird groups of the South Pacific. However, speciation rates are much higher among the sexually dimorphic, polygynous birds of paradise (33 species) than among manucodes (5 species).

Wilkesia hobdyi

Madia sativa (tarweed)

Dubautia menziesii

Argyroxiphium sandwicense

23.17 Rapid Evolution among Hawaiian Silverswords The Hawaiian silverswords, three closely related genera of the sunflower family, are believed to have descended from a single common ancestor (a plant similar to the tarweed; upper right) that colonized Hawaii from the Pacific coast of North America. The four plants shown here are more closely related than they appear to be based on their morphology.

in the mountains. Despite their extraordinary morphological diversification, the silverswords are genetically very similar.

The Hawaiian silverswords are more diverse in size and shape than the mainland tarweeds because the tarweed ancestors first arrived on islands that harbored very few plant species. In particular, there were few trees and shrubs, because such large-seeded plants rarely disperse to oceanic islands. Trees and shrubs have evolved from nonwoody ancestors on many oceanic islands. On the mainland, however, tarweeds live in ecological communities that contain many tree and shrub species in lineages with long evolutionary histories. In those environments, opportunities to exploit the "tree" way of life had already been preempted.

23.4 RECAP

Dispersal ability, dietary specialization, and mechanisms of sexual selection affect rates of speciation. Speciation rates in plants can depend on mechanisms of pollination. Open ecological niches present opportunities for evolutionary radiations.

- Explain how pollinator specialization in plants and sexual selection in animals can increase rates of speciation. **See pp. 493–494**

- Why do adaptive radiations often occur when a founder species invades an isolated geographic area? **See p. 494**

The processes described in this chapter, operating over billions of years, have produced a world in which life is organized into millions of species, each adapted to live in a particular environment and to use environmental resources in a particular way. In the next chapter we consider how species evolve at the level of their genes and genomes.

CHAPTER SUMMARY

23.1 What Are Species?

- **Speciation** is the process by which one species splits into two or more daughter species, which thereafter evolve as distinct lineages.

- The **biological species concept** distinguishes species on the basis of **reproductive isolation**.

- The **morphological species concept** distinguishes species on the basis of physical similarities; it often underestimates or over-estimates the actual number of reproductively isolated species.

- The **lineage species concept** recognizes evolutionarily inde-pendent lineages as species, allowing biologists to consider species over evolutionary time.

23.2 How Do New Species Arise?

- Speciation usually results from the interruption of gene flow within a population.

- The Dobzhansky-Muller model describes how reproductive iso-lation can develop between two descendant species. **Review Figure 23.2**

- **Allopatric speciation**, which results when populations are sepa-rated by a physical barrier, is the dominant mode of speciation. This type of speciation may follow from founder events, in which some members of a population cross a barrier and found a new, isolated population. **Review Figure 23.5, ANIMATED TUTORIAL 23.1**

- **Sympatric speciation** results when the genomes of two groups diverge in the absence of physical isolation. It can result from disruptive selection for two or more distinct microhabitats.

- Sympatric speciation can occur within two generations via **polyploidy**, an increase in the number of chromosomes sets. Polyploidy may arise from chromosome duplications within a species (**autopolyploidy**) or from hybridization that results in combining the chromosomes of two species (**allopolyploidy**). **Review Figure 23.8**

SEE ANIMATED TUTORIAL 23.2

23.3 What Happens When Newly Formed Species Come Together?

- **Prezygotic barriers** to reproduction operate before fertilization; **postzygotic barriers** to reproduction operate after fertilization. Prezygotic barriers may be favored by natural selection if postzygotic barriers are incomplete. **Review Figures 23.9, 23.11**

- **Hybrid zones** may form when previously separated populations come into contact and reproductive isolation is incomplete. **Review Figure 23.14**

23.4 Why Do Rates of Speciation Vary?

- Dispersal ability, dietary specialization, type of pollination, and sexual selection all influence speciation rates. **Review Figure 23.15**

SEE WEB ACTIVITY 23.1 for a concept review of this chapter.

SELF-QUIZ

1. The biological species concept defines a species as a group of
   a. actually interbreeding natural populations that are repro-ductively isolated from other such groups.
   b. potentially interbreeding natural populations that are reproductively isolated from other such groups.
   c. actually or potentially interbreeding natural populations that are reproductively isolated from other such groups.
   d. actually or potentially interbreeding natural populations that are reproductively connected to other such groups.
   e. actually interbreeding natural populations that are repro-ductively connected to other such groups.

2. Which of the following is *not* a condition expected to favor allopatric speciation?
   a. Continents drift apart and separate previously connected lineages.
   b. A mountain range separates formerly connected popula-tions.
   c. Different environments on two sides of a barrier cause populations to diverge.
   d. The range of a species is separated by loss of intermedi-ate habitat.
   e. Tetraploid individuals arise in one part of the range of a species.

3. Finches speciated in the Galápagos Islands because
   a. the Galápagos Islands are not far from the mainland.
   b. the Galápagos Islands are thought to promote sympatric speciation in birds.
   c. hybridization across different island populations of finch-es led to high levels of polyploidy.

   d. the islands of the Galápagos archipelago are sufficiently isolated from one another that there is little migration among them.
   e. the islands of the Galápagos archipelago are close enough to one another that there is considerable migra-tion among them.

4. Which of the following is *not* a potential prezygotic repro-ductive barrier?
   a. Temporal segregation of breeding seasons
   b. Differences in chemicals that attract mates
   c. Hybrid infertility
   d. Spatial segregation of mating sites
   e. Sperm that cannot penetrate an egg

5. A common means of sympatric speciation is
   a. polyploidy.
   b. hybrid infertility.
   c. temporal segregation of breeding seasons.
   d. spatial segregation of mating sites.
   e. imposition of a geographic barrier.

6. Narrow hybrid zones may persist for long times because
   a. hybrids are always at a disadvantage.
   b. hybrids have an advantage only in narrow zones.
   c. hybrid individuals never move far from their birthplaces.
   d. individuals that move into the zone have not previously encountered individuals of the other species, so rein-forcement of reproductive barriers has not occurred.
   e. Narrow hybrid zones are artifacts because biologists gen-erally restrict their studies to contact zones between species.

7. Which statement about speciation is *not* true?
   a. It always takes thousands of years.
   b. Reproductive isolation may develop slowly between diverging lineages.
   c. Among animals, it usually requires a physical barrier.
   d. Among plants, it often happens as a result of polyploidy.
   e. It has produced the millions of species living today.

8. Which of the following is often associated with higher rates of speciation?
   a. Sexually dimorphic compared with sexually monomorphic birds
   b. Insects with specialized diets compared with insects with generalized diets
   c. Species with low dispersal ability compared with species with high dispersal ability
   d. Plants with animal pollination compared with plants with wind pollination
   e. All of the above

9. Evolutionary radiations
   a. happen often on continents but rarely on island archipelagoes.
   b. characterize birds and plants but not other groups of organisms.
   c. have happened on continents as well as on islands.
   d. require major reorganizations of the genome.
   e. never happen in species-poor environments.

10. Speciation is an important component of evolution because it
   a. generates the variation on which natural selection acts.
   b. generates the variation on which genetic drift and mutations act.
   c. enabled Charles Darwin to perceive the mechanisms of evolution.
   d. generates the high extinction rates that drive evolutionary change.
   e. has resulted in a world with millions of species, each adapted for a particular way of life.

FOR DISCUSSION

1. The North American snow goose has two distinct color forms, blue and white. Matings between the two color forms are common. However, blue individuals pair with blue individuals and white individuals pair with white individuals much more frequently than would be expected by chance. Suppose that blue and white snow geese are equally frequent in a population, and that 75 percent of all mated pairs consist of two individuals of the same color. What would you conclude about speciation processes in these geese? What if 100 percent of pairs were the same color?

2. Suppose pairs of snow geese of mixed colors were found only in a narrow zone within the broad Arctic breeding range of the geese, with blue geese found on one side and white geese found on the other side of this narrow zone. Would your answer to Question 1 remain the same?

3. Although many butterfly species are divided into local populations among which there is little gene flow, these species often show relatively little morphological variation among populations. Describe the studies you would conduct to determine what maintains this morphological similarity.

4. Evolutionary radiations are common and easily studied on oceanic islands. In what types of *mainland* situations would you expect to find major evolutionary radiations? Why?

5. Fruit flies of the genus *Drosophila* are distributed worldwide, but 30 to 40 percent of all the species in the genus are found on the Hawaiian Islands (which comprise far less than 1% of Earth's total land area). What might account for this distribution pattern?

6. What factors can cause extinction rates to exceed speciation rates in a clade? Name some clades in which human activities are increasing extinction rates without increasing speciation rates.

7. If it is true that natural selection does not directly favor lower viability of hybrids, why is it that hybrid individuals so often have lowered viability?

ADDITIONAL INVESTIGATION

In the two *Aquilegia* species shown in Figure 23.12, the orientation of the flowers and the length of flower spurs are associated with the respective pollinator species (hummingbirds and hawkmoths). Columbine flowers vary in other ways as well; for example, they differ in color, and probably in odor. What experiments could you design to determine the traits that various pollinators use to distinguish among the flowers of different columbine species?

WORKING WITH DATA (GO TO yourBioPortal.com)

Examining Evidence for Reinforcement of Prezygotic Barriers
In this exercise based on Figure 23.13, you will examine some of the data collected by Don Levin to study reinforcement of prezygotic reproductive barriers in *Phlox*. You will also critique the study design of the experiment, and consider alternative explanations for the results.

Shocking evolution

Some fishes, including the famous electric eel of Central and South America, can produce high-voltage discharges of electricity (up to 650 volts) that they use to stun their prey. A variety of other fish species are known to produce somewhat weaker electric discharges. Most of these latter species live in murky water where visual cues are limited; they use electric signals to locate (but not to stun) their prey. Electric signals also allow them to communicate with other individuals of their own species.

Electric organs have evolved independently in several fish lineages. How did these organs evolve? Let's consider first the physical basis of the electrical signal. *Voltage-gated sodium channels* are large proteins that underlie the generation and propagation of rapid electrical signals in nerve, muscle, and heart tissues (see Chapter 45). Electric signals are transmitted along nerves to muscles as the sodium channels embedded in cell membranes are stimulated to open. These channels control the concentration of positively charged sodium ions (Na^+) on the inside relative to the outside of cells, resulting in an electric charge that is transmitted across the surface of the muscle, leading to muscular contraction.

Most vertebrates have a number of different copies of the genes encoding the several proteins that make up the sodium channel. These copies arose through a series of *gene duplications* in the distant past of vertebrate genome evolution. Such duplications allowed for the specialization of protein function, making it possible for different sodium channels to exist in different types of tissue. In the case of electric fishes, one of the sodium channel genes ordinarily expressed in muscle diverged and a new functional protein evolved. Changes in a relatively small number of nucleotide positions in the gene resulted in modified sodium channels, allowing the development of a new organ with a unique function—the generation of externally transmitted electric energy.

The "living battery" electric organ differs from skeletal muscle in important ways. The organ is composed of many *electrocytes*, each of which is a derived muscle cell capable of producing a small electric charge. Electrocytes are stacked in series, much like the plates in a car battery. Rather than producing muscle contraction and movement, however, the organ generates an electric discharge. This signal is species-specific, which allows intraspecific communication and also serves as an isolating mechanism between species (see Chapter 23).

The repeated evolution of electric organs from muscle tissue is facilitated by relatively simple molecular changes

An Electric Fish The elephant-nose fish (*Gnathonemus petersi*), a river-dwelling species from West Africa, is one of many fishes in which weakly discharging electric organs have evolved via modifications in sodium channel proteins.

A High-Voltage Electric Fish This torpedo ray can put out as much as 220 volts of electricity. So far this particular species remains unidentified; it has been found only in a single bay among the islands of Komodo National Park, Indonesia.

in certain genes, changes that result in major functional changes to sodium channels. Gene duplication facilitates the process, since "extra" genes allow for such specialization in protein function. Finally, interspecific differences in sodium channel function result from additional changes in nucleotide sequences of the respective genes. These small differences allow different species of fishes to use different communication signals, which improves intraspecific communication while reducing interspecific interference.

The evolution of sodium channels is just one example of how an understanding of the evolution of genes and genomes helps biologists understand the diversity of life on Earth. Molecular investigations also allow biologists to observe the process of evolution directly in the laboratory, and to use evolutionary principles to produce new molecules with useful functions.

IN THIS CHAPTER we will see how molecular biologists infer both the patterns and the causes of molecular evolution from studies of nucleic acids and proteins. We will explore how the functions of molecules change, how genomes change in size, and where new genes come from. Finally, we will explore some practical applications of molecular evolution for producing new molecules with novel functions.

24.1 How Are Genomes Used to Study Evolution?

An organism's **genome** is the full set of genes it contains, as well as any noncoding regions of the DNA (or in the case of some viruses, RNA). Most of the genes of eukaryotic organisms are found on chromosomes in the nucleus, but genes are also present in chloroplasts and mitochondria. In organisms that reproduce sexually, both males and females transmit nuclear genes, but mitochondrial and chloroplast genes usually are transmitted only via the cytoplasm of one of the two gametes (usually from the female parent).

Genomes must be replicated to be transmitted from parents to offspring. DNA replication does not occur without error, however. Mistakes in DNA replication—mutations—provide much of the raw material for evolutionary change. Mutations are essential for the long-term survival of life, because they are the initial source of the genetic variation that permits organisms to evolve in response to changes in their environment.

A particular copy of a gene will not be passed on to successive generations unless an individual with that copy survives and reproduces. Therefore, the capacity to cooperate with different combinations of other genes is likely to increase the probability that a particular allele will become fixed in a population. Moreover, the degree and timing of a gene's expression are affected by its location in the genome. For these reasons, the genes of an individual organism can be viewed as interacting members of a group, among which there are divisions of labor but also strong interdependencies.

A genome, then, is not simply a random collection of genes in random order along chromosomes. Rather, it is a complex set of integrated genes, regulatory sequences, and structural elements, as well as vast stretches of noncoding DNA that may have little direct function. The positions of genes, as well as their sequences, are subject to evolutionary change, as are the extent and location of noncoding DNA. All of these changes can affect the phenotype of an organism. Biologists have now sequenced the complete genomes of a large number of organisms, including humans. This information is helping us to understand how and why organisms differ, how they function, and how they have evolved.

Evolution of genomes results in biological diversity

The field of **molecular evolution** investigates the mechanisms and consequences of the evolution of macromolecules. Molecular evolutionists study relationships between the structures of genes and proteins and the functions of organisms. They also examine molecular variation to reconstruct evolutionary history and to study the mechanisms and consequences of evolution. The molecules of special interest to molecular evolutionists are nucleic acids (DNA and RNA) and proteins. Students of this field ask questions such as: What does molecular variation tell us about a gene's function? Why do the genomes of different organisms vary in size? What evolutionary forces shape patterns of variation among genomes? And a crucial question from an evolutionary perspective, How do genomes acquire new functions? Investigations into the evolution of particular nucleic acids and proteins are instrumental in reconstructing the evolutionary histories of genes and in determining which organisms carry them. Ultimately, molecular evolutionists hope to explain the molecular basis of biological diversity.

The evolution of nucleic acids and proteins depends on genetic variation introduced by mutations. One of several ways in which genes evolve is by means of *nucleotide substitutions*. In genes that encode proteins, nucleotide substitutions sometimes result in amino acid replacements that can change the charge, the structure (secondary or tertiary), and other chemical and physical properties of the encoded protein. Phenotypic changes in the protein often affect the way that protein functions in the organism.

Evolutionary changes in genes and proteins can be identified by comparing the nucleotide or amino acid sequences of different organisms. The longer two sequences have been evolving separately, the more differences they accumulate (bearing in mind that different genes in the same species evolve at different rates). Determining when changes in nucleotide or amino acid sequences occurred is a first step toward inferring their causes. Knowledge of the pattern and rate of evolutionary change in a given macromolecule is useful in reconstructing the evolutionary history of groups of organisms.

To compare genes or proteins across different organisms, biologists need a way to identify homologous parts of molecules. (Recall from Section 22.1 that *homologous* features are shared by two or more species and have been inherited from a common ancestor.) Homologous parts of a protein can be traced to homologous amino acid sequences. And, since nucleotide sequences encode amino acid sequences, the concept of homology extends down to the level of individual nucleotide positions. Therefore, one of the first steps in studying the evolution of genes or proteins is to align homologous positions in the nucleotide or amino acid sequence of interest.

Genes and proteins are compared through sequence alignment

Once the DNA or amino acid sequences of molecules from different organisms have been determined, they can be compared. Homologous positions can be identified only if we first pinpoint the locations of deletions

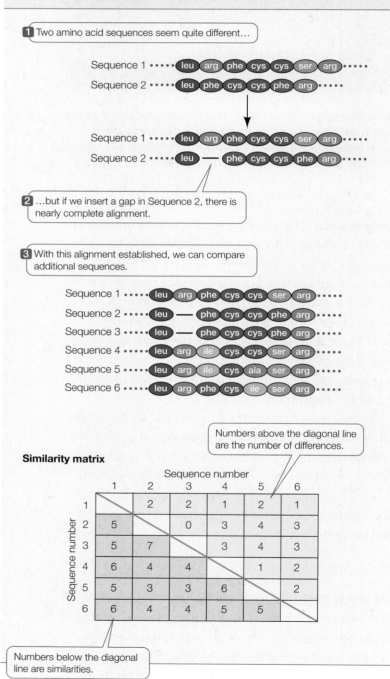

TOOLS FOR INVESTIGATING LIFE

24.1 Amino Acid Sequence Alignment

Amino acid sequence alignment is a way of arranging protein sequences to identify regions of homology between the sequences. Gaps are inserted between the amino acid residues to align similar residues in columns. Differences and similarities between each pair of aligned sequences are then summarized in a similarity matrix. Homologous DNA sequences can be aligned in a similar manner.

1 Two amino acid sequences seem quite different…

Sequence 1 ····· leu arg phe cys cys ser arg ·····
Sequence 2 ····· leu phe cys cys phe arg ·····

Sequence 1 ····· leu arg phe cys cys ser arg ·····
Sequence 2 ····· leu — phe cys cys phe arg ·····

2 …but if we insert a gap in Sequence 2, there is nearly complete alignment.

3 With this alignment established, we can compare additional sequences.

Sequence 1 ····· leu arg phe cys cys ser arg ·····
Sequence 2 ····· leu — phe cys cys phe arg ·····
Sequence 3 ····· leu — phe cys cys phe arg ·····
Sequence 4 ····· leu arg ile cys cys ser arg ·····
Sequence 5 ····· leu arg ile cys ala ser arg ·····
Sequence 6 ····· leu arg phe cys ile ser arg ·····

Numbers above the diagonal line are the number of differences.

Similarity matrix

						Sequence number
		1	2	3	4	5	6
Sequence number	1		2	2	1	2	1
	2	5		0	3	4	3
	3	5	7		3	4	3
	4	6	4	4		1	2
	5	5	3	3	6		2
	6	6	4	4	5	5	

Numbers below the diagonal line are similarities.

yourBioPortal.com
GO TO **Web Activity 24.1** • Amino Acid Sequence Alignment

and insertions that have occurred in the molecules of interest in the time since the organisms diverged from a common ancestor. A simple hypothetical example illustrates this **sequence alignment** technique. In **Figure 24.1** we compare two amino acid sequences (1 and 2) from homologous proteins in different organisms. The two sequences at first appear to differ in both the number and identity of their amino acids, but if we insert a gap after the first amino acid in Sequence 2 (after leucine), similarities in the two sequences become more obvious. This gap represents the occurrence of one of two evolutionary events: an insertion of an amino acid in the longer protein, or a deletion of an amino acid in the shorter protein.

Having adjusted for this insertion or deletion event, we can see that the two sequences differ by only one amino acid at position 6 (serine or phenylalanine). Adding a single gap—that is, identifying a deletion or an insertion—*aligns* these sequences. Longer sequences and those that have diverged more extensively require more elaborate adjustments. Explicit models (incorporated into computer algorithms) have been developed to account for the relative probabilities of deletions, insertions, and particular amino acid replacements.

Having aligned the sequences, we can compare them by counting the number of nucleotides or amino acids that differ between them. Summing the number of similar and different amino acids in each pair of sequences allows us to construct a **similarity matrix**, which gives us a measure of the minimum number of changes that have occurred during the divergence between each pair of organisms (see Figure 24.1).

---

yourBioPortal.com

GO TO Web Activity 24.2 • Similarity Matrix Construction

Models of sequence evolution are used to calculate evolutionary divergence

The sequence comparison procedure illustrated in Figure 24.1 gives a simple count of the number of differences and similarities between the proteins of two species. In the context of two aligned DNA sequences, we can count the number of differences at homologous nucleotide positions, and this count indicates the minimum number of nucleotide changes that must have occurred since the two sequences diverged from a common ancestral sequence.

Although it is useful in determining a *minimum* number of changes between two sequences, the count provided by sequence alignment almost certainly underestimates the actual number of changes that have occurred since the sequences diverged. Any given change counted in a similarity matrix of DNA sequences may result from multiple substitution events

that occurred at a given nucleotide position over time. As illustrated in **Figure 24.2**, any of the following events may have occurred at a given nucleotide position that would not be revealed by a simple count of similarities and differences between two DNA sequences:

- *Multiple substitutions.* More than one change occurs at a given position between the ancestral sequence and at least one of the observed sequences.

- *Coincident substitutions.* At a given position, different substitutions occur between the ancestral sequence and each observed sequence.

- *Parallel substitutions.* The same substitution occurs independently between the ancestral sequence and each observed sequence.

- *Back substitutions* (also called *reversions*). In a variation on multiple substitutions, after a change at a given position, a subsequent substitution changes the position back to the ancestral state.

To correct for undercounting of substitutions, molecular evolutionists have developed mathematical models that describe how DNA (and protein) sequences evolve. These models take into account the relative rates of change from one nucleotide to another; for example, *transitions* (changes between the two purines, A ↔ G, or between the pyrimidines, C ↔ T) are more frequent than *transversions* (a purine is replaced by a pyrimidine, or vice versa). Models also include parameters such as the different rates of substitution across different parts of a gene and the proportions of each nucleotide present in a given sequence. Once such parameters have been estimated, the model

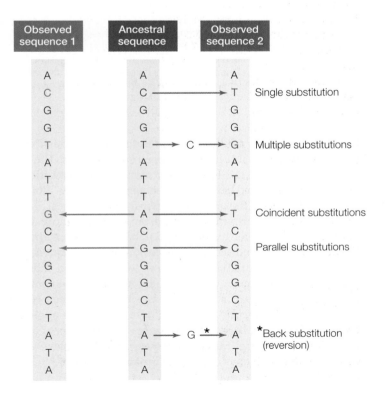

24.2 Multiple Substitutions Are Not Reflected in Pairwise Sequence Comparisons Two observed sequences are descended from a common ancestral sequence (center) through a series of substitutions. Although the two observed sequences differ in only three nucleotide differences (colored letters), these three differences resulted from a total of nine substitutions (arrows).

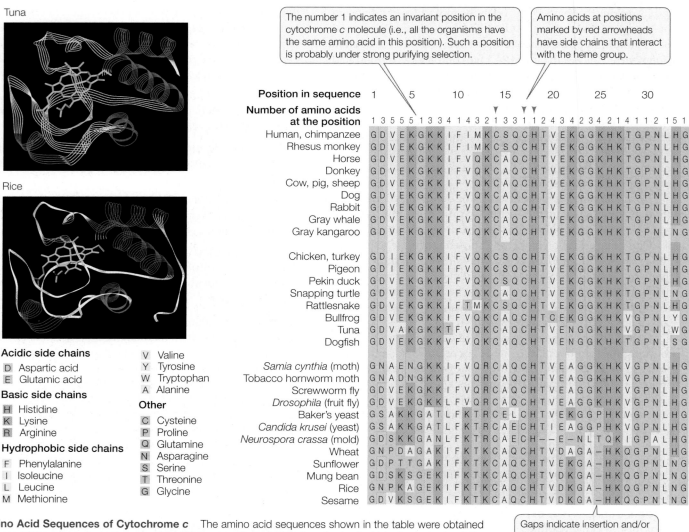

24.3 Amino Acid Sequences of Cytochrome c The amino acid sequences shown in the table were obtained from analyses of the enzyme cytochrome c from 33 species of plants, fungi, and animals. Note the lack of variation across the sequences at positions 70–80, suggesting that this region is under strong stabilizing selection and that changing its amino acid sequence would impair the protein's function. The computer graphics at the upper left are created from these sequences and show the three-dimensional structures of tuna and rice cytochrome c. Alpha helixes are in red, and the molecule's heme group is shown in yellow.

is used to correct for multiple substitutions, coincident substitutions, parallel substitutions, and back substitutions. The revised estimate accounts for the *total* number of substitutions likely to have occurred between two sequences, which is almost always greater than the observed number of differences.

As sequence information becomes available for more and more genes in an ever-expanding database, sequence alignments can be extended across multiple homologous sequences, and the minimum number of insertions, deletions, and substitutions can be summed across homologous genes of an entire group of organisms. Similar databases have also been constructed for homologous proteins. **Figure 24.3** shows aligned data for cytochrome c protein sequences in 33 species of animals, plants, and fungi. Such information is used extensively in determining evolutionary relationships among species.

Experimental studies examine molecular evolution directly

Although molecular evolutionists are often interested in naturally evolved genes and proteins, molecular and phenotypic evolution can also be observed directly in the laboratory. Increasingly, evolutionary biologists are studying evolution experimentally. Because substitution rates are related to generation time rather than to absolute time, most of these experiments use unicellular organisms or viruses with short generations. Viruses, bacteria, and unicellular eukaryotes (such as the yeasts) can be cultured in large populations in the laboratory, and many of these organisms can evolve rapidly. In the case of some RNA viruses, the natural substitution rate may be as high as 1 substitution per 1,000 nucleotides per generation. Therefore, in a virus of a few thousand nucleotides, one

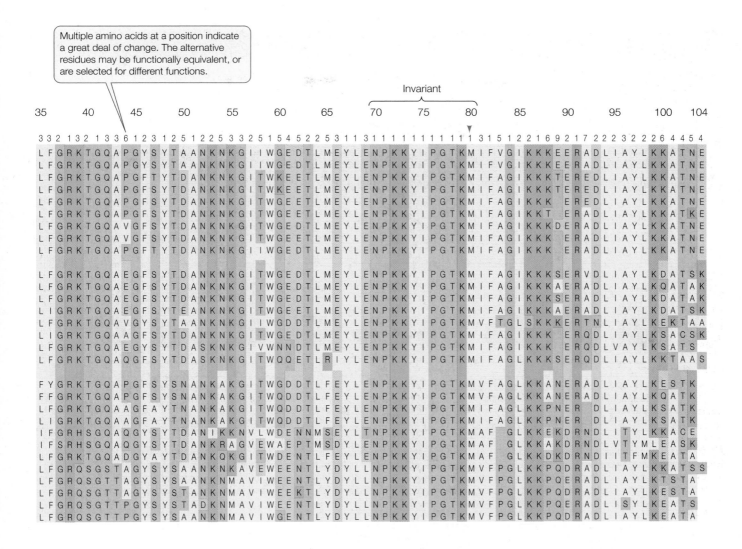

Multiple amino acids at a position indicate a great deal of change. The alternative residues may be functionally equivalent, or are selected for different functions.

or more substitutions are expected (on average) every generation, and these changes can easily be determined by sequencing the entire genome (because of its small size). Generation time may be only tens of minutes (rather than years or decades, as in humans), so biologists can directly observe substantial molecular evolution in a controlled population over the course of days, weeks, or months.

An example of an experimental evolutionary study is shown in **Figure 24.4**. Paul Rainey and Michael Travisano wanted to examine a potential cause of adaptive radiations, which are a major source of biological diversity (see Section 23.4). For instance, near the beginning of the Cenozoic era, mammals rapidly diversified into species as diverse as elephants, moles, whales, and bats. While Rainey and Travisano clearly couldn't experimentally manipulate mammals over many millions of years, they could test the idea that heterogeneous environments with unoccupied niches lead to adaptive radiation by experimentally manipulating a bacterial lineage.

Rainey and Travisano inoculated several flasks containing culture medium with the same strain of the bacterium *Pseudomonas fluorescens*. They then shook some of the cultures

to maintain a constantly uniform environment. Others they left alone (static cultures), allowing them to develop a spatially distinct structure. In the static cultures, the environment on the surface film of the medium differed from that on the walls of the flasks and from parts of the culture not touching any surfaces.

When the cultures were started, the ancestral phenotype of the bacterium produced a smooth colony, which the investigators called a "smooth morph." In just a few days, however, the static cultures consistently and independently developed two other morphs: a "wrinkly spreader" and a "fuzzy spreader." The researchers determined that the two new morphs had a genetic basis and were adaptively superior in some of the environments found within the static cultures. For example, the "wrinkly spreader" cells adhered firmly to one another as well as to surfaces, and thus were able to form a mat across the surface of the medium, where they could compete successfully for oxygen.

DNA sequencing of the genomes of these morphs showed that the same phenotypes had evolved repeatedly, and that many different substitutions could produce the same phenotypes. The homogeneous shaken cultures, in contrast, showed no evolution in phenotype. The same mutations occurred in the

INVESTIGATING LIFE

24.4 Evolution in a Heterogeneous Environment

Rainey and Travisano cultured the rapidly evolving bacterium *Pseudomonas fluorescens* in homogeneous and heterogeneous environments to examine the relationship between phenotypic diversity and environmental variability.

HYPOTHESIS Heterogeneous environments are more conducive to the evolution of phenotypic diversity than are homogeneous environments.

METHOD One colony of *Pseudomonas fluorescens* (all of a single genotype) is used to inoculate many replicate cultures.

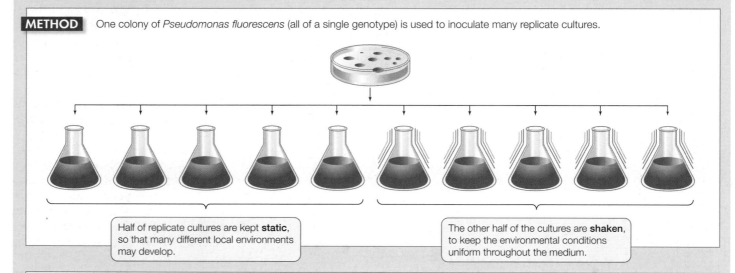

Half of replicate cultures are kept **static**, so that many different local environments may develop.

The other half of the cultures are **shaken**, to keep the environmental conditions uniform throughout the medium.

RESULTS In the shaken flasks, the ancestral morph persisted; the uniform environment did not result in morphological diversification. In the static flasks, two new morphotypes regularly arose, each adapted to a different local environment. Molecular analysis revealed that the mutations that produce these phenotypes arose in both shaken and static cultures, but the mutations did not persist in the uniform (shaken) environment because the phenotypes they produced were selectively disadvantageous under homogeneous conditions.

Smooth morph (ancestral)

"Wrinkly spreader"

"Fuzzy spreader"

CONCLUSION Phenotypic change and diversification are enhanced in a heterogeneous environment.

FURTHER INVESTIGATION: Do you think the two evolved phenotypes could compete in the homogeneous environment if they were introduced after having become successfully established in the heterogeneous environment? How would you test your hypothesis?

Go to **yourBioPortal.com** for original citations, discussions, and relevant links for INVESTIGATING LIFE figures.

shaken cultures but did not persist, because the novel phenotypes they produced were selectively disadvantageous (i.e., less fit) under the "shaken" environmental conditions.

Experimental molecular evolutionary studies are used for a wide variety of purposes and have greatly expanded the ability of evolutionary biologists to test evolutionary concepts and principles. Biologists now routinely study evolution in the laboratory and, as we will see later in this chapter, use in vitro evolutionary techniques to produce novel molecules that perform new functions with industrial and pharmaceutical uses.

(A)

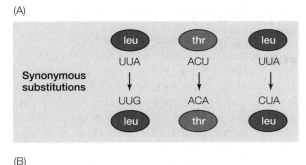

(B)

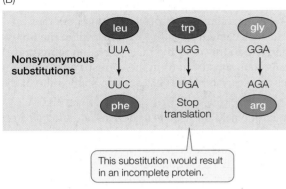

This substitution would result in an incomplete protein.

24.5 When One Nucleotide Does or Doesn't Make a Difference
(A) Synonymous substitutions do not change the amino acid specified and do not affect protein function; such substitutions are less likely to be subject to natural selection. (B) Nonsynonymous substitutions do change the amino acid sequence and are likely to have an effect (often deleterious) on protein function; such substitutions are targets for natural selection.

24.1 RECAP

The genomes of all organisms evolve over time, as can be detected by direct observation in the laboratory, as well as by aligning genes and proteins between species. Experimental studies of molecular evolution allow biologists to study many processes of evolution directly under controlled conditions.

- How do biologists align nucleotide and amino acid sequences they wish to compare, and how do they calculate the minimum number of changes that have occurred between pairs of aligned sequences? See pp. 500–501 and Figure 24.1

- Explain why a simple count of nucleotide differences between two sequences underestimates the actual number of nucleotide substitutions since the sequences diverged. See Figure 24.2

We have seen that molecular evolutionists can directly observe the evolution of genomes over time, and can compare the genomes of different organisms and reconstruct the changes that have occurred during their evolution. Let's turn now to the question of how genomes change and examine some of the consequences of those changes.

24.2 What Do Genomes Reveal About Evolutionary Processes?

A *mutation*, as we saw in Chapter 15, is any change in the genetic material. A nucleotide substitution is one type of mutation. Many nucleotide substitutions have no effect on phenotype, even if the change occurs in a gene that encodes a protein, because most amino acids are specified by more than one codon (see Figure 14.6). A substitution that does not change the encoded amino acid is known as a **synonymous substitution** or **silent substitution** (**Figure 24.5A**). Synonymous substitutions do not affect the functioning of a protein (although they may have other effects, such as changes in mRNA stability or translation rates; see Section 14.5), and are therefore less likely to be influenced by natural selection.

A nucleotide substitution that *does* change the amino acid sequence encoded by a gene is known as a **nonsynonymous substitution**, also known as a *missense substitution* (**Figure 24.5B**). In general, nonsynonymous substitutions are more likely to be deleterious to the organism. But not every amino acid replacement alters a protein's shape and charge (and hence its functional properties). Therefore, some nonsynonymous substitutions may also be selectively neutral, or nearly so. Conversely, an amino acid replacement that confers an advantage to the organism would result in positive selection for the corresponding nonsynonymous substitution.

The rate of nonsynonymous nucleotide substitutions in several mammalian protein-coding genes is about 3×10^{-9} substitutions per position per year. Synonymous substitutions in these genes have occurred about five times more frequently than nonsynonymous substitutions. In other words, *substitution rates are highest at nucleotide positions that do not change the amino acid being expressed* (**Figure 24.6**). The rate of substitution is even higher in **pseudogenes**, which are duplicate copies of genes that are no longer functional.

As we saw Chapter 21, most natural populations harbor far more genetic variation than we would expect to find if genetic

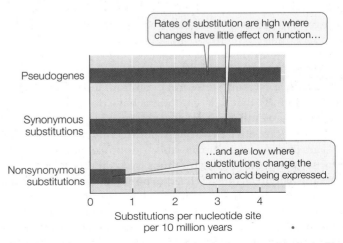

Rates of substitution are high where changes have little effect on function...

...and are low where substitutions change the amino acid being expressed.

24.6 Rates of Substitution Differ Rates of nonsynonymous substitution typically are much lower than rates of synonymous substitution and the substitution rate in pseudogenes. This pattern reflects differing levels of functional constraints.

variation were influenced by natural selection alone. This discovery, combined with the knowledge that many mutations do not change molecular function, stimulated the development of the *neutral theory* of molecular evolution.

Much of evolution is neutral

In 1968, Motoo Kimura proposed the *neutral theory of molecular evolution*. Kimura suggested that, at the molecular level, the majority of variants we observe in most populations are selectively neutral; that is, they confer neither an advantage nor a disadvantage on their bearers. Therefore, these neutral variants accumulate through genetic drift rather than through positive selection.

The rate of fixation of neutral mutations by genetic drift is independent of population size. To see why this is so, consider a population of size N and a neutral mutation rate μ (mu) per gamete per generation at a locus. The number of new mutations would be, on average, $\mu \times 2N$, because $2N$ gene copies are available to mutate in a population of diploid organisms. The probability that a given mutation will be fixed by drift alone is its frequency, which equals $1/(2N)$ for a newly arisen mutation. We can multiply these two terms to get the rate of fixation of neutral mutations in a given population of N individuals:

$$2N\mu \frac{1}{2N} = \mu$$

Therefore, the rate of fixation of neutral mutations depends only on the neutral mutation rate μ and is independent of population size. A given mutation is more likely to appear in a large population than in a small one, but any mutation that does appear is more likely to become fixed in a small population. These two influences of population size cancel each other out.

Therefore, the rate of fixation of neutral mutations is equal to the mutation rate. As long as the underlying mutation rate is constant, macromolecules evolving in different populations should diverge from one another in neutral changes at a constant rate. Empirically, the rate of evolution of particular genes and proteins is often relatively constant over time, and therefore can be used as a "molecular clock." As we described in Section 22.3, molecular clocks can be used to calculate evolutionary divergence times between species.

Although much of the genetic variation we observe in populations is the result of neutral evolution, the neutral theory does not imply that most mutations have no effect on the organism. Many mutations are never observed in populations because they are lethal or strongly detrimental to the organism and are thus quickly removed from the population through natural selection. Similarly, mutations that confer a selective advantage tend to be quickly fixed in populations, so they do not result in variation at the population level either. Nonetheless, if we compare homologous proteins from different populations or species, some amino acid positions will remain constant under purifying selection, others will vary through neutral genetic drift, and still others will differ between species as a result of positive selection for change. How can these evolutionary processes be distinguished?

Positive and purifying selection can be detected in the genome

As we have just seen, substitutions in a protein-coding gene can be either synonymous or nonsynonymous, depending on whether they change the resulting amino acid sequence of the protein. The relative rates of synonymous and nonsynonymous substitutions are expected to differ in regions of genes that are evolving neutrally, under positive selection for change, or staying unchanged under purifying selection.

- If a given amino acid in a protein can be one of many alternatives (without changing the protein's function), then an amino acid replacement is *neutral* with respect to the fitness of an organism. In this case, the rates of synonymous and nonsynonymous substitutions in the corresponding DNA sequences are expected to be very similar, so the ratio of the two rates would be close to 1.

- If a given amino acid position is under *positive selection for change*, the rate of nonsynonymous substitutions is expected to exceed the rate of synonymous substitutions in the corresponding DNA sequences.

- If a given amino acid position is under *purifying selection*, then the rate of synonymous substitutions in the corresponding DNA sequences is expected to be much higher than the rate of nonsynonymous substitutions.

By comparing the gene sequences that encode proteins from many species, scientists can determine the history and timing of synonymous and nonsynonymous substitutions. This information can be mapped on a phylogenetic tree, as we saw in Chapter 22.

Regions of genes that are evolving under neutral, purifying, or positive selection can be identified by comparing the nature and rates of substitutions across the phylogenetic tree. The evolution of lysozyme illustrates how and why particular positions of a gene sequence might be under different modes of selection.

The enzyme lysozyme (see Figure 3.8) is found in almost all animals. It is produced in the tears, saliva, and milk of mammals and in the albumen (whites) of bird eggs. Lysozyme digests the cell walls of bacteria, rupturing and killing them. As a result, it plays an important role as a first line of defense against invading bacteria. Most animals defend themselves against bacteria by digesting them, which is probably why most animals have lysozyme. Some animals also use lysozyme in the digestion of food.

Among mammals, a mode of digestion called *foregut fermentation* has evolved twice. In mammals with this mode of digestion, the foregut—the posterior esophagus and/or the stomach—has been converted into a chamber in which bacteria break down ingested plant matter by fermentation. Foregut fermenters can obtain nutrients from the otherwise indigestible cellulose that makes up a large proportion of plant tissue. Foregut fermentation evolved independently in ruminants (a group of hoofed mammals that includes cattle) and in certain leaf-eating monkeys, such as langurs. We know these evolutionary events were independent, because both langurs and ruminants have close relatives that are not foregut fermenters.

In both foregut-fermenting lineages, the enzyme lysozyme has been modified to play a new, nondefensive role. This lysozyme ruptures some of the bacteria that live in the foregut, releasing nutrients metabolized by the bacteria, which the mammal then absorbs. How many changes in the lysozyme molecule were needed to allow it to perform this function amid the digestive enzymes and acidic conditions of the mammalian foregut? To answer this question, molecular evolutionists compared the lysozyme-coding sequences in foregut fermenters with those in several of their nonfermenting relatives. They determined which amino acids differed and which were shared among the species (**Figure 24.7A**), as well as the rates of synonymous and nonsynonymous substitutions in lysozyme genes across the evolutionary history of the sampled species.

For many of the amino acid positions of lysozyme, the rate of synonymous substitutions (in the corresponding gene) is much higher than the rate of nonsynonymous substitutions. This observation indicates that many of the amino acids that make up lysozyme are evolving under purifying selection. In other words, there is selection against change in the protein at these positions, and the observed amino acids must therefore be critical for lysozyme function. At other positions, several different amino acids function equally well, and the corresponding regions of the genes have similar rates of synonymous and nonsynonymous substitutions. The most striking finding is that amino acid replacements in lysozyme happened at a much higher rate in the lineage leading to langurs than in any other primates. The high rate of nonsynonymous substitutions in the langur lysozyme gene shows that lysozyme went through a period of rapid change in adapting to the stomachs of langurs. Moreover, the lysozymes of langurs and cattle share five amino acid replacements, all of which lie on the surface of the lysozyme molecule, well away from the enzyme's active site. Several of these shared replacements involve changes from arginine to lysine, which makes the proteins more resistant to attack by the stomach enzyme pepsin. By understanding the functional significance of amino acid replacements, molecular evolutionists can explain the observed changes in amino acid sequences in terms of changes in the functioning of the protein.

A large body of fossil, morphological, and molecular evidence shows that langurs and cattle do not share a recent common ancestor. However, langur and ruminant lysozymes share several amino acids that neither mammal shares with the lysozymes of its own closer relatives. The lysozymes of these two mammals have undergone *evolutionary convergence* at some amino acid positions despite their very different ancestry. The amino acids they share give these lysozymes the ability to lyse the bacteria that ferment plant material in the foregut.

The hoatzin, an unusual leaf-eating South American bird and the only known avian foregut fermenter, offers another remarkable story of the convergent evolution of lysozyme (**Figure 24.7B**). Many birds have an enlarged esophageal chamber called a *crop*. The crop of the hoatzin contains lysozyme and bacteria and acts as a fermenting chamber. Many of the amino acid replacements that occurred in the adaptation of hoatzin crop lysozyme are identical to those that evolved in ruminants and langurs. Thus, even though the hoatzin and foregut-fermenting mammals have not shared a common ancestor in hundreds of millions of years, they have all evolved similar adaptations in their lysozymes that enable them to recover nutrients from their fermenting bacteria.

Genome size and organization also evolve

We know that genome size varies tremendously among organisms. Across broad taxonomic categories, there is some correlation between genome size and organismal complexity. The

24.7 Convergent Molecular Evolution of Lysozyme   (A) The number of amino acid differences in the lysozymes of several pairs of mammals are shown above the diagonal line; the number of convergent similarities between these same pairs are shown below the diagonal. The two foregut-fermenting species share convergent amino acid replacements related to this digestive adaptation. (B) The hoatzin—the only known foregut-fermenting bird species—has been evolving independently from mammals for hundreds of millions of years but has independently evolved similar modifications to lysozyme.

(A)

Semnopithecus sp.

The lysozymes of langurs and cattle are convergent for 5 amino acid residues, indicative of the independent evolution of foregut fermentation in these two species.

	Langur	Baboon	Human	Rat	Cattle	Horse
Langur		14	18	38	32	65
Baboon	0		14	33	39	65
Human	0	1		37	41	64
Rat	0	0	0		55	64
Cattle	5	0	0	0		71
Horse	0	0	0	0	1	

(B) *Opisthocomus hoazin*

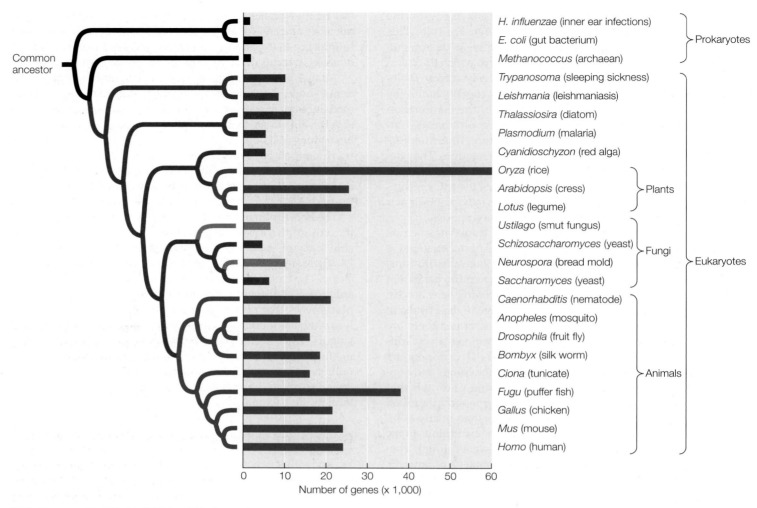

24.8 Genome Size Varies Widely This figure shows the number of genes from a sample of organisms with fully sequenced genomes, arranged by their evolutionary relationships. Bacteria and archaea typically have fewer genes than most eukaryotes. Among eukaryotes, multicellular organisms with tissue organization (plants and animals; blue branches) have more genes than single-celled organisms (red branches) or multicellular organisms that lack pronounced tissue organization (green branches).

genome of the tiny bacterium *Mycoplasma genitalium* has only 470 genes. *Rickettsia prowazekii*, the bacterium that causes typhus, has 634 genes. *Homo sapiens*, by contrast, has about 23,000 protein-coding genes. **Figure 24.8** shows the number of genes from a sample of organisms with fully sequenced genomes, arranged by their evolutionary relationships. As this figure reveals, however, a larger genome does not always indicate greater complexity (compare rice to the other plants, for example). It is not surprising that more complex genetic instructions are needed for building and maintaining a large, multicellular organism than a small, single-celled bacterium. What is surprising is that some organisms, such as lungfishes, some salamanders, and lilies, have about 40 times as much DNA as humans do. Structurally, a lungfish or a lily is not 40 times more complex than a human. So why does genome size vary so much?

Differences in genome size are not so great if we take into account only the portion of DNA that actually encodes RNAs or proteins. The organisms with the largest total amounts of nuclear DNA (some ferns and flowering plants) have 80,000 times as much DNA as do the bacteria with the smallest genomes, but no species has more than about 100 times as many protein-coding genes as a bacterium. Therefore, much of the variation in genome size lies not in the number of functional genes but in the amount of noncoding DNA (**Figure 24.9**).

Why do the cells of most eukaryotic organisms have so much noncoding DNA? Does this noncoding DNA have a function, or is it "junk"? Although some of this DNA does not appear to have a direct function, it can alter the expression of the surrounding genes. The degree or timing of gene expression can be changed dramatically depending on the gene's position relative to noncoding sequences. Other regions of noncoding DNA consist of pseudogenes that are simply carried in the genome because the cost of doing so is very small. These pseudogenes may become the raw material for the evolution of new genes with novel functions. Some noncoding sequences function in maintaining chromosomal structure. Still others consist of parasitic transposable elements that spread through populations because they reproduce faster than the host genome.

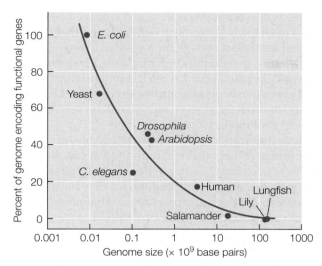

24.9 A Large Proportion of DNA Is Noncoding Most of the DNA of bacteria and yeasts encodes RNAs or proteins, but a large percentage of the DNA of multicellular species is noncoding.

Investigators can use retrotransposons to estimate the rates at which species lose DNA. Retrotransposons are transposable elements (see Figure 17.5) that copy themselves through an RNA intermediate. The most common type of retrotransposon carries duplicated sequences at each end, called long terminal repeats, or LTRs. Occasionally, LTRs recombine in the host genome, so that the DNA between them is excised. When this happens, one recombined LTR is left behind. The number of such "orphaned" LTRs in a genome is a measure of how many retrotransposons have been lost. By comparing the number of LTRs in the genomes of Hawaiian crickets (*Laupala*) and fruit flies (*Drosophila*), investigators found that *Laupala* loses DNA more than 40 times more slowly than does *Drosophila*. Therefore, it is not surprising that the genome of *Laupala* is much larger than that of *Drosophila*.

Why do species differ so greatly in the rate at which they gain or lose apparently functionless DNA? One hypothesis is that genome size is related to the rate at which the organism develops, which may be under selection pressure. Large genomes can slow down the rate of development and thus alter the relative timing of expression of particular genes. As discussed in Section 20.2, changes in the timing of gene expression (*heterochrony*) can produce major changes in phenotype. Thus, although some noncoding DNA sequences may have no direct function, they may still affect the development of the organism.

Another hypothesis is that the proportion of noncoding DNA is related primarily to population size. Noncoding sequences that are only slightly deleterious to the organism are likely to be purged by selection most efficiently in species with large population sizes. In species with small populations, the effects of genetic drift can overwhelm selection against noncoding sequences that have small deleterious consequences. Therefore, selection against the accumulation of noncoding sequences is most effective in species with large populations, so such species (such as bacteria or yeasts) have relatively little noncoding DNA compared with species with small populations (see Figure 24.9).

24.2 RECAP

By examining the relative rates of synonymous and nonsynonymous substitutions in genes across evolutionary history, biologists can distinguish the evolutionary mechanisms acting on individual genes. Neutral theory provides an explanation for the relatively constant rate of molecular change seen in many species.

- Describe how the ratio of synonymous to nonsynonymous substitutions can be used to determine whether a particular gene region is evolving neutrally, under positive selection, or under stabilizing selection. See pp. 505–506 and Figure 24.6

- Contrast two hypotheses for the wide diversity of genome sizes among different organisms. See pp. 508–509

We have examined some of the ways that biologists can use genomes to study the molecular mechanisms of evolution. But how do organisms gain new functions through time?

24.3 How Do Genomes Gain and Maintain Functions?

As we noted in the previous section, most multicellular organisms have many more genes than do most unicellular species. But multicellular organisms evolved from unicellular ancestors. Therefore, some mechanisms must exist that result in the increase of gene numbers within genomes over evolutionary time. There are two primary ways to accomplish this increase: genes can be transferred from other species, or genes can be duplicated within species.

Lateral gene transfer can result in the gain of new functions

In Chapter 22, we noted that the tree of life is usually visualized as a branching diagram, with each lineage dividing into two (or more) lineages through time, from one common ancestor to the millions of species that are alive today. Chapter 23 described how, in the process of speciation, ancestral lineages divide into descendant lineages, and it is those speciation events that the tree of life captures. However, there are also processes of **lateral gene transfer**, which allow individual genes, organelles, or fragments of genomes to move horizontally from one lineage to another. Some species may pick up fragments of DNA directly from the environment. Other genes may be picked up in a viral genome and then transferred to a new host when the virus becomes integrated in the new host's genome. Hybridization between species also results in the transfer of large numbers of genes.

Lateral gene transfer can be highly advantageous to the species that incorporates novel genes from a distant relative. Genes that confer antibiotic resistance, for example, are commonly transferred among different species of bacteria. Lateral gene transfer is another way, in addition to mutation and recombination, that species can increase their genetic variability. Genetic variability then provides the raw material on which selection acts, resulting in evolution.

A phylogenetic tree constructed from a single laterally transferred genome fragment is likely to reflect only that transfer event, rather than the overall organismal phylogeny (see Section 26.3). Most biologists prefer to build trees from large samples of genes or their products, so that the underlying species tree (as well as any lateral gene transfer events) can be reconstructed. The depiction of lateral gene transfer events on the underlying species tree are known as *reticulations* on the phylogenetic tree.

The degree to which lateral gene transfer events occur in various parts of the tree of life is a matter of considerable current investigation and debate. Lateral gene transfer appears to be relatively uncommon among most eukaryote lineages, although the two major endosymbioses that gave rise to mitochondria and chloroplasts can be viewed as lateral transfers of entire bacteria genomes to the eukaryote lineage. Some groups of eukaryotes, most notably some plants, are subject to relatively high levels of hybridization among closely related species. Hybridization leads to the exchange of many genes among recently separated lineages of plants. The greatest degree of lateral transfer, however, appears to occur among species of bacteria. Many bacteria genes have been transferred repeatedly among lineages of bacteria, to the point that relationships among the bacteria species are often hard to decipher. Nonetheless, the broad relationships of the major groups of bacteria can still be determined (as we will discuss in Part Seven of this book). Lateral transfer of genes makes it difficult to identify the boundaries of bacteria species, which is one reason why fewer bacteria species have been named than are known to exist.

Most new functions arise following gene duplication

Gene duplication is yet another way in which genomes can acquire new functions. When a gene is duplicated, one copy of that gene is potentially freed from having to perform its original function. The identical copies of a duplicated gene can have any one of four different fates:

- Both copies of the gene may retain their original function (which can result in a change in the amount of gene product that is produced by the organism).

- Both copies of the gene may retain the ability to produce the original gene product, but the expression of the genes may diverge in different tissues or at different times of development.

- One copy of the gene may be incapacitated by the accumulation of deleterious substitutions and become a functionless pseudogene.

- One copy of the gene may retain its original function while the second copy accumulates enough substitutions that it can perform a different function.

How often do gene duplications arise, and which of these four outcomes is most likely? Investigators have found that rates of gene duplication are fast enough for a yeast or *Drosophila* population to acquire several hundred duplicate genes over the course of a million years. They have also found that most of the duplicated genes in these organisms are very young. Many extra genes are lost from a genome within 10 million years (which is rapid on an evolutionary time scale).

Many gene duplications affect only one or a few genes at a time, but entire genomes are often duplicated in polyploid organisms (including many plants). When all the genes are duplicated, there are massive opportunities for new functions to evolve. That is exactly what appears to have happened in the evolution of vertebrates. The genomes of the jawed vertebrates appear to have four diploid sets of many major genes, which leads biologists to believe that two genome-wide duplication events occurred in the ancestor of these species. These duplications have allowed considerable specialization of individual vertebrate genes, many of which are now highly tissue-specific in their expression. A good example is the duplication of sodium channel genes, which allowed the evolution of the electric organs of electric fishes described in the opening of this chapter.

Although many extra genes disappear rapidly, some duplication events lead to the evolution of genes with new functions. Several successive rounds of duplication and mutation may result in a **gene family**, a group of homologous genes with related functions, often arrayed in tandem along a chromosome. An example of this process is provided by the *globin gene family* (see Figure 17.10). The globins were among the first proteins to be sequenced and compared. Comparisons of their amino acid sequences strongly suggest that the different globins arose via gene duplications. These comparisons can also tell us how long the globins have been evolving separately, because differences among these proteins have accumulated with time.

Hemoglobin, a tetramer (four-subunit molecule) consisting of two α-globin and two β-globin polypeptide chains, carries oxygen in blood. Myoglobin, a monomer, is the primary O_2 storage protein in muscle. Myoglobin's affinity for O_2 is much higher than that of hemoglobin, but hemoglobin has evolved to be more diversified in its role. Hemoglobin binds O_2 in the lungs or gills, where the O_2 concentration is relatively high, transports it to deep body tissues, where the O_2 concentration is low, and releases it in those tissues. With its more complex tetrameric structure, hemoglobin is able to carry four molecules of O_2, as well as hydrogen ions and carbon dioxide, in the blood.

To estimate the time of the globin gene duplication that gave rise to the α- and β-globin gene clusters, we can create a phylogenetic tree of the gene sequences that encode the various globins (**Figure 24.10**). The rate of molecular evolution of globin genes has been estimated from other studies, using the divergence times of groups of vertebrates that are well documented in the fossil record. These studies indicate an average rate of divergence for globin genes of about 1 nucleotide substitution every 2 million years. Applying this rate to the gene tree, the two globin gene clusters are estimated to have split about 450 million years ago.

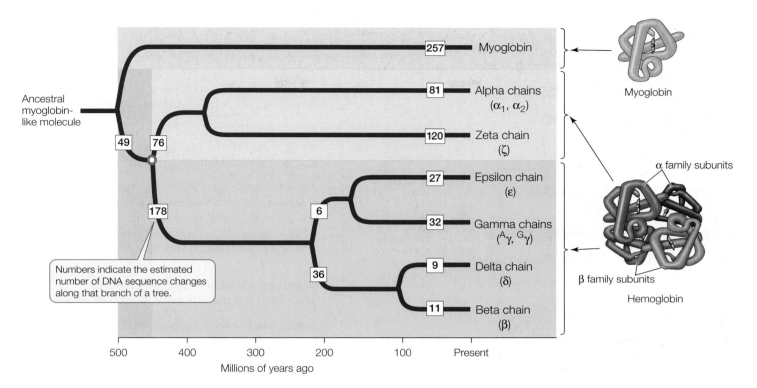

24.10 A Globin Family Gene Tree This gene tree suggests that the α-globin and β-globin gene clusters diverged about 450 million years ago (open circle), soon after the origin of the vertebrates.

───── **yourBioPortal.com** ─────

GO TO **Web Activity 24.3 • Gene Tree Construction**

Some gene families evolve through concerted evolution

Although the members of the globin gene family have diversified in form and function, the members of many other gene families do not evolve independently of one another. For instance, almost all organisms have many copies (up to thousands) of the ribosomal RNA genes. *Ribosomal RNA* (rRNA) is the principal structural element of ribosomes and, as such, has a primary role in protein synthesis. Every living species needs to synthesize proteins, often in large amounts (especially during early development). Having many copies of the rRNA genes ensures that organisms can rapidly produce many ribosomes and thereby maintain a high rate of protein synthesis.

Like all portions of the genome, ribosomal RNA genes evolve, and differences accumulate in the rRNA genes of different species. But within any one species, the multiple copies of rRNA genes are very similar, both structurally and functionally. This similarity makes sense, because ideally every ribosome in a species should synthesize proteins in the same way. In other words, within a given species, the multiple copies of these rRNA genes are evolving in concert with one another, a phenomenon called **concerted evolution**.

How does concerted evolution occur? There must be one or more mechanisms to cause a substitution in one copy to spread to other copies in a species so that all of the copies remain similar. In fact, two different mechanisms appear to be responsi-

ble for concerted evolution. The first of these is *unequal crossing over*. When DNA is replicated during meiosis in a diploid species, the homologous chromosome pairs align and recombine by crossing over (see Section 11.4).

In the case of highly repeated genes, however, it is easy for genes to become displaced in alignment, since so many copies of the same genes are present in the repeats (**Figure 24.11A**). The end result is that one chromosome will gain extra copies of the repeat and the other chromosome will have fewer copies of the repeat. If a new substitution arises in one copy of the repeat, it can spread to new copies (or be eliminated) through unequal crossing over. Thus, over time, a novel substitution will either become fixed or lost entirely from the repeat. In either case, all the copies of the repeat will remain very similar to one another.

The second mechanism that produces concerted evolution is *biased gene conversion*. This mechanism can be much faster than unequal crossing over, and has been shown to be the primary mechanism for concerted evolution of rRNA genes. DNA strands break often, and are repaired by the DNA repair systems of cells (see Section 13.4). At many times during the cell cycle, the genes for ribosomal RNA are clustered close together. If damage occurs to one of the genes, a copy of the rRNA gene on another chromosome may be used to repair the damaged copy, and the sequence that is used as a template can thereby replace the original sequence (**Figure 24.11B**). In many cases, this repair system appears to be biased in favor of using particular sequences as templates for repair, and thus the favored sequence rapidly spreads across all copies of the gene. In this way, changes may appear in a single copy and then rapidly spread to all the other copies.

Regardless of the mechanism responsible, the net result of concerted evolution is that the copies of a highly repeated gene do not evolve independently of one another. Mutations still oc-

(A) Unequal crossing over

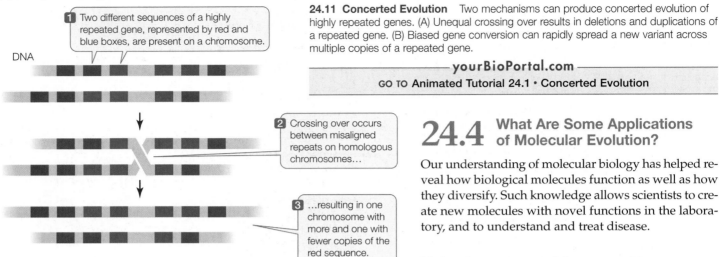

1 Two different sequences of a highly repeated gene, represented by red and blue boxes, are present on a chromosome.

DNA

2 Crossing over occurs between misaligned repeats on homologous chromosomes…

3 …resulting in one chromosome with more and one with fewer copies of the red sequence.

(B) Biased gene conversion

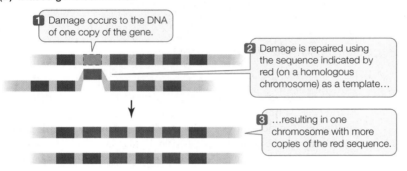

1 Damage occurs to the DNA of one copy of the gene.

2 Damage is repaired using the sequence indicated by red (on a homologous chromosome) as a template…

3 …resulting in one chromosome with more copies of the red sequence.

24.11 Concerted Evolution Two mechanisms can produce concerted evolution of highly repeated genes. (A) Unequal crossing over results in deletions and duplications of a repeated gene. (B) Biased gene conversion can rapidly spread a new variant across multiple copies of a repeated gene.

---
yourBioPortal.com

GO TO **Animated Tutorial 24.1 • Concerted Evolution**

24.4 What Are Some Applications of Molecular Evolution?

Our understanding of molecular biology has helped reveal how biological molecules function as well as how they diversify. Such knowledge allows scientists to create new molecules with novel functions in the laboratory, and to understand and treat disease.

Molecular sequence data are used to determine the evolutionary history of genes

A **gene tree** shows the evolutionary relationships of a single gene in different species or of the members of a gene family (as in Figure 24.10). The methods for constructing a gene tree are the same as those we described in Section 22.2 for building phylogenetic trees of species. The process involves identifying differences between genes and using those differences to reconstruct the evolutionary history of the genes. Gene trees are often used to infer phylogenetic trees of species, but the two types of trees are not necessarily equivalent. Processes such as gene duplication can give rise to differences between the phylogenetic trees of genes and species. From a gene tree, biologists can reconstruct the history and timing of gene duplication events and learn how gene diversification has resulted in the evolution of new protein functions.

All of the genes of a particular gene family have similar sequences because they have a common ancestry. As we discussed in Section 22.1, features that are similar as a result of common ancestry are referred to as *homologs* of one another. When discussing gene trees, however, we usually need to distinguish between two forms of homology. Genes found in different species, and whose divergence we can trace to the speciation events that gave rise to various species, are called **orthologs**. Genes in the same or different species that are related through gene duplication events are called **paralogs**. When we examine a gene tree, the questions we wish to address determine whether we should compare orthologous or paralogous genes. If we wish to reconstruct the evolutionary history of the species that contain the genes, then our comparison should be restricted to orthologs (because they will reflect the history of speciation events). If we are interested in the changes in function that have resulted from gene duplication events, however, then the appropriate comparison is among paralogs (because they will reflect the history of gene duplication events). If our focus is on the diversification of a gene family through both processes, then we will want to include both paralogs and orthologs in our analysis.

Figure 24.12 depicts a gene tree for the members of a gene family called *engrailed* (its members encode transcription factors

cur, but once they arise in one copy, they either spread rapidly across all the copies or are lost from the genome completely. This process allows the products of each copy to remain similar through time in both sequence and function.

24.3 RECAP

Gene duplication can lead to the evolution of new functions. Lateral gene transfer can result in the spread of genetic functions between distantly related species. Some highly repeated genes evolve in concert, which maintains uniform functionality.

- Explain the potential advantages of lateral gene transfer. **See pp. 509–510**

- What are four possible outcomes of gene duplication? **See p. 510**

- Describe the pattern of concerted evolution among highly repeated genes and the mechanisms that lead to concerted evolution. **See p. 511 and Figure 24.11**

We have seen how the principles and methods of molecular evolution have opened new vistas in evolutionary science. Now let's consider some of the practical applications of this field.

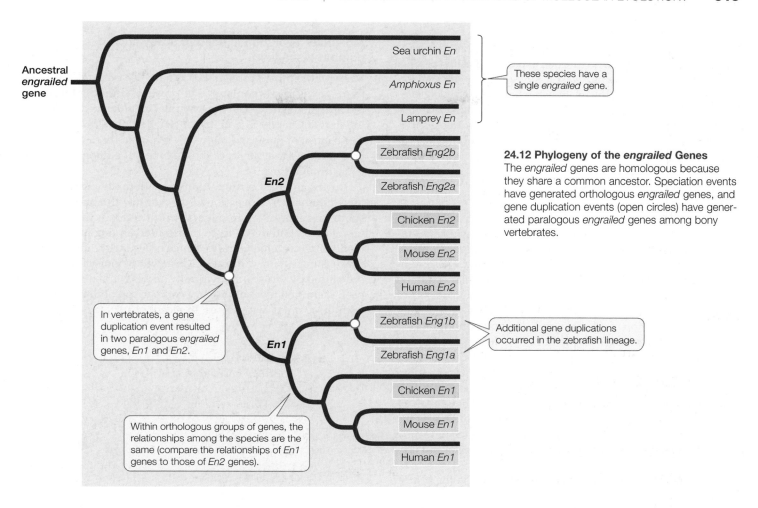

Sea urchin *En*

Amphioxus En

Lamprey *En*

These species have a single *engrailed* gene.

Ancestral *engrailed* gene

Zebrafish *Eng2b*

Zebrafish *Eng2a*

En2

Chicken *En2*

Mouse *En2*

Human *En2*

Zebrafish *Eng1b*

Additional gene duplications occurred in the zebrafish lineage.

Zebrafish *Eng1a*

En1

Chicken *En1*

Mouse *En1*

Human *En1*

In vertebrates, a gene duplication event resulted in two paralogous *engrailed* genes, *En1* and *En2*.

Within orthologous groups of genes, the relationships among the species are the same (compare the relationships of *En1* genes to those of *En2* genes).

24.12 Phylogeny of the *engrailed* Genes
The *engrailed* genes are homologous because they share a common ancestor. Speciation events have generated orthologous *engrailed* genes, and gene duplication events (open circles) have generated paralogous *engrailed* genes among bony vertebrates.

that regulate development). At least three gene duplications have occurred in this family, resulting in up to four different *engrailed* genes (*En*) in some vertebrate species (such as the zebrafish). All of the *engrailed* genes are homologs because they have a common ancestor. Gene duplication events have generated paralogous *engrailed* genes in some lineages of vertebrates. We could compare the orthologous sequences of the *En1* group of genes to reconstruct the history of the bony vertebrates (i.e., all the species in Figure 24.12 except the lamprey), or we could use the orthologous sequences of the *En2* group of genes and expect the same answer (because there is only one history of the underlying speciation events). All bony vertebrates have both groups of *engrailed* genes because the two groups arose from a gene duplication event in the common ancestor of bony vertebrates. If we wanted to focus on the diversification that occurred as a result of this duplication, then the appropriate comparison would be between the paralogous genes of the *En1* versus *En2* groups.

Gene evolution is used to study protein function

Earlier in this chapter we discussed the ways in which biologists can detect regions of genes that are under positive selection for change. What are the practical uses of this information? Consider the evolution of the family of gated sodium channel genes, which we introduced in the opening of this chapter. Sodium channels have many functions, including the control of nerve

impulses in the nervous system. Sodium channels can become blocked by various toxins, such as the tetrodotoxin that is present in puffer fishes and many other animals. A human who eats the tissues of a puffer fish that contain tetrodotoxin can become paralyzed and die, because the tetrodotoxin blocks sodium channels and prevents nerves and muscles from functioning.

Puffer fish themselves have sodium channels; so why doesn't the tetrodotoxin cause paralysis in a puffer fish? The sodium channels of puffer fish (and other animals that sequester tetrodotoxin) have evolved to become resistant to the toxin. Nucleotide substitutions in the puffer fish genome have resulted in changes to the proteins that make up sodium channels, and those changes prevent tetrodotoxin from binding to the sodium channel pore and blocking it. Several different substitutions that result in tetrodotoxin resistance have evolved in the various duplicated sodium channel genes of the many species of puffer fish.

Many other changes that have nothing to do with the evolution of tetrodotoxin resistance have occurred in these genes as well. Biologists who study the function of sodium channels can learn a great deal about how the channels work (and about neurological diseases that are caused by mutations in the sodium channel genes) by understanding which changes have been selected for tetrodotoxin resistance. They can do this by comparing the rates of synonymous and nonsynonymous substitutions across the genes in various lineages that have evolved tetrodotoxin resistance. In a similar manner, molecular evolutionary principles are used to understand function and diversification of function in many other proteins.

As biologists studied the relationship between selection, evolution, and function in macromolecules, they realized that molecular evolution could be used in a controlled laboratory environment to produce new molecules with novel and useful functions. Thus were born the applications of in vitro evolution.

In vitro evolution produces new molecules

Living organisms produce thousands of compounds that humans have found useful. The search for such naturally occurring compounds, which can be used for pharmaceutical, agricultural, or industrial purposes, has been termed *bioprospecting*. These compounds are the result of millions of years of molecular evolution across millions of species of living organisms. Yet biologists can imagine molecules that could have evolved but have not, lacking the right combination of selection pressures and opportunities.

For instance, we might like to have a molecule that binds a particular environmental contaminant so that it can be easily isolated and extracted from the environment. But if the environmental contaminant is synthetic (not produced naturally), then it is unlikely that any living organism would have evolved a molecule with the function we desire. This problem was the inspiration for the field of **in vitro evolution**, in which new molecules are produced in the laboratory to perform novel and useful functions.

The principles of in vitro evolution are based on the principles of molecular evolution that we have learned from the natural world. Consider the evolution of a new RNA molecule that was produced in the laboratory using the principles of mutation and selection. This molecule's intended function was to join two other RNA molecules (acting as a ribozyme with a function similar to that of the naturally occurring DNA ligase described in Section 13.3, but for RNA molecules). The process started with a large pool of random RNA sequences (10^{15} different sequences, each about 300 nucleotides long), which were then selected for any ligase activity (**Figure 24.13**). None were very effective ribozymes for ligase activity, but some were slightly better than others. The best of the ribozymes were selected and reverse-transcribed into cDNA (using the enzyme reverse transcriptase). The cDNA molecules were then amplified using the polymerase chain reaction (PCR; see Figure 13.22).

PCR amplification is not perfect, and it introduced many new mutations into the pool of sequences. These sequences were then transcribed back into RNA molecules using RNA polymerase, and the process was repeated. The ligase activity of the RNAs evolved quickly; after 10 rounds of in vitro evolution, it had increased by about 7 million times (see Figure 24.13). Similar techniques have since been used to create a wide variety of molecules with novel enzymatic and binding functions.

Molecular evolution is used to study and combat diseases

Many of the most problematic human diseases are caused by living, evolving organisms that present a moving target for modern medicine. Recall the example of influenza described at the start of Chapter 21 and that of HIV described in Chapter 22. The control of these and many other human diseases depends on techniques that can track the evolution of pathogenic organisms through time.

During the past century, transportation advances have allowed humans to move around the world with unprecedented speed and increasing frequency. Unfortunately, this mobility has allowed pathogens to be transmitted among human populations at much higher rates, which has led to the global emergence of many "new" diseases. Most of these emerging diseases are caused by viruses, and virtually all new viral diseases have been identified by evolutionary comparison of their genomes with those of known viruses. In recent years, for example, ro-

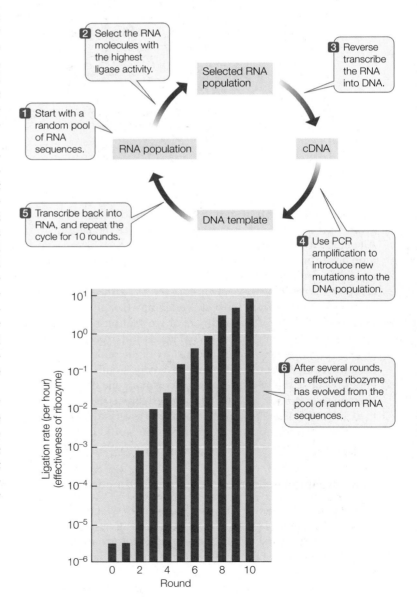

24.13 In Vitro Evolution Starting with a large pool of random RNA sequences, Bartel and Szostak produced a new ribozyme through rounds of mutation and selection for the ability to ligate RNA sequences.

dent-borne hantaviruses have been identified as the source of widespread respiratory illnesses, and the virus (and its host) that causes Sudden Acute Respiratory Syndrome (SARS) has been identified using evolutionary comparisons of genes. Studies of the origins, the timing of emergence, and the global diversity of many human pathogens depend on the principles of molecular evolution, as do the efforts to develop and use effective vaccines against these pathogens. For example, the techniques to develop polio vaccines, as well as the methods used to track their effectiveness in human populations, rely on molecular evolutionary approaches.

In the future, molecular evolution will become even more critical to the identification of human (and other) diseases. Once biologists have collected data on the genomes of enough organisms, it will be possible to identify an infection by sequencing a portion of the infecting organism's genome and comparing this sequence with other sequences on an evolutionary tree. At present, it is difficult to identify many common viral infections (those that cause "colds," for instance). As genomic databases and evolutionary trees increase, however, automated methods of sequencing and rapid phylogenetic comparison of the sequences will allow us to identify and treat a much wider array of human illnesses.

24.4 RECAP

Molecular evolutionary studies have provided biologists with new tools to understand the functions of macromolecules and how those functions can change through time. Molecular evolution is used to develop synthetic molecules for industrial and pharmaceutical uses and to identify and combat human diseases.

- Why might a biologist limit a particular investigation to orthologous (as opposed to paralogous) genes? See pp. 512–513

- Explain how gene evolution can be used to study protein function. See pp. 513–514

- Describe the process of in vitro evolution. See p. 514 and Figure 24.13

Now that we have discussed how organisms and biological molecules evolve, we are ready to consider the evolutionary history of the Earth. Chapter 25 describes the long-term evolutionary changes that have given rise to all of life's diversity.

CHAPTER SUMMARY

24.1 How Are Genomes Used to Study Evolution?
SEE WEB ACTIVITY 24.1

- The field of **molecular evolution** concerns relationships between the structures of genes and proteins and the functions of organisms.

- A **genome** is an organism's full set of genes and noncoding DNA. In eukaryotes, the genome includes genetic material in the nucleus of the cell as well as in mitochondria and chloroplasts (where present).

- Nucleotide substitutions may or may not result in amino acid replacements in the encoded proteins.

- The estimated number of substitutions between sequences can be calculated from a **similarity matrix** using models of sequence evolution that account for changes that cannot be observed directly. Review Figure 24.1, **WEB ACTIVITY 24.2**

- The concept of homology (similarity that results from common ancestry) extends down to the level of particular positions in nucleotide or amino acid sequences. **Sequence alignments** from different organisms allow us to compare the sequences and identify homologous positions. Review Figure 24.3

24.2 What Do Genomes Reveal about Evolutionary Processes?

- **Nonsynonymous substitutions** of nucleotides result in amino acid replacements in proteins, but **synonymous substitutions** do not. Review Figure 24.5

- Rates of synonymous substitution are typically higher than rates of nonsynonymous substitution in protein-coding genes (a result of stabilizing selection). Review Figure 24.6

- Much of the molecular change in nucleotide sequences is a result of **neutral evolution**. The rate of fixation of neutral mutations is independent of population size and is equal to the mutation rate.

- Positive selection for change in a protein-coding gene may be detected by a higher rate of nonsynonymous versus synonymous substitutions.

- Genome size evolves by the addition or deletion of genes and noncoding DNA. The total size of genomes varies much more widely across multicellular species than does the number of functional genes. Review Figures 24.8 and 24.9

- Even though many noncoding regions of the genome may not have direct functions, these regions can affect the phenotype of an organism by influencing gene expression.

- Functionless **pseudogenes** can serve as the raw material for the evolution of new genes.

24.3 How Do Genomes Gain and Maintain Functions?

- Lateral gene transfer can result in the rapid acquisition of new functions from distantly related species.

- **Gene duplications** can result in increased production of the gene's product, in pseudogenes, or in new gene functions. Several rounds of gene duplication can give rise to multiple genes with related functions, known as a **gene family**.

- Some highly repeated genes evolve by **concerted evolution**: multiple copies within an organism maintain high similarity, while the genes continue to diverge between species. **SEE ANIMATED TUTORIAL 24.1**

24.4 What Are Some Applications of Molecular Evolution?
SEE WEB ACTIVITY 24.3

- **Gene trees** describe the evolutionary history of particular genes or gene families.
- **Orthologs** are genes that are related through speciation events, whereas **paralogs** are genes that are related through gene duplication events. Review Figure 24.12

- Protein function can be studied by examining gene evolution. Detection of positive selection can be used to identify molecular changes that have resulted in functional changes.
- **In vitro evolution** is used to produce synthetic molecules with particular desired functions. Review Figure 24.13
- Many diseases are identified, studied, and combated through molecular evolutionary investigations.

SELF-QUIZ

1. A higher rate of synonymous relative to nonsynonymous substitutions in a protein-coding gene is expected under
   a. purifying selection.
   b. positive selection.
   c. neutral evolution.
   d. concerted evolution.
   e. none of the above

2. Before nucleotide and amino acid sequences can be compared in an evolutionary framework, they must be aligned to account for
   a. deletions and insertions.
   b. selection and neutrality.
   c. parallelisms and convergences.
   d. gene families.
   e. all of the above

3. Models of nucleotide sequence evolution, developed by biologists to estimate sequence divergence, include parameters that account for
   a. substitution rates between different nucleotides.
   b. differences in substitution rates across different positions in a gene.
   c. differences in nucleotide frequencies.
   d. all of the above
   e. none of the above

4. The rate of fixation of neutral mutations is
   a. independent of population size.
   b. higher in small populations than in large populations.
   c. higher in large populations than in small populations.
   d. slower than the rate of fixation of deleterious mutations.
   e. none of the above

5. Genome size differs widely among different multicellular species of eukaryotes. What is the greatest contributing cause for these differences?
   a. The number of protein-coding genes
   b. The amount of noncoding DNA
   c. The number of duplicated genes
   d. The degree of concerted evolution
   e. The amount of positive selection for change in protein-coding genes

6. Which of the following is *not* true of concerted evolution?
   a. Concerted evolution refers to the nonindependent evolution of some repeated genes within a species.
   b. Unequal crossing over may produce concerted evolution.
   c. Biased gene conversion may produce concerted evolution.
   d. Ribosomal RNA genes are an example of a gene family that has undergone concerted evolution.
   e. Concerted evolution results in divergence of members of a gene family within an organism.

7. When a gene is duplicated, which of the following may occur?
   a. Production of the gene's product may increase.
   b. The two copies may become expressed in different tissues.
   c. One copy of the gene may accumulate deleterious substitutions and become functionless.
   d. The two copies may diverge and acquire different functions.
   e. All of the above

8. Paralogous genes are genes that trace back to a common
   a. speciation event.
   b. substitution event.
   c. insertion event.
   d. deletion event.
   e. duplication event.

9. Which of the following is true of in vitro evolution?
   a. In vitro evolution refers to bioprospecting for naturally occurring macromolecules.
   b. In vitro evolution can produce new molecular sequences not known from nature.
   c. In vitro evolution can only produce new proteins.
   d. In vitro evolution only selects for changes that were present in the starting pool of molecules, and does not introduce any new mutations.
   e. All of the above

10. Which of the following is true of the use of molecular evolutionary studies of human disease?
    a. Molecular evolutionary studies are useful for identifying many diseases.
    b. Molecular evolutionary studies are often used to determine the origin of emerging diseases.
    c. Molecular evolutionary studies are important for developing vaccines against diseases.
    d. Molecular evolutionary studies are used to track the effectiveness of polio vaccines in human populations.
    e. All of the above

FOR DISCUSSION

1. Rates of evolutionary change differ among different molecules, and different species differ widely in generation times and population sizes. How does this variation limit how and in what ways we can use the concept of a molecular clock to help us answer questions about the evolution of both molecules and organisms?

2. One hypothesis proposed to explain the existence of large amounts of noncoding DNA is that the cost of maintaining that DNA is so small that natural selection is too weak to reduce it. How could you test this hypothesis against the hypothesis that genome size is functionally related to developmental rate?

3. If fossil evidence and molecular evidence disagree on the date of a major lineage split, which of the two kinds of evidence would you favor? Why?

4. Scientists can produce and release into the wild genetically modified mosquitoes that are unable to harbor and transmit malarial parasites. What ethical issues need to be discussed before such releases are permitted?

ADDITIONAL INVESTIGATION

Over evolutionary history, many groups of organisms that inhabit caves have lost the organs of sight. For instance, although surface-dwelling crayfishes have functional eyes, several crayfish species that are restricted to underground habitats lack eyes. Opsins are a group of light-sensitive proteins known to have an important function in vision (see Chapter 46), and opsin genes are expressed in eye tissues. Opsin genes are present in the genomes of eyeless, cave-dwelling crayfish. Two alternative hypotheses are (1) the opsin genes are no longer experiencing purifying selection (because there is no longer selection for function in vision); or (2) the opsin genes are experiencing selection for a function other than vision. How would you investigate these alternatives using the sequences of the opsin genes in various species of crayfishes?

When hawk-sized dragonflies ruled the air

Almost anyone who has spent time around fresh water ponds is familiar with dragonflies. Their hovering flight, bright colors, and transparent wings stimulate our visual senses on bright summer afternoons as they fly about their business of devouring mosquitoes, mating, and laying their eggs. The largest dragonflies alive today have wingspans that can be covered by a human hand. Three hundred million years ago, however, dragonflies such as *Meganeuropsis permiana* had wingspans of more than 70 centimeters—well over 2 feet, matching or exceeding the wingspans of many modern birds of prey—and were the largest flying predators on Earth.

No flying insects alive today are anywhere near this size. But during the Carboniferous and Permian geological periods, between 350 and 250 million years ago, many groups of flying insects contained gigantic members. *Meganeuropsis* probably ate huge mayflies and other giant flying insects that shared their home in the Permian swamps. These enormous insects were themselves eaten by giant amphibians.

None of the giant flying insects or amphibians of that time would be able to survive on Earth today. The oxygen concentrations in Earth's atmosphere were about 50 percent higher then than they are now, and those high oxygen levels are thought to have been necessary to support giant insects and their huge amphibian predators.

Paleontologists have uncovered fossils of *Meganeuropsis permiana* in the rocks of Kansas. How do we know the age of these fossils, and how can we know how much oxygen that long-vanished atmosphere contained? The *stratigraphic* layering of the rocks allows us to tell their ages relative to each other, but it does not by itself indicate a given layer's absolute age.

One of the remarkable achievements of twentieth-century scientists was to develop sophisticated techniques that use the decay rates of various radioisotopes, changes in Earth's magnetic field, and the ratios of certain molecules to infer conditions and events in the remote past and to date them accurately. It is those methods that allow us to age the fossils of *Meganeuropsis* and to calculate the concentration of oxygen in Earth's atmosphere at the time.

The development of the science of biology is intimately linked to changing concepts of

Giant Dragonflies *Meganeuropsis permiana*, shown here in a reconstruction from fossils, dwarfed modern dragonflies (shown in the inset at the same scale) in size. Otherwise, however, the Permian giant was quite similar to modern dragonflies in general appearance.

Younger Rocks Lie on Top of Older Rocks In the Grand Canyon, the Colorado River cut through and exposed many strata of ancient rocks. The oldest rocks visible here formed about 540 million years ago. The youngest, at the top, are about 500 million years old. Knowing the ages of rock strata allows scientists to date the fossils found in each stratum.

time, especially of the age of Earth. About 150 years ago, geologists first provided solid evidence that Earth is ancient; before 1850, most people believed it was no more than a few thousand years old. For many more years, physicists continued to underestimate Earth's age, until an understanding of radioactive decay was developed. Today we know that Earth is about 4.5 billion years old and that life has existed on it for about 3.8 billion of those years. That means human civilizations have occupied Earth for less than 0.0003 percent of the history of life. Discovering what happened before humans were around is an ongoing and exciting area of science.

IN THIS CHAPTER we will examine how biologists assign dates to events in the distant evolutionary past, and how such dating allows us to review the major changes in physical conditions on Earth during the past 4 billion years. We will then look at how these changes in physical conditions have influenced the major patterns in the evolution of life, and describe how scientists organize our knowledge of biological diversity based on the relationships among species.

25.1 How Do Scientists Date Ancient Events?

Many evolutionary changes happen rapidly enough to be studied directly and manipulated experimentally. Plant and animal breeding by agriculturalists and insects' evolution of resistance to pesticides are examples of rapid, short-term evolution. Other changes, such as the appearance of new species and evolutionary lineages, usually take place over much longer time frames.

To understand the long-term patterns of evolutionary change, we must think in time frames spanning many millions of years, and consider events and conditions very different from those we observe today. Earth of the distant past was so unlike the present that it seams like a foreign planet inhabited by strange organisms. The continents were not where they are today, and climates were sometimes dramatically different from those of today.

Fossils—the preserved remains of ancient organisms—can tell us a great deal about the body form, or *morphology*, of organisms that lived long ago, as well as how and where they lived. Fossils provide a direct record of evolution. But to understand patterns of evolutionary change, we must also understand how Earth has changed over time.

Earth's history is largely recorded in its rocks. We cannot tell the ages of rocks just by looking at them, but we can determine the ages of rocks relative to one another. The first person to formally recognize that this could be done was the seventeenth-century Danish physician Nicolaus Steno. Steno realized that in undisturbed **sedimentary rocks** (rocks formed by the accumulation of grains on the bottom of bodies of water), the oldest layers of rock, or **strata** (singular *stratum*), lie at the bottom; thus successively higher strata are progressively younger.

Geologists, particularly the eighteenth-century English scientist William Smith, subsequently combined Steno's insight with their observations of fossils contained in sedimentary rocks. They concluded that:

- Fossils of similar organisms are found in widely separated places on Earth.
- Certain fossils are always found in younger rocks, and certain other fossils in older rocks.
- Organisms found in higher, more recent strata are more similar to modern organisms than are those found in lower, more ancient strata.

These patterns revealed much about the relative ages of sedimentary rocks as well as patterns in the evolution of life. But the geologists still could not tell how old the rocks were. A method of dating rocks did not become available until after radioactivity was discovered at the beginning of the twentieth century.

Radioisotopes provide a way to date rocks

Radioactive isotopes of atoms (see Section 2.1) decay in a predictable pattern over long time periods. During each successive time interval, known as a **half-life**, half of the remaining radioactive material of the radioisotope decays to become a different, stable isotope (**Figure 25.1A**).

To use a radioisotope to date a past event, we must know or estimate the concentration of the isotope at the time of that event. In the case of carbon, the production of new carbon-14 (^{14}C) in the upper atmosphere (by the reaction of neutrons with nitrogen-14) just balances the natural radioactive decay of ^{14}C to ^{14}N. Therefore, the ratio of ^{14}C to its stable isotope, carbon-12 (^{12}C), is relatively constant in living organisms and their environment. As soon as an organism dies, however, it ceases to exchange carbon compounds with its environment. Its decaying ^{14}C is no longer replenished, and the ratio of ^{14}C to ^{12}C in its remains decreases through time. Paleontologists can use the ratio of ^{14}C to ^{12}C in fossil material to date fossils that are less than 50,000 years old (and thus the sedimentary rocks that contain those fossils). If fossils are older than that, so little ^{14}C remains that the limits of detection using this particular isotope are reached.

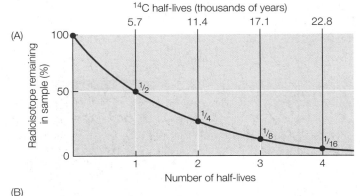

(B)

Radioisotope	Half-life (years)	Decay product	Useful dating range (years)
Carbon-14 (^{14}C)	5,700	Nitrogen-14 (^{14}N)	100 – 50,000
Potassium-40 (^{40}K)	1.3 billion	Argon-40 (^{40}Ar)	10 million – 4.5 billion
Uranium-238 (^{238}U)	4.5 billion	Lead 206 (^{206}Pb)	10 million – 4.5 billion

25.1 Radioactive Isotopes Allow Us to Date Ancient Rocks   The decay of radioactive "parent" atoms into stable "daughter" isotopes happens at a steady rate known as a half-life. (A) The graph demonstrates the principle of half-life using carbon-14 (^{14}C) as an example. (B) Radioisotopes have different characteristic half-lives that allow us to measure how much time has elapsed since the rocks containing them were laid down.

TABLE 25.1

Earth's Geological History

RELATIVE TIME SPAN	ERA	PERIOD	ONSET	MAJOR PHYSICAL CHANGES ON EARTH
	Cenozoic	Quaternary	2.6 mya	Cold/dry climate; repeated glaciations
		Tertiary	65 mya	Continents near current positions; climate cools
	Mesozoic	Cretaceous	145 mya	Northern continents attached; Gondwana begins to drift apart; meteorite strikes Yucatán Peninsula
		Jurassic	200 mya	Two large continents form: Laurasia (north) and Gondwana (south); climate warm
		Triassic	251 mya	Pangaea begins to slowly drift apart; hot/humid climate
	Paleozoic	Permian	297 mya	Extensive lowland swamps; O_2 levels 50% higher than present; by end of period continents aggregate to form Pangaea, and O_2 levels begin to drop rapidly
		Carboniferous	359 mya	Climate cools; marked latitudinal climate gradients
		Devonian	416 mya	Continents collide at end of period; meteorite probably strikes Earth
		Silurian	444 mya	Sea levels rise; two large land masses emerge; hot/humid climate
		Ordovician	488 mya	Massive glaciation, sea level drops 50 meters
		Cambrian	542 mya	O_2 levels approach current levels
Precambrian			900 mya	O_2 level at >5% of current level
			1.5 bya	O_2 level at >1% of current level
	Precambrian		3.8 bya	O_2 first appears in atmosphere
			4.5 bya	

Note: mya, million years ago; bya, billion years ago.

Radioisotope dating methods have been expanded and refined

Sedimentary rocks are formed from materials that existed for varying lengths of time before being transported, sometimes over long distances, to the site of their deposition. Therefore, the inorganic isotopes in a sedimentary rock do not contain reliable information about the date of its formation. Dating rocks more ancient than 50,000 years requires estimating isotope concentrations in *igneous* rocks—rocks formed when molten material cools. To date older sedimentary rocks, geologists search for places where sedimentary rocks show igneous intrusions of volcanic ash or lava flows.

A preliminary estimate of the age of an igneous rock determines which isotope is used to date it (**Figure 25.1B**). The decay of potassium-40 (which has a half-life of 1.3 billion years) to argon-40 has been used to date many of the ancient events in the evolution of life. Fossils in the adjacent sedimentary rock that are similar to those in other rocks of known ages provide additional clues.

Radioisotope dating of rocks, combined with fossil analysis, is the most powerful method of determining geological age. But in places where sedimentary rocks do not contain suitable ig-

neous intrusions and few fossils are present, paleontologists turn to other methods.

One method, known as **paleomagnetic dating**, relates the ages of rocks to patterns in Earth's magnetism, which change over time. Earth's magnetic poles move and occasionally reverse themselves. Because both sedimentary and igneous rocks preserve a record of Earth's magnetic field at the time they were formed, paleomagnetism helps determine the ages of those rocks. Other dating methods use information about continental drift, sea level changes, and molecular clocks (the last of which is described in Section 22.3).

Using these methods, geologists divided the history of life into *eras*, which in turn are subdivided into *periods* (**Table 25.1**). The boundaries between these time frames are based on striking differences scientists have observed in the assemblages of fossil organisms contained in successive layers of rocks. Geologists defined and named these divisions before they were able to establish the ages of fossils, adding and refining the time scales as new methods for geological dating were developed.

MAJOR EVENTS IN THE HISTORY OF LIFE
Humans evolve; many large mammals become extinct
Diversification of birds, mammals, flowering plants, and insects
Dinosaurs continue to diversify; mass extinction at end of period (≈76% of species disappear)
Diverse dinosaurs; radiation of ray-finned fishes; first fossils of flowering plants
Early dinosaurs; first mammals; marine invertebrates diversify; mass extinction at end of period (≈65% of species disappear)
Reptiles diversify; giant amphibians and flying insects present; mass extinction at end of period (≈96% of species disappear)
Extensive "fern" forests; first reptiles; insects diversify
Fishes diversify; first insects and amphibians; mass extinction at end of period (≈75% of species disappear)
Jawless fishes diversify; first ray-finned fishes; plants and animals colonize land
Mass extinction at end of period (≈75% of species disappear)
Rapid diversification of multicellular animals; diverse photosynthetic protists
Ediacaran fauna; earliest fossils of multicellular animals
Eukaryotes evolve
Origin of life; prokaryotes flourish

25.1 RECAP

Fossils in sedimentary rocks enabled geologists to determine the relative ages of organisms, but absolute dating was not possible until the discovery of radioactivity. Geologists divide the history of life into eras and periods, based on assemblages of fossil organisms found in successive layers of rocks.

- What observations about fossils suggested to geologists that they could be used to determine the relative ages of rocks? See p. 519
- How is the rate of decay of radioisotopes used to estimate the absolute ages of rocks? See p. 520 and Figure 25.1

The scale at the left of Table 25.1 gives a relative sense of geological time, especially the vast expanse of the Precambrian era, during which early life evolved amid stupendous physical changes of Earth and its atmosphere. During the Precambrian to Cambrian transition, an "explosion" of new life forms took place as representatives of many of the major multicellular groups of life evolved. Earth continued to undergo massive physical changes that influenced the evolution of life, and these events and important milestones are listed in the table. In the next two sections we'll discuss the most important of these changes.

25.2 How Have Earth's Continents and Climates Changed over Time?

The globes and maps that adorn our walls, shelves, and books give an impression of a static Earth. It would be easy for us to assume that the continents have always been where they are, but we would be wrong. The idea that Earth's land masses have changed position over the millennia, and that they continue to do so, was first put forth in 1912 by the German meteorologist

25.2 Plate Tectonics and Continental Drift The heat of Earth's core generates convection currents (arrows) in the magma that push the lithospheric plates, along with the land masses lying on them, together or apart. When lithospheric plates collide, one often slides under the other. The resulting seismic activity can create mountains and deep rift valleys (the latter known as trenches when they occur under ocean basins).

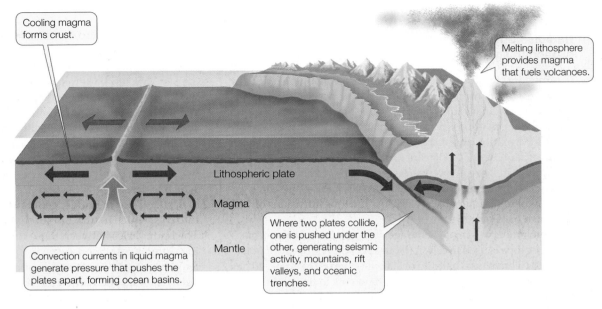

Cooling magma forms crust.

Melting lithosphere provides magma that fuels volcanoes.

Lithospheric plate

Magma

Mantle

Convection currents in liquid magma generate pressure that pushes the plates apart, forming ocean basins.

Where two plates collide, one is pushed under the other, generating seismic activity, mountains, rift valleys, and oceanic trenches.

and geophysicist Alfred Wegener. His book *The Origin of Continents and Oceans* was initially met with skepticism and resistance. By the 1960s, however, physical evidence and increased understanding of the geophysics of **plate tectonics**—the study of movement of major land masses—had convinced virtually all geologists of the reality of Wegener's vision.

Earth's crust consists of several solid plates approximately 40 kilometers thick, which collectively make up the lithosphere. The lithospheric plates float on a fluid layer of molten rock, or magma (**Figure 25.2**). Heat produced by radioactive decay deep in Earth's core sets up convection currents in the fluid magma, which then rises and exerts tremendous pressure on the solid plates. When the pressure of the rising magma pushes plates apart, ocean basins may form between them. When plates are pushed together, they either move sideways past each other or one plate slides under the other, pushing up mountain ranges and carving deep rift valleys. When they occur under the water of ocean basins, rift valleys are known as trenches. The

movement of the lithospheric plates and the continents they contain is known as **continental drift**.

We now know that at times the drifting of the plates has brought continents together and at other times has pushed them apart (these movements are depicted in Figure 25.12). The positions and sizes of the continents influence oceanic circulation patterns, global climates, and sea levels. Major drops in sea level have usually been accompanied by massive extinctions—particularly of marine organisms, which could not survive the exposure of vast areas of the continental shelves and the disappearance of the shallow seas that covered them (**Figure 25.3**).

yourBioPortal.com

GO TO **Animated Tutorial 25.1 • Evolution of the Continents**

25.3 Sea Levels Have Changed Repeatedly Most mass extinctions of marine organisms (indicated by asterisks) have coincided with periods of low sea levels.

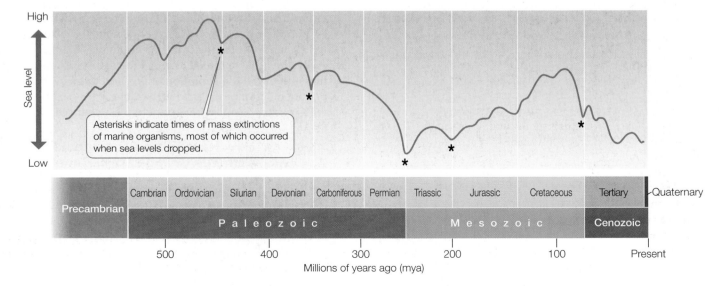

Asterisks indicate times of mass extinctions of marine organisms, most of which occurred when sea levels dropped.

High

Sea level

Low

| Precambrian | Cambrian | Ordovician | Silurian | Devonian | Carboniferous | Permian | Triassic | Jurassic | Cretaceous | Tertiary | Quaternary |

Paleozoic   Mesozoic   Cenozoic

500   400   300   200   100   Present

Millions of years ago (mya)

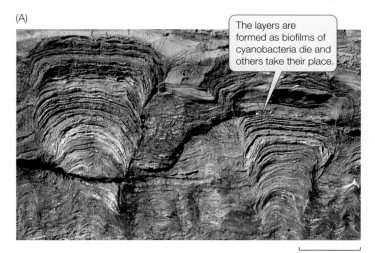

The layers are formed as biofilms of cyanobacteria die and others take their place.

(A)

12 cm

(B)

Living cyanobacteria are found in the upper parts of these structures.

30 cm

25.4 Stromatolites   (A) A vertical section through a fossil stromatolite. (B) These rocklike structures are living stromatolites that thrive in the very salty waters of Shark Bay in western Australia. Layers of cyanobacteria are found in the uppermost parts of the structures.

stromatolites, which are abundantly preserved in the fossil record. Cyanobacteria are still forming stromatolites today in a few very salty places on Earth (**Figure 25.4**). Cyanobacteria liberated enough O_2 to open the way for the evolution of oxidation reactions as the energy source for the synthesis of ATP (see Section 9.1).

The evolution of life thus irrevocably changed the physical nature of Earth. Those physical changes, in turn, influenced the evolution of life. When it first appeared in the atmosphere, O_2 was poisonous to the anaerobic prokaryotes that inhabited Earth at the time. Over millennia, however, prokaryotes that evolved the ability to metabolize O_2 not only survived but gained several advantages. Aerobic metabolism proceeds more rapidly and harvests energy more efficiently than anaerobic metabolism (see Section 9.4), and organisms with aerobic metabolism replaced anaerobes in most of Earth's environments.

An atmosphere rich in O_2 also made possible larger cells and more complex organisms. Small unicellular aquatic organisms can obtain enough O_2 by simple diffusion even when O_2 concentrations are very low. Larger unicellular organisms have lower surface area-to-volume ratios (see Figure 5.2); to obtain enough O_2 by simple diffusion, they must live in an environment with a relatively high oxygen concentration. Bacteria can thrive on 1 percent of the current atmospheric O_2 levels; eukaryotic cells require levels that are at least 2–3 percent of current concentrations. (For concentrations of dissolved O_2 in the oceans to reach these levels, much higher atmospheric concentrations were needed.)

Probably because it took many millions of years for Earth to develop an oxygenated atmosphere, only unicellular prokaryotes lived on Earth for more than 2 billion years. About 1.5 bya, atmospheric O_2 concentrations became high enough for large eukaryotic cells to flourish (**Figure 25.5**). Further increases in atmospheric O_2 levels 750 to 570 million years ago (mya) enabled several groups of multicellular organisms to evolve.

Oxygen concentrations in Earth's atmosphere have changed over time

As the continents have moved over Earth's surface, the world has experienced other physical changes, including large increases and decreases in atmospheric oxygen. The atmosphere of early Earth probably contained little or no free oxygen gas (O_2). The increase in atmospheric O_2 came in two big steps more than a billion years apart. The first step occurred at least 2.4 billion years ago (bya), when certain bacteria evolved the ability to use water as the source of hydrogen ions for photosynthesis. By chemically splitting H_2O, these bacteria generated atmospheric O_2 as a waste product. They also made electrons available for reducing CO_2 to form organic compounds (see Section 10.3).

One group of O_2-generating bacteria, the *cyanobacteria*, formed rocklike structures called

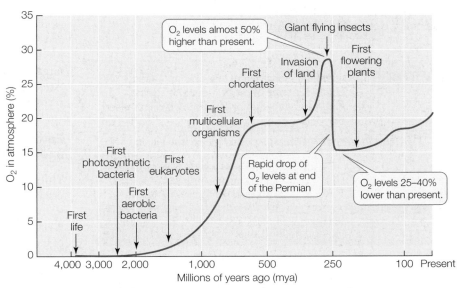

25.5 Larger Cells, Larger Organisms Need More Oxygen   Changes in oxygen concentrations have strongly influenced, and been influenced by, the evolution of life. (Note that the horizontal axis of the graph is on a logarithmic scale.)

INVESTIGATING LIFE

25.6 Rising Oxygen Levels and Body Size in Insects

In this experiment, flies were raised under hyperbaric conditions (increased atmospheric pressure), thus increasing the partial pressure of O_2 in a manner that simulated the greater levels of atmospheric O_2 characteristic of the Carboniferous and Permian. Robert Dudley asked if flies raised in hyperbaric conditions would grow larger than their normal counterparts.

HYPOTHESIS Under increased atmospheric pressure, the increased partial pressure of O_2 will allow directional selection for increased body size in flying insects.

METHOD

1. Divide a population of fruit flies (*Drosophila melanogaster*) into two lines.
2. Raise one line (the control) at current atmospheric oxygen conditions. Raise the experimental line in hyperbaric conditions (increased partial pressure of O_2, simulating increased atmospheric oxygen concentrations). Continue for 5 generations.
3. Raise the F_6 offspring of both lines under identical environmental conditions.
4. Weigh all the F_6 individuals and test for statistical differences in the average body mass of the flies in each population.

RESULTS The average body mass of F_6 individuals of both sexes in the experimental line was significantly ($p < 0.0001$) greater than that of insects in the control line.

CONCLUSION In at least some flying insects, increased concentrations of oxygen could lead to a long-term evolutionary trend toward increased body size.

FURTHER INVESTIGATION: How would you confirm that the change in average body size is related to increased partial pressure of O_2, and not to other aspects of overall increased atmospheric pressure?

Go to **yourBioPortal.com** for original citations, discussions, and relevant links for all INVESTIGATING LIFE figures.

O_2 concentrations increased again during the Carboniferous and Permian periods because of the evolution of large vascular plants in the expansive lowland swamps that existed then (see

Table 25.1). These swamps resulted in extensive burial of plant debris from vascular plants, which led to the formation of Earth's vast coal deposits. As the buried organic material was not subject to oxidation, and the living plants were producing large quantities of O_2, atmospheric O_2 increased to concentrations that have not been reached again in Earth's history (see Figure 25.5). As mentioned in the opening of this chapter, high concentrations of atmospheric O_2 allowed the evolution of giant flying insects and amphibians that could not survive in today's atmosphere. The drying of the lowland swamps at the end of the Permian reduced global organic burial, and also the production of atmospheric O_2, so O_2 concentrations dropped rapidly. Over the past 200 million years, with the diversification of flowering plants, O_2 concentrations have again increased, but not to the levels that characterized the Carboniferous and Permian periods.

Biologists have conducted experiments that demonstrate the changing selective pressures that can accompany changes in O_2 levels. In experimental conditions, an increase in O_2 concentration can be simulated by increasing atmospheric pressure in a hyperbaric chamber. Increasing atmospheric pressure increases the *partial pressure of oxygen* (see Chapter 49) in a manner that simulates an increase in O_2 concentration at normal atmospheric pressure. When lines of fruit flies (*Drosophila*) are raised in artificial hyperbaric atmospheres (which have higher partial pressure of O_2), they quickly evolve larger body sizes over just a few generations (**Figure 25.6**). The current levels of atmospheric O_2 appear to constrain body size evolution of these flying insects; increases in O_2 appear to relax these constraints. This demonstrates that the *stabilizing selection* on body size at present O_2 concentrations can quickly switch to *directional selection* (see Section 21.3) for a change in body size in response to a change in O_2 levels. Directional selection over a period of millions of years would be sufficient to account for giant insects such as *Meganeuropsis*, described at the beginning of this chapter.

Many physical conditions on Earth have oscillated in response to the planet's internal processes, such as volcanic activity and continental drift. Extraterrestrial events, such as collisions with meteorites, have also left their mark. In some cases, as we saw earlier and will see again in this chapter, changing physical parameters caused **mass extinctions**, during which a large proportion of the species living at the time disappeared. After each mass extinction, the diversity of life rebounded, but recovery took millions of years.

Earth's climate has shifted between hot/humid and cold/dry conditions

Through much of its history, Earth's climate was considerably warmer than it is today, and temperatures decreased more gradually toward the poles. At other times, Earth was colder than it is today. Large areas were covered with glaciers near the end of the Precambrian and Ordovician, and during parts of the Carboniferous and Permian periods. These cold periods were separated by long periods of milder climates (**Figure 25.7**). Because we are living in one of the colder periods in Earth's history, it is difficult for us to imagine the mild climates that were found at high latitudes during much of the history of life. During the

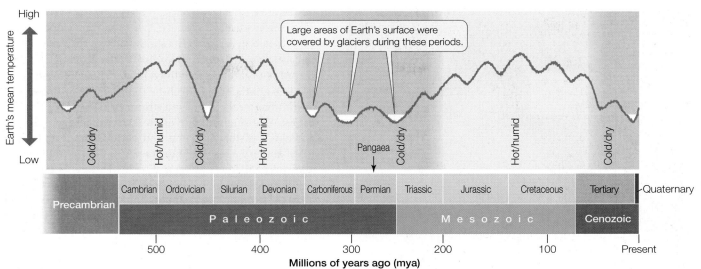

25.7 Hot/Humid and Cold/Dry Conditions Have Alternated over Earth's History Throughout Earth's history, periods of cold climates and glaciations (white depressions) have been separated by long periods of milder climates.

Quaternary period there has been a series of glacial advances, interspersed with warmer interglacial intervals during which the glaciers retreated.

"Weather" refers to daily events, such as individual storms. "Climate" refers to long-term average expectations of the various seasons at a given location. Weather often changes rapidly; climates typically change slowly. Major climatic shifts have taken place over periods as short as 5,000 to 10,000 years, primarily as a result of changes in Earth's orbit around the sun. A few climatic shifts have been even more rapid. For example, during one Quaternary interglacial period, the ice-locked Antarctic Ocean became nearly ice-free in less than 100 years. Such rapid changes are usually caused by sudden shifts in ocean currents. Some climate changes have been so rapid that the extinctions caused by them appear to be nearly "instantaneous" in the fossil record.

We are currently living in a time of rapid climate change thought to be caused by a buildup of atmospheric CO_2, primarily from the burning of fossil fuels. We are reversing the process of organic burial that occurred (especially) in the Carboniferous and Permian, but we are doing so over a few hundred years rather than the many millions of years over which these deposits accumulated. The current rate of increase of atmospheric CO_2 is unprecedented in Earth's history. A doubling of the atmospheric CO_2 concentration—which may happen during the current century—is expected to increase the average temperature of Earth, change rainfall patterns, melt glaciers and ice caps, and raise sea level. The possible consequences of such climate changes are discussed in Chapters 58 and 59.

Volcanoes have occasionally changed the history of life

Most volcanic eruptions produce only local or short-lived effects, but a few large volcanic eruptions have had major consequences for life. When Krakatoa erupted in Indonesia in 1883,

it ejected more than 25 cubic kilometers of ash and rock, as well as large quantities of sulphur dioxide gas (SO_2). The SO_2 was ejected into the stratosphere and then moved by high-level winds around the planet. This led to high concentrations of sulphurous acid (H_2SO_3) in high-level clouds, which meant less sunlight got through to Earth's surface. Global temperatures dropped by 1.2°C in the year following the eruption, and global weather patterns showed strong effects for another 5 years. This was all the result of a single volcanic eruption. The collision of continents during the Permian period (about 275 mya) formed a single, gigantic land mass (Pangaea) and caused many massive volcanic eruptions. These eruptions resulted in considerable blockage of sunlight, contributing to the glaciations of that time (see Figure 25.7). Massive volcanic eruptions occurred again as the continents drifted apart during the late Triassic and at the end of the Cretaceous.

Extraterrestrial events have triggered changes on Earth

At least 30 meteorites between the sizes of baseballs and soccer balls hit Earth each year. Collisions with large meteorites or comets are rare, but such collisions have probably been responsible for several mass extinctions. Several types of evidence tell us about these collisions. Their craters, and the dramatically disfigured rocks that resulted from their impact, are found in many places. Geologists have also discovered compounds in these rocks that contain helium and argon with isotope ratios characteristic of meteorites, which are very different from the ratios found elsewhere on Earth.

A meteorite caused or contributed to a mass extinction at the end of the Cretaceous period (about 65 mya). The first clue that a meteorite was responsible came from the abnormally high concentrations of the element iridium in a thin layer separating rocks deposited during the Cretaceous from those deposited during the Tertiary (**Figure 25.8**). Iridium is abundant in some meteorites, but it is exceedingly rare on Earth's surface. Scientists discovered a circular crater 180 kilometers in diameter buried beneath the northern coast of the Yucatán Peninsula of

Iridium-rich layer at the
Cretaceous-Tertiary
(K/T) boundary

25.8 Evidence of a Meteorite Impact   The white layers of rock are
Cretaceous in age; the layers at the upper left were deposited in the
Tertiary. Between the two is a thin, dark layer of clay that contains large
amounts of iridium, a metal common in some meteorites but rare on
Earth. Its high concentration in sediments deposited about 65 million
years ago suggests the impact of a large meteorite.

Mexico. When it collided with Earth, the meteorite released energy equivalent to that of 100 million megatons of high explo-
sives, creating great tsunamis. A massive plume of debris
swelled to a diameter of up to 200 kilometers, spread around
Earth, and descended. The descending debris heated the atmosphere to several hundred degrees, ignited massive fires, and
blocked the sun, preventing plants from photosynthesizing. The
settling debris formed the iridium-rich layer. About a billion
tons of soot, which has a composition that matches smoke from
forest fires, was also deposited. Many fossil species (particularly
dinosaurs) that are found in Cretaceous rocks are not found in
the Tertiary rocks of the next layer.

25.2 RECAP

Conditions on Earth have changed dramatically over
time. Changes in atmospheric concentrations of O_2
and in Earth's climate have had major effects on biological evolution. Continental drift, volcanic eruptions, and large meteorite strikes have contributed
to climatic changes during Earth's history.

- Describe how increases in atmospheric concentrations
  of O_2 affected the evolution of multicellular organisms.
  See pp. 523–524 and Figure 25.5

- How have volcanic eruptions and meteorite strikes influenced the course of life's evolution? See p. 525

The many dramatic physical events of Earth's history have influenced the nature and timing of evolutionary changes among
Earth's living organisms. We now will look more closely at some
of the major events that characterize the history of life on Earth.

25.3 What Are the Major Events in Life's History?

Life first evolved on Earth about 3.8 bya. By about 1.5 bya, eukaryotic organisms had evolved (see Table 25.1). The fossil
record of organisms that lived prior to 550 mya is fragmentary,
but it is good enough to show that the total number of species
and individuals increased dramatically in late Precambrian
times. As discussed above, pre-Darwinian geologists divided
geological history into eras and periods based on their distinct
fossil assemblages. Biologists refer to the assemblage of all organisms of all kinds living at a particular time or place as a **biota**.
All of the plants living at a particular time or place are its **flora**;
all of the animals are its **fauna**. Table 25.1 describes some of the
physical and biological changes, such as mass extinctions and
dramatic increases in the diversity of major groups of organisms, associated with each unit of time.

About 300,000 species of fossil organisms have been described, and the number steadily grows. The number of named
species, however, is only a tiny fraction of the species that have
ever lived. We do not know how many species lived in the past,
but we have ways of making reasonable estimates. Of the present-day biota, nearly 1.8 million species have been named. The
actual number of living species is probably well over 10 million,
and possibly much higher, because many species have not yet
been discovered and described by biologists. So the number of
described fossil species is only about 3 percent of the estimated
minimum number of living species. Life has existed on Earth
for about 3.8 billion years. Many species last only a few million years before undergoing speciation or going extinct; therefore, Earth's biota must have turned over many times during
geological history. So the total number of species that have lived
over evolutionary time must vastly exceed the number living
today. Why have only about 300,000 of these tens of millions
of species been described from fossils to date?

Several processes contribute to the paucity of fossils

Only a tiny fraction of organisms ever become fossils, and only
a tiny fraction of fossils are ever discovered by paleontologists.
Most organisms live and die in oxygen-rich environments in
which they quickly decompose. They are not likely to become
fossils unless they are transported by wind or water to sites that
lack oxygen, where decomposition proceeds slowly or not at
all. Furthermore, geological processes often transform rocks, de-
stroying the fossils they contain, and many fossil-bearing rocks
are deeply buried and inaccessible. Paleontologists have studied only a tiny fraction of the sites that contain fossils, but they
find and describe many new ones every year.

The fossil record is most complete for marine animals that
had hard skeletons (which resist decomposition). Among the
nine major animal groups with hard-shelled members, approximately 200,000 species have been described from fossils—
roughly twice the number of living marine species in these
same groups. Paleontologists lean heavily on these groups in
their interpretations of the evolution of life. Insects and spiders
are also relatively well represented in the fossil record, because
they are numerically abundant and have hard exoskeletons

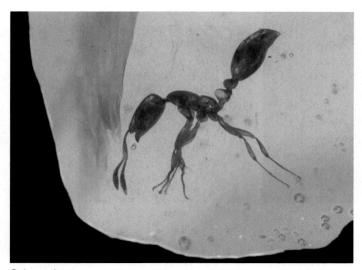

Solenopsis sp.

25.9 Insect Fossils Chunks of amber—fossilized tree resin—often contain insects that were preserved when they were trapped in the sticky resin. This fire ant fossil is some 30 million years old.

(**Figure 25.9**). The fossil record, though incomplete, is good enough to document clearly the factual history of the evolution of life.

By combining information about geological changes during Earth's history with evidence from the fossil record, scientists have composed portraits of what Earth and its inhabitants may have looked like at different times. We know in general where the continents were and how life changed over time, but many of the details are poorly known, especially for events in the more remote past.

Precambrian life was small and aquatic

For most of its history, life was confined to the oceans, and all organisms were small. Over the long ages of the **Precambrian era**—more than 3 billion years—the shallow seas slowly be-

gan to teem with life. For most of the Precambrian, life consisted of microscopic prokaryotes; eukaryotes evolved about two-thirds of the way through the era (**Figure 25.10**). Unicellular eukaryotes and small multicellular animals fed on floating photosynthetic microorganisms. Small floating organisms, known collectively as *plankton*, were strained from the water and eaten by slightly larger *filter-feeding* animals. Other animals ingested sediments on the seafloor and digested the remains of organisms within them. By the late Precambrian (630–542 mya), many kinds of multicellular soft-bodied animals had evolved. Some of them were very different from any animals living today, and may be members of groups that have no living descendants (**Figure 25.11**).

Life expanded rapidly during the Cambrian period

The Cambrian period (542–488 mya) marks the beginning of the Paleozoic era. The oxygen concentration in the Cambrian atmosphere was approaching its current level, and the land masses had come together to form several large continents. A geologically rapid diversification of life took place that is sometimes referred to as the **Cambrian explosion** (although in fact it began before the Cambrian, and the "explosion" took millions of years). Several of the major groups of animals that have species living today first evolved during the Cambrian. An overview of the continental and biotic shifts that characterized the Cambrian and subsequent periods is shown in **Figure 25.12** on the following pages.

For the most part, fossils tell us only about the hard parts of organisms, but in three known Cambrian fossil beds—the Burgess Shale in British Columbia, Sirius Passet in northern Greenland, and the Chengjiang site in southern China—the soft parts of many animals were preserved. Crustacean arthropods (crabs, shrimps, and their relatives) are the most diverse group

25.10 A Sense of Life's Time The top timeline shows the 4.5 billion year history of life on Earth. Most of this time is accounted for by the Precambrian, a 3.4 billion year era that saw the origin of life and the evolution of cells, photosynthesis, and multicellularity. The final 600 million years are expanded in the second timeline and detailed in Figure 25.12.

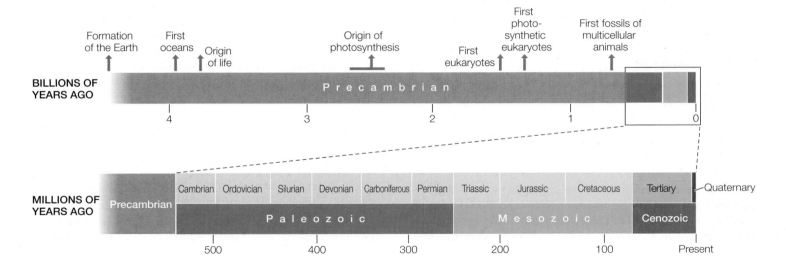

Spriggina floundersi   *Mawsonites spriggi*   *Dickinsonia costata*

25.11 Precambrian Life   These fossils of soft-bodied invertebrates, excavated at Ediacara in southern Australia, were formed about 600 million years ago. Very different from later life forms, they illustrate the diversity of life at the end of the Precambrian era.

in the Chinese fauna; some of them were large carnivores. Multicellular diversity was largely or completely aquatic during the Cambrian. If there was life on land at this time, it was probably restricted to microbial organisms.

Many groups of organisms that arose during the Cambrian later diversified

Geologists divide the remainder of the Paleozoic era into the Ordovician, Silurian, Devonian, Carboniferous, and Permian periods. Each period is characterized by the diversification of specific groups of organisms. Mass extinctions marked the ends of the Ordovician, Devonian, and Permian.

THE ORDOVICIAN (488–444 MYA)   During the Ordovician period, the continents, which were located primarily in the Southern Hemisphere, still lacked multicellular plants. Evolutionary radiation of marine organisms was spectacular during the early Ordovician, especially among animals, such as brachiopods and mollusks, that lived on the seafloor and filtered small prey from the water. At the end of the Ordovician, as massive glaciers formed over the southern continents, sea levels dropped about 50 meters and ocean temperatures dropped. About 75 percent of the animal species became extinct, probably because of these major environmental changes.

THE SILURIAN (444–416 MYA)   During the Silurian period, the continents began to merge together. Marine life rebounded from the mass extinction at the end of the Ordovician. Animals able to swim in open water and feed above the ocean bottom appeared for the first time. Jawless fishes diversified, and the first ray-finned fishes evolved. The tropical sea was uninterrupted by land barriers, and most marine organisms were widely distributed. On land, the first vascular plants evolved late in the Silurian (about 420 mya). The first terrestrial arthropods—scorpions and millipedes—evolved at about the same time.

THE DEVONIAN (416–359 MYA)   Rates of evolutionary change accelerated in many groups of organisms during the Devonian period. The major land masses continued to move slowly toward each other. In the oceans there were great evolutionary radiations of corals and of shelled, squidlike cephalopod mollusks. Fishes diversified as jawed forms replaced jawless ones and as heavy armor gave way to the less rigid outer coverings of modern fishes.

Terrestrial communities changed dramatically during the Devonian. Club mosses, horsetails, and tree ferns became common; some attained the size of large trees. Their roots accelerated the weathering of rocks, resulting in the development of the first forest soils. The ancestors of gymnosperms—the first plants to produce seeds—appeared in the Devonian. The first known fossils of centipedes, spiders, mites, and insects date to this period, and fishlike amphibians began to occupy the land.

A massive extinction of about 75 percent of all marine species marked the end of the Devonian. Paleontologists are uncertain about its cause, but two large meteorites that collided with Earth at that time (one in present-day Nevada and the other in western Australia) may have been responsible, or at least a contributing factor. The continued coalescence of the continents, with the corresponding reduction in continental shelves, may have also contributed to this mass extinction event.

THE CARBONIFEROUS (359–297 MYA)   Large glaciers formed over high-latitude portions of the southern land masses during the Carboniferous period, but extensive swamp forests grew on the tropical continents. These forests were not made up of the kinds of trees we know today, but were dominated by giant tree ferns and horsetails with small leaves. Fossilized remains of those forests formed the coal we now mine for energy. In the seas,

25.12 A Brief History of Multicellular Life   The geologically rapid "explosion" of life during the Cambrian saw the rise of animal groups that have representatives surviving today. The following three pages depict life's history from the Cambrian forward. Movements of the major continents during the past half-billion years are shown in the maps of Earth, and associated biotas for each time period are depicted. The artists' reconstructions are based on fossils such as those shown in the photographs.

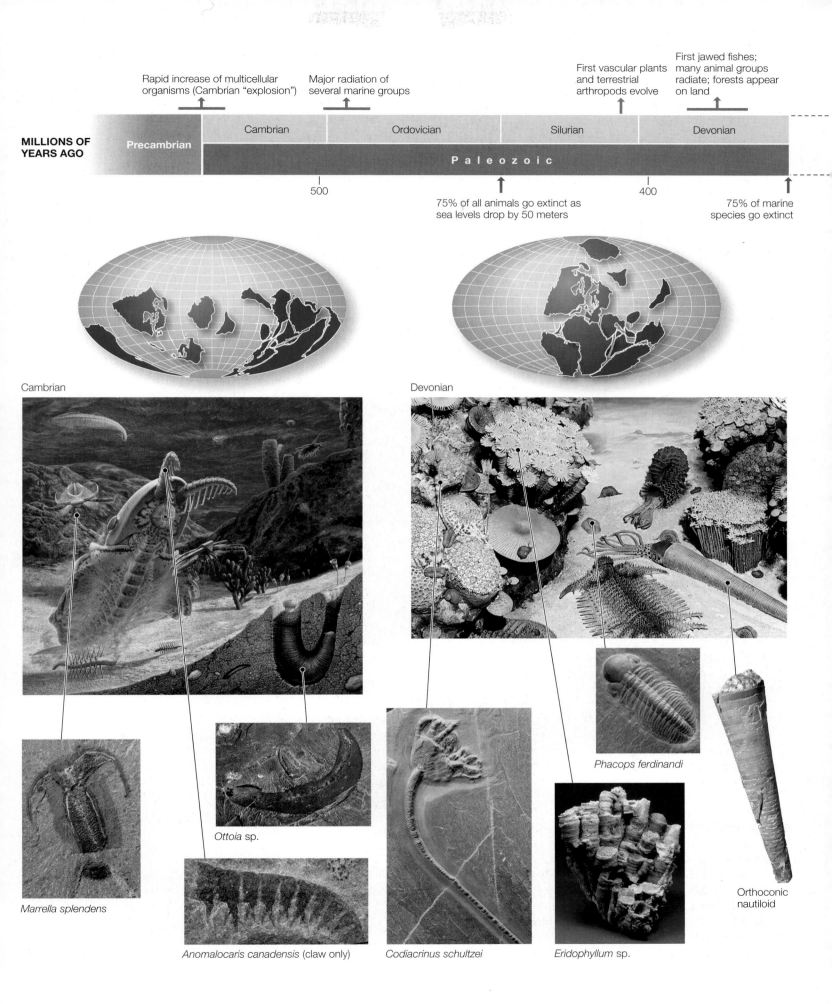

MILLIONS OF YEARS AGO

Rapid increase of multicellular organisms (Cambrian "explosion")

Major radiation of several marine groups

First vascular plants and terrestrial arthropods evolve

First jawed fishes; many animal groups radiate; forests appear on land

Precambrian | Cambrian | Ordovician | Silurian | Devonian

Paleozoic

500

75% of all animals go extinct as sea levels drop by 50 meters

400

75% of marine species go extinct

Cambrian

Devonian

Marrella splendens

Ottoia sp.

Anomalocaris canadensis (claw only)

Codiacrinus schultzei

Eridophyllum sp.

Phacops ferdinandi

Orthoconic nautiloid

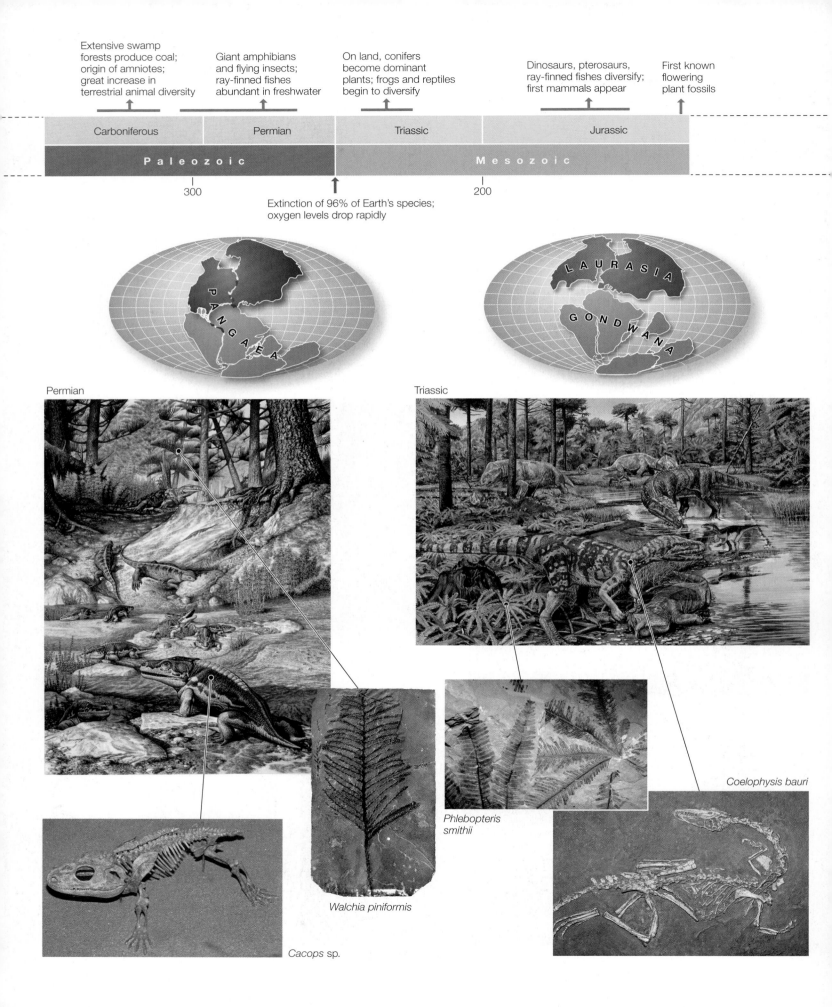

Extensive swamp forests produce coal; origin of amniotes; great increase in terrestrial animal diversity

Giant amphibians and flying insects; ray-finned fishes abundant in freshwater

On land, conifers become dominant plants; frogs and reptiles begin to diversify

Dinosaurs, pterosaurs, ray-finned fishes diversify; first mammals appear

First known flowering plant fossils

| Carboniferous | Permian | Triassic | Jurassic |

Paleozoic | Mesozoic

300

200

Extinction of 96% of Earth's species; oxygen levels drop rapidly

PANGAEA

LAURASIA
GONDWANA

Permian

Triassic

Walchia piniformis

Cacops sp.

Phlebopteris smithii

Coelophysis bauri

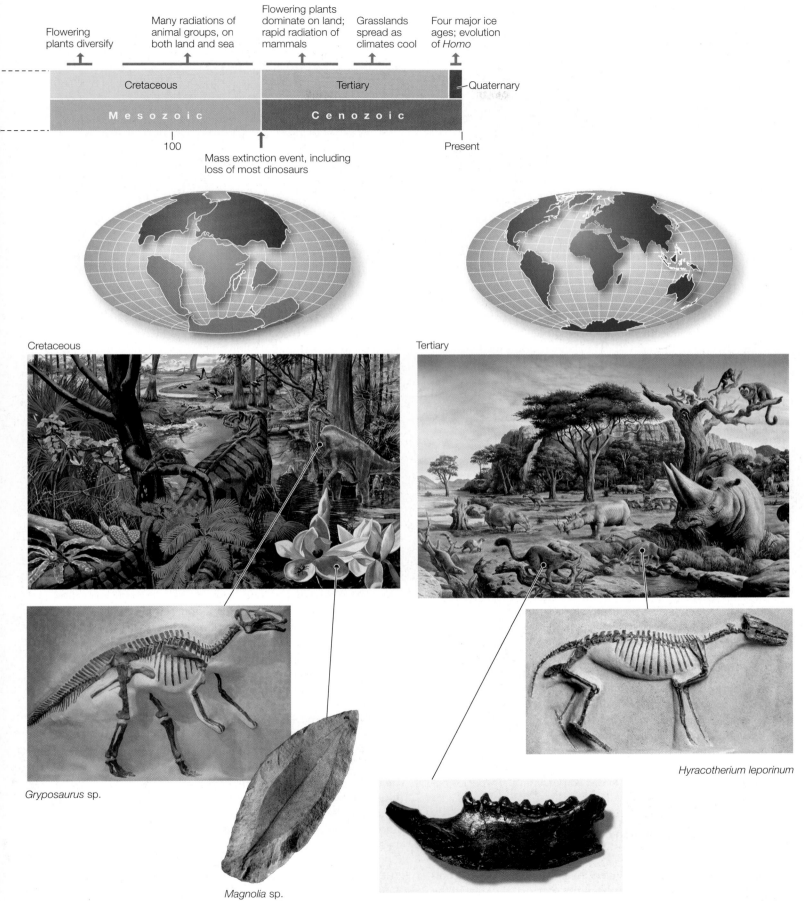

Flowering plants diversify

Many radiations of animal groups, on both land and sea

Flowering plants dominate on land; rapid radiation of mammals

Grasslands spread as climates cool

Four major ice ages; evolution of *Homo*

Cretaceous

Tertiary

Quaternary

M e s o z o i c

C e n o z o i c

100

Present

Mass extinction event, including loss of most dinosaurs

Cretaceous

Tertiary

Gryposaurus sp.

Magnolia sp.

Plesiadapis fodinatus (jaw)

Hyracotherium leporinum

25.13 Evidence of Insect Diversification The margins of this fossil fern leaf from the Carboniferous have been chewed by insects.

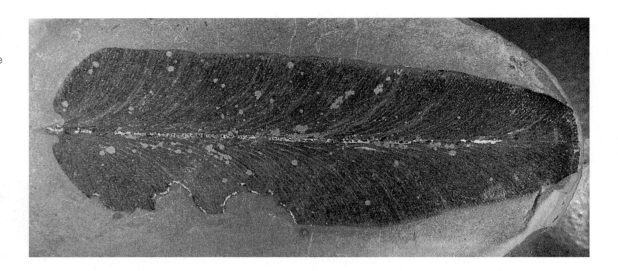

crinoids (sea lilies and feather stars) reached their greatest diversity, forming "meadows" on the seafloor.

The diversity of terrestrial animals increased greatly during the Carboniferous. Snails, scorpions, centipedes, and insects were abundant and diverse. Insects evolved wings, becoming the first animals to fly. Flight gave herbivorous insects easy access to tall plants; plant fossils from this period show evidence of chewing by insects (**Figure 25.13**). The terrestrial vertebrate lineage split, and amphibians became larger and better adapted to terrestrial existence, while the sister lineage led to the **amniotes**, vertebrates with well-protected eggs that can be laid in dry places.

THE PERMIAN (297–251 MYA) During the Permian period, the continents coalesced completely into the supercontinent **Pangaea**. Permian rocks contain representatives of many major groups of insects we know today. By the end of the period, the reptiles split from a second amniote lineage (which would lead to the mammals). Ray-finned fishes became common in the fresh waters of Pangaea.

Conditions for life deteriorated toward the end of the Permian. Massive volcanic eruptions resulted in outpourings of lava that covered large areas of Earth. The ash and gases produced by the volcanoes blocked sunlight and cooled the climate, resulting in the largest glaciers in Earth's history. Atmospheric oxygen concentrations gradually dropped from about 30 to 15 percent. At such low concentrations, most animals would have been unable to survive at elevations above 500 meters; thus about half of the land area would have been uninhabitable at the end of the Permian. The combination of these changes resulted in the most drastic mass extinction event in Earth's history. Scientists estimate that about 96 percent of all species became extinct at the end of the Permian.

Geographic differentiation increased during the Mesozoic era

The few organisms that survived the Permian mass extinction found themselves in a relatively empty world at the start of the Mesozoic era (251 mya). As Pangaea slowly began to break apart, the oceans rose and once again flooded the continental shelves, forming huge, shallow inland seas. Atmospheric oxygen concentrations gradually rose. Life once again proliferated and diversified, but different groups of organisms came to the fore. The three groups of phytoplankton (floating photosynthetic organisms) that dominate today's oceans—dinoflagellates, coccolithophores, and diatoms—became ecologically important at this time; their remains are the primary origin of the world's oil deposits. Seed-bearing plants replaced the trees that had ruled the Permian forests.

The Mesozoic era is divided into three periods: the Triassic, Jurassic, and Cretaceous. The Triassic and Cretaceous were terminated by mass extinctions, probably caused by meteorite impacts.

THE TRIASSIC (251–200 MYA) Pangaea began to break apart during the Triassic period. Many invertebrate groups became more species-rich, and many burrowing animals evolved from groups living on the surfaces of seafloor sediments. On land, conifers and seed ferns were the dominant trees. The first frogs and turtles appeared. A great radiation of reptiles began, which eventually gave rise to crocodilians, dinosaurs, and birds. The end of the Triassic was marked by a mass extinction that eliminated about 65 percent of the species on Earth.

THE JURASSIC (200–145 MYA) During the Jurassic period, Pangaea became fully divided into two large continents: **Laurasia** drifted northward and **Gondwana** drifted south. Ray-finned fishes rapidly diversified in the oceans. The first lizards appeared, and flying reptiles (pterosaurs) evolved. Most of the large terrestrial predators and herbivores of the period were dinosaurs. Several groups of mammals made their first appearance, and the earliest known fossils of flowering plants are from late in this period.

THE CRETACEOUS (145–65 MYA) By the early Cretaceous period, Laurasia and Gondwana had begun to break apart into the continents we know today. A continuous sea encircled the tropics. Sea levels were high, and Earth was warm and humid. Life proliferated both on land and in the oceans. Marine invertebrates increased in diversity and in number of species. On land, the reptile radiation continued as dinosaurs diversified further and the first snakes appeared. Early in the Cretaceous, flowering plants began the radiation that led to their current dominance of the land. By the end of the period, many groups of mammals had evolved. Most early mammals were small, but one species recently discovered in China, *Repenomamus giganticus*, was large enough to capture and eat young dinosaurs.

As described in Section 25.2, another meteorite-caused mass extinction took place at the end of the Cretaceous (the impact site was near the present day Yucatán Peninsula of Mexico). In the seas, many planktonic organisms and bottom-dwelling invertebrates became extinct. On land, almost all animals larger than about 25 kilograms in body weight became extinct. Many species of insects died out, perhaps because the growth of their food plants was greatly reduced following the impact. Some species in northern North America and Eurasia survived in areas that were not subjected to the devastating fires that engulfed most low-latitude regions.

Modern biota evolved during the Cenozoic era

By the early Cenozoic era (65 mya), the positions of the continents resembled those of today, but Australia was still attached to Antarctica, and the Atlantic Ocean was much narrower. The Cenozoic was characterized by an extensive radiation of mammals, but other groups were also undergoing important changes.

Flowering plants diversified extensively and came to dominate world forests, except in the coolest regions, where the forests were composed primarily of gymnosperms. Mutations of two genes in one group of plants (the legumes) allowed them to use atmospheric nitrogen directly by forming symbioses with a few species of nitrogen-fixing bacteria (see Section 36.4). The evolution of this symbiosis between certain early Cenozoic plants and these specialized bacteria was the first "green revolution" and dramatically increased the amount of nitrogen available for terrestrial plant growth; the symbiosis remains fundamental to the ecological base of life as we know it today.

The Cenozoic era is divided into the Tertiary and the Quaternary periods. Because both the fossil record and our subsequent knowledge of evolutionary history become more extensive as we approach our own time, paleontologists have subdivided these two periods into *epochs* (**Table 25.2**).

THE TERTIARY (65–2.6 MYA) During the Tertiary period, Australia began its northward drift. By 20 mya it had nearly reached its current position. The early Tertiary was a hot and humid time, and the ranges of many plants shifted latitudinally. The tropics were probably too hot for rainforests and were clothed in low-lying vegetation instead. In the middle of the Tertiary, however, Earth's climate became considerably cooler and drier. Many lineages of flowering plants evolved herbaceous (non-woody) forms, and grasslands spread over much of Earth.

TABLE 25.2
Subdivisions of the Cenozoic Era

PERIOD	EPOCH	ONSET (MYA)
Quaternary	Holocene[a]	0.01 (~10,000 years ago)
	Pleistocene	2.6
Tertiary	Pliocene	5.3
	Miocene	23
	Oligocene	34
	Eocene	55.8
	Paleocene	65

[a]The Holocene is also known as the Recent.

By the start of the Cenozoic era, invertebrate faunas had already come to resemble those of today. It is among the terrestrial vertebrates that evolutionary changes during the Tertiary were most rapid. Frogs, snakes, lizards, birds, and mammals all underwent extensive radiations during this period. Three waves of mammals dispersed from Asia to North America across one of the several land bridges that have intermittently connected the two continents during the past 55 million years. Rodents, marsupials, primates, and hoofed mammals appeared in North America for the first time.

THE QUATERNARY (2.6 MYA TO PRESENT) We are living in the Quaternary period. It is subdivided into two epochs, the Pleistocene and the Holocene (the Holocene also being known as the Recent).

The Pleistocene was a time of drastic cooling and climate fluctuations. During 4 major and about 20 minor "ice ages," massive glaciers spread across the continents, and the ranges of animal and plant populations shifted toward the equator. The last of these glaciers retreated from temperate latitudes less than 15,000 years ago. Organisms are still adjusting to these changes. Many high-latitude ecological communities have occupied their current locations for no more than a few thousand years.

It was during the Pleistocene that divergence within one group of mammals, the primates, resulted in the evolution of the hominoid lineage. Subsequent hominoid radiation eventually led to the species *Homo sapiens*—modern humans (see Section 33.5). Many large bird and mammal species became extinct in Australia and in the Americas when *H. sapiens* arrived on those continents about 45,000 and 15,000 years ago, respectively. Many paleontologists believe these extinctions were probably the result of hunting and other influences of *Homo sapiens*.

The tree of life is used to reconstruct evolutionary events

The fossil record reveals broad patterns in life's evolution. To reconstruct major events in the history of life, biologists also rely on the phylogenetic information in the tree of life (see Chapter 22 and the Tree of Life Appendix). We can use phylogeny (in combination with the paleontological record) to reconstruct the timing of such major events as the acquisition of mitochondria in the ancestral eukaryotic cell, the several independent origins of multicellularity, and the movement of life onto dry land. We can also follow major changes in the genomes of organisms, and even reconstruct many gene sequences of species that are long extinct (see Chapter 24).

Changes to the physical environment on Earth have clearly influenced the great diversity in living organisms we see on the planet today. To study the evolution of that diversity, biologists examine the evolutionary relationships among species. Deciphering these relationships is an important step in understanding how life has diversified on Earth. Part Seven of this book explores the major groups of life and the different solutions that have evolved for major functions such as reproduction, energy acquisition, dispersal, and escape from predation.

---

yourBioPortal.com

GO TO **The Interactive Tree of Life**

25.3 RECAP

Life evolved in the Precambrian oceans. It diversified as atmospheric oxygen approached its current level and the continents came together to form several large land masses. Numerous climate changes and rearrangements of the continents, as well as meteorite impacts, contributed to five major mass extinctions.

- Why have so few of the multitudes of organisms that have existed over millennia become fossilized? **See pp. 526–527**

- What do we mean when we refer to the "Cambrian explosion"? **See p. 527**

- In what ways has continental drift affected the evolution of life on Earth? **See Figure 25.12**

CHAPTER SUMMARY

25.1 How Do Scientists Date Ancient Events?

- The relative ages of organisms can be determined by the dating of fossils and the **strata** of **sedimentary rocks** in which they are found.

- Paleontologists use a variety of radioisotopes with different **half-lives** to date events at different times in the remote past. Review Figure 25.1

- Geologists divide the history of life into eras and periods, based on major differences in the fossil assemblages found in successive layers of rocks. Review Table 25.1

25.2 How Have Earth's Continents and Climates Changed over Time?

- Earth's crust consists of solid lithospheric plates that float on fluid magma. **Continental drift** caused by convection currents in the magma moves these plates and the continents that lie on top of them. Review Figure 25.2, **ANIMATED TUTORIAL 25.1**

- Conditions on Earth have changed dramatically over time. Increases in atmospheric oxygen and changes in Earth's climate have greatly influenced the evolution of life on Earth. Review Figures 25.5 and 25.7

- Oxygen-generating cyanobacteria liberated enough O_2 to open the door to oxidation reactions in metabolic pathways. The aerobic prokaryotes were able to harvest more energy than anaerobic organisms and began to proliferate. Increases in atmospheric O_2 levels supported the evolution of large eukaryotic cells.

- Major physical events on Earth, such as the collision of continents that formed the supercontinent **Pangaea**, have affected Earth's surface, climate, and atmosphere. In addition, extraterrestrial events such as meteorite strikes created sudden and dramatic environmental shifts. All of these changes have affected the history of life.

25.3 What Are The Major Events in Life's History?

- Paleontologists use fossils and evidence of geological changes to determine what Earth and its **biota** may have looked like at different times.

- During most of its history, life was confined to the oceans. Multicellular life diversified extensively during the **Cambrian explosion**. Review Figure 25.11

- The periods of the Paleozoic era were each characterized by the diversification of specific groups of organisms. **Amniotes**—vertebrates whose eggs can be laid in dry places—first appeared during the Carboniferous period.

- During the Mesozoic era, distinct terrestrial biotas evolved on each continent.

- Five episodes of **mass extinction** punctuated the history of life in the Paleozoic and Mesozoic eras.

- Earth's **flora** has been dominated by flowering plants since the Cenozoic era.

- Phylogenetic trees help reconstruct the timing of evolutionary events and clarify relationships among modern species.

SEE WEB ACTIVITY 25.1 for a concept review of this chapter.

SELF-QUIZ

1. Which of the following is *not* true of the giant flying dragonflies of the Carboniferous and Permian?
   a. Some species grew to have wing spans as wide or wider than many modern birds of prey.
   b. They were the largest flying predators of the time.
   c. Such large flying insects could exist because of the higher concentrations of atmospheric oxygen compared to the present.
   d. Their predators were giant reptiles.
   e. Fossils of one large species, *Meganeurosis permiana*, have been found in the Permian rocks of Kansas.

2. In undisturbed strata of sedimentary rock, the oldest rocks
   a. lie at the top.
   b. lie at the bottom.
   c. are in the middle.
   d. are distributed among the strata of younger rocks.
   e. none of the above

3. ^{14}C can be used to determine the ages of fossil organisms because
   a. all organisms contain many carbon compounds.
   b. ^{14}C has a regular rate of decay to ^{14}N.
   c. the ratio of ^{14}C to ^{12}C in living organisms is always the same as that in the atmosphere.
   d. the production of new ^{14}C in the atmosphere just balances the natural radioactive decay of ^{14}C.
   e. all of the above

4. The concentration of oxygen in the Earth's atmosphere
   a. has increased steadily through time.
   b. has decreased steadily through time.
   c. has been both higher and lower in the past than at present.
   d. was lower during most of the Permian than at present.
   e. was at its highest levels in the Cambrian.

5. The total of all species of organisms in a given region is known as the region's
   a. biota.
   b. flora.
   c. fauna.
   d. flora and fauna.
   e. biogeography.

6. The coal beds we now mine for energy are largely the remains of
   a. plants that grew in swamps during the Carboniferous period.
   b. algae that grew in marshes during the Devonian period.
   c. giant insects and amphibians of the Permian period.
   d. plants that grew in the oceans during the Carboniferous period.
   e. none of the above

7. The mass extinction at the end of the Ordovician period was probably caused by
   a. the collision of Earth with a large meteorite.
   b. massive volcanic eruptions.
   c. massive glaciation on the southern continents and associated climatic changes.
   d. the uniting of all continents to form Pangaea.
   e. changes in Earth's orbit.

8. The cause of the mass extinction at the end of the Mesozoic era probably was
   a. continental drift.
   b. the collision of Earth with a large meteorite.
   c. changes in Earth's orbit.
   d. massive glaciation.
   e. changes in the salt concentration of the oceans.

9. Which of the following times was marked by the largest mass extinction of life in the history of Earth?
   a. The end of the Cretaceous
   b. The end of the Devonian
   c. The end of the Permian
   d. The end of the Triassic
   e. The end of the Silurian

10. Paleontologists have subdivided the Cenozoic era into epochs because
    a. *Homo sapiens* evolved at the start of the Cenozoic.
    b. the continents had achieved their present positions.
    c. the number of species stopped increasing at this time.
    d. our knowledge of the evolutionary events of the Cenozoic is more extensive than for other eras.
    e. starting with the Cenozoic, the fossil record is no longer a necessary source of information about evolutionary relationships.

FOR DISCUSSION

1. Some groups of organisms have evolved to contain large numbers of species; other groups have produced only a few species. Is it meaningful to consider the former groups more successful than the latter? What does the word "success" mean in evolution?

2. Scientists date ancient events using a variety of methods, but nobody was present to witness or record those events. Accepting those dates requires us to understand the accuracy and appropriateness of indirect measurement techniques. What other basic scientific concepts are also based on the results of indirect measurement techniques?

3. Why is it useful to be able to date past events absolutely as well as relatively?

4. If we are living during one of the cooler periods in Earth's history, why should we be concerned about human activities that are thought to contribute to global climate warming?

5. What conditions may have favored the evolution of multicellular groups of organisms near the end of the Precambrian?

6. In what ways do endosymbiotic events (such as the origin of mitochondria and chloroplasts) complicate the classification of the major groups of life?

ADDITIONAL INVESTIGATION

The experiment in Figure 25.6 showed that body size of insects may evolve quickly following changes in atmospheric oxygen concentrations. What other experiments could you devise to test the effects of changing atmospheric oxygen?

26 Bacteria and Archaea: The Prokaryotic Domains

Life on a strange planet

It must have been quite a shock when Thomas "Grif" Taylor's Antarctic exploration team first spotted Blood Falls in 1911. The blood-red falls were certainly a surprise in the snowy, icy terrain. What could possibly cause a red waterfall in Antarctica?

A few million years ago, the Taylor Glacier (which now bears the explorer's name) moved above a pool of salty water, trapping the pool under 400 meters of ice. The harsh environment in the resulting enclosed subglacial sea could hardly seem more hostile to life. It is extremely cold; there is no light and virtually no oxygen; and salt concentrations are several times higher than seawater. In short, it is not a place one might expect to find a diverse living ecosystem.

Some water is able to seep out of this subglacial sea. This water is stained a dark, rusty red, and it spills from the head of Taylor Glacier to form Blood Falls. Taylor speculated that red algae might account for the red coloration, but in the 1960s geologist Robert Black discovered that the water's color arises from iron oxides that come from the underlying bedrock. With the methods then available, biologists could not detect any living organisms in the cold, saline, iron-rich water.

A half-century later, biologists were better equipped to study microscopic life in strange places. By then it was also possible to amplify and sequence genes from single microbes, and to place these gene sequences into the framework of the tree of life to identify and classify the microbes. Microbiologist Jill Mikucki and her colleagues used these techniques on water samples from Blood Falls, and reported in 2009 that the falls contain an unusual ecosystem of at least 17 different species of bacteria. The bacteria survive by metabolizing minute amounts of organic matter trapped in the subglacial sea, using sulfate and iron ions as catalysts and electron acceptors.

The presence of living organisms in Blood Falls confirms that it is hard to find a place on or even near the surface of the Earth that does not contain populations of prokaryotes. There are prokaryotes in volcanic vents, in the clouds, in environments as acidic as battery acid or as alkaline as household ammonia. There are species that can survive below the freezing point and above the boiling point

A Splash of Color in a Frozen World of White Antarctica's Blood Falls is the outflow of a subglacial sea that contains an unusual ecosystem of bacteria that rely on sulfate and iron ions for metabolism.

Prokaryotes Can Take the Heat Entire ecosystems of prokaryotes create the beauty of Morning Glory Pool, a hot spring in Yellowstone National Park. Cyanobacteria impart the "morning glory blue" color. Archaea live in the intensely hot regions of the pool's interior.

of water. There are more prokaryotes living on and inside our bodies than we have human cells. The prokaryotes are masters of metabolic ingenuity, having developed more ways to obtain energy from the environment than the eukaryotes have. They have been around much longer than other organisms, too.

Prokaryotes are by far the most numerous organisms on Earth. Late in the twentieth century, it became apparent to microbiologists that all prokaryotes are not most closely related to one another. Two prokaryotic lineages diverged early in life's evolution: Bacteria and Archaea. An early merging between members of these two groups is thought to have given rise to the eukaryotic lineage, which includes humans.

IN THIS CHAPTER we will discuss the distribution of prokaryotes and examine their remarkable metabolic diversity. We will describe the difficulties involved in determining evolutionary relationships among the prokaryotes and will survey the surprising diversity of organisms in each domain. We will discuss how prokaryotes can have enormous influence on their environments. Finally, we will discuss the evolutionary origin and diversity of viruses and their relationship to the rest of life.

26.1 How Did the Living World Begin to Diversify?

You may think that you have little in common with unicellular prokaryotes. But multicellular eukaryotes like yourself actually share many attributes with Bacteria and Archaea. For example, all three of you:

- conduct glycolysis
- use DNA as the genetic material that encodes proteins
- produce those proteins by transcription and translation using a similar genetic code
- replicate DNA semiconservatively
- have plasma membranes and ribosomes in abundance

These features support the conclusion that all living organisms are related to one another. If life had multiple origins, there would be little reason to expect all organisms to use overwhelmingly similar genetic codes or to share structures as unique as ribosomes. Furthermore, similarities in DNA sequences of universal genes (such as those that encode the structural components of ribosomes) confirm the monophyly of life.

Despite the commonalities found across all three domains, major differences have evolved as well. Let's first distinguish between Eukarya and the two prokaryotic domains. Note that "domain" is a subjective term used for the largest groups of life. There is no objective definition of a domain, any more than there is of a kingdom or a family.

The three domains differ in significant ways

Prokaryotic cells differ from eukaryotic cells in three important ways:

- *Prokaryotic cells lack a cytoskeleton and a nucleus, so they do not divide by mitosis.* Instead, after replicating their DNA (see Figure 11.2), prokaryotic cells divide by their own method, *binary fission.*
- *The organization of the genetic material differs.* The DNA of the prokaryotic cell is not organized within a membrane-enclosed nucleus. DNA molecules in prokaryotes (both bacteria and archaea) are often circular. Many (but not all) prokaryotes have only one main chromosome and are effectively haploid, although many have additional smaller DNA molecules, called *plasmids*, as well (see Section 12.6).

TABLE 26.1
The Three Domains of Life on Earth

CHARACTERISTIC	BACTERIA	DOMAIN ARCHAEA	EUKARYA
Membrane-enclosed nucleus	Absent	Absent	Present
Membrane-enclosed organelles	Absent	Absent	Present
Peptidoglycan in cell wall	Present	Absent	Absent
Membrane lipids	Ester-linked	Ether-linked	Ester-linked
	Unbranched	Branched	Unbranched
Ribosomes[a]	70S	70S	80S
Initiator tRNA	Formylmethionine	Methionine	Methionine
Operons	Yes	Yes	No
Plasmids	Yes	Yes	Rare
RNA polymerases	One	One[b]	Three
Ribosomes sensitive to chloramphenicol and streptomycin	Yes	No	No
Ribosomes sensitive to diphtheria toxin	No	Yes	Yes

[a]70S ribosomes are smaller than 80S ribosomes.
[b]Archaeal RNA polymerase is similar to eukaryotic polymerases.

- *Prokaryotes have none of the membrane-enclosed cytoplasmic organelles—mitochondria, Golgi apparatus, and others—that are found in most eukaryotes.* However, the cytoplasm of a prokaryotic cell may contain a variety of infoldings of the plasma membrane and photosynthetic membrane systems not found in eukaryotes.

A glance at **Table 26.1** will show you that there are also major differences (most of which cannot be seen even under an electron microscope) between the two prokaryotic domains. In some ways archaea are more like eukaryotes; in other ways they are more like bacteria. (Note that we use lowercase when referring to the members of these domains and uppercase when referring to the domains themselves.) The structures of prokaryotic and eukaryotic cells are compared in Chapter 5. The basic unit of an *archaeon* (the term for a single archaeal organism) or *bacterium* (a single bacterial organism) is the prokaryotic cell. Each single-celled organism contains a full complement of genetic and protein-synthesizing systems, including DNA, RNA, and all the enzymes needed to transcribe and translate the genetic information into proteins. The prokaryotic cell also contains at least one system for generating the ATP it needs.

Genetic studies clearly indicate that all three domains had a single common ancestor. For a major portion of their genome, eukaryotes share a more recent common ancestor with Archaea than they do with Bacteria (**Figure 26.1**). However, the mitochondria of eukaryotes (as well as the chloroplasts of photosynthetic eukaryotes, such as plants) originated through the endosymbiosis of a bacterium, as described in Section 5.5. Some biologists prefer to view the origin of eukaryotes as a fusion of two equal partners (one ancestor that was related to modern archaea, and another that was more closely related to modern bacteria). Others view the divergence of the early eukaryotes from the archaea as a separate and earlier event than the later endosymbiosis of the bacterium (the origin of mitochondria). In either case, some genes of eukaryotes are more closely related to those of archaea, while others are more closely related to those of bacteria. The tree of life therefore contains some merging of lineages, as well as the predominant diverging of lineages.

The common ancestor of all three domains had DNA as its genetic material, and its machinery for transcription and translation produced RNAs and proteins, respectively. This ancestor probably had a circular chromosome.

Three shapes are particularly common among the bacteria: spheres, rods, and curved or helical forms (**Figure 26.2**). A spherical bacterium is called a *coccus* (plural *cocci*). Cocci may live singly or may associate in two- or three-dimensional arrays

yourBioPortal.com
GO TO **Animated Tutorial 26.1 • The Evolution of the Three Domains**

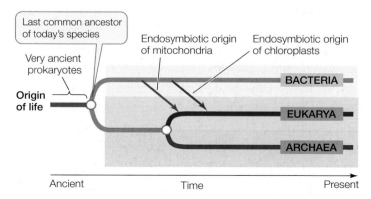

26.1 The Three Domains of the Living World All three domains share a common prokaryotic ancestor.

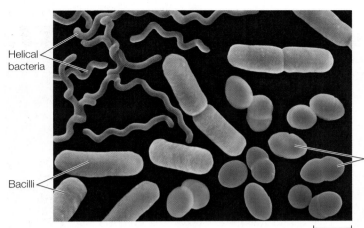

Helical bacteria

Bacilli

Cocci

0.50 μm

26.2 Bacterial Cell Shapes This composite, colorized micrograph shows the three common types of bacterial morphology. Spherical cells are called cocci; the acid-producing cocci shown here in green are a species of *Enterococcus* from the mammalian gut. The rod-shaped bacilli (orange) are represented by *Escherichia coli*, also a resident of the gut. *Leptospira interrogans* is a helical (spiral) bacterium and a human pathogen.

26.2 What Are Some Keys to the Success of Prokaryotes?

If success is measured by numbers of individuals, the prokaryotes are the most successful organisms on Earth. Individual bacteria and archaea in the oceans number more than 3×10^{28}. This stunning number is perhaps 100 million times as great as the number of stars in the visible universe. In fact, the bacteria living in your intestinal tract outnumber all the humans who have ever lived.

Prokaryotes are unicellular organisms, although many form multicellular colonies that contain many individual cells. These multicellular associations are not cases of true multicellular organisms, however, because each individual cell is fully viable and independent. These associations arise as cells adhere to one another after reproducing by binary fission. Associations in the form of chains are called **filaments**. Some filaments become enclosed in delicate tubular sheaths.

Prokaryotes generally form complex communities

Prokaryotic cells and their associations do not usually live in isolation. Rather, they live in communities of many different species of organisms, often including microscopic eukaryotes. (Microscopic organisms are often collectively referred to as *microbes*.) Some microbial communities form layers in sediments, and others form clumps a meter or more in diameter. While some microbial communities are harmful to humans, others provide important services. They help us digest our food, break down municipal waste, and recycle organic matter in the environment.

Many microbial communities tend to form dense **biofilms**. Upon contacting a solid surface, the cells secrete a gel-like sticky polysaccharide matrix that then traps other cells (**Figure 26.3**). Once this biofilm forms, it is difficult to kill the cells. Pathogenic (disease-causing) bacteria are difficult for the immune system—and modern medicine—to combat once they form a biofilm. For example, the film may be impermeable to antibiotics. Worse, some drugs stimulate the bacteria in a biofilm to lay down more matrix, making the film even more impermeable.

Biofilms form on contact lenses, on artificial joint replacements, and on just about any available surface. They foul metal pipes and cause corrosion, a major problem in steam-driven electricity generation plants. The stain on our teeth that we call dental plaque is also a biofilm. Fossil stromatolites—large, rocky structures made up of alternating layers of fossilized microbial biofilm and calcium carbonate—are the oldest remnants of early life on

as chains, plates, blocks, or clusters of cells. A rod-shaped bacterium is called a *bacillus* (plural *bacilli*). The spiral form (like a corkscrew), or *helix* (plural *helices*), is the third main bacterial shape. Bacilli and helices may be single, form chains, or gather in regular clusters. Among the other bacterial shapes are long filaments and branched filaments.

Less is known about the shapes of archaea because many of these organisms have never been seen. Many archaea are known only from samples of DNA from the environment, as we describe in Section 26.4. However, the morphology of some species is known, including cocci, bacilli, and even triangular and square-shaped species; the latter grow on surfaces, arranged like sheets of postage stamps.

Archaea, Bacteria, and Eukarya are all products of billions of years of mutation, natural selection, and genetic drift, and they are all well adapted to present-day environments. None is "primitive." Their last common ancestor probably lived 2 to 3 billion years ago. The earliest prokaryotic fossils date back at least 3.5 billion years, and they indicate that there was considerable diversity among the prokaryotes even during the earliest days of life.

26.1 RECAP

Bacteria and archaea are highly divergent from each other and are only distantly related on the tree of life. Eukaryotes received ancient evolutionary contributions from both of these prokaryotic lineages.

- What are the principal differences between the prokaryotes and the eukaryotes? See pp. 537–538 and Table 26.1
- Why don't we group Bacteria and Archaea together in a single domain? See p. 538 and Table 26.1

The prokaryotes were alone on Earth for a very long time, adapting to new environments and to changes in existing environments. They have survived to this day, in massive numbers and incredible diversity, and they are found everywhere.

(A)

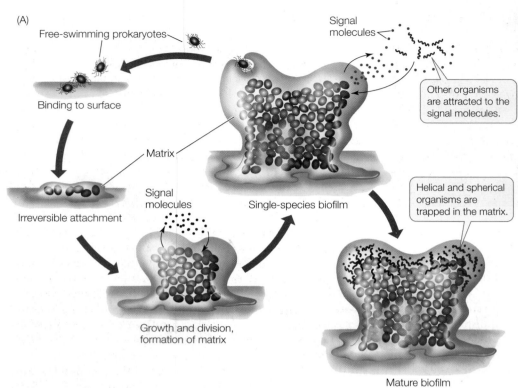

Free-swimming prokaryotes

Binding to surface

Irreversible attachment

Matrix

Signal
molecules

Growth and division,
formation of matrix

Single-species biofilm

Signal
molecules

Other organisms
are attracted to the
signal molecules.

Helical and spherical
organisms are trapped in the matrix.

Mature biofilm

(B)

100 µm

26.3 Forming a Biofilm (A) Free-swimming bacteria and archaea readily attach themselves to surfaces and form films stabilized and protected by a surrounding matrix. Once the population size is large enough, the developing biofilm can send chemical signals that attract other microorganisms. (B) Scanning electron micrography reveals a biofilm of plaque on a used toothbrush bristle. The matrix of dental plaque consists of proteins from both bacterial secretions and saliva.

Earth (see Figure 25.4). Stromatolites still form today in some parts of the world.

Biofilms are the subject of much current research. For example, some biologists are studying the chemical signals that bacteria in biofilms use to communicate with one another. By blocking the signals that lead to the production of the matrix polysaccharides, researchers may be able to prevent biofilms from forming.

A team of bioengineers and chemical engineers recently devised a sophisticated technique that enables them to monitor biofilm development in extremely small populations of bacteria, cell by cell. They developed a tiny chip housing six separate growth chambers, or "microchemostats" (**Figure 26.4**). The techniques of *microfluidics* use microscopic tubes and computer-controlled valves to direct fluid flow through complex "plumbing circuits" in the growth chambers.

TOOLS FOR INVESTIGATING LIFE

26.4 The Microchemostat

Using techniques from microfluidic engineering, biologists can monitor the dynamics of extremely small bacterial populations. The photograph shows six microchemostats on a chip. Each of the six is equipped with input ports for growth and flushing media, and a number of output ports (diagram). Tiny valves, controlled by a computer, direct flow. Samples are removed through the output ports and are analyzed to record changes in the bacterial population.

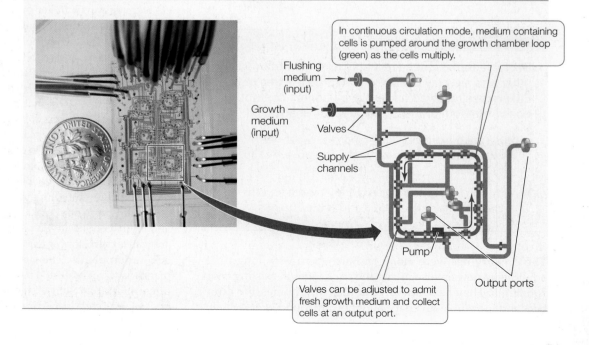

In continuous circulation mode, medium containing cells is pumped around the growth chamber loop (green) as the cells multiply.

Flushing medium (input)

Growth medium (input)

Valves

Supply channels

Pump

Output ports

Valves can be adjusted to admit fresh growth medium and collect cells at an output port.

Prokaryotes have distinctive cell walls

Many prokaryotes have a thick and relatively stiff cell wall. It is quite different from the cell walls of land plants and algae, which contain cellulose and other polysaccharides, and from those of fungi, which contain chitin. The cell walls of almost all bacteria contain **peptidoglycan** (a cross-linked polymer of amino sugars), which produces a meshlike structure around the cell. Archaeal cell walls are of differing types, but most contain significant amounts of protein. One group of archaea has *pseudopeptidoglycan* in its cell wall; as you can probably guess from the prefix *pseudo*, pseudopeptidoglycan is similar to, but distinctly different from, the peptidoglycan of bacteria. The monomers making up pseudopeptidoglycan differ from and are differently linked than those of peptidoglycan. Peptidoglycan is a substance unique to bacteria; its absence from the walls of archaea is a key difference between the two prokaryotic domains.

To appreciate the complexity of some bacterial cell walls, consider the reactions of bacteria to a simple staining process. A test called the **Gram stain** separates most types of bacteria into two distinct groups, Gram-positive and Gram-negative. A smear of cells on a microscope slide is soaked in a violet dye and treated with iodine; it is then washed with alcohol and counter-stained with a red dye (safranine). **Gram-positive bacteria** retain the violet dye and appear blue to purple (**Figure 26.5A**). The alcohol washes the violet stain out of Gram-negative cells; these cells then pick up the safranine counterstain, so **Gram-negative bacteria** appear pink to red (**Figure 26.5B**).

For most bacteria, the Gram-staining results are determined by the chemical structure of the cell wall. A Gram-negative cell wall usually has a thin peptidoglycan layer, and outside the peptidoglycan layer the cell is surrounded by a second, outer membrane quite distinct in chemical makeup from the plasma membrane (see Figure 26.5B). Between the inner (plasma) and outer membranes of Gram-negative bacteria is a *periplasmic space*. This space contains proteins that are important in digesting some materials, transporting others, and detecting chemical gradients in the environment.

A Gram-positive cell wall usually has about five times as much peptidoglycan as a Gram-negative wall. This thick peptidoglycan layer is a meshwork that may serve some of the same purposes as the periplasmic space of the Gram-negative cell wall.

The consequences of the different features of prokaryotic cell walls are numerous and relate to the disease-causing characteristics of some bacteria. Indeed, the cell wall is a favorite target in medical combat against pathogenic bacteria because it has no counterpart in eukaryotic cells. Antibiotics such as penicillin and ampicillin, as well as other agents that specifically interfere with the synthesis of peptidoglycan-containing cell walls, tend to have little, if any, effect on the cells of humans and other eukaryotes.

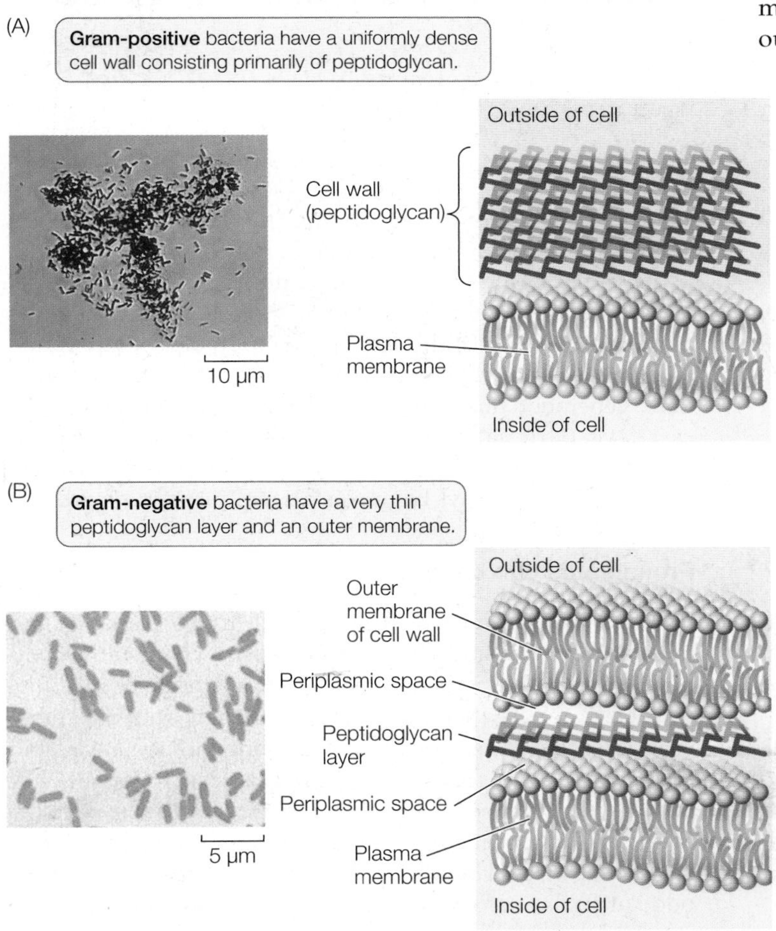

(A) Gram-positive bacteria have a uniformly dense cell wall consisting primarily of peptidoglycan.

Cell wall (peptidoglycan)

Outside of cell

Plasma membrane

Inside of cell

10 μm

(B) Gram-negative bacteria have a very thin peptidoglycan layer and an outer membrane.

Outside of cell

Outer membrane of cell wall

Periplasmic space

Peptidoglycan layer

Periplasmic space

Plasma membrane

Inside of cell

5 μm

26.5 The Gram Stain and the Bacterial Cell Wall When treated with Gram stain, the cell walls of different bacteria react in one of two ways. (A) Gram-positive bacteria have a thick peptidoglycan cell wall that retains the violet dye and appears deep blue or purple. (B) Gram-negative bacteria have a thin peptidoglycan layer that does not retain the violet dye but picks up the counterstain and appears pink to red.

Prokaryotes have distinctive modes of locomotion

Although many prokaryotes cannot move, others are *motile*. These organisms move by one of several means. Some helical bacteria, called *spirochetes*, use a corkscrew-like motion made possible by modified flagella, called *axial filaments*, running along the axis of the cell beneath the outer membrane (**Figure 26.6A**). Many cyanobacteria and a few other groups of bacteria use various poorly understood gliding mechanisms, including rolling. Various aquatic prokaryotes, including some cyanobacteria, can move slowly up and down in the water by adjusting the amount of gas in *gas vesicles* (**Figure 26.6B**). By far the most common type of locomotion in prokaryotes, however, is that driven by flagella.

Prokaryotic **flagella** are slender filaments that extend singly or in tufts from one or both ends of the cell or are distributed all around it (**Figure 26.7**). A prokaryotic flagellum consists of a single fibril made of the protein *fla-*

─── **yourBioPortal.com** ───
GO TO Web Activity 26.1 • Gram Stain and Bacteria

(A)

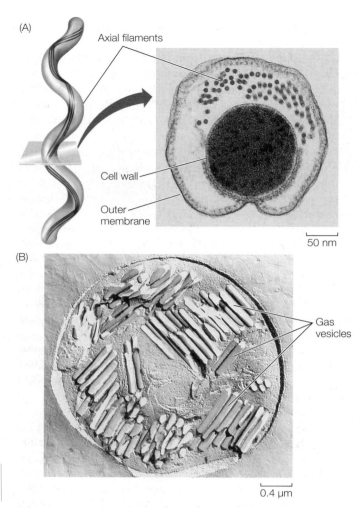

(B)

26.6 **Structures Associated with Prokaryote Motility** (A) A spirochete from the gut of a termite, seen in cross section, shows the axial filaments used to produce a corkscrew-like motion. (B) Gas vesicles in a cyanobacterium, visualized by the freeze-fracture technique.

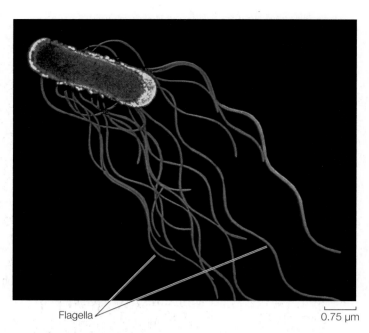

26.7 **Some Prokaryotes Use Flagella for Locomotion** Multiple flagella propel this *Salmonella* bacillus.

gellin, projecting from the cell surface, plus a hook and basal body responsible for motion (see Figure 5.5). In contrast, the flagellum of eukaryotes is enclosed by the plasma membrane and usually contains a circle of nine pairs of microtubules surrounding two central microtubules, all containing the protein *tubulin*, along with many other associated proteins. The prokaryotic flagellum rotates about its base, much like a propeller, rather than beating in a whiplike manner, as a eukaryotic flagellum or cilium does.

Prokaryotes reproduce asexually, but genetic recombination can occur

Prokaryotes reproduce by binary fission, an asexual process (see Figure 11.2). Recall, however, that there are also processes—transformation, conjugation, and transduction—that allow the exchange of genetic information between some prokaryotes without reproduction occurring. So prokaryotes can exchange and recombine their DNA with other individuals (this is sex in

the genetic sense of the word), but this genetic exchange is not directly linked to reproduction as it is in most eukaryotes.

If conditions are favorable, some prokaryotes can multiply very rapidly. The shortest known prokaryote generation times are about 10 minutes, although these rapid rates of replication usually are not maintained for long. Under less optimal conditions, generation times often extend to many hours or even several days. Bacteria living deep in Earth's crust may suspend their growth for more than a century without dividing, then multiply for a few days before once again suspending growth.

Prokaryotes can communicate

Prokaryotes can send and receive signals from one another and from other organisms. One communication channel they employ is chemical. Another is physical, with light as the medium.

Bacteria release chemical substances that are sensed by other bacteria of the same species. They can announce their availability for conjugation, for example, by means of such signals. They can also monitor the density of their population. As the density of bacteria in a particular region increases, the concentration of a chemical signal builds up. When the bacteria sense that their population has become sufficiently dense, they can commence activities that smaller densities could not manage, such as forming a biofilm (see Figure 26.3). This density-sensing technique is called **quorum sensing**.

Like fireflies and many other organisms, some bacteria can emit light by a process called **bioluminescence**. A complex, enzyme-catalyzed reaction requiring ATP causes the emission of light but not heat. Often such bacteria luminesce only when a quorum has been sensed. The bioluminescent spots present in some deep-sea fishes are produced by colonies of biolumines-

26.2 | WHAT ARE SOME KEYS TO THE SUCCESS OF PROKARYOTES? **543**

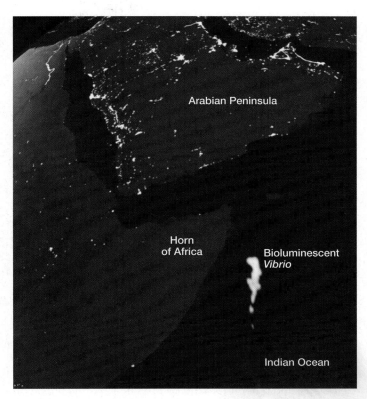

26.8 Bioluminescent Bacteria Seen from Space In this satellite photo, legions of bioluminescent *Vibrio harveyi* form a glowing patch thousands of square kilometers in area in the Indian Ocean, off the Horn of Africa. Compare their blue glow with the white light of cities in eastern Africa and the Middle East.

cent bacteria. On land, some soil-dwelling bioluminescent bacteria produce eerily glowing patches of ground at night.

How is bioluminescence useful to a prokaryote? One fairly well understood case is that of some bacteria of the genus *Vibrio*. These bacteria can live freely, but they truly thrive inside the guts of fish. Inside the fish, they may attach to food particles and then can be expelled as waste along with particulate matter. Reproducing on the particles, a bacteria population increases until a glowing particle attracts another fish, which ingests the bacteria along with the particle—giving the bacteria a new home and food source for a while. In this case, *Vibrio* are both communicating with another species and enhancing their own nutritional status. In the Indian Ocean off the eastern coast of Africa, *Vibrio* sometimes concentrate over such a large area (several thousand square kilometers) that their bioluminescence is visible from space (**Figure 26.8**).

Prokaryotes have amazingly diverse metabolic pathways

Bacteria and archaea outdo the eukaryotes in terms of metabolic diversity. Although much more diverse in size and shape, eukaryotes draw on fewer metabolic mechanisms for their energy needs. In fact, much of the eukaryotes' energy metabolism is carried out in organelles—mitochondria and chloro-

plasts—that are endosymbiotic descendants of bacteria, as described in Section 5.5.

The long evolutionary history of bacteria and archaea, during which they have had time to explore a wide variety of habitats, has led to the extraordinary diversity of their metabolic "lifestyles"—their use or nonuse of oxygen, their energy sources, their sources of carbon atoms, and the materials they release as waste products.

ANAEROBIC VERSUS AEROBIC METABOLISM Some prokaryotes can live only by anaerobic metabolism because molecular oxygen is poisonous to them. These oxygen-sensitive organisms are called **obligate anaerobes**. Other prokaryotes can shift their metabolism between anaerobic and aerobic modes (see Chapter 9) and thus are called **facultative anaerobes**. Many facultative anaerobes alternate between anaerobic metabolism (such as fermentation) and cellular respiration as conditions dictate. **Aerotolerant anaerobes** cannot conduct cellular respiration but are not damaged by oxygen when it is present. By definition, an anaerobe does not use oxygen as an electron acceptor for its respiration.

At the other extreme from the obligate anaerobes, some prokaryotes are **obligate aerobes**, unable to survive for extended periods in the *absence* of oxygen. They require oxygen for cellular respiration.

NUTRITIONAL CATEGORIES All living organisms face the same nutritional challenges: they must synthesize energy-rich compounds such as ATP to power their life-sustaining metabolic reactions, and they must obtain carbon atoms to build their own organic molecules. Biologists recognize four broad nutritional categories of organisms: photoautotrophs, photoheterotrophs, chemolithotrophs, and chemoheterotrophs. Prokaryotes are represented in all four groups (**Table 26.2**).

Photoautotrophs perform photosynthesis. They use light as their energy source and carbon dioxide (CO_2) as their carbon source. Like green plants and other photosynthetic eukaryotes, the cyanobacteria, a group of photoautotrophic bacteria, use chlorophyll *a* as their key photosynthetic pigment and produce

TABLE 26.2		
How Organisms Obtain Their Energy and Carbon		
NUTRITIONAL CATEGORY	ENERGY SOURCE	CARBON SOURCE
Photoautotrophs (found in all three domains)	Light	Carbon dioxide
Photoheterotrophs (some bacteria)	Light	Organic compounds
Chemolithotrophs (some bacteria, many archaea)	Inorganic substances	Carbon dioxide
Chemoheterotrophs (found in all three domains)	Organic compounds	Organic compounds

oxygen gas (O_2) as a by-product of noncyclic electron transport (see Section 10.1).

There are other photosynthetic groups among the bacteria, but these use *bacteriochlorophyll* as their key photosynthetic pigment, and they do not release O_2. Indeed, some of these photosynthesizers produce particles of pure sulfur, because hydrogen sulfide (H_2S) rather than H_2O is their electron donor for photophosphorylation (see Section 10.2). Bacteriochlorophyll molecules absorb light of longer wavelengths than the chlorophyll molecules used by all other photosynthesizing organisms. As a result, bacteria using this pigment can grow in water under fairly dense layers of algae, using light of wavelengths that are not absorbed by the algae (**Figure 26.9**).

Photoheterotrophs use light as their energy source but must obtain their carbon atoms from organic compounds made by other organisms. Their "food" consists of organic compounds such as carbohydrates, fatty acids, and alcohols. For example, compounds released from plant roots (as in rice paddies) or from decomposing photosynthetic bacteria in hot springs are taken up by photoheterotrophs and metabolized to form building blocks for other compounds; sunlight provides the necessary ATP through photophosphorylation. The purple nonsulfur bacteria, among others, are photoheterotrophs.

Chemolithotrophs (also called chemoautotrophs) obtain their energy by oxidizing inorganic substances, and they use some of that energy to fix CO_2. Some chemolithotrophs use reactions identical to those of the typical photosynthetic cycle, but others use alternative pathways to fix CO_2. Some bacteria oxidize ammonia or nitrite ions to form nitrate ions. Others oxidize hydrogen gas, hydrogen sulfide, sulfur, and other materials. Many archaea are chemolithotrophs.

Deep-sea hydrothermal vent ecosystems are dependent on chemolithotrophic prokaryotes that are incorporated into large communities of crabs, mollusks, and giant worms, all living at a depth of 2,500 meters—below any hint of sunlight. These bacteria obtain energy by oxidizing hydrogen sulfide and other substances released in the near-boiling water flowing from volcanic vents in the ocean floor.

Finally, **chemoheterotrophs** obtain both energy and carbon atoms from one or more complex organic compounds that have been synthesized by other organisms. Most known bacteria and archaea are chemoheterotrophs—as are all animals and fungi and many protists.

NITROGEN AND SULFUR METABOLISM Key metabolic reactions in many prokaryotes involve nitrogen or sulfur. For example, some bacteria carry out respiratory electron transport without using oxygen as an electron acceptor. These organisms use oxidized inorganic ions such as nitrate, nitrite, or sulfate as electron acceptors. Examples include the **denitrifiers**, bacteria that release nitrogen to the atmosphere as nitrogen gas (N_2). These normally aerobic bacteria, mostly species of the genera *Bacillus* and *Pseudomonas*, use nitrate (NO_3^-) as an electron acceptor in place of oxygen if they are kept under anaerobic conditions:

$$2\ NO_3^- + 10\ e^- + 12\ H^+ \rightarrow N_2 + 6\ H_2O$$

Nitrogen fixers convert atmospheric nitrogen gas into a chemical form (ammonia) usable by the nitrogen fixers themselves as well as by other organisms, especially land plants:

$$N_2 + 6\ H \rightarrow 2\ NH_3$$

All organisms require nitrogen in order to build proteins, nucleic acids, and other important compounds. Nitrogen fixation is thus vital to life as we know it. This all-important biochemical process is carried out by a wide variety of archaea and bacteria (including cyanobacteria) but by no other organisms, so we depend on these prokaryotes for our very existence. We describe the details of nitrogen fixation in Chapter 36.

Ammonia is oxidized to nitrate in soil and in seawater by chemolithotrophic bacteria called **nitrifiers**. Bacteria of two genera, *Nitrosomonas* and *Nitrosococcus*, convert ammonia to nitrite ions (NO_2^-), and *Nitrobacter* oxidizes nitrite to nitrate (NO_3^-).

What do the nitrifiers get out of these reactions? Their metabolism is powered by the energy released by the oxidation of ammonia or nitrite. For example, by passing the electrons from nitrite through an electron transport chain (see Section 9.3), *Nitrobacter* can make ATP, and using some of this ATP, can also make NADH. With this ATP and NADH, the bacterium can convert CO_2 and H_2O to glucose.

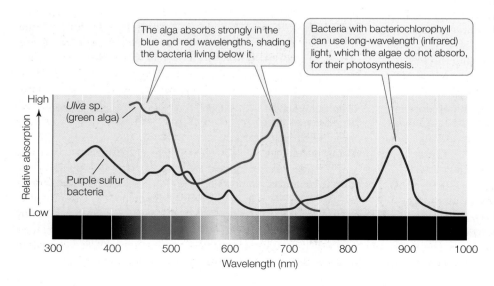

The alga absorbs strongly in the blue and red wavelengths, shading the bacteria living below it.

Bacteria with bacteriochlorophyll can use long-wavelength (infrared) light, which the algae do not absorb, for their photosynthesis.

Ulva sp. (green alga)

Purple sulfur bacteria

High

Low

Relative absorption

300   400   500   600   700   800   900   1000

Wavelength (nm)

26.9 Bacteriochlorophyll Absorbs Long-Wavelength Light The chlorophyll in *Ulva*, a green alga, absorbs no light of wavelengths longer than 750 nm. Purple sulfur bacteria, which contain bacteriochlorophyll, can conduct photosynthesis using longer wavelengths.

Prokaryotes have established themselves everywhere on Earth. They may form communities called biofilms that coat materials with a gel-like matrix. Prokaryotes have distinctive cell walls and modes of locomotion, communication, reproduction, and nutrition.

- How do biofilms form and why are they of special interest to researchers? See pp. 539–540 and Figure 26.3
- Describe bacterial cell wall architecture. See p. 541 and Figure 26.5
- How are the four nutritional categories of prokaryotes distinguished? See pp. 543–544 and Table 26.2
- Explain why nitrogen metabolism in the prokaryotes is vital to other organisms. See p. 544

We noted earlier that only recently have scientists appreciated the huge distinctions between Bacteria and Archaea. How do researchers approach the classification of organisms they can't even see?

26.3 How Can We Resolve Prokaryote Phylogeny?

As detailed in Chapter 22, classification schemes serve three primary purposes: to identify organisms, to reveal evolutionary relationships, and to provide universal names. Classifying bacteria and archaea is of particular importance to humans because scientists and medical technologists must be able to identify bacteria quickly and accurately; when the bacteria are pathogenic, lives may depend on it. In addition, many emerging biotechnologies (see Chapter 18) depend on a thorough knowledge of prokaryote biochemistry, and understanding an organism's phylogeny allows biologists to make predictions about the distribution of biochemical processes across the wide diversity of prokaryotes.

The small size of prokaryotes has hindered our study of their phylogeny

Until about 300 years ago, nobody had even *seen* an individual prokaryote; these organisms remained invisible to humans until the invention of the first simple microscope. Prokaryotes are so small that even the best light microscopes don't reveal much about them. It took the advanced microscopic equipment and techniques of the twentieth century (see Figure 5.3) to open up the microbial world.

Until recently, taxonomists based prokaryote classification on observable phenotypic characters such as shape, color, motility, nutritional requirements, antibiotic sensitivity, and reaction to the Gram stain. When biologists learned how to grow bacteria in pure culture on nutrient media, they learned a great deal about the genetics, nutrition, and metabolism of those species that could be cultured. However, these species represent prob-

ably less than 1 percent of living prokaryote species. Furthermore, this work provided little insight into how prokaryotic organisms evolved—a question of great interest to microbiologists and evolutionary biologists. Only recently have systematists developed the appropriate tools to produce classification schemes that make sense in evolutionary terms.

The nucleotide sequences of prokaryotes reveal their evolutionary relationships

Analyses of nucleotide sequences of ribosomal RNA (rRNA) genes provided the first comprehensive evidence of evolutionary relationships among prokaryotes. For several reasons, rRNA is particularly useful for evolutionary studies of living organisms:

- rRNA is evolutionarily ancient, as it was found in the common ancestor of life.
- No free-living organism lacks rRNA, so rRNA genes can be compared throughout the tree of life.
- rRNA plays a critical role in translation in all organisms, so *lateral transfer* of rRNA genes among distantly related species is unlikely.
- rRNA has evolved slowly enough that gene sequences can be aligned and analyzed among even distantly related species.

Comparisons of rRNA genes from a great many organisms have revealed the probable phylogenetic relationships from throughout the tree of life. Databases such as GenBank contain rRNA gene sequences from hundreds of thousands of species—more than any other gene sequences.

Although these data are helpful, it is clear that even distantly related prokaryotes sometimes exchange genetic material. In some groups of prokaryotes, analyses of multiple gene sequences have suggested several different phylogenetic patterns. How could such differences among different gene sequences arise?

Lateral gene transfer can lead to discordant gene trees

As noted earlier, prokaryotes reproduce by binary fission. If we could follow these divisions back through evolutionary time, we would be tracing the path of the complete tree of life for bacteria and archaea. This underlying tree of relationships, represented in highly abbreviated form in Appendix A, is called the *organismal* (or *species*) *tree*. Because whole genomes are replicated during asexual binary fission divisions, we expect phylogenetic trees constructed from most gene sequences to reflect these same relationships (see Chapter 22).

From early in evolution to the present day, however, some genes have been moving "sideways" from one prokaryotic species to another, a phenomenon known as **lateral gene transfer**. Mechanisms of lateral gene transfer include transfer by plasmids and viruses and uptake of DNA from the environment by transformation. Lateral gene transfers are well documented, especially among closely related species; some have been documented even across the three primary domains of life.

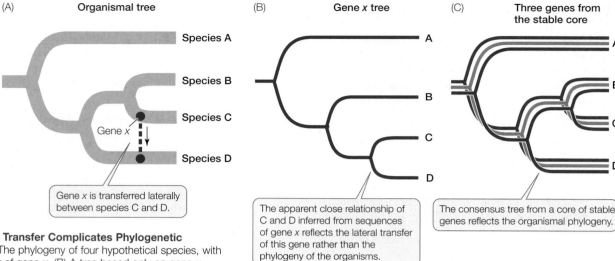

(A) Organismal tree

Species A

Species B

Species C

Gene *x*

Species D

Gene *x* is transferred laterally between species C and D.

(B) Gene *x* tree

A

B

C

D

The apparent close relationship of C and D inferred from sequences of gene *x* reflects the lateral transfer of this gene rather than the phylogeny of the organisms.

(C) Three genes from the stable core

A

B

C

D

The consensus tree from a core of stable genes reflects the organismal phylogeny.

26.10 Lateral Gene Transfer Complicates Phylogenetic Relationships (A) The phylogeny of four hypothetical species, with a lateral gene transfer of gene *x*. (B) A tree based only on gene *x* shows the phylogeny of the laterally transferred gene rather than the organismal phylogeny. (C) In many cases, a "stable core" of prokaryote genes can be used to reconstruct the organismal phylogeny of prokaryotes.

Consider, for example, the genome of *Thermotoga maritima*, a bacterium that can survive extremely high temperatures. In comparing the 1,869 gene sequences of *T. maritima* against sequences for the same proteins in other species, investigators found that some of this bacterium's genes have their closest relationships not with those of other bacterial species, but with the genes of archaeal species that live in similar environments.

When genes involved in lateral transfer events are sequenced and analyzed phylogenetically, the resulting individual gene trees will not match the organismal phylogeny in every respect (**Figure 26.10**). Individual gene trees will vary because the history of lateral gene transfer events is different for each gene. Biologists reconstruct the underlying organismal phylogeny by comparing multiple genes (to produce a *consensus tree*), or by concentrating on genes that are unlikely to be involved in lateral gene transfer events. For example, genes that are involved in fundamental cell processes (such as the rRNA genes discussed above) are unlikely to be replaced by the same genes from other species, since functional, locally adapted copies of these genes are already present.

What kinds of genes are most likely to be involved in lateral gene transfer? Genes that result in a new, adaptive function that will convey higher fitness to a recipient species are most likely to be transferred repeatedly among species. For example, genes that produce antibiotic resistance are often transferred on plasmids among many bacterial species, especially under the strong selective conditions of antibiotic medication by humans. This selection for antibiotic resistance is why informed physicians are now more careful in prescribing antibiotics. Improper or frequent use of antibiotics can lead to selection for resistant strains of bacteria, which are then much harder to treat effectively.

It is debatable whether lateral gene transfer has seriously complicated our attempts to resolve the tree of prokaryotic life.

Recent work suggests that it has not—while it complicates studies in some individual species, it need not present problems at higher levels. It is now possible to make nucleotide sequence comparisons involving entire genomes, and these studies are revealing a *stable core* of crucial genes that are uncomplicated by lateral gene transfer. Gene trees based on this stable core more accurately reveal relationships of the organismal phylogeny (see Figure 26.10). The problem remains, however, that only a very small proportion of the prokaryotic world has been described and studied.

The great majority of prokaryote species have never been studied

Most prokaryotes have defied all attempts to grow them in pure culture, causing biologists to wonder how many species, and possibly even important clades, we might be missing. A window onto this problem was opened with the introduction of a new way to look at nucleic acid sequences. Unable to work with the whole genome of a single species, biologists instead examine sequences in individual genes collected from a random sample of the environment.

Norman Pace of the University of Colorado isolated individual rRNA gene sequences from extracts of environmental samples such as soil and seawater. Comparing such sequences with previously known ones revealed an extraordinary number of new sequences, implying that they came from previously unrecognized species. Biologists have described only about 10,000 species of bacteria and only a few hundred species of archaea (see Figure 1.10). The results of Pace's and similar studies suggest that there may be millions, perhaps hundreds of millions, of prokaryote species on Earth. Other biologists put the estimate much lower, and argue that the high dispersal ability of many bacterial species greatly reduces local endemism (geographically restricted species). Only the magnitude of these estimates differ, however; all sides agree that we have just begun to uncover Earth's bacterial and archaeal diversity.

26.3 RECAP

The study of prokaryote phylogeny and diversity has been inhibited by the organisms' small size, our inability to grow some of them in pure culture, and lateral gene transfer. However, nucleotide sequences of essential genes are providing a much clearer picture of bacterial and archaeal evolutionary relationships.

- How did biologists classify bacteria before it became possible to determine nucleotide sequences? **See p. 545**

- Explain why nucleotide sequences of rRNA genes are useful for evolutionary studies. **See p. 545**

- How does lateral gene transfer complicate evolutionary studies? **See p. 545–546 and Figure 26.10**

With the advent of sequencing techniques, biologists have made rapid progress in understanding the phylogeny of prokaryotes. In the next section, we identify the characteristics and life history of the major groups.

26.4 What Are the Major Known Groups of Prokaryotes?

Here we use a widely accepted classification scheme that has considerable support from nucleotide sequence data. More than a dozen major clades have been proposed under this scheme, just a few of which we discuss here. We pay the closest attention to six groups that have received the most study: the spirochetes, chlamydias, high-GC Gram-positives, cyanobacteria, low-GC Gram-positives, and proteobacteria (**Figure 26.11**). First, however, a few words about the origins of the prokaryotes are in order.

Several of the earliest branching lineages of bacteria and archaea are **thermophiles** (Greek, "heat-lovers"). This observation is in line with the hypothesis that the first living organisms were thermophiles, given that most environments on early Earth were much hotter than those of today. While additional evidence continues to support this hypothesis, some researchers believe that the various thermophilic groups evolved more recently than did the lineages leading to the spirochetes and chlamydias.

Spirochetes move by means of axial filaments

Spirochetes are Gram-negative, motile, chemoheterotrophic bacteria characterized by unique structures called axial filaments, which are modified flagella running through the periplasmic space (see Figure 26.6A). The cell body is a long cylinder coiled into a helix (**Figure 26.12**). The axial filaments begin at either end of the cell and overlap in the middle. Protein motors connect the axial filaments to the cell wall, enabling rotation of these structures as they do in other prokaryotic flagella. Many spirochetes live in humans as parasites; a few are pathogens, including those that cause syphilis and Lyme disease. Others live free in mud or water.

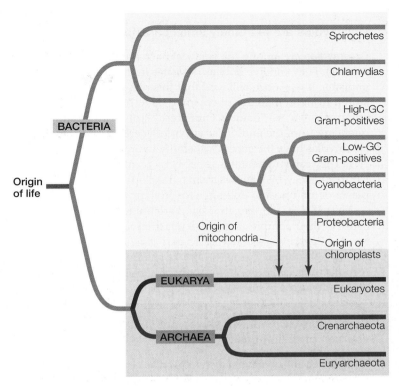

26.11 Two Domains: A Brief Overview This abridged summary classification of the domains Bacteria and Archaea shows their relationships to each other and to Eukarya. The relationships among the many clades of bacteria, not all of which are listed here, are incompletely resolved at this time.

Treponema pallidum   200 nm

26.12 A Spirochete This corkscrew-shaped bacterium causes syphilis in humans.

Chlamydias are extremely small parasites

Chlamydias are among the smallest bacteria (0.2–1.5 μm in diameter). They can live only as parasites in the cells of other organisms. It was once believed that this obligate parasitism resulted from an inability of chlamydias to produce ATP—that chlamydias were "energy parasites." However, genome sequencing from the end of the twentieth century indicates that chlamydias have the genetic capability to produce at least some ATP. They can augment this capacity by using an enzyme called a translocase, which allows them to take up ATP from the cytoplasm of their host in exchange for ADP from their own cells.

These tiny, Gram-negative cocci are unique prokaryotes because of their complex life cycle, which involves two different forms of cells, *elementary bodies* and *reticulate bodies* (**Figure 26.13**). In humans, various strains of chlamydias cause eye infections (especially trachoma), sexually transmitted diseases, and some forms of pneumonia.

Some high-GC Gram-positives are valuable sources of antibiotics

High-GC Gram-positives, also known as *actinobacteria*, derive their name from the relatively high ratio of G-C to A-T nucleotide base pairs in their DNA. These bacteria develop an elaborately branched system of filaments (**Figure 26.14**) and can resemble the filamentous growth habit of fungi, albeit at a reduced scale. Some high-GC Gram-positives reproduce by forming chains of spores at the tips of the filaments. In species that do not form spores, the branched, filamentous growth ceases, and the structure breaks up into typical cocci or bacilli, which then reproduce by binary fission.

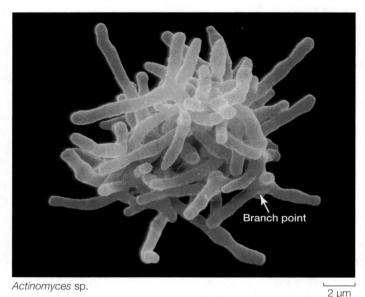

Actinomyces sp.

2 μm

26.14 Filaments of a High-GC Gram-Positive The branching filaments seen in this scanning electron micrograph are typical of this medically important bacterial group.

The high-GC Gram-positives include several medically important bacteria. *Mycobacterium tuberculosis* causes tuberculosis, which kills 3 million people each year. Genetic data suggest that this bacterium arose 3 million years ago in East Africa, making it the oldest known human bacterial affliction. *Streptomyces* produce streptomycin as well as hundreds of other antibiotics. We derive most of our antibiotics from members of the high-GC Gram-positives.

Cyanobacteria are important photoautotrophs

Cyanobacteria, sometimes called *blue-green bacteria* because of their pigmentation, are photoautotrophs that require only water, nitrogen gas, oxygen, a few mineral elements, light, and carbon dioxide to survive. They use chlorophyll *a* for photosynthesis and release oxygen gas; many species also fix nitrogen. Their photosynthesis was the basis of the "oxygen revolution" that transformed Earth's atmosphere (see Section 25.3).

Cyanobacteria carry out the same type of photosynthesis that is characteristic of eukaryotic photosynthesizers. They contain elaborate and highly organized internal membrane systems called *photosynthetic lamellae*. The chloroplasts of photosynthetic eukaryotes are derived from an endosymbiotic cyanobacterium.

Cyanobacteria may live free as single cells or associate in colonies. Depending on the species and on growth conditions, colonies may range from flat sheets one cell thick to filaments to spherical balls of cells.

Some filamentous colonies of cyanobacteria differentiate into three cell types: vegetative cells, spores, and heterocysts (**Figure 26.15**). **Vegetative cells** photosynthesize, **spores** are resting stages that can survive harsh environmental conditions and eventually develop into new filaments, and **heterocysts** are cells specialized for nitrogen fixation. All of the known cyanobacteria with heterocysts fix nitrogen. Heterocysts also have a role in reproduction: when filaments break apart to reproduce, the heterocyst may serve as a breaking point.

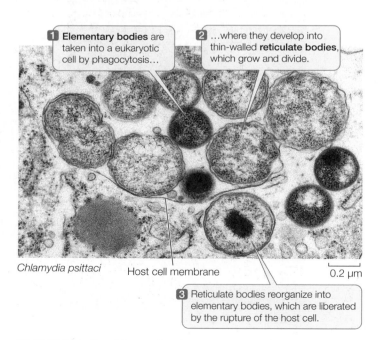

1 **Elementary bodies** are taken into a eukaryotic cell by phagocytosis…

2 …where they develop into thin-walled **reticulate bodies**, which grow and divide.

Chlamydia psittaci   Host cell membrane   0.2 μm

3 Reticulate bodies reorganize into elementary bodies, which are liberated by the rupture of the host cell.

26.13 Chlamydias Change Form during their Life Cycle
Elementary bodies and reticulate bodies are the two major phases of the chlamydia life cycle.

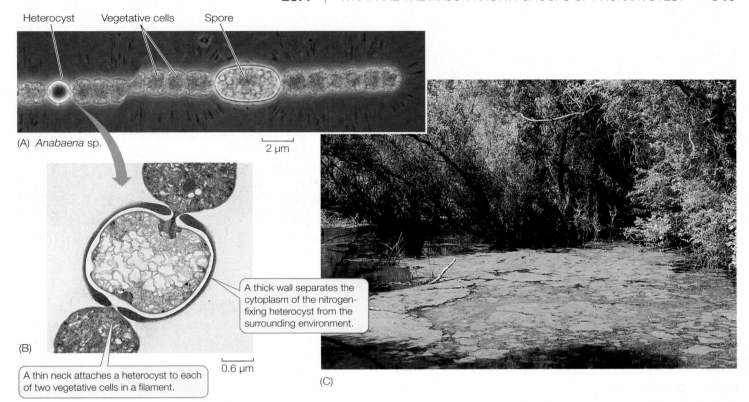

Heterocyst Vegetative cells Spore

(A) *Anabaena* sp.

2 µm

A thick wall separates the cytoplasm of the nitrogen-fixing heterocyst from the surrounding environment.

(B)

0.6 µm

A thin neck attaches a heterocyst to each of two vegetative cells in a filament.

(C)

26.15 Cyanobacteria (A) *Anabaena* is a genus of cyanobacteria that form filamentous colonies containing three cell types. (B) Heterocysts are specialized for nitrogen fixation and serve as a breaking point when filaments reproduce. (C) Cyanobacteria appear in enormous numbers in some environments. This California pond has experienced eutrophication: phosphorus and other nutrients generated by human activity have accumulated, feeding an immense green mat (commonly referred to as "pond scum") that is made up of several species of free-living cyanobacteria.

The low-GC Gram-positives include the smallest cellular organisms

As their name suggests, the **low-GC Gram-positives** have a lower ratio of G-C to A-T nucleotide base pairs than do the high-GC Gram-positives. Some of the low-GC Gram-positives are in fact Gram-negative, and some have no cell wall at all. Despite these differences among the various species, phylogenetic analyses of DNA sequences support the monophyly of this clade.

Some low-GC Gram-positives can produce heat-resistant resting structures called **endospores** (**Figure 26.16**). When a key nutrient such as nitrogen or carbon becomes scarce, the bacterium replicates its DNA and encapsulates one copy, along with some of its cytoplasm, in a tough cell wall heavily thickened with peptidoglycan and surrounded by a spore coat. The parent cell then breaks down, releasing the endospore. Endospore production is not a reproductive process; the endospore merely replaces the parent cell. The endospore, however, can survive harsh environmental conditions that would kill the parent cell, such as high or low temperatures or drought, because it is *dormant*—its normal activity is suspended. Later, if it encounters favorable conditions, the endospore becomes metabolically active and divides, forming new cells that are like the parent cells.

Some endospores can be reactivated after more than 1,000 years of dormancy. There are even credible claims of reactivation of *Bacillus* endospores after millions of years. Members of this endospore-forming group of low-GC Gram-positives include the many species of *Clostridium* and *Bacillus*.

Dormant endospores of *Bacillus anthracis* are the source of an *exotoxin* (see page 555) that causes anthrax. The spores germinate when they sense specific molecules (macrophages; see Chapter 42) in the cytoplasm of mammalian blood cells. However, endospores of other, nonpathogenic *Bacillus* species do not germi-

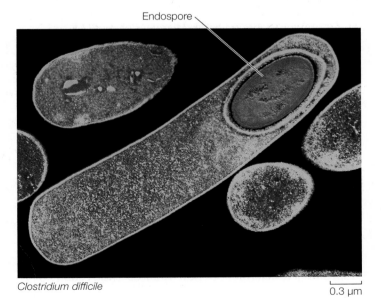

Endospore

Clostridium difficile

0.3 µm

26.16 A Structure for Waiting Out Bad Times This low-GC Gram-positive bacterium, which can cause severe colitis in humans, produces endospores as resistant resting structures.

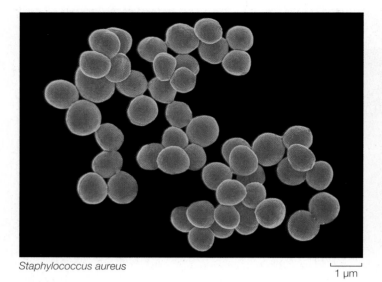

Staphylococcus aureus

1 μm

26.17 Low-GC Gram-Positives "Grape clusters" are the usual arrangement of staphylococci.

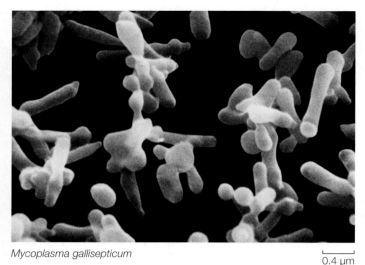

Mycoplasma gallisepticum

0.4 μm

26.18 The Tiniest Cells Containing only about one-fifth as much DNA as *E. coli*, mycoplasmas are the smallest known bacteria.

nate in this environment. *B. anthracis* has been used as a bioterrorism agent because large quantities of its endospores are relatively easy to transport and spread among human populations, where they may be inhaled or ingested.

The genus *Staphylococcus*—the **staphylococci**—includes low-GC Gram-positives that are abundant on the human body surface; they are responsible for boils and many other skin problems (**Figure 26.17**). *Staphylococcus aureus* is the best-known human pathogen in this genus; it is found in 20 to 40 percent of normal adults (and in 50 to 70 percent of hospitalized adults). In addition to skin diseases, it can cause respiratory, intestinal, and wound infections.

Another interesting group of low-GC Gram-positives, the **mycoplasmas**, lack cell walls, although some have a stiffening material outside the plasma membrane. The mycoplasmas include the smallest cellular creatures known (**Figure 26.18**)—even smaller than chlamydias. The smallest mycoplasmas capable of multiplying by fission have a diameter of about 0.2 μm. They are small in another crucial sense as well: they have less than half as much DNA as most other prokaryotes. It has been speculated that the amount of DNA in a mycoplasma, which codes for fewer than 500 proteins, may be the minimum amount required to encode the essential properties of a living cell.

The proteobacteria are a large and diverse group

By far the largest group of bacteria, in terms of number of described species, is the **proteobacteria**. The proteobacteria include many species of Gram-negative, bacteriochlorophyll-containing, sulfur-using photoautotrophs—as well as dramatically diverse bacteria that bear no phenotypic resemblance to the photoautotrophic species. Genetic and morphological evidence indicates that the mitochondria of eukaryotes were derived from a proteobacterium by endosymbiosis (see Section 27.2).

No characteristic demonstrates the diversity of the proteobacteria more clearly than their metabolic pathways (**Figure 26.19**). There are five groups of proteobacteria: alpha, beta, gamma, delta, and epsilon. The common ancestor of all the pro-

teobacteria was a photoautotroph. Early in evolution, two groups of proteobacteria lost their ability to photosynthesize and have been chemotrophs ever since. The other three groups still have photoautotrophic members, but in each group some evolutionary lines have abandoned photoautotrophy and taken up other modes of nutrition. There are chemolithotrophs and chemoheterotrophs in all three groups. Why? One possibility is that each of the trends shown in Figure 26.19 was an evolution-

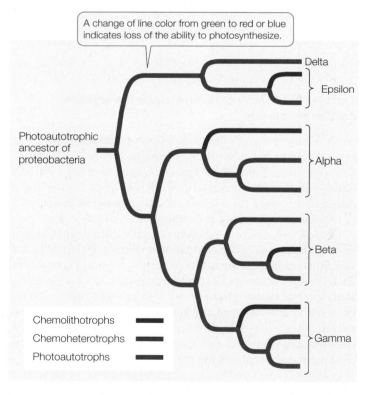

A change of line color from green to red or blue indicates loss of the ability to photosynthesize.

Photoautotrophic ancestor of proteobacteria

Delta
Epsilon
Alpha
Beta
Gamma

Chemolithotrophs
Chemoheterotrophs
Photoautotrophs

26.19 Modes of Nutrition in the Proteobacteria The common ancestor of all proteobacteria was probably a photoautotroph. As they encountered new environments, the delta and epsilon proteobacteria lost the ability to photosynthesize. In the other three groups, some evolutionary lineages became chemolithotrophs or chemoheterotrophs.

ary response to selective pressures encountered as these bacteria colonized new habitats that presented new challenges and opportunities. Lateral gene transfer may have played a role in these responses.

Among the proteobacteria are some nitrogen-fixing genera, such as *Rhizobium* (see Figure 36.9), and other bacteria that contribute to the global nitrogen and sulfur cycles. *Escherichia coli*, one of the most studied organisms on Earth, is a proteobacterium. So, too, are many of the most famous human pathogens, such as *Yersinia pestis* (which causes bubonic plague), *Vibrio cholerae* (cholera), and *Salmonella typhimurium* (gastrointestinal disease).

Although fungi cause most plant diseases, and viruses cause others, about 200 known plant diseases are of bacterial origin. *Crown gall*, with its characteristic tumors (**Figure 26.20**), is one of the most striking. The causal agent of crown gall is *Agrobacterium tumefaciens*, a proteobacterium that harbors a plasmid used in recombinant DNA studies as a vehicle for inserting genes into new plant hosts (see Section 18.2).

We have discussed six clades of bacteria in some detail. Other bacterial clades are well known, and there are probably dozens more waiting to be discovered. This estimate is conservative because so few bacteria have been cultured and studied in the laboratory.

Archaea differ in several important ways from bacteria

The separation of Archaea from Bacteria and Eukarya was originally based on phylogenetic relationships determined from sequences of rRNA genes. This conclusion was supported when biologists sequenced the first archaeal genome. It consisted of 1,738 genes, more than half of which were unlike any genes ever found in the other two domains. Archaea are well known for living in extreme habitats such as those with high salinity (salt content), low oxygen concentrations, high temperatures, or high or low pH (**Figure 26.21**). However, many archaea are not extremeophiles but live in moderate habitats; they are common in

26.20 Crown Gall A crown gall, the type of tumor shown here growing on the stem of a bushy shrub, is caused by the proteobacterium *Agrobacterium tumefaciens*.

INVESTIGATING LIFE

26.21 What Is the Highest Temperature Compatible with Life?
Can any organism thrive at temperatures above 120°C? This is the temperature used for sterilization, known to destroy all previously described organisms. Kazem Kashefi and Derek Lovley isolated an unidentified prokaryote from water samples taken near a hydrothermal vent and found it survived and even grew at 121°C. The organism was dubbed "Strain 121," and its gene sequencing results indicate that it is an archaeal species.

HYPOTHESIS Some prokaryotes survive and even multiply at temperatures above the 120°C threshold of sterilization.

METHOD
1. Seal samples of unidentified, iron-reducing, thermal vent prokaryotes in tubes with a medium containing Fe^{3+} as an electron acceptor. Control tubes contain Fe^{3+} but no organisms.
2. Hold both tubes in a sterilizer at 121°C for 10 hours. If the iron-reducing organisms are metabolically active, they will reduce the Fe^{3+} to Fe^{2+} (as magnetite, which can be detected with a magnet).
3. Isolate any surviving organisms and test for growth at various temperatures.

RESULTS

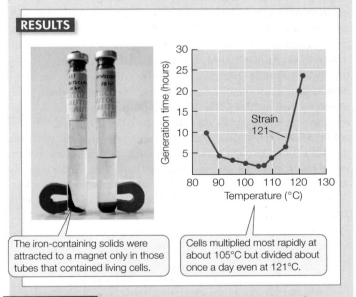

The iron-containing solids were attracted to a magnet only in those tubes that contained living cells.

Cells multiplied most rapidly at about 105°C but divided about once a day even at 121°C.

CONCLUSION Some prokaryotic organisms can survive and grow at temperatures above the previously defined sterilization limit.

FURTHER INVESTIGATION: Note that Strain 121 did not grow during a 2-hour exposure to 130°C, but it did not die, either. How would you demonstrate that it was still alive?

Go to **yourBioPortal.com** for original citations, discussions, and relevant links for all INVESTIGATING LIFE figures.

soil, for example. Perhaps the largest number of archaea live in the ocean depths.

One current classification scheme divides Archaea into two principal groups, **Euryarchaeota** and **Crenarchaeota**. Less is known about two more recently discovered groups, **Korarchaeota** and **Nanoarchaeota**. In fact, we know relatively little about the phylogeny of archaea, in part because the study of archaea is still in its early stages.

Two characteristics shared by all archaea are the absence of peptidoglycan in their cell walls and the presence of lipids of distinctive composition in their cell membranes (see Table 26.1). The unusual lipids in the membranes of archaea are found in all archaea and in no bacteria or eukaryotes. Most bacterial and eukaryotic membrane lipids contain unbranched long-chain fatty acids connected to glycerol molecules by *ester linkages*:

$$\begin{array}{c} \text{O} \quad\quad \text{H} \\ \| \quad\quad\, | \\ -\text{C}-\text{O}-\text{C}- \\ | \\ \text{H} \end{array}$$

In contrast, some archaeal membrane lipids contain long-chain hydrocarbons connected to glycerol molecules by *ether linkages*:

$$\begin{array}{c} \text{H} \quad\quad \text{H} \\ | \quad\quad\, | \\ -\text{C}-\text{O}-\text{C}- \\ | \quad\quad\, | \\ \text{H} \quad\quad \text{H} \end{array}$$

These ether linkages are a synapomorphy of archaea. In addition, the long-chain hydrocarbons of archaea are branched. One class of these lipids, with hydrocarbon chains 40 carbon atoms in length, contains glycerol at *both* ends of the hydrocarbons (**Figure 26.22**). This *lipid monolayer* structure, unique to archaea, still fits in a biological membrane because the lipids are twice as long as the typical lipids in the bilayers of other membranes. Lipid monolayers and bilayers are both found among the archaea. The effects, if any, of these structural features on membrane performance are unknown. In spite of this striking difference in their membrane lipids, the membranes seen in all three domains have similar overall structures, dimensions, and functions.

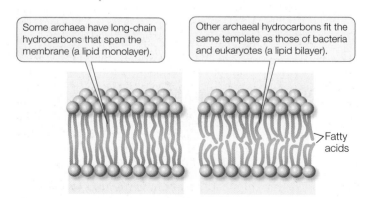

Some archaea have long-chain hydrocarbons that span the membrane (a lipid monolayer).

Other archaeal hydrocarbons fit the same template as those of bacteria and eukaryotes (a lipid bilayer).

Fatty acids

26.22 Membrane Architecture in Archaea   The long-chain hydrocarbons of many archaeal membranes have glycerol molecules at both ends, so that the membranes consist of a lipid monolayer. In contrast, the membranes of other archaea, bacteria, and eukaryotes consist of a lipid bilayer.

Most Crenarchaeota live in hot and/or acidic places

Most known Crenarchaeota are either thermophilic (heat loving), acidophilic (acid loving), or both. Members of the genus *Sulfolobus* live in hot sulfur springs at temperatures of 70°C to 75°C. They become metabolically inactive at 55°C (131°F). Hot sulfur springs are also extremely acidic. *Sulfolobus* grows best in the range from pH 2 to pH 3, but some members of this genus readily tolerate pH values as low as 0.9. Most acidophilic thermophiles maintain an internal pH of 5.5 to 7 (close to neutral) in spite of their acidic environment. These and other thermophiles thrive where very few other organisms can even survive. The archaea living in the volcanic vent shown in the opening of this chapter are examples of such thermophiles.

Euryarchaeota are found in surprising places

Some species of Euryarchaeota share the property of producing methane (CH_4) by reducing carbon dioxide. All of these methanogens are obligate anaerobes, and methane production is the key step in their energy metabolism. Comparison of rRNA gene sequences has revealed a close evolutionary relationship among these methanogenic species, which were previously assigned to several different bacterial groups.

Methanogens release approximately 2 billion tons of methane gas into Earth's atmosphere each year, accounting for 80 to 90 percent of the methane in the atmosphere, including the methane produced in some mammalian digestive systems. Approximately a third of this methane comes from methanogens living in the guts of grazing herbivores such as cattle, sheep, and deer, and another large fraction comes from the methanogens that live in the guts of termites and cockroaches. Methane is increasing in Earth's atmosphere by about 1 percent per year and contributes to the greenhouse effect. Part of the increase is due to increases in cattle and rice farming and the methanogens associated with both.

One methanogen, *Methanopyrus*, lives on the ocean bottom near hot hydrothermal vents. *Methanopyrus* can survive and grow at 122°C. It grows best at 98°C and not at all at temperatures below 84°C.

Another group of Euryarchaeota, the **extreme halophiles** (salt lovers), lives exclusively in very salty environments, such as the water of Blood Falls described in the opening of this chapter. Because they contain pink carotenoid pigments, halophiles are sometimes easy to see (**Figure 26.23**). Halophiles grow in the Dead Sea and in brines of all types: the reddish pink spots that can occur on pickled fish are colonies of halophilic archaea. Few other organisms can live in the saltiest of the homes that the extreme halophiles occupy; most would "dry" to death, losing too much water to the hypertonic environment. Extreme halophiles have been found in lakes with pH values as high as 11.5—the most alkaline environment inhabited by living organisms, and almost as alkaline as household ammonia.

Some of the extreme halophiles have a unique system for trapping light energy and using it to form ATP—without using any form of chlorophyll—when oxygen is in short supply. They use the pigment *retinal* (also found in the vertebrate eye) com-

26.23 Extreme Halophiles Commercial seawater evaporating ponds (these are in San Francisco Bay) are home to salt-loving archaea, easily visible here because of their carotenoid pigments.

bined with a protein to form a light-absorbing molecule called *bacteriorhodopsin*, and they form ATP by a chemiosmotic mechanism of the kind described in Figure 9.9.

Another member of the Euryarchaeota, *Thermoplasma*, has no cell wall. It is thermophilic and acidophilic, its metabolism is aerobic, and it lives in coal deposits. It has the smallest genome among the archaea, and among the smallest (along with the mycoplasmas) of any free-living organism—1,100,000 base pairs.

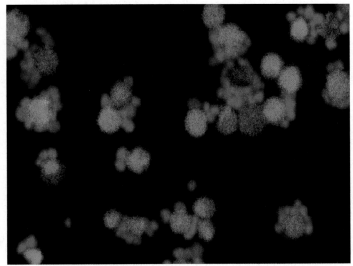

1 μm

26.24 A Nanoarchaeote Growing in Mixed Culture with a Crenarchaeote *Nanoarchaeum equitans* (red), discovered living near deep-ocean hydrothermal vents, is the only representative of the nanoarchaeote group so far discovered. This tiny organism lives attached to cells of the crenarchaeote *Ignicoccus* (green). For this confocal laser micrograph, the two species were visually differentiated by fluorescent dye "tags" that are specific to their separate gene sequences.

Korarchaeota and Nanoarchaeota are less well known

The Korarchaeota are known only by evidence derived from DNA isolated directly from hot springs. No korarchaeote has been successfully grown in pure culture.

Another distinctive archaeal lineage has been discovered at a deep-sea hydrothermal vent off the coast of Iceland. It is the first representative of a group christened Nanoarchaeota because of their minute size. This organism lives attached to cells of *Ignicoccus*, a crenarchaeote. Because of their association, the two species can be grown together in culture (**Figure 26.24**).

26.4 RECAP

Bacteria and Archaea are highly diverse groups that survive in almost every imaginable habitat on Earth. Many can survive and even thrive in habitats where no eukaryotes can live, including extremely hot, acidic, or alkaline conditions.

- Can you explain how metabolic diversity could have become so great in the proteobacteria? **See pp. 550–551 and Figure 26.19**

- What makes the membranes of archaea unique? **See p. 552 and Figure 26.22**

Because prokaryotes have so many different metabolic and nutritional capabilities, and because they can live in so many environments, it is reasonable to expect that they affect their environments in many ways. As we are about to see, prokaryotes directly affect humans—in ways both beneficial and harmful.

26.5 How Do Prokaryotes Affect Their Environments?

Prokaryotes live in and exploit all kinds of environments and are part of all ecosystems. In this section we examine the roles of prokaryotes that live in soils, in water, and even in other organisms, where they may exist in a neutral, beneficial, or parasitic relationship with their host's tissues. The roles of some prokaryotes living in extreme environments have yet to be determined.

Remember that in spite of our frequent mention of prokaryotes as human pathogens, only a small minority of the known prokaryotic species are pathogenic. Many more prokaryotes play positive roles in our lives and in the biosphere. We make direct use of many bacteria and a few archaea in such diverse applications as cheese production, sewage treatment, and the industrial production of an amazing variety of antibiotics, vitamins, organic solvents, and other chemicals.

Prokaryotes are important players in element cycling

Many prokaryotes are decomposers—organisms that metabolize organic compounds in dead organisms and other organic material and return the products to the environment as inorganic substances. Prokaryotes, along with fungi, return tremendous quantities of organic carbon to the atmosphere as carbon dioxide, thus carrying out a key step in the carbon cycle. Prokaryotic decomposers also return inorganic nitrogen and sulfur to the environment.

Animals depend on plants and other photosynthetic organisms for their food, directly or indirectly. But plants depend on other organisms—prokaryotes—for their own nutrition. The extent and diversity of life on Earth would not be possible without nitrogen fixation by prokaryotes. Nitrifiers are crucial to the biosphere because they convert the products of nitrogen fixation into nitrate ions, the form of nitrogen most easily used by many plants (see Figure 36.11). Plants, in turn, are the source of nitrogen compounds for animals and fungi. Denitrifiers also play a key role in keeping the nitrogen cycle going. Without denitrifiers, which convert nitrate ions back into nitrogen gas, all forms of nitrogen would leach from the soil and end up in lakes and oceans, making life on land impossible. Other prokaryotes—both bacteria and archaea—contribute to a similar cycle of sulfur.

In the ancient past, the cyanobacteria had an equally dramatic effect on life: their photosynthesis generated oxygen, converting Earth's atmosphere from an anaerobic to an aerobic environment (see Section 25.3). A major result was the wholesale loss of obligate anaerobic species that could not tolerate the oxygen generated by the cyanobacteria. Only those anaerobes that adapted to aerobic conditions or colonized environments that remained anaerobic survived. However, this transformation to aerobic environments made possible the evolution of cellular respiration and the subsequent explosion of eukaryotic life.

Prokaryotes live on and in other organisms

Prokaryotes work together with eukaryotes in many ways. As we have seen, mitochondria and chloroplasts are descended from what were once free-living bacteria. Much later in evolutionary history, some plants became associated with bacteria to form cooperative nitrogen-fixing nodules on their roots (see Figure 36.9).

Many animals harbor a variety of bacteria and archaea in their digestive tracts. Cattle depend on prokaryotes to perform important steps in digestion. Like most animals, cattle cannot produce cellulase, the enzyme needed to start the digestion of the cellulose that makes up the bulk of their plant food. However, bacteria living in a special section of the gut, called the rumen, produce enough cellulase to process the daily diet for the cattle.

Humans use some of the metabolic products—especially vitamins B_{12} and K—of bacteria living in our large intestine. These and other bacteria and archaea line our intestines with a dense biofilm that is in intimate contact with the mucosal lining of the gut. This biofilm facilitates nutrient transfer from the intestine into the body and induces immunity to the gut contents. The biofilm in the gut is a major part of an "organ" consisting of prokaryotes that is essential to our health. Its makeup varies from time to time and from region to region of the intestinal tract, and it has a complex ecology that scientists have just begun to explore in detail—including the possibility that the species composition of an individual's prokaryote gut fauna may contribute to obesity (or the resistance to it).

We are heavily populated inside and out by bacteria. A 2009 study of bacteria that live on human skin identified more than 1,000 species living on the outside of our bodies, and many of these are thought to be critical to maintaining the skin's health. Although only a small percentage of bacterial species are agents of disease, popular notions of bacteria as "germs" and fear of the consequences of infection arouse our curiosity about those few.

A small minority of bacteria are pathogens

The late nineteenth century was a productive era in the history of medicine—a time when bacteriologists, chemists, and physicians proved that many diseases are caused by microbial agents. During this time the German physician Robert Koch laid down a set of four rules for establishing that a particular microorganism causes a particular disease:

- The microorganism is always found in individuals with the disease.
- The microorganism can be taken from the host and grown in pure culture.
- A sample of the culture produces the same disease when injected into a new, healthy host.
- The newly infected host yields a new, pure culture of microorganisms identical to those obtained in the second step.

These rules, called **Koch's postulates**, were very important in a time when it was not widely understood that microorganisms cause disease. Although medical science today has more powerful diagnostic tools, the postulates remain useful on occasion. For example, physicians were taken aback in the 1990s when stomach ulcers—long accepted and treated as the result of excess stomach acid—were shown by Koch's postulates to be caused by the bacterium *Helicobacter pylori* (see Figure 51.14).

Only a tiny percentage of all prokaryotes are pathogens, and of those that are known, all are in the domain Bacteria. For an organism to be a successful pathogen, it must:

- arrive at the body surface of a potential host;
- enter the host's body;
- evade the host's defenses;
- multiply inside the host; and finally
- infect a new host.

Failure to successfully complete any of these steps ends the reproductive career of a pathogenic organism. However, in spite of the many defenses available to potential hosts (see Chapter 42), some bacteria are very successful pathogens.

For the host, the consequences of a bacterial infection depend on several factors. One is the **invasiveness** of the pathogen—its ability to multiply in the host's body. Another is its **toxigenicity**—its ability to produce toxins, chemical substances that are harmful to the host's tissues. *Corynebacterium diphtheriae*, the agent that causes diphtheria, has low invasiveness and multiplies only in the throat, but its toxigenicity is so great that the entire body is affected. In contrast, *Bacillus anthracis*, which causes anthrax (a disease primarily of cattle and sheep, but which is also sometimes fatal in humans), has low toxigenicity but is so invasive that the entire bloodstream ultimately teems with the bacteria.

There are two general types of bacterial toxins: exotoxins and endotoxins. **Endotoxins** are released when certain Gram-negative bacteria grow or lyse (burst). Endotoxins are lipopolysaccharides (complexes consisting of a polysaccharide and a lipid component) that form part of the outer bacterial membrane. Endotoxins are rarely fatal; they normally cause fever, vomiting, and diarrhea. Among the endotoxin producers are some strains of the gamma-proteobacteria *Salmonella* and *Escherichia*.

Exotoxins are usually soluble proteins released by living, multiplying bacteria, and they may travel throughout the host's body. They are highly toxic—often fatal—to the host, but they do not produce fevers. Human diseases induced by bacterial exotoxins include tetanus (*Clostridium tetani*), cholera (*Vibrio cholerae*), and bubonic plague (*Yersinia pestis*). Anthrax results from three exotoxins produced by *Bacillus anthracis*. Botulism is caused by exotoxins produced by *Clostridium botulinum* that are among the most poisonous ever discovered. The lethal dose of the botulinum A exotoxin for humans is about one-millionth of a gram (1 μg). Nonetheless, much smaller doses of this exotoxin, marketed under various trade names, are used to treat muscle spasms and also for cosmetic purposes (temporary wrinkle reduction in skin).

Pathogenic bacteria are often surprisingly difficult to combat, even with today's arsenal of antibiotics. One source of this difficulty is the ability of prokaryotes to form biofilms.

26.5 RECAP

Prokaryotes play key roles in the cycling of Earth's elements. Many prokaryotes are beneficial and even necessary to other forms of life; others are pathogens.

- Describe the roles of bacteria in the nitrogen cycle. See p. 554

- What are some of the challenges facing a pathogen? See p. 554

Before moving on to discuss the diversity of eukaryotic life, it is appropriate to consider how viruses are related to the rest of life. Although they are not cellular, viruses are numerically among the most abundant organisms on Earth. Their effects on other organisms are enormous. Where did viruses come from, and how do they fit into the tree of life? Biologists are still working to answer these questions.

26.6 Where Do Viruses Fit into the Tree of Life?

Some biologists do not think of viruses as living organisms, primarily because they are not cellular and must depend on cellular organisms for basic life functions such as replication and metabolism. But viruses are derived from the cells of other living organisms. They use the same essential forms of genetic storage and transmission as do cellular organisms. Viruses infect all cellular forms of life, including bacteria, archaea, and eukaryotes. They replicate, mutate, evolve, and interact with other organisms, often causing serious diseases when they infect their hosts. They are also numerically among the most abundant organisms on the planet. And, finally, viruses clearly evolve independently of other organisms, so it is almost impossible not to treat them as a part of life.

Several factors make virus phylogeny difficult to resolve. The tiny size of many viral genomes restricts the phylogenetic analyses that can be conducted to relate viruses to cellular organisms. The rapid mutational rate, which results in rapid evolution of viral genomes, tends to cloud evolutionary relationships across long periods of time. There are no known viral fossils (viruses are too small and delicate to fossilize), so the paleontological record offers no clues as to viral origins. Finally, viruses are highly diverse (see Figure 26.25), and several lines of evidence support the hypothesis that viruses have evolved repeatedly within each of the major groups of life.

Many RNA viruses probably represent escaped genomic components

Although viruses are now obligate parasites of cellular species, they may once have been cellular components involved in basic cellular functions—that is, they may be "escaped" components of cellular life that now evolve independently of their hosts.

NEGATIVE-SENSE SINGLE-STRANDED RNA VIRUSES A case in point is a class of viruses whose genome is composed of single-stranded RNA that is the complement (negative-sense) of the mRNA needed for protein translation. Many of these negative-sense single-stranded RNA viruses have only a few genes, including an RNA-dependent RNA polymerase that allows them to make mRNA from their negative-sense RNA genome. Modern cellular organisms cannot generate mRNA in this manner (at least in the absence of viral infections), but scientists speculate that single-stranded RNA genomes may have been common in the distant past, before DNA became the primary molecule for genetic information storage.

A self-replicating RNA polymerase gene that begins to replicate independently of a cellular genome could conceivably acquire a few additional protein-coding genes through recombination with its host's DNA. If one or more of these genes were to foster the development of a protein coat, the virus might then survive outside the host and infect new hosts. It is believed that this scenario has been repeated many times independently across the tree of life, given that many of the negative-sense single-stranded RNA viruses that infect organisms from bacteria to humans are not closely related to one another. In other words, negative-sense single-stranded RNA viruses do not represent a distinct taxonomic group, but exemplify a particular process of cellular escape that probably happened many different times.

Examples of familiar negative-sense single-stranded RNA viruses include the viruses that cause measles, mumps, rabies, and influenza (**Figure 26.25A**).

(A)

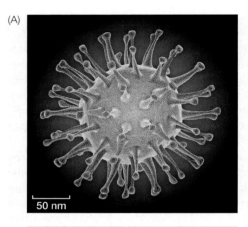

50 nm

A negative-sense single-stranded RNA virus: Influenza virus H5N1, the "bird flu" virus. Surface view.

(B)

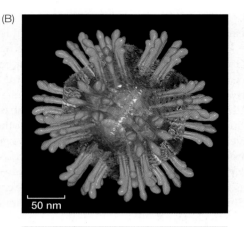

50 nm

A positive-sense single-stranded RNA virus: Hepatitis C virus, the cause of a human liver disease. Surface view.

(C)

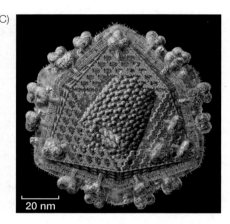

20 nm

An RNA retrovirus: One of the human immunodeficiency viruses (HIV) that causes AIDS. Cutaway view.

(D)

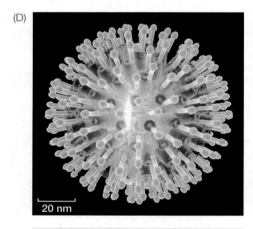

20 nm

A double-stranded DNA virus: One of the many herpes viruses (Herpesviridae). Different herpes viruses are responsible for many human infections, including chicken pox, shingles, cold sores. and genital herpes (HSV1/2). Surface view.

(E)

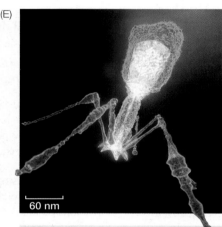

60 nm

A double-stranded DNA virus: Bacteriophage T4. Viruses that infect bacteria are referred to as bacteriophage (or simply phage). T4 attaches leglike fibers to the outside of its host cell and injects its DNA into the cytoplasm through its "tail" (pink structure in this rendition).

(F)

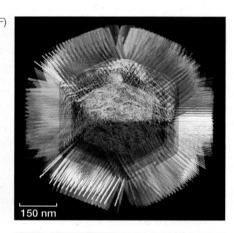

150 nm

A double-stranded DNA mimivirus: This *Acanthamoeba polyphaga* mimivirus (APMV) has the largest diameter of all known viruses and a genome larger than some prokaryote genomes. Cutaway view.

26.25 Viruses Are Diverse Relatively small genomes and rapid evolutionary rates make it difficult to reconstruct phylogenetic relationships among some classes of viruses. Instead, viruses are classified largely by general characteristics of their genomes. The images here are computer artists' reconstructions based on cryoelectron micrographs.

POSITIVE-SENSE SINGLE-STRANDED RNA VIRUSES The genome of another type of single-stranded RNA viruses is composed of positive-sense RNA. Positive-sense genomes are already set for translation; unlike in negative-sense RNA, no replication of the genome into the complement strand is needed before protein translation can take place. Positive-sense single-stranded RNA viruses are the most abundant and diverse class of viruses (**Figure 26.25B**). Most of the viruses that cause diseases in crop plants are in this group. When they infect plants, these viruses kill patches of cells in the leaves or stems of the plants, leaving live cells amid a patchwork of discolored dead plant tissue (giving them the name of *mosaic* or *mottle viruses*; **Figure 26.26**). Other viruses in this group infect bacteria, fungi, and animals. Human diseases caused by positive-sense single-stranded RNA viruses include polio, hepatitis C, and the common cold. As is true of the other functionally defined groups of viruses, these viruses appear to have evolved multiple times across the tree of life from different groups of cellular ancestors.

RNA RETROVIRUSES The RNA retroviruses are best known as the group that includes the human immunodeficiency viruses (HIV; **Figure 26.25C**). Like the previous two categories of viruses, RNA retroviruses have genomes composed of single-stranded RNA and likely evolved as escaped cellular components.

Yellow areas are dead leaf cells, killed by the mosaic virus.

26.26 Mosaic Viruses Are a Problem for Agriculture Mosaic or "mottle" viruses are the most abundant and diverse class of viruses. This leaf is from an apple tree infected with mosaic virus.

Retroviruses are so named because they regenerate themselves by reverse transcription. When the retroviruses enter the nucleus of their vertebrate host, viral reverse transcriptase produces cDNA from the viral RNA genome and then replicates the single-stranded cDNA to produce double-stranded DNA. Another virally encoded enzyme called integrase catalyzes the integration of the new piece of double-stranded viral DNA into the host's genome. The viral genome is then replicated along with the host cell's DNA; the integrated retroviral DNA is known as a *provirus*. Many components of cellular species (such as retrotransposons; see Section 17.3) resemble components of retroviruses.

Retroviruses are only known to infect vertebrates, although genomic elements that resemble portions of these viruses are a component of the genomes of a wide variety of organisms including bacteria, plants, and many animals. Several retroviruses are associated with the development of various forms of cancer, as cells infected with these viruses are likely to undergo uncontrolled replication.

DOUBLE-STRANDED RNA VIRUSES Double-stranded RNA viruses may have evolved repeatedly from single-stranded RNA ancestors—or perhaps vice versa. These viruses, which are not closely related to one another, infect organisms from throughout the tree of life. For example, many important plant diseases are caused by double-stranded RNA viruses, whereas other viruses of this type cause many cases of infant diarrhea in humans.

Some DNA viruses may have evolved from reduced cellular organisms

Another class of viruses is composed of those that have a double-stranded DNA genome (**Figure 26.25D–F**). This group is also almost certainly polyphyletic (with many independent origins). Many of the common phage that infect bacteria (*bacteriophage*) are double-stranded DNA viruses, as are the viruses that cause smallpox and herpes in humans.

Some biologists think that at least some of the DNA viruses may represent highly reduced parasitic organisms that have lost their cellular structure as well as their ability to survive as free-living species. For example, some of the largest DNA viruses are the mimiviruses (see Figure 26.25F), which have a genome in excess of a million base pairs of DNA. This is double the genome size of some parasitic bacteria such as *Mycoplasma genitalium*. Phylogenetic analyses of these DNA viruses suggest that they have evolved repeatedly from cellular organisms. Furthermore, recombination among different viruses may have allowed the exchange of various genetic modules, further complicating the history and origins of viruses.

26.6 RECAP

Viruses are highly diverse and appear to have evolved independently from many different groups of cellular organisms from throughout the tree of life. Some viruses appear to have evolved from escaped components of cellular organisms, whereas other viruses may have evolved from parasitic cellular ancestors.

- What are some the reasons that it is difficult to place viruses precisely within the phylogeny of cellular organisms? See p. 555
- Explain the two main hypotheses of viral origins. See p. 555–557

It appears that the enormous diversity of viruses is, at least in part, a result of their multiple origins from many different cellular organisms. Some may have arisen as escaped genetic components, while others may represent reduced and specialized parasites. Some appear to be derived from bacterial species, others to be derived from eukaryotic organisms. It may be best to view viruses as spinoffs from the various branches on the tree of life—sometimes evolving independently of cellular genomes, sometimes recombining with them. One way to think of viruses is as the "bark" on the tree of life: certainly an important component all across the tree, but not quite like the main branches.

CHAPTER SUMMARY

26.1 How Did the Living World Begin to Diversify?

- Two of life's three domains, Bacteria and Archaea, are prokaryotic. They are distinguished from Eukarya in several ways, including their lack of membrane-enclosed organelles. **Review Table 26.1**

- Eukaryotes are related to both Archaea and Bacteria, and appear to have formed through a merging of members of these two lineages. The last common ancestor of all three domains probably lived 2 to 3 billion years ago. **Review Figure 26.1, ANIMATED TUTORIAL 26.1**

- Prokaryotes may be single or may form chains or clusters. The three most common bacterial morphologies are **cocci** (spheres), **bacilli** (rods), and **helices** (spirals). **Review Figure 26.2**

- The cells of some bacteria aggregate, forming **filaments** and other structures.

26.2 What Are Some Keys to the Success of Prokaryotes?

- Prokaryotes are the most numerous organisms on Earth. They occur in an enormous variety of habitats, including inside other organisms and deep in Earth's crust.

- Prokaryotes form complex communities, some of which become dense films called **biofilms**. **Review Figure 26.3**

- Most prokaryotes have cell walls containing molecules important in transport, digestion, and sensing the environment. Almost all bacterial cell walls contain **peptidoglycan**. The cell walls of archaea lack peptidoglycan and instead include pseudopeptidoglycan or proteins. **Review Figure 26.5, WEB ACTIVITY 26.1**

- Bacteria can be placed into two groups by the **Gram stain**. **Gram-negative bacteria** have a periplasmic space between the inner (plasma) and outer membranes. In **Gram-positive bacteria**, the cell wall usually has about five times as much peptidoglycan as a Gram-negative wall.

- Prokaryotes move by a variety of means, including axial filaments, gas vesicles, and **flagella**. **Review Figures 26.6 and 26.7**

- Prokaryotes reproduce asexually but may undergo genetic recombination. Reproduction and genetic recombination are not directly linked in prokaryotes.

- Prokaryote metabolism is very diverse. Some prokaryotes are anaerobic, others are aerobic, and yet others can shift between these modes. Prokaryotes are classified as **photoautotrophs**, **photoheterotrophs**, **chemolithotrophs**, or **chemoheterotrophs**. **Review Table 26.2**

- The metabolic pathways of some prokaryotes involve sulfur or nitrogen. **Nitrogen fixers** convert nitrogen gas into a form organisms can metabolize.

26.3 How Can We Resolve Prokaryote Phylogeny?

- Early attempts to classify prokaryotes were hampered by the organisms' small size and difficulties growing them in pure culture. Phylogenetic classification of prokaryotes is now based on nucleotide sequences of rRNA and other slowly evolving genes.

- Although **lateral gene transfer** has occurred throughout prokaryotic evolutionary history, elucidation of prokaryote phylogeny is still possible. **Review Figure 26.10**

- Only a tiny percentage of all prokaryote species have been described.

26.4 What Are the Major Known Groups of Prokaryotes?

- Several major taxonomic groups of prokaryotes have been recognized. The members of a prokaryote clade often differ profoundly from one another. **Review Figures 26.11 and 26.19**

- Of the major taxonomic groups of Bacteria, the **proteobacteria** embrace the largest number of species. Other important groups include the **cyanobacteria**, **spirochetes**, **chlamydias**, and **low-GC Gram-positives**. Some **high-GC Gram positives** produce important antibiotics.

- The low-GC Gram-positives include the **mycoplasmas**, which are among the smallest cellular organisms ever discovered.

- Ether linkages in the branched long-chain hydrocarbons of cell membranes are a synapomorphy of archaea.

- The best-studied groups of Archaea are **Euryarchaeota** and **Crenarchaeota**. Some species of Euryarchaeota are methanogens; extreme halophiles are also found among the Euryarchaeota. Most known Crenarchaeota are both thermophilic and acidophilic.

26.5 How Do Prokaryotes Affect their Environments?

- Prokaryotes play key roles in the cycling of elements such as nitrogen, oxygen, sulfur, and carbon. One such role is as decomposers of dead organisms.

- Nitrogen-fixing bacteria fix the nitrogen needed by all other organisms. **Nitrifiers** convert that nitrogen into forms that can be used by plants, and **denitrifiers** ensure that nitrogen is returned to the atmosphere.

- Oxygen production by early photosynthetic cyanobacteria reconfigured Earth's atmosphere, which made aerobic forms of life possible.

- Prokaryotes inhabiting the guts of many animals help them digest their food.

- **Koch's postulates** establish the criteria by which an organism may be classified as a pathogen. Relatively few bacteria—and no archaea—are known to be pathogens.

26.6 Where Do Viruses Fit Into the Tree of Life?

- Viruses have evolved many times from many different groups of cellular organisms. They do not represent a single taxonomic group.

- Some viruses are probably derived from escaped genetic elements of cellular species; others are thought to have evolved as highly reduced parasites.

- Viruses are categorized by the nature of their genomes. **Review Figure 26.25**

SELF-QUIZ

1. Most prokaryotes
   a. are agents of disease.
   b. lack ribosomes.
   c. evolved from the most ancient eukaryotes.
   d. lack a cell wall.
   e. are chemoheterotrophs.

2. The division of the living world into three domains
   a. is based on the number of cells in organisms of each group.
   b. is based mostly on the major morphological differences between archaea and bacteria.
   c. emphasizes the greater importance of eukaryotes.
   d. was proposed by the early microscopists.
   e. is based on phylogenetic relationships determined from nucleotide sequences of rRNA and other genes.

3. Which statement about archaeal genomes is *true*?
   a. They are typically organized in a circular chromosome, like bacterial genomes.
   b. They include no rRNA genes.
   c. They are always much smaller than bacterial genomes.
   d. They are housed in the nucleus of the archaeal cell.
   e. No archaeal genome has yet been sequenced.

4. Which statement about nitrogen metabolism is *not* true?
   a. Certain prokaryotes reduce atmospheric N_2 to ammonia.
   b. Some nitrifiers are soil bacteria.
   c. Denitrifiers are obligate anaerobes.
   d. Nitrifiers obtain energy by oxidizing ammonia and nitrite.
   e. Without nitrifiers, terrestrial organisms would lack a nitrogen supply.

5. All photosynthetic bacteria
   a. use chlorophyll *a* as their photosynthetic pigment.
   b. use bacteriochlorophyll as their photosynthetic pigment.
   c. release oxygen gas.
   d. produce particles of sulfur.
   e. are photoautotrophs.

6. Gram-negative bacteria
   a. appear blue to purple following Gram staining.
   b. appear pink to red following Gram staining.
   c. are all either bacilli or cocci.
   d. contain no peptidoglycan in their cell walls.
   e. are all photosynthetic.

7. Endospores
   a. are produced by viruses.
   b. are reproductive structures.
   c. are very delicate and easily killed.
   d. are resting structures.
   e. lack cell walls.

8. Chlamydias
   a. are among the smallest archaea.
   b. live on the surface of human skin.
   c. are never pathogenic to humans.
   d. live only as parasites in the cells of other organisms.
   e. have a very simple life cycle.

9. Archaea
   a. have cytoskeletons.
   b. have distinctive lipids in their plasma membranes.
   c. survive only at moderate temperatures and near neutrality.
   d. all produce methane.
   e. have substantial amounts of peptidoglycan in their cell walls.

10. Genetic evidence suggests that viruses
    a. are most closely related to Bacteria.
    b. are most closely related to Archaea.
    c. are most closely related to Eukarya.
    d. have evolved multiple times from many different cellular species.
    e. evolved from the fusion of a bacterial and an archaeal species.

FOR DISCUSSION

1. Why do systematic biologists find rRNA sequence data more useful than data on metabolism or cell structure for classifying prokaryotes?

2. How can lateral gene transfer mislead studies about prokaryote phylogeny? How could this issue be addressed?

3. Differentiate among the members of the following sets of related terms:
   a. prokaryotic/eukaryotic
   b. obligate anaerobe/facultative anaerobe/obligate aerobe
   c. photoautotroph/photoheterotroph/chemolithotroph/chemoheterotroph
   d. Gram-positive/Gram-negative

4. Why are the endospores of low-GC Gram-positives not considered to be reproductive structures?

5. Originally, the cyanobacteria were called "blue-green algae" and were not grouped with the bacteria. Suggest several reasons for this (abandoned) tendency to separate the cyanobacteria from the bacteria. Why are the cyanobacteria now grouped with the other bacteria?

6. The high-GC Gram-positives are of great commercial interest. Why?

7. Thermophiles are of great interest to molecular biologists and biochemists. Why? What practical concerns might motivate that interest?

8. How can biologists discuss the Korarchaeota when they have never seen one?

9. Do you consider viruses to be living organisms? Why or why not?

ADDITIONAL INVESTIGATION

Kashefi and Lovley (see Figure 26.21) were able to grow an unnamed archaeal species at temperatures above 120°C only because they used Fe^{3+} as an electron acceptor—no other electron acceptor they tried allowed growth. How might you explore the same or other high-temperature environments for other hyperthermophilic organisms not detected by Kashefi and Lovley using Fe^{3+}?

The Origin and Diversification of Eukaryotes

How a microbial eukaryote may have changed the course of science

Charles Darwin suffered throughout his adult life from a debilitating illness that was never conclusively diagnosed. Whatever the cause of this illness, it seems to have contributed to the determination with which Darwin pursued his studies. In his autobiography, Darwin wrote that his "chief enjoyment and sole employment throughout life has been scientific work; and the excitement from such work makes me for the time forget, or drives quite away, my daily discomfort." To Darwin, science was a pleasure that provided a distraction from his pain.

One hypothesis is that Darwin suffered from Chagas' disease. Caused by the trypanosome *Trypanosoma cruzi*, one of many microbial eukaryotes, this disease currently affects 16 to 18 million people, primarily in Central and South America. The trypanosome is transmitted to people by trypanosome-infected assassin bugs, which bite to suck blood from their victims and then often defecate near the wound. When people crush the insect or scratch its bite, they can become infected with the trypanosome, which is present in the assassin bug's feces. That Darwin may have been infected in this way is suggested by a journal entry he made during his voyage around the world on the HMS *Beagle*. On March 25, 1835, while visiting Argentina, Darwin wrote:

At night I experienced an attack (for it deserves no less a name) of the Benchuca … *the great black bug of the Pampas. It is most disgusting to feel soft wingless insects, about an inch long, crawling over one's body. Before sucking they are quite thin but afterwards they become round and bloated with blood, and in this state they are easily crushed.*

Darwin's entry describes the experience of being bitten by one of the assassin bugs that carry *Trypanosoma cruzi*, so clearly Darwin had the opportunity to become infected. In fact, he even experimented with the insects to learn more about their habits. "[I]f a finger was presented, the bold insect would immediately protrude its sucker, make a charge, and if allowed, draw blood," he wrote in his journal.

There is neither a vaccine nor any effective drug treatment for Chagas'

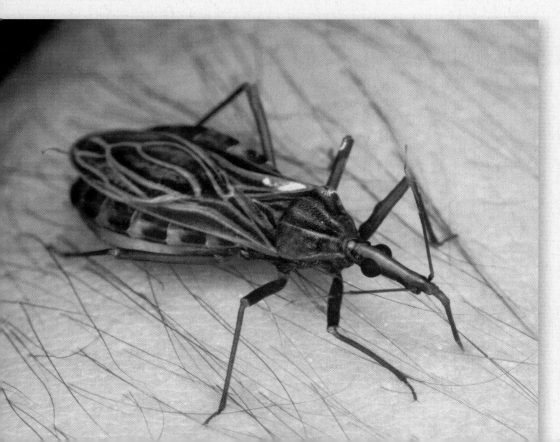

An Assassin with a Deadly Cargo
This member of the insect genus *Rhodnius* is an "assassin bug," one of several species that transmits *Trypanosoma cruzi*, the causative agent of Chagas' disease in humans.

Thugs in a Huddle Sometimes trypanosomes, like these *Leishmania major*, form clusters held together by a tangle of mucilage secreted around their flagella; nobody is sure yet why they behave this way.

disease, which kills more than 40,000 people a year. Other trypanosome species cause other debilitating diseases, including leishmaniasis (60,000 deaths/year) and African sleeping sickness (50,000 deaths/year). Since the infective organism in each of these diseases is a eukaryote like us, most of the drugs that are toxic to the trypanosomes are often toxic to humans as well. Although the treatment of bacterial diseases often capitalizes on the many differences between prokaryotes and eukaryotes in order to target only the bacteria, the diseases caused by microbial eukaryotes usually are, like Chagas' disease, much more difficult to treat.

IN THIS CHAPTER we describe the origin and early diversification of the eukaryotes and the complexity achieved by some single cells. We then explore some of the diversity of microbial eukaryote body forms and adaptations and present the developing current view of the evolutionary relationships among the major eukaryote groups.

27.1 How Did the Eukaryotic Cell Arise?

Many members of the domain Eukarya are familiar to us. We have no problem recognizing trees, mushrooms, and insects as plants, fungi, and animals, respectively. However, a dazzling assortment of other eukaryotes—mostly microscopic organisms—do not fit into any of these three groups. Eukaryotes that are not plants, animals, or fungi have traditionally been "dropped" into the category **protists**. Phylogenetic analyses, however, are clear and consistent in showing that many of the groups that fall under the rubric of "protists" in fact are not closely related to one another, but are *paraphyletic* (see Figure 22.14). Thus the word "protist" does not describe a formal taxonomic group, but is a convenience term—a shorthand way of saying "all the eukaryotes that are not plants, animals, or fungi."

The diversity of protists is reflected in both morphology and phylogeny

In terms of their evolutionary relationships, as well as in many aspects of their biology, protists are more diverse than any of the three better known eukaryote groups. Some protists are motile, while others do not move; some are photosynthetic, others heterotrophic; most are unicellular, but some are multicellular (**Table 27.1**). Most are microscopic, but a few are huge: giant kelps, for example, can grow to be longer than a football field. We refer to the unicellular species of protists as *microbial eukaryotes*, but you should keep in mind that there are large, multicellular protists as well.

The phylogeny of the major eukaryote lineages remains a subject of research and debate. Some groups of protists are closely related to animals and fungi, whereas others are closely related to the land plants, and still others are only distantly related to any of these familiar eukaryotes (**Figure 27.1**).

Cellular features support the monophyly of eukaryotes

Eukaryotic cells differ in many ways from prokaryotic cells, and these unique characters of eukaryotes lead us to conclude that eukaryotes are monophyletic. In other words, there was a single eukaryotic ancestor which diversified into the many different lineages of protists, as well as plants, animals, and fungi. Given the nature of evolutionary processes, the many synapomorphies of eukaryotes undoubtedly did not arise simultaneously. We can make some reasonable inferences about the most important events that led to the evolution of a new cell type, bearing in mind that the global environment underwent an

27.1 Major Eukaryote Groups in an Evolutionary Context This tree shows a current hypothesis for the evolutionary relationships among major groups of eukaryotes. The dashed lines indicate clades for which the evidence is weak or disputed. The root of the tree is uncertain.

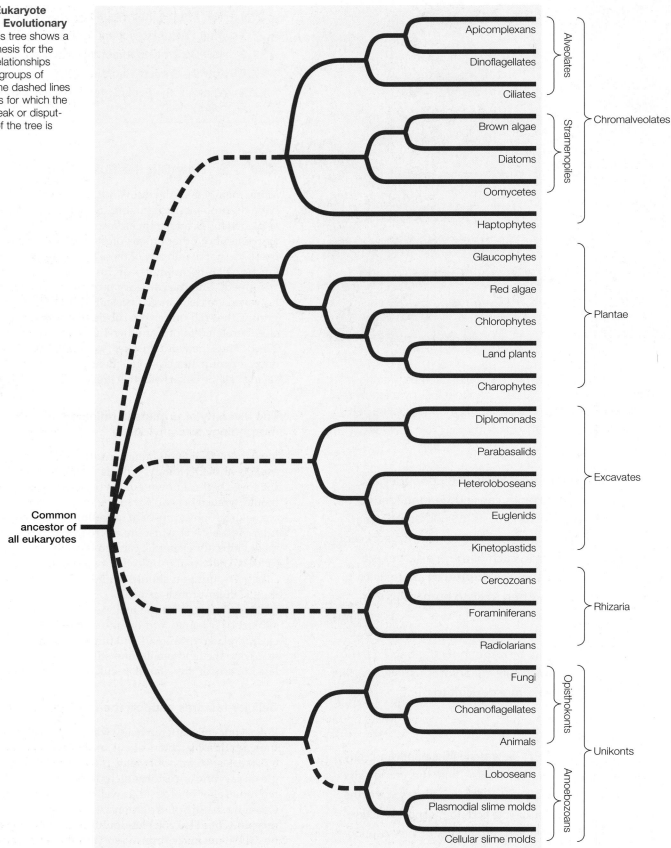

TABLE 27.1

Major Eukaryote Clades

CLADE	ATTRIBUTES	EXAMPLE (GENUS)
Chromalveolates		
Haptophytes	Unicellular, often with calcium carbonate scales	*Emiliania*
Alveolates	Sac-like structures beneath plasma membrane	
Apicomplexans	Apical complex for penetration of host	*Plasmodium*
Dinoflagellates	Pigments give golden-brown color	*Gonyaulax*
Ciliates	Cilia; two types of nuclei	*Paramecium*
Stramenopiles	Hairy and smooth flagella	
Brown algae	Multicellular; marine; photosynthetic	*Macrocystis*
Diatoms	Unicellular; photosynthetic; two-part cell walls	*Thalassiosira*
Oomycetes	Mostly coenocytic; heterotrophic	*Saprolegnia*
Plantae		
Glaucophytes	Peptidoglycan in chloroplasts	*Cyanophora*
Red algae	No flagella; chlorophyll *a* and *c*; phycoerythrin	*Chondrus*
Chlorophytes	Chlorophyll *a* and *b*	*Ulva*
*Land plants (Chs. 28–29)	Chlorophyll *a* and *b*; protected embryo	*Ginkgo*
Charophytes	Chlorophyll *a* and *b*; mitotic spindle oriented as in land plants	*Chara*
Excavates		
Diplomonads	No mitochondria; two nuclei; flagella	*Giardia*
Parabasalids	No mitochondria; flagella and undulating membrane	*Trichomonas*
Heteroloboseans	Can transform between amoeboid and flagellate stages	*Naegleria*
Euglenids	Flagella; spiral strips of protein support cell surface	*Euglena*
Kinetoplastids	Kinetoplast within mitochondrion	*Trypanosoma*
Rhizaria		
Cercozoans	Threadlike pseudopods	*Cercomonas*
Foraminiferans	Long, branched pseudopods; calcium carbonate shells	*Globigerina*
Radiolarians	Glassy endoskeleton; thin, stiff pseudopods	*Astrolithium*
Unikonts		
Opisthokonts	Single, posterior flagellum	
*Fungi (Ch. 30)	Heterotrophs that feed by absorption	*Penicillium*
Choanoflagellates	Resemble sponge cells; heterotrophic; with flagella	*Choanoeca*
*Animals (Chs. 31–33)	Heterotrophs that feed by ingestion	*Drosophila*
Amoebozoans	Amoebas with lobe-shaped pseudopods	
Loboseans	Feed individually	*Amoeba*
Plasmodial slime molds	Form coenocytic feeding bodies	*Physarum*
Cellular slime molds	Cells retain their identity in pseudoplasmodium	*Dictyostelium*

*Clades marked with an asterisk are made up of multicellular organisms and are discussed in the chapters indicated. All other groups listed are treated here as *protists*.

enormous change—from low to high availability of free oxygen—during the course of these events (see Section 25.3). Keep in mind, however, that these inferences, although reasonable and grounded, are still conjectural; the hypothesis we pursue here is one of a few that biologists are currently considering. We describe here a prominent theory on the origin of the eukaryotic cell as a framework for thinking about the challenging question of eukaryotic origins.

The modern eukaryotic cell arose in several steps

Several events preceded the origin of the modern eukaryotic cell:

- The origin of a flexible cell surface
- The origin of a cytoskeleton
- The origin of a nuclear envelope, which enclosed a genome organized into chromosomes
- The appearance of *digestive vacuoles*
- The acquisition of certain organelles via *endosymbiosis*

RAMIFICATIONS OF A FLEXIBLE CELL SURFACE

Many fossil prokaryotes look like rods, and we presume that these ancient organisms, like most present-day prokaryotic cells, had firm cell walls. The first step toward the eukaryotic condition was the loss of the cell wall by an ancestral prokaryotic cell. This wall-less condition is present in some present-day bacteria, although many others have developed new types of cell walls. Let's consider the possibilities open to a flexible cell without a wall.

First, think of cell size. As a cell grows larger, its surface area-to-volume ratio decreases (see Figure 5.2). Unless the surface area can be increased, the cell volume will reach an upper limit. If the cell's surface is flexible, it can fold inward and elaborate itself, creating more surface area for gas and nutrient exchange (**Figure 27.2**).

With a surface flexible enough to allow infolding, the cell can exchange materials with its environment rapidly enough to sustain a larger volume and more rapid metabolism (Figure 27.2, steps 1–2). Furthermore, a flexible surface can pinch off bits of the environment, bringing them into the cell by endocytosis.

CHANGES IN CELL STRUCTURE AND FUNCTION

Other early steps in the evolution of the eukaryotic cell are likely to have included three advances: the formation of ribosome-studded internal membranes, some of which surrounded the DNA; the appearance of a cytoskeleton; and the evolution of digestive vacuoles (Figure 27.2, steps 3–7).

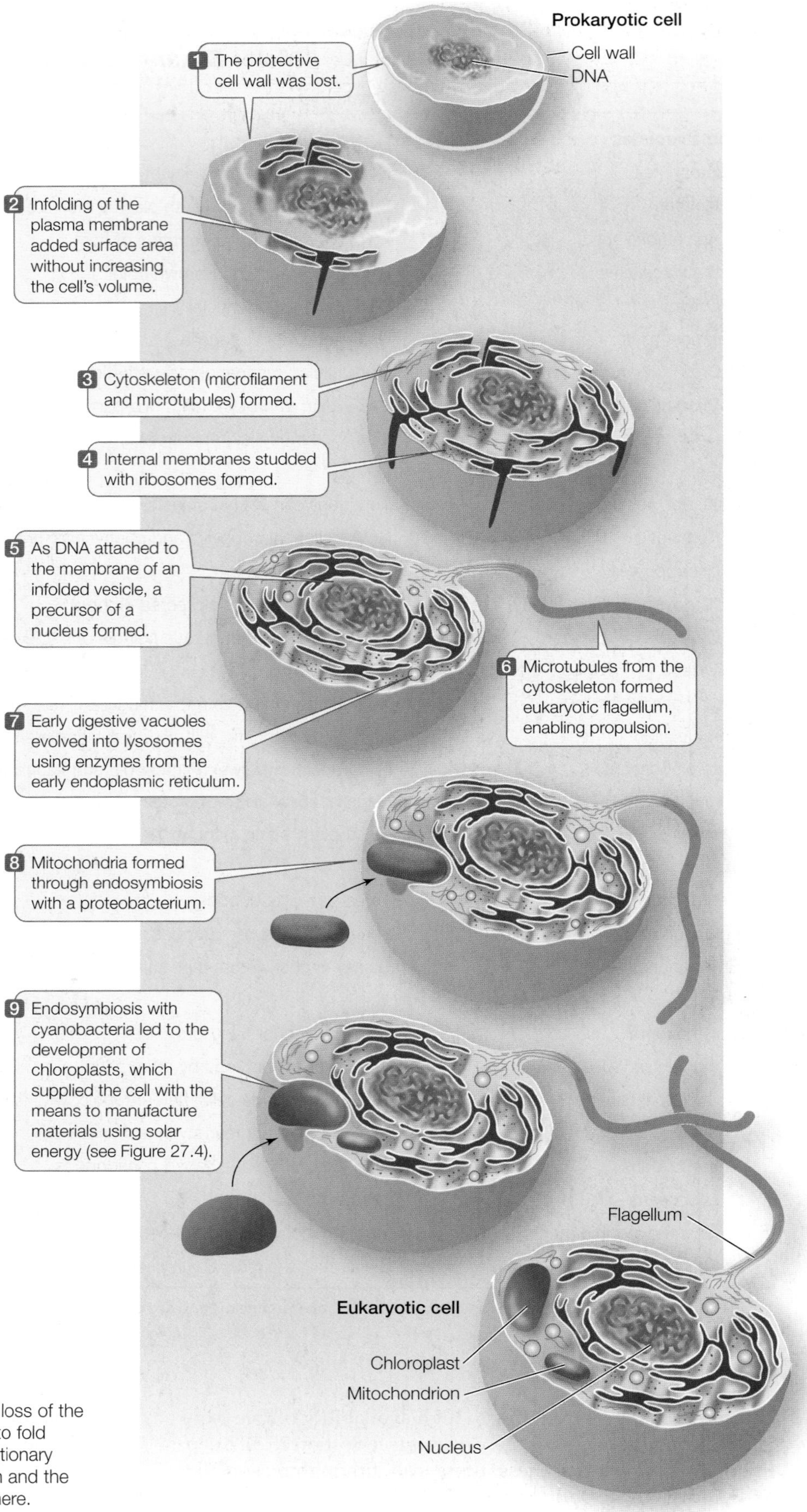

Prokaryotic cell

1 The protective cell wall was lost.

Cell wall
DNA

2 Infolding of the plasma membrane added surface area without increasing the cell's volume.

3 Cytoskeleton (microfilament and microtubules) formed.

4 Internal membranes studded with ribosomes formed.

5 As DNA attached to the membrane of an infolded vesicle, a precursor of a nucleus formed.

6 Microtubules from the cytoskeleton formed eukaryotic flagellum, enabling propulsion.

7 Early digestive vacuoles evolved into lysosomes using enzymes from the early endoplasmic reticulum.

8 Mitochondria formed through endosymbiosis with a proteobacterium.

9 Endosymbiosis with cyanobacteria led to the development of chloroplasts, which supplied the cell with the means to manufacture materials using solar energy (see Figure 27.4).

Flagellum

Eukaryotic cell

Chloroplast
Mitochondrion
Nucleus

27.2 From Prokaryotic Cell to Eukaryotic Cell   The loss of the rigid prokaryotic cell wall allowed the plasma membrane to fold inward and create more surface area. One possible evolutionary sequence, which includes the formation of a cytoskeleton and the enclosure of genetic material into the nucleus, is shown here.

A cytoskeleton composed of microfilaments and microtubules would support the cell and allow it to manage changes in shape, to distribute daughter chromosomes, and to move materials from one part of the now much larger cell to other parts. The presence of microtubules in the cytoskeleton could have evolved in some cells to give rise to the characteristic eukaryotic flagellum. The origin of the cytoskeleton is becoming clearer, as homologs of the genes that encode many cytoskeletal proteins have been found in modern prokaryotes.

The DNA of a prokaryotic cell is attached to a site on its plasma membrane. If that region of the plasma membrane were to fold into the cell, the first step would be taken toward the evolution of a *nucleus*, a primary feature of the eukaryotic cell.

From an intermediate kind of cell, the next step was probably *phagocytosis*—the ability to engulf and digest other cells. The cytoskeleton and nuclear envelope appeared early in the eukaryote lineage. Early eukaryotes may also have had an associated endoplasmic reticulum and Golgi apparatus, and perhaps one or more flagella of the eukaryotic type.

ENDOSYMBIOSIS AND ORGANELLES At the same time the processes outlined above were taking place, cyanobacteria were generating oxygen gas as a product of photosynthesis. The increasing O_2 levels in the atmosphere had disastrous consequences because most organisms of the time (archaea and bacteria) were unable to tolerate the newly oxidizing environment. But some prokaryotes managed to cope with these changes, and—fortunately for us—so did some of the new phagocytic eukaryotes.

At about this time, endosymbiosis might have come into play (Figure 27.2, steps 8–9). Recall that the theory of endosymbiosis proposes that certain organelles are the descendants of prokaryotes engulfed, but not digested, by ancient eukaryotic cells (see Section 5.5). One crucial endosymbiotic event in the history of the Eukarya was the incorporation of a proteobacterium that evolved into the mitochondrion. Initially, the new organelle's primary function was probably to detoxify O_2 by reducing it to water. Later, this reduction became coupled with the formation of ATP in cellular respiration (see Chapter 9). Upon completion of this step, the essential modern eukaryotic cell was complete.

Some important eukaryotes are the result of yet another endosymbiotic step, the incorporation of a prokaryote related to today's cyanobacteria, which became the chloroplast.

Chloroplasts are a study in endosymbiosis

Eukaryotes in several different groups possess chloroplasts, and groups with chloroplasts appear in several distantly related clades. Some of these groups differ in the photosynthetic pigments their chloroplasts contain. And we'll see that not all chloroplasts have a pair of surrounding membranes—in some microbial eukaryotes, they are surrounded by *three or more* membranes. We now view these observations in terms of a remarkable series of endosymbioses, supported by extensive evidence from electron microscopy and nucleic acid sequence comparisons.

All chloroplasts trace their ancestry back to the engulfment of one cyanobacterium by a larger eukaryotic cell (**Figure 27.3A**).

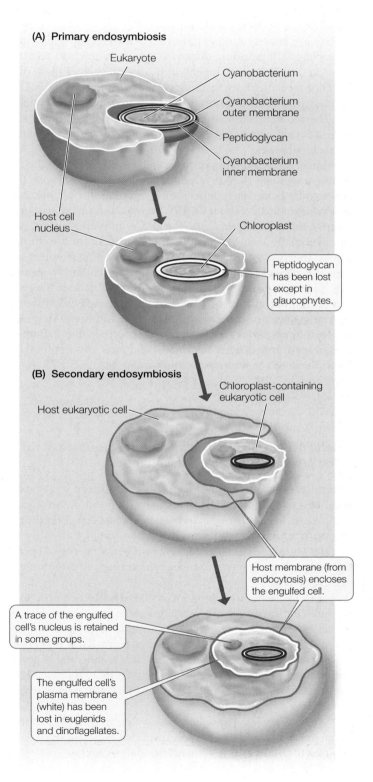

(A) Primary endosymbiosis

Eukaryote

Cyanobacterium

Cyanobacterium outer membrane

Peptidoglycan

Cyanobacterium inner membrane

Host cell nucleus

Chloroplast

Peptidoglycan has been lost except in glaucophytes.

(B) Secondary endosymbiosis

Chloroplast-containing eukaryotic cell

Host eukaryotic cell

Host membrane (from endocytosis) encloses the engulfed cell.

A trace of the engulfed cell's nucleus is retained in some groups.

The engulfed cell's plasma membrane (white) has been lost in euglenids and dinoflagellates.

27.3 Endosymbiotic Events in the "Family Tree" of Chloroplasts (A) A single instance of primary endosymbiosis ultimately gave rise to all of today's chloroplasts. A eukaryotic cell engulfed a cyanobacterium but did not digest it. (B) Secondary endosymbiosis—the uptake and retention of a chloroplast-containing cell by another eukaryotic cell—took place several times, independent of each other.

This event, the step that gave rise to the photosynthetic eukaryotes, is known as **primary endosymbiosis**. The cyanobacterium, a Gram-negative bacterium, had both an inner and outer membrane. Thus the original chloroplasts had two surrounding membranes—the inner and outer membranes of the cyanobacterium. Remnants of the peptidoglycan-containing wall of the bacterium are present in the form of a bit of peptidoglycan between the chloroplast membranes of *glaucophytes*, the first microbial eukaryote group to branch off following primary endosymbiosis of the cyanobacterium.

Primary endosymbiosis gave rise to the chloroplasts of the "green algae" (including *chlorophytes* and *charophytes*) and the *red algae*. Studies of phylogeny indicate that each of these distinct lineages trace back to a single primary endosymbiosis. The photosynthetic land plants arose later from a green algal ancestor. The red algal chloroplast retains certain pigments of the original cyanobacterial endosymbiont that are absent in green algal chloroplasts.

Almost all remaining photosynthetic eukaryotes are the result of additional rounds of endosymbiosis. For example, the photosynthetic *euglenids* derived their chloroplasts from **secondary endosymbiosis (Figure 27.3B)**. Their ancestor took up a unicellular chlorophyte, retaining the endosymbiont's chloroplast and eventually losing the rest of the constituents of the chlorophyte. This history explains why the photosynthetic euglenids have the same photosynthetic pigments as the chlorophytes and land plants. It also accounts for the third membrane of the euglenoid chloroplast, which is derived from the euglenid's plasma membrane (as a result of endocytosis). An additional round of endosymbiosis (**tertiary endosymbiosis**) occurred when a dinoflagellate apparently lost its chloroplast and took up a haptophyte (itself the result of secondary endosymbiosis).

─── **yourBioPortal.com** ───
GO TO **Animated Tutorial 27.1 • Family Tree of Chloroplasts**

Lateral gene transfer accounts for the presence of some prokaryotic genes in eukaryotes

Several uncertainties remain about the origins of eukaryotic cells. Lateral gene transfer (see Section 26.4 and Figure 26.10) complicates the study of eukaryote origins, just as it complicates the study of relationships among prokaryotes. Lateral gene transfer accounts for the increasing numbers of genes of bacterial origin that are being found in eukaryotes by ongoing genetic analyses.

An endosymbiotic origin of mitochondria and chloroplasts accounts for the presence of bacterial genes encoding enzymes for energy metabolism (respiration and photosynthesis) in eukaryotes, but it does not explain the presence of some other bacterial genes. The eukaryotic genome clearly is a mixture of genes with different origins. A recent suggestion is that Eukarya might have arisen from the mutualistic fusion of a Gram-negative bacterium and an archaean.

Many interesting ideas about eukaryotic origins await additional data and analysis. We can expect that these questions and others will eventually yield to additional research.

27.1 RECAP

The modern eukaryotic cell likely arose from an ancestral prokaryote in several steps, including endosymbiosis. Primary endosymbiosis involves the engulfment of a bacterium by another free-living organism; the endosymbiosis of a bacterium probably gave rise to mitochondria, whereas the chloroplasts of photosynthesizing eukaryotes are thought to have originated in the endosymbiosis of a cyanobacterium.

- Explain why protists are described as paraphyletic rather than monophyletic. See p. 561, Table 27.1, and Figure 27.1

- Why was the development of a flexible cell surface a key event in eukaryotic cell history? See p. 564

- Identify some of the probable events involved in the evolution of the eukaryotic cell from a prokaryotic cell. See pp. 564–565 and Figure 27.2

- What is the difference between primary and secondary endosymbiosis? See p. 566 and Figure 27.3

Having considered some of the known and suspected steps that led from the prokaryotic to the eukaryotic condition, let's now see what use the early eukaryotes made of their new features.

27.2 What Features Account for Protist Diversity?

The eukaryotic cell possesses some very useful features (detailed in Chapter 5). The cytoskeleton allows for various means of locomotion and also manages the controlled movement of cellular constituents (notably the mitotic and meiotic chromosomes). The specialized organelles of eukaryotes support a variety of activities. Given these tools, eukaryotes have been able to explore many environments and have exploited a variety of nutrient sources.

Protists occupy many different niches

Most protists are aquatic. Some live in marine environments, others in fresh water, and still others in the body fluids of host organisms. Many aquatic protists are plankton: free-floating aquatic organisms. The *slime molds* inhabit damp soil, animal feces, and the moist, decaying bark of rotting trees. Other microbial protists also live in soil water, and some of them contribute to the global nitrogen cycle by preying on soil bacteria and recycling their nitrogen compounds into nitrates.

Many metabolic lifestyles are found among the protists. Some protists are photosynthetic autotrophs, some are heterotrophs, and some switch with ease between the autotrophic and heterotrophic modes of nutrition. Some of the heterotrophs ingest their food; others, including many parasites, absorb nutrients from their environment.

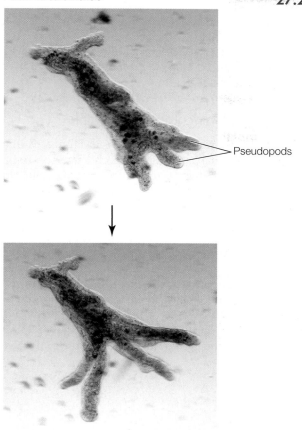

Pseudopods

120 μm

27.4 An Amoeba in Motion Its flowing pseudopods are constantly changing shape as this "chaos amoeba" moves and feeds.

Two general terms are sometimes used to designate two broad categories of protists: the *protozoans* and *algae*. These are convenience terms used to describe categories of distantly related species that have some similar attributes. The term *protozoans* refers to various groups of protists, formerly classified as animals, that often survive by ingesting other species. Likewise, the term *algae* (singular *alga*) refers to several groups of photosynthetic protists. Neither of these terms designate formal taxonomic groups, however.

Protists have diverse means of locomotion

Although a few protist groups consist entirely of nonmotile organisms, most groups include organisms that move by amoeboid motion, by ciliary action, or by means of flagella. Each of these types of motion is based on activities of the cytoskeleton.

In *amoeboid motion*, the cell forms **pseudopods** ("false feet") which are extensions of its constantly changing cell shape. Cells such as the one shown in **Figure 27.4** simply extend a pseudopod and then flow into it. Regions of the cytoplasm alternate between a more liquid state and a stiffer state,

and a network of cytoskeletal microfilaments squeezes the more liquid cytoplasm forward.

The proteins of the eukaryotic cytoskeleton form microtubules which allowed the evolution of different means of locomotion. *Cilia* are tiny, hairlike organelles that beat in a coordinated fashion to move the cell forward or backward. Some ciliated organisms can change direction rapidly in response to their environment. A eukaryotic flagellum moves like a whip; some flagella *push* the cell forward, others *pull* the cell forward. Cilia and eukaryotic flagella are identical in cross section, with a "9 + 2" arrangement of microtubules (see Figure 5.20); they differ only in length.

Protists employ vacuoles in several ways

Most unicellular organisms are microscopic. As noted above, an important reason that cells are small is that they need enough membrane surface area in relation to their volume to support the exchange of materials required for their existence. Some relatively large unicellular eukaryotes minimize this problem by having membrane-enclosed vacuoles of various types that increase their effective surface area.

Organisms living in fresh water are hypertonic to their environment (see Section 6.3). Many freshwater protists such as *Paramecium* address this problem by means of specialized vacuoles that excrete the excess water they constantly take in by osmosis. Members of several groups have such **contractile vacuoles**. The excess water collects in the contractile vacuoles, which then expel the water from the cell (**Figure 27.5**).

A **digestive vacuole** is a second important type of vacuole found in *Paramecium* and many other protists. These organisms engulf solid food by endocytosis, forming a vacuole within which the food is digested. Smaller vesicles containing digested food pinch away from the digestive vacuole and enter the cy-

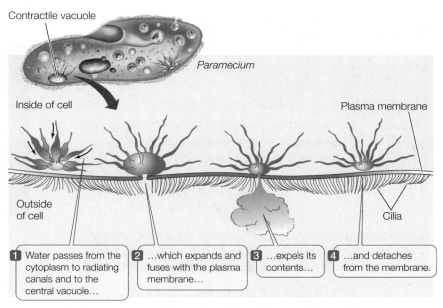

Contractile vacuole

Paramecium

Inside of cell

Plasma membrane

Outside of cell

Cilia

1 Water passes from the cytoplasm to radiating canals and to the central vacuole…

2 …which expands and fuses with the plasma membrane…

3 …expels its contents…

4 …and detaches from the membrane.

27.5 Contractile Vacuoles Bail Out Excess Water Contractile vacuoles remove the water that constantly enters freshwater protists by osmosis.

INVESTIGATING LIFE

27.6 The Role of Vacuoles in Ciliate Digestion

An experiment with the ciliate protist *Paramecium* demonstrates the function of food vacuoles. Given that an acidic environment is known to aid digestion in many organisms, does this microbial eukaryote use acid to obtain nutrients?

HYPOTHESIS The food vacuoles of *Paramecium* produce an acidic environment that allows the organism to digest food particles.

METHOD

1. Feed *Paramecium* yeast cells stained with Congo red, a dye that is red at neutral or basic pH but turns green at acidic pH.
2. Under a light microscope, observe the formation and degradation of food vacuoles within the *Paramecium*. Note time and sequence of color (i.e., acid level) changes.

RESULTS

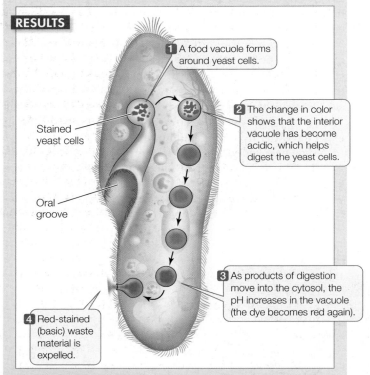

1 A food vacuole forms around yeast cells.

Stained yeast cells

2 The change in color shows that the interior vacuole has become acidic, which helps digest the yeast cells.

Oral groove

3 As products of digestion move into the cytosol, the pH increases in the vacuole (the dye becomes red again).

4 Red-stained (basic) waste material is expelled.

CONCLUSION Acidification of food vacuoles assists digestion in this ciliate protist.

FURTHER INVESTIGATION: How might you determine whether the acid level changes in *Paramecium's* food vacuoles are the result of enzymes?

Go to **yourBioPortal.com** for original citations, discussions, and relevant links for all INVESTIGATING LIFE figures.

──── **yourBioPortal.com** ────

GO TO **Animated Tutorial 27.2** • **Food Vacuoles Handle Digestion and Excretion**

toplasm. These tiny vesicles provide a large surface area across which the products of digestion may be absorbed by the rest of the cell (**Figure 27.6**).

The cell surfaces of protists are diverse

A few protists, such as the amoeba in Figure 27.4, are surrounded by only a plasma membrane, but most have stiffer surfaces that maintain the structural integrity of the cell. Many have cell walls, which are often complex in structure, outside the plasma membrane. Other protists that lack cell walls have a variety of ways of strengthening their surfaces.

Paramecium has proteins in its cell surface—known as a *pellicle* in this genus (see Figure 27.18)—that make it flexible but resilient. Other groups have external "shells," which the organism either produces itself or makes from bits of sand and thickenings immediately beneath the plasma membrane, as some amoebas do (**Figure 27.7A**). The complex cell walls of diatoms are glassy, based on silica (silicon dioxide; **Figure 27.7B**). Biologists recently measured, at a microscopic scale, the forces

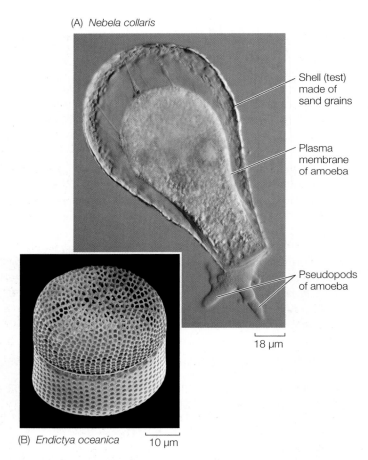

(A) *Nebela collaris*

Shell (test) made of sand grains

Plasma membrane of amoeba

Pseudopods of amoeba

18 μm

(B) *Endictya oceanica*   10 μm

27.7 Cell Surfaces among Microbial Eukaryotes (A) This testate amoeba has built a lightbulb-shaped shell, or test, by gluing sand grains together. Its pseudopods extend through the single aperture in the test (compare with Figure 27.4). (B) Scanning electron micrography reveals the intricate patterning of the silica-dense cell walls of this diatom. These spectacular unicellular, photosynthetic eukaryotes dominate the aquatic phytoplankton community (see also Figure 27.19).

needed to break single, living diatoms, and discovered that the glassy cell walls are exceptionally strong. Evolution of these walls by natural selection may have given diatoms an enhanced defense against predators, and thus an edge over competitors.

27.2 RECAP

Protists are diverse in their habitat, nutrition, locomotion, and body form. Some protists move by use of pseudopods, cilia, or flagella. Protists may have cell walls or external coverings that provide structural support and protection.

- Can you explain the roles of the cytoskeleton in the locomotion of protists? See p. 567

- Do you understand the operation of contractile and food vacuoles in *Paramecium*? See p. 567 and Figures 27.5 and 27.6

The diversity of body form, habitat, nutrition, and locomotion found among the protists reflects the diversity of avenues pursued during the early evolution of eukaryotes. Protists have an enormous effect on biotic and physical environments.

27.3 How Do Protists Affect the World Around Them?

Protists are extremely diverse, and their effects on other organisms and on the physical environment are almost as diverse. Some protists are food for marine animals, while others poison the sea; some are packaged as nutritional supplements, and some are pathogens; the remains of some form the sands of many modern beaches, and others are a major source of today's ever more expensive crude oil. Many protists are constituents of the plankton. Photosynthetic members of the plankton are called **phytoplankton**.

A single eukaryote clade, the *diatoms*, is responsible for about a fifth of all photosynthetic carbon fixation on Earth—about the same amount of photosynthesis performed by all of Earth's rainforests. These spectacular unicellular organisms (see Figures 27.7B and Figure 27.19) are the predominant members of the phytoplankton, but other protist clades also include important phytoplanktonic species that contribute heavily to global photosynthesis. Like green plants on land, the phytoplankton serve as a gateway for energy from the sun into the living world; in other words, they are *primary producers*. In turn, they are eaten by heterotrophs, including animals and many other protists. Those consumers are, in turn, eaten by other consumers. Most aquatic heterotrophs (with the exception of some species existing in the deep ocean) depend on the photosynthesis performed by phytoplankton.

Some protists are endosymbionts

As we have described, endosymbiosis is the condition in which two organisms live together, one inside the other. Endosymbiosis is common among the microbial protists, many of which live within the cells of animals. Members of the *dinoflagellates* are common symbionts in both animals and in other protists; most but not all dinoflagellate endosymbiont species are photosynthetic. Some dinoflagellates live endosymbiotically in the cells of corals, contributing products of their photosynthesis to the partnership. The importance to the coral is demonstrated when the dinoflagellates die as a result of changing environmental conditions, a phenomenon known as coral bleaching (**Figure 27.8A**); the coral is ultimately damaged or destroyed when its nutrient supply is reduced.

Many *radiolarians* also harbor photosynthetic endosymbionts (**Figure 27.8B**). As a result, the radiolarians, which are not photosynthetic themselves, appear greenish or golden, depending on the type of endosymbiont they contain. This arrangement is often mutually beneficial: the radiolarian can make use of the organic

(A)

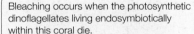
Bleaching occurs when the photosynthetic dinoflagellates living endosymbiotically within this coral die.

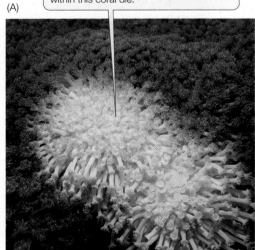
Goniopora sp.

(B)

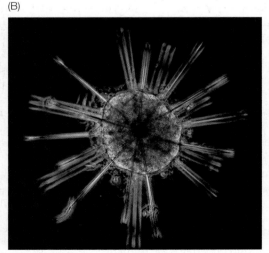
Astrolithium sp.

250 μm

27.8 Dinoflagellate Endosymbionts are Photosynthesizers (A) Some corals lose their chief nutritional source when their photosynthetic endosymbionts die, often as a result of changing environmental conditions such as warming water. (B) Dinoflagellates live endosymbiotically inside a radiolarian (another protist), providing organic nutrients for the radiolarian and imparting the golden brown pigmentation seen at the center of its glassy skeleton.

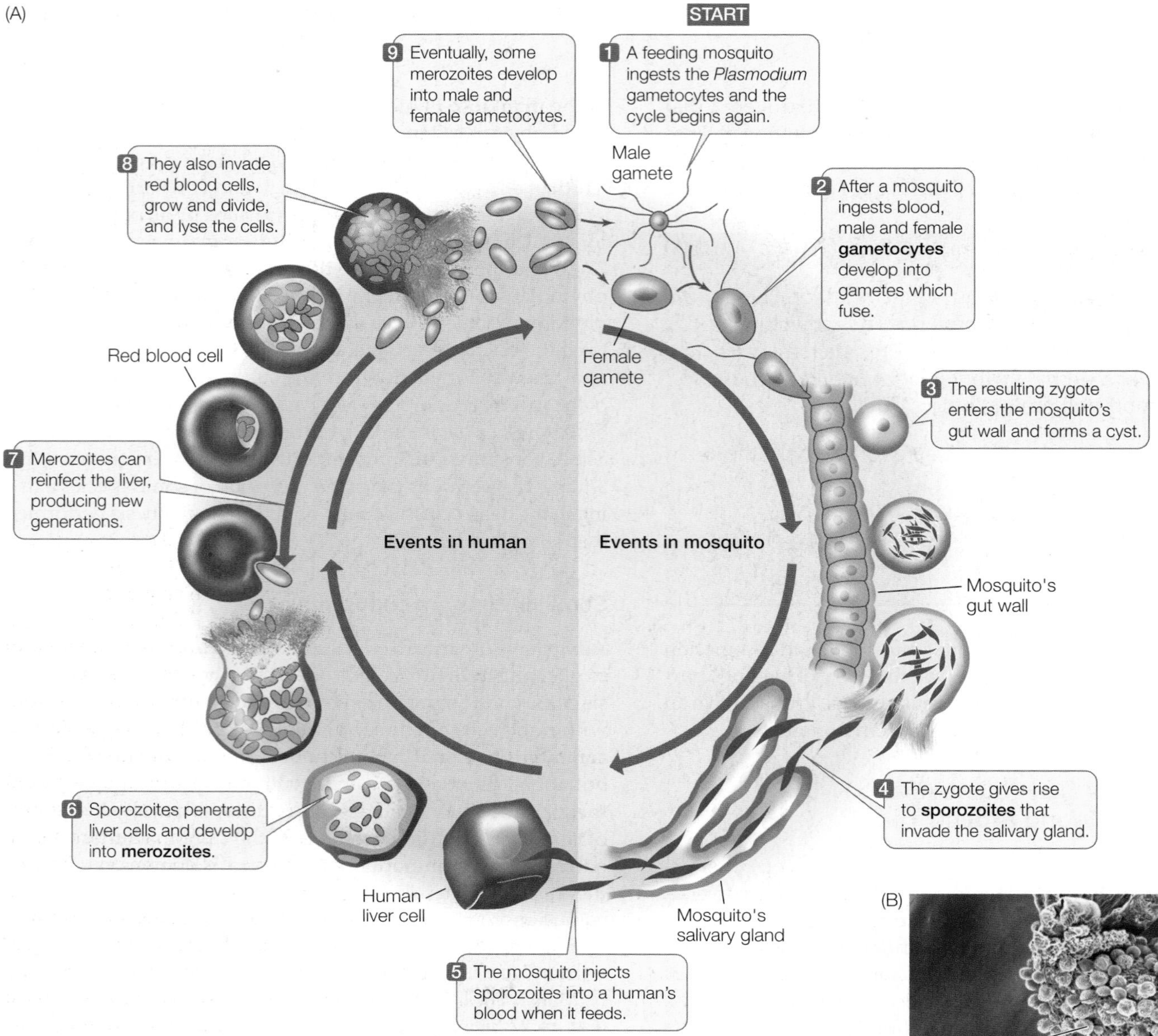

(A)

START

9 Eventually, some merozoites develop into male and female gametocytes.

1 A feeding mosquito ingests the *Plasmodium* gametocytes and the cycle begins again.

8 They also invade red blood cells, grow and divide, and lyse the cells.

Male gamete

2 After a mosquito ingests blood, male and female **gametocytes** develop into gametes which fuse.

Red blood cell

Female gamete

3 The resulting zygote enters the mosquito's gut wall and forms a cyst.

7 Merozoites can reinfect the liver, producing new generations.

Events in human

Events in mosquito

Mosquito's gut wall

6 Sporozoites penetrate liver cells and develop into **merozoites**.

4 The zygote gives rise to **sporozoites** that invade the salivary gland.

Human liver cell

Mosquito's salivary gland

5 The mosquito injects sporozoites into a human's blood when it feeds.

(B)

Cysts

Mosquito's gut wall

170 μm

nutrients produced by its photosynthetic guest, and the guest may in turn make use of metabolites made by the host or receive physical protection. In some cases, the guest is exploited for its photosynthetic products while receiving little or no benefit itself.

Some microbial protists are deadly

The best-known pathogenic microbial protists are members of the genus *Plasmodium*, a highly specialized group of *apicomplexans* that spend part of their life cycle as parasites in human red blood cells, where they are the cause of malaria (**Figure 27.9**). In terms of the number of people affected, malaria is one of the world's three most serious infectious diseases; it kills about 880,000 people each year, out of 250 million infected individuals. On average, someone dies from malaria every 36 seconds—usually in sub-Saharan Africa, although malaria occurs in more than 100 countries.

Female mosquitoes of the genus *Anopheles* transmit *Plasmodium* to humans. The parasite enters the human circulatory system when an infected *Anopheles* mosquito penetrates the human skin in search of blood. The parasites find their way to cells in

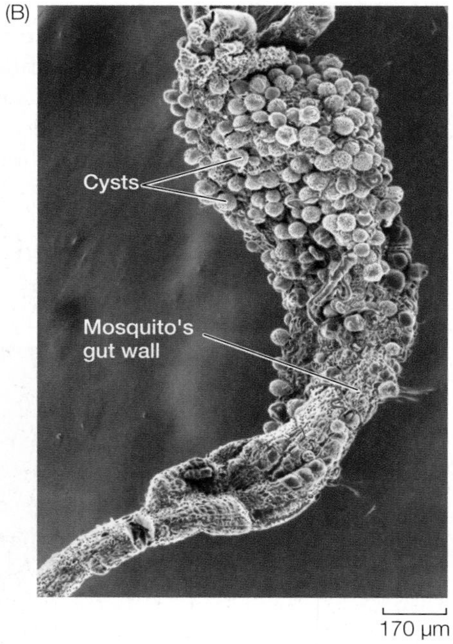

27.9 Life Cycle of the Malarial Parasite (A) Like many parasitic species, the apicomplexan *Plasmodium falciparum* has a complex life cycle, part of which is spent in mosquitoes of the genus *Anopheles* and part in humans. The sexual phase (gamete fusion) of this life cycle takes place in the insect, and the zygote is the only diploid stage. (B) Encysted *Plasmodium* zygotes (artificially colored blue) cover the stomach wall of a mosquito. Invasive sporozoites will hatch from the cysts and be transmitted to a human, in whom the parasite causes malaria.

(B)

A coccolithophore's scales reflect sunlight.

(A)

Emiliania huxleyi

0.9 µm

27.10 Chromalveolates Can Bloom in the Oceans (A) By reproducing in astronomical numbers, the dinoflagellate *Gonyaulax tamarensis* can cause toxic red tides, such as this one along the coast of Baja California. (B) Massive blooms of this coccolithophore, a tiny haptophyte, can reduce the amount of sunlight able to penetrate to the waters below.

trypanosomes cause sleeping sickness, leishmaniasis, and Chagas' disease. The genomes of all three of the trypanosomes responsible for these diseases have now been sequenced.

Some *chromalveolates*, including diatoms, dinoflagellates, and haptophytes, reproduce in enormous numbers in warm and somewhat stagnant waters. The result can be a "red tide," so called because of the reddish color of the sea that results from the dinoflagellates' pigments (**Figure 27.10A**). During a dinoflagellate red tide, the concentration of cells may reach 60 million per liter of ocean water. Some red tide species produce a potent nerve toxin that can kill tons of fish. The genus *Gonyaulax* produces a toxin that can accumulate in shellfish in amounts that, although not fatal to the shellfish, may kill a person who eats the shellfish.

The haptophyte *Emiliania huxleyi* is one of the smallest unicellular protists, but it can form tremendous blooms in ocean waters. This *coccolithophore* ("sphere of stone") has an armored coating that makes the surface water more reflective (**Figure 27.10B**). This reflectivity cools the deeper layers of water below the bloom by reducing the amount of sunlight that penetrates. At the same time, it is possible that *E. huxleyi* contributes to global warming, because its metabolism increases the amount of dissolved CO_2 in ocean waters.

We continue to rely on the products of ancient marine protists

Diatoms are often lovely to look at, but their importance to us goes far beyond aesthetics. They store oil as an energy reserve and to help them float at the correct depth in the ocean. Over millions of years, diatoms have died and sunk to the ocean floor, ultimately undergoing chemical changes and becoming a major source of petroleum and natural gas, two of our most important energy supplies and political concerns.

Other marine protists have also contributed to today's world. Some *foraminiferans*, for example, secrete shells of calcium carbonate. After they reproduce (by mitosis and cytokinesis), the daughter cells abandon the parent shell and make new shells of their own. The discarded shells of ancient foraminiferans make up extensive limestone deposits in various parts of the world, forming a layer hundreds to thousands of meters deep over mil-

the liver and the lymphatic system, change their form, multiply, and reenter the bloodstream, attacking red blood cells.

The parasites multiply inside the red blood cells, which then burst, releasing new swarms of parasites. If another *Anopheles* bites the victim, the mosquito takes in *Plasmodium* cells along with blood. Some of the ingested cells are gametes that formed in human cells. The gametes unite in the mosquito, forming zygotes that lodge in the mosquito's gut, divide several times, and move into its salivary glands, from which they can be passed on to another human host. Thus *Plasmodium* is an extracellular parasite in the mosquito vector and an intracellular parasite in the human host.

Plasmodium has proved to be a singularly difficult pathogen to attack. The complex *Plasmodium* life cycle is best broken by the removal of stagnant water, in which mosquitoes breed. Using insecticides to reduce the *Anopheles* population can be effective, but the benefits must be weighed against the ecological, economic, and health risks posed by the insecticides themselves.

The genomes of one malarial parasite, *Plasmodium falciparum*, and one of the mosquitoes that transmits malaria, *Anopheles gambiae*, have been sequenced. These advances should lead to a better understanding of the biology of malaria and to the development of drugs, vaccines, or other means of dealing with this pathogen or its insect vectors. The opening story of Chapter 30 describes a novel weapon against malaria: mosquito netting containing fungi that attack mosquitoes.

Some *kinetoplastids* are human pathogens, such as the trypanosomes discussed at the opening of this chapter. Recall that

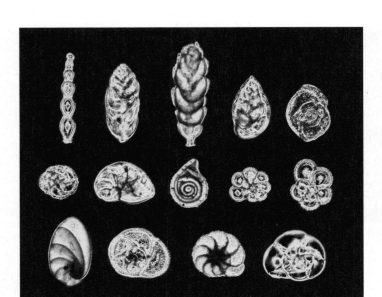

250 μm

27.11 Foraminiferan Shells Are Building Blocks   Some foraminiferan shells are made of calcium carbonate that has been mineralized on an organic matrix external to the cell. Over millions of years, the remains of foraminiferans have formed limestone deposits and sandy beaches. Several species are shown in this micrograph.

27.4 How Do Protists Reproduce?

Although most protists engage in both asexual and sexual reproduction, sexual reproduction has yet to be observed in some groups. As we will see, some protists separate the acts of sex and reproduction, so that the two are not directly linked.

Asexual reproductive processes found among the protists include:

- *Binary fission:* Equal splitting of one cell into two, with mitosis followed by cytokinesis
- *Multiple fission:* Splitting of one cell into multiple (i.e., more than two) cells
- *Budding:* The outgrowth of a new cell from the surface of an old one
- *Spores:* The formation of specialized cells that are capable of developing into new organisms

Sexual reproduction also occurs among the protists, and it takes various forms. In some protists, as in animals, the gametes are the only haploid cells. In others, the zygote is the only diploid cell. In still others, both diploid and haploid cells undergo mitosis, giving rise to *alternation of generations* (the alternation of multicellular diploid and haploid life stages).

Some protists have reproduction without sex, and sex without reproduction

Members of the genus *Paramecium* are *ciliates,* a protist group characterized by the possession of two types of nuclei in a single cell—commonly one *macronucleus* and from one to several *micronuclei.* The micronuclei, which are typical eukaryotic nuclei, are essential for genetic recombination. The macronucleus is derived from micronuclei. Each macronucleus contains many copies of the genetic information, packaged in units containing very few genes each. The macronuclear DNA is transcribed and translated to regulate the life of the cell. In asexual reproduction, all of the nuclei are copied before the cell divides.

Paramecia also have an elaborate sexual behavior called **conjugation**, in which two paramecia line up tightly against each other and fuse in the oral groove region of the body. Nuclear material is extensively reorganized and exchanged over the next several hours (**Figure 27.12**). Each cell ends up with two haploid micronuclei, one of its own and one from the other cell, which fuse to form a new diploid micronucleus. A new macronucleus develops from the micronucleus through a series of dramatic chromosomal rearrangements. The exchange of nuclei is fully reciprocal—each of the two paramecia gives and receives an equal amount of DNA. The two organisms then separate and go their own ways, each equipped with new combinations of alleles.

lions of square kilometers of ocean bottom. Foraminiferan shells also make up much of the sand of some beaches. A single gram of such sand may contain as many as 50,000 foraminiferan shells and shell fragments.

The shells of individual foraminiferans are easily preserved as fossils in marine sediments. The shells of foraminiferan species have distinctive shapes (**Figure 27.11**), and each geological period has a distinctive assemblage of foraminiferan species. For this reason, and because they are so abundant, the remains of foraminiferans are especially valuable in classifying and dating sedimentary rocks, as well as in oil prospecting. Analyses of foraminiferan shells are also used in determining the global temperatures prevalent at the time of their existence.

27.3 RECAP

Protists have many effects, both positive and negative, on other organisms and on global ecosystems. Some species are primary producers, many are endosymbionts, and some are pathogens. They are among the most important producers of petroleum products, and they are important for producing and dating sedimentary rock formations.

- Can you describe the role of female mosquitoes of the genus *Anopheles* in the transmission of malaria? **See pp. 570–571 and Figure 27.9**

- Do you understand the role of dinoflagellates in the very different phenomena of coral bleaching and red tides? **See pp. 569 and 571**

This section has presented a brief overview of the many ways protists interact with one another and with other species. Next we examine their diverse forms of reproduction.

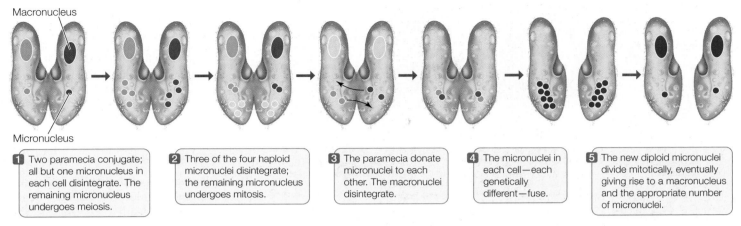

1 Two paramecia conjugate; all but one micronucleus in each cell disintegrate. The remaining micronucleus undergoes meiosis.

2 Three of the four haploid micronuclei disintegrate; the remaining micronucleus undergoes mitosis.

3 The paramecia donate micronuclei to each other. The macronuclei disintegrate.

4 The micronuclei in each cell—each genetically different—fuse.

5 The new diploid micronuclei divide mitotically, eventually giving rise to a macronucleus and the appropriate number of micronuclei.

27.12 Paramecia Achieve Genetic Recombination by Conjugating
The exchange of micronuclei by two conjugating *Paramecium* individuals results in genetic recombination. After conjugation, the cells separate and continue their lives as two individuals.

Conjugation in *Paramecium* is a *sexual* process of genetic recombination, but it is not a *reproductive* process. Two cells begin the process and two cells are there at the end, so no new cells are created. As a rule, each asexual clone of paramecia must periodically conjugate. Experiments have shown that if some species are not permitted to conjugate, the clones can live through no more than approximately 350 cell divisions before they die out.

Some protist life cycles feature alternation of generations

Alternation of generations is a type of life cycle found in many multicellular protists, land plants, and some fungi. The term describes a life cycle in which a multicellular, diploid, spore-producing organism gives rise to a multicellular, haploid, gamete-producing organism. When two haploid gametes fuse (*fertilization*, or *syngamy*), a diploid organism is formed (**Figure 27.13**). The haploid organism, the diploid organism, or both may also reproduce asexually.

The two alternating generations (spore-producing and gamete-producing) differ genetically (one has diploid cells, the other haploid cells), but they may or may not differ morphologically. In **heteromorphic** alternation of generations, the two generations differ morphologically; in **isomorphic** alternation of generations, they do not, despite their genetic difference.

Examples of both heteromorphic and isomorphic alternation of generations are found in both brown algae and green algae. As we discuss the life cycles of land plants and multicellular photosynthetic protists, we will use the terms **sporophyte** ("spore plant") and **gametophyte** ("gamete plant") to refer to the multicellular diploid and haploid generations, respectively.

Gametes are not produced by meiosis because the gametophyte generation is already haploid. Instead, specialized cells of the diploid sporophyte, called **sporocytes**, divide meiotically to produce four haploid spores. The spores may eventually germinate and divide mitotically to produce multicellular haploid gametophytes, which produce gametes by mitosis and cytokinesis.

Gametes, unlike spores, can produce new organisms only by fusing with other gametes. The fusion of two gametes produces a diploid zygote, which then undergoes mitotic divisions to produce a diploid organism: the sporophyte generation. The sporocytes of the sporophyte generation then undergo meiosis and produce haploid spores, starting the cycle anew.

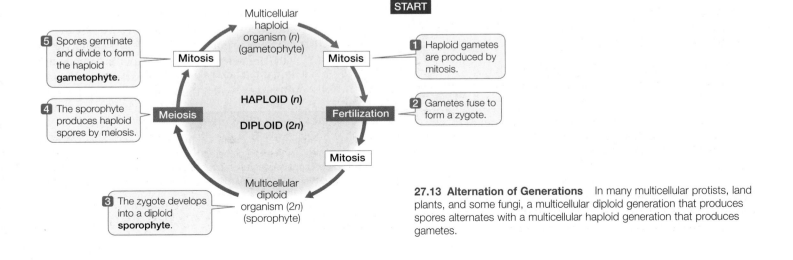

27.13 Alternation of Generations In many multicellular protists, land plants, and some fungi, a multicellular diploid generation that produces spores alternates with a multicellular haploid generation that produces gametes.

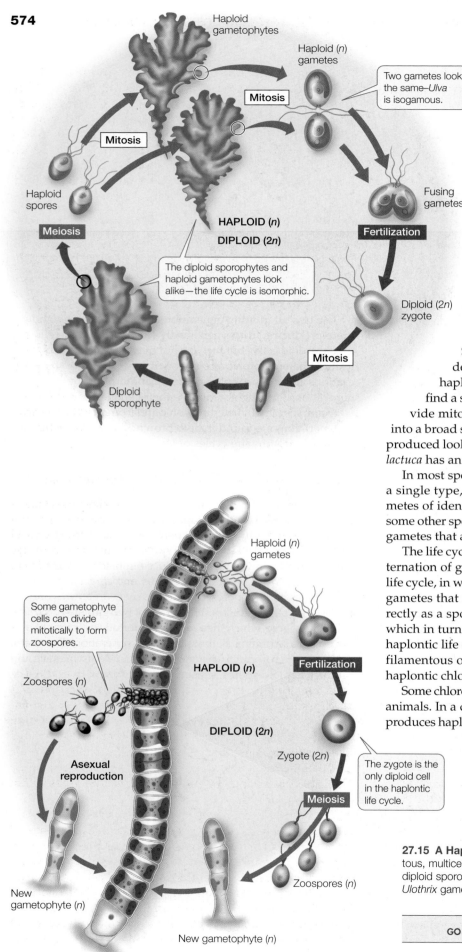

Haploid gametophytes

Haploid (*n*) gametes

Mitosis

Two gametes look the same—*Ulva* is isogamous.

Mitosis

Haploid spores

Fusing gametes

HAPLOID (*n*)

DIPLOID (2*n*)

Fertilization

Meiosis

The diploid sporophytes and haploid gametophytes look alike—the life cycle is isomorphic.

Diploid (2*n*) zygote

Diploid sporophyte

Mitosis

Some gametophyte cells can divide mitotically to form zoospores.

Haploid (*n*) gametes

Zoospores (*n*)

HAPLOID (*n*)

Fertilization

Asexual reproduction

DIPLOID (2*n*)

Zygote (2*n*)

The zygote is the only diploid cell in the haplontic life cycle.

Meiosis

New gametophyte (*n*)

Zoospores (*n*)

New gametophyte (*n*)

27.14 An Isomorphic Life Cycle The sexual life cycle of sea lettuce (*Ulva lactuca*) is an example of isomorphic alternation of generations.

yourBioPortal.com
GO TO Web Activity 27.1 • An Isomorphic Life Cycle

Chlorophytes provide examples of several life cycles

The major features of protist life cycles can all be seen in the chlorophytes. Let's begin with the common sea lettuce, *Ulva lactuca*. Like many chlorophytes, sea lettuce exhibits alternation of generations. The diploid sporophyte of this common multicellular seashore organism is a broad sheet only two cells thick. Some of its cells (sporocytes) differentiate and undergo meiosis and cytokinesis, producing motile haploid spores. These *zoospores* swim away, and some find a suitable place to settle. The zoospores begin to divide mitotically, producing a thin filament that develops into a broad sheet only two cells thick. The gametophyte thus produced looks just like the sporophyte—in other words, *Ulva lactuca* has an isomorphic life cycle (**Figure 27.14**).

In most species of *Ulva*, the gametes are structurally of just a single type, making those species **isogamous**—having gametes of identical appearance. Other chlorophytes, including some other species of *Ulva*, are **anisogamous**—they have female gametes that are distinctly larger than the male gametes.

The life cycles of many other chlorophytes do not feature alternation of generations. Some chlorophytes have a **haplontic** life cycle, in which a multicellular haploid individual produces gametes that fuse to form a zygote. The zygote functions directly as a sporocyte, undergoing meiosis to produce spores, which in turn produce a new haploid individual. In the entire haplontic life cycle, only one cell—the zygote—is diploid. The filamentous organisms of the genus *Ulothrix* are examples of haplontic chlorophytes (**Figure 27.15**).

Some chlorophytes have a **diplontic** life cycle like that of many animals. In a diplontic life cycle, meiosis of diploid sporocytes produces haploid gametes directly; the gametes fuse, and the resulting diploid zygote divides mitotically to form a new multicellular diploid sporophyte. In such organisms, all cells except the gametes are diploid. Between these two extremes are chlorophytes in which the gametophyte and sporo-

27.15 A Haplontic Life Cycle In the life cycle of *Ulothrix*, a filamentous, multicellular haploid gametophyte generation alternates with a diploid sporophyte generation consisting of a single cell (the zygote). *Ulothrix* gametophytes can also reproduce asexually.

yourBioPortal.com
GO TO Web Activity 27.2 • A Haplontic Life Cycle

phyte generations are both multicellular, but one generation (usually the sporophyte) is much larger and more prominent than the other.

The life cycles of some protists require more than one host species

The trypanosome diseases discussed at the opening of this chapter share a striking feature with malaria: in each case, the eukaryote pathogen completes part of its life cycle in the human host and part in an insect (see Figure 27.9). Many other protist life cycles require the participation of two different host species.

What might be the advantage of a life cycle with two hosts? This remains an intriguing question. It may be relevant that in the human pathogens described above, the sexual phase of the organism's life cycle—the fusion of gametes into a zygote—takes place in the insect host. Could this imply that the human host is nothing but a copying machine for the products of sexual reproduction in the insect host?

27.4 RECAP

Protists reproduce both asexually and sexually, although sex may occur independently of reproduction in some species. Some multicellular protists exhibit alternation of generations, alternating between multicellular haploid and diploid life stages. Parasitic protists may have complex life cycles in which they infect more than one host species.

- Why is conjugation between paramecia considered a sexual process but not a reproductive process? See p. 573 and Figure 27.12

- Can you explain the difference between the diplontic human life cycle and a life cycle with alternation of generations? See pp. 573–574

The success of protists' diverse adaptations for nutrition, locomotion, and reproduction is evident from the abundance and diversity of eukaryotes living today. In the next section we survey that diversity.

27.5 What Are the Evolutionary Relationships among Eukaryotes?

Biologists used to classify the various groups of protists largely on the basis of life history and reproductive features. However, scientists using electron microscopy and gene sequencing have revealed many new patterns of evolutionary relatedness. Analyses of slowly evolving gene sequences are making it possible to explore evolutionary relationships among eukaryotes in ever greater detail and with greater confidence. Today we recognize great diversity among the many distantly related protist clades, whose members have explored a great variety of lifestyles.

Most eukaryotes can be classified in one of five major hypothesized clades: chromalveolates, Plantae, excavates, rhizaria,

or unikonts (see Figure 27.1 and Table 27.1). As we will see, some of these groups consist of organisms with enormously diverse body forms and nutritional lifestyles. The phylogenetic relationships among the major groups of eukaryotes comprise an active area of study, and new data from genomes are rapidly changing our understanding of the evolution of these species. These relationships also help us understand how the major multicellular eukaryotic groups (brown algae, plants, fungi, and animals) originated from the microbial eukaryotes.

CHROMALVEOLATES

We begin our tour of protist groups with the **chromalveolates**, a group of photosynthetic organisms, usually with cellulose in their cell walls, that includes the haptophytes, alveolates and stramenopiles. The monophyly of the chromalveolates is not yet well established. The **haptophytes** are unicellular organisms with flagella; many are "armored" with elaborate scales (see Figure 27.10B). The alveolates and stramenopiles are large and diverse clades that we explore here in greater detail.

Alveolates have sacs under their plasma membrane

The synapomorphy that characterizes the **alveolate** clade is the possession of sacs called *alveoli* just below their plasma membranes. The alveoli may play a role in supporting the cell surface. These organisms are all unicellular, but they are diverse in body form. The alveolate groups we consider in detail here are the dinoflagellates, apicomplexans, and ciliates.

DINOFLAGELLATES The **dinoflagellates** are of great ecological, evolutionary, and morphological interest. Most dinoflagellates are marine, and they are important primary photosynthetic producers of organic matter in the oceans. A distinctive mixture of photosynthetic and accessory pigments gives dinoflagellate chloroplasts a golden brown color. (Section 27.1 describes the endosymbiotic events that gave rise to dinoflagellates with different numbers of membranes surrounding their chloroplasts.) Some are photosynthetic endosymbionts living in the cells of other organisms, including invertebrate animals (such as corals) and other marine protists (see Figure 27.8). Some are nonphotosynthetic and live as parasites within other marine organisms.

Dinoflagellates have a distinctive appearance. They are unicellular and generally have two flagella, one in an equatorial groove around the cell, the other starting near the same point as the first and passing down a longitudinal groove before extending into the surrounding medium (**Figure 27.16**). Some dinoflagellates can take on different forms, including amoeboid ones, depending on environmental conditions. It has been claimed that the dinoflagellate *Pfiesteria piscicida* can occur in at least two dozen distinct forms, although this claim is highly con-

Peridinium sp.

Longitudinal groove

Equatorial groove

15 μm

27.16 A Dinoflagellate The dinoflagellates are an important group of alveolates. Most of them are photosynthetic and are a crucial component of the world's phytoplankton. They are often endosymbiotic (see Figure 27.8) and can be the agents of deadly ocean "blooms" (see Figure 27.10A).

troversial. In any case, this remarkable dinoflagellate is harmful to fish and can, when present in great numbers, both stun and feed on them.

APICOMPLEXANS The exclusively parasitic **apicomplexans** derive their name from the *apical complex*, a mass of organelles contained in the *apical* end (the tip) of a cell. These organelles help the apicomplexan invade its host's tissues. For example, the apical complex enables the merozoites and sporozoites of *Plasmodium*, the causative agent of malaria, to enter their target cells in the human body.

Like many obligate parasites, apicomplexans have elaborate life cycles featuring asexual and sexual reproduction by a series of very dissimilar life stages. Often these stages are associated with two different types of host organisms, as is the case with *Plasmodium*. The apicomplexan *Toxoplasma* alternates between cats and rats to complete its life cycle. A rat infected with *Toxoplasma* loses its fear of cats, making it more likely to be eaten by, and thus transfer the parasite to, a cat.

Apicomplexans lack contractile vacuoles. They contain a much-reduced chloroplast that no longer has a photosynthetic function (derived, like all chromalveolate chloroplasts, from secondary endosymbiosis of a red alga). This chloroplast might be a target for a future antimalarial drug.

CILIATES The **ciliates** are so named because they characteristically have numerous hairlike cilia shorter than, but otherwise identical to, eukaryotic flagella. This group is noteworthy for its diversity and ecological importance (**Figure 27.17**). Almost all ciliates are heterotrophic (although a few contain photosynthetic endosymbionts), and they are much more complex in body form than are most other unicellular eukaryotes. The definitive characteristic of ciliates is the possession of two types of nuclei (as seen in the paramecia in Figure 27.12).

Paramecium, a frequently studied ciliate genus, exemplifies the complex structure and behavior of ciliates (**Figure 27.18**). The slipper-shaped cell is covered by an elaborate pellicle, a structure composed principally of an outer membrane and an inner layer of closely packed, membrane-enclosed sacs (the alveoli) that surround the bases of the cilia. Defensive organelles

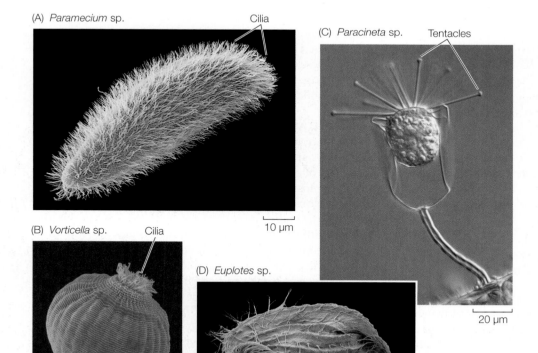

(A) *Paramecium* sp.   Cilia

(B) *Vorticella* sp.   Cilia

(C) *Paracineta* sp.   Tentacles

(D) *Euplotes* sp.

10 μm

20 μm

10 μm

Oral groove   Rows of fused cilia   25 μm

27.17 Diversity among the Ciliates (A) A free-swimming organism, this paramecium belongs to a ciliate group whose members have many cilia of uniform length. (B) Members of this group have cilia on their mouthparts. (C) In this group, cilia are replaced by tentacles as development proceeds. (D) Some of the cilia in *Euplotes* are grouped into flat sheets that sweep food particles into the oral groove.

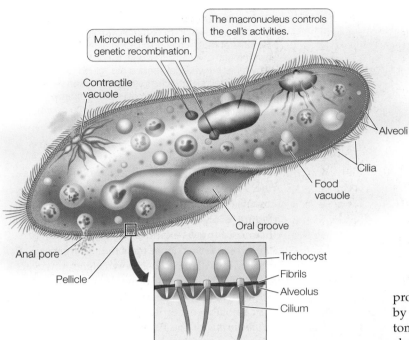

27.18 Anatomy of *Paramecium* This diagram shows the complex structure of a typical paramecium, detailing the pellicle (outer covering) with its trichocysts and alveoli.

───────── **yourBioPortal.com** ─────────
GO TO Web Activity 27.3 • Anatomy of *Paramecium*

called *trichocysts* are also present in the pellicle. In response to a threat, a microscopic explosion expels the trichocysts in a few milliseconds, and they emerge as sharp darts, driven forward at the tip of a long, expanding filament.

The cilia provide a form of locomotion that is generally more precise than locomotion by flagella or pseudopods. A paramecium can coordinate the beating of its cilia to propel itself either forward or backward in a spiraling manner. It can also back off swiftly when it encounters a barrier or a negative stimulus. The coordination of ciliary beating is probably the result of a differential distribution of ion channels in the plasma membrane near the two ends of the cell.

Stramenopiles have two unequal flagella, one with hairs

A morphological synapomorphy shared by most **stramenopiles** is the possession of rows of tubular hairs on the longer of their two flagella. Some stramenopiles lack flagella, but they are descended from ancestors that possessed flagella. The stramenopiles include the diatoms and the brown algae, which are photosynthetic, and the oomycetes and slime nets, which are not. Most golden algae are photosynthetic, but nearly all of them become heterotrophic when light intensity is limited or when there is a plentiful food supply; some even feed on diatoms or bacteria. The slime nets (not to be confused with slime molds) are unicellular organisms that produce networks of filaments along which the cells move.

DIATOMS All of the **diatoms** are unicellular, although some species associate in filaments. Many have sufficient carotenoids in their chloroplasts to give them a yellow or brownish color. All make carbohydrates and oils as photosynthetic storage products. Diatoms lack flagella except in male gametes.

Architectural magnificence on a microscopic scale is the hallmark of the diatoms. As mentioned earlier, almost all diatoms deposit silica (hydrated silicon dioxide) in their cell walls. The cell wall is constructed in two pieces, with the top overlapping the bottom like the top and bottom of a petri plate. The silica-impregnated walls have intricate patterns unique to each species (**Figure 27.19**). Despite their remarkable morphological diversity, all diatoms are symmetrical—either bilaterally (with "right" and "left" halves) or radially (with the type of symmetry possessed by a circle).

Diatoms reproduce both sexually and asexually. Asexual reproduction is by binary fission and is somewhat constrained by the stiff, silica-containing cell wall. Both the top and the bottom of the "petri plate" become tops of new "plates" without changing appreciably in size; as a result, the new cell made from the former bottom is smaller than the parent cell. If this process continued indefinitely, one cell line would simply vanish, but sexual reproduction largely solves this potential problem. Gametes are formed, shed their cell walls, and fuse. The resulting zygote then increases substantially in size before a new cell wall is laid down.

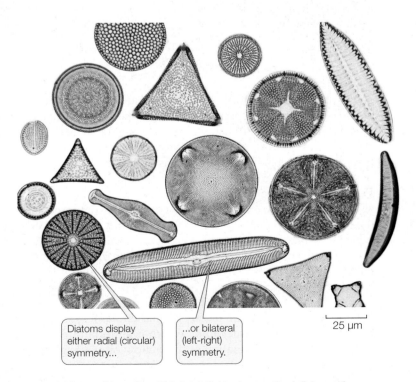

27.19 Diatom Diversity This brightfield micrograph of diatoms shows a variety of species-specific forms. Diatoms are photosynthesizers and dominant components of the world's phytoplankton.

(A) *Cystoseira usneoides*

(B) *Ectocarpus* sp.

60 µm

(C) *Postelsia palmaeformis*

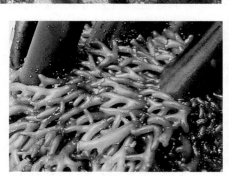

(D) *Postelsia palmaeformis*

27.20 Brown Algae (A) This seaweed illustrates the filamentous growth form of the brown algae. (B) Filaments of the microscopic brown alga *Ectocarpus* seen through a light microscope. (C) Sea palms and many other brown algal species are "glued" to the substratum by tough, branched structures called holdfasts. (D) Sea palms exemplify the leaflike form of brown algae. Holdfasts allow them to withstand the pounding of the surf.

The brown algae are almost exclusively marine. They are composed either of branched filaments (**Figure 27.20A,B**) or of leaflike growths (**Figure 27.20D**). Some float in the open ocean; the most famous example is the genus *Sargassum*, which forms dense mats in the Sargasso Sea in the mid-Atlantic. Most brown algae, however, are attached to rocks near the shore. A few thrive only where they are regularly exposed to heavy surf; a notable example is the sea palm *Postelsia palmaeformis* of the Pacific coast. All of the attached forms develop a specialized structure, called a *holdfast*, that literally glues them to the rocks (**Figure 27.20C**). The "glue" of the holdfast is *alginic acid*, a gummy polymer of sugar acids found in the walls of many brown algal cells. In addition to its function in holdfasts, alginic acid cements algal cells and filaments together, and is harvested and used by humans as an emulsifier in ice cream, cosmetics, and other products.

Some brown algae differentiate extensively into specialized organs. Some, like the sea palm, have stemlike stalks and leaflike blades. Some develop gas-filled cavities or bladders that serve as floats. In addition to organ differentiation, the larger brown algae also exhibit considerable tissue differentiation. Most of the giant kelps have photosynthetic filaments only in the outermost regions of their stalks and blades. Within the stalks and blades lie filaments of tubular cells that closely resemble the nutrient-conducting tissue of land plants. Called *trumpet cells* because they have flared ends, these tubes rapidly conduct the products of photosynthesis through the body of the organism.

Diatoms are found in all the oceans and are frequently present in great numbers, making them major photosynthetic producers in coastal waters. Diatoms are also common in fresh water and even occur on the wet surfaces of terrestrial mosses. They are also the dominant organisms in the dense "blooms" of plankton that occasionally appear in the open ocean.

Because the silica-containing walls of dead diatom cells resist decomposition, certain sedimentary rocks are composed almost entirely of diatom skeletons that sank to the seafloor over time. Diatomaceous earth, which is obtained from such rocks, has many industrial uses, such as insulation, filtration, and metal polishing. It has also been used as an "Earth-friendly" insecticide that clogs the tracheae (breathing structures) of insects.

BROWN ALGAE The **brown algae** obtain their namesake color from the carotenoid *fucoxanthin*, which is abundant in their chloroplasts. The combination of this yellow-orange pigment with the green of chlorophylls *a* and *c* yields a brownish tinge. All brown algae are multicellular, and some are extremely large. Giant kelps, such as those of the genus *Macrocystis*, may be up to 60 meters long.

OOMYCETES A nonphotosynthetic stramenopile group called the **oomycetes** consists in large part of the water molds and their terrestrial relatives, such as the downy mildews. Water molds are filamentous and stationary, and they are *absorptive heterotrophs*—that is, they secrete enzymes that digest large food molecules into smaller molecules that the water mold can absorb. If you have seen a whitish, cottony mold growing on dead fish or dead insects in water, it was probably a water mold of the common genus *Saprolegnia* (**Figure 27.21**).

Don't be misled by the "mycete" in the name of this group. That term means "fungus," and it is there because these organisms were once classified as fungi. However, we now know that the oomycetes are more distantly related to the fungi than

Saprolegnia sp.

27.21 An Oomycete The filaments of a water mold radiate from the carcass of an insect.

are many other eukaryote groups, including ourselves (see Figure 27.1), and the similarity of oomycetes to fungi is only superficial. For example, the cell walls of oomycetes are typically made of cellulose, whereas those of fungi are made of chitin.

Some oomycetes are **coenocytes**, which means they have many nuclei enclosed in a single plasma membrane. Their filaments have no cross-walls to separate the many nuclei into discrete cells. Their cytoplasm is continuous throughout the body of the organism, and there is no single structural unit with a single nucleus, except in certain reproductive stages. A distinguishing feature of the oomycetes is their flagellated reproductive cells. Oomycetes are diploid throughout most of their life cycle.

The water molds, such as *Saprolegnia*, are all aquatic and **saprobic**—meaning they feed on dead organic matter. Some other oomycetes are terrestrial. Although most of

the terrestrial oomycetes are harmless or helpful decomposers of dead matter, a few are serious plant parasites that attack crops such as avocados, grapes, and potatoes.

Although their presumed chromalveolate ancestors had chloroplasts and were photosynthetic, the oomycetes lack chloroplasts.

PLANTAE

Plantae contains mostly photosynthetic species and consists of several major clades, including glaucophytes, red algae, chlorophytes, charophytes, and the land plants, all of which probably trace their chloroplasts back to a single incidence of endosymbiosis (see Section 27.1). It is for this reason that the small clade known as the **glaucophytes**, unicellular organisms that live in fresh water, is of particular interest.

Glaucophytes
Red algae
Chlorophytes
Land plants
Charophytes

The glaucophytes were likely the first group to diverge after the primary endosymbiosis event. Their chloroplast is unique in containing a small amount of peptidoglycan between its inner and outer membranes—the same arrangement as that found in cyanobacteria (see Figure 27.3A). The presence of peptidoglycan, the characteristic cell wall component of bacteria, suggests that this feature has been retained in glaucophytes but lost in the other Plantae groups.

Red algae have a distinctive accessory photosynthetic pigment

Almost all **red algae** are multicellular (**Figure 27.22**). Their characteristic color is a result of the accessory photosynthetic pigment *phycoerythrin*, which is found in relatively large amounts in the chloroplasts of many species. In addition to phycoerythrin, red algae contain phycocyanin, carotenoids, and chlorophyll *a*.

The red algae include species that grow in the shallowest tide pools as well as the photosynthesizers found deepest in the ocean (as deep as 260 meters if nutrient conditions are right and the water is clear enough to permit light to penetrate). Very few red algae inhabit fresh water. Most grow attached to a substratum by a holdfast.

In a sense, the red algae are misnamed. They have the capacity to change the relative amounts of their

(B) *Calliarthron tuberculatum*

(A) *Ceramium* sp.

1.5 mm

7.5 mm

27.22 Red Algae (A) Differential contrast light microscopy reveals the rich red color of the pigment phycoerythrin. (B) Coralline red alga, named for its coral-like appearance.

various photosynthetic pigments depending on the light conditions where they are growing. Thus the leaflike *Chondrus crispus*, a common North Atlantic red alga, may appear bright green when it is growing at or near the surface of the water and deep red when growing at greater depths. The ratio of pigments present depends to a remarkable degree on the intensity of the light that reaches the alga. In deep water, where the light is dimmest, the alga accumulates large amounts of phycoerythrin. The algae in deep water have as much chlorophyll as the green ones near the surface, but the accumulated phycoerythrin makes them look red.

In addition to being the only photosynthetic eukaryotes with phycoerythrin among their pigments, the red algae have two other distinctive characteristics:

- They store the products of photosynthesis in the form of *floridean starch*, which is composed of very small, branched chains of approximately 15 glucose monomers.

- They produce no motile, flagellated cells at any stage of their life cycle. The male gametes lack cell walls and are slightly amoeboid; the female gametes are completely immobile.

Some red algal species enhance the formation of coral reefs (see Figure 27.22B). Like coral animals, they possess the biochemical machinery for secreting calcium carbonate, which they deposit both in and around their cell walls. After the deaths of corals and algae, the calcium carbonate persists, sometimes forming substantial rocky masses.

Some red algae produce large amounts of mucilaginous polysaccharide substances, which contain the sugar galactose with a sulfate group attached. This material readily forms solid gels and is the source of agar, a substance widely used in the laboratory for making a solid aqueous medium on which tissue cultures and many microorganisms can be grown.

The distinctive chloroplasts of the photosynthetic chromalveolates are derived by secondary endosymbiosis of a red alga, as discussed in Section 27.1.

Chlorophytes, charophytes, and land plants contain chlorophylls a and b

One major clade of "green algae" is the **chlorophytes**. A sister group to the chlorophytes contains another green algal clade—the **charophytes**, or **Charales**—along with the land plants (see Section 28.1). The green algae share several characters that distinguish them from other protists: like the land plants, they contain chlorophylls *a* and *b*, and their reserve of photosynthetic products is stored as starch in chloroplasts. Through secondary endosymbiosis, a chlorophyte became the chloroplast of the euglenids.

There are more than 17,000 species of chlorophytes. Most are aquatic—some are marine, though more are freshwater forms—but others are terrestrial, living in moist environments. The chlorophytes range in size from microscopic unicellular forms to multicellular forms many centimeters long.

The chlorophytes display an incredible variety of shapes and body forms. *Chlamydomonas* is an example of the simplest type: unicellular and flagellated. Surprisingly large and well-formed

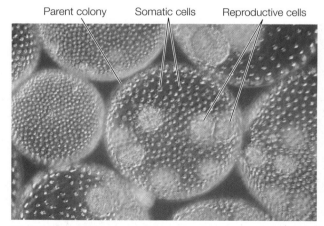

Parent colony   Somatic cells   Reproductive cells

(A) *Volvox* sp.   120 μm

(B) *Ulva lactuca*   3 cm

27.23 Chlorophytes (A) *Volvox* colonies are precisely spaced arrangements of cells. Specialized reproductive cells produce daughter colonies, which will eventually release new individuals. (B) Sea lettuce grows in ocean tidewaters.

colonies of cells are found in such freshwater groups as the genus *Volvox* (**Figure 27.23A**). The cells in these colonies are not differentiated into specialized tissues and organs, as in land plants and animals, but the colonies show vividly how the preliminary step of this great evolutionary innovation might have been taken. In *Volvox*, the origins of cell specialization can be seen in certain cells in the colony that are specialized for reproduction.

While *Volvox* is colonial and spherical, *Oedogonium* is multicellular and filamentous, and each of its cells has only one nucleus. *Cladophora* is multicellular, but each cell is multinucleate. *Bryopsis* is tubular and coenocytic, forming cross-walls only when reproductive structures form. *Acetabularia* is a single giant, uninucleate cell a few centimeters in length that becomes multinucleate only at the end of its reproductive stage. *Ulva lactuca* is a thin, membranous sheet a few centimeters across; its distinctive appearance justifies its common name of sea lettuce (**Figure 27.23B**).

As mentioned above, the chlorophytes are the largest clade of green algae, but there are other green algal clades as well. Those clades are branches of a clade that also includes the land plants, which are described in the next chapter.

EXCAVATES

Excavates include a number of diverse clades, several of which lack mitochondria. This absence of mitochondria once led to the view that these groups might represent early diverging groups of eukaryotes that diversified before the evolution of mitochondria. However, the absence of mitochondria seems to be a derived condition, judging in part from the presence of nuclear genes normally associated with mitochondria. Ancestors of these organisms probably possessed mitochondria that were lost or reduced in the course of evolution. The existence of such organisms today shows that eukaryotic life is feasible without mitochondria, and for that reason, these groups are the focus of much attention.

Diplomonads and parabasalids are excavates that lack mitochondria

The **diplomonads** and **parabasalids**, all of which are unicellular, are distinctive in their lack of mitochondria. *Giardia lamblia*, a diplomonad, is a familiar parasite that contaminates water supplies and causes the intestinal disease giardiasis (**Figure 27.24A**). This tiny organism contains two nuclei bounded by nuclear envelopes, and it has a cytoskeleton and multiple flagella.

Trichomonas vaginalis is a parabasalid responsible for a sexually transmitted disease in humans (**Figure 27.24B**). Infection of the male urethra, where it may occur without symptoms, is less common than infection of the vagina. In addition to flagella and a cytoskeleton, the parabasalids have undulating membranes that also contribute to the cell's locomotion.

Heteroloboseans alternate between amoeboid forms and forms with flagella

The amoeboid body form appears in several protist groups—including the **loboseans** and **heteroloboseans**—that are only distantly related to one another. These groups belong, respectively, to the unikonts and excavates. Amoebas of the free-living heterolobosean genus *Naegleria*, some of which can enter humans and cause a fatal disease of the nervous system, usually have a two-stage life cycle, in which one stage has amoeboid cells and the other flagellated cells.

Euglenids and kinetoplastids have distinctive mitochondria and flagella

The euglenids and kinetoplastids, both of which are excavates, together constitute a clade of unicellular organisms with flagella. Their mitochondria contain distinctive, disc-shaped cristae, and their flagella contain a crystalline rod not found in other organisms. They reproduce primarily asexually by binary fission.

EUGLENIDS The **euglenids** possess flagella arising from a pocket at the anterior end of the cell. Spiraling strips of proteins under their plasma membranes control the cell's shape. Some members

(A) *Giardia* sp.

(B) *Trichomonas vaginalis*

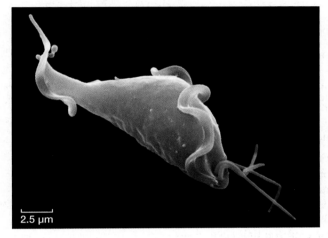

27.24 Some Excavate Groups Lack Mitochondria (A) *Giardia*, a diplomonad, has flagella and two nuclei. (B) *Trichomonas*, a parabasalid, has flagella and undulating membranes. Neither of these organisms possesses mitochondria.

of the group are photosynthetic. Euglenids used to be claimed by the zoologists as animals and by the botanists as plants.

Figure 27.25 depicts a cell of the genus *Euglena*. Like most other euglenids, this common freshwater organism has a complex cell structure. It propels itself through the water with

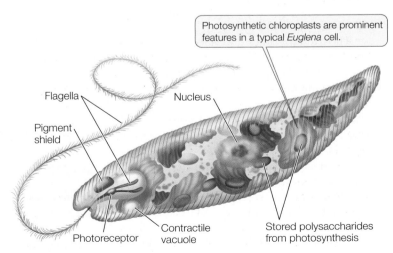

Photosynthetic chloroplasts are prominent features in a typical *Euglena* cell.

Flagella

Nucleus

Pigment shield

Photoreceptor

Contractile vacuole

Stored polysaccharides from photosynthesis

27.25 A Photosynthetic Euglenid Several *Euglena* species possess flagella. In this species, the second flagellum is rudimentary.

TABLE 27.2
A Comparison of Three Kinetoplastid Trypanosomes

	Trypanosoma brucei	*Trypanosoma cruzi*	*Leishmania major*
Human disease	Sleeping sickness	Chagas' disease	Leishmaniasis
Insect vector	Tsetse fly	Assassin bug	Sand fly
Vaccine or effective cure	None	None	None
Strategy for survival	Changes surface recognition molecules frequently	Causes changes in surface recognition molecules on host cell	Reduces effectiveness of macrophage hosts
Site in human body	Bloodstream; attacks nerve tissue in final stages	Enters cells, especially muscle cells	Enters cells, primarily macrophages
Approximate number of deaths per year	50,000	43,000	60,000

the longer of its two flagella, which may also serve as an anchor to hold the organism in place. The second flagellum is often rudimentary.

Euglenids have diverse nutritional requirements. Many species are always heterotrophic. Other species are fully autotrophic in sunlight, using chloroplasts to synthesize organic compounds through photosynthesis. The chloroplasts of euglenids are surrounded by three membranes as a result of secondary endosymbiosis (see Figure 27.3B). When kept in the dark, these euglenids lose their photosynthetic pigment and begin to feed exclusively on dissolved organic material in the water around them. Such a "bleached" *Euglena* resynthesizes its photosynthetic pigment when it is returned to the light and becomes autotrophic again. But *Euglena* cells treated with certain antibiotics or mutagens lose their photosynthetic pigment completely; neither they nor their descendants are ever autotrophs again. However, those descendants function well as heterotrophs.

KINETOPLASTIDS The **kinetoplastids** are unicellular parasites with two flagella and a single, large mitochondrion. That mitochondrion contains a *kinetoplast*—a unique structure housing multiple circular DNA molecules and associated proteins. Some of these DNA molecules encode "guides" that edit messenger RNA in the mitochondrion.

The trypanosomes described at the opening of this chapter are kinetoplastids. They are able to change their cell surface recognition molecules frequently, which allows them to evade our best attempts to kill them and eradicate the diseases they cause (**Table 27.2**).

RHIZARIA

The primary groups of **Rhizaria** are unicellular aquatic eukaryotes. Foraminiferans, radiolarians, and cercozoans typically have long, thin pseudopodia that contrast with the broader, lobelike pseudopodia of the familiar amoebas. These groups have contributed to ocean sediments, some of which have become terrestrial features in the course of geological history.

The **cercozoans** are a diverse group, with many forms and habitats. Some are amoeboid, while others have flagella. Some are aquatic; others live in soil. One group of cercozoans possesses chloroplasts derived from a green alga by secondary endosymbiosis—and that chloroplast contains a trace of the green alga's nucleus.

Foraminiferans have created vast limestone deposits

Some **foraminiferans** secrete external shells of calcium carbonate (see Figure 27.11), which over time have accumulated to produce much of the world's limestone. Some foraminiferans live as plankton; others live on the seafloor. Living foraminiferans have been found at the deepest point in the world's oceans—10,896 meters down in the Challenger Deep, in the western Pacific. At that depth, however, they cannot secrete a normal shell because the surrounding water is too poor in calcium carbonate.

Long, threadlike, branched pseudopods reach out through numerous microscopic apertures in the shell and interconnect to create a sticky, reticulated net, which planktonic foraminifera use to catch smaller plankton. The pseudopods provide locomotion in some species.

Radiolarians have thin, stiff pseudopods

The **radiolarians** are recognizable by their thin, stiff pseudopods, which are reinforced by microtubules. These pseudopods:

- greatly increase the surface area of the cell for exchange of materials with the environment
- help the cell float in its marine environment

Found exclusively in marine environments, radiolarians are immediately recognizable by their distinctive radial symmetry (see Figure 27.8B). Almost all radiolarian species secrete glassy *endoskeletons* (internal skeletons). A central capsule lies within the cytoplasm. The skeletons of the different species are as varied as snowflakes, and many have elaborate geometric designs (**Figure 27.26**). A few radiolarians are among the largest of the unicellular eukaryotes, measuring several millimeters across.

Podocyrtis cothurnata

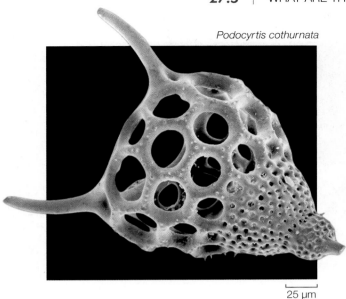

25 μm

27.26 A Radiolarian's Glass House Radiolarians secrete intricate glassy skeletons such as the one shown here. A living radiolarian with its endosymbionts is shown in Figure 27.8B.

(A) *Codosiga botrytis*

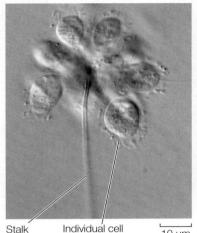

Stalk   Individual cell

10 μm

(B) *Choanoeca* sp.

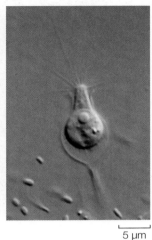

5 μm

27.27 A Link to the Animals Choanoflagellates are the sister group to the animals. (A) The formation of colonies by unicellular organisms, as in this choanoflagellate species, is one route to the evolution of multicellularity. (B) A solitary choanoflagellate illustrates the similarity of this microbial protist group to a cell type present in the multicellular sponges (see Figure 31.7).

UNIKONTS

We now consider the **unikonts**, a large clade that may be close to the root of the eukaryote tree. The name *unikont* derives from the Greek for "single cone," and in these eukaryotes, the flagella, if present, are single. Unikonts consist of two major groups, the *opisthokonts* (which include animals and fungi) and the *amoebozoans* (see Figure 27.1).

Fungi
Choanoflagellates
Animals
Opisthokonts

Loboseans
Plasmodial slime molds
Cellular slime molds
Amoebozoans

A morphological synapomorphy of the opisthokonts is that their flagellum, if present, is posterior, as in animal sperm. The flagella of all other eukaryotes are anterior. In addition to the animals and fungi, opisthokonts also include the choanoflagellates. Fungi and animals are discussed in Chapters 30–33. The choanoflagellates, or collar flagellates, are sister to the animals, and the animal–choanoflagellate clade is sister to the fungi.

Some choanoflagellates are colonial (**Figure 27.27A**). They bear a striking resemblance to the most characteristic type of cell found in the sponges (compare **Figure 27.27B** with Figure 31.7).

Amoebozoans use lobe-shaped pseudopods for locomotion

The lobe-shaped pseudopods used by amoebozoans are a hallmark of the amoeboid body plan. The amoebozoan pseudopod differs in form and function from the slender pseudopods of rhizaria. We consider three amoebozoan groups here: the loboseans and two clades of slime molds.

LOBOSEANS A lobosean, such as the *Peloxima carolinensis* shown in Figure 27.4, consists of a single cell. Unlike the cells of the slime molds, loboseans live independently of one another and

do not aggregate. A lobosean feeds on small organisms and particles of organic matter by phagocytosis, engulfing them with its pseudopods. Many loboseans are adapted for life on the bottoms of lakes, ponds, and other bodies of water. Their creeping locomotion and their manner of engulfing food particles fit them for life close to a relatively rich supply of sedentary organisms or organic particles. Most loboseans exist as predators, parasites, or scavengers.

Some loboseans are shelled, living in casings of sand grains glued together (see Figure 27.7A). Others have shells secreted by the organism itself.

SLIME MOLDS The two major groups of *slime molds* share only general characteristics. All are motile, all ingest particulate food by endocytosis, and all form spores on erect structures called *fruiting bodies*. They undergo striking changes in organization during their life cycles, and one stage consists of isolated cells that take up food particles by endocytosis. Some slime molds may cover areas of 1 meter or more in diameter while in their less aggregated stage. Such a large slime mold may weigh more than 50 grams. Slime molds of both types favor cool, moist habitats, primarily in forests. They range from colorless to brilliant yellow and orange.

PLASMODIAL SLIME MOLDS If the nucleus of an amoeba began rapid mitotic division, accompanied by a tremendous increase in cytoplasm and organelles but no cytokinesis, the resulting organism might resemble the multinucleate mass of a plasmodial slime mold. During its vegetative (feeding) phase, a plasmodial slime mold is a wall-less mass of cytoplasm with numerous diploid nuclei. This mass streams very slowly over its substratum in a remarkable network of strands called a *plas-*

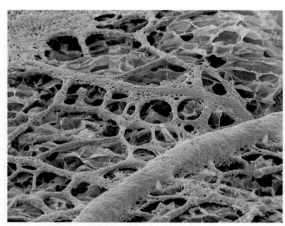

(A) *Physarum polycephalum*

0.1 mm

(B) *Physarum polycephalum*

0.25 mm

27.28 Plasmodial Slime Molds (A) The slime mold *Physarum* is most often seen in its plasmodial form, covering rocks, decaying logs, and other objects as it engulfs bacteria and other food items from the surface. (B) Fruiting structures of *Physarum*.

modium. The plasmodium of such a slime mold is another example of a coenocyte, with many nuclei enclosed in a single plasma membrane. The outer cytoplasm of the plasmodium (closest to the environment) is normally less fluid than the interior cytoplasm and thus provides some structural rigidity (**Figure 27.28A**).

Plasmodial slime molds provide a dramatic example of movement by cytoplasmic streaming. The outer cytoplasmic region of the plasmodium becomes more fluid in places, and cytoplasm rushes into those areas, stretching the plasmodium. This streaming somehow reverses its direction every few minutes as cytoplasm rushes into a new area and drains away from an older one, moving the plasmodium over its substratum. Sometimes an entire wave of plasmodium moves across the substratum, leaving strands behind. Microfilaments and a contractile protein called *myxomyosin* interact to produce the streaming movement. As it moves, the plasmodium engulfs food particles by endocytosis—predominantly bacteria, yeasts, spores of fungi, and other small organisms, as well as decaying animal and plant remains.

A plasmodial slime mold can grow almost indefinitely in its plasmodial stage, as long as the food supply is adequate and other conditions, such as moisture and pH, are favorable. However, one of two things can happen if conditions become unfavorable. First, the plasmodium can form an irregular mass of hardened cell-like components called a *sclerotium.* This resting structure rapidly becomes a plasmodium again when favorable conditions are restored.

Alternatively, the plasmodium can transform itself into spore-bearing fruiting structures (**Figure 27.28B**). These stalked or branched structures rise from heaped masses of plasmodium. They derive their rigidity from walls that form and thicken between their nuclei. The diploid nuclei of the plasmodium divide by meiosis as the fruiting structure develops. One or more knobs, called *sporangia*, develop on the end of the stalk.

Within a sporangium, haploid nuclei become surrounded by walls and form spores. Eventually, as the fruiting body dries, it sheds its spores.

The spores germinate into wall-less, haploid cells called *swarm cells,* which can either divide mitotically to produce more haploid swarm cells or function as gametes. Swarm cells can live as separate individual cells that move by means of flagella or pseudopods, or they can become walled and resistant resting cysts when conditions are unfavorable; when conditions improve again, the cysts release swarm cells. Two swarm cells can also fuse to form a diploid zygote, which divides by mitosis (but without a wall forming between the nuclei) and thus forms a new, coenocytic plasmodium.

CELLULAR SLIME MOLDS Whereas the plasmodium is the basic vegetative (feeding, nonreproductive) unit of the plasmodial slime molds, an amoeboid cell is the vegetative unit of the cellular slime molds. Large numbers of cells called *myxamoebas,* which have single haploid nuclei, engulf bacteria and other food particles by endocytosis and reproduce by mitosis and fission. This simple life cycle stage, consisting of swarms of independent, isolated cells, can persist indefinitely as long as food and moisture are available.

When conditions become unfavorable, the cellular slime molds aggregate and form fruiting structures, as do their plasmodial counterparts. The individual myxamoebas aggregate into a mass called a *slug* or *pseudoplasmodium* (**Figure 27.29**). Unlike the true plasmodium of the plasmodial slime molds, this structure is not simply a giant sheet of cytoplasm with many nuclei; the individual myxamoebas retain their plasma membranes and, therefore, their identity.

A slug may migrate over its substratum for several hours before becoming motionless and reorganizing to construct a delicate, stalked fruiting structure. Cells at the top of the fruiting structure develop into thick-walled spores, which are eventually released. Later, under favorable conditions, the spores germinate, releasing myxamoebas.

The cycle from myxamoebas through slug and spores to new myxamoebas is asexual. Cellular slime molds also have a sex-

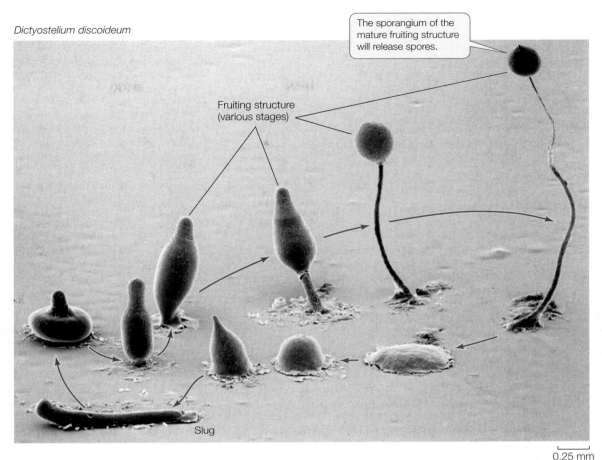

Dictyostelium discoideum

The sporangium of the mature fruiting structure will release spores.

Fruiting structure (various stages)

Slug

0.25 mm

27.29 A Cellular Slime Mold The life cycle of the slime mold *Dictyostelium* is shown here in a composite micrograph.

ual cycle, in which two myxamoebas fuse. The product of this fusion develops into a spherical structure that ultimately germinates, releasing new haploid myxamoebas.

In subsequent chapters we explore the three major groups of multicellular eukaryotes. Chapters 28 and 29 describe the origin and diversification of the plants, Chapter 30 presents the fungi, and Chapters 31–33 describe the animals. All three of these groups arose from protist ancestors.

CHAPTER SUMMARY

27.1 How Did the Eukaryotic Cell Arise?

- The modern eukaryotic cell evolved from an ancestral prokaryote. Probable early events in this evolution include the loss of the cell wall and infolding of the plasma membrane. Such infolding probably led to the segregation of the genetic material in a membrane-enclosed nucleus. The development of a cytoskeleton gave the evolving cell increasing control over its shape and distribution of daughter chromosomes. **Review Figure 27.2**
- Some organelles were acquired by endosymbiosis. Mitochondria evolved from a proteobacterium.
- **Primary endosymbiosis** of a eukaryote and a cyanobacterium gave rise to the chloroplasts, beginning with those of glaucophytes, red algae, green algae, and land plants. **Secondary** and **tertiary endosymbiosis** of these chloroplast-containing eukaryotes within other eukaryotes gave rise to the chloroplasts of euglenids, dinoflagellates, and other groups. **Review Figure 27.3, ANIMATED TUTORIAL 27.1**

27.2 What Features Account for Protist Diversity?

- **Protists** are a diverse, paraphyletic assemblage of mostly unicellular eukaryotes. **Review Figure 27.1**
- Some protists are photosynthetic autotrophs, some are heterotrophs, and some are both.
- The cytoskeleton allows for various means of locomotion. Most protists are motile, moving by amoeboid motion with **pseudopods** or by means of cilia or flagella.
- Some protist cells contain **contractile vacuoles** that pump out excess water, or **digestive vacuoles** where food is digested. **Review Figures 27.5 and 27.6, ANIMATED TUTORIAL 27.2**
- Many protists have protective cell surfaces such as cell walls, external "shells," or shells constructed from sand.

27.3 How Do Protists Affect the World Around Them?

- The diatoms, part of the plankton, are responsible for about a fifth of all the photosynthetic carbon fixation on Earth.

- **Phytoplankton** are the primary producers in the marine environment.
- Endosymbiosis is common among protists and often helpful to both partners. Review Figure 27.8
- Pathogenic protists include species of *Plasmodium* and trypanosomes. Review Figure 27.9

27.4 How Do Protists Reproduce?

- Most protists reproduce both asexually and sexually.
- **Conjugation** in paramecia is a sexual process but not a reproductive one. Review Figure 27.12
- **Alternation of generations**, which includes a multicellular diploid phase and a multicellular haploid phase, is a feature of many protist life cycles. Review Figure 27.13
- Alternation of generations may be **heteromorphic** or **isomorphic**. The alternating generations are the (diploid) **sporophyte** and (haploid) **gametophyte**. Specialized cells of the sporophyte, called **sporocytes**, divide meiotically to produce haploid spores.
- Depending on whether their gametes look identical or dissimilar, species are termed **isogamous** or **anisogamous**.
- In a **diplontic** life cycle, the gametes are the only haploid cells. In a **haplontic** life cycle, the zygote is the only diploid cell. Review Figures 27.14 and 27.15, **WEB ACTIVITIES 27.1 AND 27.2**
- Some protist life cycles involve more than one host species.

27.5 What Are the Evolutionary Relationships Among Eukaryotes?

- Most eukaryotes can be classified in one of five major clades: chromalveolates, Plantae, excavates, rhizaria, or unikonts. Review Table 27.1 and Figure 27.1
- The **chromalveolates** include the haptophytes, alveolates, and stramenopiles.
- **Alveolates** are unicellular organisms with sacs (alveoli) beneath their plasma membranes. Alveolate clades include the marine **dinoflagellates**, the parasitic **apicomplexans**, and the diverse, highly motile **ciliates**. **REVIEW WEB ACTIVITY 27.3**

- **Stramenopiles** typically have two flagella of unequal length, the longer one bearing rows of tubular hairs. Included among the stramenopiles are the unicellular **diatoms**, the multicellular **brown algae**, and the nonphotosynthetic **oomycetes**, including the water molds and downy mildews.
- Clades in the **Plantae** include the **glaucophytes**, **red algae**, **chlorophytes**, **charophytes**, and the land plants. All are photosynthetic and contain chloroplasts. Glaucophyte chloroplasts contain peptidoglycan between their inner and outer membranes.
- The **excavates** include the diplomonads, parabasalids, heteroloboseans, euglenids, and kinetoplastids. The **diplomonads** and **parabasalids** lack mitochondria, having apparently lost them during their evolution. **Heteroloboseans** are amoebas with a two-stage life cycle. **Euglenids** are often photosynthetic and have anterior flagella and strips of protein that support their cell surface. **Kinetoplastids** have a single, large mitochondrion in which mitochondrial mRNA is edited.
- **Rhizaria** are unicellular and aquatic; most are amoeboid. This group includes the **foraminiferans**, whose shells have contributed to great limestone deposits; the **radiolarians** with thin, stiff pseudopods and glassy endoskeletons; and the **cercozoans**, which take many forms and live in diverse habitats.
- The **unikonts** encompass organisms with single flagella on their flagellated cells (if any). They can be divided into two subgroups, the opisthokonts and the amoebozoans.
- In the **opisthokonts**, the flagellum (when present) is posterior. The opisthokont subgroups are the fungi, choanoflagellates, and animals. **Choanoflagellates** resemble the cells of sponges and are sister to the animal clade.
- The **amoebozoans** move by means of lobe-shaped pseudopodia. They comprise the **loboseans**, plasmodial slime molds, and cellular slime molds. A lobosean consists of a single cell; these cells do not aggregate. Plasmodial slime molds are amoebozoans whose feeding phase is coenocytic. In the feeding phase, movement is by cytoplasmic streaming. In cellular slime molds, the individual cells maintain their identity at all times but aggregate to form fruiting bodies.

SELF-QUIZ

1. Microbial eukaryotes with flagella
   a. appear in several clades.
   b. are all algae.
   c. all have pseudopods.
   d. are all colonial.
   e. are never pathogenic.

2. Which statement about eukaryotic phytoplankton is *not* true?
   a. Some are important primary producers.
   b. Some contributed to the formation of petroleum.
   c. Some form toxic "red tides."
   d. Some are food for marine animals.
   e. They constitute a clade.

3. Apicomplexans
   a. possess flagella.
   b. possess a glassy shell.
   c. are all parasitic.
   d. are algae.
   e. include the trypanosomes that cause sleeping sickness.

4. The ciliates
   a. move by means of short flagella.
   b. use amoeboid movement.
   c. include *Plasmodium*, the agent of malaria.
   d. possess both a macronucleus and micronuclei.
   e. are autotrophic.

5. The chloroplasts of photosynthetic protists
   a. are structurally identical.
   b. gave rise to mitochondria.
   c. are all descended from a once free-living cyanobacterium.
   d. all have exactly two surrounding membranes.
   e. are all descended from a once free-living red alga.

6. Which statement about the brown algae is *not* true?
   a. They are all multicellular.
   b. They use the same photosynthetic pigments as do land plants.
   c. They are almost exclusively marine.
   d. A few are many meters in length.
   e. They are stramenopiles.
7. Which statement about the chlorophytes is *not* true?
   a. They use the same photosynthetic pigments as do land plants.
   b. Some are unicellular.
   c. Some are multicellular.
   d. All are microscopic in size.
   e. They display a great diversity of life cycles.
8. The red algae
   a. are mostly unicellular.
   b. are mostly marine.
   c. owe their red color to a special form of chlorophyll.
   d. have flagella on their gametes.
   e. are all heterotrophic.
9. The plasmodial slime molds
   a. form a plasmodium that is a coenocyte.
   b. lack fruiting bodies.
   c. consist of large numbers of myxamoebas.
   d. consist at times of a mass called a pseudoplasmodium.
   e. possess flagella.
10. The cellular slime molds
   a. possess apical complexes.
   b. lack fruiting bodies.
   c. form a plasmodium that is a coenocyte.
   d. have haploid myxamoebas.
   e. possess flagella.

FOR DISCUSSION

1. For each type of organism below, give a single characteristic that may be used to differentiate it from the other, related organism(s) named in parentheses.
   a. Foraminiferans (radiolarians)
   b. *Euglena* (*Volvox*)
   c. *Trypanosoma* (*Giardia*)
   d. Plasmodial slime molds (cellular slime molds)
2. In what sense are sex and reproduction independent of each other in the ciliates? What does that suggest about the role of sex in biology?
3. Why are dinoflagellates and apicomplexans placed in one group of microbial eukaryotes and brown algae and oomycetes in another?
4. Unlike many protists, apicomplexans lack contractile vacuoles. Why don't apicomplexans need a contractile vacuole?
5. Giant seaweeds (mostly brown algae) have "floats" that aid in keeping their fronds suspended at or near the surface of the water. Why is it important that the fronds be suspended in this way?
6. Why are algal pigments so much more diverse than those of land plants?
7. Consider the chloroplasts of chlorophytes, euglenids, and red algae. For each of these groups, indicate how many membranes surround their chloroplasts, and offer a reasonable explanation in each case. Why do some dinoflagellates have more membranes around their chloroplasts than other dinoflagellates?

ADDITIONAL INVESTIGATION

Mitochondrial, chloroplast, and nuclear genomes of eukaryotes each contain ribosomal RNA genes. If the endosymbiotic theory for the origin of mitochondria and chloroplasts is correct, would you expect the ribosomal genes of these organelles to be more closely related to the nuclear ribosomal RNA genes, or to ribosomal RNA genes in proteobacteria and cyanobacteria? How would you test the endosymbiotic theory for the origin of these organelles using a phylogenetic analysis of ribosomal RNA genes?

What surprises lurk in a rock?

John William Dawson, later Sir William Dawson, was an outspoken critic of Charles Darwin and evolutionary theory. Ironically, Dawson made one of the first contributions to our understanding of the early evolution of plants. While surveying the geology of Nova Scotia, this Canadian scientist found a puzzling fossil fragment on the Gaspé Peninsula. It was 1859—the very year in which Darwin published *The Origin of Species*.

What Dawson found was the remains of a remarkable plant, which he named *Psilophyton*, meaning "naked plant." The fossilized plant appeared to have no roots and no leaves, and its stem grew both below and above the ground. The aboveground part of the stem, which branched and ended in spore cases, was about 50 centimeters tall. Could this Devonian fossil plant represent one of the stages in the momentous transition of plants from an aquatic existence to the forests that cover the land today? Dawson presented *Psilophyton* and other finds in a major lecture he delivered in 1870, but his audience was unreceptive, with his fellow scientists chuckling that *Psilophyton* grew nowhere but in Dawson's imagination. His published drawing of the fossil plant was regarded as an amusing curiosity.

Decades later, Dawson's interpretation of *Psilophyton* was vindicated by the work of two British botanists who studied plant fossils. The year 1915 found Robert Kidston and William Lang in the hills near Rhynie, Scotland, where they discovered evidence that a marsh had existed there 400 million years ago. In the intervening hundreds of millions of years, that Devonian marsh had become a flinty, fossil-laden rock called chert. Although the rock now lay in a hilly site some 50 kilometers from the sea, its original location must have been near the shore.

The most startling feature of the Rhynie chert was the great abundance of fossils of a small plant with spore cases but neither roots nor leaves—a plant, in fact, that looked strikingly like Dawson's "imaginary" *Psilophyton*! Kidston and Lang described their find in detail and gave it the genus name *Rhynia* in honor of the site of its discovery.

Making It on Land An artist's reconstruction of the earliest vascular plant ancestors, based on the Rhynie chert fossil bed in Scotland. The plants in this scene are rhyniophytes. Heavily mineralized water from hot springs and geysers may have been partly responsible for the remarkable preservation of plant fossils in the Rhynie bed.

Green Mansions Seedless plants—mosses and ferns—dominate the understory of this pristine rainforest on the island of Maui, Hawaii. The only seed plants to be seen are the trees, which are almost obscured by mosses and ferns.

Rhynia is just one of several ancient groups of plants that lack seeds and other features of more modern plants. The glory days of the seedless plants are long past, but many of them—most notably the mosses and ferns—are still abundant. We rely on these surviving plants, as well as on fossils, to help us understand the evolution of land plants from aquatic algae growing at the edge of a sea or marsh. Much of the fossilized history, and the great bulk of the mass of seedless plants, was preserved as coal, which today is an economically important substance that we burn as fuel and to produce electricity. What a library of botanical history the fossils in the world's coal beds provide!

IN THIS CHAPTER we will see how plants first invaded land, how land plants evolved, and how plant clades diversified, resulting in plants ever better equipped to face the challenges of terrestrial environments. The descriptions here will concentrate on those land plants that lack seeds. Chapter 29 considers the seed plants that dominate the terrestrial scene today.

28.1 How Did the Land Plants Arise?

The question "Where did plants come from?" embraces two questions: "What were the ancestors of plants?" and "Where did those ancestors live?" We consider both questions in this section.

The **land plants** are monophyletic: all land plants descend from a single common ancestor and form a branch of the evolutionary tree of life. Although we refer to this clade of mostly terrestrial species as land plants, some of them actually live in shallow water. One of the key shared derived traits, or *synapomorphies*, of the land plants is development from an embryo protected by tissues of the parent plant. For this reason, land plants are sometimes called **embryophytes** (*phyton*, "plant"). Land plants retain the derived features they share with the "green algae" described in Chapter 27: the use of chlorophylls *a* and *b* in photosynthesis, and the use of starch as a photosynthetic storage product. Both land plants and green algae have cellulose in their cell walls.

There are several ways to define "plant" and still refer to a clade (**Figure 28.1**; see also Figure 27.1). Throughout this book and in everyday language, the unmodified common name "plants" usually refers to the land plants. Some biologists, however, use the term "plants" to include land plants plus some closely related groups of green algae; this monophyletic group is also known as the **streptophytes**. The addition of the remainder of the green algae to the streptophytes results in a more inclusive clade, commonly called the **green plants**, which encompasses all the groups that possess chlorophyll *b*. Green plants, streptophytes, and land plants each have been called "the plant kingdom" by different authorities; others take an even broader view and include red algae and glaucophytes as "plants." To avoid confusion in this chapter, we will use modifying terms ("land plants" or "green plants," for example) to refer to the various clades shown in Figure 28.1.

There are ten major groups of land plants

The land plants that exist today fall naturally into ten major clades (**Table 28.1**). Members of seven of those clades possess well-developed vascular systems that transport materials throughout the plant body. We call these seven groups, collectively, the **vascular plants**, or **tracheophytes**, because they all possess fluid-conducting cells called **tracheids**. Taken together, the seven groups of vascular plants constitute a clade.

The remaining three clades (liverworts, hornworts, and mosses) lack tracheids. These three groups are sometimes collectively called *bryophytes*, but in this text we reserve that term for the largest monophyletic group—the mosses—and refer to these three clades collectively as **nonvascular land plants**.

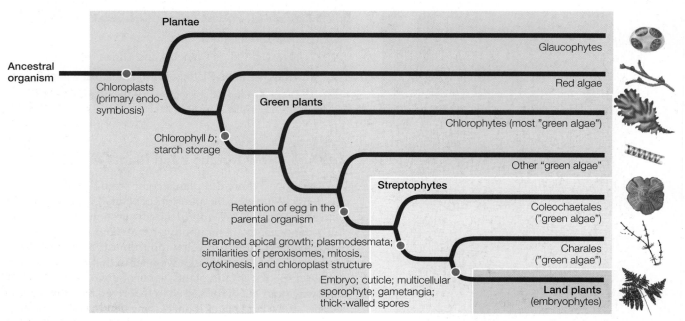

28.1 What Is a Plant? In its broadest definition, the term "plant" includes the green plants, red algae, and glaucophytes—all the groups descended from a common ancestor with primary chloroplasts. Some biologists restrict the term to the green plants (those with chlorophyll *b*) or, even more narrowly, to the land plants (embryophytes).

Note, however, that *these three groups do not form a clade*. Some nonvascular land plants have conducting cells, but none have tracheids.

Where did the land plants come from? To determine which living organisms are most closely related to land plants, biologists have considered several of the synapomorphies of land plants and looked for their origins in various other groups.

The land plants arose from a green algal clade

Several microscopic structural features, backed by clear-cut evidence from molecular studies, indicate that the closest relatives of the land plants are two groups of aquatic green algae, the **Coleochaetales** and the **Charales** (see Figure 28.1). Both of these algal groups retain their eggs in the parental organism, as do land plants. Of these two candidates, Charales is thought to be the sister-group of land plants, based on the following synapomorphies:

- Plasmodesmata join the cytoplasm of adjacent cells (see Figure 34.6)
- Growth is branching and apical (from the tip)

TABLE 28.1
Classification of Land Plants

GROUP	COMMON NAME	CHARACTERISTICS
NONVASCULAR LAND PLANTS		
Hepatophyta	Liverworts	No filamentous stage; gametophyte flat
Anthocerophyta	Hornworts	Embedded archegonia; sporophyte grows basally (from the ground)
Bryophyta	Mosses	Filamentous stage; sporophyte grows apically (from the tip)
VASCULAR PLANTS		
Lycopodiophyta	Lycophytes: Club mosses and allies	Microphylls in spirals; sporangia in leaf axils
Monilophyta	Horsetails, whisk ferns, ferns	Differentiation between main stem and side branches (overtopping growth)
SEED PLANTS		
Gymnosperms		
Cycadophyta	Cycads	Compound leaves; swimming sperm; seeds on modified leaves
Ginkgophyta	Ginkgo	Deciduous; fan-shaped leaves; swimming sperm
Gnetophyta	Gnetophytes	Vessels in vascular tissue; opposite, simple leaves
Coniferophyta	Conifers	Seeds in cones; needlelike or scalelike leaves
Angiosperms	Flowering plants	Endosperm; carpels; gametophytes much reduced; seeds within fruit

(A) *Chara vulgaris* (stonewort)

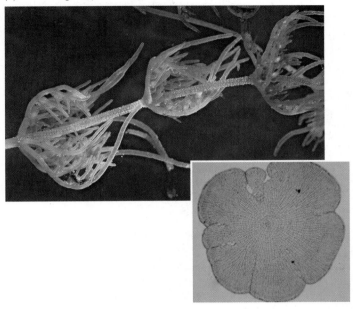

(B) *Coleochaete* sp.

28.2 The Closest Relatives of Land Plants (A) The land plants probably evolved from a common ancestor shared with the Charales, a green algal group. (B) This species of Coleochaete is a representative of the Coleochaetales, the sister group of Charales plus land plants.

- Similar peroxisome contents, mechanics of mitosis and cytokinesis, and chloroplast structure

Both of these algal groups, however, have some features that are similar to groups of land plants. The Charales, as represented by stoneworts of the genus *Chara* (**Figure 28.2A**), exhibit the branching growth form found among most land plants. But the flattened growth form of many members of Coleochaetales, as represented by the genus *Coleochaete* (**Figure 28.2B**), is more like the growth form of basal land plants such as liverworts.

28.1 RECAP

Land plants are photosynthetic organisms that develop from embryos protected by parental plant tissue.

- Explain the different possible uses of the term "plant." See p. 589 and Figure 28.1

- What is the key difference between the vascular plants and the other three clades of land plants? See pp. 589–590 and Table 28.1

- What evidence supports the phylogenetic relationship between land plants and Charales? See p. 590

The green algal ancestors of the land plants lived at the margins of ponds or marshes, ringing them with a green mat. It was from such a marginal habitat, which was sometimes wet and sometimes dry, that early plants made the transition onto land.

28.2 How Did Plants Colonize and Thrive on Land?

Land plants, or their immediate ancestors in those ancient green mats, first appeared in the terrestrial environment between 400 and 500 million years ago. How did they survive in an environment that differed so dramatically from the aquatic environment of their ancestors? While the water essential for life is everywhere in the aquatic environment, water is difficult to obtain and retain in the terrestrial environment.

Adaptations to life on land distinguish land plants from green algae

No longer bathed in fluid, organisms on land faced potentially lethal desiccation (drying). Large terrestrial organisms had to develop ways to transport water to body parts distant from the source. And whereas water provides aquatic organisms with support against gravity, a plant living on land must either have some other support system or sprawl unsupported on the ground. A land plant must also use different mechanisms for dispersing its gametes and progeny than its aquatic relatives, which can simply release them into the water. Survival on land required numerous adaptations. The first colonists—the *nonvascular land plants*—met at least some of these challenges.

Most of the characteristics that distinguish land plants from green algae are evolutionary adaptations to life on land:

- The *cuticle*, a waxy covering that retards water loss

- *Stomata*, small closable openings in leaves and stems that are used to regulate gas exchange (stomata are not present in liverworts)

- *Gametangia*, multicellular organs that enclose plant gametes and prevent them from drying out

- *Embryos*, young plants contained within a protective structure

- Certain *pigments* that afford protection against the mutagenic ultraviolet radiation that bathes the terrestrial environment

- Thick *spore walls* containing a polymer (called sporopollenin) that protects the spores from desiccation and resists decay

- A *mutually beneficial association with a fungus* that promotes nutrient uptake from the soil

The **cuticle** may be the most important—and earliest—of these features. Composed of several unique waxy lipids (see Section 3.4) that coat the leaves and stems of land plants, the cuticle has several functions, the most obvious and important of which is to keep water from evaporating from the plant body.

As ancient plants colonized land, they modified the terrestrial environment by contributing to the formation of soil. Acid secreted by plants helps break down rock, and the organic compounds produced by the breakdown of dead plants contribute to soil structure. Such effects are repeated today as plants grow in new areas.

Nonvascular land plants usually live where water is readily available

Living species of liverworts, mosses, and hornworts are thought to be similar in many ways to the earliest land plants. Most of these plants grow in dense mats, usually in moist habitats. Even the largest of these species are only about half a meter tall, and most are only a few centimeters tall or long. Why have they not evolved to be taller? The probable answer is that they lack an efficient vascular system for conducting water and minerals from the soil to distant parts of the plant body.

The nonvascular land plants lack the leaves, stems, and roots that characterize the vascular plants, although they have structures analogous to each. Their growth pattern allows water to move through the mats of plants by capillary action. They have leaflike structures that readily catch and hold any water that splashes onto them. They are small enough that minerals can be distributed throughout their bodies by diffusion. As in all land plants, layers of maternal tissue protect their embryos from desiccation. Nonvascular land plants also have a cuticle, although it is often very thin (or even absent in some species) and thus is not highly effective in retarding water loss.

Most nonvascular land plants live on the soil or on vascular plants, but some grow on bare rock, dead and fallen tree trunks, and even on buildings. The ability to grow on such marginal surfaces results from a mutualistic association with the glomeromycetes, a clade of fungi. The earliest association of land plants and fungi dates back at least 460 million years. This mutualism probably facilitated the absorption of water and minerals, especially phosphorus, from the first soils.

Nonvascular land plants are widely distributed over six continents and even exist (albeit very locally) on the coast of the seventh, Antarctica. They are well adapted to their environments. Most are terrestrial. Although a few species live in fresh water, these aquatic forms are descended from terrestrial ones. None live in the oceans.

Land plants and green algae differ not just in structure but also in their life cycles. Differences in the life cycles of land plants and those of their ancestors are partially a function of the relative dependence of these respective plant groups on a water supply for reproduction.

Life cycles of land plants feature alternation of generations

A universal feature of the life cycles of land plants is alternation of generations (**Figure 28.3**). Recall from Section 27.4 the two hallmarks of alternation of generations:

- The life cycle includes both a multicellular diploid stage and a multicellular haploid stage.
- Gametes are produced by mitosis, not by meiosis. Meiosis produces spores that develop into multicellular haploid organisms.

If we begin looking at the land plant life cycle at the single-cell stage—the diploid zygote—then the first phase of the cycle is the formation, by mitosis and cytokinesis, of a multicellular em-

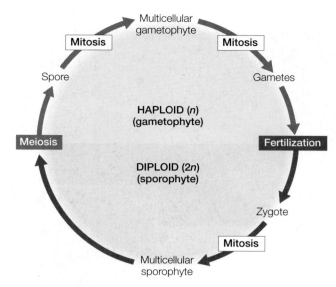

28.3 Alternation of Generations in Plants   A multicellular diploid sporophyte generation produces spores by meiosis and alternates with a multicellular haploid gametophyte generation that produces gametes by mitosis.

bryo, which eventually grows into a mature diploid plant. This multicellular diploid plant is the **sporophyte** ("spore plant").

Cells contained within **sporangia** (singular *sporangium*) of the sporophyte undergo meiosis to produce haploid, unicellular spores. By mitosis and cytokinesis, a spore forms a haploid plant. This multicellular haploid plant, called the **gametophyte** ("gamete plant"), produces haploid gametes by mitosis. The fusion of two gametes (*syngamy*, or *fertilization*) forms a single diploid cell—the zygote—and the cycle is repeated (**Figure 28.4**).

The *sporophyte generation* extends from the zygote through the adult multicellular diploid plant and sporangium formation; the *gametophyte generation* extends from the spore through the adult multicellular haploid plant to the gametes. The transitions between the generations are accomplished by fertilization and meiosis. In all land plants, the sporophyte and gametophyte differ genetically: the sporophyte has diploid cells, and the gametophyte has haploid cells.

There is a trend toward reduction of the gametophyte generation in plant evolution. In the nonvascular land plants, the gametophyte is larger, longer-lived, and more self-sufficient than the sporophyte. In those groups that appeared later in plant evolution, however, the sporophyte generation is the larger, longer-lived, and more self-sufficient one. In the seed plants, this evolutionary trend has led to a condition in which water is not required for the sperm to reach the egg.

The sporophytes of nonvascular land plants are dependent on gametophytes

In nonvascular land plants, the conspicuous green structure visible to the naked eye is the gametophyte (see Figure 28.4). This is in contrast to vascular plants, such as ferns and seed plants, in which the familiar forms are sporophytes. The gametophyte of liverworts, hornworts, and mosses is photosynthetic and

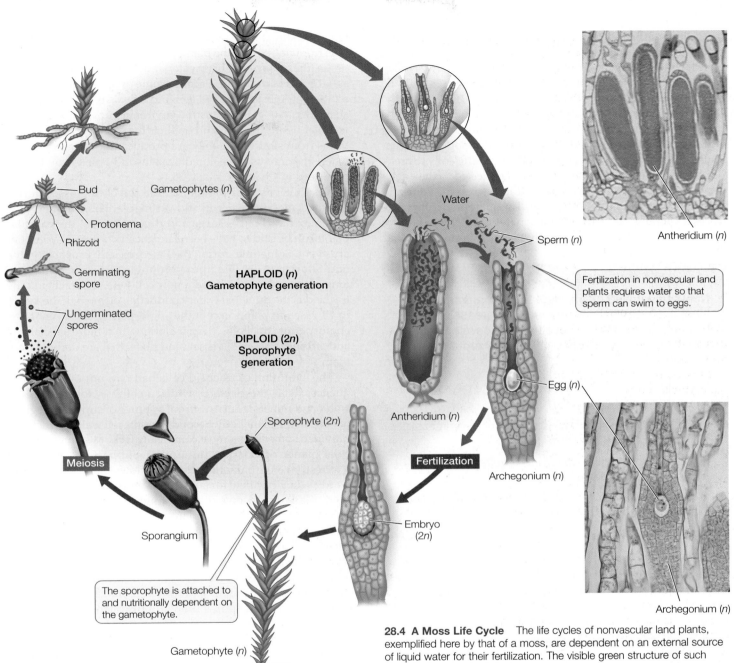

Bud

Protonema

Rhizoid

Gametophytes (*n*)

Germinating spore

Ungerminated spores

HAPLOID (*n*)
Gametophyte generation

DIPLOID (2*n*)
Sporophyte generation

Water

Sperm (*n*)

Antheridium (*n*)

Fertilization in nonvascular land plants requires water so that sperm can swim to eggs.

Egg (*n*)

Antheridium (*n*)

Meiosis

Sporophyte (2*n*)

Fertilization

Archegonium (*n*)

Sporangium

Embryo (2*n*)

The sporophyte is attached to and nutritionally dependent on the gametophyte.

Gametophyte (*n*)

Archegonium (*n*)

28.4 A Moss Life Cycle The life cycles of nonvascular land plants, exemplified here by that of a moss, are dependent on an external source of liquid water for their fertilization. The visible green structure of such plants is the gametophyte.

──yourBioPortal.com──
GO TO **Animated Tutorial 28.1 • Life Cycle of a Moss**

therefore nutritionally independent; the sporophyte may or may not be photosynthetic, but it is always nutritionally dependent on the gametophyte and remains permanently attached to it.

In nonvascular land plants, a sporophyte produces unicellular haploid spores as products of meiosis within a sporangium. A spore germinates, giving rise to a multicellular haploid gametophyte whose cells contain chloroplasts and are thus photosynthetic. Eventually gametes form within specialized sex organs, the **gametangia**. The **archegonium** is a multicellular, flask-shaped female sex organ with a long neck and a swollen base, which produces a single egg. The **antheridium** is a male sex organ in which sperm, each bearing two flagella, are produced in large numbers (see the insets in Figure 28.4). Both archegonia and antheridia are produced on the same individ-

ual, so each individual can have both male and female reproductive structures. Fertilization is often between adjacent individuals, however, which helps maintain genetic diversity in the population.

Once released from the antheridium, the sperm must swim or be splashed by raindrops to a nearby archegonium on the same or a neighboring plant—a constraint that reflects the aquatic origins of the nonvascular land plants' ancestors. The sperm are aided in this task by chemical attractants released by the egg or the archegonium. Before sperm can enter the archegonium, certain cells in the neck of the archegonium must break down, leaving a water-filled canal through which the sperm swim to complete their journey. Note that *all of these events require liquid water.*

On arrival at the egg, the nucleus of a sperm fuses with the egg nucleus to form a diploid zygote. Mitotic divisions of the zygote produce a multicellular, diploid sporophyte embryo. The base of the archegonium grows and surrounds the embryo, which protects the embryo during its early development. Eventually, the developing sporophyte elongates sufficiently to break out of the archegonium, but it remains connected to the gametophyte by a "foot" that is embedded in the parent tissue and absorbs water and nutrients from it. The sporophyte then produces a sporangium, within which meiotic divisions produce spores and thus the next gametophyte generation.

28.2 RECAP

Nonvascular land plants were the first plants to inhabit a terrestrial environment. The transition to land required numerous evolutionary developments, including the cuticle, gametangia, and protected embryos.

- Describe several adaptations of plants to the terrestrial environment. See p. 591

- Explain what is meant by alternation of generations. See p. 592 and Figure 28.3

- Why do nonvascular land plants require liquid water for fertilization? See pp. 592–594

- In the nonvascular land plants, how is the sporophyte dependent on the gametophyte? See pp. 593–594

Further adaptations to the terrestrial environment appeared as plants continued to evolve. One of the most important of these later adaptations was the appearance of vascular tissues.

28.3 What Features Distinguish the Vascular Plants?

The first land plants were nonvascular, lacking both water-conducting and food-conducting tissues. These plants did not leave a very complete fossil record, however, because most nonvascular plant species readily decompose, leaving little or nothing to fossilize.

That the nonvascular land plants evolved tens of millions of years before the earliest vascular plants is supported by extensive phylogenetic and molecular-clock analyses of DNA sequences. The first plants possessing vascular tissue (characterized by fluid-conducting tracheid cells) arose much later (**Figure 28.5**).

Vascular tissues transport water and dissolved materials

Vascular plants differ from the other land plants in crucial ways, one of which is the possession of a well-developed **vascular sys-**tem consisting of tissues specialized for the transport of materials from one part of the plant to another. One type of vascular tissue, the **xylem**, conducts water and minerals from the soil to aerial parts of the plant. Because some of its cell walls contain a stiffening substance called *lignin*, xylem also provides support against gravity in the terrestrial environment. The other type of vascular tissue, the **phloem**, conducts the products of photosynthesis from sites where they are produced or released to sites where they are used or stored. (Xylem and phloem are further discussed in Chapters 34 and 35.)

Familiar vascular plants include the club mosses, ferns, conifers, and angiosperms (flowering plants). Although they are an extraordinarily large and diverse group, the vascular plants can be said to have been launched by a single evolutionary event. Sometime during the Paleozoic era, probably in the mid-Silurian (430 Mya), the sporophyte generation of a now long-extinct plant produced a new cell type, the tracheid. The tracheid is the principal water-conducting element of the xylem in all vascular plants except the angiosperms, and even in the angiosperms, tracheids persist along with a more specialized and efficient system of vessels and fibers that are derived from them.

The evolution of tracheid cells had two important consequences. First, these cells provided a pathway for transport of water and mineral nutrients from a source of supply to regions of need in the plant body. Second, the stiff cell walls of tracheids provided something almost completely lacking among nonvascular plants: rigid structural support. Support is important in a terrestrial environment because it allows plants to grow upward as they compete for sunlight to power photosynthesis. A taller plant can receive direct sunlight and photosynthesize more readily than a shorter plant, whose leaves may be shaded by the taller one. Increased height also improves the dispersal of spores. Thus tracheids set the stage for the complete and permanent invasion of land by plants.

The vascular plants featured another evolutionary novelty: a branching, independent sporophyte. A branching sporophyte can produce more spores than an unbranched body, and it can develop in complex ways. The sporophyte of a vascular plant is nutritionally independent of the gametophyte at maturity. Among the vascular plants, the sporophyte is the large and obvious plant that one normally pays attention to in nature. In contrast, the sporophyte of nonvascular land plants is attached to, dependent on, and usually much smaller than the gametophyte.

The present-day evolutionary descendants of the early vascular plants include the lycophytes, monilophytes, and seed plants (see Figure 28.5; the vascular plant groups are bracketed on the right). Two types of life cycles are seen in the vascular plants: one that involves seeds and another that does not. As we discuss in greater detail in the next chapter, a **seed** consists of a plant embryo, together with a food source, surrounded by a protective coat. The life cycles of the club mosses, ferns, whisk ferns, and horsetails do not involve seeds. We will describe these seedless vascular plant groups in detail after taking a closer look at vascular plant evolution. The major groups of seed plants are described in Chapter 29.

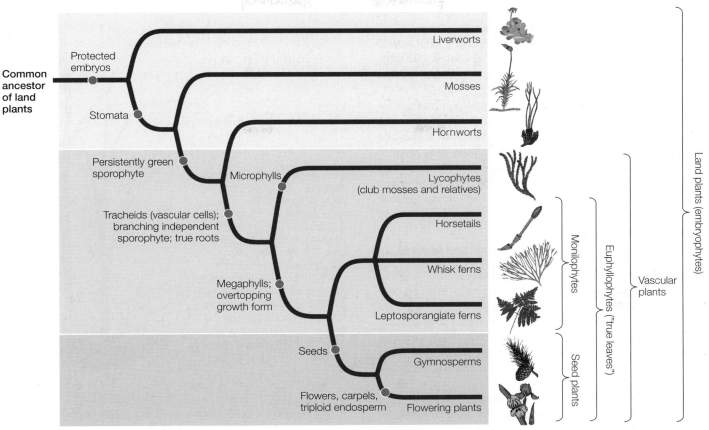

28.5 The Evolution of Plants Three key characteristics that emerged during plant evolution—protected embryos, vascular tissues, and seeds— are adaptations to life in a terrestrial environment.

Vascular plants have been evolving for almost half a billion years

The evolution of an effective cuticle and protective layers for the gametangia (archegonia and antheridia) helped make the first vascular plants successful, as did the initial absence of herbivores (plant-eating animals) on land. By the late Silurian period (about 425 Mya), vascular plants were being preserved as fossils that we can study today. During the Silurian, the largest vascular plants were only a few centimeters tall, yet fossils uncovered in Wales in 2004 give clear evidence for the earliest known wildfire, which burned vigorously even in the Silurian atmosphere, which had 14 percent less oxygen than today's atmosphere (see Figure 25.6). The small plants must have been abundant to sustain fire in such an atmosphere. Their proliferation made the terrestrial environment more hospitable to animals. Amphibians and insects arrived on land soon after land plants became established.

Trees of various kinds appeared in the Devonian period and dominated the landscape of the Carboniferous period (359–297 Mya). Forests of lycophytes (club mosses) up to 40 meters tall, along with horsetails and tree ferns, flourished in the tropical swamps of what would become

North America and Europe (**Figure 28.6**). Plant parts from those forests sank in the swamps and were gradually covered by sediment. Over millions of years, as the buried plant material was subjected to intense pressure and elevated temperatures, it was

28.6 Reconstruction of an Ancient Forest This Carboniferous forest once thrived in what is now Michigan. The "trees" to the left and in the background are lycophytes of the genus *Lepidodendron*; abundant ferns are visible to the right. The plant in the foreground is a relative of the modern horsetails.

transformed into coal. Today that coal provides over half of our electricity (and contributes to air pollution and global warming). The world's coal deposits, although huge, are not infinite, and they cannot be renewed because the conditions that created coal no longer exist.

In the subsequent Permian period, the continents came together to form a single gigantic land mass called Pangaea. The continental interior became warmer and drier, but late in the period glaciation was extensive. The 200-million-year reign of the lycophyte–fern forests came to an end as they were replaced by forests of seed plants (gymnosperms), which were prevalent until a different group of seed plants (angiosperms) overtook the landscape about 65 million years ago.

The earliest vascular plants lacked roots and leaves

The earliest known vascular plants belonged to the now-extinct group called **rhyniophytes**. The rhyniophytes were one of a very few types of vascular plants in the Silurian period. The landscape at that time probably consisted of bare ground, with stands of rhyniophytes in low-lying moist areas. Early versions of the structural features of all the other vascular plant groups appeared in the rhyniophytes of that time. These shared features strengthen the case for the origin of all vascular plants from a common nonvascular plant ancestor.

The beginning of this chapter described the discovery of some important fossils in Devonian rocks near Rhynie, Scotland. The preservation of these plants was remarkable, considering that the rocks were more than 395 million years old. These fossil plants had a simple vascular system of phloem and xylem, but not all had the tracheids characteristic of today's vascular plants.

These plants also lacked roots. Like most modern ferns and lycophytes, they were apparently anchored in the soil by horizontal portions of stem, called **rhizomes**, which bore water-absorbing unicellular filaments called **rhizoids**. These rhizomes also bore aerial branches, and sporangia—homologous to the sporangia of mosses—were found at the tips of those branches. Their branching pattern was *dichotomous*; that is, the apex (tip) of the shoot divided to produce two equivalent new branches, each pair diverging at approximately the same angle from the original stem (**Figure 28.7**). Scattered fragments of such plants had been found earlier, but never in such profusion or as well preserved as those discovered by Kidston and Lang.

Although they were apparently ancestral to the other vascular plant groups, the rhyniophytes themselves are long gone. None of their fossils appear anywhere after the Devonian period.

The vascular plants branched out

The **lycophytes** (club mosses and their relatives) first appeared in the Silurian period. The **monilophytes** (ferns and fern allies) appeared during the Devonian period. These two groups, both still with us today, arose from rhyniophyte-like ancestors. Important new features of the vascular plants are found in these groups, such as true roots, true leaves, and a differentiation between two types of spores. The monilophytes and seed plants constitute a clade called the **euphyllophytes** (*eu*, "true"; *phyllos*, "leaf").

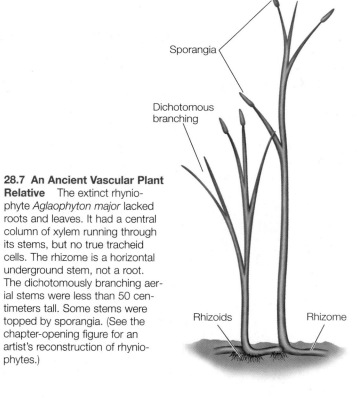

28.7 An Ancient Vascular Plant Relative The extinct rhyniophyte *Aglaophyton major* lacked roots and leaves. It had a central column of xylem running through its stems, but no true tracheid cells. The rhizome is a horizontal underground stem, not a root. The dichotomously branching aerial stems were less than 50 centimeters tall. Some stems were topped by sporangia. (See the chapter-opening figure for an artist's reconstruction of rhyniophytes.)

An important synapomorphy of the euphyllophytes is **overtopping**, a growth pattern in which one branch differentiates from and grows beyond the others. Overtopping growth would have given these plants an advantage in the competition for light for photosynthesis, enabling them to shade their dichotomously growing competitors. And, as we'll see, the overtopping growth of the euphyllophytes enabled a new type of leaf to evolve.

Roots may have evolved from branches

The rhyniophytes had only rhizoids arising from a rhizome with which to gather water and minerals. How, then, did subsequent groups of vascular plants come to have the complex roots we see today?

It is probable that roots had their evolutionary origins as a branch, either of a rhizome or of the aboveground portion of a stem. That branch presumably penetrated the soil and branched further. The underground portion could anchor the plant firmly, and even in this primitive condition, it could absorb water and minerals. The discovery of several fossil plants from the Devonian period, all having horizontal stems (rhizomes) with both underground and aerial branches, supports this hypothesis.

Underground and aboveground branches, growing in sharply different environments, were subjected to very different selection pressures during the succeeding millions of years. Thus the two parts of the plant body—the aboveground shoot system and the underground root system—diverged in structure and evolved distinct internal and external anatomies. In spite of these differences, scientists believe that the root and shoot systems of vascular plants are homologous—that they were once part of the same organ.

nium containing a single egg. The male organ is an antheridium, producing many sperm. Such plants, which bear a single type of spore, are said to be **homosporous** (Figure 28.10A).

A system with two distinct types of spores evolved somewhat later. Plants of this type are said to be **heterosporous** (Figure 28.10B). In heterospory, one type of spore—the **megaspore**—develops into a specifically female gametophyte (a *megagametophyte*) that produces only eggs. The other type, the **microspore**, is smaller and develops into a male gametophyte (a *microgametophyte*) that produces only sperm. The sporophyte produces megaspores in small numbers in *megasporangia*, and microspores in large numbers in *microsporangia*. Heterospory affects not only the spores and the gametophyte but also the sporophyte plant itself, which must develop two types of sporangia.

The most ancient vascular plants were all homosporous, but heterospory evidently evolved several times in the early descendants of the rhyniophytes. The fact that heterospory evolved repeatedly suggests that it affords selective advantages. Subsequent evolution in the land plants featured ever greater specialization of the heterosporous condition.

28.3 RECAP

A new type of cell, the tracheid, marked the origin of the vascular plants. Later evolutionary events included the appearance of roots and leaves.

- How do the vascular tissues xylem and phloem serve the vascular plants? **See p. 594**

- Describe the difference between the two leaf types (microphylls and megaphylls). **See p. 597 and Figure 28.8**

- Explain the concept of heterospory. **See p. 599 and Figure 28.10**

The liverworts, hornworts, mosses, lycophytes, and monilophytes have come a long way from the aquatic environment to meet the challenges of life on dry land. Let's look at the diversity within these groups.

28.4 What Are the Major Clades of Seedless Plants?

Three principal clades of living land plants lack tracheids; these nonvascular land plants are the liverworts, hornworts, and mosses. The structure and growth pattern of the sporophyte differ among the three groups. There are four principal clades of seedless vascular plants—lycophytes (club mosses and their relatives), horsetails, whisk ferns, and leptosporangiate ferns. Most plants known as "ferns" are leptosporangiate ferns, although there are also small groups called "ferns" that are more closely related to horsetails and whisk ferns.

Liverworts may be the most ancient surviving plant clade

There are about 9,000 species of **liverworts** (Hepatophyta, "liver-plants"). Most liverworts have leafy gametophytes (**Figure 28.11A**). Some have *thalloid* gametophytes—green, leaflike layers that lie flat on the ground (**Figure 28.11B**). The simplest liverwort gametophytes, however, are flat plates of cells, a centimeter or so long, that produce antheridia or archegonia on their upper surfaces and rhizoids on their lower surfaces.

Liverwort sporophytes are shorter than those of mosses and hornworts, rarely exceeding a few millimeters. The liverwort sporophyte has a stalk that connects sporangium and foot. In most species, the stalk elongates by expansion of cells throughout its length. This elongation raises the sporangium above ground level, allowing the spores to be dispersed more widely. The sporangia of liverworts are simple: a globular sporangium wall surrounds a mass of spores. In some species of liverworts, spores are not released by the sporophyte until the surrounding sporangium wall rots. In other liverworts, however, the

```
┌─ Liverworts
│
│  ┌─ Mosses
└──┤
   │  ┌─ Hornworts
   └──┤
      └─ Vascular plants
```

28.11 Liverwort Structures Liverworts display various characteristic structures. (A) The gametophyte of a leafy liverwort. (B) Gametophytes of a thalloid liverwort. (C) This thalloid liverwort bears archegonia in the structures that look like bunches of bananas. It also bears cups containing gemmae.

These cups contain gemmae—small, lens-shaped outgrowths of the plant body, each capable of developing into a new plant.

The banana-like structures bear archegonia.

(A) *Bazzania trilobata*

(B) *Marchantia* sp.

(C) *Marchantia* sp.

spores are thrown from the sporangium by structures that shorten and compress a "spring" as they dry out. When the stress becomes sufficient, the compressed spring snaps back to its resting position, throwing spores in all directions.

Among the most familiar thalloid liverworts are species of the genus *Marchantia*. *Marchantia* is easily recognized by the characteristic structures on which its male and female gametophytes bear their antheridia and archegonia (**Figure 28.11C**). Like most liverworts, *Marchantia* also reproduces asexually by simple fragmentation of the gametophyte. *Marchantia* and some other liverworts and mosses also reproduce asexually by means of *gemmae* (singular *gemma*), which are lens-shaped clumps of cells. In a few liverworts, the gemmae are held in structures called *gemmae cups*, which promote dispersal of the gemmae by raindrops (see Figure 28.11C).

Water- and sugar-transport mechanisms first emerged in the mosses

The most familiar of the nonvascular land plants are the **mosses** (**Bryophyta**). These hardy little plants, of which there are about 15,000 species, are found in almost every terrestrial environment. They are often found on damp, cool ground, where they form thick mats (**Figure 28.12**). The mosses are the sister lineage to the vascular plants plus the hornworts (see Figure 28.5).

The mosses, along with the hornworts and vascular plants, share an advance over the liverwort clade in their adaptation to life on land: they have stomata, which are important for both water and gas exchange. Stomata are a shared derived trait of mosses and all other land plants except liverworts.

In mosses, the gametophyte begins its development following spore germination as a branched, filamentous structure called a *protonema* (see Figure 28.4). Although the protonema looks a bit like a filamentous green alga, this structure is unique to the mosses. Some of the filaments contain chloroplasts and are photosynthetic; others, called rhizoids, are nonphotosynthetic and anchor the protonema to the substratum. After a period of linear growth, cells close to the tips of the photosynthetic filaments divide rapidly in three dimensions to form *buds*. The buds eventually develop a distinct tip, or apex, and produce the familiar leafy moss shoot with leaflike structures arranged spirally. These leafy shoots produce antheridia or archegonia (see Figure 28.4).

Some moss gametophytes are so large they could not transport enough water solely by diffusion. Gametophytes and sporophytes of many mosses contain a type of cell called a *hydroid*, which dies and leaves a tiny channel through which water can travel. The hydroid is functionally similar to the tracheid, the characteristic water-conducting cell of the vascular plants, but it lacks lignin and the cell-wall structure found in tracheids. The possession of hydroids and of a limited system for transport of sugar by some mosses (via cells called *leptoids*) shows that the term "nonvascular plant" is somewhat misleading when applied to mosses. Despite their simple system of internal transport, however, the mosses are not vascular plants because they lack true xylem and phloem.

(A)

Sporophytes

Gametophytes

(B) *Polytrichum* sp.

28.12 Mosses Grow in Dense Mats (A) Dense layers of moss carpet a field of solidified volcanic lava in Iceland. (B) A close-up view of moss growing on a forest floor in Michigan.

Mosses of the genus *Sphagnum* (**Figure 28.13A**) often grow in cool, swampy places, where the plants begin to decompose in the water after they die. Rapidly growing upper layers of moss compress the deeper-lying, decomposing layers. Partially decomposed plant matter is called *peat*. In some parts of the world, people derive the majority of their fuel from peat bogs (**Figure 28.13B**). *Sphagnum*-dominated peatlands cover an area approximately half the size of the United States—more than 1 percent of Earth's surface. Millions of years ago, continued compression of peat composed primarily of other seedless plants gave rise to coal.

Hornworts have distinctive chloroplasts and sporophytes without stalks

The approximately 100 species of **hornworts** comprise the group Anthocerophyta ("horn plants"), so named because their sporophytes look like little horns (**Figure 28.14**). Hornworts appear at first glance to be liverworts with very simple gametophytes. Their gametophytes are flat plates of cells a few cells thick.

Hornworts have two characteristics that distinguish them from liverworts and mosses. First, the cells of hornworts each

(A)

— Sporophyte

— Gametophyte

Sphagnum sp.

(B)

28.13 Sphagnum Moss (A) *Sphagnum* bogs are extremely dense growths of the moss shown here. These bogs can cover large areas in temperate climates. (B) A farmer mines a bog for peat, a fossil fuel formed from decomposing *Sphagnum* mosses.

The sporophytes of hornworts can reach 20 cm in height.

Gametophytes are flat plates a few cells thick.

Anthoceros sp.

28.14 A Hornwort The sporophytes of many hornworts resemble little horns.

contain a single large, platelike chloroplast, whereas the cells of the other two groups contain numerous small, lens-shaped chloroplasts. Second, of the sporophytes in all three groups, those of the hornworts come closest to being capable of growth without a set limit. Liverwort and moss sporophytes have a stalk that stops growing as the sporangium matures, so elongation of the sporophyte is strictly limited. The hornwort sporophyte, however, has no stalk. Instead, a basal region of the sporangium remains capable of indefinite cell division, continuously producing new spore-bearing tissue above. The sporophytes of some hornworts growing in mild and continuously moist conditions can become as tall as 20 centimeters. Eventually, however, the sporophyte's growth is limited by the lack of a transport system.

Hornworts have evolved a symbiotic relationship that promotes their growth by providing them with greater access to nitrogen, which is often a limiting resource. Hornworts have internal cavities filled with mucilage; these cavities are often populated by cyanobacteria that convert atmospheric nitrogen gas into a form usable by their host plant.

We present the hornworts here as sister to the vascular plants, but this is only one possible interpretation of the current data. The hornworts and vascular plants are united by DNA sequence analyses, and they also share a persistently green sporophyte. The exact evolutionary position of the hornworts is still unclear, and in some morphological analyses they are placed as the sister group to the mosses plus the vascular plants (the two groups that express apical cell division).

Some vascular plants have vascular tissue but not seeds

The earliest vascular plants did not have seeds, and several major clades of seedless vascular plants have survived to the present. These plants have a large, independent sporophyte and a small gametophyte that is independent of the sporophyte. The gametophytes of the surviving seedless vascular plants are rarely more than 2 centimeters long and are short-lived, whereas their sporophytes are often highly visible and long-lived; the sporophyte of a tree fern, for example, may be up to 20 meters tall and live for many years.

The most prominent resting stage in the life cycle of the seedless vascular plants is the single-celled spore. A spore may "rest" for some time before developing further. This feature makes the life cycle of seedless vascular plants similar to those of the fungi, green algae, and nonvascular land plants, but not, as we will see in the next chapter, to that of the seed plants. Like nonvascular land plants, seedless vascular plants must have an aqueous environment for at least one stage of their life cycle because fertilization is accomplished by flagellated, swimming sperm.

The leptosporangiate ferns are the most abundant and diverse group of seedless vascular plants today, but the club mosses and horsetails were once dominant elements of Earth's vegetation. A fourth group, the whisk ferns, contains only two genera. Let's look at the characteristics of these four groups and at some of the evolutionary advances that appeared in them.

The lycophytes are sister to the other vascular plants

The club mosses and their relatives, the spike mosses and quillworts, together called **lycophytes**, are the sister-group of the remaining vascular plants. There are relatively few surviving species of lycophytes—just over 1,200.

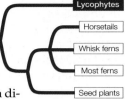

The lycophytes have roots that branch dichotomously. The arrangement of vascular tissue in their stems is simpler than in the other vascular plants. They bear only microphylls, and these simple leaves are arranged spirally on the stem. Growth in club mosses comes entirely from apical cell division, and branching in the stems is also dichotomous, by a division of the apical cluster of dividing cells.

The sporangia of many club mosses are aggregated in cone-like structures called *strobili* (singular *strobilus*; **Figure 28.15**). The strobilus of a club moss is a cluster of spore-bearing leaves inserted on an axis (linear supporting structure). Other club mosses lack strobili and bear their sporangia on (or adjacent to) the upper surfaces of leaves called *sporophylls*. This placement contrasts with the terminal sporangia of the rhyniophytes. There are both homosporous and heterosporous species of club mosses.

Although they are only a minor element of present-day vegetation, the lycophytes are one of two groups that appear to have been the dominant vegetation during the Carboniferous period. One type of coal (cannel coal) is formed almost entirely from fossilized spores of the tree lycophyte *Lepidodendron*—which gives us an idea of the abundance of this genus in the forests of that time (see Figure 28.6). Other major elements of Carboniferous vegetation included horsetails and ferns.

Horsetails, whisk ferns, and ferns constitute a clade

Once thought to be only distantly related, the horsetails, whisk ferns, and ferns form a clade, the monilophytes, or "ferns and fern allies." Within that clade, the whisk ferns and the horsetails are both monophyletic; the ferns are not. However, most ferns do belong to a single clade, the *leptosporangiate ferns*. Other small groups that are commonly called ferns are actually more closely related to horsetails or whisk ferns. In the monilophytes—as in all seed plants—there is differentiation between the main stem and side branches. This pattern contrasts with the dichotomous branching characteristic of the lycophytes and rhyniophytes, in which each split gives rise to two branches of similar size (see Figure 28.7).

HORSETAILS Today there are only about 15 species of **horsetails**, all in the genus *Equisetum*. Horsetails have been sometimes called "scouring rushes" because rough silica deposits found in their cell walls made them useful for cleaning. They have true roots that branch irregularly. Horsetails have a large sporophyte and a small gametophyte, both independent.

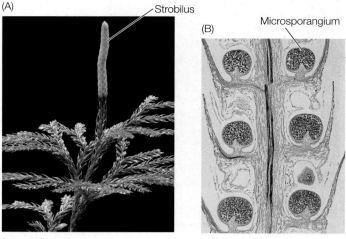

Lycopodium obscurum

28.15 Club Mosses   (A) A strobilus is visible at the tip of this club moss. Club mosses have microphylls arranged spirally on their stems. (B) A thin section through a strobilus of a club moss, showing microsporangia.

The small leaves of horsetails are reduced megaphylls and form in distinct whorls (circles) around the stem (**Figure 28.16**). Growth in horsetails originates to a large extent from discs of dividing cells just above each whorl of leaves, so each segment of the stem grows from its base. Such basal growth is uncommon in plants, although it is found in the grasses, a major group of flowering plants, as well as in the hornworts.

Equisetum pratense   *Equisetum arvense*

28.16 Horsetails   (A) Horsetails have a distinctive growth pattern in which the stem grows in segments above each whorl of leaves. These are fertile shoots, with sporangia-bearing structures at the apex. (B) Scanning electron micrograph of horsetail sporangia.

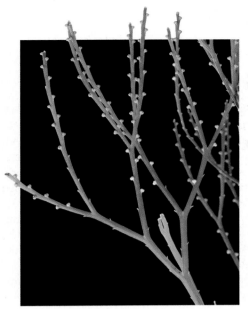

Psilotum flaccidum

28.17 A Whisk Fern DNA sequence data reveals that *Psilotum*—once considered to be a surviving rhyniophyte—is in fact a much more recent genus that arose from a fernlike ancestor. Whisk ferns are now grouped with the monilophytes.

WHISK FERNS There once was some disagreement about whether the rhyniophytes are entirely extinct. The confusion arose because of the existence today of about 15 species in two genera of rootless, spore-bearing plants, *Psilotum* and *Tmesipteris*, collectively called the **whisk ferns**. *Psilotum* (**Figure 28.17**) has minute scales instead of true leaves, but plants of the genus *Tmesipteris* have flattened photosynthetic organs—reduced megaphylls—with well-developed vascular tissue. Are these two genera the living relics of the rhyniophytes, or do they have more recent origins?

The whisk ferns once were thought to be evolutionarily ancient descendants of anatomically simple ancestors. That hypothesis was weakened by an enormous hole in the fossil record between the rhyniophytes, which apparently became extinct more than 300 million years ago, and *Psilotum* and *Tmesipteris*, which are modern plants. DNA sequence data finally settled the question in favor of a more recent origin of the whisk ferns from fernlike ancestors. The whisk ferns are a clade of highly specialized plants that evolved fairly recently from anatomically more complex ancestors by loss or reduction of megaphylls and true roots. Whisk fern gametophytes live below the surface of the ground and lack chlorophyll. They depend on fungal partners for their nutrition.

LEPTOSPORANGIATE FERNS The first **leptosporangiate ferns** appeared during the Devonian period; today this group comprises more than 12,000 species. The sporangia of leptosporangiate ferns are borne on a stalk and have walls only one cell thick. The name "leptosporangiate" refers to these thin-walled sporangia (*lepton*, "thin"). Two other small groups that are also called ferns are actually more closely related to horsetails and whisk ferns, even though they are superficially similar to the leptosporangiate ferns.

The sporophytes of ferns, like those of the seed plants, have true roots, stems, and leaves. Ferns are characterized by large leaves with branching vascular strands (**Figure 28.18A**). During its development, the fern leaf unfurls from a tightly coiled "fiddlehead" (**Figure 28.18B**). Some fern leaves become climbing organs and may grow to be as long as 30 meters. A few species have small leaves as a result of evolutionary reduction, but even these small leaves have more than one vascular strand, and are thus megaphylls (**Figure 28.18C**).

(A)

Dicksonia sp.

Blechnum discolor (crown ferns)

(B)

(C)

Marsilea sp.

Salvinia sp.

28.18 Fern Leaves Take Many Forms (A) Tree ferns and crown ferns dominate this forest on Stewart Island, New Zealand. (B) The "fiddlehead" (developing leaf) of a common forest fern will unfurl and expand, giving rise to a complex adult leaf such as that of a crown fern. (C) The leaves of two species of water ferns.

THE FERN LIFE CYCLE In all ferns, *spore mother cells* inside the sporangia undergo meiosis to form haploid spores. Once shed, the spores may be blown great distances by the wind and eventually germinate to form independent gametophytes far from the parent sporophyte. A case in point is Old World climbing fern (*Lygodium microphyllum*), which is currently spreading disastrously through the Florida Everglades, choking off the growth of other plants. This rapid spread is testimony to the effectiveness of windborne spores. Another example is the remarkable diversity of ferns that have spread through the isolated Hawaiian Islands.

Fern gametophytes have the potential to produce both antheridia and archegonia, although not necessarily at the same time or on the same gametophyte. Sperm swim through water to archegonia—often to those on other gametophytes—where they unite with an egg. The resulting zygote develops into a new sporophyte embryo. The young sporophyte sprouts a root and can thus grow independently of the gametophyte. In the alternating generations of a fern, the gametophyte is small, delicate, and short-lived, but the sporophyte can be very large and can sometimes survive for hundreds of years (**Figure 28.19**).

Because they require liquid water for the transport of the male gametes to the female gametes (as do all other plant groups discussed in this chapter), most ferns inhabit shaded, moist woodlands and swamps. Tree ferns can reach heights of 20 meters. Tree ferns are not as rigid as woody plants, and they have poorly developed root systems. Thus they do not grow in sites exposed directly to strong winds, but rather in ravines or beneath trees in forests. The sporangia of ferns typically are found on the undersurfaces of the leaves, sometimes covering the entire undersurface and sometimes only at the edges. In most species, the sporangia are found in clusters called *sori* (singular *sorus*).

Most ferns are homosporous. However, two groups of aquatic ferns, the Marsileaceae and the Salviniaceae (see Figure 28.18C),

yourBioPortal.com
GO TO Web Activity 28.3 • The Fern Life Cycle

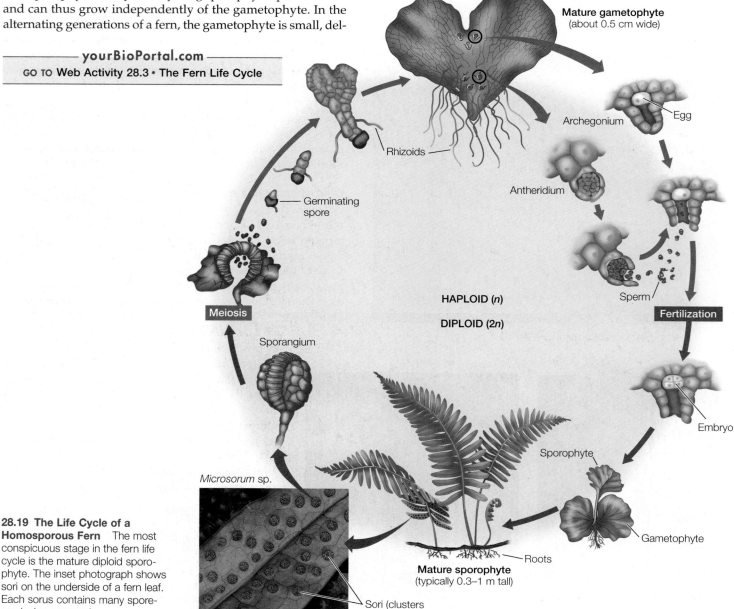

Mature gametophyte
(about 0.5 cm wide)

Archegonium
Egg

Antheridium

Rhizoids

Germinating spore

Sperm

Meiosis

Fertilization

Sporangium

HAPLOID (*n*)

DIPLOID (*2n*)

Embryo

Sporophyte

Microsorum sp.

Gametophyte

Roots

Mature sporophyte
(typically 0.3–1 m tall)

Sori (clusters of sporangia)

28.19 The Life Cycle of a Homosporous Fern The most conspicuous stage in the fern life cycle is the mature diploid sporophyte. The inset photograph shows sori on the underside of a fern leaf. Each sorus contains many spore-producing sporangia.

are derived from a common ancestor in which heterospory evolved. The megaspores and microspores of these plants (which germinate to produce female and male gametophytes, respectively) are produced in different sporangia (megasporangia and microsporangia), and the microspores are always much smaller and greater in number than the megaspores.

A few genera of ferns produce a tuberous, fleshy gametophyte instead of the characteristic flattened, photosynthetic structure produced by most ferns. These tuberous gametophytes depend on a mutualistic fungus for nutrition. In some genera, even the sporophyte embryo must become associated with the fungus before its development can proceed. Chapter 30 discusses many other important plant–fungus mutualisms.

The seedless vascular plants, and especially the ferns, were long considered an evolutionary cul-de-sac—that is, a group with great diversity in the fossil record but less diversity in the present. However, recent DNA-based research has suggested that the diversification of today's ferns took place much more recently than previously thought. The expansion of seed plants and their dominance of forests actually predates the diversification of extant ferns, which presumably took advantage of the new environments created by those forests.

28.4 RECAP

Three clades of land plants lack true vascular systems (liverworts, mosses, and hornworts). The seedless vascular plants include the club mosses, horsetails, whisk ferns, and leptosporangiate ferns.

- Describe the different branching patterns of lycophytes and monilophytes. **See pp. 602–603**
- Why were whisk ferns once thought to be close relatives of the rhyniophytes? **See p. 603**
- Why do most ferns live in moist, shady areas? **See p. 604**

All of the groups described in this chapter—the nonvascular land plants and the seedless vascular plants—require water at a key stage in their life cycles, because their sperm can reach their eggs only by means of liquid water. In addition, the vascular plant groups we have discussed thus far all disperse by spores. The next chapter describes the *seed plants* that dominate Earth's vegetation today, and in which the seed affords new sporophytes protection unavailable to the progeny of seedless plants.

CHAPTER SUMMARY

28.1 How Did the Land Plants Arise?

- **Land plants**, sometimes referred to as **embryophytes**, are photosynthetic eukaryotes that develop from embryos protected by parental tissue. **Review Figure 28.1**
- **Streptophytes** include the land plants and certain green algae. **Green plants** include the streptophytes and the remaining green algae.
- Land plants arose from an aquatic green algal ancestor related to today's **Charales**.
- The **vascular plants** have well developed water-conducting tissues with cells including **tracheids**; the three major groups of **nonvascular land plants** do not. **Review Table 28.1**

28.2 How Did Plants Colonize and Thrive on Land?

- The acquisition of a **cuticle**, **stomata**, **gametangia**, a protected embryo, protective pigments, thick spore walls with a protective polymer, and a mutualistic association with a fungus are all adaptations to terrestrial life.
- All land plant life cycles feature alternation of generations, in which a multicellular diploid **sporophyte** alternates with a multicellular haploid **gametophyte**. **Review Figure 28.3**
- Spores form in **sporangia**; gametes form in gametangia. In seedless land plants, the female and male gametangia are, respectively, an **archegonium** and an **antheridium**.
- In liverworts, hornworts, and mosses, the sporophyte is smaller than the gametophyte and depends on it for water and nutrition. **Review Figure 28.4, ANIMATED TUTORIAL 28.1**

28.3 What Features Distinguish the Vascular Plants?

- A **vascular system** consisting of **xylem** and **phloem** conducts water, minerals, and products of photosynthesis through the bodies of vascular plants.

- Among living vascular plant groups, the **lycophytes** (club mosses and relatives) have only small, simple leaves (**microphylls**). Larger, more complex leaves (**megaphylls**) are found in **monilophytes** (horsetails, ferns, and allies) and seed plants. These latter two groups comprise the **euphyllophytes**. **Review Figure 28.5**
- In vascular plants, the sporophyte is larger than and independent of the gametophyte.
- The **rhyniophytes**, the earliest vascular plants, are known to us only in fossil form. They lacked roots and leaves but possessed **rhizomes** and **rhizoids**. **Review Figure 28.7**
- Roots may have evolved either from rhizomes or from branches. Microphylls probably evolved from sterile sporangia, and megaphylls may have resulted from the flattening and reduction of an **overtopping** branching stem system. **Review Figure 28.8**
- Many seedless vascular plants are **homosporous**, but **heterospory**—the production of distinct **megaspores** and **microspores**—evolved several times. Megaspores develop into megagametophytes; microspores develop into microgametophytes. **Review Figure 28.10, WEB ACTIVITIES 28.1 and 28.2**

28.4 What Are the Major Clades of Seedless Plants?

- The nonvascular land plant clades are the **liverworts**, the **mosses**, and the **hornworts**. The seedless vascular plant groups are the **lycophytes** (club mosses and their relatives) and the monilophytes (**horsetails**, **whisk ferns**, and **leptosporangiate ferns**). **Review Figure 28.5**
- Mosses, hornworts, and vascular plants all have surface pores (stomata) in their leaves.
- The gametophyte of ferns is small, delicate, and short-lived, whereas the sporophyte of ferns is typically much larger and longer lived. **Review Figure 28.19, WEB ACTIVITY 28.3**

SELF-QUIZ

1. Land plants differ from photosynthetic protists in that only the plants
   a. are photosynthetic.
   b. are multicellular.
   c. possess chloroplasts.
   d. have multicellular embryos protected by the parent.
   e. are eukaryotic.

2. Which statement about alternation of generations in land plants is *not* true?
   a. The gametophyte and sporophyte differ in appearance.
   b. Meiosis occurs in sporangia.
   c. Gametes are always produced by meiosis.
   d. The zygote is the first cell of the sporophyte generation.
   e. The gametophyte and sporophyte differ in chromosome number.

3. Which statement is *not* evidence for the origin of land plants from the green algae?
   a. Some green algae have multicellular sporophytes and multicellular gametophytes.
   b. Both plants and green algae have cellulose in their cell walls.
   c. The two groups have the same photosynthetic pigments.
   d. Both plants and green algae produce starch as their principal storage carbohydrate.
   e. All green algae produce large, stationary eggs.

4. Liverworts, mosses, and hornworts
   a. lack a sporophyte generation.
   b. grow in dense masses, allowing capillary movement of water.
   c. possess xylem and phloem.
   d. possess true leaves.
   e. possess true roots.

5. Which statement is *not* true of the mosses?
   a. The sporophyte is dependent on the gametophyte.
   b. Sperm are produced in archegonia.
   c. There are more species of mosses than of liverworts and hornworts combined.
   d. The sporophyte grows by apical cell division.
   e. Mosses are probably sister to the vascular plants plus hornworts.

6. Megaphylls
   a. probably evolved only once.
   b. are found in all the vascular plant groups.
   c. probably arose from sterile sporangia.
   d. are the characteristic leaves of club mosses.
   e. are the characteristic leaves of horsetails and ferns.

7. The rhyniophytes
   a. lacked tracheids.
   b. possessed true roots.
   c. possessed sporangia at the tips of stems.
   d. possessed leaves.
   e. lacked branching stems.

8. Club mosses and horsetails
   a. have larger gametophytes than sporophytes.
   b. possess small leaves.
   c. are represented today primarily by trees.
   d. have never been a dominant part of the vegetation.
   e. produce fruits.

9. Which statement about ferns is *not* true?
   a. The sporophyte is larger than the gametophyte.
   b. Most are heterosporous.
   c. The young sporophyte can grow independently of the gametophyte.
   d. The leaf is a megaphyll.
   e. The gametophytes produce archegonia and antheridia.

10. The leptosporangiate ferns
    a. are not a monophyletic group.
    b. have sporangia with walls more than one cell thick.
    c. constitute a minority of all ferns.
    d. are monilophytes.
    e. produce seeds.

FOR DISCUSSION

1. Mosses and ferns share a common trait that makes water droplets a necessity for sexual reproduction. What is that trait?

2. Are the mosses well adapted to terrestrial life? Explain your answer.

3. Ferns display a dominant sporophyte generation (with large leaves). Describe the major advance in anatomy that enables most ferns to grow much larger than mosses.

4. What features distinguish club mosses from horsetails? What features distinguish these groups from rhyniophytes? From ferns?

5. Why did some botanists once believe that the whisk ferns should be classified together with the rhyniophytes?

6. Contrast microphylls with megaphylls in terms of structure, evolutionary origin, and occurrence among plants.

ADDITIONAL INVESTIGATION

The findings of Osborne and co-workers on the evolution of megaphylls (see Figure 28.9) support the concept that large megaphylls became common only after the atmospheric CO_2 level had dropped, so that more stomata were produced, allow- ing water to evaporate and cool larger leaves. How might you extend that work to confirm the involvement of temperature as a factor limiting leaf size?

29 The Evolution of Seed Plants

A seed from biblical times germinates after 2,000 years

The fruit of the Judean date palm was once much prized. The prophet Muhammad admired its nutritional and medicinal properties; in the Koran, it is associated with heaven and described as a symbol of goodness. The Judean date was the source of the "honey" in the biblical "land of milk and honey." Today that ancient strain of date is extinct. Or is it?

Around two thousand years ago, at the beginning of the Common Era, a seed developed in a fruit on a Judean date palm. The fruit that contained that seed found its way to a storeroom in the fortress Masada in Judea. In 73 C.E. almost a thousand Jewish Zealots involved in a religious revolt against Rome fled to this refuge with their families. Roman legions followed, and the ensuing siege lasted more than two years. In the end, rather than be killed or enslaved by the Roman soldiers, the Zealots are said to have killed themselves and their families in a dramatic mass suicide.

Twenty centuries later, archeologists working in Masada discovered the date seed that was long ago stored in the Masada fortress and confirmed its age. The previous record for seed survival and germination was 1,300 years, held by lotus seeds that recently germinated under the care of scientists in China. But botanist Elaine Solowey succeeded in making the 2,000-year-old date seed germinate! The resulting seedling (nicknamed Methuselah) has continued to thrive and grow. Perhaps the ancient Egyptians knew what they were doing when they placed date seeds in the tombs of the Pharaohs as symbols of immortality.

Seeds are important structures for the survival of plants. They protect the plant embryo within from environmental extremes through what may be a long and stressful resting period—in the case of the Judean date, many centuries in a harsh desert. Seeds of the coconut palm remain dormant for years as they float across vast expanses of ocean, finally washing up on a distant shore, where they germinate and grow. Such hardiness is one of the properties that have contributed to making seed plants the predominant plants on Earth. All of today's forests are dominated by seed plants.

So will the seedling growing under Dr. Solowey's care serve as the parent of a new population of Judean dates, thus resurrecting that genotype from extinction? Unfortunately, it cannot do so alone, because date trees are of two different sexes. However, it may be possible to germinate

A Refuge King Herod of Judea fortified Masada and stocked it with water and food—including Judean dates.

The Hardy Seed This coconut seed arrived on a beach, where it germinated successfully. The evolution of seeds was a major factor in the eventual dominance of the seed plants.

other Judean dates, and/or to cross the tree with Moroccan, Egyptian, or Iraqi varieties of date palms. Although these latter varieties differ from Judean dates at almost half of the gene loci that have been examined, crossing these varieties may introduce many beneficial alleles from Judean dates back into modern date plantations.

Today, humans are collecting and storing seeds from many species in seed banks (see Chapter 34). Many plant species or cultivars that would otherwise be lost to extinction are thus being preserved for restoration projects or other uses by future generations.

IN THIS CHAPTER we will describe the defining characteristics of the seed plants as a group. Living seed plants include the gymnosperms, which produce seeds not protected by ovary or fruit tissue, and the angiosperms, which are characterized by flowers and fruits. We will cover some of the unsolved problems in seed plant evolution, concluding with a survey of the diversity of living seed plants.

29.1 How Did Seed Plants Become Today's Dominant Vegetation?

By the late Devonian period, more than 360 million years ago, Earth was home to a great variety of land plants, many of which are discussed in Chapter 28. The land plants shared the hot, humid terrestrial environment with insects, spiders, centipedes, and fishlike amphibians (early tetrapods). The plants and animals affected one another, acting as agents of natural selection.

In the Devonian an innovation had appeared: some plants developed extensively thickened woody stems, which resulted from the proliferation of xylem. This type of growth in the diameter of stems and roots is called **secondary growth**. Among the first plants with this adaptation were seedless vascular plants called *progymnosperms*, all species of which are now extinct.

The earliest fossil evidence of seed plants is found in late Devonian rocks. Like the progymnosperms, these *seed ferns* were woody. They possessed fernlike foliage but had seeds attached to their leaves. By the end of the Permian, other groups of seed plants became dominant (**Figure 29.1**).

The living seed plants fall into two major groups, the **gymnosperms** (such as pines and cycads) and the **angiosperms** (flowering plants). There are several competing phylogenetic hypotheses regarding the major groups of gymnosperms relative to the angiosperms, but **Figure 29.2** shows the most widely supported relationships. All living gymnosperms and many angiosperms show secondary growth. The life cycles of all seed plants also share distinctive features, as we are about to see.

Features of the seed plant life cycle protect gametes and embryos

Section 28.2 describes a trend in plant evolution: the sporophyte became less dependent on the gametophyte, which became smaller in relation to the sporophyte. This trend continued with the appearance of the seed plants, whose gametophyte generation is reduced even further than it is in the ferns (**Figure 29.3**). The haploid gametophyte develops partly or entirely while attached to and nutritionally dependent on the diploid sporophyte.

Among the seed plants, only the earliest groups of gymnosperms (such as cycads and ginkgos) had swimming sperm. Later groups of gymnosperms and the angiosperms evolved other means of bringing eggs and sperm together. The culmination of this striking evolutionary trend in seed plants was independence from the liquid water that earlier plants needed to assist sperm in reaching the egg. The advent of the seed gave seed plants the opportunity to colonize drier areas and spread over the terrestrial environment.

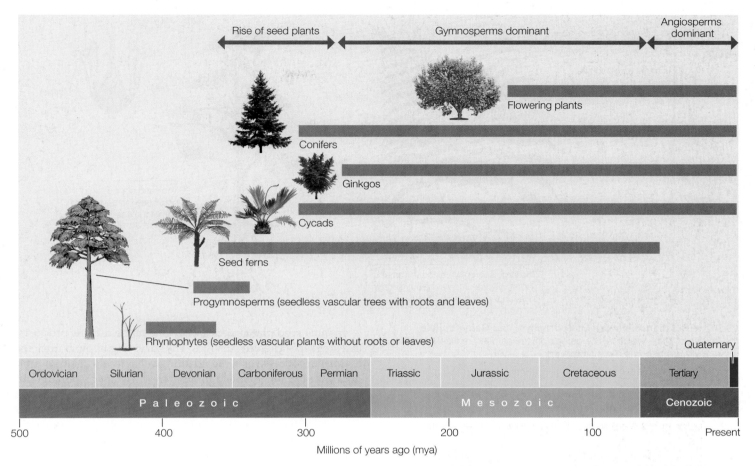

29.1 The Fossil Record of Seed Plants and Some of Their Extinct Seedless Relatives Woody growth evolved in the seedless progymnosperms. The now-extinct seed ferns had woody growth, fernlike foliage, and seeds attached to their leaves. New lineages of seed plants arose during the Carboniferous, but the earliest known fossils of flowering plants are from the late Jurassic.

Seed plants are heterosporous (see Figure 28.9B); that is, they produce two types of spores, one that becomes the male gametophyte and one that becomes the female gametophyte. They form separate microsporangia and megasporangia on structures that are grouped on short axes, such as the stamens and pistils of an angioperm flower.

Within the microsporangium, the meiotic products are microspores, which divide mitotically within the spore wall one or a few times to form a multicellular male gametophyte called a **pollen grain**. Pollen grains are released from the microsporangium to be distributed by wind or by an animal pollinator (**Figure 29.4**). The wall of the pollen grain contains *sporopollenin*, the most chemically resistant biological compound known, which protects the pollen grain against dehydration and chemical damage—another advantage in terms of survival in the terrestrial environment. Recall that sporopollenin in spore walls contributed to the successful colonization of the terrestrial environment by the earliest land plants.

In contrast to the pollen grains produced by the microporangia, the megaspores of seed plants are not shed. Instead, they develop into female gametophytes within the megasporangia. These megagametophytes are dependent on the sporophyte for food and water.

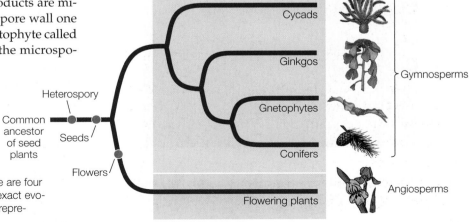

29.2 The Major Groups of Living Seed Plants There are four groups of gymnosperms and one of angiosperms. Their exact evolutionary relationship is still uncertain, but this cladogram represents one current interpretation.

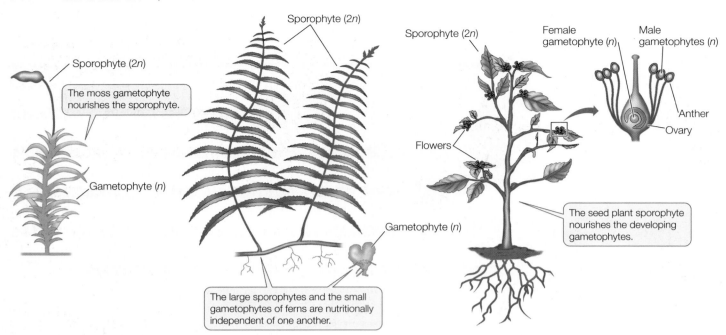

Sporophyte (2n)

The moss gametophyte nourishes the sporophyte.

Gametophyte (n)

Sporophyte (2n)

Gametophyte (n)

The large sporophytes and the small gametophytes of ferns are nutritionally independent of one another.

Sporophyte (2n)

Female gametophyte (n)

Male gametophytes (n)

Flowers

Anther

Ovary

The seed plant sporophyte nourishes the developing gametophytes.

29.3 The Relationship between Sporophyte and Gametophyte In the course of plant evolution, the gametophyte has been reduced and the sporophyte has become more prominent.

Betula pendula

29.4 Pollen Grains Pollen grains are the male gametophytes of seed plants. The pollen of this silver birch is dispersed by the wind, and grains may land near the female gametophytes of the same or other silver birch trees.

In most seed plant species, only one of the meiotic products in a megasporangium survives. The surviving haploid nucleus divides mitotically, and the resulting cells divide again to produce a multicellular female gametophyte. The megasporangium is surrounded by sterile sporophytic structures, which form an **integument** that protects the megasporangium and its contents. Together, the megasporangium and integument constitute the **ovule**, which will develop into a seed after fertilization (**Figure 29.5**).

The arrival of a pollen grain at an appropriate landing point, close to a female gametophyte on a sporophyte of the same species, is called **pollination**. A pollen grain that reaches this point develops further. It produces a slender **pollen tube** that elongates and digests its way toward the megagametophyte (see Figure 29.5). When the tip of the pollen tube reaches the megagametophyte, sperm are released from the tube and fertilization occurs.

The resulting diploid zygote divides repeatedly, forming an embryonic sporophyte. After a period of embryonic development, growth is temporarily suspended (the embryo enters a *dormant* stage). The end product at this stage is a multicellular **seed**.

The seed is a complex, well-protected package

A seed contains tissues from three generations. A *seed coat* develops from the integument—tissues of the diploid sporophyte parent that surround the megasporangium. Within the megasporangium is the haploid female gametophytic tissue from the next generation, which contains a supply of nutrients for the developing embryo. (This tissue is fairly extensive in most gymnosperm seeds. In angiosperm seeds it is greatly reduced, and nutrition for the embryo is supplied by a tissue called *endosperm*, which is described below.) In the center of the seed is the third generation, the embryo of the new diploid sporophyte.

The seed of a gymnosperm or an angiosperm is a well-protected resting stage. The seeds of some species may remain dormant but stay *viable* (capable of growth and development) for

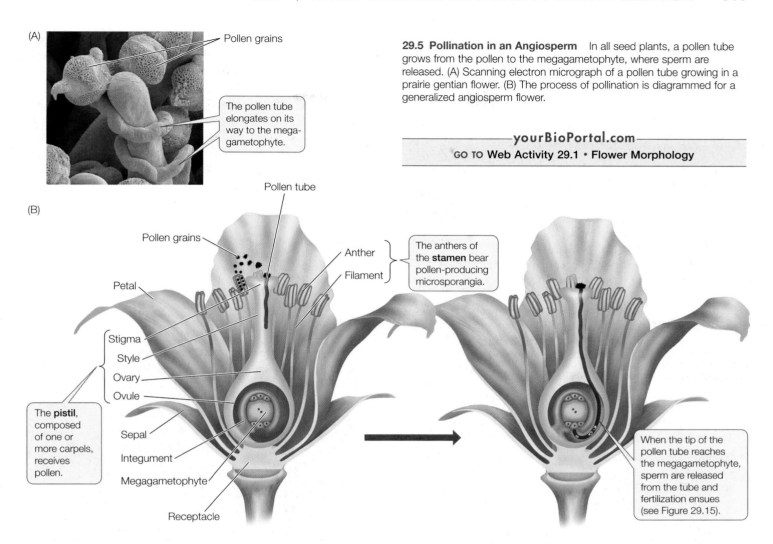

(A)

Pollen grains

The pollen tube elongates on its way to the mega-gametophyte.

29.5 Pollination in an Angiosperm   In all seed plants, a pollen tube grows from the pollen to the megagametophyte, where sperm are released. (A) Scanning electron micrograph of a pollen tube growing in a prairie gentian flower. (B) The process of pollination is diagrammed for a generalized angiosperm flower.

yourBioPortal.com
GO TO **Web Activity 29.1 • Flower Morphology**

(B)

Pollen tube

Pollen grains

Petal

Stigma
Style
Ovary
Ovule

The **pistil**, composed of one or more carpels, receives pollen.

Sepal

Integument

Megagametophyte

Receptacle

Anther
Filament

The anthers of the **stamen** bear pollen-producing microsporangia.

When the tip of the pollen tube reaches the megagametophyte, sperm are released from the tube and fertilization ensues (see Figure 29.15).

many years, germinating only when conditions are favorable for the growth of the sporophyte, as happened with the 2,000-year-old Judean date seed mentioned at the beginning of this chapter. In contrast, the embryos of seedless plants develop directly into sporophytes, which either survive or die, depending on environmental conditions. Spores of some seedless plants may remain dormant and viable for long periods of time, but seeds provide a more secure and lasting dormant stage.

During the dormant stage, the seed coat protects the embryo from excessive drying and may also protect it against potential predators that would otherwise eat the embryo and its nutrient reserves. Many seeds have structural adaptations that promote their dispersal by wind or, more often, by animals. When the young sporophyte resumes growth, it draws on the food reserves in the seed. The possession of seeds is a major reason for the enormous evolutionary success of the seed plants, which are the dominant life forms of most modern terrestrial floras.

A change in anatomy enabled seed plants to grow to great heights

The most ancient seed plants produced **wood**—extensively proliferated xylem—which gave them the support to grow taller than other plants around them, thus capturing more light for

photosynthesis. The younger portion of wood is well adapted for water transport, whereas older wood becomes clogged with resins or other materials. Although no longer functional in transport, the older wood continues to provide support for the plant.

Not all seed plants are woody. In the course of seed plant evolution, many seed plants lost the woody growth habit; however, other advantageous attributes helped them become established in an astonishing variety of places.

29.1 RECAP

Pollen, seeds, and wood are major evolutionary innovations of the seed plants. Protection of the gametes and embryos is a hallmark of seed plants.

- Distinguish between the roles of the megagametophyte and the pollen grain. **See p. 609**

- Explain the importance of pollen in freeing seed plants from dependence on liquid water. **See pp. 609–610 and Figure 29.5**

- What are some of the advantages afforded by seeds? By wood? **See pp. 610–611**

The seed ferns have long been extinct, but the surviving seed plants have been remarkable successes. Next we examine the gymnosperms, the next group of plants to dominate terrestrial environments.

29.2 What Are the Major Groups of Gymnosperms?

The gymnosperms are seed plants that do not form flowers. Gymnosperms (which means "naked-seeded") are so named because their ovules and seeds are not protected by ovary or fruit tissue. Although there are probably fewer than 1,200 species of living gymnosperms, these plants are second only to the angiosperms in their dominance of the terrestrial environment.

> Cycads
> Ginkgos
> Gnetophytes
> Conifers
> Angiosperms

Although the modern gymnosperms are probably a clade, their monophyly has not been established beyond a doubt. The four major groups of living gymnosperms bear little superficial resemblance to one another.

- The **cycads** (**Cycadophyta**) are palmlike plants of the tropics and subtropics, growing as tall as 20 meters (**Figure 29.6A**). Of the present-day gymnosperms, the cycads are probably the earliest-diverging clade. There are about 300 species. Their tissues are often highly toxic to humans if ingested.

- **Ginkgos** (**Ginkgophyta**), which were common during the Mesozoic era, are represented today by a single genus and species: *Ginkgo biloba*, the maidenhair tree (**Figure 29.6B**).

There are both male (microsporangiate) and female (megasporangiate) maidenhair trees. The difference is determined by X and Y sex chromosomes, as in humans; few other plants have distinct sex chromosomes.

- **Gnetophytes** (**Gnetophyta**) number about 90 species in three very different genera, which share certain characteristics analogous to ones found in the angiosperms, as we will see. One of the gnetophytes is *Welwitschia* (**Figure 29.6C**), a long-lived desert plant with just two straplike leaves that sprawl on the sand and can grow as long as 3 meters.

29.6 Diversity among the Gymnosperms   (A) Many cycads have growth forms that resemble both ferns and palms, but cycads are not closely related to either. (B) The characteristic fleshy seed coat and broad leaves of the maidenhair tree. (C) A gnetophyte growing in the Namib Desert of Africa. Straplike leaves grow throughout the life of the plant, breaking and splitting as they grow. (D) Conifers, like this giant sequoia growing in Sequoia National Park, California, dominate many modern forests.

(A) *Encephalartos villosus*

(B) *Ginkgo biloba*

(C) *Welwitschia mirabilis*

(D) *Sequoiadendron giganteum*

- **Conifers** (**Coniferophyta**) are by far the most abundant of the gymnosperms. There are about 700 species of these cone-bearing plants, including the pines and redwoods (**Figure 29.6D**).

With the exception of the gnetophytes, the living gymnosperm groups have only tracheids as water-conducting and support cells within the xylem. In most angiosperms, cells called vessel elements and fibers (specialized for water conduction and support, respectively) are found alongside tracheids. While the gymnosperm water-transport and support system may thus seem somewhat less efficient than that of the angiosperms, it serves some of the largest trees known. The coastal redwoods of California are the tallest gymnosperms; the largest are well over 100 meters tall.

During the Permian, as environments became warmer and dryer, the conifers and cycads flourished. Gymnosperm forests changed over time as the gymnosperm groups evolved. Gymnosperms dominated the Mesozoic era, during which the continents drifted apart and large dinosaurs walked the Earth. Gymnosperms were the principal trees in all forests until about 65 million years ago, and even down to the present day, conifers are the dominant trees in many forests, especially of higher latitudes and altitudes. The oldest living single organism on Earth today is a gymnosperm in California—a bristlecone pine that germinated about 4,800 years ago, at about the time the ancient Egyptians were just starting to develop writing.

Conifers have cones but no motile gametes

The great Douglas fir and cedar forests found in the northwestern United States and the massive boreal forests of pine, fir, and spruce of the northern regions of Eurasia and North America, as well as on the upper slopes of mountain ranges everywhere, rank among the great vegetation formations of the world. All these trees belong to one group of gymnosperms, Coniferophyta—the conifers, or cone-bearers.

Male and female **cones** contain the reproductive structures

of conifers. The female (seed-bearing) cone is known as a **megastrobilus** (plural: *megastrobili*); this is the familiar woody cone of pine trees. The seeds in a megastrobilus are protected by a tight cluster of woody scales, which are modifications of branches extending from a central axis (**Figure 29.7A**). The typically much smaller male (pollen-bearing) cone is known as a **microstrobilus**. The microstrobilus is typically herbaceous rather than woody, as its scales are composed of modified leaves, beneath which are the pollen-bearing microsporangia (**Figure 29.7B**).

We will use the life cycle of a pine to illustrate reproduction in gymnosperms (**Figure 29.8**). The production of male gametophytes in the form of pollen grains frees the plant completely from its dependence on liquid water for fertilization. Wind, rather than water, assists conifer pollen grains in their first stage of travel from the strobilus to the female gametophyte inside a cone (see Figure 29.4). The pollen tube provides the sperm with the means for the last stage of travel by elongating through maternal sporophytic tissue, as diagrammed for an angiosperm in Figure 29.5. When it reaches the female gametophyte, it releases two sperm, one of which degenerates after the other unites with an egg. Union of sperm and egg results in a zygote; mitotic di-

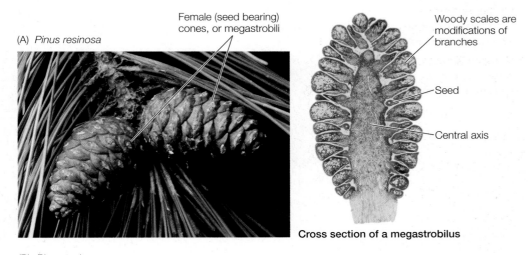

(A) *Pinus resinosa*

Female (seed bearing) cones, or megastrobili

Woody scales are modifications of branches

Seed

Central axis

Cross section of a megastrobilus

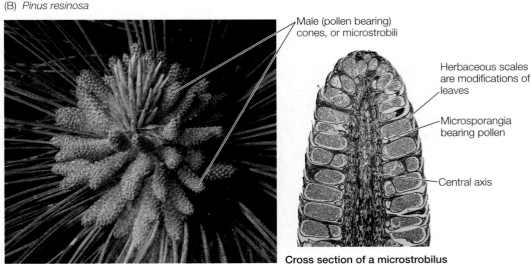

(B) *Pinus resinosa*

Male (pollen bearing) cones, or microstrobili

Herbaceous scales are modifications of leaves

Microsporangia bearing pollen

Central axis

Cross section of a microstrobilus

29.7 Female and Male Cones
(A) The scales of female cones (megastrobili) are modified branches. (B) The spore-bearing structures in male cones (microstrobili) are modified leaves.

visions and further development of the zygote result in an embryo.

─────── yourBioPortal.com ───────
GO TO Web Activity 29.2 and Animated Tutorial 29.1 •
Life Cycle of a Conifer

The megasporangium, in which the female gametophyte will form, is enclosed in a layer of sporophytic tissue—the integument—that will eventually develop into the seed coat that protects the embryo. The integument, the megasporangium inside it, and the tissue attaching it to the maternal sporophyte constitute the ovule. The pollen grain enters through a small opening in the integument at the tip of the ovule, the **micropyle**.

Most conifer ovules (which will develop into seeds after fertilization) are borne exposed on the upper surfaces of the mod-

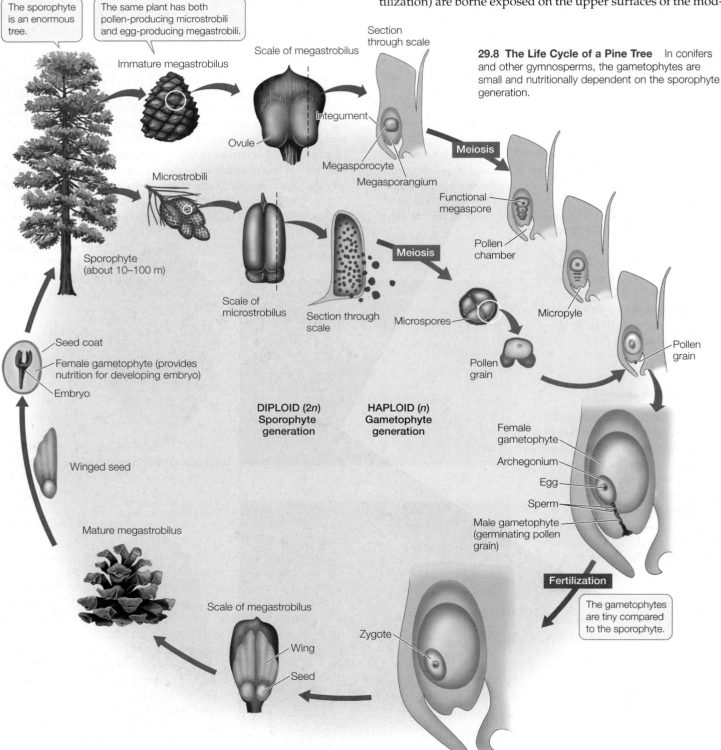

29.8 The Life Cycle of a Pine Tree In conifers and other gymnosperms, the gametophytes are small and nutritionally dependent on the sporophyte generation.

The sporophyte is an enormous tree.

The same plant has both pollen-producing microstrobili and egg-producing megastrobili.

Immature megastrobilus

Scale of megastrobilus

Section through scale

Integument

Ovule

Megasporocyte

Megasporangium

Meiosis

Microstrobili

Functional megaspore

Pollen chamber

Sporophyte (about 10–100 m)

Scale of microstrobilus

Section through scale

Meiosis

Microspores

Micropyle

Pollen grain

Seed coat

Female gametophyte (provides nutrition for developing embryo)

Embryo

Pollen grain

Pollen grain

DIPLOID (2n) Sporophyte generation

HAPLOID (n) Gametophyte generation

Female gametophyte

Archegonium

Egg

Sperm

Male gametophyte (germinating pollen grain)

Winged seed

Mature megastrobilus

Fertilization

The gametophytes are tiny compared to the sporophyte.

Scale of megastrobilus

Zygote

Wing

Seed

29.9 From Devastation, New Life Emerges   A stand of lodgepole pines in Yellowstone National Park. The mature trees were destroyed by a forest fire in 1988. The fire released large numbers of lodgepole pine seeds from the cones, and now many young lodgepole pine trees are emerging in the burn area.

ified branches that form the scales of the cone (megastrobilus). The only protection of the ovules comes from the scales, which are tightly pressed against one another within the cone. Some pines, such as the lodgepole pine, have such tightly closed cones that only fire suffices to split them open and release the seeds. A fire devastated lodgepole pine forests in Yellowstone National Park in 1988 but also released large numbers of seeds from the cones. As a result, large numbers of lodgepole pine seedlings are now emerging in the burn area (**Figure 29.9**).

About half of all conifer species have soft, fleshy modifications of cones that envelop their seeds; examples are the fruit-like cones or "berries" of juniper and yew. Animals may eat these tissues and then disperse the seeds in their feces, often carrying them considerable distances from the parent plant.

29.2 RECAP

Living gymnosperms include cycads, ginkgos, gnetophytes, and conifers, all of which are woody and have seeds that are not protected by ovaries.

- Explain the different functions of a megastrobilus and a microstrobilus. **See p. 613 and Figure 29.8**

- What is the role of the integument in a gymnosperm seed? **See p. 614 and Figure 29.8**

- Do you understand how fire can be necessary for the survival of some species? **See p. 615**

The "berries" on some gymnosperms (such as juniper and yew) are not true fruits but rather are fleshy cones. As we will see, true fruits are the ripened ovaries of plants. Ovaries are absent in gymnosperms but are a characteristic of the plant group that is dominant today: the angiosperms. Let's look at other ways in which angiosperms differ from gymnosperms.

29.3 What Features Contributed to the Success of the Angiosperms?

The oldest fossil evidence of angiosperms dates back to the late Jurassic period, about 150 million years ago (see Figure 29.1). The angiosperms radiated explosively in the Tertiary (beginning 65 million years ago) and became the dominant plant life on Earth. Today there are more than a quarter million species of angiosperms.

Cycads
Ginkgos
Gnetophytes
Conifers
Angiosperms

The female gametophyte of the angiosperms is even more reduced than that of the gymnosperms, usually consisting of just seven cells. Thus the angiosperms represent the current extreme of the trend we have traced throughout the evolution of the vascular plants: the sporophyte generation becomes larger and more independent of the gametophyte, while the gametophyte generation becomes smaller and more dependent on the sporophyte. What else sets the angiosperms apart from other plants?

The major synapomorphies (shared derived traits) that characterize the angiosperms include:

- Double fertilization
- Production of a nutritive tissue called the endosperm
- Ovules and seeds enclosed in a carpel
- Flowers
- Fruits
- Phloem with companion cells
- Reduced gametophytes

In the angiosperms, pollination consists of the arrival of a microgametophyte—a pollen grain—on a receptive surface in a flower (the *stigma*). As in the gymnosperms, pollination is the first in a series of events that result in the formation of a seed. The next is the growth of a pollen tube extending to the megagametophyte (see Figure 29.5). The third event is a fertilization process that, in detail, is unique to the angiosperms.

Double fertilization is often considered the single most reliable distinguishing characteristic of the angiosperms. *Two* male gametes, contained in a single microgametophyte, participate in fertilization events in the megagametophyte of an angiosperm. The nucleus of one sperm combines with that of the egg to produce a diploid zygote, the first cell of the sporophyte generation. In most angiosperms, the other sperm nucleus combines with two other haploid nuclei of the female gametophyte to form a *triploid* (3n) nucleus (further discussed below; see Figure 29.15). That nucleus, in turn, divides to form a triploid tissue, the **endosperm**, that nourishes the embryonic sporophyte during its early development. This process, in which two fertilization events take place, is known as **double fertilization**.

Double fertilization occurs in nearly all present-day angiosperms. We are not sure when and how it evolved because there is no known fossil evidence on this point. It may have first resulted in two diploid embryos, as it does in the three existing genera of the gnetophytes.

The name *angiosperm* ("enclosed seed") is drawn from another distinctive characteristic of these plants: the ovules and seeds are enclosed in a modified leaf called a **carpel**. Besides protecting the ovules and seeds, the carpel often interacts with incoming pollen to prevent self-pollination, thus favoring cross-pollination and increasing genetic diversity. Of course, the most obvious diagnostic feature of angiosperms is that they have **flowers**. Production of **fruits** is another of their unique characteristics. As we will see, both flowers and fruits afford major advantages to angiosperms.

Most angiosperms are distinguished by the possession of specialized water-transporting cells called **vessel elements** in their xylem (see Chapter 34). These cells are broad in diameter and connect without obstruction, allowing easy water movement. A second distinctive cell type in angiosperm xylem is the **fiber**, which plays an important role in supporting the plant body. Angiosperm phloem possesses another unique cell type, called a **companion cell**. Like the gymnosperms, woody angiosperms show secondary growth, producing secondary xylem and secondary phloem and growing in diameter.

In the remainder of this section we examine the structure and function of flowers, evolutionary trends in flower structure, the functions of pollen and fruits, and the angiosperm life cycle.

The sexual structures of angiosperms are flowers

If you examine any familiar flower, you will notice that the outer parts look somewhat like leaves. In fact, all the parts of a flower *are* modified leaves.

We showed a generalized flower (for which there is no exact counterpart in nature) in Figure 29.5. The structures bearing microsporangia are called **stamens**. Each stamen is composed of a **filament** bearing an **anther** that contains pollen-producing microsporangia. The structures bearing megasporangia are the carpels. A structure composed of one carpel or two or more fused carpels is called a **pistil**. The swollen base of the pistil, containing one or more ovules (each containing a megasporangium surrounded by its protective integument), is called the **ovary**. The apical stalk of the pistil is the **style**, and the terminal surface that receives pollen grains is the **stigma**.

In addition, a flower often has several specialized sterile (non-spore-bearing) leaves. The inner ones are called **petals** (collectively, the **corolla**) and the outer ones **sepals** (collectively, the **calyx**). The corolla and calyx (collectively, the *perianth*) can be quite showy and often play roles in attracting animal pollinators to the flower. The calyx more commonly protects the immature flower in bud. From base to apex, the sepals, petals, stamens, and carpels (which are referred to as the *floral organs*; see Figure 19.14) are usually positioned in circular arrangements or whorls and attached to a central stalk called the **receptacle**.

The generalized flower in Figure 29.5 has functional megasporangia and microsporangia; such flowers are referred to as

perfect (or hermaphroditic). Many angiosperms produce two types of flowers, one with only megasporangia and the other with only microsporangia. Consequently, either the stamens or the carpels are nonfunctional or absent in a given flower, and the flower is referred to as **imperfect**.

Species such as corn or birch, in which both megasporangiate (female) and microsporangiate (male) flowers occur on the same plant, are said to be **monoecious** (meaning "one-housed"—but, it must be added, one house with separate rooms). Complete separation is the rule in some other angiosperm species, such as willows and date palms; in these species, a given plant produces either flowers with stamens or flowers with carpels, but never both. Such species are said to be **dioecious** ("two-housed").

Flowers come in an astonishing variety of forms—just think of some of the flowers you recognize. The generalized flower in Figure 29.5 has distinct petals and sepals arranged in distinct whorls. In nature, however, petals and sepals sometimes are indistinguishable. Such appendages are called **tepals**. In other flowers, petals, sepals, or tepals are completely absent.

Flowers may be single, or they may be grouped together to form an **inflorescence**. Different families of flowering plants have characteristic types of inflorescences, such as the compound umbels of the carrot family (**Figure 29.10A**), the heads of the aster family (**Figure 29.10B**), and the spikes of many grasses (**Figure 29.10C**).

Flower structure has evolved over time

The flowers of the most basal clades of angiosperms have a large and variable number of tepals (or sepals and petals), carpels, and stamens (**Figure 29.11A**). Evolutionary change within the angiosperms has included some striking modifications of this early condition: reductions in the number of each type of floral organ to a fixed number, differentiation of petals from sepals, and changes in symmetry from radial (as in a lily or magnolia) to bilateral (as in a sweet pea or orchid), often accompanied by an extensive fusion of parts (**Figure 29.11B**).

According to one theory, the first carpels to evolve were leaves with marginal sporangia, folded but incompletely closed. Early in angiosperm evolution, the carpels fused and became progressively more buried in receptacle tissue, forming the ovary (**Figure 29.12A**). In some flowers, the other floral organs are attached at the top of the ovary, rather than at the bottom as in Figure 29.5. The stamens of the most ancient flowers may have appeared leaflike (**Figure 29.12B**), little resembling those of the generalized flower in Figure 29.5.

Why do so many flowers have pistils with long styles and anthers with long filaments? Natural selection has favored length in both of these structures, probably because length increases the likelihood of successful pollination. Long filaments may bring the anthers into contact with insect bodies, or they may place the anthers in a better position to catch the wind. Similar arguments apply to long styles.

A perfect flower represents a compromise of sorts. On the one hand, in attracting a pollinating bird or insect, the plant is attending to both its female and male functions with a single

29.10 Inflorescences (A) The inflorescence of bishop's goutweed (a member of the carrot family) is a compound umbel. Each umbel bears flowers on stalks that arise from a common center. (B) Sunflowers are members of the aster family; their inflorescence is a head. In a head, each of the long, petal-like structures is a ray flower; the central portion of the head consists of dozens to hundreds of disc flowers. (C) Grasses such as this fountain grass have inflorescences called spikes.

(A) *Magnolia watsonii*

(B) *Paphiopedilum maudiae*

29.11 Flower Form and Evolution (A) A magnolia flower shows the major features of early flowers: it is radially symmetrical, and the individual tepals, carpels, and stamens are separate, numerous, and attached at their bases. (B) Orchids, such as this ladyslipper, have a bilaterally symmetrical structure that evolved much later.

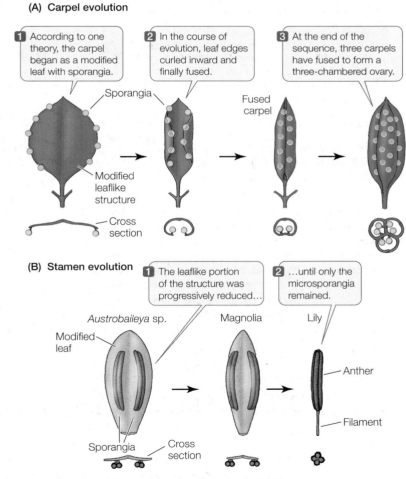

(A) Carpel evolution

1 According to one theory, the carpel began as a modified leaf with sporangia.

2 In the course of evolution, leaf edges curled inward and finally fused.

3 At the end of the sequence, three carpels have fused to form a three-chambered ovary.

Sporangia

Fused carpel

Modified leaflike structure

Cross section

(B) Stamen evolution

1 The leaflike portion of the structure was progressively reduced...

2 ...until only the microsporangia remained.

Austrobaileya sp.

Magnolia

Lily

Modified leaf

Anther

Sporangia

Cross section

Filament

29.12 Carpels and Stamens Evolved from Leaflike Structures
(A) Possible stages in the evolution of a carpel from a more leaflike structure.
(B) The stamens of three modern plants show the various stages in the evolution of that organ. It is *not* implied that these species evolved from each other; their structures simply illustrate the possible stages.

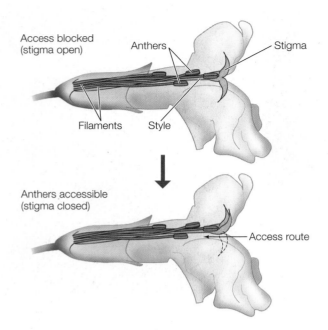

29.13 An Unusual Solution to Selfing Both long stamens and long styles facilitate cross pollination, but if these male and female structures are too close to each other, the likelihood of (disadvantageous) self-pollination increases. In *Mimulus aurantiacus*, initially the stigma is open, blocking access to the anthers. A hummingbird's touch as it deposits pollen on the stigma causes one lobe of the stigma to retract, creating a path to the anthers and allowing pollen dispersal.

flower type, whereas plants with imperfect flowers must create that attraction twice—once for each type of flower. On the other hand, the perfect flower can favor self-pollination, which is usually disadvantageous. Another potential problem is that the female and male functions might interfere with each other—for example, the stigma might be so placed as to make it difficult for pollinators to reach the anthers, thus reducing the export of pollen to other flowers.

Might there be a way around these problems? One solution is seen in the bush monkeyflower (*Mimulus aurantiacus*), which is pollinated by hummingbirds and has a stigma that initially serves as a screen, hiding the anthers (**Figure 29.13**). Once a hummingbird touches the stigma, one of the stigma's two lobes folds, so that subsequent hummingbirds pick up pollen from the previously screened anthers. The first bird transfers pollen from another plant to the stigma, eventually leading to fertilization. Later visitors pick up pollen from the now-accessible anthers, fulfilling the flower's male function. **Figure 29.14** describes the experiment that revealed the function of this mechanism.

Angiosperms have coevolved with animals

Whereas many gymnosperms are wind-pollinated, most angiosperms are animal-pollinated. The many different pollination symbioses between plants and animals are vital to both par-

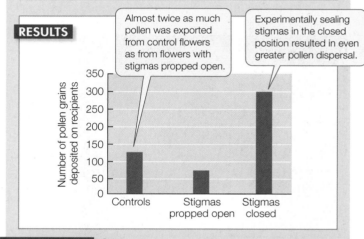

INVESTIGATING LIFE

29.14 Stigma Behavior in *Mimulus* Flowers
Elizabeth Fetscher's experiments showed that the unusual stigma retraction mechanism in monkeyflowers (see Figure 29.13) enhances the dispersal of pollen from the anthers.

HYPOTHESIS Stigma responses in *M. aurantiacus* increase the likelihood that an individual's pollen will be exported once pollen from another individual has been deposited on the stigma.

METHOD
1. Set up experimental arrays such that only one flower in each array can donate pollen. In control arrays, the pollen donor styles function normally. In another set of arrays, donor stigmas are artificially sealed closed; in a third set, the pollen donor stigmas are permanently propped open.
2. Allow hummingbirds to visit the arrays, then count the pollen grains from each donor on the next flower visited.

RESULTS

Almost twice as much pollen was exported from control flowers as from flowers with stigmas propped open.

Experimentally sealing stigmas in the closed position resulted in even greater pollen dispersal.

CONCLUSION Stigma responses enhance the male function of the flower (dispersal of pollen) once the female function (receipt of pollen) has been performed.

FURTHER INVESTIGATION: How might you test how this mechanism affects self-pollination of the flower?

Go to **yourBioPortal.com** for original citations, discussions, and relevant links for all INVESTIGATING LIFE figures.

ties. This fascinating coevolutionary field is covered in more detail in Chapter 56, but we mention some aspects here.

Many flowers entice animals to visit them by providing food rewards. Some flowers produce a sugary fluid called nectar, and the pollen grains themselves sometimes serve as food for animals. In the process of visiting flowers to obtain nectar or pollen, animals often carry pollen from one flower to another or from one plant to another. Thus, in their quest for food, the animals contribute to the genetic diversity of the plant population. In-

29.15 The Life Cycle of an Angiosperm The formation of a triploid endosperm is one of the features that distinguishes the angiosperms from the gymnosperms. One sperm nucleus fertilizes the egg to form the zygote, while the other combines with the two polar nuclei to form the endosperm.

yourBioPortal.com
GO TO Animated Tutorial 29.2 •
Life Cycle of an Angiosperm

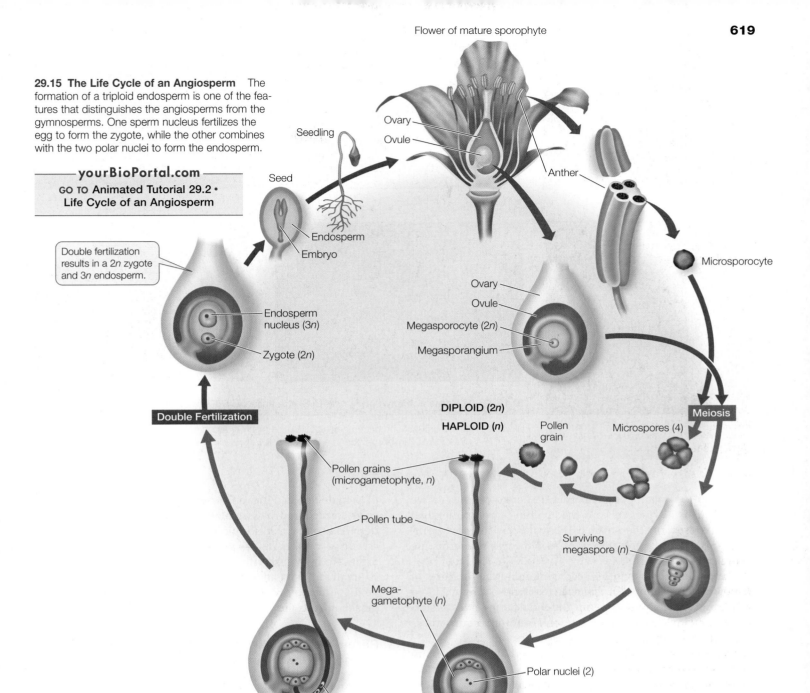

Flower of mature sporophyte

Seedling

Seed

Ovary

Ovule

Anther

Microsporocyte

Endosperm

Embryo

Double fertilization results in a 2*n* zygote and 3*n* endosperm.

Endosperm nucleus (3*n*)

Zygote (2*n*)

Ovary

Ovule

Megasporocyte (2*n*)

Megasporangium

DIPLOID (2*n*)

HAPLOID (*n*)

Meiosis

Pollen grain

Microspores (4)

Double Fertilization

Pollen grains (microgametophyte, *n*)

Pollen tube

Surviving megaspore (*n*)

Mega-gametophyte (*n*)

Polar nuclei (2)

Tube cell nucleus

Sperm (2)

Egg

sects, especially bees, are among the most important pollinators; birds and some species of bats are also major pollinators.

For more than 150 million years, angiosperms and their animal pollinators have coevolved in the terrestrial environment. The animals have affected the evolution of the plants, and the plants have affected the evolution of the animals. Flower structure has become incredibly diverse under these selection pressures. Some of the products of coevolution are highly specific; for example, some yucca species are pollinated by only one species of moth. Pollination by just one or a few animal species provides a plant species with a reliable mechanism for transferring pollen from one of its members to another.

Most plant–pollinator interactions are much less specific; that is, many different animal species pollinate the same plant species, and the same animal species pollinate many different plant species. However, even these less specific interactions have developed some specialization. Bird-pollinated flowers are often red and odorless. Many insect-pollinated flowers have characteristic odors, and bee-pollinated flowers may have conspicuous markings, or *nectar guides*, that may be visible only in the ultraviolet region of the spectrum, where bees have better vision than in the red region (see Table 56.1 and Figure 56.10).

The angiosperm life cycle features double fertilization

The angiosperm life cycle is considered in detail in Chapter 38, but let's look at it briefly here, comparing **Figure 29.15** with the conifer life cycle shown in Figure 29.8.

29.16 Fruits Come in Many Forms (A–C) Simple fruits. (A) The single-seeded drupe of the fleshy-fruited plum is dispersed by animals. (B) Each macadamia seed is covered by a hard, woody fruit that allows it to survive drought and other hardships. (C) The highly reduced fruits of dandelions are dispersed by wind. (D) A multiple fruit (pineapple). (E) An aggregate fruit (raspberry). (F) An accessory fruit (strawberry).

Like all seed plants, angiosperms are heterosporous. As we have seen, their ovules are contained within carpels rather than being exposed on the surfaces of scales, as in most gymnosperms. The male gametophytes, as in the gymnosperms, are pollen grains.

The ovule develops into a seed containing the products of the double fertilization that characterizes angiosperms: a diploid zygote and a triploid endosperm. The endosperm serves as storage tissue for starch or lipids, proteins, and other substances that will be needed by the developing embryo.

The zygote develops into an embryo, which consists of an embryonic axis (the "backbone" that will become a stem and a root) and one or two **cotyledons**, or seed leaves. The cotyledons have three different possible fates in different plants. In many, they serve as absorptive organs that take up and digest the endosperm. In others, they enlarge and become photosynthetic when the seed germinates. Often they play both roles.

Angiosperms produce fruits

The ovary of a flower (together with the seeds it contains) develops into a fruit after fertilization. The fruit protects the seeds and can also promote seed dispersal by becoming attached to or being eaten by an animal. Other fruits are adapted for dispersal by wind or water. A fruit may consist only of the mature ovary and its seeds, or it may include other parts of the flower or structures associated with it (**Figure 29.16**). A *simple fruit*, such as a plum, is one that develops from a single carpel or several united carpels. A raspberry is an example of an *aggregate fruit*—one that develops from several separate carpels

of a single flower. Pineapples and figs are examples of *multiple fruits*, formed from a cluster of flowers (an inflorescence). Fruits derived from parts in addition to the carpel and seeds are called *accessory fruits*; examples are apples, pears, and strawberries. Fruits are not necessarily fleshy; they can be hard and woody or small and have modified structures that allow the seeds to be dispersed by wind. The development, ripening, and dispersal of fruits are considered in Chapters 37 and 38.

Recent analyses have revealed the oldest split among the angiosperms

Figure 29.17 shows the relationships among the major angiosperm clades. The two largest clades— the **monocots** and the **eudicots**—include the great majority of angiosperm species. The monocots are so called because they have a single embryonic cotyledon; the eudicots have two. (Chapter 34 describes other differences between these two groups.)

Some familiar angiosperms belong to other clades, including the water lilies, star anise and its relatives, and the magnoliid complex (**Figure 29.18**). The magnoliids are less numerous than the monocots and eudicots, but they include many familiar and useful plants such as magnolias, avocados, cinnamon, and black pepper.

The root of the evolutionary tree of flowering plants was once a matter of great controversy. A fundamental issue was identifying the group that is sister to the remaining angiosperms, and the magnoliid clade was a leading candidate for the postion. At the close of the twentieth century, however, an impressive convergence of molecular and morpholog-

29.17 Evolutionary Relationships among the Angiosperms Recent analyses of many angiosperm genes have clarified the relationships among the major groups.

Common ancestor of angiosperms

Carpels; endosperm; seeds in fruit; reduced gametophytes; double fertilization; flowers; phloem with companion cells

Transitional tracheid vessel elements

Vessel elements

Carpels fused by tissue connection

Perianth of two whorls

Single cotyledon

Pollen with three grooves

Amborella

Water lilies

Star anise

Magnoliids

Monocots

Eudicots

(A) *Amborella trichopoda*

(B) *Nymphaea* sp.

(C) *Illicium anisatum*

(D) *Piper nigrum*

(E) *Aristolochia macrophylla*

(F) *Persea americana*

29.18 Monocots and Eudicots Are Not the Only Surviving Angiosperms (A) *Amborella*, a shrub, is the sister to the remaining extant angiosperms. Of interest in this photo of a female flower is the pair of false anthers (male flower parts), possibly serving to lure insects that are searching for pollen. (B) The water lily clade is the next clade to diverge after *Amborella*. (C) Star anise and its relatives belong to another basal clade. (D–F) The largest clade other than the monocots and eudicots is the magnoliid complex, represented here by (D) black pepper, (E) Dutchman's pipe, and (F) avocado. The magnolia in Figure 29.11A is another magnoliid.

ical evidence led to the conclusion that the sister group to the remaining flowering plants is a clade that today consists of a single species of the genus *Amborella* (see Figure 29.18A). This woody shrub, with cream-colored flowers, lives only on New Caledonia, an island in the South Pacific. Its 5 to 8 carpels are in a spiral arrangement, and it has 30 to 100 stamens. The xylem of *Amborella* lacks vessel elements, which evolved after this deepest split in the angiosperm evolutionary tree.

29.3 RECAP

The synapomorphies of angiosperms include double fertilization, endosperm (triploid in most species), flowers, fruit, and distinctive cells in their xylem and phloem. The largest angiosperm clades are the monocots and the eudicots.

- Explain the difference between pollination and fertilization. See pp. 615–616

- Give some examples of how animals have affected the evolution of the angiosperms. See pp. 618–619

- What are the respective roles of the two sperm in double fertilization? See pp. 619–620 and Figure 29.15

- What are the different functions of flowers, fruits, and seeds?

The remarkable diversity of the seed plants has been shaped in part by the different environments in which these and other plants have evolved. In turn, land plants—and seed plants in particular—affect their environments.

29.4 How Do Plants Support Our World?

Once life moved onto land, it was plants that shaped the environment, and today's environment is dominated by angiosperms. Representatives of the two largest angiosperm clades are everywhere. The monocots (**Figure 29.19**) include grasses, cattails, lilies, orchids, and palms. The eudicots (**Figure 29.20**) include the vast majority of familiar seed plants, including most herbs (i.e., nonwoody plants), vines, trees, and shrubs. Among the eudicots are such diverse plants as oaks, willows, beans, snapdragons, and sunflowers.

Plants make profound contributions to ecosystem services—processes by which the environment maintains resources that benefit humans. These benefits include the effects of plants on soil, water, the atmosphere, and the climate. Plants produce oxygen and remove carbon dioxide from the atmosphere, as well as play important roles in forming and renewing the fertility of soils. Plant roots help hold soil in place, providing a defense against erosion by wind and water. They also moderate local climate in various ways, such as by increasing humidity, providing shade, and blocking wind.

Seed plants are our primary food source

Plants are **primary producers**; that is, their photosynthesis traps energy and carbon, making those resources available not only for their own needs but also for the herbivores and omnivores that consume them, for the carnivores and omnivores that eat the herbivores, and for the prokaryotes and fungi that complete the food chain. The earliest steps in human civilization involved cultivating angiosperms to provide a reliable food supply.

Today, twelve species of seed plants stand between the human race and starvation: rice, coconut, wheat, corn (also called maize), potato, sweet potato, cassava (also called tapioca or man-

(A) *Phoenix dactylifera*

(B) *Triticum aestivum*

(C) *Hemerocallis* sp.

29.19 Monocots   (A) Palms are among the few monocot trees. Date palms are a major food source in some areas of the world. (B) Grasses such as this cultivated wheat and the fountain grass in Figure 29.10C are monocots. (C) Monocots include popular garden flowers such as these daylilies. Orchids (Figure 29.11B) are also highly sought-after monocot flowers.

(C) *Crataegus viridis*

(B) *Mimosa nuttallii*

29.20 Eudicots (A) The cactus family is a large group of eudicots, with about 1,500 species in the Americas. This black lace cactus bears large pink flowers for a brief period of the year. (B) This sensitive briar is a legume ("bean and pea family"), an economically important plant group with a large number of species. (C) The green hawthorn is a small eudicot tree and a member of the family that includes roses.

ioc), sugarcane, sugar beet, soybean, common bean (*Phaseolus vulgaris*), and banana. Hundreds of other seed plants are cultivated for food, but none rank with these twelve in importance.

Indeed, more than half of the world's human population derives the bulk of its food energy from the seeds of a single plant: rice, *Oryza sativa*. Rice is particularly important in the Far East, where it has been cultivated for more than 8,000 years. People also use rice straw in many ways, such as thatching for roofs, food and bedding for livestock, and clothing. Rice hulls, too, have many uses, ranging from fertilizer to fuel.

Another vitally important angiosperm is the coconut (*Cocos nucifera*), whose seed is illustrated at the opening of this chapter. In some cultures of the coastal tropics, this monocot tree is known as the "tree of life." Every aboveground part of the plant—its seeds, fruit, stem (trunk), leaves, buds, and even its sap—is of use and value to humans. Millions of people get most of their protein from the "meat" of coconut seeds, and the seed's "milk" is vital in areas where water is scarce or unfit to drink.

Seed plants have been sources of medicine since ancient times

One of the oldest human professions is that of medicine man or shaman—a person who cures others with medicines derived from seed plants. It is claimed that a legendary Chinese emperor around 2700 B.C.E. knew some 365 medicinal plants. Although we also use medicines derived from fungi, lichens, and actinobacteria, seed plants are the source of many of our medications, just a few of which are shown in **Table 29.1**. Even in synthetic pharmaceuticals, the chemical structures of active ingredients are often based on the biochemistry of substances isolated from plants.

How are plant-based medicines discovered? These days many are found by systematic testing of tremendous numbers of plants from all over the world, a process that began in the 1960s. One example is taxol, an important anticancer drug. Among the myriad plant samples that had been tested by 1962, extracts of the bark of Pacific yew (*Taxus brevifolia*) showed antitumor activity in tests against rodent tumors. The active ingredient, taxol, was isolated in 1971 and tested against human cancers in 1977. After another 16 years, the U.S. Food and Drug Administration approved it for human use, and taxol is now widely used in treating breast and ovarian cancers as well as several other types of cancers.

TABLE 29.1		
Some Medicinal Plants and Their Products		
PRODUCT	**PLANT SOURCE**	**MEDICAL APPLICATION**
Atropine	Belladonna	Dilating pupils for eye examination
Bromelain	Pineapple stem	Controlling tissue inflammation
Digitalin	Foxglove	Strengthening heart muscle contraction
Ephedrine	*Ephedra*	Easing nasal congestion
Menthol	Japanese mint	Relief of coughing
Morphine	Opium poppy	Relief of pain
Quinine	*Cinchona* bark	Treatment of malaria
Taxol	Pacific yew	Treatment of ovarian and breast cancers
Tubocurarine	Curare plant	Muscle relaxant in surgery
Vincristine	Periwinkle	Treatment of leukemia and lymphoma

Widespread screening of plant samples eventually was deemphasized in favor of a purely chemical approach. Using automation and miniaturization, pharmaceutical laboratories generate vast numbers of compounds that are screened just as plant materials were screened in the search for taxol and other plant-based medicines. Now, however, the plant screening is getting renewed interest. Both approaches are based on trial and error.

The other leading source of medicinal plants is work by *ethnobotanists*, who study how people use and view plants in their local environments. This work proceeds all over the globe today. An older example is the discovery of quinine as a treatment for malaria. In the sixteenth century, Spanish priests in Peru became aware that the native population used the bark of local *Cinchona* trees to treat fevers. The priests successfully used the bark to treat malaria. Word of the medicine spread to Europe, where it is said to have been in use in Rome by 1631. The active ingredient of *Cinchona* bark—quinine—was identified in 1820, and quinine remained the standard malarial remedy well into the twentieth century.

CHAPTER SUMMARY

29.1 How Did Seed Plants Become Today's Dominant Vegetation?

- Fossils of woody seed ferns are the earliest evidence of vascular seed plants. The surviving groups of seed plants are the **gymnosperms** and **angiosperms**.
- All seed plants are heterosporous, and their gametophytes are much smaller than (and dependent on) their sporophytes. **Review Figure 29.3**
- An **ovule** consists of the seed plant megagametophyte and the **integument** that protects it. The ovule develops into a **seed**.
- **Pollen grains**, the microgametophytes, do not require liquid water to perform their functions. Following **pollination**, a **pollen tube** emerges from the pollen grain and elongates to deliver gametes to the megagametophyte. **Review Figure 29.5, WEB ACTIVITY 29.1**
- Seeds are well protected, and they are often capable of long periods of dormancy, germinating when conditions are favorable.

29.2 What Are the Major Groups of Gymnosperms?

- The gymnosperms produce seeds that are not protected by ovaries.
- The major gymnosperm groups are the **cycads**, **ginkgos**, **gnetophytes**, and **conifers**. **Review Figure 29.2.**
- The megaspores of conifers are produced in woody **cones** called **megastrobili**, and microspores are produced in herbaceous cones called **microstrobili**. Pollen reaches the megagametophyte by way of the **micropyle**, an opening in the integument of the ovule. **Review Figure 29.8, WEB ACTIVITY 29.2, ANIMATED TUTORIAL 29.1**

29.3 What Features Contributed to the Success of the Angiosperms?

- Only angiosperms have **flowers** and **fruits**.
- The ovules and seeds of angiosperms are enclosed in and protected by **carpels**. **Review Figure 29.12**

- Angiosperms have **double fertilization**, resulting in the production of a zygote and **endosperm** (which is triploid in most species). **Review Figure 29.15, ANIMATED TUTORIAL 29.2**
- The xylem and phloem of angiosperms are more complex and efficient than those of the gymnosperms. **Vessel elements** in the xylem of angiosperms function in water transport. **Fibers** in angiosperm xylem play an important role in structural support.
- The floral organs, from the apex to the base of the flower, are the **pistil, stamens, petals**, and **sepals**. Stamens bear microsporangia in **anthers**. The pistil (consisting of one or more carpels) includes an **ovary** containing ovules. The **stigma** is the receptive surface of the pistil. The floral organs are borne on the **receptacle**. **Review Figure 29.5**
- A flower with both megasporangia and microsporangia is **perfect**; all other flowers are **imperfect**. Flowers may be grouped to form an **inflorescence**. Flowers may be pollinated by wind or animals.
- A **monoecious** species has megasporangiate and microsporangiate flowers on the same plant. A **dioecious** species is one in which megasporangiate and microsporangiate flowers occur on different plants.
- The most species-rich angiosperm clades are the **monocots** and the **eudicots**. The magnoliids form the sister group to the monocots and eudicots. **Review Figure 29.17**
- The oldest evolutionary split among the angiosperms is between the single species in the genus *Amborella* and all the remaining flowering plants.

29.4 How Do Plants Support Our World?

- Plants provide **ecosystem services** that affect soil, water, air quality, and climate.
- Plants are **primary producers** and as such are the foundation of the entire terrestrial food web.
- Plants provide many important medicinal products. **Review Table 29.1**

SELF-QUIZ

1. Which of the following statements about seed plants is *true*?
   a. Seeds are produced only by flowering plants (angiosperms).
   b. The sporophyte generation is more reduced in seed plants than in the ferns.
   c. The gametophytes of seed plants are independent of the sporophytes.
   d. All seed plant species are heterosporous.
   e. The zygote of seed plants divides repeatedly to form the gametophyte.

2. The gymnosperms
   a. dominate all land masses today.
   b. have never dominated land masses.
   c. have secondary growth.
   d. all have vessel elements.
   e. lack sporangia.

3. Conifers
   a. produce ovules in microstrobili and pollen in megastrobili.
   b. depend on liquid water for fertilization.
   c. have triploid endosperm.
   d. have pollen tubes that release two sperm.
   e. have vessel elements.

4. Most angiosperms
   a. have seeds enclosed in a carpel.
   b. produce triploid endosperm by the union of two eggs and one sperm.
   c. lack secondary growth.
   d. bear two kinds of cones.
   e. lack flowers.

5. Which statement about flowers is *not* true?
   a. Pollen is produced in the anthers.
   b. Pollen is received on the stigma.
   c. An inflorescence is a cluster of flowers.
   d. A species having female and male flowers on the same plant is dioecious.
   e. A flower with both megasporangia and microsporangia is said to be perfect.

6. Which statement about fruits is *not* true?
   a. They develop from ovaries.
   b. They may include other parts of the flower.
   c. A multiple fruit develops from several carpels of a single flower.
   d. They are produced only by angiosperms.
   e. A cherry is a simple fruit.

7. Which statement is *not* true of angiosperm pollen?
   a. It is the male gamete.
   b. It is haploid.
   c. It produces a long tube.
   d. It interacts with the carpel.
   e. It is produced in microsporangia.

8. Which statement is *not* true of carpels?
   a. They are thought to have evolved from leaves.
   b. They bear megasporangia.
   c. They may fuse to form a pistil.
   d. They are floral organs.
   e. They are absent in perfect flowers.

9. *Amborella*
   a. was the first flowering plant.
   b. belongs to the first gymnosperm clade.
   c. is the sister group of all other living angiosperms.
   d. is a eudicot.
   e. has vessel elements in its xylem.

10. The eudicots
   a. include many herbs, vines, shrubs, and trees.
   b. and the monocots are the only extant angiosperm clades.
   c. are not a clade.
   d. include the magnolias.
   e. include orchids and palm trees.

FOR DISCUSSION

1. In most seed plant species, only one of the products of meiosis in the megasporangium survives. How might this be advantageous?

2. Suggest an explanation for the great success of the angiosperms in occupying terrestrial habitats.

3. In many locales, large gymnosperms predominate over large angiosperms. Under what conditions might gymnosperms have the advantage, and why?

4. Not all flowers possess all of the following floral organs: sepals, petals, stamens, and carpels. Which floral organ or organs do you think might be found in the flowers that have the smallest number of floral organ types? Discuss the possibilities, both for a single flower and for a species.

5. The origin of the angiosperms has long been "an abominable mystery," as Charles Darwin once put it. The earliest known angiosperm fossils are from the late Jurassic, but fossils of their sister-group, the gymnosperms, are known from as early as the late Carboniferous (about 150 million years earlier). Given that these two sister-groups are thought to have arisen at the same time from a single split in the seed plant tree, what might explain the lack of earlier angiosperms fossils?

ADDITIONAL INVESTIGATION

The flower of a particular species of orchid has a long, spurlike tube into which a pollinating insect can insert its proboscis to suck nectar. The spurs are of different lengths in different habitats, apparently correlated with the lengths of the probosces of the local pollinators. How might you test the hypothesis that this correlation increases reproductive success, in terms of pollen transfer to the flower?

Fungi: Recyclers, Pathogens, Parasites, and Plant Partners

A fungus battles witchweed

More than 300 million Africans in 25 countries are suffering because their crops have been invaded by witchweed (*Striga*), a parasitic flowering plant. Witchweed has attacked more than two-thirds of the sorghum, corn, and millet crops in sub-Saharan Africa. Reduced crop yields cost an estimated U.S. $7 billion each year.

A team of Canadian scientists set out to find a biological solution to the *Striga* problem. Their strategy was to look for an organism that would destroy witchweed in the fields. They succeeded in isolating a strain of a fungus—the mold *Fusarium oxysporum*—that has several outstanding properties. First, it grows on *Striga* and kills a high percentage of these parasitic plants. Second, the fungus does

not attack the crop plants that *Striga* parasitizes. And finally, *F. oxysporum* is not toxic to humans. In subsequent fieldwork, the scientists established specific techniques for applying *F. oxysporum* to witchweed. Farmers who apply the fungus to their crops are rewarded by greatly increased crop yields as *Striga* is held in check.

It may be possible to repeat the *Striga* story—the use of a fungus to wipe out a particular type of flowering plant—in a very different context. A different strain of *F. oxysporum* preferentially attacks coca plants, the source of cocaine. A controversial proposal to use *F. oxysporum* to wipe out the coca plantations in Andean South America and other parts of the world has been proposed; however, the specificity of the fungus to infect only coca is not clear. Establishing the degree of host specificity is crucial, because some naturally occurring strains of *F. oxysporum* attack important crops in various parts of the world. Introducing this plant pathogen on a widespread basis could have unintended consequences for many non-target plant species.

Fungi can also be used to battle animal pests. Research teams reported in 2005 that two fungi, *Beauveria bassiana* and *Metarhizium anisopliae*, killed malaria-carrying mosquitoes when applied to mosquito netting. Certain fungi are already used against other insect pests, notably termites and aphids. In

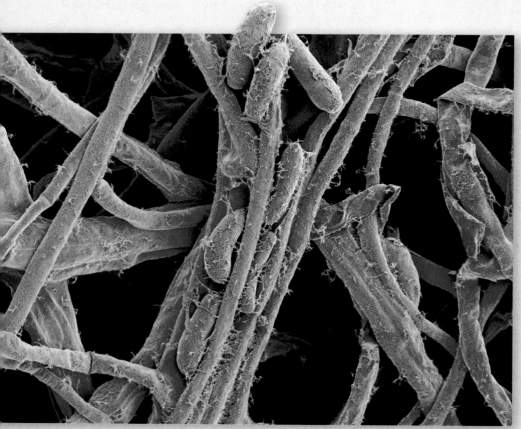

Pathogenic Fungus, Parasitic Plant
The fungus *Fusarium oxysporum* is a potent pathogen of witchweed (*Striga*), a parasitic plant that attacks crops. The fungal spores are shown in blue; fungal filaments are in tan. Both colors were added to enhance this scanning electron micrograph.

An Alien Meal The tropical fungus whose fruiting body is growing on the stalk projecting from this ant's carcass developed internally in the host, from a spore ingested by the ant.

research conducted thus far, no insect pests have yet been known to develop resistance to fungi (as they can to DDT and other pesticides).

Of course, fungi don't need human help to find organisms to grow on. Fungal spores ingested by suitable animal hosts such as ants can germinate in the host's gut and develop into internal parasites. Absorbing nutrients from their unwitting hosts, some ingested fungi may eventually kill the host, producing new spores to infect new hosts as they do so.

Fungi interact with other organisms in many different ways, some of which are beneficial and some harmful to the other organisms. As we begin our study, recall that the fungi are more closely related to animals than to plants. That means molds and mushrooms are more closely related to you than they are to the plants discussed in Chapter 29.

IN THIS CHAPTER we will see that fungi differ from other eukaryotes in some very interesting ways. We will explore the diversity of body forms, reproductive structures, and life cycles that have evolved among six major groups of fungi. We will also examine the mutually beneficial associations of certain fungi with other organisms.

30.1 What Is a Fungus?

Modern fungi are believed to have evolved from a unicellular protist ancestor that had a flagellum. The probable common ancestor of the animals was also a flagellated microbial eukaryote that may have been similar to the existing choanoflagellates (see Figure 27.27). Current evidence suggests that today's choanoflagellates, fungi, and animals share a common ancestor not shared by other eukaryotes, and thus the three lineages are often grouped together as the *opisthokonts*. Synapomorphies that distinguish the fungi among the opisthokonts include absorptive heterotrophy and the presence of chitin in their cell walls (**Figure 30.1**). Molecular sequences of many genes also support these relationships among the opisthokonts. Thus, fungi represent one of four large, independent evolutionary origins of multicellular organisms (plants, brown algae, and animals are the other three).

The fungi live by **absorptive heterotrophy**: they secrete digestive enzymes outside their bodies to break down large food molecules in the environment, then absorb the breakdown products through the plasma membranes of their cells. Absorptive heterotrophy is successful in virtually every conceivable environment. Many fungi are **saprobes**, which absorb nutrients from dead organic matter. Others are **parasites**, which absorb nutrients from living hosts (such as the ant shown at the opening of this chapter). Still others are **mutualists** living in intimate associations with other organisms that benefit both partners.

We will discuss six major groups of fungi (**Table 30.1**): microsporidia, chytrids, zygospore fungi (Zygomycota), arbuscular mycorrhizal fungi (Glomeromycota), sac fungi (Ascomycota), and club fungi (Basidiomycota). The chytrids and zygospore fungi are not thought to represent monophyletic groups, but instead consist of several distantly related lineages that retain some ancestral features. Nonetheless, these groupings represent convenient categories for a general introduction to the fungi. The clades that are thought to be monophyletic within these two paraphyletic groupings are listed in Table 30.1. Major fungal groups were originally defined by their methods and structures for sexual reproduction and also, to a lesser extent, by other morphological differences. More recently, evidence from DNA analyses has established the placement of microsporidia among the fungi, the paraphyly of chytrids and zygospore fungi, the independence of arbuscular mycorrhizal fungi from the other fungal groups, and the monophyly of sac fungi and club fungi (**Figure 30.2**). The chytrids are almost all aquatic, but the other groups are mostly terrestrial.

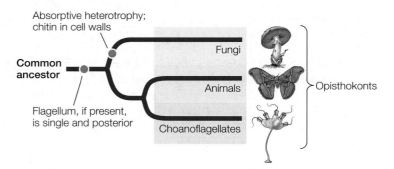

30.1 Fungi in Evolutionary Context Absorptive heterotrophy and the presence of chitin in their cell walls distinguish the fungi from other opisthokonts.

Unicellular fungi are known as yeasts

Most fungi are multicellular, but single-celled species are found in most of the fungal groups. Unicellular forms of zygospore fungi, sac fungi, and club fungi are called **yeasts** (**Figure 30.3**). Some fungi that have yeast stages also include a filamentous stage. Yeasts live in liquid or moist environments and absorb nutrients directly across their cell surfaces. The term "yeast" does not refer to a single taxonomic group of organisms but rather to a lifestyle that has evolved multiple times. The ease with which many yeast species are cultured, combined with their rapid growth rates, has made them ideal model organisms for study in the laboratory. They present many of the same advantages to laboratory investigators as do many bacteria, but since they are eukaryotes, they have genomic structure and cells that are much more like those of humans and other eukaryotes.

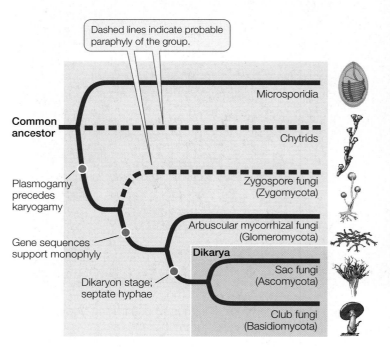

30.2 Phylogeny of the Fungi Microsporidia are reduced, parasitic fungi whose relationships among the fungi are uncertain. They may be the sister group of most other fungi or more closely related to particular groups of chytrids or zygospore fungi. The dashed lines indicate that chytrids and zygospore fungi are thought to be paraphyletic; the relationships of the lineages within these two informal groups (see Table 30.1) are not yet well resolved. The sac fungi and club fungi together form the clade Dikarya.

yourBioPortal.com
GO TO **Web Activity 30.1 • Fungal Phylogeny**

TABLE 30.1
Classification of the Fungi

GROUP	COMMON NAME	FEATURES
Microsporidia	Microsporidia	Intracellular parasites of animals; greatly reduced, among smallest eukayotes known; polar tube used to infect hosts
Chytrids (paraphyletic)[a] Chytridiomycota Neocallimastigomycota Blastocladiomycota	Chytrids	Mostly aquatic and microscopic; zoospores have flagella
Zygomycota (paraphyletic)[a] Entomophthoromycotina Kickxellomycotina Mucoromycotina Zoopagomycotina	Zygospore fungi	Reproductive structure is a unicellular zygospore with many diploid nuclei in a zygosporangium; no regularly occurring septa; usually no fleshy fruiting body
Glomeromycota	Arbuscular mycorrhizal fungi	Form arbuscular mycorrhizae on plant roots; only asexual reproduction is known
Ascomycota	Sac fungi	Sexual reproductive saclike structure known as an ascus, which contains haploid ascospores; perforated septa; dikaryon
Basidiomycota	Club fungi	Sexual reproductive structure is a basidium, a swollen cell at the tip of a specialized hypha that supports haploid basidiospores; perforated septa; dikaryon

[a]The formally named groups within the chytrids and Zygomycota are each thought to be monophyletic, but their relationships to one another (and to Microsporidia) are not yet well resolved.

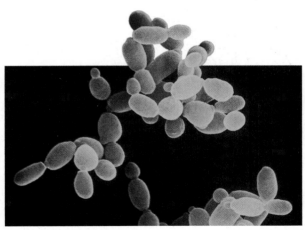

Saccharomyces cerevisiae

10 μm

30.3 Yeasts Are Unicellular Fungi Unicellular species of the zygospore fungi, sac fungi, and club fungi are known as yeasts. Many yeasts reproduce by budding—mitosis followed by asymmetrical cell division—as those shown here are doing.

The body of a multicellular fungus is composed of hyphae

The body of a multicellular fungus is called a **mycelium** (plural *mycelia*). A mycelium is composed of a mass of individual tubular filaments called **hyphae** (singular *hypha*; **Figure 30.4A,B**). The cell walls of the hyphae are greatly strengthened by microscopic fibrils of *chitin*, a nitrogen-containing polysaccharide. In some species of fungi, the hyphae are subdivided into cell-like compartments by *incomplete* cross-walls called **septa** (singular *septum*); these hyphae are referred to as **septate**. Septa do not completely close off compartments in the hyphae. Gaps in the septa known as *pores* allow organelles—sometimes even nuclei—to move in a controlled way between compartments (**Figure 30.4C**). In other species of fungi, the hyphae lack septa but may contain hundreds of nuclei; these hyphae are referred to as **coenocytic**. The coenocytic condition results from repeated nuclear divisions without cytokinesis.

Certain modified hyphae, called *rhizoids*, anchor some fungi to their substratum (the dead organism or other matter on which they feed). These rhizoids are not homologous to the rhizoids of plants, and they are not specialized to absorb nutrients and water. Parasitic fungi, however, may possess modified hyphae that take up nutrients from their host.

The total hyphal growth of a fungal mycelium (not the growth of an individual hypha) may exceed 1 kilometer a day! The hyphae may be widely dispersed to forage for nutrients over a large area, or they may clump together in a cottony mass to exploit a rich nutrient source. In some members of certain fungal groups, when sexual spores are produced, portions of the mycelium become reorganized into a reproductive *fruiting body*, such as a mushroom. The mycelial mass is often far larger than the mushroom alone. The mycelium of one individual fungus in Michigan covers 15 hectares underground and weighs

(A)

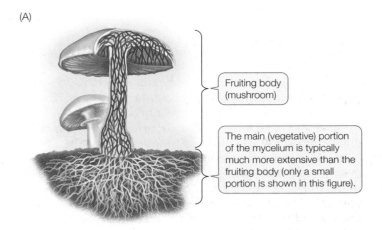

Fruiting body (mushroom)

The main (vegetative) portion of the mycelium is typically much more extensive than the fruiting body (only a small portion is shown in this figure).

(B) Xylem of wood   Fungal hyphae

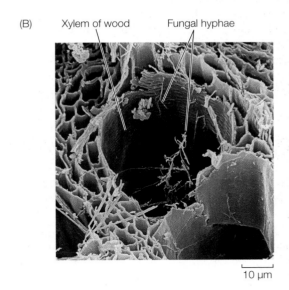

10 μm

(C)

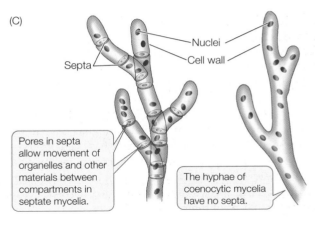

Nuclei

Cell wall

Septa

Pores in septa allow movement of organelles and other materials between compartments in septate mycelia.

The hyphae of coenocytic mycelia have no septa.

30.4 Mycelia Are Made Up of Hyphae (A) The fruiting body of a club fungus is transient, but the filamentous, nutrient-absorbing mycelium can be long-lived and cover large areas. (B) The minute individual hyphae of fungal mycelia can penetrate small spaces. In this artificially colored micrograph, hyphae (yellow structures) of a dry-rot fungus are penetrating the xylem cavities and other woody tissues of a log. (C) The hyphae of septate fungal species are divided into organelle-containing compartments by porous septa, while coenocytic hyphae have no septa.

more than a blue whale. Aboveground, this individual is evident only as isolated clumps of mushrooms.

Fungi are in intimate contact with their environment

The filamentous hyphae of a fungus give it a unique relationship with its physical environment. The fungal mycelium has an enormous surface area-to-volume ratio compared with that of most large multicellular organisms. This large ratio is a marvelous adaptation for absorptive heterotrophy. Throughout the mycelium (except in fruiting structures), all of the hyphae are very close to their environmental food source.

The downside of the great surface area-to-volume ratio of the mycelium is its tendency to lose water rapidly in a dry environment. Thus fungi are most common in moist environments. You have probably observed the tendency of molds, toadstools, and other fungi to appear in damp places.

Another characteristic of some fungi is a tolerance for highly hypertonic environments (those with a solute concentration higher than their own; see Section 6.3). Many fungi are more resilient than bacteria in hypertonic surroundings. Jelly in the refrigerator, for example, will not become a growth medium for bacteria because it is too hypertonic to those organisms, but it may eventually harbor mold colonies. This presence of fungi in the refrigerator illustrates yet another trait of many fungi: tolerance of temperature extremes. Many fungi tolerate temperatures as low as –6°C, and some tolerate temperatures above 50°C.

Fungi reproduce both sexually and asexually

Both asexual and sexual reproduction occur among the fungi (**Figure 30.5**). Asexual reproduction takes several forms:

- The production of (usually) haploid spores within structures called *sporangia*
- The production of haploid spores (not enclosed in sporangia) at the tips of hyphae; such spores are called *conidia* (Greek *konis*, "dust")
- Cell division by unicellular fungi—either a relatively equal division of one cell into two (*fission*) or an asymmetrical division in which a smaller daughter cell is produced (*budding*)
- Simple breakage of the mycelium

Asexual reproduction in fungi can be spectacular in terms of spore quantity. A 2.5-centimeter colony of *Penicillium*, the mold that produces the antibiotic penicillin, can produce as many as 400 million conidia. The air we breathe contains as many as 10,000 fungal spores per cubic meter.

Sexual reproduction is rare (or even unknown) in some groups of fungi but common in others. Sexual reproduction may not occur, or it may occur so rarely that biologists have never observed it. Species in which no sexual stage has been observed were once placed in a separate taxonomic group, because the sexual life cycle was considered necessary for classifying fungi. Now, however, these species can be related to other species of fungi through analysis of their DNA sequences.

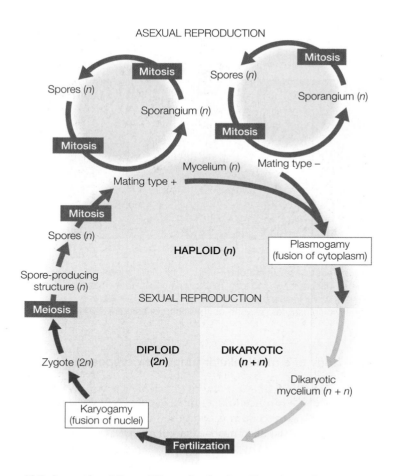

30.5 Asexual and Sexual Reproduction in a Fungal Life Cycle
Environmental conditions may determine which mode of reproduction takes place at a given time. In the sexual phase, many fungi are characterized not by male and female individuals but by genetically distinct mating types (top of figure).

When it does occur, sexual reproduction in many fungi features an interesting twist. There is often no morphological distinction between female and male structures, or between female and male individuals. Rather, there is a genetically determined distinction between two *or more* **mating types**. Individuals of the same mating type cannot mate with one another, but they can mate with individuals of another mating type within the same species, thus preventing self-fertilization. Individuals of different mating types differ genetically but are often visually and behaviorally indistinguishable. Many protists also have mating type systems.

Fungi reproduce sexually when hyphae (or, in one fungal group, motile cells) of different mating types meet and fuse. In many fungi, the zygote nuclei formed by sexual reproduction are the only diploid nuclei in the life cycle. These nuclei undergo meiosis, producing haploid nuclei that become incorporated into spores. Haploid fungal spores, whether produced sexually in this manner or asexually, germinate, and their nuclei divide mitotically to produce haploid hyphae.

We will discuss reproduction in fungi in more detail in Section 30.3.

Fungi are important components of healthy ecosystems. Fungi interact with other organisms in many ways, some of which are harmful and some beneficial to other organisms.

30.2 How Do Fungi Interact with Other Organisms?

Without the fungi, our planet would be very different. Picture Earth with only a few stunted plants and watery environments choked with the remains of dead organisms. Fungi do much of Earth's garbage disposal. Fungi that absorb nutrients from dead organisms not only help clean up the landscape and form soil but also play a key role in recycling mineral elements. Furthermore, the colonization of the terrestrial environment was made possible in large part by associations fungi formed with other organisms.

Saprobic fungi are critical to the planetary carbon cycle

Saprobic fungi (those that consume nonliving organic matter), along with bacteria, are the major decomposers on Earth, contributing to decay and thus to recycling the elements used by living things. In forests, for example, the mycelia of fungi absorb nutrients from fallen trees, thus decomposing their wood. Fungi are the principal decomposers of cellulose and lignin, the main components of plant cell walls (most bacteria cannot break down these materials). Other fungi produce enzymes that decompose keratin and thus break down animal structures such as hair and nails.

Were it not for the fungal decomposers, Earth's carbon cycle would fail: great quantities of carbon atoms would remain trapped forever on forest floors and elsewhere (see Chapter 58). Instead, those carbon atoms are returned to the atmosphere in the form of respiratory CO_2, available for photosynthesis by plants.

In fact, there was a time when populations of saprobic fungi declined significantly. During the Carboniferous period, plants in the vast tropical swamps died and began to form peat (see Section 28.4). Peat formation led to acidification of the swamps; that acidity, in turn, drastically reduced the fungal population. The result? With the decomposers largely absent, large quantities of peat remained on the swamp floor and over time were converted into coal.

In contrast to their decline during the Carboniferous, fungi did very well at the end of the Permian, a quarter of a billion years ago, when the aggregation of continents produced volcanic eruptions that triggered a planetwide extinction event of many other organisms (see Chapter 25). The fossil record shows that even though 96 percent of all species became extinct, fungi flourished, demonstrating both their hardiness and their role in recycling the elements in the dead bodies of plants and animals.

Many saprobic fungi can be grown on artificial media, and it is relatively easy to perform experiments to determine their exact nutritional requirements. Sugars are the favored source of carbon for saprobic fungi. Most fungi obtain nitrogen from proteins or the products of protein breakdown. Many fungi can use nitrate (NO_3^-) or ammonium (NH_4^+) ions as their sole source of nitrogen. No known fungus can get its nitrogen directly from inorganic nitrogen gas as can some bacteria and plant–bacteria associations (that is, fungi cannot "fix" nitrogen; see Section 36.4). Nutritional studies also reveal that most fungi are unable to synthesize certain vitamins and must absorb them from their environment. However, fungi can synthesize some vitamins that animals cannot. Like all organisms, fungi also require some mineral elements.

What happens when a fungus faces a dwindling food supply? A common strategy is to reproduce rapidly and abundantly. When conditions are good, fungi produce great quantities of spores, but the rate of spore production is commonly even higher when nutrient supplies go down. The spores may then remain dormant until conditions improve, or may be dispersed to areas where nutrient supplies are higher.

Not only are fungal spores abundant in number, but they are extremely tiny and easily spread by wind or water (**Figure 30.6**). This virtually assures that the individual that produced them

Lycoperdon pyriforme

30.6 Spores Galore Puffballs (a type of club fungus) disperse trillions of spores in great bursts. Few of the spores travel very far, however; some 99 percent of them fall within 100 meters of the parent puffball.

will have many progeny, which may be scattered over great distances. No wonder we find fungi just about everywhere.

Fungi may engage in parasitic and predatory interactions

Whereas saprobic fungi obtain their energy, carbon, and nitrogen directly from dead organic matter, other species of fungi obtain their nutrition from parasitic—and even predatory—interactions.

PARASITIC FUNGI Biologists distinguish between two classes of parasitic fungi, based on the degree of dependence on their host species. *Facultative* parasites can attack living organisms but can also grow by themselves, including on artificial media. *Obligate* parasites can grow only on their specific living host, usually a plant species. Because their growth depends on a living host, obligate parasites have specialized nutritional requirements.

The filamentous structure of fungal hyphae is especially well suited to a life of absorbing nutrients from plants. The slender hyphae of a parasitic fungus can invade a plant through stomata, through wounds, or in some cases, by direct penetration of epidermal cell walls (**Figure 30.7A**). Once inside the plant, the hyphae branch out to expand the mycelium. Some hyphae produce **haustoria**, branching projections that push through cell walls into living plant cells, absorbing the nutrients within those cells. The haustoria do not break through the plant cell plasma membranes inside the cell walls; they simply invaginate into the membranes, with the plasma membrane fitting them like a glove (**Figure 30.7B**). Fruiting structures may form, either within the plant body or on its surface in a symbiotic relationship that is usually not lethal to the plant. Some parasitic fungi, however, are *pathogenic*, sickening or even killing the hosts from which they derive nutrition.

PATHOGENIC FUNGI Although most human diseases are caused by bacteria or viruses, fungal pathogens are a major cause of death among people with compromised immune systems. Most people with AIDS die of fungal diseases, such as the pneumonia caused by *Pneumocystis jirovecii* or incurable diarrhea caused by other fungi. *Candida albicans* and certain other yeasts also cause severe diseases, such as esophagitis (which impairs swallowing), in individuals with AIDS and in individuals taking immunosuppressive drugs. Fungal diseases are a growing international health problem, requiring vigorous research. Our limited understanding of the basic biology of these fungi still hampers our ability to treat the diseases they cause. Various fungi cause other, less threatening human diseases, such as ringworm and athlete's foot.

The worldwide decline of amphibian species has been linked to the spread of a chytrid fungus, *Batrachochytrium dendrobatidis*. Genetic analyses indicate that the fungus populations attacking amphibian populations around the world are genetically almost identical, which suggests a recent introduction of the fungus across the globe. This chytrid appears to be endemic to southern Africa, and its spread around the world may have initiated in the 1930s with exports of the African clawed frog (*Xenopus laevis*), which was once widely used in human pregnancy tests.

Fungi are by far the most important plant pathogens, causing crop losses amounting to billions of dollars. Bacteria and viruses are less important as plant pathogens. Major fungal diseases of crop plants include black stem rust of wheat and other diseases of wheat, corn, and oats. The agent of black stem rust is *Puccinia graminis*, which has a complicated life cycle that involves two plant hosts (wheat and barberry). In an epidemic in 1935, *P. graminis* was responsible for the loss of about one-fourth of the entire wheat crop in Canada and the United States. However, as we saw at the beginning of this chapter, pathogenic fungi that kill certain weed species can be a boon to agriculture.

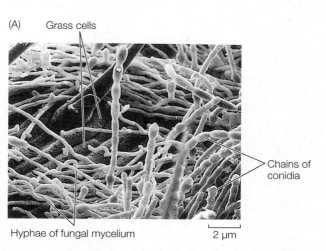

(A) Grass cells

Chains of conidia

Hyphae of fungal mycelium

2 μm

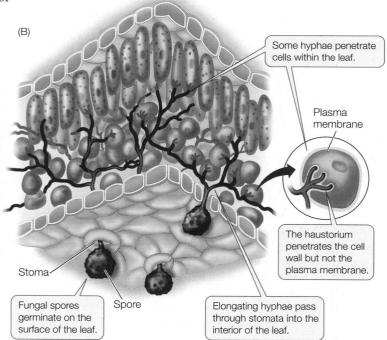

(B)

Some hyphae penetrate cells within the leaf.

Plasma membrane

The haustorium penetrates the cell wall but not the plasma membrane.

Stoma

Fungal spores germinate on the surface of the leaf.

Spore

Elongating hyphae pass through stomata into the interior of the leaf.

30.7 Attacks on a Leaf (A) The thin, light-colored structures in the micrograph are hyphae of the parasitic fungus *Blumeria graminis* growing on the dark surface of the leaf of a grass. (B) Haustoria are fungal hyphae that push into the living cells of plants, from which they absorb nutrients.

Nematode   Fungal hyphae

20 μm

30.8 Some Fungi Are Predators   A nematode is trapped by hyphal rings of the soil-dwelling fungus *Arthrobotrys dactyloides*.

PREDATORY FUNGI   Some fungi have adaptations that enable them to function as active predators, trapping nearby microscopic protists or animals. The most common predatory strategy seen in fungi is to secrete sticky substances from the hyphae so that passing organisms stick tightly to them. The hyphae then quickly invade the prey, growing and branching within it, spreading through its body, absorbing nutrients, and eventually killing it.

A more dramatic adaptation for predation is the constricting ring formed by some species of *Arthrobotrys*, *Dactylaria*, and *Dactylella* (**Figure 30.8**). All of these fungi grow in soil. When nematodes (tiny roundworms) are present in the soil, these fungi form three-celled rings with a diameter that just fits a nematode. A nematode crawling through one of these rings stimulates the fungus, causing the cells of the ring to swell and trap the worm. Fungal hyphae quickly invade and digest the unlucky victim.

Some fungi engage in relationships beneficial to both partners

Certain kinds of relationships between fungi and other organisms have nutritional consequences for both partners. Two of these relationships are highly specific and are **symbiotic** (the partners live in close, permanent contact with one another) as well as **mutualistic** (the relationship benefits both partners; see Chapter 56).

Lichens are associations of a fungus with a cyanobacterium, a unicellular photosynthetic alga, or both. **Mycorrhizae** (singular *mycorrhiza*) are associations between fungi and the roots of plants. In these associations, the fungus obtains organic compounds from its photosynthetic partner and provides it with minerals and water in return, so that the partner's nutrition is also promoted. In fact, many plants grow very poorly without their fungal partners.

LICHENS   A lichen is not a single organism but rather a meshwork of two radically different organisms: a fungus and a photosynthetic microorganism. Together the organisms constituting a lichen can survive some of the harshest environments on Earth. The biota of Antarctica, for example, features more than a hundred times as many species of lichens as of plants.

In spite of their hardiness, lichens are highly sensitive to air pollution because they are unable to excrete any toxic substances they absorb. This sensitivity means that lichens are good biological indicators of air pollution levels. It also explains why they are not commonly found in heavily industrialized regions or large cities.

The fungal components of most lichens are sac fungi (Ascomycota), which also include various cup fungi, yeasts, and molds such as the *Fusarium* mentioned at the beginning of the chapter. The photosynthetic component of a lichen is most often a unicellular green alga, but it can be a cyanobacterium, or can include both. Relatively little experimental work has focused on lichens, perhaps because they grow so slowly—typically less than 1 centimeter in a year.

There are nearly 30,000 described "species" of lichens, each of which is assigned the name of its fungal component. These fungal components may constitute as many as 20 percent of all fungal species. Some of these fungi are able to grow independently without a photosynthetic partner, but most have never been observed in nature other than in a lichen association.

Lichens are found in all sorts of exposed habitats: on tree bark, on open soil, and on bare rock. Reindeer moss (not a moss at all, but the lichen *Cladonia subtenuis*) covers vast areas in Arctic, sub-Arctic, and boreal regions, where it is an important part of the diets of reindeer and other large mammals. Lichens come in various forms and colors. *Crustose* (crustlike) lichens look like colored powder dusted over their substratum (**Figure 30.9A**); *foliose* (leafy) and *fruticose* (shrubby) lichens may have complex forms (**Figure 30.9B,C**).

30.9 Lichen Body Forms   Lichens fall into three principal classes based on their body form. (A) Two crustose lichens are growing on the surface of exposed rock. (B) Foliose lichens have a leafy appearance. (C) The brown and orange growth is of "shrubby" fruticose lichens.

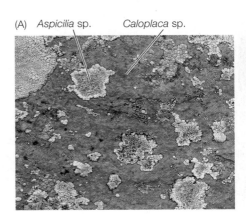

(A) *Aspicilia* sp.   *Caloplaca* sp.

(B) *Parmotrema* sp.

(C) *Teloschistes exilis*

The most widely held interpretation of the lichen relationship is that it is a mutually beneficial symbiosis (a *mutualism*). The hyphae of the fungal mycelium are tightly pressed against the algal or cyanobacterial cells and sometimes even invade them without breaching the plasma membrane (as we described earlier for haustoria in parasitic fungi of plants; see Figure 30.7). The bacterial or algal cells not only survive these indignities but continue their growth and photosynthesis. In fact, the algal cells in a lichen "leak" photosynthetic products at a greater rate than do similar cells growing on their own, and photosynthetic cells from lichens grow more rapidly on their own than when associated with a fungus. On this basis, we could consider lichen fungi to be parasitic on their photosynthetic partners. In many places where lichens grow, however, the photosynthetic cells would not grow at all on their own.

Lichens can reproduce simply by fragmentation of the vegetative body (the *thallus*), or by means of specialized structures called *soredia* (singular *soredium*). Soredia consist of one or a few photosynthetic cells bound by fungal hyphae. The soredia become detached from the lichen, are dispersed by air currents, and upon arriving at a favorable location, develop into a new lichen thallus. Alternatively, the fungal partner may go through its sexual cycle, producing haploid spores. When these spores are discharged, however, they disperse alone, unaccompanied by the photosynthetic partner.

Visible in a cross section of a typical foliose lichen are a tight upper region of fungal hyphae, a layer of photosynthetic cyanobacteria or algae, a looser hyphal layer, and finally hyphal rhizoids that attach the entire structure to its substratum (**Figure 30.10**). The meshwork of fungal hyphae takes up some nutrients needed by the photosynthetic cells and provides a suitably moist environment for them by holding water tenaciously. The fungi derive fixed carbon from the photosynthetic products of the algal or cyanobacterial cells.

Lichens are often the first colonists on new areas of bare rock. They get most of the nutrients they need from the air and rainwater, augmented by minerals absorbed from dust. A lichen begins to grow shortly after a rain, as it begins to dry. As it grows, the lichen acidifies its environment slightly, and this acidity contributes to the slow breakdown of rocks, an early step in soil formation. After further drying, the lichen's photosynthesis ceases. The water content of the lichen may drop to less than 10 percent of its dry weight, at which point it becomes highly insensitive to extremes of temperature.

MYCORRHIZAE Many vascular plants depend on a symbiotic association with fungi. Unassisted, the root hairs of such plants often do not take up enough water or minerals to sustain growth. However, their roots usually do become infected with fungi, forming an association called a mycorrhiza. Mycorrhizae are of two types, based on whether or not the fungal hyphae penetrate the plant cell walls.

In *ectomycorrhizae*, the fungus wraps around the root, and its mass is often as great as that of the root

A **soredium** consists of one or a few photosynthetic cells surrounded by fungal hyphae.

Soredia detach from the parent lichen and travel in air currents, founding new lichens when they settle in a suitable environment.

Upper layer of hyphae

Photosynthetic cell layer

Loose layer of hyphae

Lower level of hyphal rhizoids

30.10 Lichen Anatomy Cross section showing the layers of a foliose lichen and the release of soredia.

itself (**Figure 30.11A**). The fungal hyphae wrap around individual cells in the root but do not penetrate the cell walls. An extensive web of hyphae penetrates the soil in the area around the root, so that up to 25 percent of the soil volume near the root may be fungal hyphae. The hyphae attached to the root increase the surface area for the absorption of water and minerals, and the mass of the mycorrhiza in the soil, like a sponge, holds water efficiently in the neighborhood of the root. Infected roots characteristically are short, swollen, and club-shaped, and they lack root hairs.

The fungal hyphae of *arbuscular mycorrhizae* enter the root and penetrate the cell wall of the root cells, forming arbuscular (treelike) structures inside the cell wall but outside the plasma membrane. These structures, like the haustoria of par-

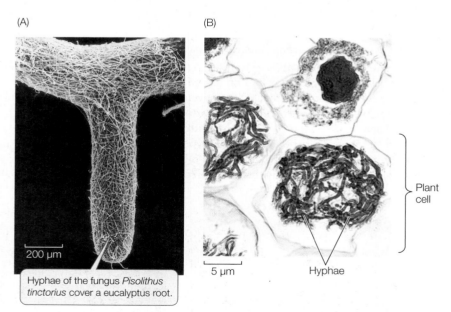

(A)

(B)

200 µm

Hyphae of the fungus *Pisolithus tinctorius* cover a eucalyptus root.

5 µm

Plant cell

Hyphae

30.11 Mycorrhizal Associations (A) Ectomycorrhizal fungi wrap themselves around a plant root, increasing the area available for absorption of water and minerals. (B) Hyphae of arbuscular mycorrhizal fungi infect the root internally and penetrate the root cell walls, branching within the cells and forming treelike (arbuscular) structures that provide the plant with nutrients. Hyphae fill much of the cell outside the nucleus and invaginate the plasma membrane without puncturing it.

asitic fungi and the contact regions of fungi and algal cells in lichens, become the primary site of exchange between plant and fungus (**Figure 30.11B**). As in the ectomycorrhizae, the fungus forms a vast web of hyphae leading from the root surface into the surrounding soil.

The mycorrhizal association is important to both partners. The fungus obtains needed organic compounds, such as sugars and amino acids, from the plant. In return, the fungus, because of its very high surface area-to-volume ratio and its ability to penetrate the fine structure of the soil, greatly increases the plant's ability to absorb water and minerals (especially phosphorus). The fungus may also provide the plant with certain growth hormones and may protect it against attack by disease-causing microorganisms. Plants that have active arbuscular mycorrhizae typically are a deeper green and may resist drought and temperature extremes better than plants of the same species that have little mycorrhizal development. Attempts to introduce some plant species to new areas have failed until a bit of soil from the native area (presumably containing the fungus necessary to establish mycorrhizae) was provided. Trees without ectomycorrhizae will not grow well in the absence of abundant nutrients and water, so the health of our forests depends on the presence of ectomycorrhizal fungi.

The partnership between plant and fungus results in a plant that is better adapted for life on land. It has been suggested that the evolution of mycorrhizae was the single most important step in the colonization of the terrestrial environment by living things. Fossils of mycorrhizal structures 460 million years old have been found. Some liverworts, which are representatives of one of the oldest lineages of terrestrial plants (see Section 28.4), form mycorrhizal associations with fungi.

Certain plants that live in nitrogen-poor habitats, such as cranberry bushes and orchids, invariably have mycorrhizae. Orchid seeds will not germinate in nature unless they are already infected by the fungus that will form their mycorrhizae. Plants that lack chlorophyll always have mycorrhizae, which they often share with the roots of green, photosynthetic plants. In effect, these plants without chlorophyll are feeding on nearby green plants, using the fungus as a bridge.

Biologists had long suspected that roots secrete a chemical signal that enables fungi to find and invade them to form arbuscular mycorrhizae. This was proved to be correct in 2005 when researchers succeeded in isolating the signaling compound. Might the compound also be used by parasitic plants to attack their host plants? Indeed it is. *Striga*, discussed at the beginning of this chapter, turns out to be one of the parasitic plants that use exactly this signal. Thus, in attracting its helper fungus, a plant may also attract a dangerous parasite. The *Striga* story is continued at the end of Chapter 36.

Endophytic fungi protect some plants from pathogens, herbivores, and stress

In a tropical rainforest, 10,000 or more fungal spores land on a single leaf each day. Some are plant pathogens, some do not attack the plant at all, and some invade the plant in a beneficial way. Fungi that live within aboveground parts of plants

without causing obvious deleterious symptoms are called **endophytic fungi**. Recent research has shown that endophytic fungi are abundant in plants in all terrestrial environments.

Grasses with endophytic fungi are more resistant to pathogens and to insect and mammalian herbivores than are grasses lacking endophytes. The fungi produce alkaloids (nitrogen-containing compounds) that are toxic to animals. The alkaloids do not harm the host plant; in fact, some plants produce alkaloids (such as nicotine) themselves. The fungal alkaloids also increase the ability of host plants to resist stress of various types, including drought (water shortage) and salty soils. Such resistance is useful in agriculture.

The role, if any, of endophytic fungi in most broad-leafed plants is unclear, however. They may convey protection against pathogens or simply occupy space within leaves, without conferring any benefit but also without doing harm. The benefit, in fact, might be all for the fungus.

30.2 RECAP

Fungi interact with other organisms in many ways, both harmful and beneficial. Lichens are mutualistic associations of a fungus with an alga and/or a cyanobacterium. Mycorrhizae are associations of fungi and the roots of plants; they are essential for the survival of most plant species.

- What is the role of fungi in Earth's carbon cycle? See p. 631

- Describe the nature and benefits of the lichen association. See pp. 633–635

- Why do plants grow better in the presence of mycorrhizal fungi? See p. 635

One of the most important criteria for assigning fungi to taxonomic groups, before molecular techniques clarified phylogenetic relationships among fungi, was the nature of their life cycles.

30.3 What Variations Exist among Fungal Life Cycles?

Different fungal groups have different life cycles. Some chytrids feature alternation of generations, a type of life cycle found in all plants and some protists. The life cycles of sac fungi and club fungi feature a unique stage called a *dikaryon*, in which a single hypha has two genetically distinct nuclei. In this section we examine the diverse life cycles of major groups of fungi.

Alternation of generations is seen among some aquatic chytrids

The alternation between multicellular haploid (n) and multicellular diploid ($2n$) generations that evolved in plants and certain protist groups (see Section 28.2) is seen in some chytrids as well (**Figure 30.12A**). Alternation of generations is not usual for the life cycles of other fungal groups. The basal chytrids, which are

(A) Chytrids

The life cycle of some aquatic chytrids features alternation of generations.

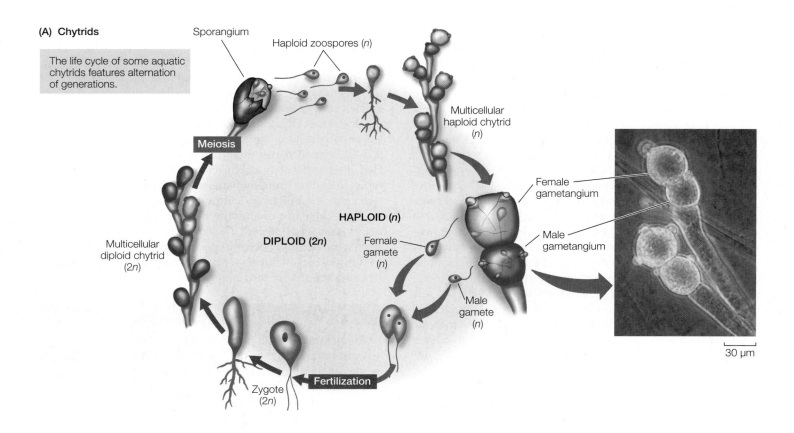

(B) Zygospore fungi (Zygomycota)

The sporangium of zygospore fungi contains haploid nuclei that are incorporated into spores.

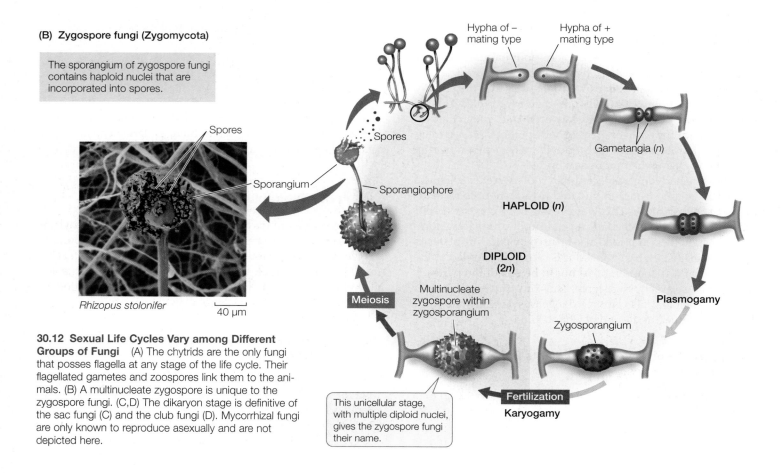

Rhizopus stolonifer

This unicellular stage, with multiple diploid nuclei, gives the zygospore fungi their name.

30.12 Sexual Life Cycles Vary among Different Groups of Fungi (A) The chytrids are the only fungi that posses flagella at any stage of the life cycle. Their flagellated gametes and zoospores link them to the animals. (B) A multinucleate zygospore is unique to the zygospore fungi. (C,D) The dikaryon stage is definitive of the sac fungi (C) and the club fungi (D). Mycorrhizal fungi are only known to reproduce asexually and are not depicted here.

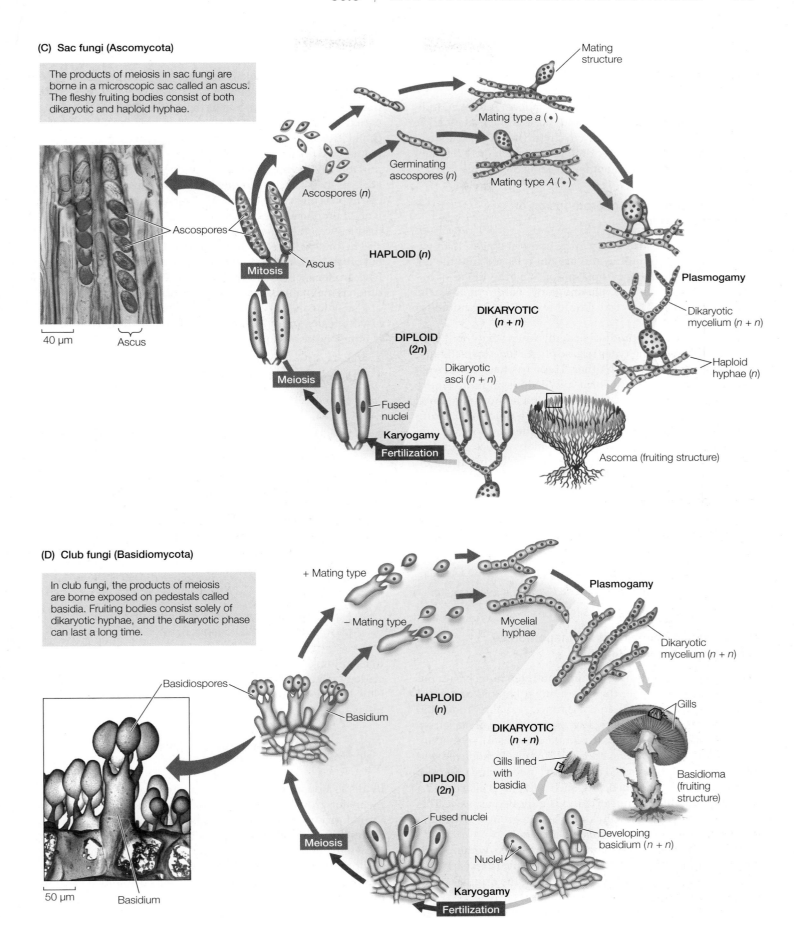

(C) Sac fungi (Ascomycota)

The products of meiosis in sac fungi are borne in a microscopic sac called an ascus. The fleshy fruiting bodies consist of both dikaryotic and haploid hyphae.

Mating structure

Mating type *a* (•)

Germinating ascospores (*n*)

Mating type *A* (•)

Ascospores (*n*)

Ascospores

HAPLOID (*n*)

Plasmogamy

Ascus

Mitosis

Dikaryotic mycelium (*n* + *n*)

DIKARYOTIC (*n* + *n*)

DIPLOID (2*n*)

Haploid hyphae (*n*)

Meiosis

Dikaryotic asci (*n* + *n*)

40 µm

Ascus

Fused nuclei

Karyogamy

Fertilization

Ascoma (fruiting structure)

(D) Club fungi (Basidiomycota)

In club fungi, the products of meiosis are borne exposed on pedestals called basidia. Fruiting bodies consist solely of dikaryotic hyphae, and the dikaryotic phase can last a long time.

+ Mating type

Plasmogamy

− Mating type

Mycelial hyphae

Dikaryotic mycelium (*n* + *n*)

Basidiospores

Basidium

HAPLOID (*n*)

Gills

DIKARYOTIC (*n* + *n*)

Gills lined with basidia

DIPLOID (2*n*)

Basidioma (fruiting structure)

Fused nuclei

Meiosis

Developing basidium (*n* + *n*)

Nuclei

50 µm

Basidium

Karyogamy

Fertilization

aquatic, possess flagellated gametes and flagellated spores. Flagella have been lost in the terrestrial fungi.

What are the consequences of alternation of generations in the chytrids? It is possible that the multicellular haploid organisms serve as a "filter" for harmful mutations. A haploid individual with such a mutation would die, and the mutant allele would not be passed to progeny. In chytrids with alternation of generations, the multicellular diploid stage includes a resistant structure capable of withstanding freezing or drying.

Terrestrial fungi have separate fusion of cytoplasms and nuclei

Although the terrestrial fungi grow in moist places, their gamete nuclei are not motile and are not released into the environment. Instead, the cytoplasms of two individuals of different mating types fuse (**plasmogamy**) before their nuclei fuse (**karyogamy**). Therefore, liquid water is not required for fertilization.

Zygospore fungi reproduce sexually when adjacent hyphae of two different mating types release pheromones, which cause them to grow toward each other. These hyphae produce gametangia, which fuse to form a zygosporangium (**Figure 30.12B**). Sometime later, the gamete nuclei now contained within the zygosporangium fuse to form a unicellular multinucleate **zygospore**, which is the basis of the name of the zygospore fungi. The zygosporangium develops a thick, multilayered wall that protects the zygospore. The highly resistant zygospore may remain dormant for months before its nuclei undergo meiosis and a *sporangiophore* sprouts, bearing a *sporangium*. The sporangium contains the products of meiosis: haploid nuclei that are incorporated into spores. These spores disperse and germinate to form a new generation of haploid hyphae.

— **yourBioPortal.com** —
GO TO Animated Tutorial 30.1 • Life Cycle of a Zygomycete

The dikaryotic condition is a synapomorphy of sac fungi and club fungi

Certain hyphae of terrestrial sac fungi and club fungi have a nuclear configuration other than the familiar haploid or diploid states (**Figures 30.12C,D**). In these fungi, sexual reproduction begins in two distinctive steps: karyogamy (fusion of nuclei) occurs long after plasmogamy (fusion of cytoplasm), so that *two genetically different haploid nuclei coexist and divide within the same hypha*. Such a hypha is called a **dikaryon** ("two nuclei"). Because the two nuclei differ genetically, such a hypha is also called a *heterokaryon* ("different nuclei") and is indicated as $n + n$. This dikaryon is a synapomorphy of these two groups, which are placed together in the clade called Dikarya.

Eventually, specialized fruiting structures form, within which the pairs of genetically dissimilar nuclei—one from each parent—fuse, giving rise to zygotes long after the original "mating." The diploid zygote nucleus undergoes meiosis, producing four haploid nuclei. The mitotic descendants of those nuclei

become spores, which germinate to give rise to the next generation—usually of hyphae.

A life cycle with a dikaryon stage has several unusual features. First, there are no gamete *cells*, only gamete *nuclei*. Second, the only true diploid structure is the zygote, although for a long period the genes of both parents are present in the dikaryon and can be expressed. In effect, the hypha is neither diploid ($2n$) nor haploid (n); rather, it is *dikaryotic* ($n + n$). A harmful recessive mutation in one nucleus may be compensated for by a normal allele on the same chromosome in the other nucleus, and dikaryotic hyphae often have characteristics that are different from their n or $2n$ products. The dikaryotic condition is perhaps the most distinctive of the genetic peculiarities of the fungi.

The dikaryotic condition is short-lived in most sac fungi but often lasts for months or even years in club fungi. Club fungi have an elegant mechanism that ensures that the dikaryotic condition is maintained as new cells are formed. One consequence of a sustained dikaryotic condition is an increased opportunity for fusions of hyphae of two different mating types before fruiting bodies are formed.

— **yourBioPortal.com** —
GO TO Web Activity 30.2 • Life Cycle of a Dikaryotic Fungus

30.3 RECAP

Some chytrid fungi exhibit alternation of generations between haploid and diploid multicellular states. The life cycle of zygospore fungi includes a resistant-spore stage with many diploid nuclei. The sac fungi and club fungi share a derived dikaryotic condition, in which two genetically different haploid nuclei coexist in the same cell.

- What is the role of the zygospore in the life cycle of zygospore fungi? See p. 638

- Explain the phenomenon of dikaryosis in terms of plasmogamy and karyogamy. See p. 638

In examining the most important properties of the fungi as a clade, we have often drawn examples from specific groups. Now let's look in more detail at the diversity and evolution of fungi.

30.4 How Have Fungi Evolved and Diversified?

In this section we examine representative species from each of the major groups of fungi (see Table 30.1). Recall that the informal groups called chytrids and zygospore fungi are not clades, but the various lineages within these informal groups shown in Table 30.1 are each thought to be monophyletic. In addition, Microsporidia, Glomeromycota (arbuscular mycorrhizal fungi), Ascomycota (sac fungi), and Basidiomycota (club fungi) are all clades, and the latter two groups form a monophyletic group called Dikarya.

Microsporidia are highly reduced, parasitic fungi

Microsporidia are unicellular fungi with walls that contain chitin. They are among the smallest eukaryotes known, with infective spores that are only 1–40 μm in diameter. About 1,500 species have been described, but many more species are thought to exist. Their relationships among the eukaryotes have puzzled biologists for many decades.

Microsporidia lack true mitochondria, although they have reduced structures known as *mitosomes* that are derived from mitochondria. Unlike mitochondria, however, mitosomes contain no DNA; the mitochondrial genome has been completely transferred to the nucleus. Because microsporidia lack mitochondria, biologists initially suspected that they represent an early lineage of eukaryotes that branched before the endosymbiotic event from which mitochondria evolved. The presence of the mitosome, however, indicates that this hypothesis is incorrect. DNA sequence analysis has confirmed that microsporidia are in fact highly reduced, parasitic fungi, although the exact placement of microsporidia among the fungal lineages is still being investigated.

Microsporidia are obligate intracellular parasites of animals, especially of insects, crustaceans, and fishes. Some species are known to infect mammals, including humans. Most infections by microsporidia cause chronic diseases in the hosts, with effects that include weight loss, reduced fertility, and shortened life span. The host cell is penetrated by a *polar tube* of the microsporidian spore, and the contents of the spore are injected into the host (**Figure 30.13**). The sporoplasm then replicates within the host cell and produces new infective spores. The life cycle of some species is complex and involves multiple hosts, whereas other species infect a single host. In some insects, the parasitic microsporida are transmitted vertically (i.e., from parent to offspring).

Chytrids are the only fungi with flagella

Chytrids include several distinct lineages of aquatic microorganisms once classified with the protists. However, morphological evidence (cell walls that consist primarily of chitin) and molecular evidence support their classification as basal fungi. In this book we use the term "chytrid" to refer to all three of the formally named clades shown as chytrids in Table 30.1, but some mycologists use this term to refer to one particular clade, the Chytridiomycota. There are fewer than 1,000 described species among the three taxonomic groups of chytrids.

Like the animals (and many other eukaryotes), most chytrids possess flagellated gametes. The retention of this character reflects the aquatic environment in which fungi first evolved. Chytrids are the only fungi that have flagella at any life cycle stage.

Chytrids may be parasitic (on organisms such as algae, mosquito larvae, and nematodes) or saprobic. Some have complex mutualistic relationships in the compound stomachs of foregut-fermenting animals such as cattle and deer. Many chytrids live in freshwater habitats or in moist soil, but some are marine. Some chytrids are unicellular, others have rhizoids (**Figure 30.14**), and still others have coenocytic hyphae. Chytrids reproduce both sexually and asexually, but they do not have a dikaryon stage.

Allomyces, a well-studied genus of chytrids, is a member of the group of chytrids (**Blastocladiomycota**; see Table 30.1) that displays alternation of generations. Both female and male gametes have flagella. The motile female gamete produces a *pheromone*, a chemical signal that attracts the swimming male gamete. The two gametes fuse, and then their nuclei fuse to form a diploid zygote that germinates to form a diploid mycelium. This is the diploid generation. Mitosis and cytokine-

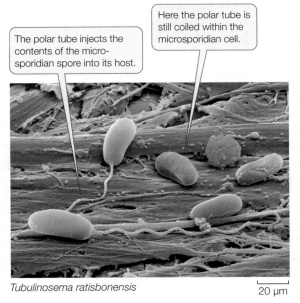

The polar tube injects the contents of the microsporidian spore into its host.

Here the polar tube is still coiled within the microsporidian cell.

Tubulinosema ratisbonensis

20 μm

30.13 Spores of Microsporidia Inject Host Cells The polar tubes of microsporidia spores transfer their contents into the host's cells. This species infects many animals, including humans.

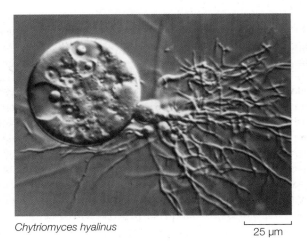

Chytriomyces hyalinus

25 μm

30.14 A Chytrid Branched rhizoids emerge from the sporangium of a mature chytrid.

sis in the zygote give rise to a small, multicellular, diploid organism, which produces numerous diploid flagellated zoospores. These diploid zoospores disperse and germinate to form more diploid organisms. Eventually, the diploid organisms produce a second kind of zoosporangium, a thick-walled resting sporangium, that can survive unfavorable conditions such as dry weather or freezing. Nuclei in the resting sporangium undergo meiosis, giving rise to haploid zoospores that are released into the water and begin the haploid stage of the life cycle.

The haploid mycelium is similar to that of the diploid mycelium, but it matures to produce female and male *gametangia* (gamete cases; see Figure 30.12A). Note that meiosis occurred in the earlier diploid generation and not immediately before gamete formation, as it does in animals. This life cycle with meiosis in spore formation is similar to the alternation of generations seen in plants.

Zygospore fungi are terrestrial saprobes, parasites, and mutualists

Zygospore fungi (Zygomycota) include four major lineages of terrestrial fungi that live on soil as saprobes, as parasites of insects and spiders, or as mutualists of other fungi and invertebrate animals. They produce no cells with flagella, and only one diploid cell—the zygote—appears in the entire life cycle. Their hyphae are coenocytic. The mycelium spreads over its substratum, growing forward by means of vegetative hyphae. Most species do not form a fleshy fruiting structure; rather, the hyphae spread in an apparently random fashion, with occasional stalked **sporangiophores** reaching up into the air (**Figure 30.15**). These reproductive structures may bear one or many sporangia.

More than a thousand species of zygospore fungi have been described. One species you may have seen is *Rhizopus stolonifer*, the black bread mold. *Rhizopus* produces many stalked sporangiophores, each bearing a single sporangium containing hundreds of minute spores (see Figure 30.12B). As in other filamentous fungi, the spore-forming structure is separated from the rest of the hypha by a wall.

Arbuscular mycorrhizal fungi form symbioses with plants

Arbuscular mycorrhizal fungi (Glomeromycota) are terrestrial fungi that associate with plant roots in a close symbiotic relationship (see Figure 30.11B). As we noted earlier in this chapter, these associations are important for most species of plants, which benefit from absorption of water and mineral nutrients through the large surface area of the fungal mycelium. Many of the fungi found in soils are arbuscular mycorrhizal fungi. Fewer than 200 species have been described, but 80 to 90 percent of all plants have associations with them. Molecular systematic studies have suggested that arbuscular mycorrhizal fungi are the sister group to Dikarya (sac fungi and club fungi), although this position is still subject to some debate.

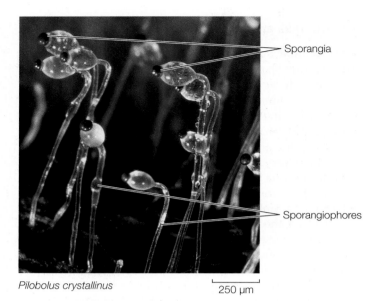

Pilobolus crystallinus

250 µm

30.15 Zygospore Fungi Produce Sporangiophores These transparent structures are sporangiophores (spore-bearing hyphae) growing on decomposing animal dung. Sporangiophores grow toward the light and end in tiny sporangia, which the filamentous sporangiophores can eject as far as 2 meters. Animals ingest those sporangia that land on grass and then disseminate the spores in their feces.

The hyphae of arbuscular mycorrhizal fungi are coenocytic. These fungi use glucose from their plant partners as their primary energy source, converting the glucose to other, fungus-specific sugars that cannot return to the plant. Arbuscular mycorrhizal fungi reproduce asexually; there is not any direct evidence that they reproduce sexually.

The next two fungal clades that we'll discuss, the sac fungi and the club fungi, are related groups with many similarities, including a dikaryon stage and septate hyphae. The two clades differ in sexual reproductive structure; in the sac fungi the sexual spores are borne inside a sac, whereas in the club fungi they are borne on a pedestal.

The sexual reproductive structure of sac fungi is the ascus

The **sac fungi** (Ascomycota) are a large and diverse group of fungi found in marine, freshwater, and terrestrial habitats. There are approximately 64,000 known species, nearly half of which are the fungal partners in lichens. The hyphae of sac fungi are segmented by more or less regularly spaced septa. A pore in each septum permits extensive movement of cytoplasm and organelles (including nuclei) from one segment to the next.

Sac fungi are distinguished by the production of sacs called **asci** (singular *ascus*), which after meiosis and cytoplasmic cleavage contain sexually produced haploid *ascospores* (see Figure 30.12C). The ascus is the characteristic sexual reproductive structure of the sac fungi. The sac fungi contain many diverse groups and in the past were divided on the basis of whether or not the asci are contained within a specialized fruiting structure known as an **ascoma** (plural *ascomata*), and on the morphology of this fruiting structure. DNA sequence analyses have resulted in a revision of these traditional groupings, however.

Some species of sac fungi are unicellular yeasts. Perhaps the best known of the 800 or so species of yeasts in this group is baker's, or brewer's, yeast (*Saccharomyces cerevisiae*; see Figure 30.3). These yeasts are among the most important domesticated fungi. *S. cerevisiae* metabolizes glucose obtained from its environment to ethanol and carbon dioxide by fermentation. It forms carbon dioxide bubbles in bread dough and gives baked bread its light texture. Although they are baked away in bread making (which produces the pleasant aroma of baking bread), the ethanol and carbon dioxide are both retained when yeast ferments grain into beer. Other yeasts live on fruits such as figs and grapes and play an important role in the making of wine. Many other yeasts are associated with insects; in the guts of some insects, yeasts provide enzymes for digestion of refractory materials, especially cellulose.

Sac fungus yeasts reproduce asexually by budding. Sexual reproduction takes place when two adjacent haploid cells of opposite mating types fuse. In some species, the resulting zygote buds to form a diploid cell population. In others, the zygote nucleus undergoes meiosis immediately; when this happens, the entire cell becomes an ascus. Depending on whether the products of meiosis then undergo mitosis, a yeast ascus contains either eight or four ascospores. The ascospores germinate to become haploid cells. The sac fungus yeasts have lost the dikaryon stage.

Most sac fungi are filamentous species, such as the cup fungi (**Figure 30.16**), in which the ascomata are cup-shaped and can be as large as several centimeters across (although most are much smaller). The inner surfaces of the cups, which are covered with a mixture of specialized hyphae and asci, produce huge numbers of spores. The edible ascomata of some species, including morels and truffles, are regarded by humans as gourmet delicacies (and can sell at prices higher than gold). The un-

derground ascomata of truffles have a strong odor that attracts mammals such as pigs, which then eat and disperse the fungus.

The sac fungi also include many of the filamentous fungi known as molds. Many of these species are parasites of flowering plants. Chestnut blight and Dutch elm disease are both caused by filamentous molds. Between its introduction to the United States in the 1890s and 1940, the chestnut blight fungus destroyed the American chestnut as a commercial species. Before the blight, this species accounted for more than half the trees in the forests of the eastern United States. Another familiar story is that of the American elm. Sometime before 1930, the Dutch elm disease fungus (first discovered in the Netherlands but native to Asia) was introduced into the United States on infected elm logs from Europe. Spreading rapidly—sometimes by way of connected root systems—the fungus destroyed great numbers of American elm trees.

Other plant pathogens among the sac fungi include the powdery mildews that infect cereal grains, lilacs, and roses, among many other plants. Mildews can be a serious problem to farmers and gardeners, and a great deal of research has focused on ways to control these agricultural pests.

Brown molds of the genus *Aspergillus* are important in some human diets. *A. tamarii* acts on soybeans in the production of soy sauce, and *A. oryzae* is used in brewing the Japanese alcoholic beverage sake. Some species of *Aspergillus* that grow on grains and on nuts such as peanuts and pecans produce extremely carcinogenic (cancer-inducing) compounds called *aflatoxins*. In the United States and most other industrialized countries, moldy grain infected with *Aspergillus* is thrown out. In Africa, where food is scarcer, the grain gets eaten, moldy or not, and causes severe health problems, including high levels of certain cancers.

Penicillium is a genus of green molds, of which some species produce the antibiotic penicillin, presumably for defense against competing bacteria. Two species, *P. camembertii* and *P. roquefortii*, are the organisms responsible for the characteristic strong flavors of Camembert and Roquefort cheeses, respectively.

The filamentous sac fungi reproduce asexually by means of conidia that form at the tips of specialized hyphae (**Figure 30.17**). Small chains of conidia are produced by the millions and can survive for weeks in nature. The conidia are what give molds their characteristic colors. *Fusarium oxysporum*, the plant pathogen mentioned at the beginning of this chapter, is a filamentous sac fungus with no known sexual stage. It produces conidia in abundance.

The sexual reproductive cycle of filamentous sac fungi includes the formation of a dikaryon, although this stage is relatively brief compared with that in club fungi. Many filamentous sac fungi form mating structures of two different mating types (types *A* and *a*; see Figure 30.12C). A nucleus from one of the mating structures (specialized hyphae) on type *a* enters the

(A) *Sarcoscypha coccinea*

(B) *Morchella* sp.

30.16 Cup Fungi (A) These brilliant red cups are the ascomata of a cup fungus. (B) Morels, which have a spongelike ascoma and a subtle flavor, are considered a culinary delicacy by humans.

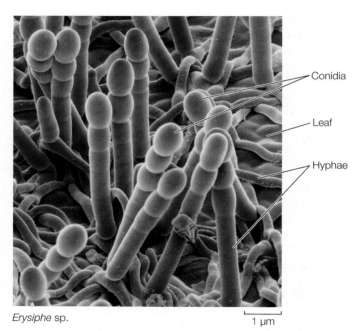

Erysiphe sp.

1 µm

30.17 Conidia   Chains of conidia are developing at the tips of specialized hyphae arising from this powdery mildew growing on a leaf.

hypha of type *A* (or vice versa). *Ascogenous* (ascus-forming) hyphae develop from the now dikaryotic hypha. The mating type *a* and *A* nuclei divide simultaneously with the host nuclei, and eventually asci form at the tips of the ascogenous hyphae. Only with the formation of asci do the nuclei finally fuse within the cell that now is defined as the ascus. Both nuclear fusion and the subsequent meiosis to produce the haploid ascospores take place within individual asci. The meiotic products are incorporated into ascospores that are ultimately released (sometimes shot off forcefully) by the ascus to begin the new haploid generation.

The sexual reproductive structure of club fungi is a basidium

Club fungi (Basidiomycota) produce some of the most spectacular fruiting structures found anywhere among the fungi. These fruiting structures, called **basidiomata** (singular, basidioma), include puffballs (see Figure 30.6), some of which may be more than half a meter in diameter, mushrooms of all kinds, and the bracket fungi often encountered on trees and fallen logs in a damp forest. About 30,000 species of club fungi have been described. These include about 4,000 species of mushrooms, including the familiar *Agaricus bisporus* you may enjoy on your pizza, as well as poisonous species such as some members of the genus *Amanita* (**Figure 30.18A**). Bracket fungi (**Figure 30.18B**) do great damage both to cut lumber and timber stands;

Zygomycota

Glomeromycota

Ascomycota

Basidiomycota

they also play an important role in returning carbon to the carbon cycle. Some of the most damaging plant pathogens are club fungi, including the rust fungi and the smut fungi that parasitize cereal grains. In contrast, other club fungi contribute to the survival of plants as fungal partners in ectomycorrhizae.

The hyphae of club fungi characteristically have septa with small, distinctive pores. The **basidium** (plural *basidia*), a swollen cell at the tip of a specialized hypha, is the characteristic sexual reproductive structure of the club fungi. In mushroom-forming club fungi, the basidia form on specialized structures known as gills or pores. The basidium is the site of nuclear fusion and meiosis and thus plays the same role in the club fungi as the ascus does in the sac fungi and the zygosporangium does in the zygospore fungi.

As seen in Figure 30.12D, after nuclei fuse in the basidium, the resulting diploid nucleus undergoes meiosis, and the four resulting haploid nuclei are incorporated into haploid *basidiospores*, which form on tiny stalks on the outside of the basidium. A single basidioma of the common bracket fungus *Ganoderma applanatum* can produce as many as 4.5 *trillion* basidiospores in one growing season. Basidiospores typically are forcibly discharged from their basidia and then germinate, giving rise to hyphae with haploid nuclei. As these hyphae grow, haploid hyphae of different mating types meet and fuse, forming dikaryotic hyphae, each cell of which contains two nuclei, one from each parent hypha. The dikaryotic mycelium grows and eventually, when triggered by rain or another environmental cue, produces a basidioma. The dikaryon stage may persist for years—some club fungi live for decades or even centuries. This pattern contrasts with the life cycle of the sac fungi, in which the dikaryon is found only in the stages leading up to formation of the asci.

30.4 RECAP

The ancestor of all fungi was probably aquatic like the chytrids, but fungi have diversified to become important components of terrestrial ecosystems. Sac fungi and club fungi (which together form the clade Dikarya) contain the largest number of species.

- Explain how the microsporidia infect the cells of their animal hosts. See p. 639 and Figure 30.13

- What feature of the chytrids suggests an aquatic ancestor to the fungi? See p. 639

- What distinguishes the fruiting bodies of sac fungi from those of club fungi? See pp. 640–642 and Figures 30.12C,D

Whether living on their own or in symbiotic associations, fungi have spread successfully over much of Earth since their origin from a protist ancestor. That ancestor also gave rise to the choanoflagellates and the animals, as we describe in Chapter 31.

30.18 Club Fungus Fruiting Structures The basidiomata of the club fungi are probably the most familiar of all fungal structures. (A) These highly poisonous mushrooms were produced by a fungus in the genus *Amanita*, which forms ecto-mycorrhizal relationships with trees. (B) This edible bracket fungus is an agent of decay on dead wood.

(A) *Amanita muscaria*

(B) *Laetiporus sulphureus*

CHAPTER SUMMARY

30.1 What Is a Fungus?

- Fungi are opisthokonts with **absorptive heterotrophy** and with chitin in their cell walls. Fungi have various nutritional modes: some are **saprobes**, others are **parasites**, and some are **mutualists**. **Yeasts** are unicellular fungi. **SEE WEB ACTIVITY 30.1**

- The body of a multicellular fungus is a **mycelium**—a meshwork of **hyphae** that may be **septate** (having **septa**) or **coenocytic**. **Review Figure 30.4**

- Fungi are tolerant of hypertonic environments, and many are tolerant of low or high temperatures.

- Many species of fungi reproduce both sexually and asexually. Sexual reproduction occurs between individuals of different **mating types**.

30.2 How Do Fungi Interact with Other Organisms?

- Saprobic fungi, as decomposers, make crucial contributions to the recycling of elements, especially carbon. Certain fungi have relationships with other organisms that are both **symbiotic** and **mutualistic**.

- Many fungi are parasitic plant pathogens, harvesting nutrients from plant cells by means of **haustoria**. **Review Figure 30.7**

- Some fungi associate with cyanobacteria and/or green algae to form **lichens**, which live on many exposed surfaces of rocks, trees, and soil. **Review Figure 30.10**

- **Mycorrhizae** are mutualistic associations of fungi with plant roots. They improve a plant's ability to take up nutrients and water. **Review Figure 30.11**

- **Endophytic fungi** live within plants and can provide protection to their hosts from herbivores.

30.3 What Variations Exist among Fungal Life Cycles?

- Some chytrids have a life cycle that includes alternation of generations. **Review Figure 30.12A**

- In the sexual reproduction of terrestrial fungi, hyphae fuse, allowing "gamete" nuclei to be transferred. **Plasmogamy** (fusion of cytoplasm) precedes **karyogamy** (fusion of nuclei).

- Zygospore fungi have a resistant-spore stage with many diploid nuclei, known as a **zygospore**. **Review Figure 30.12B, ANIMATED TUTORIAL 30.1**

- In sac fungi and club fungi, a **dikaryon** is formed. The dikaryotic ($n + n$) condition is unique to the fungi. **Review Figures 30.12C,D, WEB ACTIVITY 30.2**

30.4 How Have Fungi Evolved and Diversified?

- The relationships of microsporidia, chytrids, and zygospore fungi are not well resolved, but these groups diversified early in fungal evolution. The mycorrhizal fungi, sac fungi, and club fungi form a monophyletic group, and the latter two groups form the clade Dikarya. **Review Figure 30.2 and Table 30.1**

- **Microsporidia** are reduced, intracellular parasitic species of fungi that infect several animal groups, especially insects, crustaceans, and fishes.

- The three distinct lineages of **chytrids** have flagellated gametes. **Review Figure 30.12A**

- The four distinct lineages of **zygospore fungi** have coenocytic hyphae and a zygospore in their life cycle. Their fruiting structures are simple stalked **sporangiophores**. **Review Figure 30.12B**

- **Arbuscular mycorrhizal fungi** form symbiotic associations with plant roots. These mycorrhizae increase water and mineral uptake by the plants and provide a carbon source to the fungi. They are only known to reproduce asexually. Their hyphae are coenocytic.

- **Sac fungi** have septate hyphae; their sexual reproductive structures are **asci**. Many sac fungi are partners in lichen and endophytic associations. Filamentous sac fungi produce fleshy fruiting bodies called **ascomata**. The dikaryon stage in the sac fungus life cycle is relatively brief. **Review Figure 30.12C**

- **Club fungi** have septate hyphae. Many of the species are plant pathogens, although mushroom-forming species are more familiar to most people. Their fruiting bodies are called **basidiomata**, and their sexual reproductive structures are **basidia**. The dikaryon stage may last for years. **Review Figure 30.12D**

SELF-QUIZ

1. Which statement about fungi is *not* true?
   a. A multicellular fungus has a body called a mycelium.
   b. Hyphae are composed of individual mycelia.
   c. Many fungi tolerate highly hypertonic environments.
   d. Many fungi tolerate low temperatures.
   e. Some fungi are anchored to their substrate by rhizoids.

2. The absorptive heterotrophy of fungi is aided by
   a. dikaryon formation.
   b. spore formation.
   c. the fact that they are all parasites.
   d. their large surface area-to-volume ratio.
   e. their possession of chloroplasts.

3. Which statement about fungal nutrition is *not* true?
   a. Some fungi are active predators.
   b. Some fungi form mutualistic associations with other organisms.
   c. All fungi require mineral nutrients.
   d. Fungi can make some of the compounds that are vitamins for animals.
   e. Facultative parasites can grow only on their specific hosts.

4. Which statement about dikaryosis is *not* true?
   a. The cytoplasm of two cells fuses before their nuclei fuse.
   b. The two haploid nuclei are genetically different.
   c. The two nuclei are of the same mating type.
   d. The dikaryon stage ends when the two nuclei fuse.
   e. Not all fungi have a dikaryon stage.

5. Reproductive structures consisting of one or more photosynthetic cells surrounded by fungal hyphae are called
   a. ascospores.
   b. basidiospores.
   c. conidia.
   d. soredia.
   e. gametes.

6. Members of the zygospore fungi
   a. have hyphae without regularly occurring septa.
   b. produce motile gametes.
   c. form fleshy fruiting bodies.
   d. are haploid throughout their life cycle.
   e. have sexual reproductive structures similar to those of the sac fungi.

7. Which statement about sac fungi is *not* true?
   a. Some species are yeasts.
   b. They form reproductive structures called asci.
   c. Their hyphae are segmented by septa.
   d. Many of their species have a dikaryotic state.
   e. All have fruiting structures called ascomata.

8. Club fungi
   a. often produce fleshy fruiting structures.
   b. have hyphae without septa.
   c. have no sexual stage.
   d. produce basidia within basidiospores.
   e. form diploid basidiospores.

9. Microsporidia
   a. lack true mitochondria.
   b. are parasites of animals.
   c. contain mitosomes.
   d. are among the smallest eukaryotes known.
   e. all of the above

10. Which statement about lichens is *not* true?
    a. They can reproduce by fragmentation of the vegetative body.
    b. They are often the first colonists in a new area.
    c. They render their environment more basic (alkaline).
    d. They contribute to soil formation.
    e. They may contain less than 10 percent water by weight.

FOR DISCUSSION

1. You are shown an object that looks superficially like a pale green mushroom. Describe at least three criteria (including anatomical and chemical traits) that would enable you to tell whether the object is a piece of a plant or a piece of a fungus.

2. Differentiate among the members of the following pairs of related terms:
   a. hypha/mycelium
   b. ascus/basidium
   c. ectomycorrhiza/arbuscular mycorrhiza

3. For each type of organism listed below, give a single characteristic that may be used to differentiate it from the other, related organism(s) in parentheses.
   a. zygospore fungi (sac fungi)
   b. sac fungi (club fungi)
   c. baker's yeast (*Penicillium*)

4. Many fungi are dikaryotic during part of their life cycle. Why are dikaryons described as $n + n$ instead of $2n$?

5. If all the fungi on Earth were suddenly to die, how would the surviving organisms be affected?

6. How might the first mycorrhizae have arisen?

7. What attributes might account for the ability of lichens to withstand the intensely cold environment of Antarctica? Be specific in your answer.

8. What factors must be taken into account in using fungi to combat agricultural pests?

ADDITIONAL INVESTIGATION

We noted that lichens are highly sensitive to air pollution. How could you use lichen diversity and abundance to measure air quality? How would you expect lichen diversity to vary with respect to distance from major metropolitan areas? Would you expect prevailing wind direction to be a factor in this pattern?

31 Animal Origins and the Evolution of Body Plans

Getting back to our roots

In 1883, the zoologist Franz Schulze noticed something unusual in his Austrian laboratory: transparent, flattened organisms were crawling on the sides of his saltwater aquarium. Collected accidentally along with the sponges that were Schulze's primary interest, these organisms were unlike any previously described animals—especially since they continually changed shape as they moved.

Schulze's examination of the new organisms revealed that they were animals. But structurally they were among the simplest animals that he—or anyone else—had ever observed, with only four types of cells. He named the new species *Trichoplax adhaerens*, which means "sticky hairy plate," and argued that the new species had no close relationships with other major animal groups. For decades,

however, most biologists dismissed Schulze's findings, insisting that the transparent organisms must be larval forms of other, well-known, animals.

In the 1960s, new and more detailed studies confirmed Schultze's findings and the distinctive nature of *Tricho-plax*. Even then, this odd animal continued to be known almost exclusively from aquariums. Only in the past decade have biologists been able to locate and study natural field populations of *Trichoplax adhaerens*. A few additional closely related species have been discovered, and they are collectively known as placozoans (Greek, "flat animals").

The more biologists have studied *Trichoplax*, the odder this animal appears. It has the smallest genome of any animal studied to date. The mature stages lack body symmetry and have no mouth, gut, or nervous system. Is *Trichoplax* a relict representative of a group of animals that appeared early in animal evolution? Indeed, some recent phylogenetic analyses support the possibility that *Trichoplax* is a representative of the most divergent group of animals.

Although biologists agree on which groups of organisms are animals, the root of the animal tree is a subject of considerable investigation and debate. Traditionally, the first split is thought to have been between the sponges and all other animals, and most evidence still favors that view. Some gene sequence analyses, however, suggest that the major groups of sponges are not even each other's closest relatives, and that the glass sponges alone split with all the remaining animals (including other sponges)

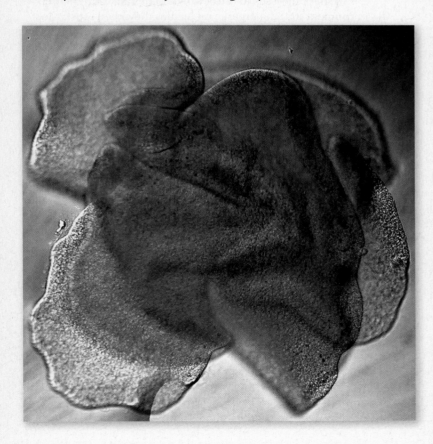

Are Placozoans at the Base of Animal Phylogeny? Is the simplicity of placozoans an ancestral feature, or did these organisms descend from ancestors with more complex body plans? Answering that question requires an understanding of animal phylogeny.

31.1 What Characteristics Distinguish the Animals?

How do we recognize an organism as an animal? That may seem obvious for many familiar animals, but less so for groups like sponges, which were once thought to be plants. Some of the general aspects that we associate with animals include:

- *Multicellularity.* In contrast to the Bacteria, Archaea, and most protists (see Chapters 26 and 27), all animals are multicellular. Animal life cycles feature complex patterns of development from a single-celled zygote into a multicellular adult.

- *Heterotrophic metabolism.* In contrast to most plants (see Chapters 28 and 29), all animals are heterotrophs. Animals are able to synthesize very few organic molecules from inorganic chemicals, so they must take in nutrients from their environment (either through their own actions, or in some cases with the aid of symbiotic species).

- *Internal digestion.* Although the fungi are also heterotrophs (see Chapter 30), animals digest their food differently. Whereas fungi rely on external digestion, most animals use internal processes to break down materials from their environment into the organic molecules they need most. Most animals ingest food into an internal gut that is continuous with the outside environment and in which digestion takes place.

- *Movement.* In contrast to the majority of plants and fungi, most animals can move. Animals must move to find food or bring food to them. Muscle tissue is unique to animals, and many animal body plans are specialized for movement.

Although these general features help us recognize animals, none is diagnostic for all animals. Some animals do not move, at least during certain life stages, and some plants and fungi do have limited movement. Some animals lack a gut. Many multicellular organisms are not animals. So what is the evidence that groups all animals together in a single clade?

Animal monophyly is supported by gene sequences and morphology

The most convincing evidence that all the organisms considered to be animals share a common ancestor comes from phylogenetic analyses of their gene sequences. Relatively few complete animal genomes are available, but many more genomes are being sequenced each year. Analyses of these genomes, as well as many individual gene sequences, have shown that all animals are indeed *monophyletic*; a currently well-supported phylogenetic tree of the animals is shown in **Figure 31.1**.

Alternative Candidates   Some studies place either sponges (left) or ctenophores (upper right) rather than placozoans as the sister group of all other multicellular animals. Most evidence favors sponges as the most divergent animal group.

at the base of the animal tree. Other genomic investigations have suggested that the ctenophores—comb jellies—may be the sister group to all other animals. Ctenophores are beautiful, transparent organisms whose sticky tentacles capture planktonic prey.

Newly acquired insights into how animal genomes are structured and have evolved play an ever more important role in understanding the relationships among the major animal groups. This chapter explores the earliest branches on the animal tree, and how a few fundamental "body plans" have been modified to yield the remarkable variety of animal forms described in this and the following two chapters.

IN THIS CHAPTER we will review the evidence that has led biologists to conclude that the animals are monophyletic and will then present the best-supported current hypotheses of animal phylogeny. We will describe how the diverse animal forms are derived from a small array of body plans. We will discuss various animal strategies for obtaining food and describe the amazingly varied life cycles of animals—how they are born, grow, disperse, and reproduce. Finally, we will describe members of several basal clades of animals.

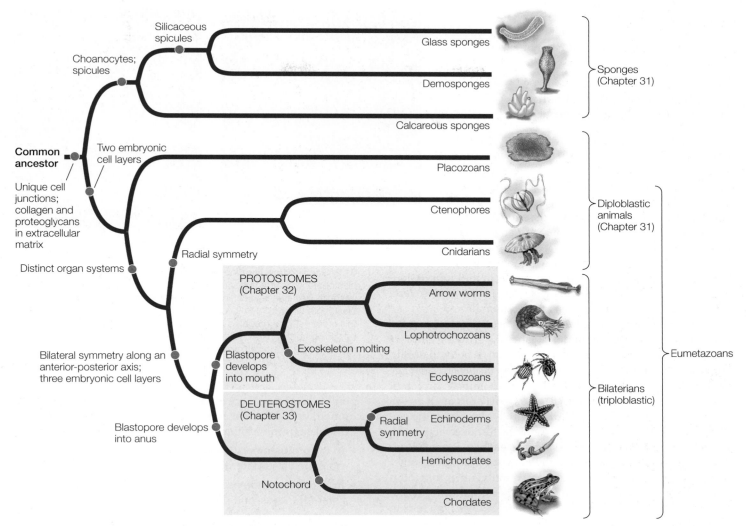

31.1 The Phylogeny of Animals This tree presents the best supported current hypotheses of evolutionary relationships among major groups of animals. The traits highlighted by red circles will be explained as you read this chapter; you should review this figure closely after you complete your reading.

─── **yourBioPortal.com** ───
GO TO **Web Activity 31.1 • Sponge and Diploblast Classification**

Although animals were considered to belong to a single clade long before gene sequencing became possible, surprisingly few morphological features are shared across all species of animals. These morphological *synapomorphies* include:

- Unique types of junctions between their cells (tight junctions, desmosomes, and gap junctions; see Figure 6.7).
- A common set of extracellular matrix molecules, including collagen and proteoglycans (see Figure 5.25).

Although some animals in a few groups lack one or another of these characteristics, it is believed that these traits were possessed by the ancestor of all animals and subsequently lost in those groups. Similarities in the organization and function of Hox and other developmental genes (see Chapter 20) provide additional

evidence of developmental mechanisms shared by a common animal ancestor. The Hox genes specify body pattern and axis formation, leading to developmental similarities across animals.

The common ancestor of animals was probably a colonial flagellated protist similar to existing colonial choanoflagellates (see Figure 27.27). The most reasonable current scenario postulates a choanoflagellate lineage in which certain cells in the colony began to be specialized—some for movement, others for nutrition, others for reproduction, and so on. Once this *functional specialization* had begun, cells could have continued to differentiate. Coordination among groups of cells could have improved by means of specific regulatory and signaling molecules that guided differentiation and migration of cells in developing embryos. Such coordinated groups of cells eventually evolved into the larger and more complex organisms that we call animals.

More than a million living animal species have been named and described, and millions of additional animal species await discovery. Clues to the evolutionary relationships among animal groups can be found in fossils, in patterns of embryonic development, in the morphology and physiology of living animals, in the structure of animal proteins, and in gene sequences.

Increasingly, studies of higher-level relationships have come to depend on genomic sequence comparisons, as genomes are ultimately the source of all inherited trait information.

A few basic developmental patterns differentiate major animal groups

Differences in patterns of embryonic development have until recently provided many of the most important clues to animal phylogeny. Analyses of gene sequences, however, are now showing that some developmental patterns are more evolutionarily variable than previously thought. We describe here the basic developmental patterns that vary among the major animal clades.

The first few cell divisions of a zygote are known as **cleavage**. In general, the number of cells in the embryo doubles with each cleavage. As described in Section 44.1, several different **cleavage patterns** exist among animals.

Cleavage patterns are influenced by the configuration of the *yolk*, the nutritive material that nourishes the growing embryo. In reptiles, for example, the presence of a large body of acellular yolk within the fertilized egg creates an *incomplete* cleavage pattern in which the dividing cells form an embryo on top of the yolk mass (see Figure 44.3B). In *echinoderms* such as sea urchins, limited yolk is evenly distributed throughout the egg cytoplasm, so cleavage is *complete*, with the fertilized egg dividing in an even pattern known as **radial cleavage**. Radial cleavage is thought to be the ancestral condition for the animals other than sponges, as it is widely distributed in the other major lineages. **Spiral cleavage**—a complicated permutation of radial cleavage—is found among many *lophotrochozoans*, including earthworms and clams. Lophotrochozoans with spiral cleavage are thus sometimes known as *spiralians.* The early branches of the *ecdysozoans* (molting animals, such as insects and nematodes) have radial cleavage, although most ecdysozoans have an idiosyncratic cleavage pattern that is neither radial nor spiral in organization (see Figure 44.3C).

Distinct layers of cells form during the early development of most animals. These cell layers differentiate into specific organs and organ systems as development continues. The embryos of **diploblastic** animals have two cell layers: an outer *ectoderm* and an inner *endoderm*. Embryos of **triploblastic** animals have, in addition to ectoderm and endoderm, a third distinct cell layer, *mesoderm*, between the ectoderm and the endoderm. The existence of three cell layers in embryos is a synapomorphy of triploblastic animals, whereas the diploblastic animals (placozoans, ctenophores, and cnidarians) exhibit the ancestral condition. Some biologists consider sponges to be diploblastic, but since they do not have clearly differentiated tissue types or embryonic cell layers, the term is not usually applied to them.

During early development in many animals, in a process known as *gastrulation*, a hollow ball one cell thick indents to form a cup-shaped structure. The opening of the cavity formed by this indentation is called the *blastopore* (**Figure 31.2**). The process of gastrulation is covered in detail in Section 44.2; the

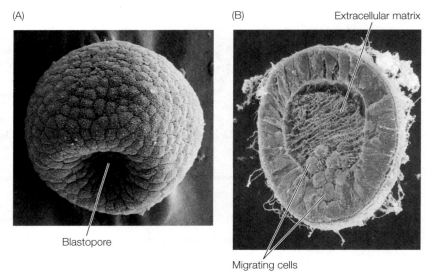

(A)

(B)

Extracellular matrix

Blastopore

Migrating cells

31.2 Gastrulation Illuminates Evolutionary Relationships (A) The blastopore is clear in this scanning electron micrograph of a sea urchin gastrula. Because sea urchins (echinoderms) are deuterostomes, this blastopore will eventually become the anal end of the animal's gut. (B) In this cross section through a later-stage sea urchin gastrula, the cells are beginning to look different from one another. The molecules of the extracellular matrix guide cell movement.

point to remember here is that the *overall pattern* of gastrulation immediately after formation of the blastopore divides the triploblastic animals into two major groups:

- In the **protostomes** (Greek, "mouth first"), the mouth arises from the blastopore, and the anus forms later.
- In the **deuterostomes** ("mouth second"), the blastopore becomes the anus, and the mouth forms later.

Although the developmental patterns of animals are more varied than suggested by this simple dichotomy, sequencing data indicate that the protostomes and deuterostomes represent distinct animal clades. Together, these two groups are known as the **bilaterians** (named for their usual bilateral symmetry), and they account for the vast majority of animal species.

31.1 RECAP

The animals are thought to be monophyletic because they share several derived traits, especially among their gene sequences. Major developmental differences also provide evidence of evolutionary relationships, although phylogenetic analyses of gene sequences have shown that the evolutionary history of these features is more complex than was once thought.

- What general features of animals distinguish this group from other living organisms? See p. 646
- Describe the difference between diploblastic and triploblastic embryos, and between protostomes and deuterostomes. See p. 648

We devote Chapter 32 to the protostomes and Chapter 33 to the deuterostomes. Later in this chapter, we describe several groups of animals that diverged before the origin of the bilaterians. We begin our exploration of animal diversity by discussing general features of animal body plans.

31.2 What Are the Features of Animal Body Plans?

The general structure of an animal, the arrangement of its organ systems, and the integrated functioning of its parts are referred to as its **body plan**. As Chapter 20 describes, the regulatory and signaling genes that govern the development of body symmetry, body cavities, segmentation, and appendages are widely shared among the different animal groups. Thus we might expect animals to share body plans. Although animal body plans vary tremendously, they can be seen as variations on four key features:

- The *symmetry* of the body
- The structure of the *body cavity*
- The *segmentation* of the body
- *External appendages* that are used for sensing, chewing, locomotion, mating, and other functions

Each of these features affects how an animal moves and interacts with its environment.

Most animals are symmetrical

The overall shape of an animal can be described by its **symmetry**. An animal is said to be *symmetrical* if it can be divided along at least one plane into similar halves. Animals that have no plane of symmetry are said to be *asymmetrical*. Placozoans and many sponges are asymmetrical, but most other animals have some kind of symmetry, which is governed by the expression of regulatory genes during development.

The simplest form of symmetry is **spherical symmetry**, in which body parts radiate out from a central point. An infinite number of planes passing through the central point can divide a spherically symmetrical organism into similar halves. Spherical symmetry is widespread among unicellular protists, but most animals possess other forms of symmetry.

In organisms with **radial symmetry**, body parts are arranged around one main axis at the body's center (**Figure 31.3A**). *Ctenophores* are radially symmetrical, as are many *cnidarians* and *echinoderms*. A perfectly radially symmetrical animal can be divided into similar halves by any plane that contains the main axis. However, most radially symmetrical animals—including the adults of echinoderms such as sea stars and sand dollars—are slightly modified so that fewer planes can divide them into identical halves. Some radially symmetrical animals are sessile (sedentary) or drift with water currents. Others move slowly but can move equally well in any direction.

Bilateral symmetry is characteristic of animals that have a distinct front end, which typically precedes the rest of the body as the animal moves. A bilaterally symmetrical animal can be

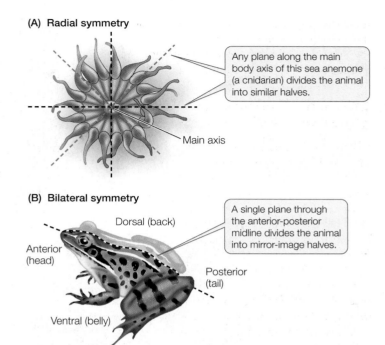

(A) Radial symmetry

Any plane along the main body axis of this sea anemone (a cnidarian) divides the animal into similar halves.

Main axis

(B) Bilateral symmetry

Dorsal (back)

A single plane through the anterior-posterior midline divides the animal into mirror-image halves.

Anterior (head)

Posterior (tail)

Ventral (belly)

31.3 Body Symmetry Most animals are either radially or bilaterally symmetrical.

divided into mirror-image (left and right) halves by a single plane that passes through the midline of its body (**Figure 31.3B**). This plane runs from the front, or **anterior**, end of the body, to its rear, or **posterior**, end.

A plane at right angles to the midline divides the body into two dissimilar sides. The back of a bilaterally symmetrical animal is its **dorsal** surface; the underside is its **ventral** surface.

Bilateral symmetry is strongly correlated with **cephalization**, which is the concentration of sensory organs and nervous tissues in a head at the anterior end of the animal. Cephalization has been evolutionarily favored because the anterior end of a bilaterally symmetrical animal typically encounters new environments first.

The structure of the body cavity influences movement

Animals can be divided into three types—*acoelomate, pseudocoelomate,* and *coelomate*—based on the presence and structure of an internal, fluid-filled **body cavity**. The structure of an animal's body cavity strongly influences the ways in which it can move.

Acoelomate animals such as flatworms lack an enclosed, fluid-filled body cavity. Instead, the space between the gut (derived from endoderm) and the muscular body wall (derived from mesoderm) is filled with masses of cells called *mesenchyme* (**Figure 31.4A**). These animals typically move by beating cilia.

Body cavities come in two types. Some lie between mesoderm and endoderm, and others are enclosed completely within mesoderm.

- **Pseudocoelomate** animals have a body cavity called a *pseudocoel*, a fluid-filled space in which many of the internal

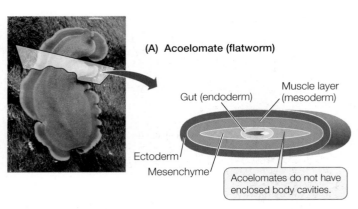

(A) Acoelomate (flatworm)

Gut (endoderm)

Muscle layer (mesoderm)

Ectoderm

Mesenchyme

Acoelomates do not have enclosed body cavities.

(B) Pseudocoelomate (roundworm)

Gut (endoderm)

Pseudocoel (cavity)

Muscle (mesoderm)

Internal organs

Ectoderm

The pseudocoel is lined with mesoderm, but no mesoderm surrounds the internal organs.

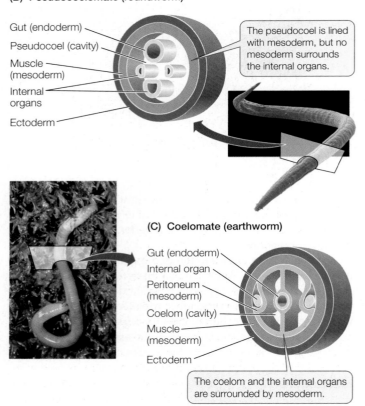

(C) Coelomate (earthworm)

Gut (endoderm)

Internal organ

Peritoneum (mesoderm)

Coelom (cavity)

Muscle (mesoderm)

Ectoderm

The coelom and the internal organs are surrounded by mesoderm.

31.4 Animal Body Cavities (A) Acoelomates do not have enclosed body cavities. (B) Pseudocoelomates have a body cavity bounded by endoderm and mesoderm. (C) Coelomates have a peritoneum surrounding the internal organs in a region bounded by mesoderm.

yourBioPortal.com
GO TO **Web Activity 31.2 • Animal Body Cavities**

organs are suspended. A pseudocoel is enclosed by muscles (mesoderm) only on its outside; there is no inner layer of mesoderm surrounding the internal organs (**Figure 31.4B**).

● **Coelomate** animals have a *coelom*, a body cavity that develops within the mesoderm. It is lined with a layer of muscular tissue called the *peritoneum*, which also surrounds the internal organs. The coelom is thus enclosed on both the inside and the outside by mesoderm (**Figure 31.4C**). A coelomate

animal has better control over the movement of the fluids in its body cavity than a pseudocoelomate animal does.

The body cavities of many animals function as **hydrostatic skeletons**. Fluids are relatively incompressible, so when the muscles surrounding them contract, fluids shift to another part of the cavity. If the body tissues around the cavity are flexible, fluids squeezed out of one region can cause some other region to expand. The moving fluids can thus move specific body parts. (You can see how a hydrostatic skeleton works by watching a snail emerge from its shell.) An animal with both *circular muscles* (encircling the body cavity) and *longitudinal muscles* (running along the length of the body) has even greater control over its movement.

In terrestrial environments, the hydrostatic function of fluid-filled body cavities applies mostly to relatively small, soft-bodied organisms. Most larger animals (as well as many smaller ones) have hard skeletons that provide protection and facilitate movement. Muscles are attached to those firm structures, which may be inside the animal or on its outer surface (in the form of a shell or cuticle).

Segmentation improves control of movement

Many animal bodies are divided into segments. **Segmentation** facilitates specialization of different body regions. It also allows an animal to alter the shape of its body in complex ways and to control its movements precisely. If an animal's body is segmented, muscles in each individual segment can change the shape of that segment independently of the others. In only a few segmented animals is the body cavity separated into discrete compartments, but even partly separated compartments allow better control of movement. As we see in Chapters 32 and 33, segmentation occurs in several groups of protostomes and deuterostomes.

In some animals, segments are not apparent externally (as with the segmented vertebrae of vertebrates). In other animals, such as annelids, similar body segments are repeated many times (**Figure 31.5A**). And in yet other animals, including most arthropods, segments are visible but differ strikingly (**Figure 31.5B**). As described in Chapter 32, the dramatic evolutionary radiation of the arthropods (including the insects, spiders, centipedes, and crustaceans) was based on changes in a segmented body plan that features muscles attached to the inner surface of an external skeleton, including a variety of external appendages that move these animals.

Appendages have many uses

Getting around under their own power is important to many animals. It allows them to obtain food, to avoid predators, and to find mates. Even some sedentary species, such as sea anemones, have larval stages that use cilia to swim, thus increasing the animal's chances of finding a suitable habitat.

Appendages that project externally from the body greatly enhance an animal's ability to move around. Many echinoderms, including sea urchins and sea stars, have myriad *tube feet* that allow them to move slowly across the substratum. Highly con-

31.5 Segmentation The body cavities of many animals are segmented. (A) All of the segments of this marine fireworm, an annelid, are similar. Its appendages are tipped with bristles (setae) that are used for locomotion and (in this species) for protection—the setae contain a noxious toxin. (B) Segmentation allows the evolution of differentiation among the segments. The segments of this scorpion, an arthropod, differ in their form, function, and the appendages they bear.

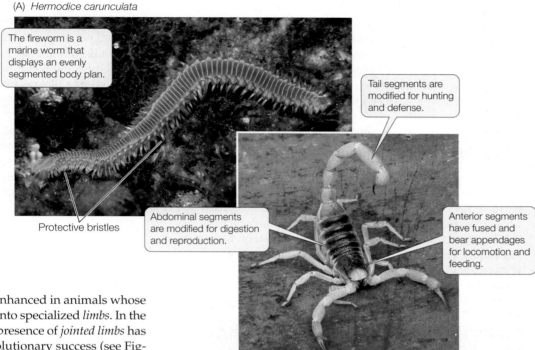

(A) *Hermodice carunculata*

The fireworm is a marine worm that displays an evenly segmented body plan.

Protective bristles

Tail segments are modified for hunting and defense.

Abdominal segments are modified for digestion and reproduction.

Anterior segments have fused and bear appendages for locomotion and feeding.

(B) *Hadrurus arizonensis*

trolled, rapid movement is greatly enhanced in animals whose appendages have become modified into specialized *limbs*. In the arthropods and the vertebrates, the presence of *jointed limbs* has been a prominent factor in their evolutionary success (see Figure 31.5B). In four independent instances—among the arthropod insects and among the vertebrate pterosaurs, birds, and bats—body plans emerged in which limbs were modified into wings, allowing these animals to take to the air.

Appendages also include many structures that are not used for locomotion. Many animals have antennae, which are specialized appendages used for sensing the environment. Other appendages (such as claws and mouth parts of many arthropods) are adaptations for capturing prey or chewing food. In some species, appendages are used for reproductive purposes, such as sperm transfer or egg incubation.

31.2 RECAP

The body plans of animals are variations on patterns of symmetry, body cavities, segmentation, and appendages. All four can affect movement and locomotion, which are important aspects of the animal way of life.

- Describe the main types of symmetry found in animals. How can an animal's symmetry influence the way it moves? See p. 649 and Figure 31.3

- Explain several ways in which body cavities and segmentation improve control over movement. See pp. 649–650

Many of the modifications to animal body plans affect ways of finding, capturing, and processing food. Evolutionary changes in symmetry, body cavities, appendages, and segmentation have played key roles in enabling animals to obtain food from their environment, as well as helping them avoid becoming food for other animals.

31.3 How Do Animals Get Their Food?

As noted in Section 31.1, animals are heterotrophs, or "other feeders." Although many animals rely on photosynthetic endosymbionts for nutrition (see Figures 5.14C and 27.8), most animals must actively obtain an outside source of nutrition, otherwise known as food. The need to locate food has favored the evolution of sensory structures that can provide animals with detailed information about their environment, as well as nervous systems that can receive, process, and coordinate that information.

To acquire food, most animals must expend energy, either to move through the environment to where food is located or to move the environment and the food it contains to them. Animals that can move from one place to another are **motile**; animals that stay in one place are **sessile**.

The principal feeding strategies that animals use fall into five broad categories:

- *Filter feeders* capture small organisms delivered to them by their environment.
- *Herbivores* eat plants or parts of plants.
- *Predators* capture and eat other animals that typically are relatively large.
- *Parasites* live in or on other, generally much larger, organisms from which they obtain energy and nutrients.
- *Detritivores* actively feed on dead organic material.

Each of these strategies can be found in many different animal groups, and none of them is limited to a single group. In addition, individuals of some species may employ more than one feeding strategy, and some animals employ different feeding strategies at different points in their life cycle. The constant and ongoing need to obtain food, the variety of nutrient sources available in any given environment, and the necessity of competing with other animals to obtain food means that a variety of feeding strategies can be found among all the major animal groups.

Filter feeders capture small prey

Air and water often contain small organisms and organic molecules that are potential food for animals. Moving air and water may carry those items to an animal that positions itself in a good location. These **filter feeders** then use some kind of straining device to filter the food from the environment. Many sessile aquatic animals rely on water currents to bring prey to them (**Figure 31.6A**).

(A) *Spirobranchus* sp.

(B) *Phoenicopterus ruber*

Motile filter feeders bring the nutrient-containing medium to them. Flamingos, for example, uses their serrated beak to filter small organisms out of the muddy mixture they pick up as they wade through shallow water (**Figure 31.6B**). Blue whales—the largest animals that have ever lived—are filter feeders that strain tiny crustaceans from the water column as they swim.

Some sessile filter feeders expend energy to move water past their food-capturing devices. Sponges, for example, bring water into their body by beating the flagella of their specialized feeding cells, called **choanocytes** (**Figure 31.7**). These flagellated feeding cells of sponges are similar in structure to protists known as choanoflagellates, which provides evidence for the close relationship of choanoflagellates to animals (see Section 27.5).

Herbivores eat plants

Animals that eat plants are called **herbivores**. An individual plant has many different structures—leaves, wood, sap, flowers, fruits, nectar, and seeds—that animals can consume. Not surprisingly, then, many different kinds of herbivores may feed on a single kind of plant, consuming different parts of the plant or eating the same part in different ways. An individual animal that is captured by a predator is likely to die, but herbivores often feed on plants without killing them.

Animals do not need to expend energy subduing and killing plants. However, they do need to digest them, and animals must expend energy to detoxify plants' defensive chemicals. Digestion can pose challenges to terrestrial herbivores because the dominant land plants tend to have several different kinds of tissues, many of which are tough or fibrous. Herbivorous animals typically have long, complex guts to accomplish the tasks involved in digesting plants (see Section 51.2).

Predators capture and subdue large prey

Predators possess features that enable them to capture and subdue relatively large animals (referred to as their **prey**). Many vertebrate predators have sensitive sensory organs that enable them to locate prey, as well as sharp teeth or claws that allow them to capture and subdue large prey (**Figure 31.8**). Predators may stalk and pursue their prey, or wait (often camouflaged) for their prey to come to them.

Another weapon of predators (as well as of prey) is toxins. We are all aware of the dangers of encountering the toxins of a venomous snake. Toxins often have both a defensive role as well as a role in the capture and sometimes the digestion of prey. Cnidarians (jellyfishes and their relatives) are one of many animal groups that use toxins to capture and subdue prey. The cnidarians' tentacles are covered with specialized cells that con-

31.6 Filter-Feeding Strategies (A) Sessile marine filter feeders such as this "Christmas tree worm," a polychaete, allow the ocean currents to bring their food—plankton—to them. (B) The greater flamingo of South America is a motile filter feeder, using its appendages (legs) to stir up mud as it wades through ocean lagoons and salty lakes. The bird then uses its beak (close-up) to strain small organisms out of the muddy mixture.

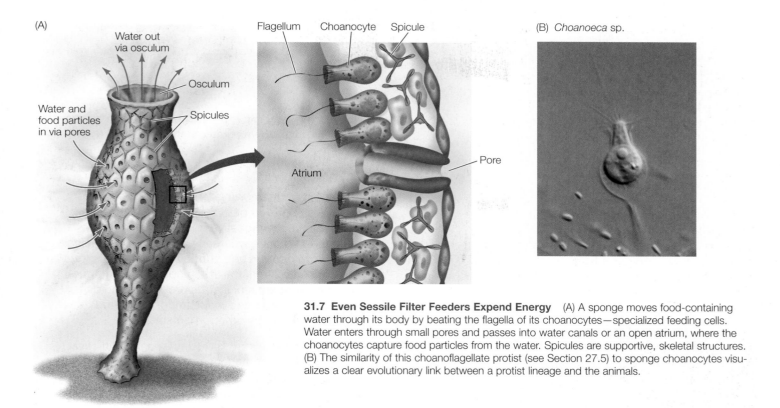

(A)

Water out via osculum

Osculum

Water and food particles in via pores

Spicules

Flagellum   Choanocyte   Spicule

Atrium

Pore

(B) *Choanoeca* sp.

31.7 Even Sessile Filter Feeders Expend Energy   (A) A sponge moves food-containing water through its body by beating the flagella of its choanocytes—specialized feeding cells. Water enters through small pores and passes into water canals or an open atrium, where the choanocytes capture food particles from the water. Spicules are supportive, skeletal structures. (B) The similarity of this choanoflagellate protist (see Section 27.5) to sponge choanocytes visualizes a clear evolutionary link between a protist lineage and the animals.

tain stinging organelles called **nematocysts**, which inject toxins into the prey (**Figure 31.9**).

Omnivores are animals, such as raccoons and humans, that eat both plants and other animals. The diet of some omnivores differs at different life stages; many songbirds, for example, eat fruit or seeds as adults but feed insects to their young.

31.8 Tooth and Claw   (A) The teeth of the predatory gray wolf are adapted for killing prey and shearing meat. (B) The appendages (legs and wings) of the bald eagle, along with its strong beak, are adaptations to the life of a predatory hunter.

Parasites live in or on other organisms

Parasites are animals that live in or on another organism—called a *host*—and obtain their nutrients from that host. Some parasites consume parts of the host itself (such as ticks that suck body fluids); others highjack nutrients the host would otherwise consume (such as tapeworms that may live in our intestines). Most animal parasites are much smaller than their hosts, and many parasites can consume parts of their host without killing it. To reside within a host, a parasite must first overcome the host's defenses. Parasites often have complex life cycles that rely on multiple hosts, as we detail in Section 31.4.

(A) *Canis lupus*

(B) *Haliaeetus leucocephalus*

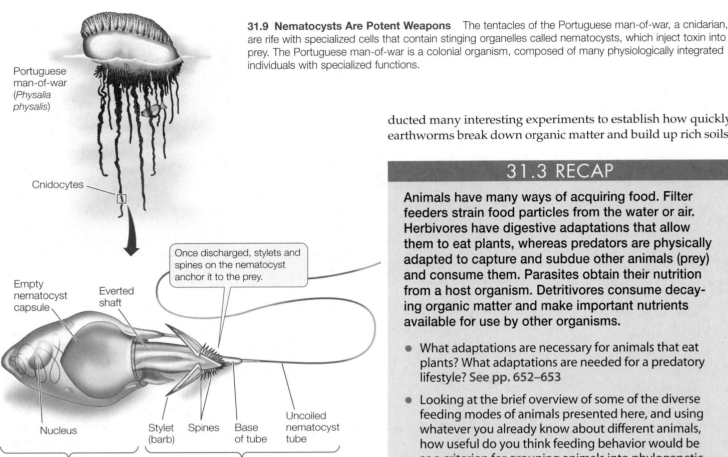

31.9 Nematocysts Are Potent Weapons The tentacles of the Portuguese man-of-war, a cnidarian, are rife with specialized cells that contain stinging organelles called nematocysts, which inject toxin into prey. The Portuguese man-of-war is a colonial organism, composed of many physiologically integrated individuals with specialized functions.

Portuguese man-of-war (*Physalia physalis*)

Cnidocytes

Once discharged, stylets and spines on the nematocyst anchor it to the prey.

Empty nematocyst capsule

Everted shaft

Nucleus

Stylet (barb)   Spines   Base of tube   Uncoiled nematocyst tube

Cnidocyte

Nematocyst

ducted many interesting experiments to establish how quickly earthworms break down organic matter and build up rich soils.

31.3 RECAP

Animals have many ways of acquiring food. Filter feeders strain food particles from the water or air. Herbivores have digestive adaptations that allow them to eat plants, whereas predators are physically adapted to capture and subdue other animals (prey) and consume them. Parasites obtain their nutrition from a host organism. Detritivores consume decaying organic matter and make important nutrients available for use by other organisms.

- What adaptations are necessary for animals that eat plants? What adaptations are needed for a predatory lifestyle? See pp. 652–653

- Looking at the brief overview of some of the diverse feeding modes of animals presented here, and using whatever you already know about different animals, how useful do you think feeding behavior would be as a criterion for grouping animals into phylogenetic categories?

Parasites that live inside their hosts are called *endoparasites*, and they are often morphologically very simple. They can often function without a digestive system because they absorb food directly from the host's gut or bodily tissues. Many flatworms are endoparasites of humans and other mammals, as described in Chapter 32.

Parasites that live outside their hosts are called *ectoparasites*, and they are generally more complex morphologically than endoparasites. Ectoparasites have digestive tracts and mouthparts that enable them to pierce the host's tissues or suck on their body fluids. Fleas and ticks are widely known ectoparasitic arthropods that many humans have unfortunately experienced.

Detritivores live off the remains of other organisms

Detritivores feed on decomposing organic matter, or *detritus*. In so doing, they perform an important ecosystem function by returning nutrients to the environment in a state that can be used by other organisms. Detrivores are common in any soil with high organic content, as well as on the ocean floors. Well-known detrivores include earthworms and other annelids, millipedes, and many insects and crustaceans.

Charles Darwin became fascinated with the action of earthworms, and wrote a book called *The Formation of Vegetable Mould Through the Action of Worms*. He was particularly impressed by the importance of earthworms in soil formation. Darwin con-

As an animal grows from a single cell into a larger, more complex adult, its body structure, its diet, and the environment in which it lives may all change. In the next section we describe some animal life cycles and discuss why they are so varied.

31.4 How Do Life Cycles Differ among Animals?

The **life cycle** of an animal encompasses its embryonic development, birth, growth to maturity, reproduction, and death. During its life an individual animal ingests food, grows, interacts with other individuals of the same and other species, and reproduces.

In some groups of animals, newborns bear many similarities to adults (a pattern called *direct development*). Newborns of most species, however, differ dramatically from adults. Consider, for example a **larva** (plural *larvae*), the immature life stage that some organisms take early in their life cycle before assuming an adult form. Some of the most striking life cycle changes are found among insects such as beetles, flies, moths, butterflies, and bees, which undergo radical changes (called *metamorphosis*) between their larval and adult stages (**Figure 31.10**). In these animals, one stage may be specialized for feeding and the other for reproduction. Adults of most moth species, for example, do not eat. In some animals species, individuals eat during all life cycle stages, but what they eat changes with the stage. For exam-

(A)

(B)

(C)

31.10 A Life Cycle with Metamorphosis (A) The larval stage (caterpillar) of the monarch butterfly (*Danaus plexippus*) is specialized for feeding. (B) The pupa is the stage during which the transformation to the adult form occurs. (C) The adult butterfly is specialized for dispersal and reproduction.

soon hatches and floats freely in the plankton, where it filters small prey from the water.

Many animals that live on the seafloor, including polychaete worms and mollusks, have a radially symmetrical larval form known as a **trochophore (Figure 31.11A)**; others, such as crustaceans, have a bilaterally symmetrical larval form called a **nauplius (Figure 31.11B)**. Both types of larvae feed for some time in the plankton before settling on a substratum and transforming into adults.

Although the main dispersal phase of many animals occurs early in the life cycle, some species that are motile as adults disperse when they are mature. A caterpillar, for example, may spend its entire larval stage feeding on a single plant, but after it metamorphoses into a flying adult—a butterfly—it may fly to and lay eggs on other plants located far from the one where it spent its caterpillar days. In some species, individuals disperse during several different life cycle stages.

No life cycle can maximize all benefits

The common saying "a jack-of-all-trades is master of none" suggests why there are constraints on the evolution of life cycles. The characteristics an animal has in any one life cycle stage may improve its performance in one activity but reduce its performance in another—a situation known as a *trade-off*. An animal that is good at filtering small food particles from the water, for example, probably cannot capture large prey. Similarly, energy devoted to building protective structures such as shells cannot be used for growth.

Some major trade-offs can be seen in animal reproduction. Some animals produce large numbers of small eggs, each with

ple, butterfly larvae, known as *caterpillars*, eat leaves and flowers, whereas most adult butterflies eat only nectar. Having different life cycle stages that are specialized for different activities may increase the efficiency with which an animal performs particular tasks.

Most animal life cycles have at least one dispersal stage

At some time during its life, an animal moves, or is moved, so few animals die exactly where they were born. Movement of organisms from a parent organism or from an existing population is called **dispersal**.

Animals that are sessile as adults typically disperse as eggs or larvae. Most sessile marine animals discharge their small eggs and sperm into the water, where fertilization takes place. A larva

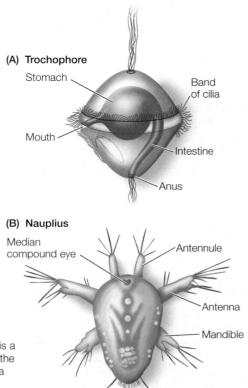

(A) Trochophore

Stomach

Band of cilia

Mouth

Intestine

Anus

(B) Nauplius

Median compound eye

Antennule

Antenna

Mandible

31.11 Planktonic Larval Forms of Marine Animals (A) The trochophore ("wheel-bearer") is a distinctive larval form found in several marine animal clades with spiral cleavage, most notably the polychaete worms and the mollusks. (B) This nauplius larva will mature into a crustacean with a segmented body and jointed appendages.

(A) *Rana sylvatica*

(B) *Pygoscelis papua*

31.12 Many Small or Few Large Allocation of energy to eggs requires trade-offs. (A) This wood frog has divided her reproductive energy among a large number of small eggs. (B) This gentoo penguin invested all of her reproductive energy in one large egg.

a small energy store (**Figure 31.12A**). Other animals produce a small number of large eggs, each with a large energy store (**Figure 31.12B**). With a fixed amount of available energy, a female animal can produce many small eggs or a few large eggs, but she cannot produce many large eggs. Thus there is a trade-off between the number of offspring produced and the energy resources each offspring receives from its mother.

The larger the energy store in an egg, the longer an offspring can develop before it must either find its own food or be fed by its parents. Birds of all species lay relatively small numbers of relatively large eggs, but incubation periods vary. In some

species, eggs hatch when the young are still helpless (**Figure 31.13A**). Such *altricial* young must be fed and cared for until they can feed themselves; parents can provide for only a small number of altricial offspring. In contrast, some bird species incubate their eggs longer, and the hatchlings are developed to a point that they are able to forage for themselves almost immediately (**Figure 31.13B**). The young of such species are called *precocial*.

Parasite life cycles evolve to facilitate dispersal and overcome host defenses

Animals that live as internal parasites are bathed in the nutritious tissues of their host or in the digested food that fills their host's digestive tract. Thus they may not need to exert much energy to obtain food, but to survive they must overcome the host's defenses. Furthermore, either they or their offspring must disperse to new hosts while their host is still living, because they die when their host dies.

The fertilized eggs of some parasites are voided with the host's feces and later ingested directly by other host individuals. Most parasite species, however, have complex life cycles involving one or more intermediate hosts and several larval stages (**Figure 31.14**). Some intermediate hosts transport individual parasites directly between other hosts. Others house and support the parasite until another host ingests it. Complex life cycles may thus facilitate the transfer of individual parasites among hosts.

Colonial organisms are composed of genetically identical, physiologically integrated individuals

Most people tend to view the distinction between individuals and populations as clear-cut. However, in several groups of animals, asexual reproduction without fission can lead to colonies

(A) *Parus caeruleus*

(B) *Branta canadensis*

31.13 Helpless or Independent (A) The altricial young of the blue tit are essentially helpless when they hatch. Their parents feed and care for them for several weeks. (B) Canada goose hatchlings are precocial, ready to swim and feed independently almost immediately after hatching.

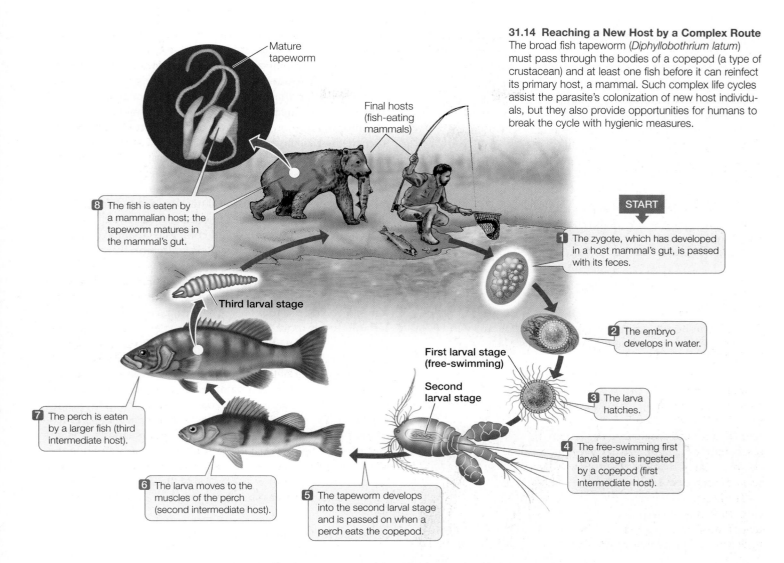

31.14 Reaching a New Host by a Complex Route
The broad fish tapeworm (*Diphyllobothrium latum*) must pass through the bodies of a copepod (a type of crustacean) and at least one fish before it can reinfect its primary host, a mammal. Such complex life cycles assist the parasite's colonization of new host individuals, but they also provide opportunities for humans to break the cycle with hygienic measures.

Mature tapeworm

Final hosts (fish-eating mammals)

START

8 The fish is eaten by a mammalian host; the tapeworm matures in the mammal's gut.

1 The zygote, which has developed in a host mammal's gut, is passed with its feces.

2 The embryo develops in water.

Third larval stage

First larval stage (free-swimming)

Second larval stage

3 The larva hatches.

7 The perch is eaten by a larger fish (third intermediate host).

4 The free-swimming first larval stage is ingested by a copepod (first intermediate host).

6 The larva moves to the muscles of the perch (second intermediate host).

5 The tapeworm develops into the second larval stage and is passed on when a perch eats the copepod.

of organisms composed of many physiologically integrated individuals, which at first appearance may look much like a single integrated organism. The individuals in a colony are clonal copies of one another, so they are genetically homogeneous. Coloniality has arisen several times among animal groups, with widely varying levels of integration and specialization among the individuals. In some species, colonies are composed of loosely connected but integrated individuals that all function alike (**Figure 31.15**). In other colonial species, the individuals may become specialized for different functions, just as different cell types in multicellular organisms have different functions. The Portuguese man-of-war (see Figure 31.9) is an example of such a colonial animal, as it is composed of many individuals of four specialized body forms, all integrated and functioning together. The individuals in the colony are themselves multicellular, however, unlike the cells of a single multicellular organism.

Diaperoecia californica

The individual animals...

...secrete a gelatinous matrix that brings the colony together.

31.15 Colonial Animals This colonial bryozoan consists of many asexually reproducing, genetically homogeneous, physiologically interacting individuals. The colony looks much like a single individual with many parts, but in fact it is many individuals acting together.

31.4 RECAP

Many animals have a larval stage that clearly differs from the adult in morphology. In some animals, the larval form is a dispersal stage; in other species, the adults are more likely to disperse than are larvae. In several groups of organisms, asexual reproduction without fission leads to coloniality.

- How do trade-offs constrain the evolution of life cycles? See pp. 655–656

- Do you understand the differences between a single multicellular organism and coloniality? See p. 657 and Figure 31.15

31.5 What Are the Major Groups of Animals?

Variations in body symmetry, body cavity structure, life cycles, patterns of development, and survival strategies differentiate millions of animal species. In the remainder of this chapter and in Chapters 32 and 33, we become acquainted with the major animal groups and learn how the general characteristics described in this chapter apply to each of them.

Table 31.1 summarizes the living members of the major animal groups. **Bilateria** is a large monophyletic group embracing all animals other than sponges, placozoans, ctenophores, and cnidarians. Some major traits that support the monophyly of bilaterians (in addition to genomic analyses) are the presence of three distinct cell layers in embryos (triploblasty) and the presence of at least seven Hox genes (see Chapters 19 and 20). Although bilateral symmetry is often viewed as a synapomorphy of bilaterians (and the trait gives the group its name), some groups of cnidarians are also bilaterally symmetrical. Recent studies have shown that the genetic basis of bilateral symmetry is the same in bilaterians and cnidarians that have bilateral symmetry, so this feature was likely present in the ancestor of these two groups.

Bilaterian animals comprise the two major categories mentioned earlier in this chapter, the protostomes and the deuterostomes (see Figure 31.1). These two groups have been evolving separately for more than 500 million years—since the early Cambrian or the late Precambrian. We describe the protostomes in Chapter 32 and the deuterostomes in Chapter 33.

The remainder of this chapter describes those animal groups that are not bilaterians. The simplest animals, the sponges, have no distinct tissue types. Placozoans have four cell types and weakly differentiated tissue layers. All other animals groups, including the bilaterians, are known as **eumetazoans**. The eumetazoans have obvious body symmetry, a gut, a nervous system, and tissues organized into distinct organs (although there have been secondary losses of some of these structures in some eumetazoans). Sponges and placozoans lack all of these features.

Sponges are loosely organized animals

Sponges are the simplest animals. Although they have some specialized cells, they have no distinct embryonic cell layers and no true organs. Early naturalists thought sponges were plants because they were sessile and lacked body symmetry.

Sponges have hard skeletal elements called **spicules**, which may be small and simple or large and complex. There are three major groups of sponges, which separated soon after the split between sponges and the rest of the animals. Members of two groups (*glass sponges* and *demosponges*) have skeletons composed of silicaceous spicules made of hydrated silicon dioxide (**Figure 31.16A,B**). These spicules are remarkable in having greater flexibility and toughness than synthetic glass rods of similar length. Members of the third group, the *calcareous sponges*, take their name

(A) *Xestospongia testudinaria*

(B) *Euplectella aspergillum*

(C) *Leucilla nuttingi*

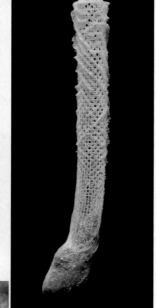

31.16 Sponge Diversity (A) The great majority of sponge species are demosponges, such as these Pacific barrel sponges. The system of pores and water canals "typical" of the sponge body plan is apparent. (B) The supporting structures of both demosponges and glass sponges are silicaceous spicules, seen here in the skeleton of a glass sponge. (C) The skeletons of calcareous sponges are made of calcium carbonate.

TABLE 31.1

Summary of Living Members of the Major Animal Groups

	APPROXIMATE NUMBER OF LIVING SPECIES DESCRIBED	MAJOR GROUPS
Sponges	9,000	Demosponges, glass sponges, calcareous sponges
Placozoans	2	
Ctenophores	150	
Cnidarians	11,000	Anthozoans: Corals, sea anemones
		Hydrozoans: Hydras and hydroids
		Scyphozoans: Jellyfishes
PROTOSTOMES		
Arrow worms	100	
Lophotrochozoans		
Bryozoans	4,500	
Flatworms	25,000	Free-living flatworms; flukes and tapeworms (all parasitic); monogeneans (ectoparasites of fishes)
Rotifers	1,800	
Ribbon worms	1,000	
Phoronids	20	
Brachiopods	335	
Annelids	16,500	Polychaetes (all marine)
		Clitellates: Earthworms, freshwater worms, leeches
Mollusks	100,000	Monoplacophorans
		Chitons
		Bivalves: Clams, oysters, mussels
		Gastropods: Snails, slugs, limpets
		Cephalopods: Squids, octopuses, nautiloids

	APPROXIMATE NUMBER OF LIVING SPECIES DESCRIBED	MAJOR GROUPS
Ecdysozoans		
Kinorhynchs	150	
Loriciferans	100	
Priapulids	16	
Horsehair worms	320	
Nematodes	25,000	
Onychophorans	150	
Tardigrades	800	
Arthropods:		
Crustaceans	52,000	Crabs, shrimps, lobsters, barnacles, copepods
Hexapods	1,000,000	Insects and relatives
Myriapods	14,000	Millipedes, centipedes
Chelicerates	98,000	Horseshoe crabs, arachnids (scorpions, harvestmen, spiders, mites, ticks)
DEUTEROSTOMES		
Echinoderms	7,000	Crinoids (sea lilies and feather stars); brittle stars; sea stars; sea daisies; sea urchins; sea cucumbers
Hemichordates	100	Acorn worms and pterobranchs
Urochordates	3,000	Ascidians (sea squirts)
Cephalochordates	30	Lancelets
Vertebrates	62,000	Hagfish; lampreys
		Cartilaginous fishes
		Ray-finned fishes
		Coelacanths; lungfishes
		Amphibians
		Reptiles (including birds)
		Mammals

from their calcium carbonate skeletons (**Figure 31.16C**). There is some question about the monophyly of sponges. Analyses of some gene sequences suggest that calcareous sponges are actually more closely related to the eumetazoans than to the other groups of sponges. However, genomic analyses that combine information from many genes do support the monophyly of sponges.

The body plan of sponges of all three groups—even large ones, which may reach a meter or more in length—is an aggregation of cells built around a water canal system. Water, along with any food particles it contains, enters the sponge by way of small pores and passes into the water canals or a central atrium, where choanocytes capture food particles (see Figure 31.7).

A skeleton of simple or branching spicules, and often a complex network of elastic fibers, supports the body of most sponges. Sponges also have an extracellular matrix, composed of collagen, adhesive glycoproteins, and other molecules, that holds the cells together. Most species are filter feeders; a few species are carnivores that trap prey on hook-shaped spicules that protrude from the body surface.

Most of the 9,000 species of sponges are marine animals; only about 50 species live in fresh water. Sponges come in a wide variety of sizes and shapes that are adapted to different movement patterns of water. Sponges living in intertidal or shallow subtidal environments with strong wave action are firmly attached to the substratum. Most sponges that live in slowly flowing water are flattened and are oriented at right angles to the direction of current flow. They intercept water and the prey it contains as it flows past them.

Sponges reproduce both sexually and asexually. In most species, a single individual produces both eggs and sperm, but individuals do not self-fertilize. Water currents carry sperm from one individual to another. Asexual reproduction is by budding and fragmentation.

Placozoans are abundant but rarely observed

As discussed in the opening of this chapter, **placozoans** are structurally very simple animals with only a few distinct cell types (see the photograph on p. 645). Individuals in the mature, asymmetrical life stage are usually observed adhering to surfaces (such as the glass of aquariums, where they were first discovered, or to rocks and other hard substrates in nature). Their structural simplicity—they have no mouth, gut, or nervous system—initially led biologists to suspect they might be the sister group of all other animals. Most phylogenetic analyses have not supported this hypothesis, however, and some aspects of the placozoans' structural simplicity may be secondarily derived. They are generally considered to have a diploblastic body plan, with upper and lower epithelial (surface) layers that sandwich a layer of contractile fiber cells.

Recent studies have found that placozoans have a pelagic (open-ocean) stage that is capable of swimming (**Figure 31.17**), but the life history of placozoans is incompletely known. Most studies have focused on the larger adherent stages that are usually found in aquariums, where they appear after being inadvertently collected with other marine organisms. The transparent nature and small size of placozoans make them very difficult to observe in nature. Nonetheless, it is known that placozoans can reproduce both asexually as well as sexually, although the details of their sexual reproduction are mostly unknown. As we

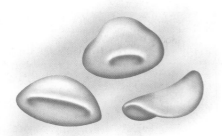

31.17 A Swimming Stage in the Life of a Placozoan Placozoans are tiny and transparent, and thus difficult to observe in nature. Recent studies have found a small, weakly swimming pelagic stage of placozoan to be abundant in many warm tropical and subtropical seas.

noted in the opening story, placozoans have mostly been studied in aquariums, although we now know that pelagic-stage placozoans are abundant in warm seas around the world.

Ctenophores are radially symmetrical and diploblastic

Ctenophores, also known as *comb jellies*, lack most of the Hox genes found in all other eumetazoans. Ctenophores have a radially symmetrical, diploblastic body plan. The two cell layers are separated by an inert, gelatinous extracellular matrix called **mesoglea**. Ctenophores have a *complete gut*: food enters through a mouth, and wastes are eliminated through two anal pores.

Ctenophores move by beating cilia rather than muscular contractions. Most of the 150 known species have eight comblike rows of cilia-bearing plates, called **ctenes** (**Figure 31.18**). The feeding tentacles of ctenophores are covered with cells that discharge adhesive material when they contact prey. After capturing its prey, a ctenophore retracts its tentacles to bring the food to its mouth. In some species, the entire surface of the body is coated with sticky mucus that captures prey. Most ctenophores

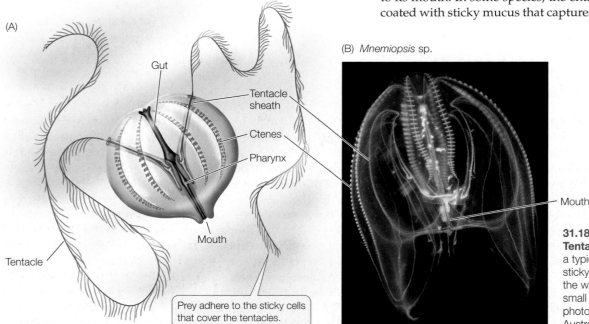

(A)

Gut

Tentacle sheath

Ctenes

Pharynx

Mouth

Tentacle

Prey adhere to the sticky cells that cover the tentacles.

(B) *Mnemiopsis* sp.

Mouth

31.18 Comb Jellies Feed with Tentacles (A) The body plan of a typical ctenophore. The long, sticky tentacles sweep through the water, efficiently harvesting small prey. (B) This comb jelly, photographed in Sydney Harbor, Australia, has short tentacles.

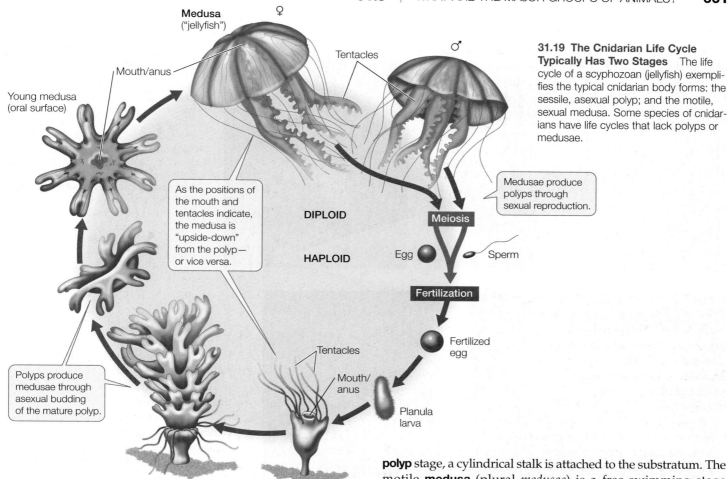

Medusa ("jellyfish") ♀

Mouth/anus

Tentacles

♂

Young medusa (oral surface)

As the positions of the mouth and tentacles indicate, the medusa is "upside-down" from the polyp—or vice versa.

31.19 The Cnidarian Life Cycle Typically Has Two Stages The life cycle of a scyphozoan (jellyfish) exemplifies the typical cnidarian body forms: the sessile, asexual polyp; and the motile, sexual medusa. Some species of cnidarians have life cycles that lack polyps or medusae.

Medusae produce polyps through sexual reproduction.

DIPLOID

Meiosis

HAPLOID

Egg   Sperm

Fertilization

Fertilized egg

Polyps produce medusae through asexual budding of the mature polyp.

Tentacles

Mouth/anus

Planula larva

Mature polyp

Polyp

eat small planktonic organisms, although some eat other ctenophores. They are common in open seas and can become abundant in protected bodies of water, where large populations of ctenophores can damage local ecosystems.

Ctenophore life cycles are uncomplicated. Gametes are released into the body cavity and then discharged through the mouth or the anal pores. Fertilization takes place in open seawater. In nearly all species, the fertilized egg develops directly into a miniature ctenophore that gradually grows into an adult.

Cnidarians are specialized carnivores

The **cnidarians** (jellyfishes, sea anemones, corals, and hydrozoans) may be the sister group of the ctenophores, although some biologists think they are more closely related to the bilaterians. The mouth of a cnidarian is connected to a blind sac called the **gastrovascular cavity** (a cnidarian thus does not have a complete gut). The gastrovascular cavity functions in digestion, circulation, and gas exchange, and it also acts as a hydrostatic skeleton. The single opening serves as both mouth and anus.

The life cycle of many cnidarians has two distinct stages, one sessile and the other motile (**Figure 31.19**), although one or the other of these stages is absent in some groups. In the sessile

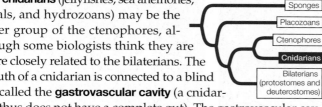

Sponges
Placozoans
Ctenophores
Cnidarians
Bilaterians (protostomes and deuterostomes)

polyp stage, a cylindrical stalk is attached to the substratum. The motile **medusa** (plural *medusae*) is a free-swimming stage shaped like a bell or an umbrella. It typically floats with its mouth and feeding tentacles facing downward. Mature polyps produce medusae by asexual budding. Medusae then reproduce sexually, producing eggs or sperm by meiosis and releasing the gametes into the water. A fertilized egg develops into a free-swimming, ciliated larva called a **planula**, which eventually settles to the bottom and develops into a polyp.

---

yourBioPortal.com

GO TO **Animated Tutorial 31.1 • Life Cycle of a Cnidarian**

Cnidarians have epithelial cells with muscle fibers whose contractions enable the animals to move, as well as simple *nerve nets* that integrate their body activities. They also have specialized structural molecules (collagen, actin, and myosin). They are specialized carnivores, using the toxin in their nematocysts to capture relatively large and complex prey (see Figure 31.9). Some cnidarians, including many corals and anemones, gain additional nutrition from photosynthetic endosymbionts that live in their tissues. Cnidarians, like ctenophores, are largely made up of inert mesoglea. They have low metabolic rates and can survive in environments where they encounter prey only infrequently.

Of the roughly 11,000 living cnidarian species, all but a few live in the oceans (**Figure 31.20**). The smallest cnidarians can hardly be seen without a microscope. The largest known jellyfish is 2.5 meters in diameter, and some colonial siphonophores

(A) *Anthopleura elegantissima*

(B) *Pteroeides* sp.

31.20 Diversity among Cnidarians
(A) The nematocyst-studded tentacles of this sea anemone from British Columbia are poised to capture large prey carried to the animal by water movement. (B) The sea pen is a colonial cnidarian that lives in soft bottom sediments and projects polyps above the substratum. (C) This jellyfish illustrates the complexity of a scyphozoan medusa. (D) The internal structure of the medusa of a North Atlantic colonial hydrozoan is visible here.

(C) *Gonionemus vertens*

(D) *Polyorchis penicillatus*

ondary polyps differentiate into feeding polyps; in some species, other secondary polyps differentiate to circulate water through the colony.

The common names of coral groups—brain corals, staghorn corals, and organ pipe corals, among others—often describe their appearance (**Figure 31.21A**). Corals are sessile and colonial. The polyps of

(which include the Portuguese man-of-war; see Figure 31.9) can reach lengths in excess of 30 meters. Here we describe three clades of cnidarians that have many species: anthozoans, scyphozoans, and hydrozoans.

ANTHOZOANS Members of the **anthozoan** clade include sea anemones, sea pens, and corals. Sea anemones (see Figure 31.20A), all of which are solitary, are widespread in both warm and cold ocean waters. Sea pens (see Figure 31.20B), by contrast, are colonial. Each colony consists of two or more different kinds of polyps. The primary polyp has a lower portion anchored in the bottom sediment and a branched upper portion that projects above the substratum. Along the upper portion, the primary polyp produces smaller secondary polyps by budding. Some of these sec-

(A) *Diploria labyrinthiformis*

31.21 Corals (A) The descriptive common name of this Caribbean coral is "brain coral." (B) Many different coral species form this reef in the Red Sea between Egypt and the Arabian Peninsula.

(B)

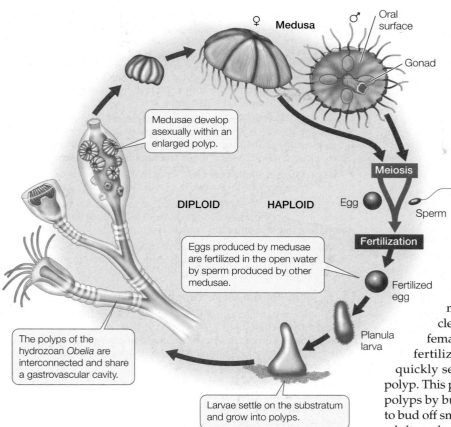

Medusa

♀

♂ Oral surface

Gonad

Medusae develop asexually within an enlarged polyp.

DIPLOID   HAPLOID

Meiosis

Egg

Sperm

Fertilization

Eggs produced by medusae are fertilized in the open water by sperm produced by other medusae.

Fertilized egg

Planula larva

The polyps of the hydrozoan *Obelia* are interconnected and share a gastrovascular cavity.

Larvae settle on the substratum and grow into polyps.

31.22 Hydrozoans Often Have Colonial Polyps
The polyps in a hydrozoan colony may differentiate to perform specialized tasks. In the species whose life cycle is diagrammed here, the medusa is the sexual reproductive stage, producing eggs and sperm in organs called gonads.

most species form a skeleton by secreting a matrix of organic molecules on which they deposit calcium carbonate, which forms the eventual skeleton of the coral colony. As the colony grows, old polyps die but their calcium carbonate skeletons remain. Living corals form a layer on top of a growing bank of skeletal remains, eventually forming chains of islands and reefs (**Figure 31.21B**). The Great Barrier Reef along the northeastern coast of Australia is a system of coral formations more than 2,000 kilometers long—about the distance from New York City to St. Louis. A single coral reef in the Red Sea has been calculated to contain more material than all the buildings in the major cities of North America combined.

Corals flourish in clear, nutrient-poor tropical waters. They grow well in such environments because unicellular photosynthetic dinoflagellates live endosymbiotically within their cells. These dinoflagellates provide the corals with products of photosynthesis; the corals, in turn, provide the dinoflagellates with nutrients and a place to live. This endosymbiotic relationship explains why reef-forming corals are restricted to clear surface waters, where light levels are high enough to support photosynthesis.

Coral reefs throughout the world are threatened both by global warming, which is raising the temperatures of shallow tropical ocean waters, and by polluted runoff from development on adjacent shorelines. Warming can lead to the loss of coral endosymbionts (*coral bleaching*; see Figure 27.8A), and an overabundance of nitrogen in runoff gives an advantage to algae, which overgrow and eventually smother the corals.

SCYPHOZOANS The several hundred species of scyphozoans are all marine. The mesoglea of their medusae is thick and firm, giving rise to their common name—jellyfishes or sea jellies. The medusa rather than the polyp dominates the life cycle of scyphozoans. An individual medusa is male or female, releasing eggs or sperm into the open sea. The fertilized egg develops into a small planula larva that quickly settles on a substratum and develops into a small polyp. This polyp feeds and grows and may produce additional polyps by budding. After a period of growth, the polyp begins to bud off small medusae, which feed, grow, and transform into adult medusae (see Figures 31.19 and 31.20C).

HYDROZOANS Hydrozoans have diverse life cycles. The polyp typically dominates the life cycle, but some species have only medusae and others have only polyps. Most hydrozoans are colonial (see Figure 31.20D). A single larval planula eventually gives rise to a colony of many polyps, all interconnected and sharing a continuous gastrovascular cavity (**Figure 31.22**). Within such a colony, some polyps have tentacles with many nematocysts; they capture prey for the colony. Other individuals lack tentacles and are unable to feed but are specialized for the asexual production of medusae. Still others are fingerlike and defend the colony with their nematocysts.

31.5 RECAP

Bilaterian animals are in one of two major clades, protostomes or deuterostomes. The non-bilaterian animals—the sponges, placozoans, ctenophores, and cnidarians—have diverse life cycles, feeding strategies, and growth forms.

- Why are sponges are considered to be animals, even though they lack the complex body structures found among most other animal groups? See pp. 658–659 and Figure 31.7

- Describe some major features of the following groups: sponges, placozoans, ctenophores, and cnidarians. See pp. 658–663 and Figures 31.16–31.19

CHAPTER SUMMARY

31.1 What Characteristics Distinguish the Animals?

- Animals share a set of derived traits not found in other groups of organisms. These traits include similarities in the sequences of many of their genes, the structure of their cell junctions, and the components of their extracellular matrix.

- Patterns of embryonic development provide clues to the evolutionary relationships among animals. **Diploblastic** animals develop two embryonic cell layers; **triploblastic** animals develop three cell layers.

- Differences in their patterns of early development characterize two major clades of triploblastic animals, the **protostomes** and the **deuterostomes**.

31.2 What Are the Features of Animal Body Plans?

- Animal **body plans** can be described in terms of **symmetry**, **body cavity** structure, **segmentation**, and type of appendages.

- A few animals have no symmetry, but most animals have either **radial symmetry** or **bilateral symmetry**. Review Figure 31.3

- Most animals with radial symmetry move slowly or not at all, whereas most animals with bilateral symmetry are able to move more rapidly. Many bilaterally symmetrical animals exhibit **cephalization**, with sensory and nervous tissues in an anterior head.

- On the basis of their body cavity structure, animals can be described as **acoelomates**, **pseudocoelomates**, or **coelomates**. Review Figure 31.4

- Segmentation, which takes many forms, improves control of movement, especially if the animal also has appendages.

31.3 How Do Animals Get Their Food?

- **Motile** animals can move to find food; **sessile** animals stay in one place and capture food by filter feeding or through interaction with endosymbionts.

- **Filter feeders** strain small organisms and organic molecules from their environment.

- **Predators** have morphological features such as sharp teeth, beaks, and claws that enable them to capture and subdue animal **prey**.

- **Herbivores** consume plants, usually without killing them.

- **Parasites** live in or on other organisms and obtain nutrition from these host individuals.

- **Detritivores** consume decaying organic matter and return nutrients to the ecosystem.

31.4 How Do Life Cycles Differ among Animals?

- The stages of an animal's **life cycle** may be specialized for different activities. An immature stage that is dramatically different from the adult stage is called a **larva**.

- Most animal life cycles have at least one **dispersal** stage, so that the animal does not die in the same place where it was born. Many sessile marine animals can be grouped by the presence of one of two distinct larval dispersal stages: **trochophore** or **nauplius**.

- Parasites have complex life cycles that may involve one or more hosts and several larval stages. Review Figure 31.14

- A characteristic of an animal or a life cycle stage may improve the animal's performance in one activity but reduce its performance in another, a situation known as a trade-off.

- Colonial organisms are composed of groups of genetically homogeneous individuals produced through clonal reproduction without subsequent fission.

31.5 What Are the Major Groups of Animals?

- All animals other than sponges, placozoans, ctenophores, and cnidarians belong to a large monophyletic group called the **Bilateria**. **Eumetazoans**, which have tissues organized into distinct organs, include all animals other than sponges and placozoans.

- **Sponges** are simple animals that lack differentiated cell layers and true organs. They have skeletons made up of silicaceous or calcareous **spicules**. They create water currents and capture food with flagellated feeding cells called choanocytes. Choanocytes are an evolutionary link between the animals and the choanoflagellate protists. Review Figure 31.7

- **Placozoans** have only a few cell types and lack true organs, although their simplicity may be secondarily derived. They are abundant in warm seas of the world, but their transparent form and small size make them difficult to observe.

- **Ctenophores** and many **cnidarians** are radially symmetrical, although some cnidarians are bilaterally symmetrical. The mature adherent stage of placozoans is asymmetrical.

- The two cell layers of ctenophores are separated by an inert extracellular matrix called **mesoglea**.

- Ctenophores move by beating fused plates of cilia called **ctenes**. Review Figure 31.18

- The life cycle of most cnidarians has two distinct stages: a sessile **polyp** stage and a motile **medusa**. A fertilized egg develops into a free-swimming larval **planula**, which settles to the bottom and develops into a polyp. Review Figures 31.19 and 31.22, **ANIMATED TUTORIAL 31.1**

SEE WEB ACTIVITIES 31.1 and 31.2 for a concept review of this chapter.

SELF-QUIZ

1. The body plan of an animal is
   a. its general structure.
   b. the integrated functioning of its parts.
   c. its general structure and the integrated functioning of its parts.
   d. its general structure and its evolutionary history.
   e. the integrated functioning of its parts and its evolutionary history.

2. A bilaterally symmetrical animal can be divided into mirror images by
   a. any plane through the midline of its body.
   b. any plane from its anterior to its posterior end.
   c. any plane from its dorsal to its ventral surface.
   d. any plane through the midline of its body from its anterior to its posterior end.
   e. a single plane through the midline of its body from its dorsal to its ventral surface.

3. Among protostomes, cleavage of the fertilized egg is
   a. delayed while the egg continues to mature.
   b. always radial.
   c. spiral, radial, or idiosyncratic.
   d. triploblastic.
   e. diploblastic.

4. Many parasites evolved complex life cycles because
   a. they are too simple to disperse readily.
   b. they are poor at recognizing new hosts.
   c. they were driven to it by host defenses.
   d. complex life cycles increase the probability of a parasite's transfer to a new host.
   e. their ancestors had complex life cycles and they simply retained them.

5. Bilateral symmetry
   a. is found only among bilaterians.
   b. is characteristic of all sponges.
   c. is characteristic of all ctenophores.
   d. is characteristic of all cnidarians.
   e. none of the above

6. In the common ancestor of protostomes and deuterostomes, the pattern of early cleavage was
   a. spiral.
   b. radial.
   c. biradial.
   d. deterministic.
   e. haphazard.

7. A fluid-filled body cavity can function as a hydrostatic skeleton because
   a. fluids are moderately compressible.
   b. fluids are highly compressible.
   c. fluids are incompressible at physiological pressures.
   d. fluids have the same density as body tissues.
   e. fluids can be moved by ciliary action.

8. Which of the following is *not* a feature that enables some animals to capture large prey?
   a. Sharp teeth
   b. Claws
   c. Toxins
   d. A filtering device
   e. Tentacles with stinging cells

9. The sponge body plan is characterized by
   a. a mouth and digestive cavity but no muscles or nerves.
   b. muscles and nerves but no mouth or digestive cavity.
   c. a mouth, digestive cavity, and spicules.
   d. muscles and spicules but no digestive cavity or nerves.
   e. the lack of a mouth, digestive cavity, muscles, or nerves.

10. The endosymbiotic dinoflagellates present in many corals
    a. provide the corals with the products of photosynthesis.
    b. allow corals to grow rapidly in clear, nutrient-poor tropical waters
    c. can be lost when environmental conditions change.
    d. all of the above
    e. none of the above

FOR DISCUSSION

1. Differentiate among the members of each of the following sets of related terms:
   a. radial symmetry/bilateral symmetry
   b. protostome/deuterostome
   c. diploblastic/triploblastic
   d. coelomate/pseudocoelomate/acoelomate

2. In this chapter we listed some of the traits shared by all animals that convince most biologists that all animals are descendants of a single common ancestral lineage. In your opinion, which of these traits provides the most compelling evidence that animals are monophyletic? If morphological and molecular data do not agree, should one type of evidence be given greater weight? If so, which one?

3. Describe some features that allow animals to capture prey that are larger and more complex than they themselves are.

4. Why is bilateral symmetry strongly associated with cephalization, the concentration of sensory organs in an anterior head?

5. How does a slow metabolic rate enable an animal to live in an unproductive environment?

ADDITIONAL INVESTIGATION

The discoveries that pelagic (swimming in open ocean) stages of placozoans are abundant in warm seas, and that the mature stages settle onto smooth surfaces, suggest how these organisms might be collected and surveyed. What sampling procedures might you use to discover whether placozoans occur at a particular location along a coast?

WORKING WITH DATA (GO TO yourBioPortal.com)

Reconstructing Animal Phylogeny In this exercise, you will analyze a subset of protein sequences that provide evidence about the evolutionary relationships of some major animal groups. Using the parsimony method described in Chapter 22, you will build a phylogenetic tree of some of the major animal lineages depicted in Figure 31.1.

Tiny parasites exert mind control

Most people have never heard of strepsipterans, and even those who *have* heard of them have probably never seen one. They don't know what they are missing! These tiny insects—there are some 600 species of them—parasitize hundreds of other insect species, including bees, wasps, ants, grasshoppers, and cockroaches. Males and females of most strepsipteran species parasitize the same host species, although in one clade males parasitize ants whereas females parasitize grasshoppers.

Strepsipteran males and females are often so different that even determining that they are members of the same species requires DNA analy-

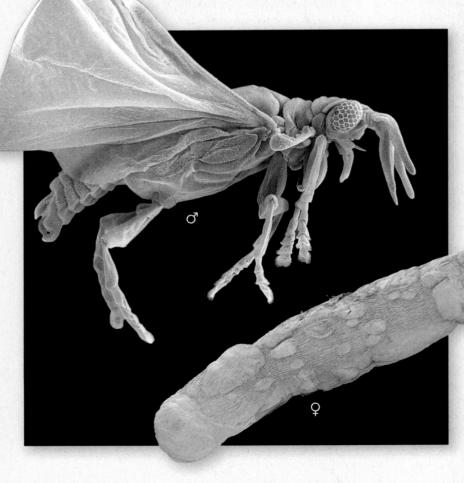

sis. They also have some of the strangest life cycles of any animals. Once grown to maturity within their hosts (whose internal organs they consume), the males of most species emerge looking like a "typical" insect. The females also consume the host from within, but they usually remain inside the host. A mature female extrudes her head and part of her body from the body of her host. The extruded body part contains an opening that receives sperm from a male. Much later, this opening becomes an exit for the strepsipteran larvae. The host insects are left dead or severely damaged and produce no offspring of their own.

Strepsipterans dramatically change the behavior of their hosts in ways that help them complete their life cycles at the expense of their hosts' reproduction. For example, when wasps—a typical host—are parasitized by strepsipterans, the parasites generate signals that induce the wasps to leave their nest and form a mating aggregation. This aggregation, however, serves the strepsipterans, not the wasps. Once the wasps aggregate, the male strepsipterans emerge from their hosts to search for and mate with female strepsipterans, whose heads are now poking out of the bodies of other wasps.

Adult male strepsipterans live only a few hours, during which they must find a female and mate. Because the protruding part of a female's body is barely visible, the males have unusually large eyes,

Same Parasite, Different Lifestyles
Strepsipterans are parasitic insects that grow to maturity within host insects. Most male strepsipterans look insectlike upon reaching maturity, when they leave their host to find and mate with a very different-appearing female strepsipteran. Females remain inside their host.

A Host Insect   Strepsipterans parasitize many different insect species; wasps of the genus *Polistes* (paper wasps) are common hosts. Three female strepsipterans (arrows) can be seen on this *P. dorsalis* individual.

and about 75 percent of their brain cells are allocated to vision. This highly developed sensory system evolved as an adaptation to help the male find a female.

Strepsipterans and their hosts are all insects, and insects account for more than half of all described species on Earth. Other protostome groups, such as mollusks, nematodes, crabs, spiders, and ticks, are also species-rich. Many protostome species are parasites. Parasites often live within their hosts and absorb nutrients through their body walls. Some parasites, including the strepsipterans, can more technically be described as *parasitoids*, which consume the host's tissues as they develop from eggs laid on or in the host's body, ultimately killing the host.

IN THIS CHAPTER we will describe the characteristics of protostome animals and describe the members of two major protostome clades, the lophotrochozoans and the ecdysozoans. We will give particular attention to the arthropods, a species-rich group of ecdysozoans with rigid exoskeletons and jointed appendages.

32.1 What Is a Protostome?

You may recall from Chapter 31 that the embryos of diploblastic animals have two cell layers: an outer ectoderm and an inner endoderm (see Section 31.1). Sometime after the origin of the diploblastic animals (the placozoans, ctenophores, and cnidarians), a third embryological germ layer evolved—the mesoderm, which lies between the ectoderm and the endoderm. Mesoderm is found in the two major triploblastic animal clades, the protostomes and the deuterostomes. If we were to judge solely on the basis of numbers, both of species and of individuals, the protostomes would emerge as by far the more successful of the two groups.

As noted in Chapter 31, the name *protostome* means "mouth first." In protostomes that have an embryonic blastopore, this opening becomes the mouth as the animal develops. In contrast, in deuterostomes ("mouth second"), the blastopore becomes the anal opening of the digestive tract (see Figure 31.2). The protostomes can be further divided into two major clades—the *lophotrochozoans* and the *ecdysozoans*—largely on the basis of DNA sequence analysis (**Figure 32.1**).

The protostomes are extremely varied, but they are all bilaterally symmetrical animals whose bodies exhibit two major derived traits:

- An anterior *brain* that surrounds the entrance to the digestive tract
- A ventral *nervous system* consisting of paired or fused longitudinal nerve cords

Other aspects of protostome body organization differ widely from group to group (**Table 32.1**). Although the common ancestor of the protostomes had a *coelom*, subsequent modifications of the coelom distinguish many protostome lineages. In at least one protostome lineage (the flatworms), the coelom has been lost (that is, the flatworms reverted to an *acoelomate* state). Some lineages are characterized by a *pseudocoelom*, which you may recall is a body cavity lined with mesoderm in which the internal organs are suspended (see Figure 31.4). In two of the most prominent protostome clades, the coelom has been highly modified:

- The *arthropods* lost the ancestral condition of the coelom over the course of evolution. Their internal body cavity has become a *hemocoel*, or "blood chamber," in which fluid from an open circulatory system bathes the internal organs before returning to blood vessels.

32.1 Phylogenetic Tree of Protostomes
Two major lineages, the lophotrochozoans and
the ecdysozoans, dominate the protostome tree.
Some small groups are not included. The phylogenetic relationships shown here are supported
mostly by genomic sequence data. Although
genomic studies are contributing greatly to our
knowledge of animal phylogeny, most species of
protostomes have yet to be studied in detail.

yourBioPortal.com
GO TO **Web Activity 32.1**
Protostome Classification

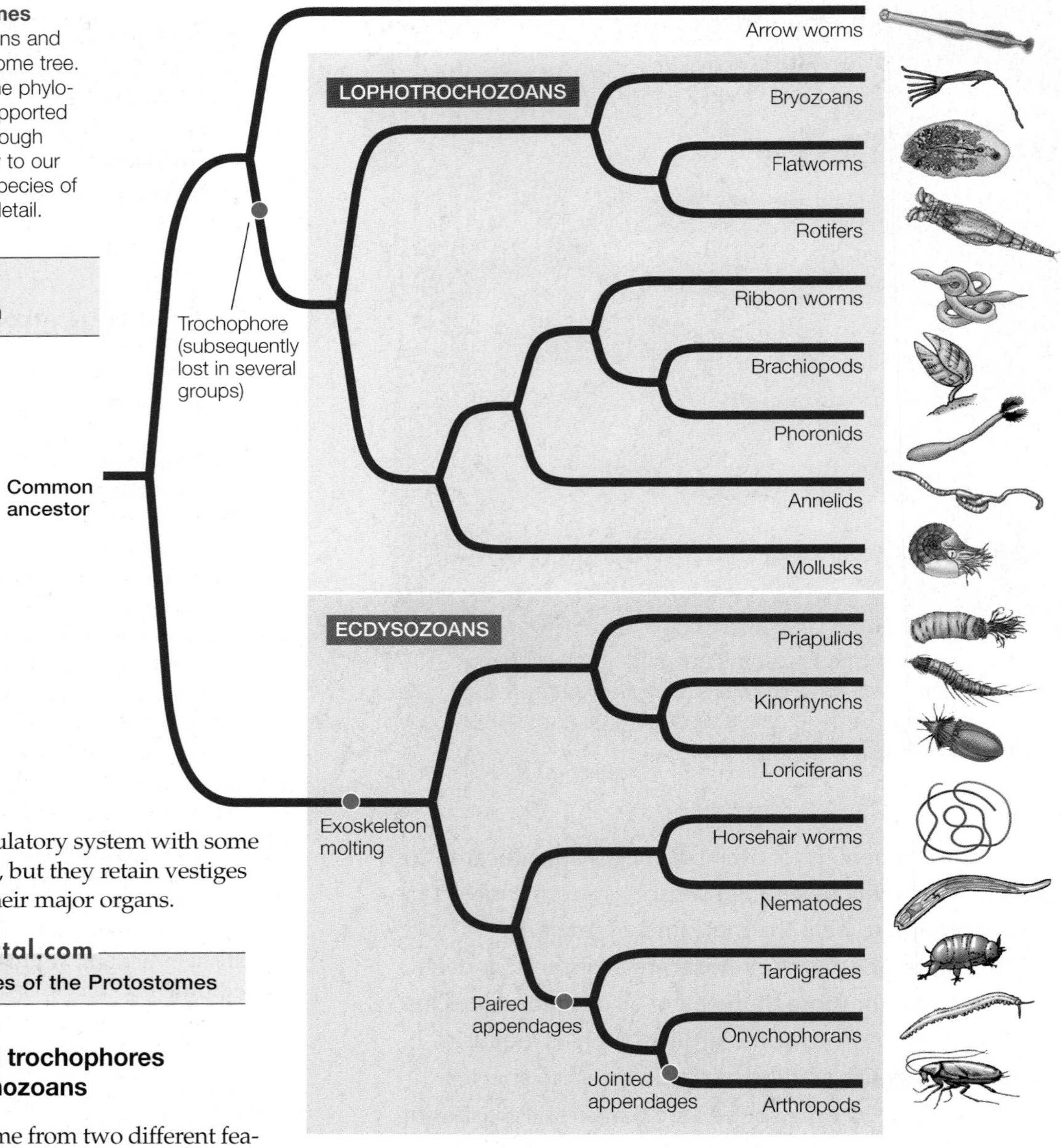

• Most *mollusks* have an open circulatory system with some
of the attributes of the hemocoel, but they retain vestiges
of an enclosed coelom around their major organs.

yourBioPortal.com
GO TO **Web Activity 32.2** • **Features of the Protostomes**

Cilia-bearing lophophores and trochophores evolved among the lophotrochozoans

Lophotrochozoans derive their name from two different features that involve cilia: a feeding structure known as a
lophophore and a free-living larva known as a *trochophore*.
Neither a lophophore nor a trochophore is universal for all
lophotrochozoans, however.

Several distantly related groups of lophotrochozoans (including bryozoans, brachiopods, and phoronids) have a **lophophore**,
a circular or U-shaped ring of ciliated, hollow tentacles around
the mouth (**Figure 32.2**). This complex structure is an organ
for both food collection and gas exchange. Biologists once
grouped taxa that have lophophores together as *lophophorates*,
but it is now clear that they are not one another's closest relatives. The lophophore appears to have evolved independently
several times, or else it is an ancestral feature of lophotrochozoans and has been lost in many groups. Nearly all animals
with a lophophore are sessile as adults. They use the tentacles
and cilia of the lophophore to capture small floating organisms
from the water. Other sessile lophotrochozoans use less well developed tentacles for the same purpose.

Some lophotrochozoans, especially in their larval form, use
cilia for locomotion. The larval form known as a **trochophore**
moves by beating a band of cilia (see Figure 31.11A). This movement of cilia also brings plankton closer to the larva, where
the plankton can be captured and ingested (similar in function
to the cilia of the lophophore). Trochophore larvae are found
among many of the major groups of lophotrochozoans, including the mollusks, annelids, ribbon worms, and bryozoans. This
larval form was probably present in the common ancestor of
lophotrochozoans, although it has been subsequently lost in several lineages.

As discussed Chapter 31, some lophotrochozoans (including
flatworms, ribbon worms, annelids, and mollusks) exhibit a
form of cleavage in early development known as *spiral cleavage*.
Some biologists group these taxa together as *spiralians,* although
phylogenetic analyses of gene sequences do not support the

TABLE 32.1
Anatomical Characteristics of Some Major Protostome Groups[a]

GROUP	BODY CAVITY	DIGESTIVE TRACT	CIRCULATORY SYSTEM
Arrow worms	Coelom	Complete	None
LOPHOTROCHOZOANS			
Flatworms	None	Dead-end sac	None
Rotifers	Pseudocoelom	Complete	None
Bryozoans	Coelom	Complete	None
Brachiopods	Coelom	Complete in most	Open
Phoronids	Coelom	Complete	Closed
Ribbon worms	Coelom	Complete	Closed
Annelids	Coelom	Complete	Closed or open
Mollusks	Reduced coelom	Complete	Open except in cephalopods
ECDYSOZOANS			
Horsehair worms	Pseudocoelom	Greatly reduced	None
Nematodes	Pseudoceolom	Complete	None
Arthropods	Hemocoel	Complete	Open

[a] Note that all protostomes have bilateral symmetry.

Bryozoans can oscillate, rotate, and retract their lophophore tentacles.

Plumatella repens

100 μm

32.2 Bryozoans Use Their Lophophore to Feed The extended lophophore dominates the anatomy of the colonial bryozoans. This species inhabits fresh water, although most bryozoans are marine. Bryozoan colonies can grow to contain more than a million individuals, all stemming from the asexual reproduction of the colony's founder.

species with spiral cleavage as monophyletic. Nonetheless, spiral cleavage may have been present in the lophotrochozoan ancestor, with several subsequent losses of this cleavage pattern.

Many lineages of lophotrochozoans are *wormlike*, which means they are bilaterally symmetrical, legless, soft-bodied, and at least several times longer than they are wide. A wormlike body form enables animals to burrow efficiently through marine sediment or soil. However, as we will describe later in this chapter, the *mollusks*—the most familiar of the lophotrochozoans to many people—have a very different body organization.

Ecdysozoans must shed their cuticles

Ecdysozoans have an external covering, or **cuticle**, which is secreted by the underlying *epidermis* (the outermost cell layer). The cuticle provides these animals with both protection and support. Once formed, however, the cuticle cannot grow. How, then, can ecdysozoans increase in size? They do so by shedding, or **molting**, the cuticle and replacing it with a new, larger one. This molting process gives the clade its name (Greek *ecdysis*, "to get out of").

A recently discovered fossil of a Cambrian soft-bodied arthropod, preserved in the process of molting, shows that molting evolved more than 500 million years ago (**Figure 32.3A**). An increasingly rich array of molecular and genetic evidence, including a set of Hox genes shared by all ecdysozoans, suggests they have a single common ancestor. Thus molting of a cuticle is a trait that may have evolved only once during animal evolution.

Before an ecdysozoan molts, a new cuticle is already forming underneath the old one. Once the old cuticle is shed, the new one expands and hardens. Until it has hardened, though, the animal is vulnerable to its enemies, both because its outer surface is easy to penetrate and because an animal with a soft cuticle moves slowly or not at all (**Figure 32.3B**).

In many ecdysozoans that have wormlike bodies, the cuticle is relatively thin and flexible; it offers the animal some protection but provides only modest body support. A thin cuticle allows the exchange of gases, minerals, and water across the body surface, but it restricts the animal to moist habitats. Many species of ecdysozoans with thin cuticles live in marine sediments from which they obtain prey, either by ingesting sediments and extracting organic material from them, or by capturing larger prey using a toothed *pharynx* (a muscular organ at the anterior end of the digestive tract). Some freshwater species absorb nutrients directly through their thin cuticles, as do parasitic species that live within their hosts. Many wormlike ecdysozoans are predators, eating protists and small animals.

The cuticles of other ecdysozoans, mainly arthopods, function as external skeletons, or **exoskeletons**. These exoskeletons are thickened by layers of protein and a strong, waterproof polysaccharide called **chitin**. An animal with a rigid, chitin-reinforced exoskeleton can neither move in a wormlike manner nor use cilia for locomotion. A hard exoskeleton also impedes the passage of oxygen and nutrients into the animal, presenting new

(A)

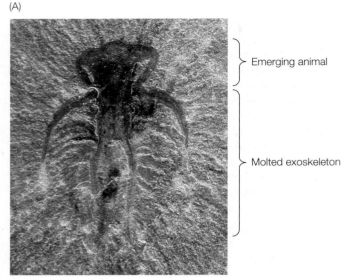

Emerging animal

Molted exoskeleton

(B) *Phrynus parvulus*

Molted exoskeleton

The newly emerged whip spider's body is still soft and vulnerable.

32.3 Molting: Past and Present (A) This 500-million-year-old fossil from the Cambrian captured an individual of a long-extinct arthropod species in the process of molting and shows that the molting process is an evolutionarily ancient trait. (B) This whip spider has just emerged from its discarded exoskelton and will be highly vulnerable until its new cuticle has hardened.

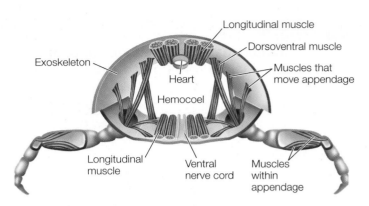

32.4 Arthropod Skeletons Are Rigid and Jointed This cross section through a thoracic segment of a generalized arthropod illustrates the arthropod body plan, which is characterized by a rigid exoskeleton with jointed appendages.

challenges in other areas besides growth. New mechanisms of locomotion and gas exchange evolved in those ecdysozoans with hard exoskeletons.

To move rapidly, an animal with a rigid exoskeleton must have body extensions that can be manipulated by muscles. Such *appendages* evolved in the late Precambrian, leading to the **arthropod** ("jointed foot") clade. Arthropod appendages exist in an amazing variety of forms. They serve many functions, including walking and swimming, gas exchange, food capture and manipulation, copulation, and sensory perception. Arthropods grasp food with their mouths and associated appendages and digest it internally. Their muscles are attached to the inside

of the exoskeleton. Each segment has muscles that operate that segment and the appendages attached to it (**Figure 32.4**).

The arthropod exoskeleton has had a profound influence on the evolution of these animals. Encasement within a rigid body covering provides support for walking on dry land, and the waterproofing provided by chitin keeps the animal from dehydrating in dry air. Aquatic arthropods were, in short, excellent candidates to invade terrestrial environments. As we will see, they did so several times.

Arrow worms retain some ancestral developmental features

Nearly all triploblastic animal groups can be readily classified as either protostomes or deuterostomes, but the evolutionary relationships of one small group, the **arrow worms**, were debated for many years. Although the early development of arrow worms seems similar to that of deuterostomes, it is now known that arrow worms merely retain developmental features that are ancestral to triploblastic animals in general. Recent studies of gene sequences clearly identify arrow worms as protostomes. There is still some question as to whether they are the closest relatives of the lophotrochozoans (as shown in Figure 32.1), or possibly the sister group of all other protostomes.

The arrow worm body is divided into three compartments: head, trunk, and tail (**Figure 32.5**). The body is transparent or translucent. Most arrow worms swim in the open sea. A few species live on the seafloor. Their abundance as fossils indicates that they were common more than 500 million years ago. The 100 or so living species of arrow worms are small enough— ranging from 3 millimeters to less than 12 centimeters in length—that their gas-exchange and waste-excretion requirements are met by diffusion through the body surface. They lack a circulatory system; wastes and nutrients are moved around the body in the coelomic fluid, which is propelled by cilia that line the coelom. There is no distinct larval stage. Miniature adults hatch directly from eggs that are fertilized internally following elaborate courtship between two hermaphroditic in-

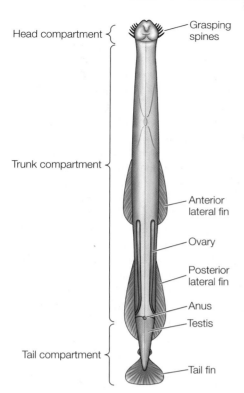

Head compartment { — Grasping spines

Trunk compartment {
— Anterior lateral fin

— Ovary

— Posterior lateral fin

— Anus

— Testis

Tail compartment {
— Tail fin

32.5 An Arrow Worm Arrow worms have a three-part body organization. Their fins and grasping spines are adaptations for a predatory lifestyle. Individuals are hermaphroditic, producing both eggs and sperm (ovary and testis).

dividuals (each arrow worm produces both male and female gametes).

Arrow worms are stabilized in the water by means of one or two pairs of lateral fins and a tail fin. They are major predators of planktonic organisms in the open ocean, ranging in size from small protists to young fish as large as the arrow worms themselves. An arrow worm typically lies motionless in the water until water movement signals the approach of prey. The arrow worm then darts forward and uses the stiff spines adjacent to its mouth to grasp its prey.

32.1 RECAP

The shared derived traits of protostomes include a blastopore that develops into a mouth, as well as the structure of the anterior brain and ventral nervous system. Several lophotrochozoan groups are characterized by a filter-feeding structure known as a lophophore and/or by cilia-bearing larvae known as trochophores. Ecdysozoans, which have a body covering known as a cuticle, must molt periodically in order to grow.

- How does an animal's body covering influence the way it breathes, feeds, and moves? See pp. 669–670

- What features made arthropods well adapted for colonizing terrestrial environments? See p. 670

32.2 What Features Distinguish the Major Groups of Lophotrochozoans?

Lophotrochozoans come in a variety of sizes and shapes, ranging from relatively simple animals with a blind gut (that is, a gut with only one opening) and no internal transport system to animals with a complete gut (having separate entrance and exit openings) and a complex internal transport system. They include some highly species-rich groups, such as flatworms, annelids, and mollusks. A number of these groups exhibit worm-like bodies, but the lophotrochozoans encompass a wide diversity of morphologies, including a few groups with external shells. Some lophotrochozoan groups have only recently been discovered by biologists.

Bryozoans live in colonies

The 4,500 species of **bryozoans** ("moss animals") are colonial animals that live in a "house" made of material secreted by the external body wall. Almost all bryozoans are marine, although a few species occur in fresh or brackish water. A bryozoan colony consists of many small (1–2 mm) individuals connected by strands of tissue along which nutrients can be moved (**Figure 32.6**). The colony is created by the asexual reproduction of its founding member, and a single colony may contain as many as 2 million individuals. Rocks in coastal regions in many parts of the world are covered with luxuriant growths of bryozoans. Some bryozoans create miniature reefs in shallow waters. In some species, the individual colony members are differentially specialized for feeding, reproduction, defense, or support.

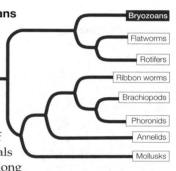

Sertella septentrionalis

32.6 A Bryozoan Colony The rigid orange tissue of this bryozoan colony connects and supplies nutrients to thousands of individual animals (see Figure 32.2).

Individual bryozoans in a colony are able to oscillate their lophophore to increase contact with prey. They can also retract it into their "house" (see Figure 32.2). Bryozoans can reproduce sexually by releasing sperm into the water, which carries the sperm to other individuals. Eggs are fertilized internally; developing embryos are brooded before they exit as larvae to seek suitable sites for attachment to the substratum.

Flatworms and rotifers are structurally diverse relatives

Flatworms and rotifers compose a structurally diverse group, and only recently have their relationships to one another been hypothesized. Before gene sequences were available for phylogenetic analysis, biologists considered the structure of the body cavity to be a critical feature in animal classification. If recent genomic studies prove correct, however, this monophyletic animal group includes both acoelomate subgroups (e.g., the flatworms) and pseudocoelomate subgroups (e.g., the rotifers), and yet the closest relatives of these two groups—the bryozoans—are coelomate. Thus, systematists today conclude that body cavity forms have undergone considerable convergence in the course of animal evolution (see Table 32.1).

Flatworms lack specialized organs for transporting oxygen to their internal tissues. Lacking a gas transport system, each cell must be near a body surface, a requirement met by the dorsoventrally flattened body form. The digestive tract of a flatworm consists of a mouth opening into a blind sac. The sac is often highly branched, forming intricate patterns that increase the surface area available for the absorption of nutrients. Some

Bryozoans

Flatworms

Rotifers

Ribbon worms

Brachiopods

Phoronids

Annelids

Mollusks

small free-living flatworms are cephalized, with a head bearing chemoreceptor organs, two simple eyes, and a tiny brain composed of anterior thickenings of the longitudinal nerve cords. Free-living flatworms glide over surfaces, powered by broad bands of cilia (**Figure 32.7A**).

Although many flatworms are free-living, most of the species are parasites. Of the parasitic species, most are internal parasites. There are also flatworms that feed externally on animal tissues (living or dead), and some graze on plants. A likely evolutionary transition was from feeding on dead organisms to feeding on the body surfaces of dying hosts to invading and consuming parts of healthy hosts.

Most of the 25,000 species of living flatworms are *tapeworms* and *flukes*; members of these two groups are internal parasites, particularly of vertebrates (**Figure 32.7B**). Because they absorb digested food from the digestive tracts of their hosts, many parasitic flatworms lack digestive tracts of their own. Some cause serious human diseases, such as schistosomiasis, which is common in parts of Asia, Africa, and South America. The species that causes this devastating disease has a complex life cycle involving both freshwater snails and mammals as hosts. Members of another flatworm group, the *monogeneans*, are external parasites of fishes and other aquatic vertebrates. The *turbellarians* include most of the free-living species.

Most **rotifers** are tiny (50–500 μm long)—smaller than some ciliate protists—but they have specialized internal organs (**Figure 32.8**). A complete gut passes from an anterior mouth to a posterior anus; the body cavity is a pseudocoel that functions as a hydrostatic skeleton. Rotifers typically propel themselves through the water by means of rapidly beating cilia rather than by muscular contraction.

The most distinctive organ of rotifers is a conspicuous ciliated organ called the *corona*, which surmounts the head of many species. Coordinated beating of the cilia sweeps particles of

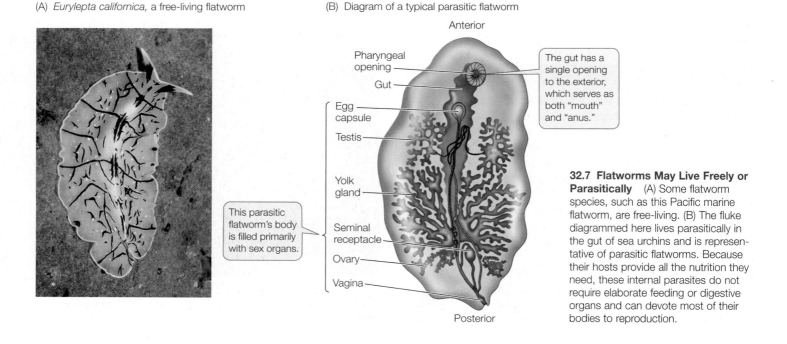

(A) *Eurylepta californica*, a free-living flatworm

This parasitic flatworm's body is filled primarily with sex organs.

(B) Diagram of a typical parasitic flatworm

Anterior

Pharyngeal opening

Gut

Egg capsule

Testis

Yolk gland

Seminal receptacle

Ovary

Vagina

The gut has a single opening to the exterior, which serves as both "mouth" and "anus."

Posterior

32.7 Flatworms May Live Freely or Parasitically (A) Some flatworm species, such as this Pacific marine flatworm, are free-living. (B) The fluke diagrammed here lives parasitically in the gut of sea urchins and is representative of parasitic flatworms. Because their hosts provide all the nutrition they need, these internal parasites do not require elaborate feeding or digestive organs and can devote most of their bodies to reproduction.

(A) *Philodina roseola*

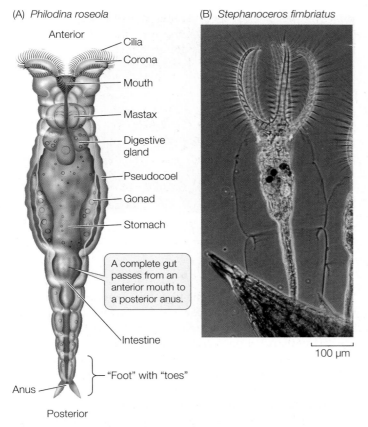

(B) *Stephanoceros fimbriatus*

A complete gut passes from an anterior mouth to a posterior anus.

100 μm

32.8 Rotifers (A) The individual diagrammed here reflects the general structure of many rotifers. (B) A micrograph reveals the internal complexity of a rotifer that has ingested photosynthetic protists. This species, one of the larger rotifers, is found in the Great Lakes of North America.

organic matter from the water into the animal's mouth and down to a complicated structure called the *mastax*, in which food is ground into small pieces. By contracting muscles around the pseudocoel, a few rotifer species that prey on protists and small animals can protrude the mastax through their mouth and seize small objects with it.

Most of the 1,800 known species of rotifers live in fresh water. Some species rest on the surfaces of mosses or lichens in a desiccated, inactive state until it rains. When rain falls, they absorb water and become mobile, feeding in the films of water that temporarily cover the plants. Most rotifers live no longer than a few weeks.

Both males and females are found in some species of rotifers, but only females are known among the *bdelloid rotifers* (the *b* in *bdelloid* is silent). Biologists have concluded that the bdelloid rotifers may have existed for tens of millions of years without regular sexual reproduction. In general, lack of genetic recombination leads to the buildup of deleterious mutations, so long-term asexual reproduction typically leads to extinction (see Section 21.4). Recent studies, however, have indicated that bdelloid rotifers may avoid this problem by picking up fragments of genes from their environment during the desiccation–rehydration cycle, which allows genetic recombination among individuals in the absence of direct sexual exchange.

Ribbon worms have a long, protrusible feeding organ

Ribbon worms (nemerteans) have simple nervous and excretory systems similar to those of flatworms. Unlike flatworms, however, they have a complete digestive tract with a mouth at one end and an anus at the other. Small ribbon worms move slowly by beating their cilia. Larger ones employ waves of muscle contraction to move over the surface of sediments or to burrow into them.

Within the body of nearly all of the 1,000 species of ribbon worms is a fluid-filled cavity called the *rhynchocoel*, within which lies a hollow, muscular *proboscis*. The proboscis, which is the worm's feeding organ, may extend much of the length of the body. Contraction of the muscles surrounding the rhynchocoel causes the proboscis to evert explosively through an anterior pore (**Figure 32.9A**). The proboscis may be armed with sharp stylets that pierce prey and discharge paralysis-causing toxins into the wound.

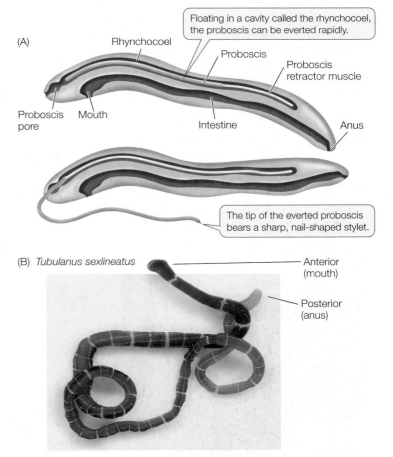

32.9 Ribbon Worms (A) The proboscis is the ribbon worm's feeding organ. (B) This large marine nemertean is found in harbors and bays along the Pacific Coast of North America. Its proboscis is not everted in this photograph.

Ribbon worms are largely marine, although there are species that live in fresh water or on land. Most species are less than 20 centimeters long, but individuals of some species reach 20 *meters* or more. Some genera feature species that are conspicuous and brightly colored (**Figure 32.9B**). Recent molecular analyses suggest that ribbon worms may be most closely related to the phoronids and brachiopods.

Phoronids and brachiopods use lophophores to extract food from the water

Recall that the bryozoans use a lophophore to feed. Phoronids and brachiopods also feed using a lophophore, but this structure may have evolved more than once among these groups. Although neither the phoronids nor the brachiopods are represented by many living species, the brachiopods (which have shells and thus leave an excellent fossil record) are known to have been much more abundant during the Paleozoic and Mesozoic eras.

> Bryozoans
> Flatworms
> Rotifers
> Ribbon worms
> **Brachiopods**
> **Phoronids**
> Annelids
> Mollusks

The 20 known species of **phoronids** are small (5–25 cm long), sessile worms that live in muddy or sandy sediments or attached to rocky substrata. Phoronids are found in marine waters, from the intertidal zone to about 400 meters deep. They secrete tubes made of chitin, within which they live (**Figure 32.10**).

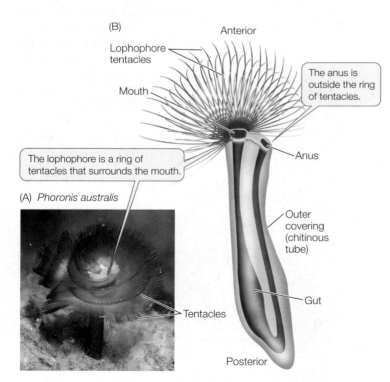

(B)

Anterior

Lophophore tentacles

Mouth

The anus is outside the ring of tentacles.

Anus

The lophophore is a ring of tentacles that surrounds the mouth.

(A) *Phoronis australis*

Outer covering (chitinous tube)

Tentacles

Gut

Posterior

32.10 Phoronids (A) The tentacles of this phoronid's lophophore form a spiral. (B) The phoronid gut is U-shaped, as seen in this generalized diagram.

Laqueus sp.

Lophophore ring

Tentacles

32.11 A Brachiopod's Lophophore The lophophore of this North Pacific brachiopod can be seen between the valves of its shell.

Their cilia drive water into the top of the lophophore, and the water exits through the narrow spaces between the tentacles. Suspended food particles are caught and transported to the mouth by ciliary action. Some species release eggs into the water, where they are fertilized, but other species produce large eggs that are fertilized internally and retained in the parent's body, where they are brooded until they hatch.

Brachiopods are solitary marine animals. They have a rigid shell that is divided into two parts connected by a ligament (**Figure 32.11**). The two halves can be pulled shut to protect the soft body. Brachiopods superficially resemble bivalve mollusks, but shells have evolved independently in the two groups. The two halves of the brachiopod shell are dorsal and ventral, rather than lateral as in bivalves. The lophophore is located within the shell. The beating of cilia on the lophophore draws water into the slightly opened shell. Food is trapped in the lophophore and directed to a ridge, along which it is transferred to the mouth. Most brachiopods are 4–6 centimeters long.

Brachiopods live attached to a solid substratum or embedded in soft sediments. Most species are attached by means of a short, flexible stalk that holds the animal above the substratum. Gases are exchanged across body surfaces, especially the tentacles of the lophophore. Most brachiopods release their gametes into the water, where they are fertilized. The larvae remain among the plankton for only a few days before they settle and develop into adults.

Brachiopods reached their peak abundance and diversity in Paleozoic and Mesozoic times. More than 26,000 fossil species have been described. Only about 335 species survive, but they remain common in some marine environments.

Annelids have segmented bodies

The bodies of **annelids** are clearly segmented. As discussed in Section 31.2, segmentation allows an animal to move different parts of its body independently of one another, giving it much better control of its movement. The earliest segmented worms, preserved as fossils from the middle Cambrian, were burrowing marine annelids.

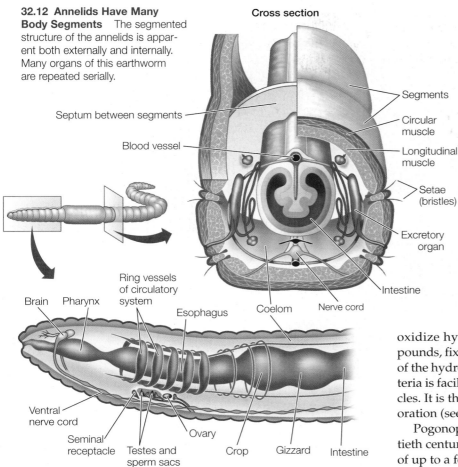

32.12 Annelids Have Many Body Segments The segmented structure of the annelids is apparent both externally and internally. Many organs of this earthworm are repeated serially.

Cross section

Septum between segments

Blood vessel

Segments

Circular muscle

Longitudinal muscle

Setae (bristles)

Excretory organ

Intestine

Brain   Pharynx

Ring vessels of circulatory system

Esophagus

Coelom   Nerve cord

Ventral nerve cord

Seminal receptacle

Testes and sperm sacs

Ovary

Crop   Gizzard   Intestine

In most large annelids, the coelom in each segment is isolated from those in other segments (**Figure 32.12**). A separate nerve center called a *ganglion* controls each segment; nerve cords that connect the ganglia coordinate their functioning. Most annelids lack a rigid external protective covering; instead, they have a thin, permeable body wall that serves as a general surface for gas exchange. Most annelids are thus restricted to moist environments because they lose body water rapidly in dry air. The approximately 16,500 described species live in marine, freshwater, and moist terrestrial environments.

Bryozoans

Flatworms

Rotifers

Ribbon worms

Brachiopods

Phoronids

Annelids

Mollusks

POLYCHAETES More than half of all annelid species are commonly known as *polychaetes* ("many hairs"), although this is a descriptive term rather than the name of a single clade. Most polychaetes are marine, and many live in burrows in soft sediments. Most of them have one or more pairs of eyes and one or more pairs of tentacles, with which they capture prey or filter food from the surrounding water, at the anterior end of the body (**Figure 32.13A**; see also Figure 31.6A). In some species, the body wall of most segments extends laterally as a series of

thin outgrowths called *parapodia*. The parapodia function in gas exchange, and some species use them to move. Stiff bristles called *setae* protrude from each parapodium, forming temporary contact with the substratum and preventing the animal from slipping backward when its muscles contract. Recent molecular studies indicate that polychaetes are paraphyletic with respect to the remaining annelids.

Members of one polychaete clade, the *pogonophorans*, secrete tubes made of chitin and other substances, in which they live (**Figure 32.13B**). Pogonophorans have lost their digestive tract (they have no mouth or gut). So how do they obtain nutrition? Part of the answer is that pogonophorans can take up dissolved organic matter directly from the sediments in which they live or from the surrounding water. Much of their nutrition, however, is provided by endosymbiotic bacteria that live in a specialized organ known as the *trophosome*. These bacteria oxidize hydrogen sulfide and other sulfur-containing compounds, fixing carbon from methane in the process. The uptake of the hydrogen sulfide, methane, and oxygen used by the bacteria is facilitated by hemoglobin in the pogonophorans' tentacles. It is this hemoglobin that gives the tentacles their red coloration (see Figure 32.13B).

Pogonophorans were not discovered until early in the twentieth century, when the first species were discovered at depths of up to a few hundred meters. In recent decades, deep-sea explorers have found them living many thousands of meters below the ocean surface. In these deep oceanic sediments, they may reach densities of many thousands per square meter. About 160 species have been described. The largest and most remarkable pogonophorans are 2 meters or more in length and live near deep-sea *hydrothermal vents*—volcanic openings in the seafloor through which hot, sulfide-rich water pours. The methane and hydrogen sulfide from these vents provide the raw materials for carbon fixation by the pogonophorans' endosymbiotic bacteria.

CLITELLATES The approximately 3,000 described species of this well-supported clade within the annelids are found in freshwater, marine, or terrestrial environments. The clitellates appear to be phylogenetically nested among various groups of polychaetes, although the exact relationships are not yet clear. There are two major groups of clitellates, the oligochaetes and the leeches.

Oligochaetes ("few hairs") have no parapodia, eyes, or anterior tentacles, and they have only four pairs of setae bundles per segment. Earthworms—the most familiar oligochaetes—burrow in and ingest soil, from which they extract food particles. All oligochaetes are *hermaphroditic*; that is, each individual is both male and female. Sperm are exchanged simultaneously between two copulating individuals (**Figure 32.13C**). Eggs and sperm are deposited in a cocoon outside the adult's body. Fertilization occurs within the cocoon after it is shed, and when development is complete, miniature worms emerge and immediately begin independent life.

(A) *Eudistylia* sp.

(C) *Lumbricus terrestris*

32.13 Diversity among the Annelids (A) "Fan worms" or "feather duster worms" are sessile marine polychaetes that grow in masses, filtering food from the water with their tentacles. This individual has been removed from its chitinous tube. (B) Pogonophorans live around hydrothermal vents deep in the ocean. Their tentacles can be seen protruding from their chitinous tubes. (C) Earthworms are hermaphroditic; when they copulate, each individual donates and receives sperm. (D) The medicinal leech has been a tool of physicians and healers for centuries. Even today, leeches have uses in modern clinical practice.

(B) *Riftia* sp.

(D) *Hirudo medicinalis*

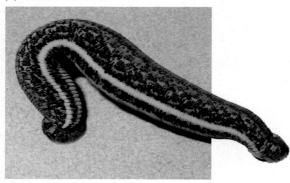

Leeches, like oligochaetes, lack parapodia and tentacles. The coelom of leeches is not divided into compartments; the coelomic space is largely filled with undifferentiated tissue. Groups of segments at each end of the body are modified to form suckers, which serve as temporary anchors that aid the leech in movement. With its posterior sucker attached to a substratum, the leech extends its body by contracting its circular muscles. The anterior sucker is then attached, the posterior one detached, and the leech shortens itself by contracting its longitudinal muscles. Leeches live in freshwater or terrestrial habitats.

A leech makes an incision in its host, from which blood flows. It can ingest so much blood in a single feeding that its body may enlarge several-fold. The leech secretes an anticoagulant into the wound that keeps the host's blood flowing. For centuries, medical practitioners employed leeches to draw blood to treat diseases they believed were caused by an excess of blood or by "bad blood." Although most leeching practices (such as inserting a leech in a person's throat to alleviate swollen tonsils) have been abandoned, *Hirudo medicinalis* (the medicinal leech; **Figure 32.13D**) is used today to reduce fluid pressure and prevent blood clotting in damaged tissues, to eliminate pools of coagulated blood, and to prevent scarring. The anticoagulants of certain other leech species that also contain anesthetics and blood vessel dilators are being studied for possible medical uses.

Mollusks have undergone a dramatic evolutionary radiation

Mollusks are the most diverse group of lophotrochozoans, both in numbers of species and in environments they occupy. Although the major groups of mollusks differ dramatically in morphology, they all share the same three major body components: a foot, a visceral mass, and a mantle (**Figure 32.14**).

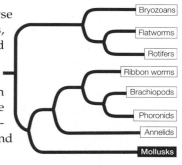

Bryozoans
Flatworms
Rotifers
Ribbon worms
Brachiopods
Phoronids
Annelids
Mollusks

- The molluscan *foot* is a large, muscular structure that originally was both an organ of locomotion and a support for the internal organs. In squids and octopuses, the foot has been modified to form arms and tentacles borne on a head with complex sensory organs. In other groups, such as clams, the foot is a burrowing organ. In some groups the foot is greatly reduced.

- The heart and the digestive, excretory, and reproductive organs are concentrated in a centralized, internal *visceral mass*.

- The *mantle* is a fold of tissue that covers the organs of the visceral mass. The mantle secretes the hard, calcareous shell that is typical of many mollusks.

In most mollusks, the mantle extends beyond the visceral mass to form a *mantle cavity*. Within this cavity lie *gills* that are used for gas exchange. When cilia on the gills beat, they create a cur-

Generalized molluscan body plan

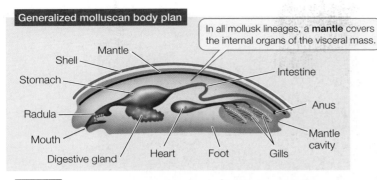

In all mollusk lineages, a **mantle** covers the internal organs of the visceral mass.

Chitons

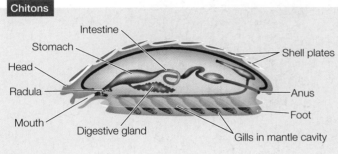

Gastropods

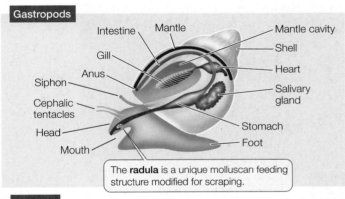

The **radula** is a unique molluscan feeding structure modified for scraping.

Bivalves

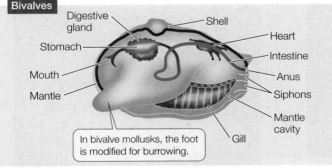

In bivalve mollusks, the foot is modified for burrowing.

Cephalopods

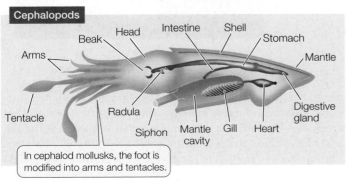

In cephalod mollusks, the foot is modified into arms and tentacles.

32.14 Organization of Molluscan Bodies The major molluscan groups display different variations on a general body plan that includes three major components: a foot, a visceral mass of internal organs, and a mantle. The mantle may secrete a calcareous shell, as in the gastropods and bivalves.

rent of water. The tissue of the gills, which is highly *vascularized* (containing many blood vessels), takes up oxygen from the water and releases carbon dioxide. Many mollusk species use their gills as filter-feeding devices, whereas others feed using a rasping structure known as the *radula* to scrape algae from rocks. In some mollusks, such as the marine cone snails, the radula has been modified into a drill or poison dart.

Molluscan blood vessels do not form a closed system. Blood and other fluids empty into a large, fluid-filled *hemocoel*, through which fluids move around the animal and deliver oxygen to the internal organs. Eventually, the fluids reenter the blood vessels and are moved by a heart.

Monoplacophorans were the most abundant mollusks during the Cambrian period, 500 million years ago, but only a few species survive today. Unlike in all other living mollusks, in monoplacophorans the gas exchange organs, muscles, and excretory pores are repeated over the length of the body.

Figure 32.15 illustrates the four major clades of living mollusks: chitons, bivalves, gastropods, and cephalopods. The species shown here are a tiny sample of the approximately 100,000 living species of mollusks.

CHITONS Eight overlapping calcareous plates, surrounded by a structure known as the girdle, protect the internal organs and muscular foot of **chitons** (**Figure 32.15A**). The chiton body is bilaterally symmetrical, and the internal organs, particularly the digestive and nervous systems, are relatively simple. Most chitons are marine omnivores that scrape algae, bryozoans, and other organisms from rocks with their sharp radula. An adult chiton spends most of its life clinging tightly to rock surfaces with its large, muscular, mucus-covered foot. It moves slowly by means of rippling waves of muscular contraction in the foot. Fertilization in most chitons takes place in the water, but in a few species fertilization is internal and embryos are brooded within the body. There are approximately 1,000 living species of chitons.

BIVALVES Clams, oysters, scallops, and mussels are all familiar **bivalves**. The 30,000 living species are found in both marine and freshwater environments. Bivalves have a very small head and a hinged, two-part shell that extends over the sides of the body as well as the top (**Figure 32.15B**). Many clams use their foot to burrow into mud and sand. Bivalves feed by taking in water through an opening called an *incurrent siphon* and filtering food from the water with their large gills, which are also the main sites of gas exchange. Water and gametes exit through the *excurrent siphon*. Fertilization takes place in open water in most species.

GASTROPODS **Gastropods** are the most species-rich and widely distributed mollusks, with nearly 70,000 living species. Snails, whelks, limpets, slugs, nudibranchs (sea slugs), and abalones

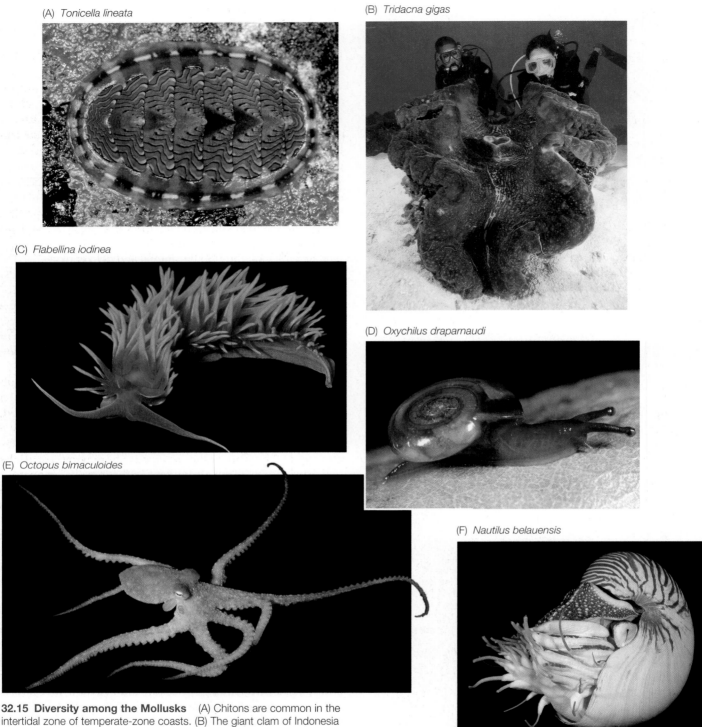

(A) *Tonicella lineata*

(B) *Tridacna gigas*

(C) *Flabellina iodinea*

(D) *Oxychilus draparnaudi*

(E) *Octopus bimaculoides*

(F) *Nautilus belauensis*

32.15 Diversity among the Mollusks (A) Chitons are common in the intertidal zone of temperate-zone coasts. (B) The giant clam of Indonesia is among the largest of the bivalve mollusks. (C) This shell-less nudibranch ("naked gills"), or sea slug, is brightly colored, a signal of its toxicity. (D) Pulmonate land snails, such as this dark-bodied glass snail, have a vascularized lung that allows them to survive in terrestrial environments. (E) Cephalopods such as the octopuses are active predators. (F) The boundaries of chambers are clearly visible on the outer surfaces of this shelled cephalopod, one of the chambered nautiluses.

are all gastropods. Most species move by gliding on their muscular foot, but in a few species—the sea butterflies and heteropods—the foot is a swimming organ with which the animal moves through open ocean waters. Nudibranchs have lost their protective shell over the course of evolution. Their sometimes brilliant coloration is *aposematic*, meaning it serves to warn potential predators of toxicity (**Figure 32.15C**). Other nudibranch species exhibit camouflaged coloration.

Shelled gastropods have one-piece shells. The only mollusks that live in terrestrial environments—land snails and slugs—are gastropods (**Figure 32.15D**). In these terrestrial species, the mantle tissue is modified into a highly vascularized lung.

CEPHALOPODS The **cephalopods**—squids, octopuses, and nautiluses—first appeared near the beginning of the Cambrian period. By the Ordovician period a variety of types were present. Today there are about 800 living species.

In the cephalopods, the excurrent siphon is modified to allow the animal to control the water content of the mantle cavity. The modification of the mantle into a device for forcibly ejecting water from the cavity through the siphon enables these animals to move rapidly by "jet propulsion" through the water. With their greatly enhanced mobility, cephalopods became the major predators in the open waters of the Devonian oceans. They remain important marine predators today.

Cephalopods capture and subdue prey with their tentacles; octopuses also use their tentacles to move over the substratum (**Figure 32.15E**). As is typical of active, rapidly moving predators, cephalopods have a head with complex sensory organs, most notably eyes that are comparable to those of vertebrates in their ability to resolve images. The head is closely associated with a large, branched foot that bears the tentacles and a siphon. The large, muscular mantle provides a solid external supporting structure. The gills hang in the mantle cavity. Many cephalopods have elaborate courtship behavior, which may involve striking color changes.

Many early cephalopods had a chambered shell divided by partitions penetrated by tubes through which gases and liquids could be moved to control the animal's buoyancy. Nautiluses (genus *Nautilus*) are the only surviving cephalopods that have such external chambered shells (**Figure 32.15F**).

32.2 RECAP

Lophotrochozoans include animals with diverse body types. Wormlike forms include some flatworms, ribbon worms, phoronids, and annelids. There has been convergent evolution of lophophores (in bryozoans versus brachiopods and phoronids) and of external two-part shell coverings (in brachiopods versus bivalve mollusks).

- Can you explain how flatworms can survive without a gas transport system? **See p. 672 and Figure 32.7**

- Do you know why most annelids are restricted to moist environments? **See p. 675**

- Briefly describe how the basic body organization of mollusks has been modified to yield a wide diversity of animals. **See pp. 676–679 and Figure 32.14**

The second of the two major protostome clades, the ecdysozoans, contains the vast majority of Earth's animal species. What evolutionary innovations led to this massive diversity?

32.3 What Features Distinguish the Major Groups of Ecdysozoans?

Many ecdysozoans are wormlike in form, although the *arthropods*, *onychophorans*, and *tardigrades* have limbs. In this section we will look at the two clades of wormlike ecdysozoans: the priapulids, kinorhynchs, and loriciferans in one group, and the horsehair worms and nematodes in the other. Section 32.4 is devoted to the most diverse ecdysozoans—the arthropods and their relatives—and the many forms their appendages take.

Several marine groups have relatively few species

Members of several species-poor groups of wormlike marine ecdysozoans—the priapulids, kinorhynchs, and loriciferans—have relatively thin cuticles that are molted periodically as the animals grow to full size. In 2004, embryos of a fossil species related to these ecdysozoans were discovered in sediments laid down in China about 500 million years ago. This remarkable discovery shows that the ancestors of these animals developed directly from an egg to the adult form, as their modern descendants do.

Priapulids
Kinorhynchs
Loriciferans
Horsehair worms
Nematodes
Tardigrades
Onychophorans
Arthropods

The 16 species of **priapulids** are cylindrical, unsegmented, wormlike animals with a three-part body plan consisting of a proboscis, trunk, and caudal appendage ("tail"). It should be clear from their appearance why they were named after the Greek fertility god Priapus (**Figure 32.16A**). Priapulids range in length from 0.5 millimeters to 20 centimeters. They burrow in fine marine sediments and prey on soft-bodied invertebrates such as polychaetes, which they capture with a toothed, muscular pharynx that they evert through their mouth and then withdraw into their body together with the grasped prey. Fertilization is external, and most species have a larval form that also lives in the mud.

About 150 species of **kinorhynchs** have been described. They live in marine sands and muds and are virtually microscopic; no kinorhynchs are longer than 1 millimeter. Their bodies are divided into 13 segments, each with a separate cuticular plate (**Figure 32.16B**). These plates are periodically molted during growth. Kinorhynchs feed by ingesting sediments through their retractable proboscis (the name means "movable snout"). They then digest the organic material found in the sediment, which may include living algae as well as dead matter. Kinorhynchs have no distinct larval stage; fertilized eggs develop directly into juveniles, which emerge from their egg cases with 11 of the 13 body segments already formed.

Loriciferans are also minute animals less than 1 millimeter long. They were not discovered until 1983. About 100 living species are known to exist, although many of these are still being described. The body is divided into a head, neck, thorax, and abdomen and is covered by six plates, from which the loriciferans get their name (Latin *lorica*, "corset"). The plates around the base of the neck bear anterior-directed spines of unknown

(A) *Priapulus caudatus*

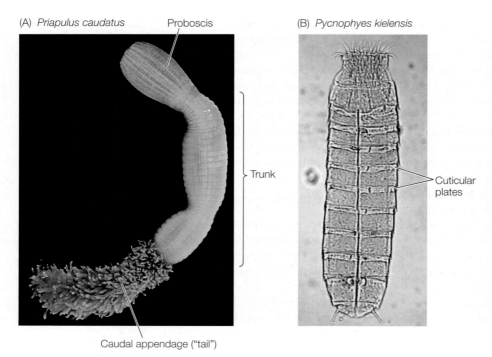

(B) *Pycnophyes kielensis*

(C) *Nanaloricus mysticus*

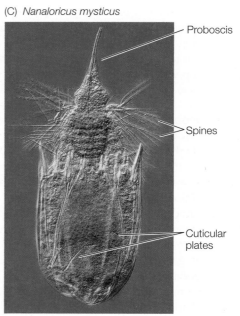

32.16 Wormlike Marine Ecdysozoans Members of these groups are marine bottom-dwellers. (A) Most priapulid species live in burrows on the ocean floor, extending the proboscis for feeding. (B) Kinorhynchs are virtually microscopic. The cuticular plates that cover their bodies are molted periodically. (C) Six cuticular plates form a "corset" around the minute loriciferan body.

function (**Figure 32.16C**). Loriciferans live in coarse marine sediments. Little is known about what they eat, but some species apparently eat bacteria.

Nematodes and their relatives are abundant and diverse

Nematodes (roundworms) have a thick, multilayered cuticle that gives their unsegmented body its shape (**Figure 32.17**). As a nematode grows, it sheds its cuticle four times. Nematodes exchange oxygen and nutrients with their environment through both the cuticle and the gut, which is only one cell layer thick. Materials are moved through the gut by rhythmic contraction of a highly muscular organ, the *pharynx*, at the worm's anterior end. Nematodes move by contracting their longitudinal muscles.

Nematodes are probably the most abundant and universally distributed of all animal groups. Many are microscopic; the largest known nematode, which reaches a length of 9 meters, is a parasite in the placentas of sperm whales. About 25,000 species have been described, but the actual number of living species may be more than a million. Countless nematodes live as scavengers in the upper layers of the soil, on the bottoms of lakes and streams, and in marine sediments. The topsoil of rich farmland may contain from 3 to 9 billion nematodes per acre. A single rotting apple may contain as many as 90,000 individuals.

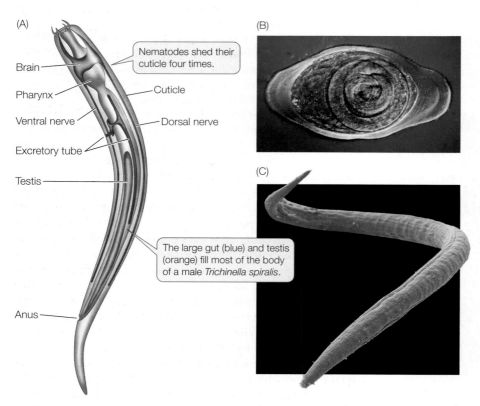

32.17 Nematodes (A) *Trichinella spiralis*, a parasitic nematode that causes trichinosis. (B) This polarized light micrograph shows a cyst of *T. spiralis* in the muscle tissue of a host. (C) This free-living nematode lives in freshwater environments.

One soil-inhabiting nematode, *Caenorhabitis elegans*, serves as a model organism in the laboratories of geneticists and developmental biologists. It is ideal for such research because it is easy to cultivate, matures in 3 days, and has a fixed number of body cells. Its genome has been completely sequenced.

Many nematodes are predators, feeding on protists and small animals (including other roundworms). Most significant to humans, however, are the many species that parasitize plants and animals. The nematodes that parasitize humans (causing serious diseases such as trichinosis, filariasis, and elephantiasis), domestic animals, and economically important plants have been studied intensively in an effort to find ways of controlling them.

The structure of parasitic nematodes is similar to that of free-living species, but the life cycles of many parasitic species have special stages that facilitate the transfer of individuals among hosts. *Trichinella spiralis*, the species that causes the human disease trichinosis, has a relatively simple life cycle. A person may become infected by eating the flesh of an animal (usually a pig) that has *Trichinella* larvae encysted in its muscles (see Figure 32.17B). The larvae are activated in the person's digestive tract, emerge from their cysts, and attach to the intestinal wall, where they feed. Later, they bore through the intestinal wall and are carried in the bloodstream to muscles, where they form new cysts. If present in great numbers, these cysts can cause severe pain or death.

About 320 species of the unsegmented **horsehair worms** have been described. As their name implies, these animals are extremely thin in diameter; horsehair worms range from a few millimeters up to a meter in length. Most adult worms live in fresh water, among leaf litter and algal mats near the edges of streams and ponds. A few species live in damp soil.

Horsehair larvae are internal parasites of freshwater crayfish and of terrestrial and aquatic insects (**Figure 32.18**). An adult

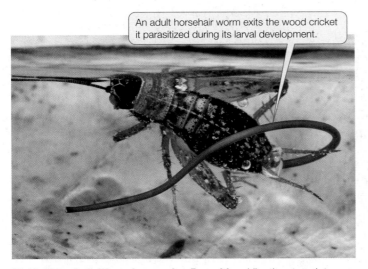

An adult horsehair worm exits the wood cricket it parasitized during its larval development.

32.18 Horsehair Worm Larvae Are Parasitic Like the strepsipteran insects described at the start of this chapter, the larvae of this horsehair worm (*Paragordius tricuspidatus*) can manipulate its host's behavior. The hatching worm causes the cricket to jump into water, where the worm will continue its life cycle as a free-living adult. The insect, having delivered its parasitic burden, drowns.

horsehair worm has no mouth, and its gut is greatly reduced and probably nonfunctional. Some species may feed only as larvae, absorbing nutrients from their hosts across the body wall. But other species continue to grow after they have left their hosts, shedding their cuticles, suggesting that adult worms may also absorb nutrients from their environment.

32.3 RECAP

Priapulids, kinorhynchs, and loriciferans are relatively small, poorly known groups of wormlike marine ecdysozoans. Horsehair worms and nematodes are also wormlike. Nematodes are probably the most abundant and universal animal group. The biology of some nematodes has been studied extensively, particularly that of *Caenorhabitis elegans*, which serves as a model laboratory organism.

- Describe at least three different ways in which nematodes have a significant impact on humans. See pp. 680–681

We now turn to the animals that not only dominate the ecdysozoan clade but are also the most diverse animals on Earth.

32.4 Why Are Arthropods So Diverse?

Arthropods and their relatives are ecdysozoans with paired appendages. Arthropods are the most diverse group of animals in numbers of species (about a million have been described, and many more remain to be discovered). Furthermore, the number of individual arthropods alive at any one time is estimated to be about 10^{18}, or a billion billion. Among the animals, only the nematodes are thought to exist in greater numbers.

Priapulids
Kinorhynchs
Loriciferans
Horsehair worms
Nematodes
Tardigrades
Onychophorans
Arthropods

Several key features have contributed to the success of the arthropods. Their bodies are segmented, and their muscles are attached to the inside of their rigid exoskeletons. Each segment has muscles that operate that segment and the jointed appendages attached to it (see Figure 32.4). Jointed appendages permit complex movements, and different appendages are specialized for different functions. Encasement of the body within a rigid exoskeleton provides the animal with support for walking in the water or on dry land and provides some protection against predators. The waterproofing provided by chitin keeps the animal from dehydrating in dry air.

Representatives of the four major arthropod groups living today are all species-rich: the *crustaceans* (including shrimps, crabs, and barnacles), *hexapods* (insects and their relatives), *myriapods* (millipedes and centipedes), and *chelicerates* (including the arachnids—spiders, scorpions, mites, and their relatives). Phylogenetic relationships among arthropod groups are currently being reexamined in light of a wealth of new information, much of it based on gene sequences. These studies show close relationships between the myriapods and chelicerates in one clade and between the crustaceans and hexapods in another, with strong support for the monophyly of arthropods as a whole.

The jointed appendages of arthropods gave the clade its name, from the Greek words *arthron*, "joint," and *podos*, "foot" or "limb." Arthropods evolved from ancestors with simple, unjointed appendages. The exact forms of those ancestors are unknown, but some arthropod relatives with segmented bodies and unjointed appendages survive today. Before we describe the modern arthropods, we will discuss those arthropod relatives, as well as an early clade that went extinct but left an important fossil record.

Arthropod relatives have fleshy, unjointed appendages

Until fairly recently, biologists debated whether the **onychophorans** (velvet worms) were more closely related to annelids or arthropods, but molecular evidence clearly links them to the latter. Indeed, with their soft, fleshy, unjointed, claw-bearing legs, onychophorans may be similar in appearance to the ancestors of arthropods (**Figure 32.19A**). The 150 species of onychophorans live in leaf litter in humid tropical environments. They have soft,

segmented bodies that are covered by a thin, flexible cuticle that contains chitin. They use their fluid-filled body cavities as hydrostatic skeletons. Fertilization is internal, and the large, yolky eggs are brooded within the body of the female.

Tardigrades (water bears) also have fleshy, unjointed legs and use their fluid-filled body cavities as hydrostatic skeletons (**Figure 32.19B**). Tardigrades are extremely small (0.1–0.5 millimeters long) and lack both a circulatory system and gas-exchange organs. The 800 known extant species live in marine sands and on temporary water films on plants. When these films dry out, the animals also lose water and shrink to small, barrel-shaped objects that can survive for at least a decade in a dormant state. Tardigardes have been found in densities as high as 2 million per square meter of moss.

Jointed appendages first appeared in the trilobites

The **trilobites** flourished in Cambrian and Ordovician seas, but they disappeared in the great Permian extinction at the close of the Paleozoic era (251 mya). Because they had heavy exoskeletons that readily fossilized, they left behind an abundant record of their existence (**Figure 32.20**). About 10,000 species have been described.

The body segmentation and appendages of trilobites followed a relatively simple, repetitive plan, but some of their

(A) *Peripatus* sp.

(B) *Echiniscus* sp.

50 µm

32.19 Arthropod Relatives with Unjointed Appendages
A) Onychophorans have unjointed legs and use the body cavity as a hydrostatic skeleton. (B) Tardigrades can be abundant on the wet surfaces of mosses and plants, and in temporary pools of water.

Acanthopyge sp.

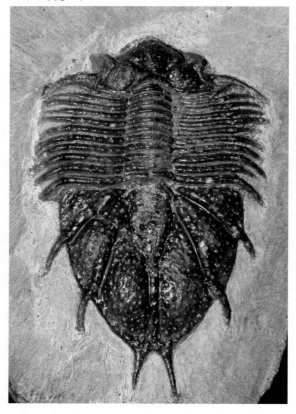

32.20 A Trilobite The relatively simple, repetitive segments of the now-extinct trilobites are illustrated by a fossil trilobite from the shallow seas of the Devonian period, some 400 million years ago.

jointed appendages were modified for different functions. This specialization of appendages is a theme in the continuing evolution of the arthropods.

Myriapods have many legs

Members of two arthropod groups, the myriapods and the chelicerates, have a body with just two regions: a head and a trunk. This contrasts with the hexapods and crustaceans, most of which have a three-part body, with a head, thorax, and abdomen.

The **myriapods** comprise the centipedes, millipedes, and their close relatives. Centipedes and millipedes have a well-formed head and a long, flexible, segmented trunk that bears many pairs of legs (**Figure 32.21**). Centipedes, which have one pair of legs per segment, prey on insects and other small animals. In millipedes, two adjacent segments are fused so that each fused segment has two pairs of legs. Millipedes scavenge and eat plants. More than 3,000 species of centipedes and 11,000 species of millipedes have been described; many more species probably remain unknown. Although most myriapods are less than a few centimeters long, some tropical species are ten times that size.

Most chelicerates have four pairs of legs

In the two-part body of **chelicerates**, the head bears two pairs of appendages modified to form mouthparts. In addition, many chelicerates have four pairs of walking legs. The 98,000 described species are usually placed in three major clades: pycnogonids, horseshoe crabs, and arachnids.

The *pycnogonids*, or sea spiders, are a poorly known group of about 1,000 marine species (**Figure 32.22A**). Most are small, with leg spans less than 1 centimeter, but some deep-sea species have leg spans up to 60 centimeters. A few pycnogonids eat algae, but most are carnivorous, eating a variety of small invertebrates.

There are four living species of *horseshoe crabs*, but many close relatives are known from fossils. Horseshoe crabs, which have changed very little morphologically during their long fossil history, have a large horseshoe-shaped covering over most of the body. They are common in shallow waters along the eastern

(A) *Endeis* sp.

(A) *Scolopendra heros*

(B) *Sigmoria trimaculata*

32.21 Myriapods (A) Centipedes have modified appendages that function as poisonous fangs for capturing active prey and one pair of legs per segment. (B) Millipedes, which are scavengers and plant eaters, have smaller jaws and legs than centipedes do. They have two pairs of legs per segment.

(B) *Limulus polyphemus*

32.22 Two Small Chelicerate Groups (A) Although they are not spiders, it is easy to see why sea spiders were given their common name. (B) This spawning aggregation of horseshoe crabs was photographed on a sandy beach of Delaware Bay. Horseshoe crabs are an ancient group that has changed very little in morphology over time; such species are sometimes referred to as "living fossils."

coast of North America and the southern and eastern coasts of Asia, where they scavenge and prey on bottom-dwelling animals. Periodically they crawl into the intertidal zone in large numbers to mate and lay eggs (**Figure 32.22B**).

Arachnids are abundant in terrestrial environments. Most arachnids have a simple life cycle in which miniature adults hatch from internally fertilized eggs and begin independent lives almost immediately. Some arachnids retain their eggs during development and give birth to live young.

The most species-rich and abundant arachnids are the spiders, scorpions, harvestmen, mites, and ticks (**Figure 32.23**). More than 50,000 described species of mites and ticks live in soil, leaf litter, mosses, and lichens, under bark, and as parasites of plants and animals. Mites are vectors for wheat and rye mosaic viruses; they cause mange in domestic animals and skin irritation in humans.

Spiders, of which 38,000 species have been described, are important terrestrial predators. Some have excellent vision that enables them to chase and seize their prey. Others spin elaborate webs made of protein threads in which they snare prey. The threads are produced by modified abdominal appendages connected to internal glands that secrete the proteins, which solidify on contact with air. The webs of different groups of spiders are strikingly varied, and this variation enables the spiders to position their snares in many different environments for many different types of prey.

Crustaceans are diverse and abundant

Crustaceans are the dominant marine arthropods today, and they are also common in freshwater and some terrestrial environments. The most familiar crustaceans are the shrimps, lobsters, crayfishes, and crabs (all *decapods*; **Figure 32.24A**) and the sow bugs (*isopods*; **Figure 32.24B**). Additional species-rich groups of crustaceans include *amphipods*, *ostracods*, and *branchiopods* (**Figure 32.24C**), all of which are found in freshwater and marine environments. (Don't confuse the branchiopods with the similarly named brachiopods, described in Section 32.2.) *Krill* are small but abundant oceanic crustaceans that are important food items for a variety of large vertebrates, including baleen whales. Also

32.23 Arachnid Diversity (A) "Tarantulas" encompass many free-living species as well as several hundred species of hairy, ground-dwelling spiders, some of which can grow to the size of a dinner plate. Their venomous bite, although painful, is not usually dangerous to humans. (B) Scorpions are nocturnal predators. (C) Harvestmen, also called daddy longlegs, are scavengers. (D) Mites include many free-living species as well as blood-sucking external parasites.

(A) *Poecilotheria metallica*

(C) *Leiobunum rotundum*

(B) *Pseudouroctonus minimus*

(D) *Brevipalpus phoenicis*

(A) *Randallia ornata*

(B) *Armadillium vulgare*

(C) *Triops longicaudatus*

(D) Cyclopoid copepod

(E) *Lepas pectinata*

32.24 Crustacean Diversity (A) This decapod crustacean, a purple sand crab, is also referred to as a "globe crab" based on its globelike body shape. (B) Isopods are sometimes referred to as "rock lice," a name derived from their preferred habitat. (C) This tadpole shrimp, a branchiopod, is common in seasonal pools of the southwestern United States. Molecular studies suggest that branchiopods may be more closely related to hexapods than to other crustaceans. (D) This minute copepod from a freshwater pond is brooding eggs. (E) Gooseneck barnacles attach to a substratum by their muscular stalks and feed by protruding and retracting their feeding appendages.

included in the crustacean clade are the small *copepods* (**Figure 32.24D**), which are also abundant in the open oceans.

Barnacles are unusual crustaceans that are sessile as adults (**Figure 32.24E**). Adult barnacles look more like mollusks than like other crustaceans, but as the zoologist Louis Agassiz remarked more than a century ago, a barnacle is "nothing more than a little shrimp-like animal, standing on its head in a limestone house and kicking food into its mouth."

Recent molecular studies suggest that crustaceans may not be monophyletic. In analyses of DNA sequences, the branchiopods (which include brine shrimp, fairy shrimp, water fleas, and tadpole shrimp; see Figure 32.24D) appear to be more closely related to the hexapods than to other crustaceans.

Most of the 52,000 described species of crustaceans have a body that is divided into three regions: head, thorax, and abdomen (**Figure 32.25**). The segments of the head are fused together, and the head bears five pairs of appendages. Each of the multiple thoracic and abdominal segments usually bears one pair of appendages. The appendages on different parts of the body are specialized for different functions, such as gas exchange, chewing, capturing food, sensing, walking, and swimming. In some cases, the appendages are branched, with different branches

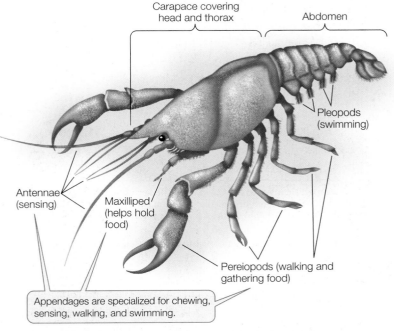

Carapace covering head and thorax

Abdomen

Pleopods (swimming)

Antennae (sensing)

Maxilliped (helps hold food)

Pereiopods (walking and gathering food)

Appendages are specialized for chewing, sensing, walking, and swimming.

32.25 Crustacean Body Plan The bodies of crustaceans are divided into three regions: the head, thorax, and abdomen. Each region bears specialized appendages, and a shell-like carapace covers the head and thorax.

TABLE 32.2
The Major Insect Groups[a]

GROUP	APPROXIMATE NUMBER OF DESCRIBED LIVING SPECIES
Jumping bristletails (Archaeognatha)	300
Silverfish (Thysanura)	370
PTERYGOTE (WINGED) INSECTS (PTERYGOTA)	
Mayflies (Ephemeroptera)	2,000
Dragonflies and damselflies (Odonata)	5,000
Neopterans (Neoptera) [b]	
Ice-crawlers (Grylloblattodea)	25
Gladiators (Mantophasmatodea)	15
Stoneflies (Plecoptera)	1,700
Webspinners (Embioptera)	300
Angel insects (Zoraptera)	30
Earwigs (Dermaptera)	1,800
Grasshoppers and crickets (Orthoptera)	20,000
Stick insects (Phasmida)	3,000
Cockroaches (Blattodea)	3,500
Termites (Isoptera)	2,750
Mantids (Mantodea)	2,300
Booklice and barklice (Psocoptera)	3,000
Thrips (Thysanoptera)	5,000
Lice (Phthiraptera)	3,100
True bugs, cicadas, aphids, leafhoppers (Hemiptera)	80,000
Holometabolous neopterans (Holometabola) [c]	
Ants, bees, wasps (Hymenoptera)	125,000
Beetles (Coleoptera)	375,000
Strepsipterans (Strepsiptera)	600
Lacewings, ant lions, dobsonflies (Neuropterida)	4,700
Scorpionflies (Mecoptera)	600
Fleas (Siphonaptera)	2,400
True flies (Diptera)	120,000
Caddisflies (Trichoptera)	5,000
Butterflies and moths (Lepidoptera)	250,000

[a] The hexapod relatives of insects include the springtails (Collembola; 3,000 spp.), two-pronged bristletails (Diplura; 600 spp.), and proturans (Protura; 10 spp.). All are wingless and have internal mouthparts.
[b] Neopteran insects can tuck their wings close to their bodies
[c] Holometabolous insects are neopterans that undergo complete metamorphosis.

serving different functions. In many species, a fold of the exoskeleton, the *carapace*, extends dorsally and laterally back from the head to cover and protect some of the other segments.

The fertilized eggs of most crustacean species are attached to the outside of the female's body, where they remain during their early development. At hatching, the young of some species are released as larvae; those of other species are released as juveniles that are similar in form to the adults. Still other species release eggs into the water or attach them to an object in the environment. The typical crustacean larva, called a *nauplius*, has three pairs of appendages and one central eye (see Figure 31.11B).

Insects are the dominant terrestrial arthropods

During the Devonian period, more than 400 million years ago, some arthropods colonized terrestrial environments. Of the several groups that successfully colonized the land, none is more prominent today than the six-legged *hexapods*: the **insects** and their relatives (**Table 32.2**). Insects are abundant and diverse in terrestrial and freshwater environments; only a few live in salt water.

Insects, like crustaceans, have a body with three regions—head, thorax, and abdomen. In addition, insects have a unique mechanism for gas exchange in air: a system of air sacs and tubular channels called *tracheae* (singular *trachea*) that extend from external openings called *spiracles* inward to tissues throughout the body. Insects have a single pair of antennae on the head and three pairs of legs attached to the thorax (**Figure 32.26**). Unlike the other arthropods, insects have no appendages growing from their abdominal segments. Insects can be distinguished from other hexapods by their external mouthparts and paired antennae that contain a motion-sensitive receptor called *Johnston's organ*.

The wingless relatives of the insects—the springtails (**Figure 32.27**), two-pronged bristletails, and proturans—are probably the most similar of living forms to insect ancestors. These relatives of insects have a simple life cycle; they hatch from eggs as miniature adults. They differ from insects in having inter-

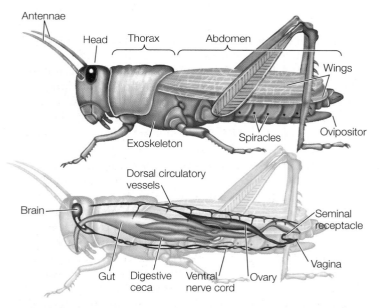

32.26 Body Plan of an Insect Insects, like crustaceans, have a three-part body plan with a head, thorax, and abdomen. The middle region, the thorax, bears three pairs of legs and, in most groups, two pairs of wings. Unlike other arthropods, the abdomen of an insect bears no appendages (see Figure 20.8).

Sminthurides aquaticus

32.27 Wingless Hexapods The wingless hexapods, such as this springtail, have a simple life cycle. They hatch looking like miniature adults, then grow by successive molts of the cuticle.

nal mouthparts. Springtails can be extremely abundant (up to 200,000 per square meter) in soil, leaf litter, and on vegetation, and are the most abundant hexapods in the world in terms of number of individuals (rather than number of species).

Insects use nearly all species of plants and many species of animals as food. Herbivorous insects can consume massive amounts of plant matter. Gypsy moth caterpillars, for example, have been known to denude entire forests of their hardwood tree leaves, and the cyclical depredations wrought by locust hordes can be of biblical proportions. Many insects are predatory, feeding on small animals or on other insects. *Detritivorous* insects such as dung beetles are important in the recycling of chemicals through ecosystems. Some insects are internal parasites of plants or animals; others suck their host's blood or consume surface body tissues.

As recently as the 1980s, most biologists thought that about half of existing insect species had been described, but today they think that the approximately 1 million species of described insects are a small fraction of the total number of living species. Why did they change their minds?

A simple but important field study published in 1988 suggested that the number of existing insect species had been significantly underestimated. Knowing that the insects of tropical forests (the most species-rich habitat on Earth) were poorly known, entomologist Terry Erwin decided to make a comprehensive sample of beetles in the canopies of a single species of tropical forest tree, *Luehea seemannii*. Erwin fogged the canopies of 19 large *L. seemannii* trees with a pesticide and collected the insects that fell from the trees in collection nets (**Figure 32.28**). His sample contained about 1,000 species of beetles—many of them undescribed—from this one species of tree. He then used a set of assumptions to estimate the total number of insect species in tropical evergreen forests. Erwin's assumptions included estimates of the number of species of host trees in these forests, the proportion of beetles that specialized on a specific species of host tree, the proportion of beetles relative to other insect groups, and the proportion of beetles that live in trees versus leaf litter, among several other factors. From this and similar studies, Erwin estimated that there may be as many as 50 million species of insects on Earth! This estimate was surpris-

ing to many biologists, since it is about 50 times higher than the known species diversity of insects.

The assumptions Erwin used to make his calculations are open to debate, and Erwin did not intend for his calculations to provide more than a rough estimate of undiscovered biodiversity. If different assumptions are used, estimates of the number of species of insects change dramatically. For example, one of Erwin's assumptions involved an estimate of the proportion of insect species thought to be *specialists*—species that specialize on a particular host plant (in this case, *Luehea seemannii*). If the proportion of specialist species was overestimated, then the total number of species was likely overestimated; if the proportion of specialists was underestimated, the total number of species may have been underestimated. Erwin's extrapolations also assumed that his samples were representative of other forest trees, and that his sampling areas were representative of other tropical forests, among many other assumptions. But even if his estimates were off by several fold, Erwin's pioneering study highlighted the fact that we live on a poorly known planet, most of whose species have yet to be named and described.

Two groups of insects—the jumping bristletails and silverfish—are wingless and have simple life cycles, like the springtails and other close insect relatives. The remaining insects are the *pterygote insects*. Pterygotes have two pairs of wings, except in

32.28 Erwin's Research Project In an important study, entomologist Terry Erwin fogged tree canopies in a Panamanian rainforest with insecticide. Using the number of species represented among the fallen insects as a base, he extrapolated to estimate the total number of insect species on Earth.

(A) *Libellula semifasciata* (B) *Brachystola magna* (C) *Lygaeus kalmii* (D) *Limnephilus* sp. (larva) (E) *Eupholus magnificus* (F) *Argema mittrei* (G) *Phaenicia sericata* (H) *Vespula vulgaris*

32.29 Diverse Winged Insects (A) Unlike most flying insects, a dragonfly cannot fold its wings over its back. (B) The plains lubber, a grasshopper, is an orthopteran. (C) Milkweed bugs are true bugs (hemipterans). The plants they feed on (*Asclepias* spp.) are toxic to many species; its coloring serves as a signal to predators that the bug also toxic. (D–H) Holometabolous insects undergo complete metamorphosis. (D) A larval caddisfly (right) emerges from its dark pupal case. (E) Coleopterans (beetles) such as this New Guinean weevil account for more than half of all holometabolous species and are the largest insect group. (F) The comet moth of Madagascar is a lepidopteran. (G) Flies such as this green bottle fly are dipterans. (H) Many hymenopteran genera, such as these yellowjacket wasps, are social.

some groups where one or both pairs of wings have been secondarily lost. Secondarily wingless groups include the parasitic lice and fleas, some beetles, and the worker individuals in many ants.

Hatchling pterygote insects do not look like adults, and they undergo substantial changes at each molt. The immature stages of insects between molts are called **instars**. As Section 31.4 describes, a substantial change that occurs between one developmental stage and another is called **metamorphosis**. If the changes between its instars are gradual, an insect is said to have **incomplete metamorphosis**. If the change between at least some instars is dramatic, an insect is said to have **complete metamorphosis**.

The most familiar examples of species with complete metamorphosis are butterflies (see Figure 31.10). The wormlike butterfly larva, called a *caterpillar*, transforms itself during a specialized phase, the **pupa**, in which larval tissues are broken down and the adult form develops. In many insects with complete metamorphosis, the different life stages are specialized for dif-

ferent environments and use different food sources. In many species, the larvae are adapted for feeding and growing, whereas the adults are specialized for reproduction and dispersal.

Pterygote insects were the first animals in evolutionary history to achieve the ability to fly. Flight opened up many new lifestyles and feeding opportunities that only the insects could exploit, and it is almost certainly one of the reasons for the remarkable numbers of insect species and individuals, and for their unparalleled evolutionary success.

The adults of most flying insects have two pairs of stiff, membranous wings attached to the thorax. True flies, however, have one pair of wings and a pair of stabilizers called *haltares*. In winged beetles, one pair of wings—the forewings—forms heavy, hardened wing covers. Flying insects are important pollinators of flowering plants.

Two groups of pterygotes, the mayflies and dragonflies (**Figure 32.29A**), cannot fold their wings against their bodies. This is the ancestral condition for pterygote insects, and these two groups are not closely related to one another. Members of these groups have predatory or herbivorous aquatic larvae that transform into flying adults after they crawl out of the water. Dragonflies (and their relatives the damselflies) are active predators as adults. In contrast, adult mayflies lack functional digestive tracts. Mayflies live only about a day, just long enough to mate and lay eggs.

All other pterygote insects—the *neopterans*—can tuck their wings out of the way upon landing and crawl into crevices and other tight places. Some neopteran groups undergo incomplete metamorphosis, so hatchlings of these insects are sufficiently similar in form to adults to be recognizable. Examples include the grasshoppers (**Figure 32.29B**), roaches, mantids, stick insects, termites, stoneflies, earwigs, thrips, true bugs (**Figure 32.29C**), aphids, cicadas, and leafhoppers. They acquire adult organ systems, such as wings and compound eyes, gradually through several juvenile instars.

More than 80 percent of all insects belong to a subgroup of the neopterans called the *holometabolous* insects (see Table 32.2),

which undergo complete metamorphosis (**Figure 32.29D**). The many species of beetles account for almost half of this group (**Figure 32.29E**). Also included are lacewings and their relatives; caddisflies; butterflies and moths (**Figure 32.29F**); sawflies; true flies (**Figure 32.29G**); and bees, wasps, and ants, some species of which display unique and highly specialized social behaviors (**Figure 32.29H**).

Molecular data suggest that insects began to diversify about 450 million years ago, about the time of the appearance of the first land plants. These early hexapods evolved in a terrestrial environment that lacked any other similar organisms, which in part accounts for their remarkable success. But the success of the insects is also due to their wings. Homologous genes control the development of insect wings and crustacean appendages, suggesting that the insect wing evolved from a dorsal branch of a crustacean-like limb (**Figure 32.30**). The dorsal limb branch of crustaceans is used for gas exchange. Thus the insect wing probably evolved from a gill-like structure that had a gas exchange function.

32.4 RECAP

All arthropods have segmented bodies. Muscles in each segment operate that segment and the appendages attached to it. These jointed, specialized appendages permit complex patterns of movement, including, in insects, the ability to fly. With flight, insects took advantage of new feeding and lifestyle opportunities, which contributed to the unparalleled evolutionary success of this group.

- What features have contributed to making arthropods among the most abundant animals on Earth, both in number of species and number of individuals? See p. 681

- Describe the difference between incomplete and complete metamorphosis. See p. 688

- Can you think of some possible reasons why there are so many species of insects? See pp. 688–689

32.30 The Origin of Insect Wings? Insect wings may have evolved from an ancestral appendage similar to that of modern crustaceans. (A) A diagram of the ancestral, multibranched arthropod limb. (B, C) The *pdm* gene, a Hox gene, is expressed throughout the dorsal limb branch and walking leg of the thoracic limb of a crayfish (B) and in the wings and legs of *Drosophila* (C).

Development of appendages in the crayfish is governed by the *pdm* gene.

Development of the insect wing is governed by the expression of the same gene.

(A) Ancestral multibranched appendage

(B) Modern crayfish

(C) *Drosophila*

Dorsal branches

■ *pdm* gene product

The protostomes encompass the majority of Earth's animal species, so it is not surprising that the different protostome groups display a huge variety of different characteristics.

An Overview of Protostome Evolution

The myriad species of protostomes encompass a staggering number of different body forms and life styles. The following aspects of protostome evolution have contributed to this enormous diversity:

- The evolution of *segmentation* permitted some groups of protostomes to move different parts of the body independently of one another. Species in some groups gradually evolved the ability to move rapidly over and through the substratum, through water, and through air.

- *Complex life cycles* with dramatic changes in form between one stage and another allow individuals of different stages to specialize on different resources.

- *Parasitism* has evolved repeatedly, and many protostome groups parasitize plants and animals.

- The evolution of *diverse feeding structures* allowed protostomes to specialize on many different food sources. Specialization on food sources undoubtedly contributed to species isolation and further diversification.

- Predation was a major selection pressure favoring the development of *hard external body coverings* (exoskeletons and

shells). Such coverings evolved independently in many lophotrochozoan and ecdysozoan groups. In addition to providing protection, these coverings became key elements in the development of new systems of locomotion.

- *Better locomotion* permitted prey to escape from predators, but also allowed predators to pursue their prey more effectively. Thus the evolution of animals has been, and continues to be, a complex "arms race" among predators and prey.

Many major evolutionary trends among the protostomes are shared by the deuterostomes, which includes the chordates, the group to which humans belong. We turn to the deuterostomes in the next chapter.

CHAPTER SUMMARY

32.1 What Is a Protostome?

- Protostomes ("mouth first") are bilaterally symmetrical animals with an anterior brain that surrounds the entrance to the digestive tract and a ventral nervous system. The blastopore of protostomes (when present in development) develops into a mouth. Protostomes comprise two major clades, the lophotrochozoans and the ecdysozoans. Review Figure 32.1, WEB ACTIVITIES 32.1 and 32.2

- **Lophotrochozoans** include a wide diversity of animals. Within this group evolved **lophophores** (a complex organ for both food collection and gas exchange), free-living **trochophore** larvae, and spiral cleavage. Some of these features were subsequently lost in some lineages.

- **Ecdysozoans** have a body covering known as the **cuticle**, which they must **molt** in order to grow. Some ecdysozoans have a relatively thin cuticle. Others, especially the **arthropods**, have a rigid cuticle reinforced with **chitin**. This rigid cuticle functions as an **exoskeleton**. New mechanisms of locomotion and gas exchange evolved among the arthropods. Review Figure 32.4

32.2 What Features Distinguish the Major Groups of Lophotrochozoans?

- Lophotrochozoans range from animals having only one entrance to the digestive tract and no oxygen transport system to animals with complete digestive tracts and complex internal transport systems.

- Lophophores, wormlike body forms, and external shells are each found in many unrelated groups of lophotrochozoans.

- The most species-rich groups of lophotrochozoans are the **flatworms**, **annelids**, and **mollusks**.

- Annelids are a diverse group of segmented worms that live in moist terrestrial and aquatic environments. Review Figure 32.12

- Mollusks underwent a dramatic evolutionary radiation based on a body plan consisting of three major components: a foot, a mantle, and a visceral mass. The four major living molluscan clades—**chitons**, **bivalves**, **gastropods**, and **cephalopods**—demonstrate the diversity that evolved from this three-part body plan. Review Figure 32.14

32.3 What Features Distinguish the Major Groups of Ecdysozoans?

- Many ecdysozoan groups are wormlike in form. Members of several species-poor groups of wormlike marine ecdysozoans—**priapulids**, **kinorhynchs**, and **loriciferans**—have thin cuticles.

- **Horsehair worms** are extremely thin; many are internal parasites as larvae.

- **Nematodes**, or roundworms, have a thick, multilayered cuticle. Nematodes are among the most abundant and universally distributed of all animal groups. Review Figure 32.17

- One major ecdysozoan clade, the arthropods, has evolved jointed, paired appendages that have a wide diversity of functions. Collectively, arthropods are the dominant animals on Earth in number of described species, and among the most abundant in number of individuals.

32.4 Why Are Arthropods So Diverse?

- Encasement within a rigid exoskeleton provides arthropods with support for walking as well as some protection from predators. The waterproofing provided by chitin keeps arthropods from dehydrating in dry air.

- Jointed appendages permit complex movement patterns. Each arthropod segment has muscles attached to the inside of the exoskeleton that operate that segment and the appendages attached to it.

- Two groups of arthropod relatives, the **onychophorans** and the **tardigrades**, have simple, unjointed appendages. **Trilobites** were early marine arthropods that disappeared in the Permian extinction.

- The bodies of **myriapods** have only two regions, a head, and a long trunk with many segments, each of which carries appendages. **Chelicerates** also have a two-part body; most chelicerates have four pairs of walking legs.

- **Crustaceans** are the dominant marine arthropods, and are also found in many freshwater and some terrestrial environments. Their segmented bodies are divided into three regions (head, thorax, and abdomen) with different, specialized appendages in each region. Review Figure 32.25

- Hexapods—**insects** and their relatives—are the dominant terrestrial arthropods. They have the same three body regions as crustaceans, but no appendages form in their abdominal segments. Wings and the ability to fly first evolved among the insects, allowing them to exploit new lifestyles. Review Figure 32.26

SELF-QUIZ

1. Members of which groups have lophophores?
   a. Phoronids, brachiopods, and nematodes
   b. Phoronids, brachiopods, and bryozoans
   c. Brachiopods, bryozoans, and flatworms
   d. Phoronids, rotifers, and bryozoans
   e. Rotifers, bryozoans, and brachiopods

2. Which of the following is *not* part of the molluscan body plan?
   a. Mantle
   b. Foot
   c. Radula
   d. Visceral mass
   e. Jointed skeleton

3. Nautiluses control their buoyancy by
   a. adjusting salt concentrations in their blood.
   b. forcibly expelling water from the mantle.
   c. pumping water and gases in and out of internal chambers.
   d. using the complex sensory organs in their heads.
   e. swimming rapidly.

4. The outer covering of ecdysozoans
   a. is always hard and rigid.
   b. is always thin and flexible.
   c. is hard and rigid in larvae but thin in adults.
   d. ranges from very thin to hard and rigid depending on the species.
   e. grows throughout life to accommodate a growing body.

5. Nematodes are abundant and diverse because
   a. they are both parasitic and free-living and eat a wide variety of foods.
   b. they are able to molt their exoskeleton.
   c. their thick cuticle enables them to move in complex ways.
   d. their body cavity is a pseudocoelom.
   e. their segmented body enables them to live in many different places.

6. The arthropod exoskeleton is composed of a
   a. mixture of several kinds of polysaccharides.
   b. mixture of several kinds of proteins.
   c. single complex polysaccharide called chitin.
   d. single complex protein called arthropodin.
   e. mixture of layers of proteins and a polysaccharide called chitin.

7. Which groups are arthropod relatives with unjointed legs?
   a. Trilobites and onychophorans
   b. Onychophorans and tardigrades
   c. Trilobites and tardigrades
   d. Onychophorans and chelicerates
   e. Tardigrades and chelicerates

8. The body plan of insects is composed of which of the three following regions?
   a. Head, abdomen, and trachea
   b. Head, abdomen, and cephalothorax
   c. Cephalothorax, abdomen, and trachea
   d. Head, thorax, and abdomen
   e. Abdomen, trachea, and mantle

9. Insects whose hatchlings are sufficiently similar in form to adults to be recognizable are said to have
   a. instars.
   b. neopterous development.
   c. accelerated development.
   d. incomplete metamorphosis.
   e. complete metamorphosis.

10. Factors that may have contributed to the remarkable evolutionary success of insects include
    a. the lack of any other similar organisms in the terrestrial environments colonized by insects.
    b. the ability to fly.
    c. complete metamorphosis.
    d. a new mechanism for delivering oxygen to their internal tissues.
    e. all of the above

FOR DISCUSSION

1. Segmentation has either arisen several times during animal evolution, or else it arose early in animal evolution and was subsequently lost multiple times. What advantages does segmentation provide? Given these advantages, why might some animals have lost their segmentation?

2. Major structural novelties have arisen only infrequently during the course of evolution. Which of the features of protostomes do you think are major evolutionary novelties?

Can you think of morphological features that may have led to major evolutionary radiations?

3. There are more described and named species of insects than of all other species on Earth combined. However, only a very few insect species live in marine environments, and those species are restricted to the intertidal zone or the ocean surface. What factors may have contributed to this lack of success by the insects in the oceans?

ADDITIONAL INVESTIGATION

If you were given funding to carry out studies to improve our understanding of Earth's biodiversity, how would you spend it? How would your investigation differ if you targeted particular major groups of animals, such as insects or nematodes? Which animal groups do you think should receive highest priority for such studies, and why?

WORKING WITH DATA (GO TO yourBioPortal.com)

Estimating the Number of Species of Insects In this exercise based on the study illustrated in Figure 32.28, you will use the data collected by Terry Erwin to estimate the number of insects in a hectare (10,000 m^2) of Panamanian rainforest, as well as to estimate the number of insect species found on Earth. You will also examine Erwin's assumptions and consider how they might be modified and tested.

33 Deuterostome Animals

Good parents, or cannibals?

Why would a frog swallow its own offspring? That may not sound like a good parenting skill, but female gastric brooding frogs of Australia shut down their digestive system to brood their tadpoles in their stomach. This provides the tadpoles with a safe haven from predators, making it much more likely that they will survive to metamorphosis. That was the case, at least, until gastric brooding frogs went extinct in the early 1980s, after humans introduced pathogenic fungi into their native range.

Not all frogs are as involved in raising their young as were gastric brooding frogs, however. Bullfrogs, for example, lay thousands of eggs each year and provide no parental care for their offspring. The eggs are fertilized by a male bullfrog and are then left to develop on their own. The eggs hatch into tadpoles and transform into tiny frogs—if they aren't eaten first; most of these offspring don't survive. In the water, the tadpoles are eaten by many species of fishes, turtles, birds, snakes, aquatic insects, and other predators. As the young tadpoles transform into frogs, they are likely to be eaten by predators hunting at the margins of the pond. Out of the tens of thousands of tadpoles an adult bullfrog produces in its lifetime, an average of only two offspring are expected to survive to reproduce themselves.

In many species of frogs, complex behaviors associated with parental care change these long odds. Rather than producing huge numbers of offspring, each with a minimal chance of surviving, some frogs invest more energy in each offspring and care for the young as they grow. This increases the chances that any one offspring will survive and reproduce—but it also means far fewer offspring can be produced, because of the greater parental investment per offspring.

Strategies for parental care include many behaviors in addition to gastric brooding. The females of many species guard a terrestrial egg clutch until it hatches. Other females carry their developing embryos or tadpoles around with them on their backs, or even in special brood pouches (as in the marsupial frogs of South America). Parental care often involves males as well. Males of many frog species guard egg masses or carry young, sometimes in unusual ways. In Darwin's frog

Eating One's Offspring? Young frogs emerge from the digestive tract of a female *Rheobatrachus silus*, one of the now-extinct gastric brooding frogs of Australia. In this unique form of gestation, eggs hatch and tadpoles develop within the protected environment of the mother's stomach.

Riding Piggyback The young of this female *Flectonotus pygmaeus* are developing within a pouch on her back. The marsupial frogs of South America take their name from another vertebrate group—marsupial mammals such as kangaroos—whose females also protect their developing young in pouches.

of southern South America, for example, the tadpoles develop within the male's vocal sacs.

Parental care can extend beyond protection to feeding the young. Some female dart-poison frogs of the tropical Americas carry each individual tadpole to one of the tiny pools of water that collect in bromeliad plants growing on the trees. The female then returns to each bromeliad "pond" and lays unfertilized eggs as food for the single tadpole developing there.

Frogs and other vertebrates constitute one of the major groups of deuterostome animals. Although there are far fewer species of deuterostomes than of protostomes, deuterostomes are of particular interest to biologists because of their importance in many ecosystems, their often complex behaviors, and their widespread use as models in developmental biology and genetics.

IN THIS CHAPTER we will introduce the deuterostomes and describe the principal animal groups: the echinoderms, hemichordates, and chordates. We will then discuss the evolution of the vertebrates within the chordates, including the features that allowed vertebrates to colonize most habitats on Earth. We will look especially closely at the primate lineage, which includes our own species.

33.1 What Is a Deuterostome?

It may surprise you to learn that both you and a sea urchin are deuterostomes. Adult sea stars, sea urchins, and sea cucumbers—the most familiar echinoderms—look so different from adult vertebrates (fishes, frogs, lizards, birds, and mammals) that it may be difficult to believe all these animals are closely related. The evidence that all deuterostomes share a common ancestor that is not shared with the protostomes includes common early developmental patterns and phylogenetic analysis of gene sequences, factors that are not apparent in the forms of the adult animals.

Historically, the deuterostomes were characterized by three early developmental patterns:

- Radial cleavage
- Formation of the mouth at the opposite end of the embryo from the blastopore (the pattern that gives the deuterostomes their name); in deuterostomes, the blastopore develops into the anus (see Figure 31.2)
- Development of a coelom from mesodermal pockets that bud off from the cavity of the gastrula rather than by splitting of the mesoderm, as occurs among protostomes

Radial cleavage is not exclusive to deuterostomes, and as noted in Chapter 31, it is now thought to be the ancestral condition for bilaterians. In fact, some of the groups now known to be protostomes were once thought to be deuterostomes because they have developmental patterns similar to those of echinoderms and chordates. The development of the blastopore into an anus may also be the ancestral condition for bilaterians, rather than a derived feature of deuterostomes. Today, the strongest support for the shared evolutionary relationships of echinoderms, hemichordates, and chordates (the groups that now compose the deuterostomes) comes from phylogenetic analyses of DNA sequences of many different genes.

Although there are far fewer species of deuterostomes than of protostomes (see Table 31.1), we have a special interest in deuterostomes, in part because we are members of that clade. The deuterostomes are also of interest because they include many large animals that strongly influence the characteristics of ecosystems. Many deuterostome species have been intensively studied in all fields of biology. Complex behaviors, such as the parenting behaviors described in the opening of this chapter, are especially well developed among some deuterostomes.

33.1 Phylogeny of the Deuterostomes The three principal groups of deuterostomes are the hemichordates, the echinoderms, and the chordates (cephalochordates, urochordates, and vertebrates). The echinoderms and the vertebrates contain most of the described species.

yourBioPortal.com
GO TO **Web Activity 33.1 • Deuterostome Phylogeny**

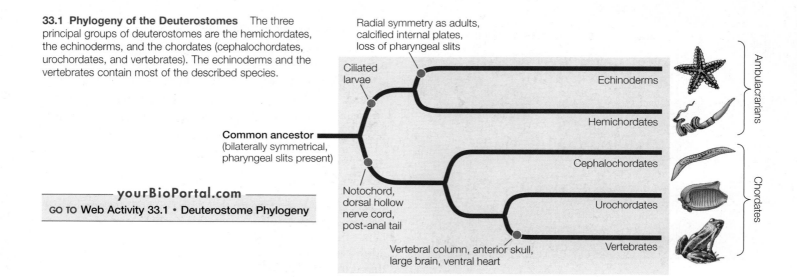

Radial symmetry as adults, calcified internal plates, loss of pharyngeal slits

Ciliated larvae

Echinoderms

Hemichordates

Ambulacrarians

Common ancestor (bilaterally symmetrical, pharyngeal slits present)

Notochord, dorsal hollow nerve cord, post-anal tail

Cephalochordates

Urochordates

Vertebral column, anterior skull, large brain, ventral heart

Vertebrates

Chordates

Living deuterostomes comprise three distinct clades (**Figure 33.1**):

- *Echinoderms*: sea stars, sea urchins, and their relatives
- *Hemichordates*: acorn worms and pterobranchs
- *Chordates*: sea squirts, lancelets, and vertebrates

All deuterostomes are triploblastic, coelomate animals (see Figure 31.4C). Skeletal support features, where present, are internal rather than external. Some species have segmented bodies, but the segments are less obvious than those of annelids and arthropods.

Scientists are learning much about the ancestors of modern deuterostomes from fossils recently discovered in 520-million-year-old rocks in China. Some of these early deuterostomes had a skeleton similar to that of a modern echinoderm, but they had pharyngeal slits and bilateral symmetry. Another group of early deuterostomes, the *yunnanozoans*, were discovered in China's Yunnan Province. These well-preserved animals had a large mouth, six pairs of external gills, and a segmented posterior body section bearing a light cuticle (**Figure 33.2**).

The features of these fossil animals support the findings from phylogenetic analyses of living species in showing that the earliest deuterostomes were bilaterally symmetrical, segmented animals with a pharynx that had slits through which water flowed. Echinoderms evolved their adult forms with unique symmetry (in which the body parts are arranged along five radial axes) much later, whereas other deuterostomes retained the ancestral bilateral symmetry.

Yunnanozoon lividum

Mouth   Esophagus   External gills   Segments

33.2 Ancestral Deuterostomes Had External Gills The extinct yunnanozoans may be ancestral to all deuterostomes. This fossil, which dates from the Cambrian, shows the six pairs of external gills and segmented posterior body that characterized these animals.

33.1 RECAP

The deuterostomes include the echinoderms, hemichordates, and chordates. The common ancestry of these groups is supported by developmental similarities and by phylogenetic analyses of DNA sequences.

- What are three developmental patterns the earliest deuterostomes had in common? See p. 693
- Why is radial cleavage no longer considered to be evidence for the monophyly of deuterostomes? See p. 693

33.2 What Are the Major Groups of Echinoderms and Hemichordates?

About 13,000 species of echinoderms in 23 major groups have been described from their fossil remains. They are probably only a small fraction of those that actually lived. Only 6 of the 23 major groups known from fossils are represented by species that survive today; many clades became extinct during the periodic mass extinctions that have occurred throughout Earth's history. Nearly all of the 7,000 extant species of echinoderms live only in marine environments. There are far fewer species of living hemichordates, with only about 100 known species.

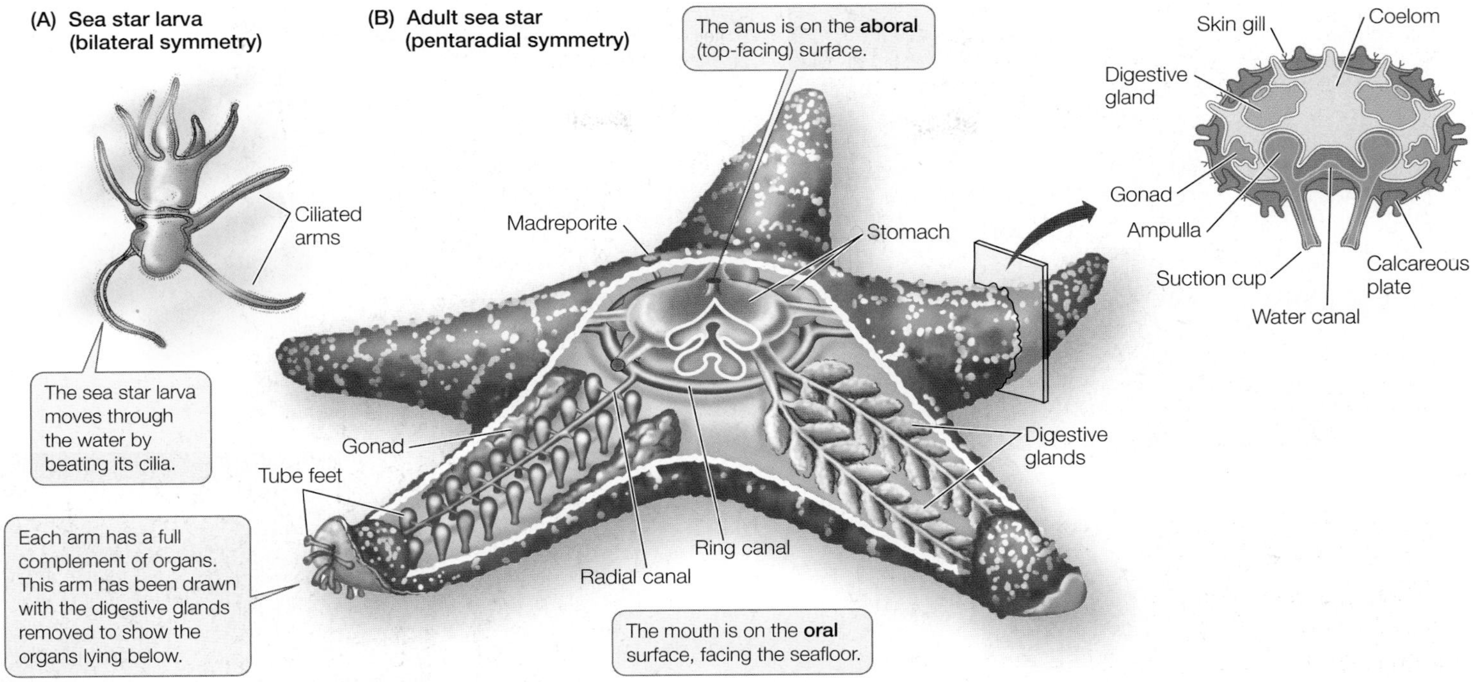

(A) Sea star larva (bilateral symmetry)

Ciliated arms

The sea star larva moves through the water by beating its cilia.

Each arm has a full complement of organs. This arm has been drawn with the digestive glands removed to show the organs lying below.

(B) Adult sea star (pentaradial symmetry)

The anus is on the **aboral** (top-facing) surface.

Madreporite

Stomach

Gonad

Tube feet

Radial canal

Ring canal

Digestive glands

The mouth is on the **oral** surface, facing the seafloor.

Skin gill

Coelom

Digestive gland

Gonad

Ampulla

Suction cup

Water canal

Calcareous plate

33.3 Echinoderms Are Bilaterally Symmetrical as Larvae but Radially Symmetrical as Adults (A) The ciliated larva of a sea star has bilateral symmetry. Hemichordates have a similar larval form. (B) The radially symmetrical adult sea star displays the canals and tube feet of the echinoderm water vascular system, as well as the calcified internal skeleton. The body's orientation is oral–aboral rather than anterior–posterior.

The echinoderms and hemichordates (together known as *ambulacrarians*) have a bilaterally symmetrical, ciliated larva (**Figure 33.3A**). Adult hemichordates also are bilaterally symmetrical. Echinoderms, however, undergo a radical change in form as they develop into adults (**Figure 33.3B**), changing from a bilaterally symmetrical larva to an adult with **pentaradial symmetry** (symmetry in five or multiples of five). As is typical of animals with radial symmetry, echinoderms have no head, and they move slowly and equally well in many directions. Rather than having an anterior–posterior (head–tail) and dorsal–ventral (back–belly) body organization, echinoderms have an *oral* side (containing the mouth) and an opposite *aboral* side (containing the anus).

Echinoderms have unique structural features

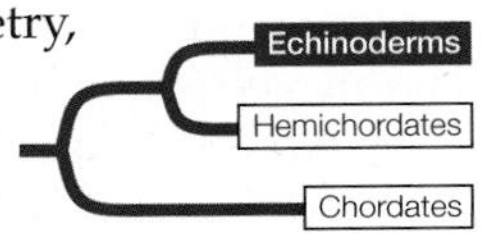

In addition to having pentaradial symmetry, adult echinoderms have two unique structural features. One is a system of calcified internal plates covered by thin layers of skin and some muscles. The calcified plates of most echinoderms are thick, and they fuse inside the entire body, forming an *internal skeleton*. The other unique feature is a **water vascular system**, a network of water-filled canals leading to extensions called **tube feet**. This system functions in gas exchange, locomotion, and feeding (see Figure 33.3B). Seawater

Echinoderms

Hemichordates

Chordates

enters the system through a perforated structure called a *madreporite*. A calcified canal leads from the madreporite to the *ring canal*, which surrounds the *esophagus* (the tube leading from the mouth to the stomach). *Radial canals* branch off from the ring canal, extending through the arms (in species that have arms) and connecting with the tube feet. Echinoderms use their tube feet in many different ways to move and to capture prey. These structural innovations have been modified in many ways, resulting in a striking array of very different animals.

Members of one major extant echinoderm clade, the *crinoids* (sea lilies and feather stars), were more abundant and species-rich 300 to 500 million years ago than they are today. There are some 80 described living sea lily species, most of which are sessile organisms attached to the substratum by a stalk. Feather stars (**Figure 33.4A**) grasp the substratum with flexible appendages that allow for limited movement. About 600 living species of feather stars have been described.

Unlike the mostly sessile crinoids, most surviving echinoderms are motile. The two main groups of motile echinoderms are the *echinozoans* (sea urchins and sea cucumbers) and *asterozoans* (sea stars and brittle stars). Sea urchins are hemispherical in shape and lack arms (**Figure 33.4B**). They are covered with spines that are attached to the underlying skeleton with ball-and-socket joints. These joints enable the spines to be moved so they can converge toward a point that has been touched. The spines vary in size and shape and can be used for locomotion; a few produce toxic substances. They provide effective protection for the urchin, as many a scuba diver has found out the hard way. Sand dollars are flattened, disc-shaped relatives of sea urchins.

Sea cucumbers also lack arms, and their bodies are oriented in an atypical manner for an echinoderm (**Figure 33.4C**). The mouth is anterior and the anus is posterior (front and rear), not oral and aboral (top and bottom) as in other echinoderms.

(A) *Oxycomanthus bennetti*

(B) *Sphaerechinus granularis*

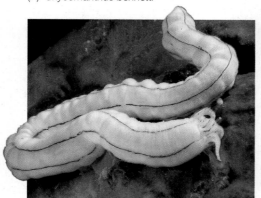

(C) *Synaptula* sp.

(D) *Asterias rubens*

(E) *Ophiothrix spiculata*

33.4 Echinoderm Diversity (A) The flexible arms of this golden feather star are clearly visible. (B) Sea urchins are important grazers on algae in the intertidal zones of the world's oceans. (C) Sea cucumbers are unique among echinderms in having an anterior–posterior rather than an oral–aboral orientation of the mouth and anus. (D) Sea stars are important predators on bivalve mollusks such as mussels and clams. Suction tips on its tube feet allow a sea star to grasp both shells of the bivalve and pull them open. (E) The arms of the brittle star are composed of hard but jointed plates.

Sea cucumbers use most of their tube feet primarily for attaching to the substratum rather than for moving.

Sea stars, popularly called starfish, are the most familiar echinoderms (**Figure 33.4D**). Their gonads and digestive organs are located in the arms, as seen in Figure 33.3B. Their tube feet serve as organs of locomotion, gas exchange, and attachment. Each tube foot of a sea star consists of an internal *ampulla* connected by a muscular tube to an external suction cup that can stick to the substratum. The tube foot is moved by expansion and contraction of the circular and longitudinal muscles of the tube. Brittle stars are similar in structure to sea stars, but their flexible arms are composed of jointed, hard plates (**Figure 33.4E**).

Echinoderms use their tube feet in a great variety of ways to capture prey. Sea lilies, for example, feed by orienting their arms in passing water currents. Food particles then strike and stick to the tube feet, which are covered with mucus-secreting

glands. The tube feet transfer these particles to grooves in the arms, where ciliary action carries the food to the mouth. Sea cucumbers capture food with their anterior tube feet, which are modified into large, feathery, sticky tentacles that can be protruded from the mouth. Periodically, a sea cucumber withdraws the tentacles, wipes off the material that has adhered to them, and digests it.

Many sea stars use their tube feet to capture large prey such as polychaetes, gastropod and bivalve mollusks, small crustaceans such as crabs, and fishes. With hundreds of tube feet acting simultaneously, a sea star can grasp a bivalve in its arms, anchor the arms with its tube feet, and by steady contraction of the muscles in its arms, gradually exhaust the muscles the bivalve uses to keep its shell closed (see Figure 33.4D). To feed on a bivalve, a sea star can push its stomach out through its mouth and then through the narrow space between the two halves of the bivalve's shell. The sea star's stomach then secretes enzymes that digest the prey.

Most sea urchins eat algae, which they catch with their tube feet from the plankton or scrape from rocks with a complex rasping structure. Most of the 2,000 species of brittle stars ingest particles from the upper layers of sediments and assimilate the organic material from them, although some species filter suspended food particles from the water and others capture small animals.

Hemichordates are wormlike marine deuterostomes

Hemichordates—acorn worms and pterobranchs—have a body organized in three major parts, consisting of a *proboscis*, a *collar* (which bears the mouth), and a *trunk* (which contains the other body parts). The 70 known species of *acorn worms* range up to 2 meters in length (**Figure 33.5A**). They live in burrows in muddy and sandy marine sediments. The digestive tract of an acorn worm consists of a mouth behind which are a muscular *pharynx* and an *intestine*. The pharynx opens to the outside through a number of *pharyngeal slits* through which water can exit. Highly vascularized tissue surrounding the pharyngeal slits serves as a gas-exchange apparatus. Acorn worms breathe by pumping water into the mouth and out through the pharyngeal slits. They capture prey with the large proboscis, which is coated with sticky mucus to which small organisms in the sediment stick.

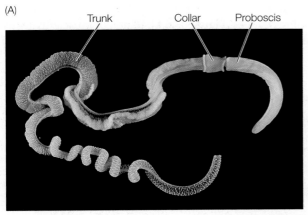

(A)

Trunk   Collar   Proboscis

Saccoglossus kowalevskii

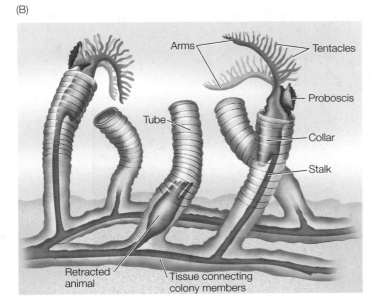

(B)

Arms — Tentacles
— Proboscis
Tube — Collar
— Stalk
Retracted animal — Tissue connecting colony members

33.5 Hemichordates (A) The proboscis of an acorn worm is modified for burrowing. (B) The structure of a colonial pterobranch.

The mucus and its attached prey are conveyed by cilia to the mouth. In the esophagus, the food-laden mucus is compacted into a ropelike mass that is moved through the digestive tract by ciliary action.

The 30 living species of *pterobranchs* are sedentary marine animals up to 12 millimeters long that live in a tube secreted by the proboscis. Some species are solitary; others form colonies of individuals joined together (**Figure 33.5B**). Behind the proboscis is a collar with anywhere from one to nine pairs of arms. The arms bear long tentacles that capture prey and function in gas exchange.

33.2 RECAP

Echinoderms have an internal skeleton of calcified plates and a unique water vascular system. Hemichordates have a bilaterally symmetrical body divided into three parts: proboscis, collar, and trunk.

- What are some of the ways that echinoderms use their tube feet to obtain food? **See p. 696**
- How do hemichordates obtain food? **See p. 697**

We have described the deuterostome groups that are most distantly related to us. Now we turn our attention to the chordates, the clade to which humans belong.

33.3 What New Features Evolved in the Chordates?

It is not obvious from examining adult animals that echinoderms and chordates share a common ancestor. The evolutionary relationships among some chordate groups are not immediately apparent, either. The features that reveal the evolutionary relationships both among the chordates and between chordates and echinoderms are seen primarily in the larvae—in other words, it is during the early developmental stages that their evolutionary relationships are evident.

There are three principal chordate clades: the **cephalochordates**, the **urochordates**, and the **vertebrates** (see Figure 33.1). There are about 3,000 living species of urochordates and 62,000 living species of vertebrates, but only about 30 living species of cephalochordates.

Adult chordates vary greatly in form, but all chordates display the following derived structures at some stage in their development (**Figure 33.6**):

- A dorsal hollow nerve cord
- A tail that extends beyond the anus
- A dorsal supporting rod called the notochord

The **notochord** is the most distinctive derived chordate trait. It is composed of a core of large cells with turgid fluid-filled vacuoles, which make it rigid but flexible. In the urochordates the notochord is lost during metamorphosis to the adult stage. In most vertebrate species, it is replaced during development by skeletal structures that provide support for the body.

(A)

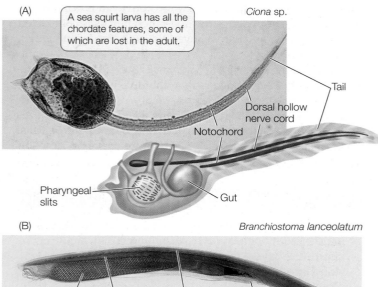

Ciona sp.

A sea squirt larva has all the chordate features, some of which are lost in the adult.

Tail

Dorsal hollow nerve cord

Notochord

Pharyngeal slits

Gut

(B)

Branchiostoma lanceolatum

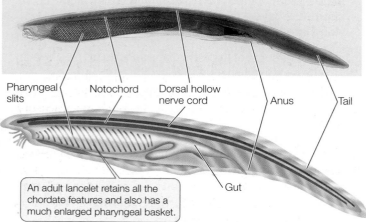

Pharyngeal slits

Notochord

Dorsal hollow nerve cord

Anus

Tail

Gut

An adult lancelet retains all the chordate features and also has a much enlarged pharyngeal basket.

33.6 The Key Features of Chordates Are Most Apparent in Early Developmental Stages The pharyngeal slits of both the urochordate sea squirt and the cephalochordate lancelet develop into a pharyngeal basket. (A) The sea squirt larva (but not the adult) has all three chordate features: dorsal hollow nerve cord, postanal tail, and a notochord. (B) All three chordate synapomorphies are retained in the adult lancelet.

The ancestral pharyngeal slits (not a derived feature of this group) are present at some developmental stage of chordates but are often lost in adults. The *pharynx*, which develops around the pharyngeal slits, functioned in chordate ancestors as the site for oxygen uptake and the elimination of carbon dioxide and water (as in acorn worms). The pharynx is much enlarged in some chordate species (as in the *pharyngeal basket* of the lancelet in Figure 33.6B) but has been lost in others.

Adults of most cephalochordates and urochordates are sessile

The 30 species of cephalochordates, or *lancelets*, are small animals that rarely exceed 5 centimeters in length. The notochord, which provides body support, extends the entire length of the body throughout their lives (see Figure 33.6B). Lancelets are found in shallow marine

Echinoderms

Hemichordates

Cephalochordates

Urochordates

Vertebrates

and brackish waters worldwide. Most of the time they lie covered in sand with their head protruding above the sediment, but they can swim. They filter prey from the water with their pharyngeal basket. During the reproductive season, the gonads of the males and females greatly enlarge. At spawning, the walls of the gonads rupture, releasing eggs and sperm into the water column, where fertilization takes place.

All members of the three major urochordate groups—the ascidians, thaliaceans, and larvaceans—are marine animals. More than 90 percent of the known species of urochordates are *ascidians* (sea squirts). Individual ascidians range in length from less than 1 millimeter to 60 centimeters. Some ascidians form colonies by asexual budding from a single founder. Colonies may measure several meters across. The baglike body of an adult ascidian is enclosed in a tough tunic, leading to its alternate name of "tunicate" (**Figure 33.7A**). The tunic is composed of proteins and a complex polysaccharide secreted by epidermal cells. The ascidian pharynx is enlarged into a pharyngeal basket that filters prey from the water passing through it.

In addition to its pharyngeal slits, an ascidian larva has a dorsal hollow nerve cord and a notochord that is restricted mostly to the tail region (see Figure 33.6A). Bands of muscle that surround the notochord provide support for the body. After a short

(A) *Clavelina dellavallei*   1 cm

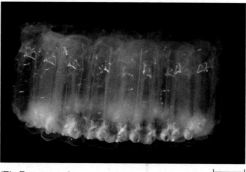

(B) *Pegea socia*   1 cm

33.7 Adult Urochordates (A) The transparent tunic and the pharyngeal basket are clearly visible in this ascidian (sea squirt, also known as a tunicate). (B) A chainlike colony of thaliaceans (salps) floats in tropical waters.

time swimming in the plankton, the larvae of most species settle on the seafloor and transform into sessile adults. The swimming, tadpolelike larvae suggest a close evolutionary relationship between ascidians and vertebrates (see Figure 22.6).

Thaliaceans (salps) can live singly or in chainlike colonies up to several meters long (**Figure 33.7B**). They float in tropical and subtropical oceans at all depths down to 1,500 meters. *Larvaceans* are solitary planktonic animals that retain their notochord and dorsal hollow nerve cord throughout their lives. Most larvaceans are less than 5 millimeters long, but some species that live near the bottom of deep ocean waters build delicate casings of mucus that may be more than a meter wide. They snare sinking organic particles (their primary food source) with elaborate filters built into their mucus "houses." When the old "house" gets clogged with excess debris, the animals build a new one.

A dorsal supporting structure replaces the notochord in vertebrates

In one chordate group, a new dorsal supporting structure evolved. The **vertebrates** take their name from the jointed, dorsal **vertebral column** that replaces the notochord during early development as their primary supporting structure (**Figure 33.8**).

> Echinoderms
> Hemichordates
> Cephalochordates
> Urochordates
> **Vertebrates**

All of the non-vertebrate deuterostomes (the hemichordates, echinoderms, cephalochordates, and urochordates) are exclusively or primarily marine. The lineage that led to the vertebrates is also thought to have evolved in the oceans, although probably in an estuarine environment (where fresh water meets salt water). Vertebrates have since radiated into marine, freshwater, terrestrial, and aerial environments worldwide.

The *hagfishes* are thought by many biologists to be the sister group to the remaining vertebrates (as shown in Figure 33.8). Hagfishes (**Figure 33.9A**) have a weak circulatory system, with three small accessory hearts (rather than a single, large heart), a partial *cranium* or skull (containing no *cerebrum* or *cerebellum*, two main regions of the brain of other vertebrates), and no jaws or stomach. They also lack separate, jointed vertebrae and have a skeleton composed of cartilage. Thus, some biologists do not consider hagfishes vertebrates, and use instead the term *craniates* to refer collectively to the hagfishes and the vertebrates. Some analyses of gene sequences suggest, however, that hagfishes may be more closely related to the vertebrate lampreys (**Figure 33.9B**); in this phylogenetic arrangement, the two groups are collectively called the *cyclostomes* ("circle mouths"). If in fact the hagfishes and lampreys do form a monophyletic group, then hagfishes must have secondarily lost many of the major vertebrate morphological features during their evolution.

The 58 known species of hagfishes are unusual marine animals that produce copious quantities of slime as a defense. They are virtually blind and rely largely on the four pairs of sensory tentacles around their mouth to detect food. Although they have no jaws, hagfishes have a tonguelike structure equipped with toothlike rasps that they can use to tear apart dead organisms and to capture their principal prey, polychaete worms. Hagfishes have direct development (no larvae), and individuals may actually change sex from year to year (from male to female and vice versa).

The nearly 50 species of lampreys either live in fresh water, or they live in coastal salt water and move into fresh water to breed. Although the lampreys and hagfishes may look super-

33.8 Phylogeny of the Living Vertebrates This phylogenetic tree shows the evolution of some of the key innovations among the major groups of vertebrates.

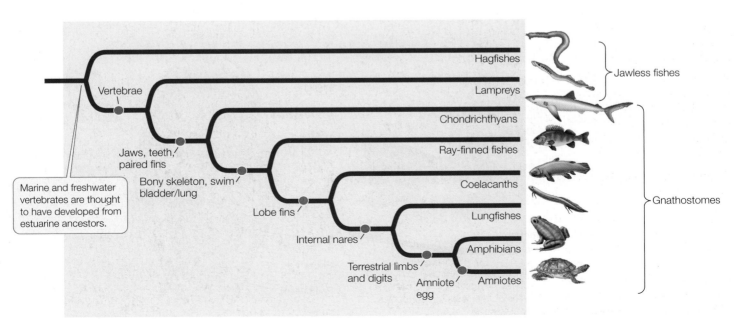

33.9 Modern Jawless Fishes   (A) Hagfishes burrow in the ocean mud, from which they extract small prey. They also scavenge on dead or dying fish. Hagfishes have degenerate eyes, which has led to their being miscalled "blind eels." (B) Sea lampreys are ectoparasites that attach to the bodies of living fish and use their large, jawless mouths to suck blood and flesh. They can survive in both fresh and salt water, as this individual attached to a salmon returning to its spawning ground will do.

(A)

Eptatretus stoutii

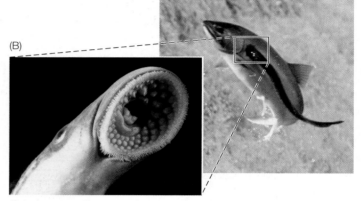

(B)

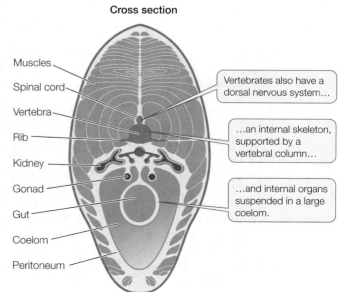

Petromyzon marinus

ficially similar (with elongate eel-like bodies and no paired fins), they differ greatly in their biology.

Lampreys have a complete braincase and distinct and separate (although rudimentary) vertebrae, all cartilaginous rather than bony. Lampreys undergo a complete metamorphosis from filter-feeding larvae known as *ammocoetes*, which are morphologically quite similar in general structure to adult lancelets. The adults of many species of lampreys are parasitic, although several lineages of lampreys evolved to become nonfeeding as adults. These nonfeeding adult lampreys survive only a few weeks after metamorphosis—just long enough to breed. In the species that are parasitic as adults, the round mouth is a rasping and sucking organ that is used to attach to their prey and rasp at the flesh (see Figure 33.9B). Some species of lampreys are critically endangered because of recent habitat changes and losses.

The vertebrate body plan can support large, active animals

Four key features characterize the vertebrates:

- An anterior *skull* with a large brain
- A rigid internal *skeleton* supported by the vertebral column
- Internal organs *suspended in a coelom*
- A well-developed *circulatory system*, driven by contractions of a ventral *heart*

This organization of the vertebrate body is exemplified by the bony fish diagrammed in **Figure 33.10**. Many kinds of jawless

fishes were found in the seas, estuaries, and fresh waters of the Devonian period, but hagfishes and lampreys are the only jawless fishes that survived beyond the Devonian. During that period, the *gnathostomes* (Greek, "jaw mouths") evolved jaws via modifications of the skeletal arches that supported the gills (**Figure 33.11A**). Jaws greatly improved feeding efficiency, as an animal with jaws can grasp, subdue, and swallow large prey.

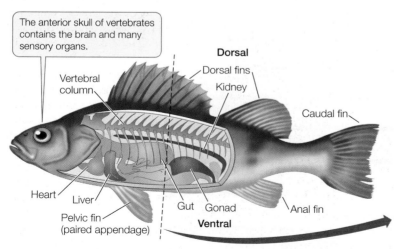

The anterior skull of vertebrates contains the brain and many sensory organs.

Dorsal

Vertebral column

Dorsal fins

Kidney

Caudal fin

Heart   Liver

Pelvic fin (paired appendage)

Gut   Gonad

Anal fin

Ventral

33.10 The Vertebrate Body Plan   A ray-finned fish is used here to illustrate the structural elements common to all vertebrates. In addition to the paired pelvic fins, these fishes have paired pectoral fins on the sides of their bodies (not seen in this cutaway view).

Cross section

Muscles

Spinal cord

Vertebra

Rib

Kidney

Gonad

Gut

Coelom

Peritoneum

Vertebrates also have a dorsal nervous system...

...an internal skeleton, supported by a vertebral column...

...and internal organs suspended in a large coelom.

(A)

Jawless fishes

Skull (cartilage)

Gill arches made of cartilage supported the gills.

Gill arches

Gill slits

Early jawed fishes
(placoderms, now extinct)

Some anterior gill arches became modified to form jaws, which at first had no teeth.

Modern jawed fishes
(cartilaginous and bony fishes)

Additional gill arches help support heavier, more efficient jaws, which in turn, support teeth.

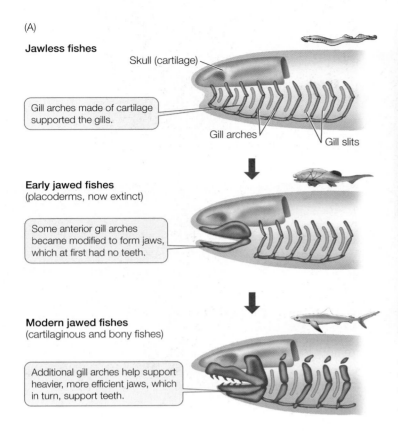

(B)

33.11 Jaws and Teeth Increased Feeding Efficiency (A) The diagrams illustrate one probable scenario for the evolution of jaws from the anterior gill arches of jawless fishes. (B) Jaws of the extinct giant shark (*Carcharodon megalodon*) display the teeth that indicate an extreme predatory lifestyle.

The earliest jaws were simple, but the evolution of *teeth* made predators more effective (**Figure 33.11B**). In predators, teeth function crucially both in grasping and in breaking up prey. In both predators and herbivores, teeth enable an animal to chew both soft and hard body parts of their food. Chewing also aids chemical digestion and improves an animal's ability to extract nutrients from its food, as we describe in Chapter 51. Vertebrates are remarkable in the diversity of their jaws and teeth.

Fins and swim bladders improved stability and control over locomotion

Paired fins stabilize the position of jawed fishes in water (and in some cases, help propel them). Most aquatic gnathostomes have a pair of pectoral fins just behind the gill slits, and a pair of pelvic fins anterior to the anal region (see Figures 33.10 and 33.12). Median dorsal and anal fins stabilize the fish, or may be used for propulsion in some species. In many fishes, the caudal fin helps propel the animal and enables it to turn rapidly.

Several groups of gnathostomes became abundant during the Devonian. Among them were the **chondrichthyans**—sharks, skates, and rays (940 living species) and chimaeras (40 living species). Like hagfishes and lampreys, these fishes have a skeleton composed entirely of firm but pliable cartilage. Their skin is flexible and leathery, sometimes bearing scales that give it the

Hagfishes
Lampreys
Cartilaginous fishes
Ray-finned fishes
Sacopterygians

consistency of sandpaper. Sharks move forward by means of lateral undulations of their body and caudal (tail) fin (**Figure 33.12A**). Skates and rays propel themselves by means of vertical undulating movements of their greatly enlarged pectoral fins (**Figure 33.12B**).

Most sharks are predators, but some feed by straining plankton from the water. Most skates and rays live on the ocean floor, where they feed on mollusks and other animals buried in the sediments. Nearly all cartilaginous fishes live in the oceans, but a few are estuarine or migrate into lakes and rivers. One group of stingrays is found in river systems of South America. The less familiar chimaeras (**Figure 33.12C**) live in deep-sea or cold waters.

In the ancestor of the bony vertebrates, gas-filled sacs supplemented the gas-exchange function of the gills by giving the animals access to atmospheric oxygen. These features enabled those fishes to live where oxygen was periodically in short supply, as it often is in freshwater environments. These lunglike sacs evolved into *swim bladders*, which are organs of buoyancy, as well as into the lungs of tetrapods. By adjusting the amount of gas in its swim bladder, a fish can control the depth at which it remains suspended in the water while expending very little energy to maintain its position.

Ray-finned fishes, and most remaining groups of vertebrates, have internal skeletons of calcified, rigid *bone* rather than flexible cartilage. The outer body surface of most species of ray-finned fishes is covered with flat, thin, lightweight scales that provide some protection or enhance movement through the wa-

(A) *Carcharhinus plumbeus*   Dorsal fin

Caudal fin   Pelvic fin   Pectoral fins

33.12 Chondrichthyans (A) Most sharks, such as this sandbar shark, are active marine predators. (B) Skates and rays, represented here by an eagle ray, feed on the ocean bottom. Their modified pectoral fins are used for propulsion; their other fins are greatly reduced. (C) A chimaera, or ratfish. These deep-sea fishes often possess modified dorsal fins that contain toxins.

(B) *Myliobatis australis*   Pectoral fins

                                        Dorsal fin

(C) *Hydrolagus colliei*   Pectoral fin   Pelvic fin

33.13 Diverse Ray-Finned Fishes (A) Eels such as this moray have the large teeth and powerful jaws typical of predatory fishes. (B) The wrasses contain more than 500 described species. Many species, such as the flame fairy wrasse seen here, inhabit coral reefs. (C) Another large ray-fin clade, the serranids, includes the sea basses and groupers. Panther groupers such as this one are endangered by the loss of Pacific coral reef habitat. (D) A unique structure that resembles a fishing lure has evolved among the anglerfishes. Deep-sea anglerfishes such as this one live below the level of light penetration; their lures are bioluminescent.

(A) *Gymnothorax favagineus*

(B) *Cirrhilabrus jordani*

(C) *Cromileptes altivelis*

(D) *Gigantactis vanhoeffeni* luring prey

ter. The gills of ray-finned fishes open into a single chamber covered by a hard flap, called an *operculum*. Movement of the operculum improves the flow of water over the gills, where gas exchange takes place.

Ray-finned fishes radiated extensively in the Tertiary. Today there are about 30,000 known living species, encompassing a remarkable variety of sizes, shapes, and lifestyles (**Figure 33.13**). The smallest are less than 1 centimeter long as adults; the largest weigh as much as 900 kilograms. Ray-finned fishes exploit nearly all types of aquatic food sources. In the oceans they filter plankton from the water, rasp algae from rocks, eat corals and other soft-bodied colonial animals, dig animals from soft sediments, and prey on virtually all kinds of other fishes. In fresh water they eat plankton, devour insects, eat fruits that fall into the water in flooded forests, and prey on other aquatic vertebrates and, occasionally, terrestrial vertebrates. Many ray-finned fishes are solitary, but in open water others form large aggregations called *schools*. Many species perform complicated behaviors to maintain schools, build nests, court and choose mates, and care for their young.

Although ray-finned fishes can readily control their position in open water using their fins and swim bladder, their eggs tend to sink. Some species produce small eggs that are buoyant enough to complete their development in the open water, but many marine fishes move to food-rich shallow waters to lay their eggs. That is why coastal waters and estuaries are so important in the life cycles of many marine fishes. Some, such as salmon, are *anadromous*; these species leave salt water when they breed, ascending rivers to spawn in freshwater streams and lakes (see Figure 33.9B).

33.3 RECAP

Chordates are characterized by a dorsal hollow nerve chord, a post-anal tail, and a dorsal supporting rod called a notochord at some point during the life cycle. Specialized structures for support (such as vertebrae), locomotion (such as fins), and feeding (such as jaws and teeth) evolved among the vertebrates, which allowed them to colonize and adapt to most of Earth's environments.

- What synapomorphies respectively characterize the chordates and the vertebrates? See pp. 697–700 and Figures 33.6 and 33.10

- How do the hagfishes differ from the lampreys in morphology and life history? Do you see why some biologists do not consider the hagfishes to be vertebrates? See p. 699

In some fishes, the lunglike sacs that gave rise to swim bladders became specialized for another purpose: breathing air. That adaptation set the stage for the vertebrates to move onto the land.

33.4 How Did Vertebrates Colonize the Land?

The evolution of lunglike sacs in fishes set the stage for the invasion of the land. Some early ray-finned fishes probably used those sacs to supplement their gills when oxygen levels in the water were low, as lungfishes and many groups of ray-finned fishes do today. But with their unjointed fins, those fishes could only flop around on land. Changes in the structure of the fins first allowed some fishes to support themselves better in shallow water and, later, to move better on land.

Jointed fins enhanced support for fishes

Two pairs of muscular, jointed fins evolved in the ancestor of the **sarcopterygians**, which include coelacanths, lungfishes, and tetrapods. Each of the jointed appendages of sarcopterygians is joined to the body by a single enlarged bone. The coelacanths flourished from the Devonian until about 65 million years ago, when they were thought to have become extinct. However, in 1938 a commercial fisherman caught a living coelacanth off South Africa. Since that time, hundreds of individuals of this extraordinary fish, *Latimeria chalumnae*, have been collected. A second species, *L. menadoensis*, was discovered in 1998 off the Indonesian island of Sulawesi. *Latimeria*, a predator of other fishes, reaches a length of about 1.8 meters and weighs up to 82 kilograms (**Figure 33.14A**). Its skeleton is composed mostly of cartilage, not bone. A cartilaginous skeleton is a derived feature in this clade because it had bony ancestors.

Lungfishes, which also have jointed fins that are connected to the body by a single enlarged bone, were important predators in shallow-water habitats in the Devonian, but most lineages died out. The six surviving species live in stagnant swamps and muddy waters in South America, Africa, and Australia (**Figure 33.14B**). Lungfishes have lungs derived from the lunglike sacs of their ancestors as well as gills. When ponds dry up, individuals of most species can burrow deep into the mud and survive for many months in an inactive state while breathing air.

It is believed that some early aquatic sarcopterygians began to use terrestrial food sources, became more fully adapted to life on land, and eventually evolved to become ancestral **tetrapods** (four-legged vertebrates).

How was the transition from an animal that swam in water to one that walked on land accomplished? Early in 2006, scientists reported the discovery of a Devonian fossil they believe represents an intermediate between the finned appendages of fishes and the limbs of terrestrial tetrapods (**Figure 33.14C**). It appears that limbs able to prop up a large fish with the front-to-rear movement necessary for walking evolved while the animals still lived in water. These limbs appear to have functioned in holding the animals upright in shallow water, perhaps even allowing them to hold their head above the water's surface. These same structures were then co-opted for movement on land, at first probably for foraging on brief trips out of water.

(A) *Latimeria chalumnae*

(B) *Protopterus annectens*

Tiktaalik's pectoral fins show some of the skeletal structures of tetrapod limbs.

(C) *Tiktaalik roseae*

33.14 The Closest Relatives of Tetrapods (A) The African coelacanth, discovered in deep waters of the Indian Ocean, represents one of two surviving species of a group that was once thought to be extinct. (B) All surviving lungfish species, such as this African lungfish, live in the Southern Hemisphere. (C) This fossil from the Devonian is believed to represent a transitional species intermediate between the finned fishes and the limbed tetrapods.

Amphibians adapted to life on land

During the Devonian, the first tetrapods arose from an aquatic ancestor. In this lineage, legs capable of movement on land evolved from the short, stubby fins of their aquatic ancestors. The basic elements of those legs have remained throughout the evolution of terrestrial vertebrates, although they have changed considerably in their form.

Most modern **amphibians** are confined to moist environments because they lose water rapidly through the skin when exposed to dry air. In addition, their eggs are enclosed within delicate membranous envelopes that cannot prevent water loss in dry conditions. In some amphibian species, adults live mostly on land but return to fresh water to lay their eggs, which are usually fertilized outside the body (**Figure 33.15**). The fertilized eggs

give rise to larvae that live in water until they undergo metamorphosis to become terrestrial adults. However, many amphibians (especially those in tropical and subtropical areas) have evolved a wide diversity of additional reproductive modes and types of parental care, as described in the opening of this chapter. Internal fertilization evolved many times among the amphibians.

Lungfishes
Amphibians
Amniotes

33.15 In and Out of the Water
Most early stages in the life cycle of many amphibians take place in water. The aquatic tadpole transforms into a terrestrial adult through metamorphosis. Some species of amphibians, however, have direct development (with no aquatic larval stage), and others are aquatic throughout life.

—— **yourBioPortal.com** ——
GO TO **Animated Tutorial 33.1** •
Life Cycle of a Frog

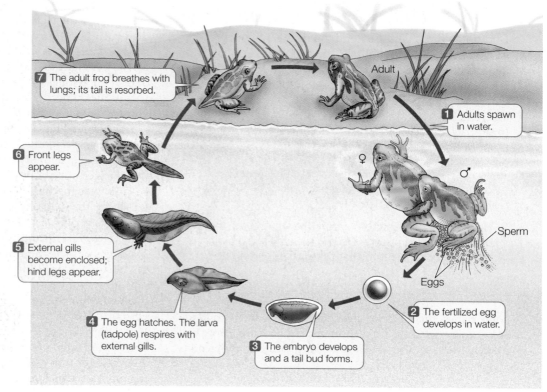

7 The adult frog breathes with lungs; its tail is resorbed.

Adult

6 Front legs appear.

1 Adults spawn in water.

♀ ♂

Sperm

5 External gills become enclosed; hind legs appear.

Eggs

4 The egg hatches. The larva (tadpole) respires with external gills.

3 The embryo develops and a tail bud forms.

2 The fertilized egg develops in water.

Many species develop directly into adultlike forms from fertilized eggs laid on land or carried by the parents. Other species of amphibians are entirely aquatic, never leaving the water at any stage of their lives, and many of these species retain a larval-like morphology.

The more than 6,500 known species of amphibians living on Earth today belong to three major groups: the wormlike, limbless, tropical, burrowing or aquatic *caecilians* (**Figure 33.16A**), the tail-less frogs and toads (collectively called *anurans;* **Figure 33.16B**), and the tailed *salamanders* (**Figure 33.16C and D**).

Anurans are most diverse in wet tropical and warm temperate regions, although a few are found at very high latitudes. There are far more anurans than any other amphibians, with about 5,800 described species and more being discovered every year. Some anurans have tough skins and other adaptations that enable them to live for long periods in very dry deserts, whereas others live in moist terrestrial and arboreal environments. Some species are completely aquatic as adults. All anurans have a very short vertebral column, with a strongly modified pelvic region that is modified for leaping, hopping, or propelling the body through water by kicking the hind legs.

The approximately 600 described species of salamanders are most diverse in temperate regions of the Northern Hemisphere, but many species are also found in cool, moist environments in the mountains of Central America, and a few species penetrate into the South American tropics. Many salamanders live in rotting logs or moist soil. One major group has lost lungs, and these species exchange gases entirely through the skin and mouth lining—body parts that all amphibians use in addition to their lungs. Through *paedomorphosis* (retention of the juvenile state; see Chapter 20), a completely aquatic lifestyle has evolved several times among the salamanders (see Figure 33.16D). Most species of salamanders have internal fertilization, which is usually achieved through the transfer of a small jellylike, sperm-embedded capsule called a *spermatophore.*

Many amphibians have complex social behaviors. Most male anurans utter loud, species-specific calls to attract females of their own species (and sometimes to defend breeding territories), and they compete for access to females that arrive at the breeding sites. Many amphibians lay large numbers of eggs, which they abandon once they are deposited and fertilized. As described in the opening of this chapter, some amphibians lay only a few eggs, which are fertilized and then guarded in a nest, or carried on the backs, in the vocal pouches, or even in the stomachs of one of the parents. A few species of frogs, salamanders, and caecilians are *viviparous,* meaning they give birth to well-developed young that have received nutrition from the female during gestation.

33.16 Diversity among the Amphibians (A) Burrowing caecilians superficially look more like worms than amphibians. (B) Male golden toads in the cloud forest of Monteverde, Costa Rica. This species has recently become extinct, one of many amphibian species to do so in the past few decades. (C) An adult barred tiger salamander. (D) This Austin blind salamander's life cycle remains aquatic; it has no adult terrestrial stage. The eyes of this cave dweller have become greatly reduced.

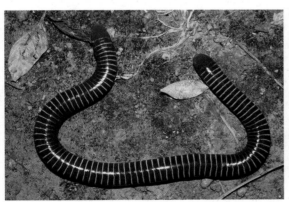

(A) *Siphonops annulatus*

(B) *Bufo periglenes*

(C) *Ambystoma mavortium*

(D) *Eurycea waterlooensis*

Amphibians are the focus of much attention today because populations of many species are declining rapidly, especially in mountainous regions of western North America, Central and South America, and northeastern Australia. Worldwide, about one-third of amphibian species are now threatened with extinction or have disappeared completely in the last few decades (as happened with the gastric brooding frogs described at the beginning of this chapter). Scientists are investigating several hypotheses to account for these population declines, including the adverse effects of habitat alteration by humans, increased solar radiation caused by destruction of Earth's ozone layer, pollution from urban and industrial areas and airborne agricultural pesticides and herbicides (see Chapter 1), and the spread of a pathogenic chytrid fungus that attacks amphibians. Scientists have documented the spread of the chytrid fungus through Central America, where many species of amphibians have become extinct (including Costa Rica's golden toad; see Figure 33.16B).

Amniotes colonized dry environments

Several key innovations contributed to the ability of members of one clade of tetrapods to exploit a wide range of terrestrial habitats. The animals that evolved these water-conserving traits are called **amniotes**.

The **amniote egg** (which gives the group its name) is relatively impermeable to water and allows the embryo to develop in a contained aqueous environment (**Figure 33.17**). The leathery or brittle, calcium-impregnated shell of the amniote egg retards evaporation of the fluids inside but permits passage of oxygen and carbon dioxide. The egg also stores large quantities of food in the form of *yolk*, allowing the embryo to attain a relatively advanced state of development before it hatches. Within the shell are *extraembryonic membranes* that protect the embryo from desiccation and assist its gas exchange and excretion of waste nitrogen.

In several different groups of amniotes, the amniote egg became modified, allowing the embryo to grow inside (and receive nutrition from) the mother. For instance, the mammalian egg lost its

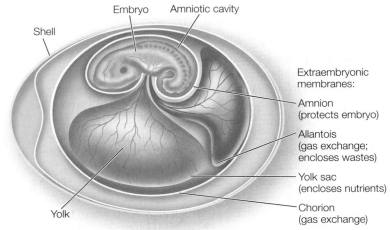

33.17 An Egg for Dry Places   The evolution of the amniote egg, with its water-retaining shell, four extraembryonic membranes, and embryo-nourishing yolk, was a major step in the colonization of the terrestrial environment.

— **yourBioPortal.com** —
GO TO **Web Activity 33.2** • **The Amniote Egg**

shell while the functions of the extraembryonic membranes were retained and expanded; we examine the roles of these membranes in detail in Section 44.4.

Other innovations evolved in the organs of terrestrial adults. A tough, impermeable skin, covered with scales or modifications of scales such as hair and feathers, greatly reduced water loss. Adaptations of the vertebrate excretory organs, the kidneys, allowed amniotes to excrete concentrated urine, ridding the body of waste nitrogen without losing a large amount of water in the process (see Chapter 52).

During the Carboniferous, amniotes split into two major groups, the mammals and reptiles (**Figure 33.18**). More than

33.18 Phylogeny of Amniotes   This tree of amniote relationships shows the primary split between mammals and reptiles. The reptile portion of the tree shows a lineage that led to the turtles, another to the lepidosaurs (snakes, lizards, and tuataras), and a third branch that includes all the archosaurs (crocodiles, several extinct groups, and the birds). There is some uncertainty in the placement of the turtle lineage; some data support a relationship between turtles and archosaurs.

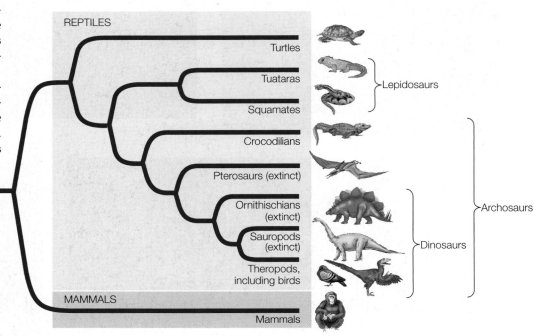

18,700 species of **reptiles** exist today, over half of which are *birds*. Birds are the only living members of the otherwise extinct *dinosaurs*, the dominant terrestrial predators of the Mesozoic.

Reptiles adapted to life in many habitats

The lineage leading to modern reptiles began to diverge from other amniotes about 250 million years ago. One reptilian group that has changed very little over the intervening millennia is the *turtles*. The dorsal and ventral bony plates of turtles form a shell into which the head and limbs can be withdrawn in many species (**Figure 33.19A**). The dorsal shell is an expansion of the ribs, and it is a mystery how the pectoral girdles evolved to be inside the ribs of turtles, making them unlike any other vertebrates. Most turtles live in aquatic environments, but several groups, such as tortoises and box turtles, are terrestrial. Sea turtles spend their entire lives at sea except when they come ashore to lay eggs. Human exploitation of sea turtles and their eggs has resulted in worldwide declines of these species, all of which are now endangered. A few species of turtles are strict herbivores or carnivores, but most species are omnivores that eat a variety of aquatic and terrestrial plants and animals.

The *lepidosaurs* constitute the second-most species-rich clade of living reptiles. This group is composed of the *squamates* (lizards, snakes, and amphisbaenians—the last a group of mostly legless, wormlike, burrowing reptiles with greatly reduced eyes) and the *tuataras*, which superficially resemble lizards but differ from them in tooth attachment and several internal anatomical features. Many species related to the tuataras lived during the Mesozoic era, but today only two species, restricted to a few islands off New Zealand, survive (**Figure 33.19B**).

The skin of a lepidosaur is covered with horny scales that greatly reduce loss of water from the body surface. These scales, however, make the skin unavailable as an organ of gas exchange. Gases are exchanged almost entirely via the lungs, which are proportionally much larger in surface area than those of amphibians. A lepidosaur forces air into and out of its lungs by bellowslike movements of its ribs. The three-chambered lepidosaur heart partially separates oxygenated blood from the lungs from deoxygenated blood returning from the body. With this type of heart, lepidosaurs can generate high blood pressure and can sustain a relatively high metabolism.

Most lizards are insectivores, but some are herbivores; a few prey on other vertebrates. The largest lizard, which grows as

33.19 Reptilian Diversity (A) Green sea turtles are widely distributed in tropical oceans. (B) This tuatara represents one of only two surviving species in a lineage that diverged long ago. (C) The leopard gecko, a desert dweller native to Afghanistan, Pakistan, and northwestern India. (D) The ringneck snake of North America is nonvenomous. It coils its tail to reveal a bright orange underbelly, which distracts potential predators from the vital head region.

(A) *Chelonia mydas*

(B) *Sphenodon punctatus*

(C) *Eublepharis macularius*

(D) *Diadophis punctatus*

long as 3 meters and can weigh more than 150 kilograms, is the predaceous Komodo dragon of the East Indies. Most lizards walk on four limbs (**Figure 33.19C**), although limblessness has evolved repeatedly in the group, especially in burrowing and grassland species. One major group of limbless squamates is the snakes (**Figure 33.19D**). All snakes are carnivores, and many can swallow objects much larger than themselves. Several snake groups evolved venom glands and the ability to inject venom rapidly into their prey.

Crocodilians and birds share their ancestry with the dinosaurs

Another reptilian group, the *archosaurs*, includes the crocodilians, dinosaurs, and birds. *Dinosaurs* rose to prominence about 215 million years ago and dominated terrestrial environments for about 150 million years; only one group of dinosaurs, the *birds*, survived the mass

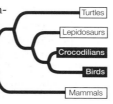

extinction event at the Cretaceous–Tertiary boundary. During the Mesozoic, most terrestrial animals more than a meter long were dinosaurs. Many were agile and could run rapidly; they had special muscles that enabled the lungs to be filled and emptied while the limbs moved. We can infer the existence of such muscles in dinosaurs from the structure of the vertebral column in fossils. Some of the largest dinosaurs weighed as much as 70,000 kilograms.

Modern *crocodilians*—crocodiles, caimans, gharials, and alligators—are confined to tropical and warm temperate environments (**Figure 33.20A**). Crocodilians spend much of their time in water, but they build nests on land or on floating piles of vegetation. The eggs are warmed by heat generated by decaying organic matter that the female places in the nest. Typically, the female guards the eggs until hatching, and she often facilitates hatching. In some species, the female continues to guard and communicate with her offspring after they hatch. All crocodilians are carnivorous. They eat vertebrates of all kinds, including large mammals.

Biologists have long accepted the phylogenetic position of birds among the reptiles, although birds clearly have many unique, derived morphological features. In addition to the strong morphological evidence for the placement of birds among the reptiles, fossil and molecular data emerging over the last few decades have provided definitive supporting evidence. Birds are thought to have emerged among the *theropods*, a group of predatory dinosaurs that share such traits as bipedal stance, hollow bones, a *furcula* ("wishbone"), elongated metatarsals with three-fingered feet, elongated forelimbs with three-fingered hands, and a pelvis that points backward.

The living bird species fall into two major groups that diverged during the late Cretaceous, about 80 to 90 million years ago, from a flying ancestor. The few modern descendants of one lineage include a group of secondarily flightless and weakly flying birds, some of which are very large. This group, called the *palaeognaths*, includes the South and Central American tinamous and several large flightless birds of the southern continents—

(A) *Crocodylus porosus*

(B) *Struthio camelus*

33.20 Archosaurs (A) Crocodiles, alligators, and their relatives live in tropical and warm temperate climates. This crocodile lives in saltwater and estuarine environments along Australia's coast. (B) Birds are the other living archosaur group, represented here by the winged but flightless ostrich.

the rheas, emu, kiwis, cassowaries, and the world's largest bird, the ostrich (**Figure 33.20B**). The second lineage, the *neognaths*, has left a much larger number of descendants, most of which have retained the ability to fly.

The evolution of feathers allowed birds to fly

Fossil dinosaurs discovered recently in early Cretaceous deposits in Liaoning Province, in northeastern China, show that the scales of some small predatory dinosaurs were highly modified to form *feathers*. The feathers of one of these dinosaurs, *Microraptor gui*, were structurally similar to those of modern birds (**Figure 33.21A**).

During the Mesozoic era, about 175 million years ago, a lineage of theropods gave rise to the birds. The oldest known avian fossil, *Archaeopteryx*, which lived about 150 million years ago, had teeth, but it was covered with feathers that are virtually identical to those of modern birds (**Figure 33.21B**). It also had well-developed wings, a long tail, and a furcula to which some of the flight muscles were probably attached. *Archaeopteryx* had clawed fingers on its forelimbs, but it also had typical perching bird claws on its hindlimbs. It probably lived in trees and shrubs and used the fingers to assist it in clambering over branches.

The evolution of feathers was a major force for diversification. Feathers are lightweight but are strong and structurally complex (**Figure 33.22**). The large quills of the flight feathers on

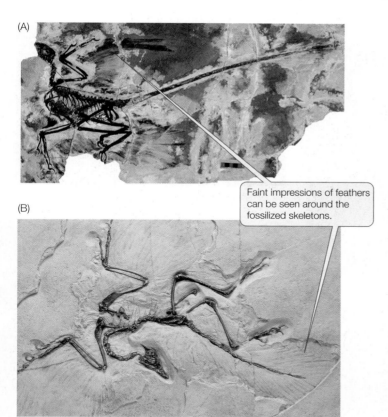

Faint impressions of feathers can be seen around the fossilized skeletons.

33.21 Mesozoic Bird Relatives Fossil remains demonstrate the evolution of birds from other dinosaurs. (A) *Microraptor gui* was a feathered dinosaur from the early Cretaceous (about 140 mya). (B) Dating from roughly the same timeframe, *Archaeopteryx* is the oldest known birdlike fossil.

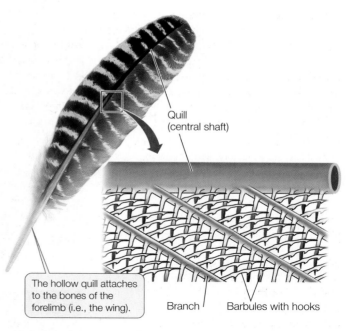

Quill (central shaft)

The hollow quill attaches to the bones of the forelimb (i.e., the wing).

Branch   Barbules with hooks

33.22 Feathers Represent a Major Evolutionary Innovation The flight feathers of birds are attached to the forelimb (wing) bones by a hollow central shaft. Fine branches with interlocking hooks and barbs radiate from the shaft, creating a strong, lightweight surface that enables flight.

the wings arise from the skin of the forelimbs to create the flying surfaces. Other strong feathers sprout like a fan from the shortened tail and serve as stabilizers during flight. The feathers that cover the body, along with an underlying layer of down feathers, provide birds with insulation that helps them to survive in virtually all of Earth's climates.

The bones of theropod dinosaurs, including birds, are hollow with internal struts that increase their strength. Hollow bones would have made early theropods lighter and more mobile; later they facilitated the evolution of flight. The sternum (breastbone) of flying birds forms a large, vertical keel to which the flight muscles are attached.

Flight is metabolically expensive. A flying bird consumes energy at a rate about 15 to 20 times faster than a running lizard of the same weight. Because birds have such high metabolic rates, they generate large amounts of heat. They control the rate of heat loss using their feathers, which may be held close to the body or elevated to alter the amount of insulation they provide. The lungs of birds allow air to flow through unidirectionally rather than by pumping air in and out (see Section 49.2). This flow-through structure of the lungs increases the efficiency of gas exchange and thereby supports an increased metabolic rate.

There are about 10,000 species of living birds, which range in size from the 150-kilogram ostrich to a tiny hummingbird

weighing only 2 grams. The teeth so prominent among other dinosaurs were secondarily lost in the ancestral birds, but birds nonetheless eat almost all types of animal and plant material. Insects and fruits are the most important dietary items for terrestrial species. Birds also eat seeds, nectar and pollen, leaves and buds, carrion, and other vertebrates. By eating the fruits and seeds of plants, birds serve as major agents of seed dispersal. **Figure 33.23** shows representatives of a few of the major groups of birds.

Mammals radiated after the extinction of dinosaurs

Small and medium-sized **mammals** coexisted with the dinosaurs for millions of years. After the non-avian dinosaurs disappeared during the mass extinction at the close of the Mesozoic era, mammals increased dramatically in numbers, diversity, and size. Today mammals range in size from tiny shrews and bats weighing only about 2 grams to the blue whale, the largest animal on Earth, which measures up to 33 meters long and can weigh as much as 160,000 kilograms. Mammals have far fewer, but more highly differentiated, teeth than do fishes, amphibians, or reptiles. Differences among mammals in the number, type, and arrangement of teeth reflect their varied diets (see Figure 51.6).

Four key features distinguish the mammals:

- *Sweat glands,* which secrete sweat that evaporates and thereby cools an animal

- *Mammary glands,* which in females secrete a nutritive fluid (milk) on which newborn individuals feed

33.23 Diversity among the Birds
(A) Perching, or passeriform, birds such as this cedar waxwing comprise the most species-rich of all bird groups. (B) This bright-plumaged male Mandarin duck is a member of the group that includes ducks, geese, and swans. (C) Great frigatebirds are among the ocean-going birds that are often found miles from shore. This male is in display mode, with an inflated pouch that advertises his presence to females. (D) This barn owl is a nighttime predator that can find prey using its sensitive auditory "sonar" system.

(B) *Aix galericulata*

- *Hair*, which provides a protective and insulating covering
- A *four-chambered heart* that completely separates the oxygenated blood coming from the lungs from the deoxygenated blood returning from the body (this last characteristic is convergent with the archosaurs, including modern birds and crocodiles)

(A) *Bombycilla cedrorum*

Mammalian eggs are fertilized within the female's body, and the embryos undergo a period of development in the female's body in an organ called the *uterus* prior to being born. Most mammals have a covering of hair (fur), which is luxuriant in some species but has been greatly reduced in others, including the cetaceans (whales and dolphins) and humans. Thick layers of insulating fat (blubber) replace hair as a heat-retention mechanism in the cetaceans; humans learned to use clothing for this purpose when they dispersed from warm tropical areas.

(D) *Tyto alba*

(C) *Frigata minor*

The approximately 5,000 species of living mammals are divided into two primary groups: the *prototherians* and the *therians*. Only five species of prototherians are known, and they are found only in Australia and New Guinea. These mammals, the duck-billed platypus and four species of echidnas, differ from other mammals in lacking a placenta, laying eggs, and having sprawling legs (**Figure 33.24**). Prototherians supply milk for their young, but they have no nipples on their mammary glands; the milk simply oozes out and is lapped off the fur by the offspring.

Reptiles
Prototherians
Marsupials
Eutherians

(A) *Tachyglossus aculeatus*

Most mammals are therians

Members of the *therian* clade are further divided into the *marsupials* and the *eutherians*. Females of most marsupial species have a ventral pouch in which they carry and feed their offspring (see Figure 33.25A). Gestation (pregnancy) in marsupials is brief; the young are born tiny but with well-developed forelimbs, with which they climb to the pouch. They attach to

33.24 Prototherians (A) The short-beaked echidna is one of the four surviving species of echidnas. (B) The duck-billed platypus is another surviving prototherian.

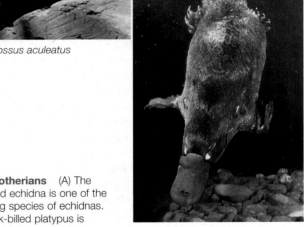

(B) *Ornithorhynchus anatinus*

33.25 Marsupials (A) Australia's eastern gray kangaroo is among the largest living marsupials. This female carries her young offspring in the characteristic marsupial pouch. (B) The carnivorous Tasmanian devil is found only in Tasmania, an island off southern Australia. (C) This arboreal opossum is a South American marsupial.

(B) *Sarcophilus harrisii*

(A) *Macropus giganteus*

(C) *Marmosa murina*

a nipple but cannot suck. The mother ejects milk into the tiny offspring until it grows large enough to suckle. Once her offspring have left the uterus, a female marsupial may become sexually receptive again. She can then carry fertilized eggs that are capable of initiating development and can replace the offspring in her pouch should something happen to them.

At one time marsupials were found on all continents, but the approximately 350 living species are now restricted to Australasia (**Figure 33.25A and B**) and the Americas (especially South America; **Figure 33.25C**). One species, the Virginia opossum, is widely distributed in North America. Marsupials radiated to become herbivores, insectivores, and carnivores, but no marsupials live in the oceans. None can fly, although some *arboreal* (tree-dwelling) marsupials are gliders. The largest living marsupials are the kangaroos of Australia, which can weigh up to 90 kilograms. Much larger marsupials existed in Australia until humans exterminated them soon after reaching that continent about 40,000 years ago.

Eutherians include the majority of mammals. Eutherians are sometimes called *placental mammals*, but this name is inappropriate because some marsupials also have placentas. Eutherians are more developed at birth than are marsupials; no external pouch houses them after they are born.

The approximately 5,000 species of living eutherians are divided into 20 major groups (**Table 33.1**). The largest group is the rodents, with about 2,300 species. Rodents are traditionally defined by the unique morphology of their teeth, which are adapted for gnawing through substances such as wood. The next largest group comprises the approximately 1,100 bat species—the flying mammals. The bats are followed by the moles and shrews, with about 430 species. The relationships of the major groups of eutherians to one another have been difficult to determine, because most of the major groups diverged in a short period of time during an explosive adaptive radiation.

Eutherians are extremely varied in their form and ecology (**Figure 33.26**). The extinction of the non-avian dinosaurs at the end of the Cretaceous may have made it possible for them to diversify and radiate into a large range of ecological *niches*. Many eutherian species grew to become quite large in size, and some

assumed the role of dominant terrestrial predators previously occupied by the large dinosaurs. Among these predators, social hunting behavior evolved in several species, including members of the canid (dog), felid (cat), and primate lineages.

Grazing and *browsing* by members of several eutherian groups helped transform the terrestrial landscape. Herds of grazing herbivores feed on open grasslands, whereas browsers feed on shrubs and trees. The effects of herbivores on plant life favored the evolution of the spines, tough leaves, and difficult-to-eat growth forms found in many plants. In turn, adaptations to the teeth and digestive systems of many herbivore lineages allowed these species to consume many plants despite such defenses—a striking example of coevolution. A large animal can survive on food of lower quality than a small animal can, and large size evolved in several groups of grazing and browsing mammals (see Figure 33.26C). The evolution of large herbivores, in turn, favored the evolution of large carnivores able to attack and overpower them.

Several lineages of terrestrial eutherians subsequently returned to the aquatic environments their ancestors had left behind (see Figure 33.26D). The completely aquatic cetaceans—whales and dolphins—evolved from artiodactyl ancestors (whales are closely related to the hippopotamuses). The seals, sea lions, and walruses also returned to the marine environment, and their limbs became modified into flippers. Weasel-like otters retain their limbs but have also returned to aquatic environments, colonizing both fresh and salt water. The manatees and dugongs colonized estuaries and shallow seas.

(A) *Erithizon dorsatum*

(B) *Artibeus lituratus*

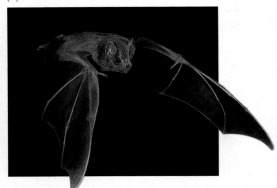

(C) *Rangifer tarandus*

33.26 Diversity among the Eutherians
(A) The North American porcupine, a large rodent covered with sharp, protective quills. Almost half of all eutherians are rodents. (B) Flight evolved in the ancestor of bats. This fruit-eating bat ranges from Central to South America. Virtually all bat species are nocturnal. (C) Large hoofed mammals are important herbivores in terrestrial environments. Although this bull is grazing by himself, caribou are usually found in huge herds. (D) Spinner dolphins are cetaceans, a cetartiodactyl group that returned to the marine environment.

(D) *Stenella longirostris*

TABLE 33.1
Major Groups of Living Eutherian Mammals

GROUP	APPROXIMATE NUMBER OF LIVING SPECIES	EXAMPLES
Gnawing mammals (Rodentia)	2,300	Rats, mice, squirrels, woodchucks, ground squirrels, beaver, capybara
Flying mammals (Chiroptera)	1,100	Bats
Soricomorph insectivores (Soricomorpha)	430	Shrews, moles
Even-toed hoofed mammals and cetaceans (Cetartiodactyla)	320	Deer, sheep, goats, cattle, antelopes, giraffes, camels, swine, hippopotamus, whales, dolphins
Carnivores (Carnivora)	290	Wolves, dogs, bears, cats, weasels, pinnipeds (seals, sea lions, walruses)
Primates (Primates)	235	Lemurs, monkeys, apes, humans
Lagomorphs (Lagomorpha)	80	Rabbits, hares, pikas
African insectivores (Afrosoricida)	50	Tenrecs, golden moles
Spiny insectivores (Erinaceomorpha)	24	Hedgehogs
Armored mammals (Cingulata)	21	Armadillos
Tree shrews (Scandentia)	20	Tree shrews
Odd-toed hoofed mammals (Perissodactyla)	20	Horses, zebras, tapirs, rhinoceroses
Long-nosed insectivores (Macroscelidea)	16	Elephant shrews
Pilosans (Pilosa)	10	Anteaters, sloths
Pholidotans (Pholidota)	8	Pangolins
Sirenians (Sirenia)	5	Manatees, dugongs
Hyracoids (Hyracoidea)	4	Hyraxes, dassies
Elephants (Proboscidea)	3	African and Indian elephants
Dermopterans (Dermoptera)	2	Flying lemurs
Aardvark (Tubulidentata)	1	Aardvark

The vertebrate colonization of dry land was facilitated by the evolution of an impermeable body covering, efficient kidneys, and the amniote egg—a structure that resists desiccation and provides an aqueous internal environment in which the embryo grows. Major amniote groups include turtles, lepidosaurs, archosaurs (crocodilians and birds), and mammals.

- In the not-too-distant past, the idea that birds were reptiles met with skepticism. Explain how fossils, morphology, and molecular evidence now support the position of birds among the reptiles. See pp. 708–709

- In reviewing the discussion of the various vertebrate groups, identify several reasons why tooth structure is such an important area of study.

The evolutionary history of one eutherian group—the primates—is of special interest to us because it includes the human lineage. The primates have been the subject of extensive research in most aspects of their biology, including behavior, ecology, physiology, and molecular biology. Let's take a closer look at the characteristics and evolutionary history of the primates.

33.5 What Traits Characterize the Primates?

The eutherian **primates** underwent extensive evolutionary radiation from an ancestral small, arboreal, insectivorous mammal. A nearly complete fossil of an early primate, *Carpolestes*, was found in Wyoming and dated at 56 million years ago; it had grasping feet with an opposable big toe that had a nail rather than a claw. Grasping limbs with opposable digits are one of the major adaptations to arboreal life that distinguish primates from other mammals.

Early in their evolutionary history, the primates split into two main clades, the prosimians and the anthropoids (**Figure 33.27**). *Prosimians*—lemurs, lorises, and their close relatives—once lived on all continents, but today they are restricted to Africa, Madagascar, and tropical Asia. All mainland prosimian species are arboreal and nocturnal (**Figure 33.28**). On the island of Madagascar, however, the site of a remarkable radiation of lemurs, there are also diurnal and terrestrial species.

A second primate lineage, the *anthropoids*—tarsiers, New World monkeys, Old World monkeys, and apes—evolved about 65 million years ago in Africa or Asia. New World monkeys diverged from Old World monkeys and apes at a slightly later date, but early enough that they might have reached South America from Africa when those two continents were still close to each other. All New World monkeys are arboreal (**Figure 33.29A**). Many of them have a long, prehensile tail with which they can grasp branches. Many Old World monkeys are arboreal as well, but a number of species are terrestrial (**Figure 33.29B**). No Old World primate has a prehensile tail.

About 35 million years ago, a lineage that led to the modern apes separated from the Old World monkeys. Between 22 and 5.5 million years ago, dozens of species of apes lived in Europe, Asia, and Africa. The Asian apes—gibbons and orangutans (**Figure 33.30A and B**)—descended from two of these ape lineages. Another extinct genus, *Dryopithecus*, is the sister group of the modern African apes—gorillas and chimpanzees (**Figure 33.30C and D**)—and of humans.

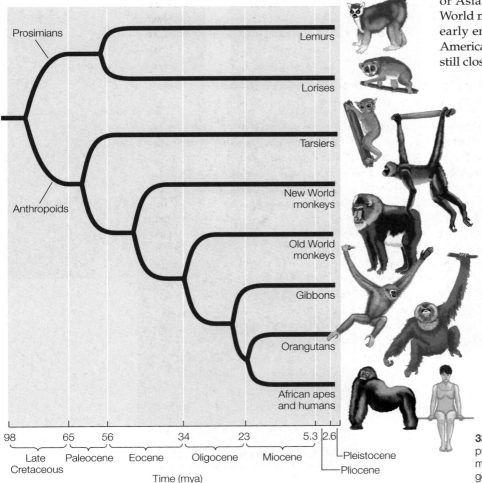

33.27 Phylogeny of the Primates The phylogeny of primates is among the best studied of any major group of mammals. This tree is based on evidence from many genes, morphology, and fossils.

Prosimians

Lemurs

Lorises

Tarsiers

New World monkeys

Anthropoids

Old World monkeys

Gibbons

Orangutans

African apes and humans

| 98 | 65 | 56 | 34 | 23 | 5.3 | 2.6 |

Late Cretaceous — Paleocene — Eocene — Oligocene — Miocene — Pleistocene — Pliocene

Time (mya)

Eulemur coronatus

(A) *Ateles geoffroyi*

(B) *Mandrillus sphinx*

33.28 A Prosimian   The crowned lemur is one of many lemur species found in Madagascar, where it is part of a unique assemblage of endemic plants and animals.

33.29 Monkeys   (A) The spider monkeys of Central America are typical of the New World monkeys, all of which are arboreal. (B) Although many Old World monkey species are arboreal, the mandrills are among the many terrestrial groups.

(A) *Hylobates lar*

(B) *Pongo pygmaeus*

(C) *Gorilla gorilla*

33.30 Apes   (A) The several genera of gibbons are all smaller in size than the other apes. Gibbons are found throughout southeastern Asia. (B) Orangutans are also native ot Asia, living in the forests of Sumatra and Borneo. (C) Gorillas—the largest apes—are restricted to humid African forests. This male is a lowland gorilla. (D) Chimpanzees, our closest relatives, are found in forested regions of Africa.

(D) *Pan troglodytes*

Human ancestors evolved bipedal locomotion

About 6 million years ago in Africa, a lineage split occurred that would lead to the chimpanzees on the one hand and to the *hominid* clade that includes modern humans and their extinct close relatives on the other.

The earliest protohominids, known as *ardipithecines*, had distinct morphological adaptations for *bipedal locomotion* (walking on two legs). Bipedal locomotion frees the forelimbs to manipulate objects and to carry them while walking. It also elevates the eyes, enabling the animal to see over tall vegetation to spot predators and prey. Bipedal locomotion is also energetically more economical than quadrupedal locomotion. All three advantages were probably important for the ardipithecines and their descendants, the australopithecines.

The first *australopithecine* skull was found in South Africa in 1924. Since then australopithecine fossils have been found at many sites in Africa. The most complete fossil skeleton of an australopithecine yet found was discovered in Ethiopia in 1974. The skeleton was approximately 3.5 million years old and was that of a young female who has since become known to the world as "Lucy." Lucy was assigned to the species *Australopithecus afarensis*, and her discovery captured worldwide interest. Fossil remains of more than 100 *A. afarensis* individuals have since been discovered, and there have been recent discoveries of fossils of other australopithecines who lived in Africa 4 to 5 million years ago.

Experts disagree over how many species are represented by australopithecine fossils, but it is clear that multiple species of hominids lived together over much of eastern Africa several million years ago (**Figure 33.31**). A lineage of larger species (weighing about 40 kilograms) is represented by *Paranthropus robustus* and *P. boisei*, both of which died out between 1 and 1.5 million years ago. Members of a smaller lineage of australopithecines gave rise to the genus *Homo*.

Early members of the genus *Homo* lived contemporaneously with *Paranthropus* in Africa for about a million years. Some 2-million-year-old fossils of an extinct species called *H. habilis* were discovered in the Olduvai Gorge, Tanzania. Other fossils of *H. habilis* have been found in Kenya and Ethiopia. Associated with the fossils are tools that these early hominids used to obtain food.

Another extinct hominid species, *Homo erectus*, evolved in Africa about 1.6 million years ago. Soon thereafter it had spread as far as eastern Asia, becoming the first hominid to leave Africa. Members of *H. erectus* were nearly as large as modern people, but their brains were smaller and they had comparatively thick skulls. The cranium, which had thick, bony walls, may have

33.31 A Phylogenetic Tree of Hominids At times in the past, more than one species of hominid lived on Earth at the same time. Originating in Africa, hominids spread to Europe and Asia multiple times. All these closely related species are now extinct, while modern *Homo sapiens* have colonized nearly every corner of the planet.

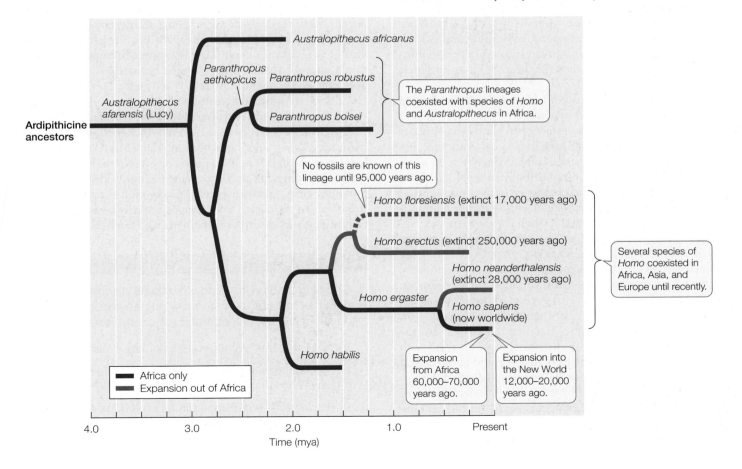

been an adaptation to protect the brain, ears, and eyes from impacts caused by a fall or a blow from a blunt object. What would have been the source of such blows? Fighting with other *H. erectus* individuals is a possible answer.

Homo erectus used fire for cooking and for hunting large animals, and made characteristic stone tools that have been found in many parts of the Old World. Populations of *H. erectus* survived until at least 250,000 years ago, although more recent fossils may also be attributable to this species. In 2004 some 18,000-year-old fossil remains of a small *Homo* were found on the island of Flores in Indonesia. Since then, numerous additional fossils of this diminutive hominid have been found on Flores, dating from 95,000 to 17,000 years ago. Many anthropologists think that this small species, named *H. floresiensis*, was most closely related to *H. erectus*.

Human brains became larger as jaws became smaller

In the hominid lineage leading to *Homo sapiens* and *H. neanderthalensis*, the brain increased rapidly in size. At the same time, the powerful jaw muscles of our ancestors dramatically decreased in size. These two changes were simultaneous, suggesting they might have been functionally correlated. A mutation in a regulatory gene that is expressed only in the head may have removed a barrier that had previously prevented this remodeling of the human cranium.

The striking enlargement of the brain relative to body size in the hominid lineage was probably favored by an increasingly complex social life. Any features that allowed group members to communicate more effectively with one another would have been valuable in cooperative hunting and gathering and for improving one's status in the complex social interactions that must have characterized early human societies, just as they do in ours today.

Several *Homo* species coexisted during the mid-Pleistocene epoch, from about 1.5 million to about 250,000 years ago. All were skilled hunters of large mammals, but plants were important components of their diets. During this period another distinctly human trait emerged: rituals and a concept of life after death. Deceased individuals were buried with tools and clothing, supplies for their presumed existence in the next world.

One species, *Homo neanderthalensis*, was widespread in Europe and Asia between about 500,000 and 28,000 years ago. Neanderthals were short, stocky, and powerfully built. Their massive skull housed a brain somewhat larger than our own. They manufactured a variety of tools and hunted large mammals, which they probably ambushed and subdued in close combat. Early modern humans (*H. sapiens*) expanded out of Africa between 70,000 and 60,000 years ago. Then about 35,000 years ago, *H. sapiens* moved into the range of *H. neanderthalensis* in Europe and western Asia, so the two species must have interacted with one another. Neanderthals abruptly disappeared about 28,000 years ago. Many anthropologists believe it is likely that the Neanderthals were exterminated by these early modern humans. Scientists have been able to isolate large parts of the genome of *H. neanderthalensis* from recent fossils and to compare it with our own. These studies suggest little or no interbreeding between the two species while they occupied the same range in Europe and western Asia.

Early modern humans made and used a variety of sophisticated tools. They created the remarkable paintings of large mammals, many of them showing scenes of hunting, found in European caves. The animals depicted were characteristic of the cold steppes and grasslands that occupied much of Europe during periods of glacial expansion. Early modern humans also spread across Asia, reaching North America perhaps as early as 20,000 years ago, although the date of their arrival in the New World is still uncertain. Within a few thousand years, they had spread southward through North America to the southern tip of South America.

Humans developed complex language and culture

As our ancestors evolved larger brains, their behavioral capabilities increased, especially the capacity for language. Most animal communication consists of a limited number of signals, which refer mostly to immediate circumstances and are associated with charged emotional states induced by those circumstances. Human language is far richer in its symbolic character than other animal vocalizations. Our words can refer to past and future times and to distant places. We are capable of learning thousands of words, many of them referring to abstract concepts. We can rearrange words to form sentences with complex meanings.

The expanded mental abilities of humans enabled the development of a complex **culture**, in which knowledge and traditions are passed along from one generation to the next by teaching and observation. Cultures can change rapidly because genetic changes are not necessary for a cultural trait to spread through a population. Cultural norms, however, are not transferred automatically and must be deliberately taught to each generation.

Cultural transmission greatly facilitated the development and use of domestic plants and animals and the resultant conversion of most human societies from ones in which food was obtained by hunting and gathering to ones in which *pastoralism* (herding large animals) and *agriculture* provided most of the food. The development of agriculture led to an increasingly sedentary life, the growth of cities, greatly expanded food supplies, rapid increases in the human population, and the appearance of occupational specializations, such as artisans, shamans, and teachers.

33.5 RECAP

Grasping limbs with opposable digits distinguish primates from other mammals. Human ancestors developed bipedal locomotion and large brains.

- What are some major trends in primate evolution?

- Describe the differences between Old World and New World monkeys. See p. 713

- Do you understand how cultural evolution differs from genetic evolution? See p. 716

CHAPTER SUMMARY

33.1 What Is a Deuterostome?

- Deuterostomes vary greatly in adult form, but based on the distinctive patterns of early development they share and on phylogenetic analyses of gene sequences, they are judged to be monophyletic. There are far fewer species of deuterostomes than of protostomes, but many deuterostomes are large and ecologically important. Review Figure 33.1, **WEB ACTIVITY 33.1**

33.2 What Are the Major Groups of Echinoderms and Hemichordates?

- Echinoderms and hemichordates both have bilaterally symmetrical, ciliated larvae.
- Adult echinoderms have **pentaradial symmetry**. Echinoderms have an internal skeleton of calcified plates and a unique **water vascular system** connected to extensions called **tube feet**. Review Figure 33.3
- Hemichordate adults are bilaterally symmetrical and have a three-part body that is divided into a proboscis, collar, and trunk. They include the acorn worms and the pterobranchs. Review Figure 33.5

33.3 What New Features Evolved In the Chordates?

- Chordates fall into three principal subgroups: **cephalochordates**, **urochordates**, and **vertebrates**.
- At some stage in their development, all chordates have a dorsal hollow nerve cord, a post-anal tail, and a **notochord**. Review Figure 33.6
- Urochordates include the ascidians (sea squirts), the larvaceans, and the thaliaceans (salps). Cephalochordates are the lancelets, which live buried in the sand of shallow marine and brackish waters.
- The vertebrate body is characterized by a rigid internal skeleton, which is supported by a **vertebral column** that replaces the notochord, internal organs suspended in a coelom, a ventral heart, and an anterior skull with a large brain. Review Figure 33.10

- The evolution of jaws from gill arches enabled individuals to grasp large prey and, together with teeth, cut them into small pieces. Review Figure 33.11
- **Chondrichthyans** have skeletons of cartilage; almost all species are marine. The skeletons of **ray-finned fishes** are made of bone; these fishes have colonized all aquatic environments.

33.4 How Did Vertebrates Colonize the Land?

- Lungs and jointed appendages enabled vertebrates to colonize the land. The earliest split in the **tetrapod** tree is between the **amphibians** and the **amniotes** (reptiles and mammals).
- Most modern amphibians are confined to moist environments because they and their eggs lose water rapidly. Review Figure 33.16, **ANIMATED TUTORIAL 33.1**
- An impermeable skin, efficient kidneys, and an egg that could resist desiccation evolved in the amniotes. Review Figure 33.17, **WEB ACTIVITY 33.2**
- The major living **reptile** groups are the turtles, the lepidosaurs (tuataras, lizards, snakes, and amphisbaenians), and the archosaurs (crocodilians and birds). Review Figure 33.18
- **Mammals** are unique among animals in supplying their young with a nutritive fluid (milk) secreted by mammary glands. There are two primary mammalian clades: the prototherians (of which there are only five species) and the species-rich therians. The therian clade is further subdivided into the marsupials and the eutherians. Review Table 33.1

33.5 What Traits Characterize the Primates?

- Grasping limbs with opposable digits distinguish **primates** from other mammals. The prosimian clade includes the lemurs and lorises; the anthropoid clade includes monkeys, apes, and humans. Review Figure 33.27
- Hominid ancestors were terrestrial primates that developed efficient bipedal locomotion. In the lineage leading to *Homo*, brains became larger as jaws became smaller; the two events may have been developmentally linked. Review Figure 33.31

SEE WEB ACTIVITY 33.3 for a concept review of this chapter.

SELF-QUIZ

1. Which of the following are *not* deuterostomes?
   a. Acorn worms
   b. Sea stars
   c. Urochordates
   d. Brachiopods
   e. Lancelets

2. The structure used by adult urochordates to capture food is a
   a. pharyngeal basket.
   b. proboscis.
   c. lophophore.
   d. mucus net.
   e. radula.

3. The pharyngeal gill slits of chordate ancestors functioned as sites for
   a. uptake of oxygen only.
   b. release of carbon dioxide only.

   c. both uptake of oxygen and release of carbon dioxide.
   d. removal of small prey from the water.
   e. forcible expulsion of water to move the animal.

4. A bony skeleton evolved in the most recent common ancestor of
   a. gnathostomes.
   b. vertebrates and hagfishes.
   c. craniates and urochordates.
   d. ray-finned fishes and sarcopterygians.
   e. amniotes and amphibians.

5. In most ray-finned fishes, lunglike sacs evolved into
   a. pharyngeal gill slits.
   b. true lungs.
   c. coelomic cavities.
   d. swim bladders.
   e. none of the above

6. Many amphibians return to water to lay their eggs because
   a. water is isotonic to egg fluids.
   b. adults must be in water while they guard their eggs.
   c. there are fewer predators in water than on land.
   d. amphibians get their nutrition from water.
   e. amphibian eggs quickly lose water and desiccate if their surroundings are dry.

7. The horny scales that cover the skin of reptiles prevent them from
   a. using their skin as an organ of gas exchange.
   b. sustaining high levels of metabolic activity.
   c. laying their eggs in water.
   d. flying.
   e. crawling into small spaces.

8. Which statement about bird feathers is *not* true?
   a. They are highly modified reptilian scales.
   b. They provide insulation for the body.
   c. They exist in two layers.
   d. They help birds fly.
   e. They are important sites of gas exchange.

9. Prototherians differ from other mammals in that they
   a. do not produce milk.
   b. lack body hair.
   c. lay eggs.
   d. live in Australia.
   e. have a pouch in which the young are raised.

10. Bipedalism is believed to have evolved in the hominid lineage because bipedal locomotion is
   a. more efficient than quadrupedal locomotion.
   b. more efficient than quadrupedal locomotion and frees the forelimbs to manipulate objects.
   c. less efficient than quadrupedal locomotion but frees the forelimbs to manipulate objects.
   d. less efficient than quadrupedal locomotion, but bipedal animals can run faster.
   e. less efficient than quadrupedal locomotion, but natural selection does not act to improve efficiency.

FOR DISCUSSION

1. In what animal groups has the ability to fly evolved? How do the structures used for flying differ among these animals?

2. Extracting suspended food from the water is a common mode of feeding among animals. Which groups contain species that extract prey from the air? Why is this mode of obtaining food so much less common than extracting prey from the water?

3. What risks and benefits are posed by large size?

4. Amphibians have survived and prospered for many millions of years, but today many species are disappearing, and populations of others are declining seriously. What features of their life histories might make amphibians especially vulnerable to the kinds of environmental changes now happening on Earth?

5. The body plan of most vertebrates is based on four appendages. What are the varied forms that these appendages take, and how are they used?

6. Compare the ways in which different animal lineages colonized the land. How were those ways influenced by the body plans of animals in the different groups?

ADDITIONAL INVESTIGATION

A mutation in the gene that encodes the myosin heavy chain (see Chapter 48) decreased the size of the jaw muscles in human ancestors. This mutation may have enabled a restructuring of the cranium and larger brain size. How could we determine when this mutation arose in the phylogenetic history of the primates?

54 Ecology and the Distribution of Life

Dung happens

In 1788, eleven ships carrying settlers from Great Britain landed on the shores of Australia (then called New South Wales), laden with all the supplies the settlers antici-pated would be needed to found a new European colony on that continent. The arriving cargo included two bulls and five cows, which would provide the foundation of an important livestock industry. The settlers, however, made a grievous error by failing to bring along dung beetles.

Dung beetles are beetle species that during their larval life consume the dung, or excrement, of vertebrates, pri-marily grazing mammals. Some species lay eggs directly into the dung pat where it was deposited; others carve up the dung pat and form the pieces into balls, which are then rolled away and buried. Australia's native dung beetles, accustomed to processing and burying the dry, fibrous dung pats of kangaroos, wombats, and other indigenous marsupial mammals, were unable to handle the copious quantities of wet manure produced by the rapidly expanding numbers of cattle and sheep. A cow can drop 10 to 12 dung pats every day and, in the absence of dung beetles to break up and bury the dung, thousands of acres of Australian pasture were rendered unusable for grazing livestock.

In addition, the cattle dung proved an exceptionally suitable habitat for the larvae of the bush fly, *Musca ve-tustissima*, so fly populations exploded. The flies became so pestiferous, flying into people's faces, that eating, drinking, and even breathing were rendered nearly im-possible in some parts of Australia at certain times of year.

In 1964, Australia's national science organization tack-led the dung problem. Entomologists traveled to Africa and other parts of the world to find dung beetle species that could process and bury cattle dung. The first such species, *Onthophagus gazella,* was imported from South Africa in 1968. Four more species arrived the next year, and by 1984, over 50 dung beetle species from around the world had been brought to Australia to take up the task of cattle dung cleanup.

Multiple species with dif-ferent habitat and climate re-quirements were needed to cover the different environ-ments across the Australian continent. The project, deemed a resounding suc-cess, not only reduced bush

***Onthophagus gazella* in Australia** Dung beetles from Africa are capable of process-ing the moist manure of intro-duced mammals such as cattle.

Freed from Bush Flies The importation of dung beetles essentially saved the livestock industries of Australia from plagues of bush flies breeding in the moist dung of cattle and other imported mammals.

fly populations, but had other desirable effects. The beetles reduced the incidence of parasitic infections in cattle by roundworms, hookworms, and other dung-breeding worms, and they increased pastureland productivity by releasing nitrogen from the dung into the soil.

The passengers and crew of the ships that brought cattle to Australia had no idea that they were carrying out a massive ecological experiment. By intent or by accident, people have been moving all kinds of organisms from their native habitats to places they have never lived before. It is now abundantly clear that introducing organisms into existing communities can have unanticipated repercussions and that even small organisms can have large effects on energy flow and nutrient cycling in ecosystems.

IN THIS CHAPTER we will introduce and define the field of ecology. We will see how the physical features of the environment, including climate and geological history, influence the distribution and abundance of organisms. Next we will introduce and describe the major biomes, biogeographic regions, and aquatic environments across which Earth's organisms are distributed.

54.1 What Is Ecology?

Ecology is the scientific investigation of interactions among organisms and between organisms and their physical environment. Ecology is a relatively new branch of the biological sciences; in fact, it did not even have a formal name until 1866. Like evolutionary biology, ecology owes its existence to Charles Darwin, who described the focus of *The Origin of Species* as "the coadaptation of organic beings to each other and to their physical conditions of life." Ernst Haeckel, a German biologist who was profoundly influenced by Darwin, constructed a new word for this new enterprise: *ecology*, from the Greek root *oikos*, "house," where "house" embraces all of an organism's environment. As Haeckel explained, "Ecology is the study of all those complex interrelations referred to by Darwin as 'the conditions of the struggle for existence.'"

Ecology provides explanations of the perceptible, palpable world. Although the *consequences* of enzyme–substrate interactions can be visible, the reactions themselves are not readily visible to the casual observer. In contrast, interactions among organisms, and between organisms and their environment, can, with persistence and ingenuity, be observed. Analysis of those observations, however, often requires additional investigation. That butterflies visit flowers has been observed, and admired aesthetically, for centuries; that butterflies perform a useful service to the flowers they visit by transporting pollen is an ecological insight less than 250 years old.

The need for sound science in making decisions about our own interactions with the environment is another important reason for studying ecology. Humans are part of the biotic environment of a tremendous variety of organisms, and our activities have profound effects on energy flow and nutrient cycling through the physical environment. Ecologists can provide objective data on the consequences of human activities. An understanding of ecology greatly improves our ability to grow food sustainably; to manage pests and diseases safely and effectively; and to deal with natural disasters such as floods and fires. The greater our understanding of ecological interactions, the greater the likelihood that we can accomplish these things without causing a cascade of unanticipated consequences for ourselves and other organisms.

Ecology is not the same as environmentalism

In defining what ecology is, it is important to emphasize what ecology is *not*. "Ecology" is sometimes equated with "environmentalism," but the two terms are not equivalent. Ecology is a science that generates knowledge about interactions in the natural world; as a field of inquiry, it is not inherently focused on human concerns. **Environmentalism** is the use of ecological knowledge, along with economics, ethics, and many other considerations, to inform both personal decisions and public policy relating to stewardship of natural resources and ecosystems.

Ecologists study biotic and abiotic components of ecosystems

From its beginnings, ecology has encompassed both the living, or **biotic**, components and the **abiotic**, or physical and chemical, components of the environment. But the field has advanced and diversified considerably since Haeckel's time. A continuous influx of new tools—mathematical models, molecular techniques, and satellite imaging, to name just a few—as well as new connections to other fields, particularly the physical sciences—have dramatically changed ecological research. Its ultimate goal, however, remains the same: to identify and understand the biotic and abiotic forces that influence the distribution and abundance of organisms.

54.1 RECAP

Ecology is the scientific investigation of interactions among organisms and between organisms and their physical environment.

- What are some reasons to consider ecology a useful scientific enterprise? See p. 1141

- How does ecology differ from environmentalism? See pp. 1142–1143

Climate is one of the abiotic factors that determine the kinds of organisms that can survive and reproduce in a particular place. Because climates set limits on the distribution and abundance of organisms, understanding Earth's climates is a good place to begin our study of ecology.

54.2 Why Do Climates Vary Geographically?

The terms *weather* and *climate* both refer to atmospheric conditions—temperature, humidity, precipitation, wind direction and velocity—but they refer to different time scales. **Weather** is the short-term state of atmospheric conditions at a particular place and time, whereas **climate** refers to the average atmospheric conditions, and the extent of their variation, at a particular place over a longer time. In other words, climate is what you expect; weather is what you get. The responses of organisms to weather are usually short-term—seeking shelter from a sudden rainstorm, for example, or shivering to keep warm when the temperature drops. Responses to climate, on the other hand, tend to be adaptations that affect physiology, morphology, and behav-

ior. These adaptations are among the forces driving speciation over evolutionary time.

Global climates are solar powered

Solar energy input, which determines air temperature, is the major determinant of climates on Earth. The intensity of solar radiation varies from place to place due to the shape of Earth and the tilt of its axis, and its orbit around the sun. The amount of solar energy reaching a given unit of Earth's surface depends primarily on the angle of the sun's rays (**Figure 54.1**). At high latitudes (areas close to the poles), sunlight strikes Earth's surface at an angle, so the incoming solar energy is distributed over a larger area (and is thus less intense) than at the Equator, where sunlight strikes Earth's surface perpendicularly. Moreover, when coming in at an angle, the sun's rays must pass through more of Earth's atmosphere, so more of their energy is absorbed and reflected before it reaches the ground. Because of this difference in solar energy input, air at the poles is colder than air at the Equator. The average temperature of the air over the course of the year decreases about 0.76°C for every degree of latitude (about 110 km) at sea level. In addition, because of the tilt of Earth's axis relative to the sun, higher latitudes experience greater variation in both day length and the angle of arriving solar energy over the course of a year, leading to more seasonal variation in temperature.

In the atmosphere closest to Earth's surface, air temperatures also decrease with elevation, so temperatures at sea level are warmer than temperatures on mountaintops at the same latitude.

Solar energy input determines atmospheric circulation patterns

When a parcel of air is warmed, it expands (that is, its molecules move farther apart) and, because it is less dense than cool air,

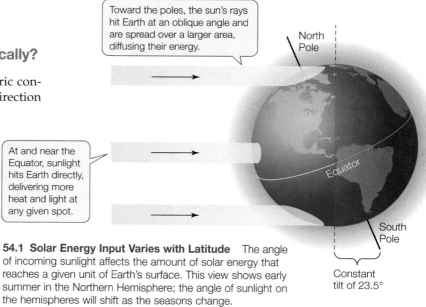

54.1 Solar Energy Input Varies with Latitude The angle of incoming sunlight affects the amount of solar energy that reaches a given unit of Earth's surface. This view shows early summer in the Northern Hemisphere; the angle of sunlight on the hemispheres will shift as the seasons change.

54.2 Air Circulation in Earth's Atmosphere Solar energy drives surface air patterns. Moist, warm air rises and releases moisture as precipitation. Cool, dry air descends, absorbing moisture from the atmosphere and surface.

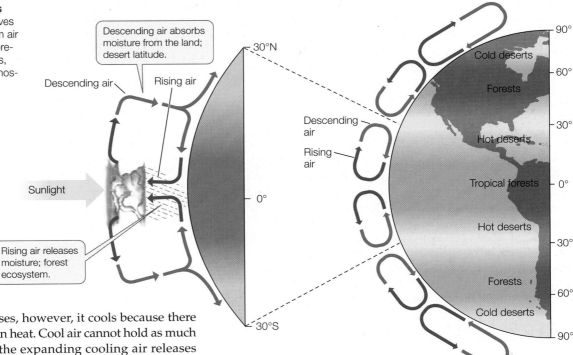

the warm air rises. As it rises, however, it cools because there are fewer molecules to retain heat. Cool air cannot hold as much moisture as warm air, so the expanding cooling air releases moisture in the form of precipitation. Air in the region surrounding the equator receives the greatest input of solar energy. As this tropical air warms, it rises, cools, and releases large amounts of rainfall. And, as this warm air rises into the atmosphere, it is replaced by surface air flowing in from the north and south (**Figure 54.2**).

The air that moves from the north and south into the equatorial tropics is replaced by air from aloft that descends at latitudes of roughly 30°N and 30°S, having traveled away from the equator high in the atmosphere. This air, which cooled and lost its moisture as it rose, now descends, warms, and *takes up* moisture rather than releasing it. Earth's great deserts, such as the Sahara of Africa, the Gobi of China, and the deserts of Australia and the American southwest are located at these latitudes.

At about 60° latitude, air rises again and moves either toward or away from the Equator. At the poles, where there is little solar energy input, air descends again. These cyclic movements of air masses are largely responsible for the prevailing wind patterns at Earth's surface.

Global oceanic circulation is driven by wind patterns

The rotation of Earth on its axis also influences prevailing winds. The velocity of rotation is fastest at the equator, where Earth's diameter is greatest, and slowest close to the poles. A stationary air mass has the same rotational velocity as does the Earth at the same latitude. As an air mass moves toward the equator, however, its rotational movement becomes slower than that of Earth beneath it. Conversely, the rotational movement of an air mass moving poleward is faster than that of the surface beneath it. Therefore, moving air masses are deflected to the right in the Northern Hemisphere and to the left in the Southern Hemisphere. Air masses moving toward the equator from the north and south veer to become the northeast and

southeast *trade winds*, respectively. Air masses moving away from the equator also veer, becoming the *westerly* winds that prevail at mid-latitudes (**Figure 54.3A**).

Incoming solar radiation also affects patterns of water movement in Earth's oceans. Solar energy heats the atmosphere and powers the wind, which in turn, through frictional drag, moves the water it blows over (**Figure 54.3B**). Thus global air circulation patterns drive the circulation patterns of surface ocean waters, known as *currents*. The trade winds, for example, cause currents to converge at the equator and move westward until they encounter a continental land mass. At that point, the currents divide, with some water moving north and some moving south along continental shores. As these currents move toward the poles, the water veers right in the Northern Hemisphere and left in the Southern Hemisphere, just as the winds do.

Ocean currents transport heat, and thus have a large effect on Earth's climates. The poleward movement of water that has warmed while in the tropics transfers large amounts of heat to high latitudes. The Gulf Stream, for example, brings warm water from the tropical Atlantic Ocean (including the Gulf of Mexico) north across the Atlantic to northern Europe, making the climate there considerably milder than that of corresponding latitudes in North America.

In addition to moving water laterally, the trade winds can also cause water to move vertically to influence climate. As trade winds pull surface water away from an area, deeper colder water can rise to replace it. An *upwelling* is an area where water from depths below 50 meters rises to mix with and replace warmer surface waters. The trade winds drive upwellings at the equator; other wind-driven currents create upwellings along coastlines.

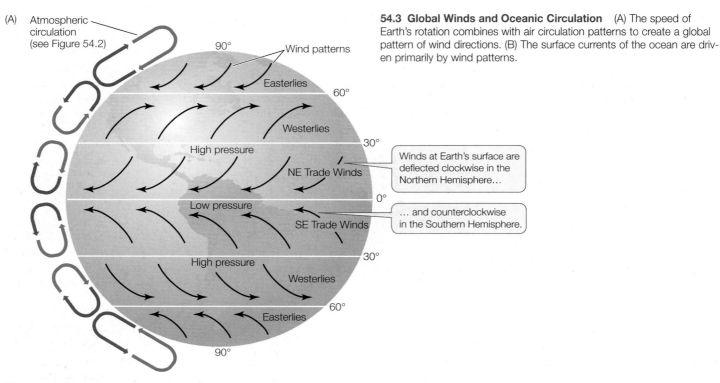

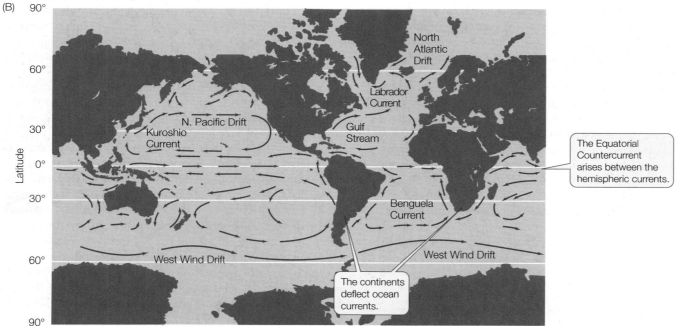

54.3 Global Winds and Oceanic Circulation (A) The speed of Earth's rotation combines with air circulation patterns to create a global pattern of wind directions. (B) The surface currents of the ocean are driven primarily by wind patterns.

Organisms adapt to climatic challenges

The climatic conditions in a region—especially temperature and water availability—act as selective agents on the organisms that live there. As a consequence, many organisms display adaptations to climatic conditions that can involve physiological, morphological, and/or behavioral specializations.

Metabolic specializations help many organisms cope with an environment that periodically becomes too hot, too cold, too dry, or short of food. These specializations are manifested in various forms of resting states that can become a programmed part of the organism's life cycle. *Estivation* (torpor or dormancy during summer), *hibernation* (torpor or dormancy during winter), and *diapause* (the invertebrate equivalent of hibernation) are such states, characterized by dramatically reduced metabolic activity and enhanced physiological resistance to the adverse conditions (see Section 40.4). These states persist until environmental signals indicate that conditions are again conducive to growth.

To deal with extremely low temperatures, cold-hardy organisms are capable of surviving by a variety of strategies, chief among which is producing antifreezes. Microbes, sponges, Arctic fishes, and many temperate-zone insects are among the organisms that produce antifreezes to lower the freezing point of their cell contents or body fluids. One of only two flowering plant species native to Antarctica, the freeze-tolerant hairgrass (*Deschampsia antarctica*) survives by producing proteins that inhibit the formation of damaging large ice crystals. In Alaska, the wood frog (*Rana sylvatica*), the only amphibian found north of the Arctic Circle, survives temperatures of –6°C for over a month with up to 65 percent of its body fluid frozen solid (**Figure 54.4A**). The frog avoids damage to its cells by allowing fluids to freeze in extracellular spaces.

Adaptation to climatic conditions is often reflected in differences in morphology. For example, some endotherms living in cold climates have proportionally rounder shapes and shorter appendages than their relatives adapted to warmer climates, which gives them a smaller surface area and allows them to conserve heat more easily (see Figure 40.18). Pigmentation can also influence rates of heat loss and gain. The larvae of some hornworm moths of the genus *Manduca* are black if they hatch at temperatures below 20°C but are green if the temperature is above that threshold. Black individuals absorb more sunlight and survive at temperatures likely to be lethal to their green conspecifics. When temperature is not an issue, the cryptic green pigmentation is favored because it provides protection from predators (**Figure 54.4B**).

Behavioral mechanisms for temperature regulation often complement physiological and morphological adaptations, particularly among ectotherms. Basking by lizards and evaporative cooling by honeybees are just two examples. Another important behavioral adaptation to climatic conditions is changing one's location to find a more suitable microclimate. A **microclimate** is a subset of climatic conditions in a small specific area that generally differ from those in the environment at large. For example, desert lizards maintain their body temperature by spending time in an underground burrow and moving about the surface only as dictated by environmental conditions (see Figure 40.10). While surface temperatures may fluctuate wildly, the microclimate of the burrow is buffered against such changes.

In addition to such localized movements, long-distance movements can be key adaptations to climatic challenges. Many organisms seek new places to live when local conditions deteriorate. If repeated seasonal changes alter an environment in predictable ways, organisms may evolve life cycles that appear to anticipate those changes. *Migration*, one response to such cyclic environmental changes is discussed in Section 53.5.

Most changes in the abiotic environment, both short-term and long-term, happen independently of anything organisms do. Storms rage, the seasons come and go, and glaciers advance and retreat uninfluenced by the organisms that must deal with their effects. Other environmental changes, however, are influenced by the activities of organisms. As we will see in subsequent chapters, many significant changes in the environment to which organisms must adapt are caused by other organisms, and many characteristics of organisms have been shaped by the history of such interactions.

54.2 RECAP

Latitudinal differences in solar energy input create patterns of atmospheric circulation, and those prevailing winds drive oceanic circulation. Organisms deal with climatic challenges by means of physiological, morphological, and behavioral adaptations.

- How does latitudinal variation in solar energy input drive global air circulation patterns? See pp. 1142–1143 and Figure 54.2

- How do global air circulation patterns drive ocean currents? See p. 1143 and Figure 54.3

The tremendous variation in Earth's climates has given rise to many different assemblages of organisms. Ecologists have found it useful to classify and name environments based on their ecological similarities.

(A) *Rana sylvatica*

(B) *Manduca quinquemaculata*

54.4 Evolutionary Adaptations to Climate (A) Wood frogs can survive frigid Alaskan winters by allowing up to 65 percent of their body fluids to freeze. (B) The larvae of some hornworm moths (genus *Manduca*) have adaptive pigmentation. The black phenotype allows the caterpillar to absorb sunlight more efficiently and survive at cooler temperatures; the cryptic green phenotype is favored if temperature is not an issue.

54.3 What Is a Biome?

A **biome** is an environment that is defined by its climatic and geographic attributes and characterized by ecologically similar organisms, particularly its dominant plants. By providing three-dimensional structure and by modifying microclimates near the ground, the dominant plants of a terrestrial environment strongly influence the lives of the other organisms living there. In contrast, aquatic environments, discussed in Section 54.5, are less dependent on plants for their structure.

The same biome may be present in several widely separated places across the globe, depending in large part on the distribution of suitable climatic conditions (**Figure 54.5**). The identities of the species in geographically separate regions may differ, but through convergent evolution in response to similar selective forces (see Section 22.1), organisms in similar biomes are likely to share many physiological, morphological, and behavioral adaptations.

The distributions of biomes are determined largely by annual patterns of temperature and rainfall. In some biomes, such as temperate deciduous forest, precipitation is relatively constant throughout the year, but temperature varies strikingly between summer and winter. In other biomes, both temperature and precipitation change seasonally. In the tropics, seasonal temperature fluctuations are small and annual cycles are dominated by wet and dry seasons. Tropical biome types are determined primarily by the length of the dry season.

The descriptions of biomes presented here are very general and fall far short of encompassing the variation that can be found within each biome. Furthermore, the boundaries between biomes tend to be arbitrary. Although sometimes an abrupt change is apparent in a landscape, more often one biome gradually merges into another. Despite these uncertainties, recognizing the major biomes of the world is useful inasmuch as these environments share ecological attributes irrespective of their location.

---

yourBioPortal.com

GO TO **Animated Tutorial 54.1 • Biomes**

54.5 The Global Distribution of Biomes The distribution of biomes is based primarily on vegetation patterns and is strongly influenced by annual patterns of temperature and precipitation.

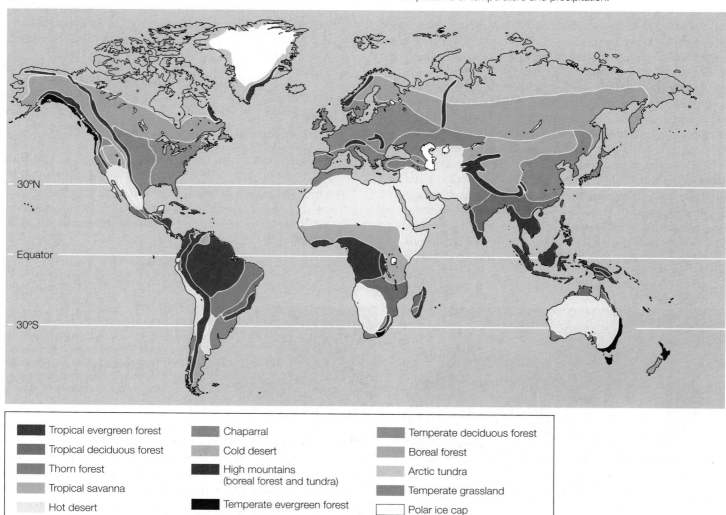

Tropical evergreen forest	Chaparral	Temperate deciduous forest
Tropical deciduous forest	Cold desert	Boreal forest
Thorn forest	High mountains (boreal forest and tundra)	Arctic tundra
Tropical savanna	Temperate evergreen forest	Temperate grassland
Hot desert		Polar ice cap

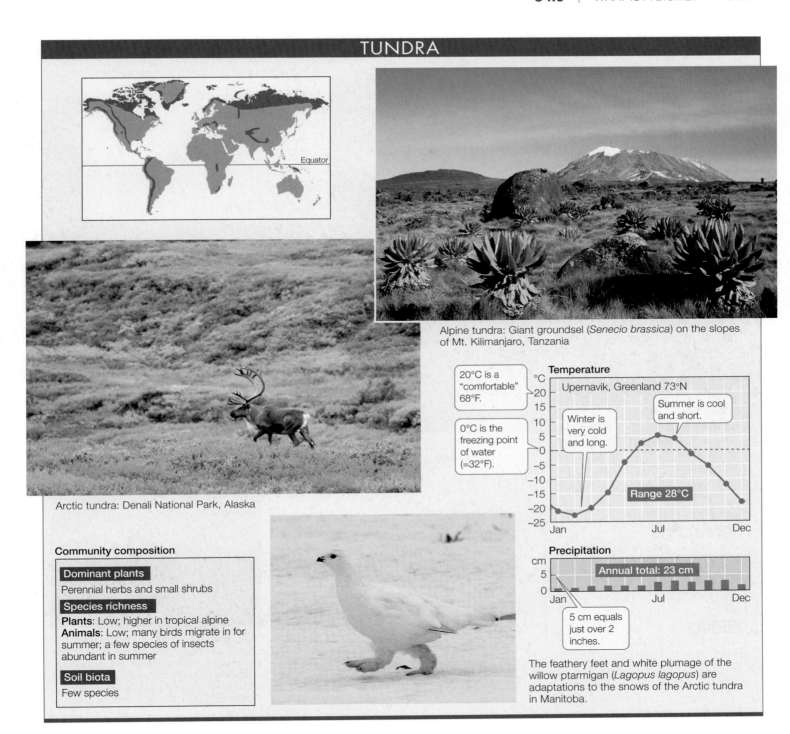

TUNDRA

Equator

Alpine tundra: Giant groundsel (*Senecio brassica*) on the slopes of Mt. Kilimanjaro, Tanzania

Arctic tundra: Denali National Park, Alaska

Community composition

Dominant plants
Perennial herbs and small shrubs

Species richness
Plants: Low; higher in tropical alpine
Animals: Low; many birds migrate in for summer; a few species of insects abundant in summer

Soil biota
Few species

20°C is a "comfortable" 68°F.

0°C is the freezing point of water (=32°F).

Temperature
Upernavik, Greenland 73°N
Winter is very cold and long.
Summer is cool and short.
Range 28°C

Precipitation
Annual total: 23 cm
5 cm equals just over 2 inches.

The feathery feet and white plumage of the willow ptarmigan (*Lagopus lagopus*) are adaptations to the snows of the Arctic tundra in Manitoba.

Tundra is found at high latitudes and high elevations

The **tundra** biome is found in the Arctic and at high elevations in mountains at all latitudes. In Arctic tundra, the vegetation consists of low-growing perennial plants and is underlain by *permafrost*—soil permeated with permanently frozen water. The top few centimeters of the soil thaw during the short summers, when the sun may be above the horizon 24 hours a day. Even though there is little precipitation, lowland Arctic tundra is wet because water cannot drain through the permafrost.

Alpine tundra is found at high elevations outside polar regions. *Tropical alpine tundra* is not underlain by permafrost, so photosynthesis and most other biological activities continue (albeit slowly) throughout the year. A variety of plant growth forms are present in tropical alpine tundra, including low-growing shrubs, rosette perennials, and grasses.

Many tundra plants have hairy leaves that trap heat; the flowers of some species move over the course of the day, tracking the sun's warmth. Most animals are either summer migrants or are dormant for much of the year. Resident birds and mammals, such as the ptarmigan and Arctic fox, have thick fur or feathers that may change color from brown in summer to white in winter.

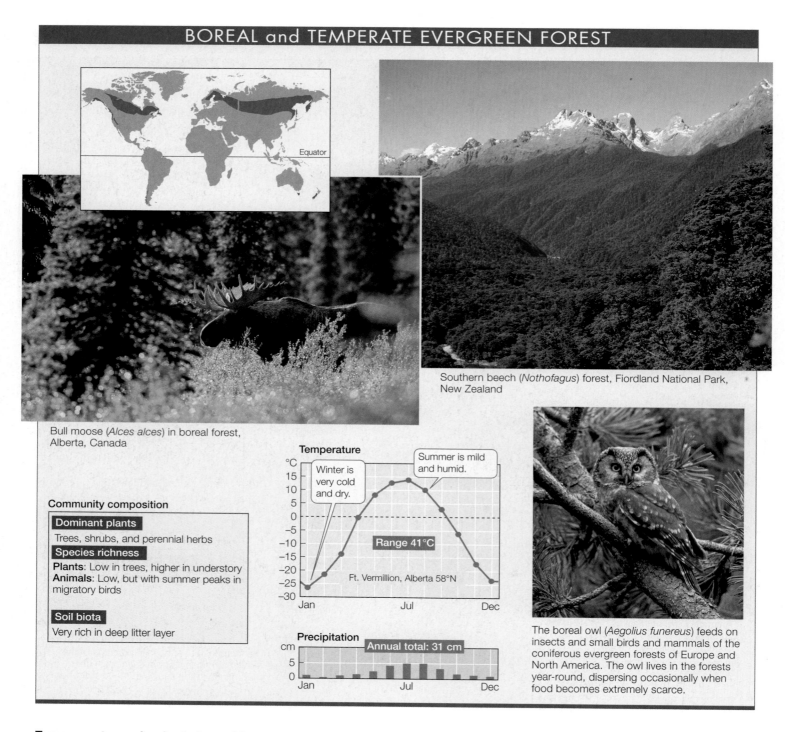

BOREAL and TEMPERATE EVERGREEN FOREST

Equator

Southern beech (*Nothofagus*) forest, Fiordland National Park, New Zealand

Bull moose (*Alces alces*) in boreal forest, Alberta, Canada

Community composition

Dominant plants
Trees, shrubs, and perennial herbs

Species richness
Plants: Low in trees, higher in understory
Animals: Low, but with summer peaks in migratory birds

Soil biota
Very rich in deep litter layer

Temperature

°C
Winter is very cold and dry.
Summer is mild and humid.
Range 41 °C
Ft. Vermillion, Alberta 58°N
Jan　Jul　Dec

Precipitation

cm
Annual total: 31 cm
Jan　Jul　Dec

The boreal owl (*Aegolius funereus*) feeds on insects and small birds and mammals of the coniferous evergreen forests of Europe and North America. The owl lives in the forests year-round, dispersing occasionally when food becomes extremely scarce.

Evergreen trees dominate boreal forests

The **boreal forest** biome (also known as *taiga*) occurs at latitudes below Arctic tundra and at elevations below alpine tundra on temperate-zone mountains. Winters are long and very cold; summers are short, although often relatively warm. The short summer favors trees with evergreen leaves that are ready to photosynthesize as soon as temperatures warm.

The evergreen forests of the Northern Hemisphere are dominated by coniferous gymnosperm species such as spruce and fir. The leaves of conifers are needlelike rather than flat; their reduced surface area cuts down on evaporative water loss. In the Southern Hemisphere forests, the dominant trees are southern beeches (*Nothofagus*), some of which are evergreen.

The dominant mammals of the boreal forest, such as moose and hares, eat leaves, but the seeds in conifer cones support a variety of rodents, birds, and insects. While some small mammals hibernate in winter, voles, lemmings, and mice remain active in snow tunnels, providing food for predators such as foxes and owls.

Temperate evergreen forests grow along the coasts of continents in both hemispheres at middle to high latitudes, where winters are mild but wet and summers are cool and dry.

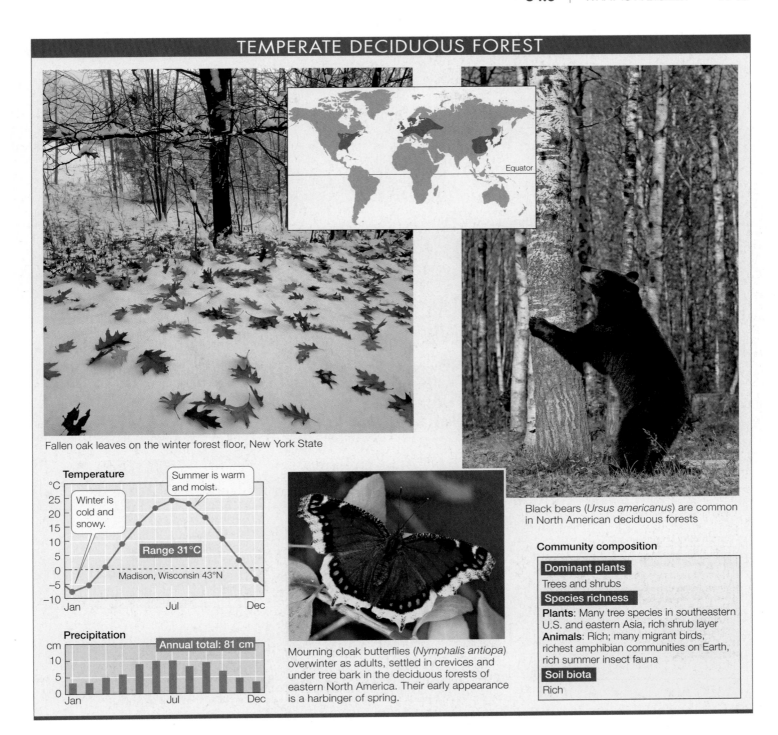

TEMPERATE DECIDUOUS FOREST

Equator

Fallen oak leaves on the winter forest floor, New York State

Temperature

°C

Winter is cold and snowy.

Summer is warm and moist.

Range 31°C

Madison, Wisconsin 43°N

Precipitation

cm

Annual total: 81 cm

Black bears (*Ursus americanus*) are common in North American deciduous forests

Mourning cloak butterflies (*Nymphalis antiopa*) overwinter as adults, settled in crevices and under tree bark in the deciduous forests of eastern North America. Their early appearance is a harbinger of spring.

Community composition

Dominant plants
Trees and shrubs

Species richness
Plants: Many tree species in southeastern U.S. and eastern Asia, rich shrub layer
Animals: Rich; many migrant birds, richest amphibian communities on Earth, rich summer insect fauna

Soil biota
Rich

Temperate deciduous forests change with the seasons

The **temperate deciduous forest** biome is found in eastern North America, eastern Asia, and Europe. Temperatures in these regions fluctuate dramatically between summer and winter, although precipitation is fairly evenly distributed throughout the year. *Deciduous* trees, which dominate these forests, lose their leaves during the cold winters and produce new leaves that photosynthesize rapidly during the warm, moist summers.

Many more tree species live here than in boreal forests. The temperate forests richest in species are those of the southern Appalachian Mountains of the United States and the ones found in eastern China and Japan—areas that were not covered by glaciers during the Pleistocene. Many plant genera are shared among the three geographically separate regions with a deciduous forest biome.

Although many animals are permanent residents of deciduous forests, some (including many birds) migrate to escape the winter cold. Others that remain through the winter acquire massive fat stores in fall and hibernate, often in underground burrows. Many resident insects pass the winter in a state of diapause, the onset of which is triggered by a decreasing number of daylight hours, a reliable predictor of the onset of winter.

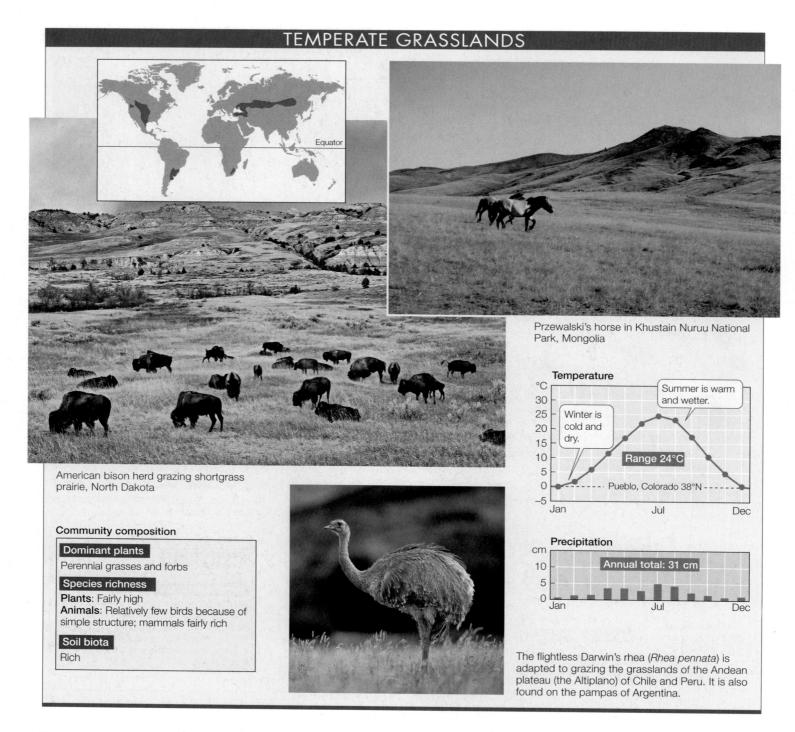

TEMPERATE GRASSLANDS

Equator

Przewalski's horse in Khustain Nuruu National Park, Mongolia

American bison herd grazing shortgrass prairie, North Dakota

Temperature

°C

Winter is cold and dry.

Summer is warm and wetter.

Range 24°C

Pueblo, Colorado 38°N

Jan   Jul   Dec

Community composition

Dominant plants
Perennial grasses and forbs

Species richness
Plants: Fairly high
Animals: Relatively few birds because of simple structure; mammals fairly rich

Soil biota
Rich

Precipitation
cm

Annual total: 31 cm

Jan   Jul   Dec

The flightless Darwin's rhea (*Rhea pennata*) is adapted to grazing the grasslands of the Andean plateau (the Altiplano) of Chile and Peru. It is also found on the pampas of Argentina.

Temperate grasslands are widespread

Temperate grasslands are found in many parts of the world, all of which are relatively dry for much of the year. Most grasslands, such as the pampas of Argentina, the veldt of South Africa, and the Great Plains of North America, have hot summers and relatively cold winters. In some grasslands, most of the precipitation falls in winter (as in California grasslands); in others, the majority falls in summer (as in the Great Plains and the Russian steppe).

Grassland vegetation is structurally simple but rich in species of perennial grasses and *forbs* (herbaceous plants other than grasses). This abundance of plant biomass supports herds of large grazing mammals. Grassland plants are adapted to grazing and to fire. They store much of their energy underground and sprout quickly after being burned or grazed. There are comparatively few trees in temperate grasslands because they cannot survive the periodic fires.

The topsoil of grasslands is usually rich and deep, and thus exceptionally well suited to growing crops, particularly grains such as corn and wheat. As a consequence, most of the world's temperate grasslands have been turned over to agriculture and no longer exist in their natural state.

HOT DESERT

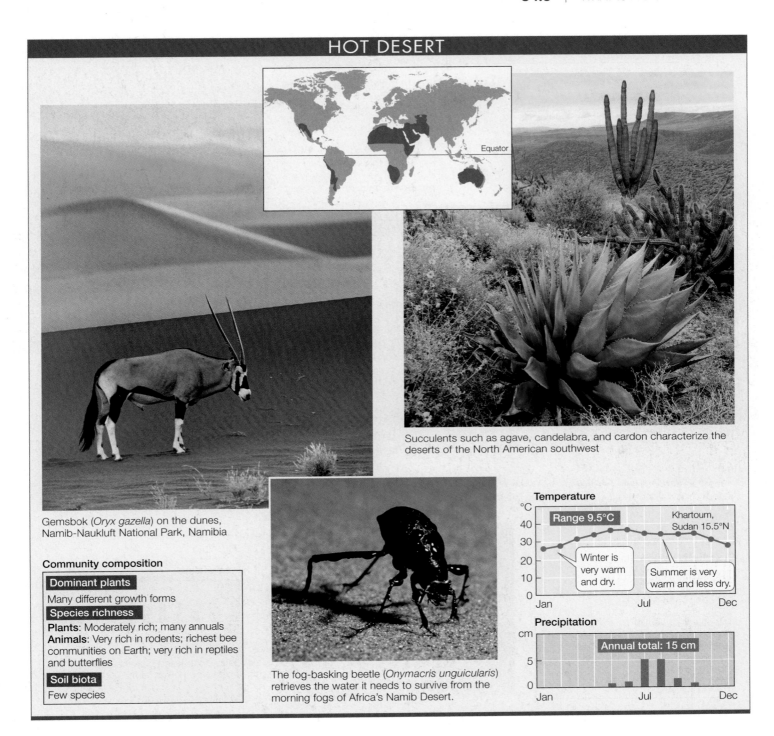

Gemsbok (*Oryx gazella*) on the dunes, Namib-Naukluft National Park, Namibia

Community composition

Dominant plants

Many different growth forms

Species richness

Plants: Moderately rich; many annuals
Animals: Very rich in rodents; richest bee communities on Earth; very rich in reptiles and butterflies

Soil biota

Few species

The fog-basking beetle (*Onymacris unguicularis*) retrieves the water it needs to survive from the morning fogs of Africa's Namib Desert.

Succulents such as agave, candelabra, and cardon characterize the deserts of the North American southwest

Temperature

Range 9.5°C

Khartoum, Sudan 15.5°N

Winter is very warm and dry.

Summer is very warm and less dry.

Precipitation

Annual total: 15 cm

Hot deserts form around 30° latitude

The **hot desert** biome is found in two belts, centered around latitudes 30°N and 30°S (see Figure 54.2). The driest regions, where rains rarely penetrate, are far from the oceans. Examples of this are in the center of Australia and the middle of the Sahara in Africa.

The structural features of desert plants reflect the need to conserve water. These include taproots that reach water far below the surface; a shallow but extensive root system that absorbs water quickly after rare rain events; and small, wax-covered leaves that open their stomata only at night. Most

perennials go dormant during dry seasons, growing quickly as soon as rains return. Their seeds tend to be heat- and drought-resistant and accumulate in a dormant state in the soil.

Small animals are inactive during the hottest part of the day, remaining in underground burrows. When the heat becomes extreme, many small invertebrates estivate. Desert mammals have physiological adaptations for conserving water, including reduced number of sweat glands and kidneys that produce highly concentrated urine. Many desert animals require no water beyond what they can extract from the carbohydrates in their food.

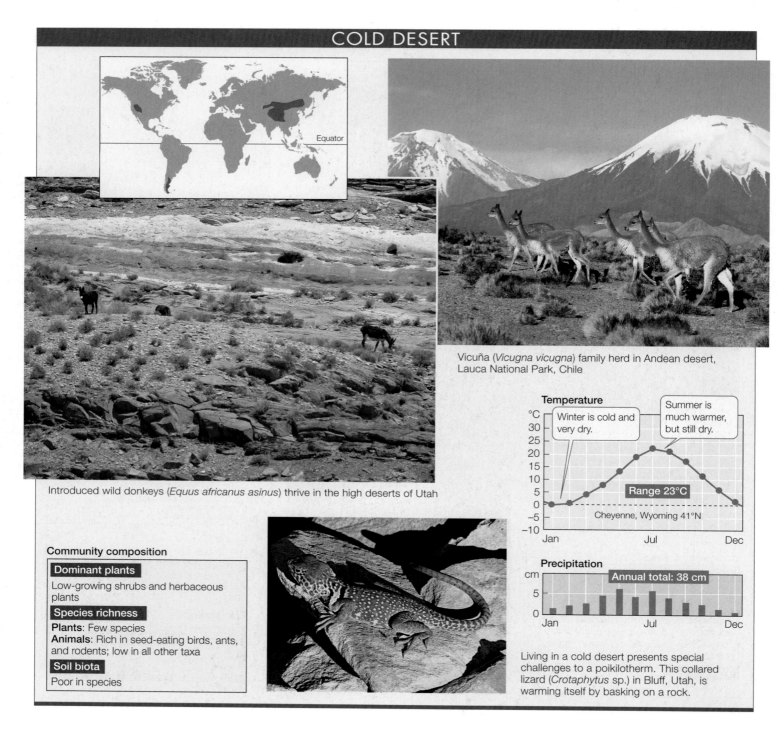

COLD DESERT

Equator

Introduced wild donkeys (*Equus africanus asinus*) thrive in the high deserts of Utah

Vicuña (*Vicugna vicugna*) family herd in Andean desert, Lauca National Park, Chile

Temperature

Winter is cold and very dry.

Summer is much warmer, but still dry.

Range 23°C

Cheyenne, Wyoming 41°N

Jan   Jul   Dec

Precipitation

cm   Annual total: 38 cm

Jan   Jul   Dec

Community composition

Dominant plants
Low-growing shrubs and herbaceous plants

Species richness
Plants: Few species
Animals: Rich in seed-eating birds, ants, and rodents; low in all other taxa

Soil biota
Poor in species

Living in a cold desert presents special challenges to a poikilotherm. This collared lizard (*Crotaphytus* sp.) in Bluff, Utah, is warming itself by basking on a rock.

Cold deserts are high and dry

The **cold desert** biome is found in dry regions at mid- to high latitudes, especially in the interiors of continents in the rain shadows of mountain ranges. Seasonal changes in temperature are great.

Cold deserts are dominated by a few species of low-growing shrubs. The surface layers of the soil are recharged with moisture in winter, and plant growth is concentrated in spring. Cold deserts are relatively poor in species diversity, but the plants that do grow there tend to produce large numbers of seeds, supporting many species of seed-eating birds, ants, and rodents. Burrowing behavior is widespread among cold desert dwellers, but, in contrast with hot desert animals, they burrow to escape cold temperatures, not excessive heat.

CHAPARRAL

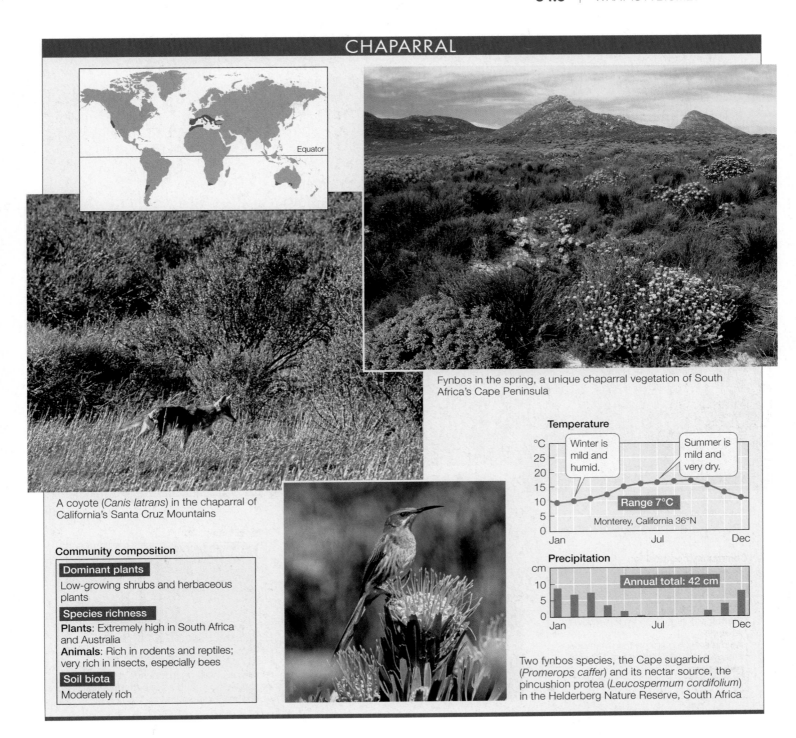

Fynbos in the spring, a unique chaparral vegetation of South Africa's Cape Peninsula

A coyote (*Canis latrans*) in the chaparral of California's Santa Cruz Mountains

Community composition

Dominant plants
Low-growing shrubs and herbaceous plants

Species richness
Plants: Extremely high in South Africa and Australia
Animals: Rich in rodents and reptiles; very rich in insects, especially bees

Soil biota
Moderately rich

Temperature

Winter is mild and humid.

Summer is mild and very dry.

Range 7°C

Monterey, California 36°N

Precipitation

Annual total: 42 cm

Two fynbos species, the Cape sugarbird (*Promerops caffer*) and its nectar source, the pincushion protea (*Leucospermum cordifolium*) in the Helderberg Nature Reserve, South Africa

Chaparral has hot, dry summers and wet, cool winters

The **chaparral** biome is found on the western sides of continents at mid-latitudes (around 40°) where cool ocean currents flow offshore. Winters in this biome are cool and wet; summers are warm and dry. Such climates are found in the Mediterranean region of Europe, coastal California, central Chile, extreme southern Africa, and southwestern Australia.

The dominant plants of chaparral vegetation are low-growing shrubs and trees with tough evergreen leaves that conserve water. The shrubs carry out most of their growth and photosynthesis in early spring, when insects are active and birds breed. Many chaparral species produce strong-smelling defensive chemicals to reduce losses of hard-to-replace foliage to herbivores. Annual plants are abundant and produce large quantities of seeds that fall onto the soil, supporting many small rodents, most of which store seeds in underground burrows. Burrowing to avoid midday heat and nocturnal foraging are strategies used by many chaparral animals. Chaparral vegetation is adapted to periodic fires; the seeds of some species do not germinate until after they have survived a fire. Many shrubs of Northern Hemisphere chaparral produce bird-dispersed fruits that ripen in late fall, when large numbers of migrant birds arrive from the north.

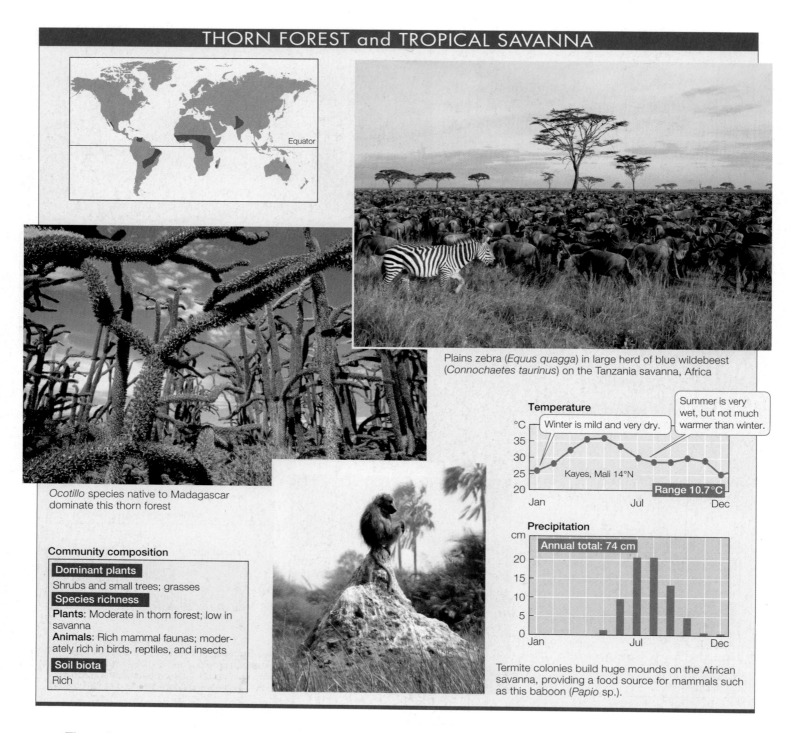

THORN FOREST and TROPICAL SAVANNA

Equator

Plains zebra (*Equus quagga*) in large herd of blue wildebeest (*Connochaetes taurinus*) on the Tanzania savanna, Africa

Ocotillo species native to Madagascar dominate this thorn forest

Temperature

Winter is mild and very dry.

Summer is very wet, but not much warmer than winter.

°C

Kayes, Mali 14°N

Range 10.7°C

Jan   Jul   Dec

Precipitation

cm

Annual total: 74 cm

Jan   Jul   Dec

Community composition

Dominant plants
Shrubs and small trees; grasses
Species richness
Plants: Moderate in thorn forest; low in savanna
Animals: Rich mammal faunas; moderately rich in birds, reptiles, and insects
Soil biota
Rich

Termite colonies build huge mounds on the African savanna, providing a food source for mammals such as this baboon (*Papio* sp.).

Thorn forests and tropical savannas have similar climates

The **thorn forest** and **tropical savanna** biomes are found at latitudes below the hot deserts of Africa, South America, and Australia. Little or no rain falls in winter, but rainfall may be heavy during summer. Thorn forests contain many plants similar to those found in hot deserts. The dominant plants are spiny shrubs and small trees, many of which drop their leaves during the long, dry winter. Trees of the genus *Acacia* are common in thorn forests and savannas worldwide. In Africa, *Andansonia* (baobob) trees are also a hallmark of this biome.

Savanna is characterized by expanses of grasses and grasslike plants with scattered individual trees. The largest tropical savannas are found in central and eastern Africa, where they are populated by herds of grazing and browsing mammals and the large carnivores that prey on them. Grazers and browsers maintain the savannas. If savanna vegetation is not grazed, browsed, or burned, it typically reverts to dense thorn forest. The migration of vast herds of African herbivores in search of "greener pastures" during the dry season is another characteristic of this impressive region.

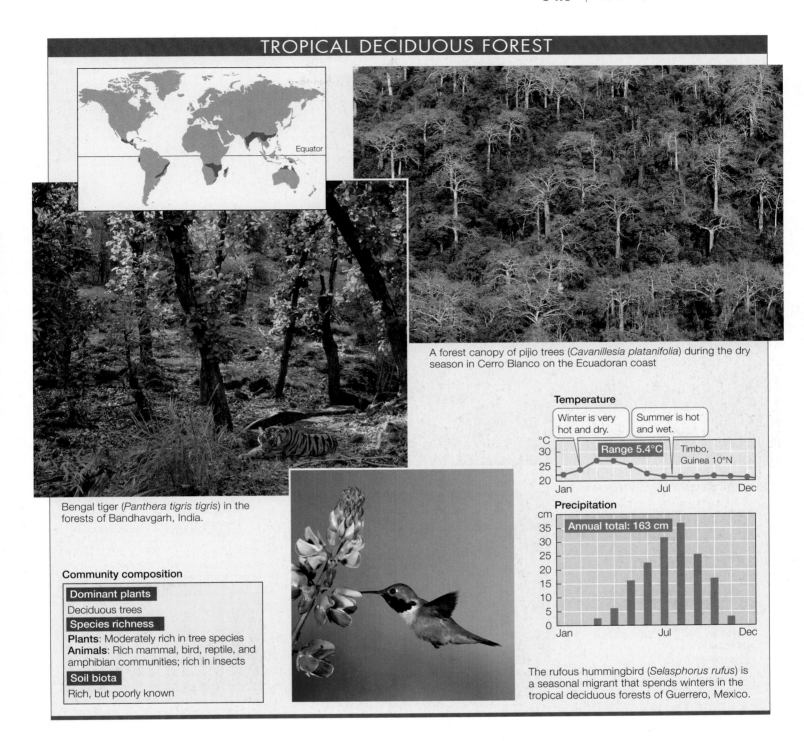

TROPICAL DECIDUOUS FOREST

Equator

A forest canopy of pijio trees (*Cavanillesia platanifolia*) during the dry season in Cerro Blanco on the Ecuadoran coast

Bengal tiger (*Panthera tigris tigris*) in the forests of Bandhavgarh, India.

Temperature

Winter is very hot and dry.

Summer is hot and wet.

Range 5.4°C

Timbo, Guinea 10°N

°C
30
25
20
Jan   Jul   Dec

Precipitation

Annual total: 163 cm

cm
35
30
25
20
15
10
5
0
Jan   Jul   Dec

The rufous hummingbird (*Selasphorus rufus*) is a seasonal migrant that spends winters in the tropical deciduous forests of Guerrero, Mexico.

Community composition

| Dominant plants |
Deciduous trees

| Species richness |
Plants: Moderately rich in tree species
Animals: Rich mammal, bird, reptile, and amphibian communities; rich in insects

| Soil biota |
Rich, but poorly known

Tropical deciduous forests occur in hot lowlands

As the length of the rainy season increases toward the Equator, the **tropical deciduous forest** biome replaces thorn forest. Tropical deciduous forests have taller trees and fewer succulent plants than thorn forests, and they support a much greater number of plant and animal species. Most of the trees, except for those growing along rivers, lose their leaves during the long, hot dry season. Activity picks up in the rainy season; many plants flower while they are still leafless.

Most plant species in this biome are pollinated by animals. In the Sierra Madre Occidental, a mountain range in the extreme southwestern United States extending into western Mexico, tropical deciduous forests are part of a "nectar corridor": a series of patches of flowering plants that provide refueling stops for long-distance migrants traveling north from overwintering sites to breeding sites in the Rocky Mountains. Among the migratory pollinators that fly this corridor are lesser long-nosed bats, white-winged doves, and rufous hummingbirds.

The soils of this biome are among the best in the tropics for agriculture because they contain more nutrients than the soils of wetter areas. As a result, most tropical deciduous forests worldwide have been cleared for agriculture and grazing.

TROPICAL EVERGREEN FOREST

Equator

The canopy of an Ecuadoran tropical forest, seen from above

A golden lion tamarin (*Leontopithecus rosalia*) in tropical forest near Rio de Janeiro, Brazil

Community composition

Dominant plants

Trees and vines

Species richness

Plants: Extremely high
Animals: Extremely high in mammals, birds, amphibians, and arthropods

Soil biota

Very rich but poorly known

The three-toed sloth (*Bradypus variegatus*) is almost totally arboreal, spending its life in the rainforest canopies of Central and South America.

Temperature

°C

The weather is warm and rainy all year.

Range 2.2°C   Iquitos, Peru 3°S

Jan   Jul   Dec

Precipitation

cm   Annual total: 262 cm

Jan   Jul   Dec

Tropical evergreen forests are rich in species

The **tropical evergreen forest** biome (commonly called the rainforest) is found in Equatorial regions where total rainfall exceeds 250 centimeters annually and the dry season lasts no longer than 2 or 3 months. With no seasons unsuitable for growth, it is the most species-rich of all biomes, with up to 500 species of trees per km². Although these forests cover less than 2 percent of Earth's surface, they are home to over half of all known species.

Along with the immense number of species they support, tropical evergreen forests have the highest overall productivity of all ecological communities. However, most mineral nutrients are tied up in the vegetation. The soils usually cannot support agriculture without massive applications of fertilizers. These forests are home to many *epiphytes*, plants that grow on other plants, deriving their nutrients and moisture from air and water rather than soil.

Tropical evergreen forests provide humans with a dazzling range of products, including fruits, nuts, medicines, fuel, pulp, and furniture wood. Many more useful species undoubtedly await discovery, as only a small proportion of this biome's species have been inventoried. The rainforests, however, are currently being deforested or converted to agriculture at a rate of almost 20 million hectares per year.

Biome distribution is not determined solely by temperature

Although warm temperatures and high precipitation support greater numbers of species than do colder, drier conditions, other climatic factors—particularly soil characteristics and the occurrence of wildfires—also influence the structure and life cycles of the dominant vegetation in an area and, consequently, the ecological attributes of the other organisms living there. For example, the plants of Australian deserts grow on extremely nutrient-poor soils and thus have difficulty growing new foliage. Such plants often protect their leaves against consumers by producing large quantities of digestibility-reducing chemicals. Such leaves are generally left uneaten, and when they senesce they drop to the ground, providing an abundance of highly flammable litter to feed the intense fires that periodically sweep across the landscape. As a result, succulent plants, which are easily killed by fires, are not found in Australia, although they are common in deserts on other continents (see p. 1151).

Major topographic features, such as mountains or large lakes, may have regional effects on temperature and precipitation. When prevailing winds bring air masses into contact with a mountain range, for example, the air rises to pass over the mountains, expanding and cooling as it does so. Thus clouds frequently form on the windward side of mountains (the side facing into the winds) and release moisture as rain or snow. On the leeward side (opposite from the direction of the winds), the now-dry air descends, warms, and once again picks up moisture. This pattern often results in a dry area called a **rain shadow** on the leeward side of a mountain range (**Figure 54.6**).

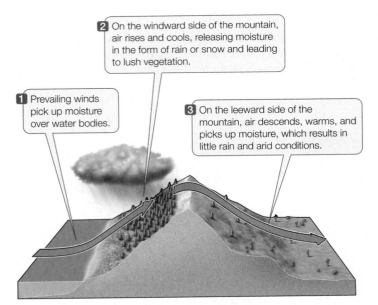

2 On the windward side of the mountain, air rises and cools, releasing moisture in the form of rain or snow and leading to lush vegetation.

1 Prevailing winds pick up moisture over water bodies.

3 On the leeward side of the mountain, air descends, warms, and picks up moisture, which results in little rain and arid conditions.

54.6 A Rain Shadow Mean annual precipitation tends to be lower on the leeward side of a mountain range than on the windward side.

yourBioPortal.com
GO TO **Animated Tutorial 54.2 • Rain Shadow**

54.3 RECAP

Ecologists recognize a number of terrestrial environment types called biomes. The geographic distribution of biomes is determined primarily by temperature and precipitation, but is also influenced by soil characteristics and fire.

- How do temperate grasslands differ from tropical savannas? See pp. 1150 and 1154
- What primary factor distinguishes a tropical biome? See pp. 1154–1156

Climatic conditions explain why chaparral vegetation is confined to the western coasts of continents at mid-latitudes, but it cannot explain why the chaparral in California and the chaparral in South Africa have no species in common. To explain the distribution of species on Earth, another aspect of the abiotic environment—geological history—must be taken into account.

54.4 What Is a Biogeographic Region?

Climate interacts with local abiotic features to influence where and how organisms live, but these are not the only factors that determine where organisms can be found. Evolutionary history—where and when groups of organisms originated and diverged—is key to determining the distributions of organisms, which in turn is profoundly influenced by geological history.

Geological history influenced the distribution of organisms

Until European naturalists traveled the globe, they had no way of knowing how organisms were distributed elsewhere. Alfred Russel Wallace, who along with Charles Darwin advanced the idea that natural selection could account for the evolution of life on Earth (see Section 21.1), was one of those global travelers. Wallace returned to England in 1862 after seven years in the Malay Archipelago, where he noticed some remarkable patterns in the distributions of organisms. For example, he described the dramatically different birds that inhabited the adjacent islands Bali and Lombok:

In Bali we have barbets, fruit-thrushes and woodpeckers; on passing over to Lombock these are seen no more, but we have an abundance of cockatoos, honeysuckers, and brush-turkeys, which are equally unknown in Bali, or any island further west. The strait here is fifteen miles wide, so that we may pass in two hours from one great division of the earth to another, differing as essentially in their animal life as Europe does from America.

Wallace pointed out that these differences could not be explained by climate or by soil characteristics, because in those respects Bali and Lombok are essentially identical.

Wallace saw that, based on the distributions of plant and animal species, he could draw a line that divided the Malay Archipelago into two distinct halves (**Figure 54.7**). He correctly de-

54.7 Wallace's Line Wallace's line corresponds to a deep-water channel between the islands of Bali and Lombok. This channel would have blocked the movement of terrestrial organisms even during the Pleistocene glaciations, when sea level was 100 meters lower than it is today.

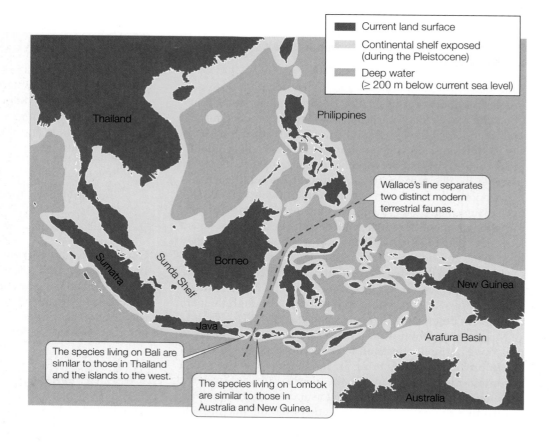

Current land surface

Continental shelf exposed (during the Pleistocene)

Deep water (≥ 200 m below current sea level)

Thailand

Philippines

Wallace's line separates two distinct modern terrestrial faunas.

Sumatra

Sunda Shelf

Borneo

New Guinea

Java

Arafura Basin

The species living on Bali are similar to those in Thailand and the islands to the west.

The species living on Lombok are similar to those in Australia and New Guinea.

Australia

duced that the dramatic differences in flora and fauna were related to the depth of the channel separating Bali and Lombok. This channel is so deep that it would have remained full of water, and thus would have been a barrier to the movement of terrestrial animals, even during the glaciations of the Pleistocene era, when sea level dropped more than 100 m and Bali and the islands to the west were connected to the Asian mainland.

With these insights, Wallace established the conceptual foundations of **biogeography**, the scientific study of the patterns of distribution of populations, species, and ecological communities across Earth. In *The Geographical Distribution of Animals*, published in 1876, he detailed the factors known at the time that influence the distributions of animals, including past glaciation, land bridges, deep ocean channels, and mountain ranges. He earned some measure of scientific immortality in that the Malay discontinuity that first piqued his curiosity is known to this day as "Wallace's line."

The biotas of different parts of the world differ enough to allow us to divide Earth into several continental-scale areas, called **biogeographic regions (Figure 54.8)**, each containing characteristic assemblages of species occupying many different biomes. The boundaries of these biogeographic regions are drawn where assemblages of species change dramatically over short distances. Wallace's line, for example, separates the Oriental biogeographic region from the Australasian biogeographic region. The biotas of biogeographic regions differ because oceans, mountains, deserts, and other barriers restrict the dispersal of organisms from one region to another. Although organisms do disperse between adjacent bio-

geographic regions, such interchanges have not been frequent or massive enough to eliminate the striking differences between them.

───── **yourBioPortal.com** ─────

GO TO **Web Activity 54.1** • Major Biogeographic Regions

Most species are confined to a single biogeographic region. A species found only within a certain region is said to be **endemic** to that region. Remote islands typically have distinctive endemic biotas because water barriers limit opportunities for

54.8 Earth's Biogeographic Regions The major biogeographic regions are separated by climatic, topographic, and/or aquatic barriers to dispersal that cause their biotas to differ strikingly from one another.

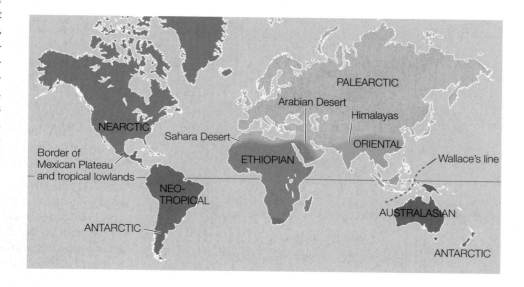

PALEARCTIC

Arabian Desert

Himalayas

NEARCTIC

Sahara Desert

ORIENTAL

Border of Mexican Plateau and tropical lowlands

ETHIOPIAN

Wallace's line

NEO-TROPICAL

AUSTRALASIAN

ANTARCTIC

ANTARCTIC

54.9 Madagascar's Endemic Species The majority of the vascular plant and vertebrate species found on the island of Madagascar are found nowhere else on Earth. Unique reptile and insect species also abound there.

Adansonia grandidieri (giant baobab tree)

Lemur catta (ring-tailed lemur) climbing *Alluaudia procera* (Madagascar ocotillo)

Chamaeleo parsonii (chameleon)

Trachelophorus giraffa (giraffe-necked weevil)

most species to colonize them. For example, nearly all the species of vascular plants and vertebrates on Madagascar, a large island 400 km off the eastern coast of Africa, are endemic to that island (**Figure 54.9**). Regions like Madagascar that are home to large numbers of endemics are classified as "biodiversity hotspots" (see Section 59.4).

Two scientific advances changed the field of biogeography

For many decades after observing that the biotas of the major biogeographic regions are strikingly different, biogeographers speculated about the causes of these differences but the field remained primarily descriptive. It was not until the second half of the twentieth century that two scientific advances transformed biogeography into a dynamic, multidisciplinary field:

the acceptance of the theory of continental drift, and the development of phylogenetic taxonomy.

CONTINENTAL DRIFT Prior to the mid-nineteenth century, most people believed that Earth was young and had changed little over time. No one at that time imagined that organisms—much less the continents themselves—had moved. But that continents can and do move was confirmed when geological evidence of continental drift was eventually discovered (see Section 25.2).

During the Triassic period, the supercontinent Pangaea broke into two great land masses, Laurasia and Gondwana (see Figure 25.12), which subsequently separated into the continents we know today. Those groups of organisms that are represented on two or more continents are believed to be ancient groups whose ancestors were distributed widely over these great landmasses before they broke apart. After the breakup, however, their de-

(A)

(B)

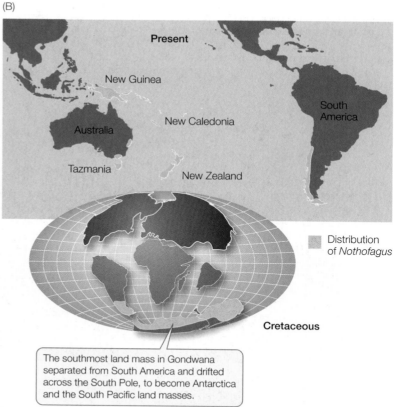

54.10 *Nothofagus* Has a Gondwanan Distribution   The modern range of southern beeches is best explained by their origin in Gondwana during the Cretaceous. Continental drift resulted in the distribution of *Nothofagus* in South America, New Zealand, and the islands of the South Pacific.

The southmost land mass in Gondwana separated from South America and drifted across the South Pole, to become Antarctica and the South Pacific land masses.

scendants evolved independently, so groups that did not originate until after that time have more discrete distributions. Thus continental drift is at least partly responsible for the existence of the seven regions shown in Figure 54.8.

Continental drift also explains certain biogeographic distributions that would otherwise be difficult to understand. For example, the southern beeches—trees of the genus *Nothofagus*—are found in both the Neotropical and the Australasian biogeographic regions. Their distribution suggests that the genus originated in Gondwana during the Cretaceous period and was geographically separated by the breakup of that land mass (**Figure 54.10**).

What evidence do we have that *Nothofagus* did not simply leapfrog from one biogeographic region to another? Fossilized *Nothofagus* pollen from 55–34 million years ago has been found in Australia, New Zealand, western Antarctica, and South America, suggesting that *Nothophagus* was once continuously distributed across a single land mass. Moreover, the modern distribution of aphid genera that feed exclusively on *Nothofagus* parallels the distribution of the trees. There are no air or water currents between Chile and New Zealand that would be likely to disperse insects, indicating that the aphids arose at a time when their host plants grew on a common land mass.

PHYLOGENETIC TAXONOMY   As we saw in Chapter 22, taxonomists have developed powerful methods of reconstructing the phylogenetic relationships among organisms. Biogeographers have adapted these methods to help them answer biogeographic questions. Biogeographers can transform phylogenetic trees into **area phylogenies** by replacing the names of the taxa on a tree with the names of the places where those taxa now live or once lived.

Suppose, for example, we wonder why zebras, which are members of the horse family (Equidae), live in Africa when the fossil record indicates that the Equidae evolved in North America. An area phylogeny of living species suggests that the ancestors of today's horses dispersed from North America to Asia, and then from Asia to Africa, and that the subsequent speciation of zebras took place entirely in Africa (**Figure 54.11**).

Biotic interchange follows fusion of land masses

Geological events not only separate land masses, but may also bring them together. When formerly separated land masses fuse, as may happen through sea level changes or continental drift, species from two different biotas may disperse into a region they had not previously inhabited. This phenomenon is called **biotic interchange**.

A great biotic interchange of mammals occurred when the Central American land bridge formed about 4 million years ago, connecting North and South America for the first time in about 65 million years. While the two continents were separated, their mammalian faunas had diverged greatly. Many of these divergent species dispersed across the newly established land bridge. Only a few South American species—the porcupine, nine-banded armadillo, and Virginia opossum—became established north of the tropical forests of Mexico. However, many North American mammals—rabbits, mice, foxes, bears, raccoons, weasels, cats, tapirs, peccaries, camels, and deer—successfully colonized South America (**Figure 54.12**).

54.11 Taxonomic Phylogeny to Area Phylogeny The conversion of a phylogenetic tree to an area phylogeny helps biogeographers explain how the current distribution of a taxon came about.

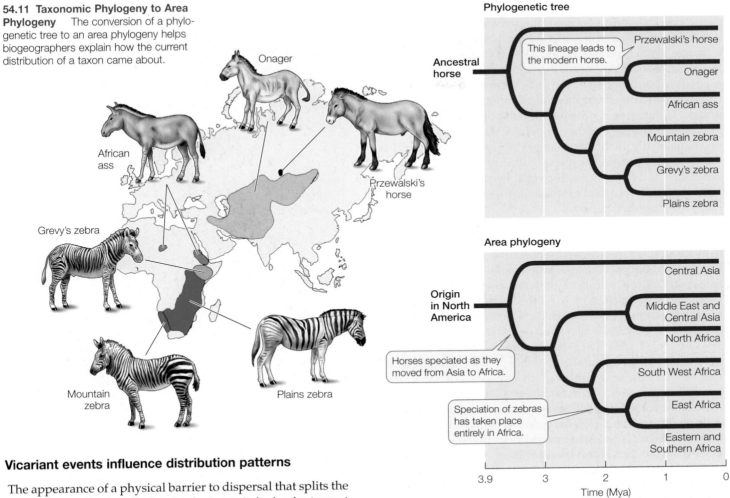

Phylogenetic tree

Onager

African ass

Grevy's zebra

Mountain zebra

Plains zebra

Przewalski's horse

Ancestral horse

This lineage leads to the modern horse.

Przewalski's horse

Onager

African ass

Mountain zebra

Grevy's zebra

Plains zebra

Area phylogeny

Origin in North America

Horses speciated as they moved from Asia to Africa.

Speciation of zebras has taken place entirely in Africa.

Central Asia

Middle East and Central Asia

North Africa

South West Africa

East Africa

Eastern and Southern Africa

3.9   3   2   1   0

Time (Mya)

Vicariant events influence distribution patterns

The appearance of a physical barrier to dispersal that splits the range of a species is called a **vicariant event**. A vicariant event divides the species into two or more discontinuous populations, even though no individuals have dispersed to new areas. If, however, members of a species cross an existing barrier and establish a new population, the discontinuous range of the species is considered to be the result of dispersal.

Given that these two major processes—vicariance and dispersal—both influence distribution patterns, how can biogeographers determine the role of each process when reconstructing the evolutionary history of a particular distribution? By

54.12 Results of a Biotic Interchange (A) The nine-banded armadillo is among only a handful of mammalian species that colonized North America from South America when the Central American land bridge formed. (B) Some species that exist today only in South America are descended from ancestors that dispersed across the land bridge from North America, including the Chacoan peccary and the guanaco (a member of the camel family).

(A) *Dasypus novemcinctus*

(B) *Catagonus wagneri*

Lama guanicoe

studying area phylogenies, a biogeographer may discover evidence suggesting that the distribution of an ancestral species was influenced by a vicariant event, such as continental drift or a change in sea level (see Figure 23.3). If that inference is correct, it is reasonable to assume that ancestral species in other lineages have been affected by the same event and that similar distribution patterns should therefore be seen in other taxonomic groups. Differences in distribution patterns among taxonomic groups may indicate that they responded differently to the same vicariant events, that they diverged at different times, or that they had very different dispersal histories. By analyzing such similarities and differences, biogeographers seek to discover the relative roles of vicariant events and dispersal in determining today's distribution patterns.

The parsimony principle, used to great effect in the reconstruction of phylogenies (see Section 22.2), can also be helpful in biogeographic studies. For example, the New Zealand flightless weevil *Lyperobius huttoni* is found in the mountains of South Island and on sea cliffs at the extreme southwestern corner of North Island (**Figure 54.13**). At first glance, its distribution might suggest that, even though this weevil cannot fly, some individuals in the distant past managed to cross Cook Strait, the 25-km body of water that separates the two islands. However, more than 60 other animal and plant species, including other flightless insects, are found on both sides of Cook Strait. Irrespective of their ability to fly, wade, or swim, it is unlikely that all 60 of these species made the same ocean crossing independently at different times over the course of their evolutionary history. In fact, geological evidence indicates that the present-day southwestern tip of North Island was once united with South Island. Thus a single vicariant event—the separation of the northern tip of South Island from the remainder of the island by the newly formed Cook Strait—could have produced the distribution pattern shared by all 60 species today.

While the concept of vicariance is easily applied in cases attributable to continental drift (due to the ease with which continental movements can be correlated with biological patterns), vicariance patterns on a smaller scale are difficult to identify unequivocally because the underlying geologic events, such as mountain uplift or glacial ice sheet movement, may have been less dramatic than continental movements. In other cases, the vicariance approach is conceptually inappropriate simply because no separation of continuous populations ever took place. There are, for example, more than 135 species of long-horned beetles endemic to the Hawaiian Islands. The islands have never been attached to any continent (they arose by volcanism from the deep ocean floor), so this distribution must have been the result of long-distance dispersal. The beetles, which are most closely related to a genus found in North and Central America, probably colonized the islands and subsequently speciated as a consequence of specializing on different host plants on different islands (see Chapter 23). Founder events must have been infrequent, because the beetles are short-lived and vulnerable to desiccation.

In fact, the present distributions of some mammalian groups provide little hint of their evolutionary origins. Canids (members of the dog family), for example, originated in North Amer-

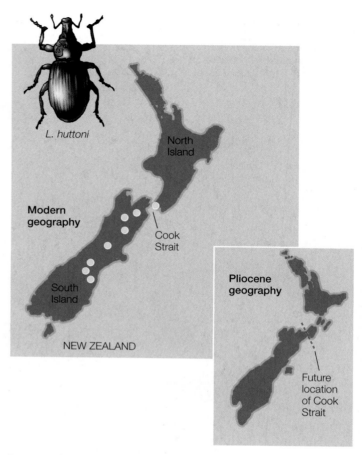

54.13 A Vicariant Distribution Explained Yellow circles indicate the current distribution of the flightless weevil *Lyperobius huttoni*. A comparison of New Zealand's present-day geography with its geography in the Pliocene, when the southern part of today's North Island was part of South Island, suggests that a vicariant event—a physical split separating populations—explains this distribution.

ica but experienced greater evolutionary success after moving into South America; today, only four canid genera are found in North America, but there are six in South America—more than on any other continent.

Humans exert a powerful influence on biogeographic patterns

One more force capable of generating unusual distribution patterns is human activity. Of the insects found both in Europe and in North America, it has been estimated that over half have been transported between the continents by humans. Some species introductions, such as those of dung beetles to Australia, have been deliberate. That introduction turned out to be beneficial, but many others have not. Species that have been introduced to new regions inadvertently include disease vectors (such as the mosquito that transmits yellow fever), household pests (including at least four cockroach species), and crop pests that can result in economic disaster. The effects of such introductions by humans is an important topic and will be discussed at length in Chapter 59.

54.4 RECAP

Biogeographers divide Earth into seven regions, each containing unique assemblages of species. Vicariant events may generate distribution patterns by splitting the ranges of species; distribution patterns may also change when species disperse across barriers.

- What determines the boundaries of Earth's major biogeographic regions? How are these boundaries different from those of the biomes described in Section 54.3? See p. 1146, p. 1158, and Figures 54.5 and 54.8

- Explain how the concepts of continental drift and phylogenetic taxonomy transformed the field of biogeography. See pp. 1159–1160

- How do vicariance and dispersal interact to generate major biogeographic patterns? See pp. 1161–1162

As described in Chapter 25, life originated in the oceans. Today the three-quarters of the planet's surface that is covered with water sustains abundant life as well.

54.5 How Is Life Distributed in Aquatic Environments?

Most of Earth's aquatic environments are oceans. However, the small percentage of the watery world that consists of freshwater lakes and streams also hosts a significant proportion of Earth's aquatic organisms.

The oceans can be divided into several life zones

Earth's oceans form one large, interconnected water mass, interrupted in places by continents. Even organisms with limited swimming abilities can travel long distances simply by floating with the great currents depicted in Figure 54.3. Even though the oceans have no physical barriers to dispersal, however, most marine organisms have restricted ranges. Water temperature, salinity, and food supply all vary spatially. Changes in water temperature, for example, can be barriers to dispersal because many marine organisms function well in only a narrow range of temperatures.

The striking physical and biological discontinuities within the oceans divide them into several distinct "life zones." These zones are identified principally by their distance from the surface, because depth determines how much light is available to sustain the photosynthetic organisms that make chemical energy available to consumer organisms. In both marine and freshwater environments, the depth of water reached by enough sunlight to support photosynthesis is called the **photic zone** (Figure 54.14). Approximately 90 percent of all aquatic life is found in the photic zone.

The **coastal zone**, extending from the shoreline to the edge of the continental shelf, is characterized by relatively shallow, well-oxygenated water and relatively stable temperatures and salinities. These conditions support high densities of phytoplankton, which in turn support some of the world's most important fisheries. Structure in coastal-zone communities can be provided by a variety of organisms. In warmer coastal waters, corals generate complex reef structures that support ecosystems rivaling the rainforest in diversity, and "forests" of the multicellular algal species (seaweeds and giant kelps) form along many coasts.

Near the shoreline, the area of the coastal zone that is affected by wave action is called the **littoral zone**. The principal autotrophs in this zone—sea grasses and algae—support a diversity of invertebrates as well as small fishes. The portion of the coastal zone lying between high- and low-tide levels is the **intertidal**, where tidal movements create highly variable conditions of temperature and salinity. Intertidal organisms, including clams, mussels, copepods, and burrowing worms, are alternately exposed to air or submerged on a regular basis.

Throughout most of the oceans, the dominant autotrophs are phytoplankton (photosynthetic floating protists; see Chapter 27). In the open ocean, referred to as the **pelagic zone**, the principal consumers of phytoplankton are zooplankton—mainly small crustaceans and larval stages of marine animals—which

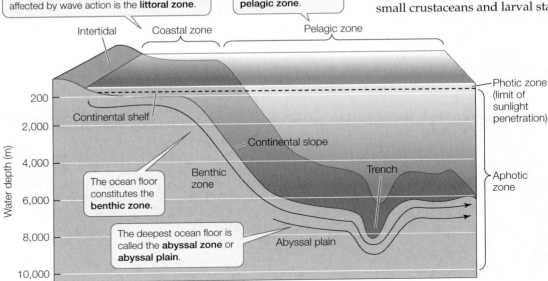

The **coastal zone** extends from the shoreline to the edge of the continental shelf. The area affected by wave action is the **littoral zone**.

The open water above the ocean floor is the **pelagic zone**.

The ocean floor constitutes the **benthic zone**.

The deepest ocean floor is called the **abyssal zone** or **abyssal plain**.

54.14 Oceanic Life Zones The ocean's life zones are in large part determined by sunlight penetration. More than 90 percent of ocean-dwelling species live in the sunlit photic zone, which comprises less than 2 percent of the open water.

in turn support many larger free-swimming vertebrate and invertebrate species.

The ocean bottom is referred to as the **benthic zone**. Many benthic organisms are adapted to life on the seafloor substrate. They include sessile animals such as sponges, bryozoans, ribbon worms, and brachiopods (see Chapters 31 and 32), as well as motile bottom feeders such as crabs and sea slugs (see Figure 32.15C).

Where water is too deep for light to penetrate, little photosynthesis can take place, and both plant and animal diversity are low. Depths reached by less than 1 percent of incoming sunlight constitute the **aphotic zone** (including the **abyssal plain** of the deep ocean floor). Many of the organisms inhabiting these regions subsist on decaying organic matter that sinks down from the photic zone. Some produce their own light by way of bioluminescent organs (see Figure 33.13D). Even deep-ocean trenches and rift valleys support ecosystems, often sustained by chemosynthetic microbes that can metabolize the nutrients in seawater without the aid of sunlight (see Section 26.2).

Freshwater environments may be rich in species

In contrast to the vast oceans, the world's freshwater environments cover less than 3 percent of Earth's surface. However, they are home to about 10 percent of all aquatic species. Freshwater environments may be found in running water (streams and rivers) or in standing water (lakes and ponds). Like the oceans, bodies of standing water can be divided into zones based on depth and light penetration, with a littoral zone close to shore characterized by warm temperatures and high species diversity, a pelagic zone with its upper layers teeming with phytoplankton, and deeper waters below, where little light penetrates and oxygen levels are low, characterized by low diversity.

Discontinuities in the ranges of aquatic organisms are seen because most animals that live in the oceans cannot survive in fresh water, and vice versa. For example, most groups of freshwater fishes cannot tolerate salt water, so they are restricted to a single continent. They can disperse only within the connected rivers and lakes of a drainage basin or in brackish coastal waters.

Estuaries have characteristics of both freshwater and marine environments

Estuaries are bodies of water that form where rivers meet the ocean, in which salt water mixes with fresh water. Having characteristics of both freshwater and marine systems, estuaries are home to many unique species, and they play an important role for other species as a conduit between marine and freshwater environments. Some salmon species that hatch in rivers, for example, spend many months in estuaries adjusting to higher salinities before swimming out to sea to grow into adults. Diversity within estuaries tends to be very high. Estuarine environments have long been an important source of human resources, not the least of which is their role in purifying groundwater. In many places around the world, however, overfishing and pollution threaten estuarine viability.

54.5 RECAP

Despite being connected, the oceans are divided into distinct regions that generate striking physical and biological discontinuities. Oceanic life zones are determined by distance from the surface, which influences how much light is available to sustain photosynthesis. Fresh waters, comprising river basins and thousands of relatively isolated lakes, are also divided into life zones according to the depth and light penetration. Estuaries are bodies of water where salt water mixes with fresh water and support many unique species.

- How does light penetration affect diversity in different life zones of the oceans? See p. 1163 and Figure 54.14

- How do estuaries link freshwater and marine systems, and why are these ecosystems so important? See p. 1164

The physical environment exerts a tremendous influence on the distribution of organisms on Earth, but to understand how and where life flourishes on Earth, it is also important to understand the influence of organisms on one another. In Chapter 55 we will consider the dynamics of populations. Chapters 56 and 57 describe how populations interact and form ecological communities. The final two chapters in the book describe global ecological patterns and emphasize the importance of ecology both to humans and to our planet as a whole.

CHAPTER SUMMARY

54.1 What Is Ecology?

- **Ecology** is the scientific investigation of interactions among organisms and between organisms and their physical environment.
- An organism's environment encompasses both **abiotic** (physical and chemical) and **biotic** components (other living organisms).

- The science of ecology generates knowledge about interactions in the natural world.
- In addition to observations of visible phenomena, ecology relies on a diversity of tools, including mathematical models, molecular techniques, and satellite imaging

- **Environmentalism** is the use of ecological knowledge to inform our decisions about the stewardship of natural resources and ecosystems.

54.2 Why Do Climates Vary Geographically?

- Weather refers to atmospheric conditions at a particular place and time. Climate is the average of atmospheric conditions, and the variation in those conditions, found in a particular place over time.
- The solar energy that reaches a given unit of Earth's surface depends primarily on the angle of the sun's rays, which in turn is a function of latitude. Latitudinal variation in air temperature drives atmospheric circulation patterns of precipitation. **See Figures 54.1 and 54.2**
- Global atmospheric circulation patterns are also driven by the Earth's rotation; these wind patterns in turn drive ocean surface currents. **Review Figure 54.3**
- Organisms respond to climatic challenges with physiological, morphological, and behavioral adaptations.

54.3 What Is a Biome?

- A **biome** is an environment that is defined by its climatic and geographic attributes and characterized by ecologically similar organisms, particularly its dominant vegetation. **Review Figure 54.5, ANIMATED TUTORIAL 54.1**
- The distribution of biomes is determined primarily by climate, but other factors, such as soil characteristics and fire, also influence vegetation. Terrestrial biomes include Arctic and alpine tundra, boreal forest, temperate evergreen and temperate deciduous forests, temperate grasslands, hot and cold deserts, chaparral, thorn forest and savanna, tropical deciduous forest, and tropical evergreen forest. **SEE ANIMATED TUTORIAL 54.2**

54.4 What Is a Biogeographic Region?

- **Biogeography** is the scientific study of the patterns of distribution of populations, species, and ecological communities.
- The boundaries of **biogeographic regions** are drawn where assemblages of species change dramatically over short distances. These boundaries generally correspond to present or past barriers to dispersal. **Review Figures 54.7 and 54.8, WEB ACTIVITY 54.1**
- A species that is found only in a certain region is said to be **endemic** to that region.
- The theory of continental drift explains some discontinuous distributions that include multiple biogeographic regions. **Review Figure 54.10**
- Biogeographers can transform phylogenetic trees into **area phylogenies** that show where organisms now live or once lived. **Review Figure 54.11**
- When formerly separated land masses come together, species from both biotas may disperse into the other region, a phenomenon known as **biotic interchange**.
- Both **vicariant events** and dispersal across boundaries generate distinctive patterns of species distributions. **Review Figure 54.13**

54.5 How Is Life Distributed in Aquatic Environments?

- Oceanic **life zones** are determined by the distance from the surface, which in turn influences how much light is available to sustain photosynthetic organisms. **Review Figure 54.14**
- Fresh waters—river basins and lakes—are also divided into life zones according to depth and light penetration.
- Estuaries are bodies of water where salt and fresh water mix. This globally small but crucial life zone supports many unique species.

SELF-QUIZ

1. Ecology and environmentalism are
   a. synonymous; the terms can be used interchangeably.
   b. differentiated by the emphasis placed by ecology on the biotic rather than the abiotic world.
   c. differentiated by the lack of utility of ecology in solving world problems.
   d. differentiated by the inherent focus of environmentalism on human concerns.
   e. both scientific fields of inquiry that generate knowledge about the natural world but use different tools.

2. A biome is
   a. a large ecological unit defined by its climatic and geographic attributes and characterized by ecologically similar organisms.
   b. a large ecological unit defined by its dominant animals.
   c. a large ecological unit defined by climate characteristics.
   d. a large ecological unit defined by the number of photosynthetic species.
   e. a large ecological unit defined by biogeochemical cycles.

3. The solar energy that reaches a given unit of Earth's surface depends primarily on
   a. the angle of the sun's rays.
   b. the moisture content of the air.
   c. the amount of cloud cover.
   d. the strength of the winds.
   e. day length.

4. Energy from the sun determines
   a. air temperature.
   b. air and wind circulation patterns.
   c. ocean surface currents.
   d. All of the above
   e. None of the above

5. Biogeography as a science began when
   a. European naturalists first traveled the globe and noted patterns in the distributions of organisms.
   b. Europeans traveled to the Middle East during the Crusades.
   c. phylogenetic methods were developed.
   d. the theory of continental drift was accepted.
   e. Charles Darwin proposed the theory of natural selection.

6. Vicariant events
   a. result from organisms dispersing across a barrier.
   b. were common in the past but are rare today.
   c. divide a species into two or more discontinuous populations.
   d. were rare in the past but are common today.
   e. caused most of today's discontinuous species ranges.

7. The oceans can be divided into life zones even though they are all connected because
   a. the rate of photosynthesis is low in the oceans.
   b. ocean currents keep organisms close to where they were born.
   c. light penetration, and hence photosynthesis, vary with depth and determine where organisms live.
   d. trade winds keep warm and cold waters separate.
   e. continents provide barriers to movement of planktonic life forms.

8. A parsimonious interpretation of a distribution pattern is one that
   a. requires the smallest number of undocumented vicariant events.
   b. requires the smallest number of undocumented dispersal events.
   c. requires the smallest total number of undocumented vicariant plus dispersal events.

   d. accords with the phylogeny of a lineage.
   e. accounts for centers of endemism.

9. The only biogeographic region that today is completely isolated from the other regions by water is
   a. the Madagascan region.
   b. the Ethiopian region.
   c. the Neotropical region.
   d. the Australasian region.
   e. the Nearctic region.

10. Which of the following characteristics is unique to Arctic tundra?
    a. Winters are long and cold.
    b. The soil is underlain by a layer of permafrost.
    c. Many small mammals hibernate in winter.
    d. Many of the plants are perennials.
    e. None of these characteristics apply to Arctic tundra.

FOR DISCUSSION

1. Discussions about the likelihood of life on other planets often focus on their proximity to a sunlike star. How different might life on Earth be if the planet's distance from the sun were considerably greater?

2. Today, by far the greatest number of species of fruit flies (genus *Drosophila*) is found in the Hawaiian Islands. Would you conclude that the genus originally evolved in Hawaii and spread to other regions? Under what circumstances do you think it is safe to conclude that a group of organisms evolved in the same region where the greatest number of species live today? (*Hint:* Review the discussion of equid phylogeny and Figure 54.11.)

3. The desert locust (*Schistocerca gregaria*), which has been a major agricultural pest for millennia, is found throughout the Middle East, western Asia, and Africa. Every other species in the genus *Schistocerca* is found in the Western Hemisphere. How could you determine whether the current distribution of species resulted from eastward dispersal (from North America to Africa) or from westward dispersal (from Africa to the Americas)?

4. Processes in nature do not always conform to the parsimony principle. Why, then, do biogeographers often use the parsimony principle to infer the histories of species and lineages?

5. A well-known legend states that Saint Patrick drove the snakes out of Ireland. Give some alternative explanations, based on biogeographic principles, for the absence of indigenous snakes in that country.

6. Most of the world's flightless birds are either nocturnal and secretive (such as the kiwi of New Zealand) or large, swift, and powerful (such as the ostrich of Africa). The exceptions are found primarily on islands, and many of these island species became extinct when humans (and their domesticated animals) arrived. What conditions on islands might permit the survival of flightless birds? Why has human colonization so often resulted in the extinction of such birds? The power of flight has been lost secondarily in representatives of many groups of birds and insects; what are some possible evolutionary advantages of flightlessness that might offset its obvious disadvantages?

ADDITIONAL INVESTIGATION

Alfred Russel Wallace observed that dramatically different birds inhabited Bali and Lombok, even though the strait that separates them is only 15 miles wide. But most birds can easily fly 15 miles, and can probably see the other island across the water. What kinds of data, in addition to those gathered by Wallace, could be obtained to determine why birds do not fly across the strait or, if indeed they do cross the water, why they fail to colonize the island on the other side?

55 Population Ecology

Reindeer games

In 1944, in the midst of World War II, the U.S. Coast Guard established a LORAN (Long-Range Aids to Navigation) tracking station on the tiny island of St. Matthew in Alaska, an isolated and otherwise unoccupied patch of tundra more than 300 km from the nearest village. As an emergency food supply for the 19 men assigned to the island, the Coast Guard brought in 29 reindeer (*Rangifer tarandus*) by barge and released them.

The reindeer thrived on the thick, lush mat of lichens that covered the 128-square-mile island. Other than the men, the island had no reindeer predators; its only other animal occupants were Arctic foxes, one species of vole, and a few ground-nesting birds. As the war wound down and the men left the island, the reindeer were left behind in an environment of plentiful food and no natural predators.

In 1957, David Klein, at the time a U.S. Fish and Wildlife biologist, visited St. Matthew with a field assistant;

together they counted more than 1,350 reindeer, most of which appeared fat and healthy. However, they also noticed areas of seriously overgrazed lichen mats. Klein did not return to the island until 1963, when he and three colleagues hitched a ride with a Coast Guard cutter. By that time, there were more than 6,000 reindeer packed in at a density of 47 per square mile. The island was covered with reindeer tracks and droppings, and the animals were distinctly smaller than the ones sighted 6 years earlier.

The winter of 1963–1964 brought punishing storms, record low temperatures, and tremendous snowfalls to St. Matthew Island. In August 1965, Coast Guard personnel reported massive reindeer deaths. Klein and two colleagues arranged a return visit in the summer of 1966, at which time they found the island littered with reindeer skeletons. The scientists could locate only 42 living reindeer, 41 of which were adult females; the lone male appeared to have deformed antlers. In a remarkably short period, the reindeer population had declined by over 99 percent. Lichens had essentially disappeared from the island, replaced almost entirely by sedges and grasses, on which reindeer cannot subsist. By 1980, the reindeer had entirely disappeared from the island.

Introducing large hoofed mammals to small islands is an inherently risky enterprise, as the experience on St. Matthew graphically illustrates. But

Reindeer Pause Part of the St. Matthew reindeer herd is seen In this photograph taken in 1963, shortly before a particularly severe winter destroyed most of this isolated population. The herd had grown exponentially for almost 40 years since being introduced to the island as a food source for soldiers during World War II.

Predators and Population Cycles Wolves are the principal non-human predators of reindeer, moose, and other large grazing mammals in northern forests and tundra. There were no wolves on St. Matthew Island, a fact largely responsible for the reindeer population's ability to grow exponentially.

such introductions do not always end in disaster. Reindeer populations introduced to the sub-Antarctic island of South Georgia almost a century ago have persisted and appear to be stable.

Why would populations of a particular species in one place explode and crash, but in another seemingly similar place remain stable over time? Understanding how and why populations change in size is more than an academic pursuit. Ecologists study how populations change over time because that knowledge is critical for understanding why some species become pests in some places and not in others, for managing sustainable harvests of economically important species, and for designing plans for conserving endangered species.

IN THIS CHAPTER we will examine how ecologists study populations and investigate how reproductive capacity and environmental resources affect the dynamics of population growth. We will identify factors that limit population densities and determine the effects of environmental variation on population dynamics. Finally, we will show how an understanding of population dynamics is applied to managing populations of importance to humans.

55.1 How Do Ecologists Study Populations?

Well before ecology became a distinct biological discipline, people engaged in population management. Whenever we grow crops or raise livestock, we are explicitly increasing populations of domesticated plants and animals. Pest control strategies aim to reduce populations of organisms whose presence we consider undesirable. Game wardens, park managers, and conservation biologists aim to maintain stable populations of fish, wildlife, and threatened or endangered species. All of these activities require an understanding of **population dynamics**: the patterns and processes of change in populations. The study of population dynamics also allows us to understand the changes in populations we make inadvertently in the course of other human activities—as when the Coast Guard introduced reindeer to St. Matthew.

A **population** consists of the individuals of a species that interact with one another within a given area at a particular time. Populations are important units for study because groups of individuals that interact in time and space have ecological characteristics that individuals do not. At any given moment, an individual organism occupies only one point in space and is a particular age and size. The members of a population, however, are distributed over space and vary in age and size.

Population density is the number of individuals per unit of area or volume. Density is a property of all populations and is a function of the processes that add individuals to the population (births and immigration, or movement of individuals into the population) and the processes that reduce the number of individuals in the population (deaths and emigration, or movement of individuals out of the population).

Populations also have a characteristic *age structure*, or distribution of individuals across age categories, and a characteristic *dispersion* pattern, or spatial distribution of individuals in the environment. These characteristics, which are constantly changing due to births, deaths, and movement, influence the stability of populations and affect the ways in which populations of one species interact with populations of other species. Thus, to study populations, ecologists need to count the individuals in a given area, determine their ages, and calculate the rates at which individuals enter and leave the population. The study of these processes is known as **demography**.

Ecologists use a variety of approaches to count and track individuals

How individuals are counted depends on the nature of the organism under study. Populations of animals, for example, can

(A)

The pattern of folds and notches on each elephant's ears is as unique as a fingerprint.

(B)

A computer chip on a bee's back logs her movements between the hive and flowers.

55.1 Identifying Individuals (A) The pattern of folds and notches on the ears of an elephant is as unique and distinctive as a fingerprint and can be used to recognize individuals in a population. (B) Worker bees in a hive are individually indistinguishable to ecologists, who have come up with ingenious methods of marking. This female honey bee sports a computer chip on her back, which not only identifies her but also logs her movements between the hive and flowers.

be more challenging to count than populations of trees. Most animals can move, so, to avoid double counting, individuals must be identified. Nevertheless, counting every tree in a forest can be logistically difficult, even though the trees are standing still.

In some species, individuals are large and distinct enough, and populations small enough, that investigators can identify all the individuals and count them. Biologists performed this type of count, called a *full census*, on the elephant population of Samburu and Buffalo Springs national reserves in Kenya. By monitoring the population for 21 months, they learned to recognize each of the 760 individual elephants, primarily by their unique and distinctive ear markings (**Figure 55.1A**). Individual recognition is impossible or impractical for most species, however, and to identify such individual organisms they must be marked in some artificial way (**Figure 55.1B**).

In most species, populations are too large and their individual members too small, too similar in appearance, and/or too

mobile for a full census to be conducted. Thus population sizes are often estimated from representative samples using statistical methods.

Population densities can be estimated from samples

Ecologists usually measure the densities of terrestrial organisms as the number of individuals per unit of area; for organisms living in soil, air, or water, the number or mass per unit of volume may be used. Ecologists obtain these measurements from sample units, then extrapolate from these samples to estimate the total population density.

Estimating population densities is easiest for sedentary organisms. Investigators need only count the individuals in a sample of representative locations and extrapolate the counts to the entire geographic range of the population. Individuals may be counted within marked and measured areas called *quadrats*. Plants are often counted along a linear *transect*: a line drawn across the population's range (often designated by a string marked at regular intervals). Any individual that touches the line is counted. By making repeated counts with either of these methods, investigators can make reasonably good estimates of the size of a population.

Counting mobile organisms is more difficult because individuals move into and out of sampling areas. In such cases, investigators may use the **mark–recapture method**. They begin by capturing, marking, and then releasing a number of individuals. Later, after the marked individuals have had time to mix with unmarked individuals in the population, another sample of individuals is captured. The proportion of individuals in the new sample that are marked can be used to estimate the total size of the population in the defined area, as follows:

$$\text{Estimated population size } N = \frac{n_1 \times n_2}{n_{1 \cap 2}}$$

where

n_1 = the total number of individuals in the first sample (captured, marked, and released)

n_2 = the total number of individuals in the second sample

$n_{1 \cap 2}$ = the number of marked individuals recaptured in second sample (the "intersection set" of samples 1 and 2)

In other words, we assume that the proportion of individuals in the second sample that were captured and marked in the first sample is about the same as the proportion of individuals in the sampling area that were captured in the first sample.

Quantifying the size and determining the density of populations are important, but these number are only a starting point for understanding population dynamics, because not all individuals contribute equally to population growth.

Populations have age structures and dispersion patterns

The **age structure** of a population—the distribution of individuals across all age groups—has a profound effect on population growth because reproductive capacity varies with age. Populations with a large proportion of young individuals have a

55.2 Changes in Age Structure Influence Population Growth The elephant population in Kidepo Valley National Park, Uganda, was monitored between 1970 and 2000. During this time, the proportion of the population achieving prime reproductive age range (15–30 years) grew considerably. Such an age structure in a population is likely to result in an increased rate of growth.

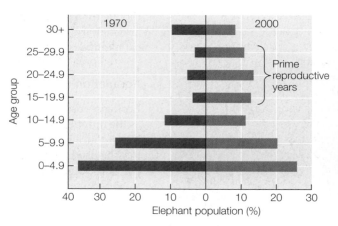

greater potential to grow than populations dominated by individuals that are beyond their peak reproductive years.

For some species, reproduction may be the province of only a tiny fraction of the entire population for only a short interval during the life cycle. Adults of the tiny insect *Clunio maritimus* (the "one-hour midge") mate, lay eggs, and die within about an hour after completing their larval development. In contrast, some vertebrates are capable of reproducing for years. **Figure 55.2** shows the results of a long-term study of the age structure of the elephant population in Kidepo Valley National Park, Uganda. Relative to 1970, the 2000 population skewed significantly toward females, particularly those over 25 years of age. These two disparate age structures were the result of years of differential mortality among young elephants due to drought, and of increased male death rates due to ivory poachers. The age structure as of 2000 portends an increase in the population's growth rate, given that female African elephants become fertile around age 10–15 and can continue producing offspring through their fifties.

Dispersion refers to the distribution of individuals in space within a population. Dispersion determines patterns of interaction among individuals and can thus have important effects on population growth. In addition, ecologists must understand the dispersion patterns of a species to choose appropriate sampling areas and methods for estimating population sizes.

Ecologists recognize three basic dispersion patterns (**Figure 55.3**):

- A *clumped* dispersion pattern occurs when the presence of one individual at any point in space increases the probability of others being near that point.

- A *regular* dispersion pattern occurs when the presence of one individual at any point in space reduces the probability of others being near that point.

- A *random* dispersion pattern occurs when there is an equal probability of an individual occupying any point in space.

(A) Clumped dispersion

(B) Regular dispersion

(C) Random dispersion

55.3 Dispersion Patterns (A) Orcas hunting in pods display a clumped dispersion pattern. (B) Nesting seabirds often stake out territories with a radius defined by their wingspans—an amount of space they can defend without leaving the nest. This results in even dispersion. (C) Dandelion seeds are dispersed by the wind in random fashion.

Spatial variation in environmental conditions strongly influences dispersion patterns. Small-scale differences in temperature, humidity, or wind speed can make particular places more or less suitable for certain organisms. Aphids, for example, cluster along the protruding veins of leaves, where they are sheltered from wind. Interactions among individuals may also bring about characteristic dispersion patterns. Social life, with the cooperation it involves, tends to promote clumped dispersion patterns, as seen in Figure 55.3A. Intraspecific competition for food, space, or mates, on the other hand, tends to space individuals apart in regular dispersion patterns (Figure 55.3B).

Changes in population size can be estimated from repeated density measurements

Ecologists can use multiple estimates of population densities made over time to estimate the rate at which a population is growing or decreasing. Over any given interval of time, the number of individuals in a population increases by the number of individuals added to the population by birth and immigration and decreases by the number of individuals lost from the population by death and emigration. This relationship is expressed mathematically in the equation

$$N_1 = N_0 + (B - D) + (I - E)$$

where

N_1 = the number of individuals at time 1

N_0 = the number of individuals at time 0

B = the number of individuals born between time 0 and time 1

D = the number that died between time 0 and time 1

I = the number that immigrated between time 0 and time 1

E = the number that emigrated between time 0 and time 1

Using this equation to estimate N_1 over multiple time intervals helps researchers estimate changes in population density.

Life tables track demographic events

The study of population dynamics requires keeping track of *demographic events* (births, deaths, immigration, and emigration) in populations and determining the *rate* (number per unit of time) at which they occur. A **life table** is a tool that ecologists use for these purposes. Life insurance companies use similar tables (called "actuarial tables") to determine how much to charge people of different ages for insurance policies. Data from life tables can be used to identify the principal *mortality factors*, or causes of death, at particular life stages, to predict future population trends, and to develop strategies for managing populations of species of commercial or ecological value.

COHORT LIFE TABLES Life tables can be constructed by a number of methods. To construct a *cohort life table*, investigators start with a **cohort**—a group of individuals born within the same time frame, or *age class*—and record their deaths until no individuals from the cohort remain alive. This type of life table is sometimes called a *horizontal life table* because it is based on data collected across the life span.

TABLE 55.1
Life Table for the 1978 Cohort of Cactus Ground Finch (*G. scandens*) on Isla Daphne

AGE CLASS (YEARS)	NUMBER ALIVE	SURVIVORSHIP[a] (l_x)	MORTALITY[b]
0–1	210	—	0.57
1–2	91	0.43	0.14
2–3	78	0.37	0.10
3–4	70	0.33	0.07
4–5	65	0.31	0.05
5–6	62	0.30	0.32
6–7	42	0.20	0.45
7–8	23	0.11	0.35
8–9	15	0.07	0.07
9–10	14	0.07	0.21
10–11	11	0.05	0.09
11–12	10	0.05	0.60
12–13	4	0.02	0.25
13	3	0.01	—

[a] Survivorship = the proportion of the original cohort (here, of 210 individuals) who survive to age x.

[b] Mortality ("death rate") = the proportion of individuals of age x who die before reaching age $x + 1$.

The age categories used in a cohort life table depend on the life cycle of the organism of interest. *Age-dependent life tables* track demographic events as a function of calendar age. *Stage-dependent life tables* track demographic events at various stages of the life cycle (e.g., eggs, larvae, pupae, and adults in insects). They are commonly used when survival and reproduction depend more on developmental stage than on calendar age, as is the case, for example, with insects and other animals that undergo metamorphosis.

Using the data in a cohort life table, investigators can calculate **mortality**: the proportion of individuals of each age class that die before reaching the next age class. They can also calculate **survivorship** (represented by the term l_x), which is the likelihood of an individual member of the cohort surviving to reach age x (**Table 55.1**).

ESTIMATING REPRODUCTIVE CAPACITY A cohort life table can also be used to track the degree to which individuals in different age categories contribute to reproduction (and hence population growth). Because only females produce offspring, life tables generally track the number of offspring produced by each female during each time period—a factor called **fecundity** (indicated in a life table by the term m_x). The portion of the life table that tracks fecundity is called a *fecundity schedule* (**Table 55.2**). Such data allows conservation biologists and other scientists to estimate a population's potential for growth (see Chapter 59).

The data shown in Tables 55.1 and 55.2 track the survivorship (l_x) and fecundity (m_x), respectively, of a cohort of cactus ground finches (*Geospiza scandens*) on Isla Daphne in the Galá-

TABLE 55.2

Fecundity Schedule for the 90 Females of the 1978 Cohort of Cactus Ground Finch (Table 55.1)

AGE CLASS (YEARS)	FECUNDITY[a] (m_x)	
0–1	0.00	
1–2	0.05	
2–3	0.67	
3–4	1.50	El Niño event;
4–5	0.66	increased rainfall
5–6	5.50	
6–7	0.69	Drought
7–8	0.00	
8–9	0.00	
9–10	2.20	
10–11	0.00	
11–12	0.00	

[a] Fecundity = number of fledglings per female per breeding season.

pagos archipelago. Peter and Rosemary Grant followed a cohort of 210 birds from 1978, when they hatched, until 1991, when only 3 individuals—all males—remained alive. All of the cactus ground finches on the island were banded so that the Grants could recognize them as individuals.

The *G. scandens* life tables show that mortality was high during the first year of life, then dropped dramatically for several years before increasing in later years. The fecundity data indicate that females of all ages breed, and breeding success does not correlate exclusively with age. Other observations of conditions on Isla Daphne revealed a correlation with rainfall, which in the Galápagos varies dramatically from year to year (see Table 55.2). These life tables and other ecological data, taken together, suggest that the survival of adult birds and the number of offspring they are able to fledge depend on food availability—that is, on cactus flower and fruit production, which are strongly correlated with rainfall. In short, life table data can be useful in separating out the many ecological factors that affect population dynamics.

Fecundity schedules vary greatly among species not only because organisms differ in the number of offspring they can produce, but also because they vary in the timing of reproduction. Female cactus ground finches begin breeding at the age of 1 or 2 years and, in favorable conditions, can fledge multiple offspring each breeding season. In contrast, female African elephants do not produce offspring until they are at least 15 years old and usually produce only one calf about every 5 years.

VERTICAL LIFE TABLES Because not all species can be easily followed through time, some life tables are constructed by sampling a population at a single time. These life tables cut across all age categories and thus are known as *vertical life tables*. One way to construct a vertical life table is to record information from a *death assemblage*, a collection of bodies or fossils of indi-

viduals that lived together in a particular place at a given time. The birth and death dates on tombstones in a cemetery, for example, can be used to construct a vertical life table for a human population and to estimate its probability of reaching different ages. A 1944 study of Dall mountain sheep (*Ovis dalli dalli*) in Mt. McKinley (now Denali) National Park, Alaska, was based on 608 sheep skulls collected throughout the park. Age at death was estimated by counting growth rings on the horns.

---
yourBioPortal.com
GO TO Working With Data • Dead Sheep Tell a Tale

SURVIVORSHIP CURVES The construction of life tables has allowed ecologists to observe common patterns, reflecting common solutions to ecological challenges, across a tremendous diversity of organisms. For example, mortality data from a life table can be used to plot a **survivorship curve**. Survivorship curves are correlated with general patterns of life history traits, which in turn suggest the mortality factors the population faces in its environment. Ecologists classify survivorship curves into three types (**Figure 55.4A**):

- *Type I*, or physiological, survivorship curves are typical of organisms that experience high overall survivorship through adulthood (such as humans and many other large mammals). Parental care and low fecundity are typical of species with this type of survivorship curve.

- *Type II*, or ecological, survivorship curves are typical of organisms faced with a constant risk of mortality at all ages (such as most birds).

- *Type III*, or maturational, survivorship curves are typical of organisms that experience low juvenile survivorship (such as most insects and annual plants). Species with this type of survivorship curve tend to produce many offspring, but provide little or no parental care.

Note that survivorship curves can differ even among close relatives within a taxonomic group (**Figure 55.4B**).

55.1 RECAP

To understand the dynamics of populations, ecologists measure population density, age structure, and dispersion patterns. Life tables can be constructed either by following a cohort of individuals through time or by recording age at death in a vertical life table.

- What are some of the ways in which population density can be measured? See p. 1169

- What kinds of information do life tables provide about a population? See pp. 1171–1172 and Tables 55.1, 55.2

- Describe the three types of survivorship curves. See p. 1171 and Figure 55.4

Survivorship and fecundity are attributes of all populations, but even within species, environmental variation across a species'

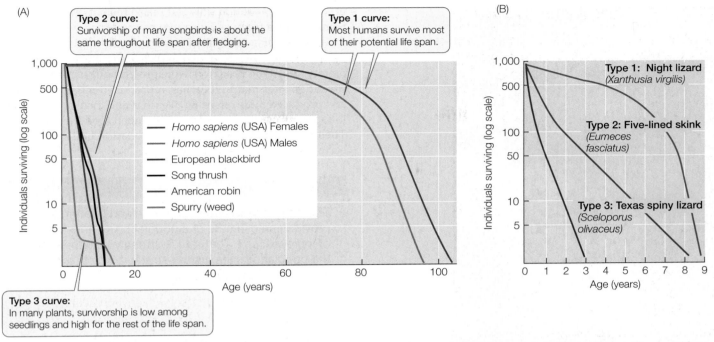

Type 2 curve:
Survivorship of many songbirds is about the same throughout life span after fledging.

Type 1 curve:
Most humans survive most of their potential life span.

Type 3 curve:
In many plants, survivorship is low among seedlings and high for the rest of the life span.

Type 1: Night lizard (*Xanthusia virgilis*)

Type 2: Five-lined skink (*Eumeces fasciatus*)

Type 3: Texas spiny lizard (*Sceloporus olivaceus*)

Legend (A):
— *Homo sapiens* (USA) Females
— *Homo sapiens* (USA) Males
— European blackbird
— Song thrush
— American robin
— Spurry (weed)

55.4 Survivorship Curves (A) Ecologists recognize three types of survivorship curves. (B) Even species that are closely related may have survivorship curves of different types, as indicated here by the survivorship curves of three species of lizards.

range influences these measures. Comparisons across populations and species reveal different patterns in life history traits, which allow organisms to cope with different environmental challenges.

55.2 How Do Environmental Conditions Affect Life Histories?

The way in which an organism partitions its time and energy among growth, maintenance, and reproduction is called its **life history strategy**. Because resources and mortality factors vary greatly among environments, life history strategies also vary dramatically. Those variations, in turn, determine how fast populations can grow.

Survivorship and fecundity determine a population's intrinsic rate of increase

To see how a population is likely to grow, ecologists can use life table data to calculate the population's per capita growth rate, also known as the **intrinsic rate of increase**, symbolized as *r*. A population's intrinsic rate of increase is the difference between the birth rate (*b*) and the death rate (*d*) per individual (*per capita*) (leaving aside, for the moment, immigration and emigration). It is expressed by the equation

$$r = b - d$$

If the birth rate is greater than the death rate, then $r > 0$, and the population is growing. If the death rate is greater than the birth

rate, then $r < 0$, and the population is declining. The equilibrium state $r = 0$ would indicate a stable population that is neither growing nor declining significantly. When applied to the human population, it is sometimes referred to as "zero population growth," with implications we will see in Section 55.5.

Life history traits vary with environmental conditions

Both survivorship (l_x) and fecundity (m_x) are highly habitat-dependent in practice, and the intrinsic rate of increase can thus change as the environment changes. The life history traits most influenced by environmental conditions include:

- age at first reproduction (generation time)
- number of broods per female (the number of times a female produces offspring)
- number of offspring per brood (the number of offspring produced each time a female reproduces)

These factors vary not only between species, but also between populations within species.

Opportunities for reproduction for some species are limited to certain locales or certain times of year, whereas other species and populations can breed continuously over their life span. Many desert wildflowers grow and flower only during the spring rainy season, and they may not be able to reproduce at all in years when rainfall is inadequate. In contrast, some tropical vines are able to flower continuously in their warm, moist rainforest environment.

Species that can reproduce multiple times over the course of their adult lives are **iteroparous** (*itero*, to repeat; *pario*, to beget). **Semelparous** species (*semel*, once), on the other hand, reproduce only once in their lives (**Figure 55.5**). Generally, semelparous species produce many more offspring in their single brood than iteroparous species do over their lifetime; semelparity is thus sometimes called "big bang" reproduction.

Orgyia antiqua

55.5 Big Bang Reproduction   Semelparous species reproduce only once and invest a great deal of energy in producing the maximum number of offspring. Female rusty tussock moths do not fly but remain with their empty cocoons, which are attached to the plants that are the caterpillar-stage food source. The stationary female lays a large number of eggs and then dies. When the eggs hatch the following spring, the larvae are surrounded by foliage they can eat.

Semelparity is typical of organisms that experience no great survival advantage upon reaching adulthood, including some fishes, many insects, and all annual plants (i.e., organisms with a type III survivorship curve). In contrast, iteroparity is typical of organisms whose survival chances increase once they reach maturity (those with a type I or II survivorship curve). For example, because environmental conditions within the nests of social insects (such as honey bees and ants) are remarkably stable, iteroparity is the rule; some queens may live 10 years or longer and reproduce over their entire adult lives.

Life history traits are influenced by interspecific interactions

Predation and other interactions among species can influence life history strategies in many ways. Some populations of guppies (*Poecilia reticulata*) in Trinidad, for example, live in streams where they are attacked and eaten by larger fish. But some streams have waterfalls that predatory fishes are unable to negotiate. Guppies that live in the predator-free areas upstream from those waterfalls have lower death rates than guppies below the falls. To see whether the risk of being eaten by a predator influenced the life history strategies of these guppies, David Reznick and his colleagues collected guppies from high-preda-

tion and low-predation sites and raised them in the laboratory. Some guppies from each group were provided with plentiful food, and others with limited food, to simulate the variation the fish would typically encounter in their home streams. In the laboratory, where no predators were present, guppies from high-predation sites matured earlier, reproduced more frequently, and produced more offspring in each brood than guppies from low-predation sites, no matter how much food they received. The investigators concluded that predation had selected for early and frequent reproduction.

55.2 RECAP

The difference between birth rate and death rate provides an estimate of a population's intrinsic rate of increase, or *r*. That rate is strongly influenced by the population's life history strategy, which in turn is highly dependent on ecological conditions.

- What life history traits vary with environmental conditions? See p. 1173

- Explain the difference between iteroparity and semelparity. See p. 1173

- How can predation affect the evolution of life history strategies? See p. 1174

A species may be rare in one location, yet elsewhere within its range it may be superabundant. A species that is rare in one locality one year may be abundant the next year. For any given species, environmental factors limit the growth of its populations in different places and at different times. Population ecologists work to identify and understand these factors.

55.3 What Factors Limit Population Densities?

What would happen if all the offspring produced by a population survived to reproduce themselves? The prospects are alarming. In 1911, L. O. Howard, then chief entomologist of the U.S. Department of Agriculture, estimated that a pair of flies beginning to reproduce in Washington, D.C., on April 15 could produce a population of 5,598,720,000,000 adults by September 10. Other entomologists took issue with Howard's calculation—they pegged the number much *higher*. Given such amazing reproductive capacities, it is clear there are forces at work that limit the growth of fly populations (and populations of every other organism).

All populations have the potential for exponential growth

As the number of individuals in a population increases, the number of new individuals added per unit of time accelerates, even if the intrinsic rate of increase remains constant. If births and deaths occur continuously and at constant rates, a graph of

(A) Elephant seals, Año Nuevo Island

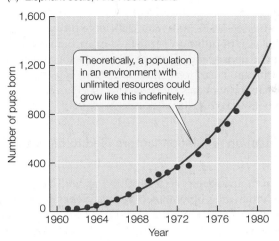

Theoretically, a population in an environment with unlimited resources could grow like this indefinitely.

(B) Reindeer, St. Matthew Island

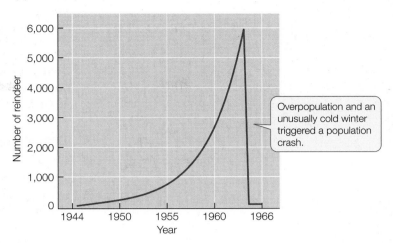

Overpopulation and an unusually cold winter triggered a population crash.

55.6 Exponential Population Growth Can Lead to a Population Crash (A) Abundant resources allow a population to grow exponentially, as did the elephant seal population on Año Nuevo Island, California. (B) The reindeer herd introduced on St. Matthew Island grew exponentially for many years despite a decreasing food supply. A single catastrophically cold winter triggered a population crash that eventually resulted in the death of the entire island reindeer population.

── **yourBioPortal.com** ──

GO TO Animated Tutorial 55.1 • Exponential Population Growth

the population size over time forms a continuous upward curve (**Figure 55.6A**). This pattern, known as **exponential growth**, can be expressed as

$$\frac{\text{change in number of individuals}}{\text{change in time}} = b - d$$

$$\frac{\Delta N}{\Delta t} = b - d$$

Using the notation of differential calculus, which is better suited to short time intervals, this equation can be expressed as

$$\frac{dN}{dt} = (b - d)N$$

Because the intrinsic rate of increase, r, is equivalent to $b - d$, the **equation for exponential growth** can be simplified to

$$\frac{dN}{dt} = rN$$

The term dN/dt is the rate of change in the size of the population over time, and the expression rN is sometimes called the *biotic potential* of the population.

For very short periods, some populations may grow at rates close to their biotic potential. During the 20 years following their introduction, the reindeer population on St. Matthew, described at the opening of this chapter, grew exponentially. When the reindeer herd was introduced to the island, it had ample habitat, abundant food, and no predators, so there was nothing to

limit the population's growth. Favorable climatic conditions also allowed the population to grow exponentially. A rapid change in those conditions—deep snow during one particularly cold winter—was a major factor leading to the population's crash (**Figure 55.6B**).

Logistic growth occurs as a population approaches its carrying capacity

No real population can maintain exponential growth for very long. As a population increases in density, the resources it requires—such as food, nest sites, or shelter—become depleted. In the absence of adequate resources to sustain more individuals, birth rates drop and death rates rise.

Any given environment has only enough resources to support a finite number of individuals of a species indefinitely. That number of individuals, referred to as the environment's **carrying capacity (K)**, is a function of its resources. The growth of a population typically slows down as its density approaches the environmental carrying capacity. A population that exhibits decreasing growth as resources become limiting will display a pattern called **logistic growth**, in which a graph of the population size over time forms an S-shaped curve. **Figure 55.7** shows this growth pattern in a laboratory population of beetles that was maintained on a constant food supply.

An S-shaped curve can be generated from the equation for exponential growth (left) by adding a term that slows the population's growth as it approaches carrying capacity:

$$K - \frac{N}{K}$$

The above term represents the *reduction in population growth caused by preemption of available resources* and is referred to as *environmental resistance*. As long as population size is less than carrying capacity (i.e., $N < K$), only a fraction of the available resources are being used. As the population approaches carrying capacity, however, the fraction of resources available for any new

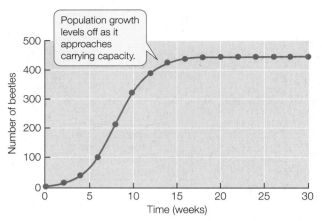

55.7 Logistic Population Growth Levels Off In an environment with limited resources, a population typically stops growing exponentially before it reaches the environmental carrying capacity (K). The data here was recorded from a laboratory population of sawtoothed grain beetles maintained on a constant food supply; it is a typical logistic growth pattern, which forms an S-shaped curve.

───────── **yourBioPortal.com** ─────────
GO TO **Web Activity 55.1 and Animated Tutorial 55.2 •**
Logistic Population Growth

individual becomes smaller. The implication is that each individual added to the population depresses population growth by an equal amount. Thus,

$$\frac{\text{rate of change in population}}{\text{size over time}} = \frac{\text{biotic potential} \times \text{environmental}}{\text{resistance}}$$

or, in mathematical terms,

$$\frac{dN}{dt} = rN\frac{K-N}{K}$$

Population growth stops when $N = K$ because at that point, $K - N = 0$, so $(K - N)/K = 0$, and thus $dN/dt = 0$.

Population growth can be limited by density-dependent or density-independent factors

When resources are limited, adding more individuals to a population runs the risk of making things worse for everyone. Factors with an effect on population size that increases in proportion to population density are called **density-dependent** regulation factors. These include:

- *Food supply.* As a population increases, it may deplete its food supply, reducing the amount of food available to each individual. Poor nutrition may then increase the death rate and/or decrease the birth rate.

- *Predators* may be attracted to areas with high densities of their prey. If predators capture a larger proportion of the prey population than they did when that population was small, the death rate of the prey population rises.

- *Pathogens* spread more easily in dense populations than in populations with fewer individuals per unit of area, resulting in a rise in the death rate.

Not all population regulation factors act in a density-dependent manner. A period of extreme cold or a hurricane that blows down most of the trees in its path may kill a large proportion of the individuals in a population regardless of its density, and thus are said to be **density-independent**. Abiotic factors (such as extreme temperatures) tend to act on populations in a density-independent manner, whereas biotic factors (such as competition for food) tend to be density-dependent.

Different population regulation factors lead to different life histories

Species vary in their capacity to reproduce, as well as in the extent to which they are vulnerable to density-dependent and density-independent mortality factors. Some of this variation in life history traits appears to result from adaptation to different habitat conditions. Generally, unpredictable habitats are associated with high fecundity and correspondingly high r, as organisms make the most of rare opportunities to reproduce; conversely, predictable habitats, where organisms have a high probability of reproductive success, are associated with low fecundity and low r.

Species whose life history strategies allow for high intrinsic rates of increase are called *r-strategists*, and species whose life history strategies allow them to persist at or near the carrying capacity (K) of their environment are called *K-strategists* (Figure 55.8). Many species display elements of both strategies.

For *r*-strategists, life is uncertain. Individuals tend to reproduce only once and to produce large numbers of offspring. They can generally tolerate a wide range of resource conditions. *K*-strategists are adapted to predictable environments, are long-lived, and reproduce several times; their smaller numbers of offspring have a high probability of surviving to adulthood. *K*-strategists tend to be more specialized in their resource use and less tolerant of variation in resource quality. That life history strategies can evolve is suggested by genetic correlations among suites of life history traits. Such genetic correlations imply either simultaneous selection on two life history traits or linkages among the genes that code for those traits. Across *Drosophila melanogaster* strains, for example, intrinsic rate of increase is positively correlated with the ability to reproduce under starvation conditions and with the ability to develop on a variety of media in the laboratory—both of which are consistent with the *r* strategy of tolerating a wide range of conditions.

Several factors explain why some species achieve higher population densities than others

Density-dependent and density-independent factors can explain how populations grow or decline, but they do not explain why some species are common whereas others are rare—that is, why the characteristic densities of species differ. Many factors explain why typical population densities vary so greatly among species, but four of them are especially influential:

- *Species that use abundant resources generally reach higher population densities than species that use scarce resources.* Thus, on average, the fruit fly *Drosophila melanogaster*, which feeds on

r-strategists	K-strategists
HABITAT Can inhabit a broad range of habitats. High tolerance for both environmental instability and low-quality resources.	**HABITAT** Specific habitat requirements, including environmental stability. Efficient users of specific and usually high-quality resources.
PHYSIOLOGY Rapid embryonic development, rapid maturation to reproductive age, small body size.	**PHYSIOLOGY** Extended embryonic development, long maturation to reproductive age, large body size.
REPRODUCTIVE STRATEGY Random mating. Reproduce once (semelparity) resulting in a large number of offspring. Little or no parental investment in each offspring.	**REPRODUCTIVE STRATEGY** Mate choice, pair bonds. Reproduce many times (iteroparity), each event producing few offspring. Large parental investment in each offspring.
SURVIVORSHIP Short life span, density-independent mortality, typically a Type 3 survivorship curve (see Figure 55.4).	**SURVIVORSHIP** Long life span, density-dependent mortality, typically a Type 1 or 2 survivorship curve (see Figure 55.4).
POPULATION FLUCTUATION Short periods of exponential population growth (r) followed by periodic or seasonal population crashes.	**POPULATION FLUCTUATION** Slowly rising population growth that stabilizes and levels off at carrying capacity (K).
EXAMPLES Dandelions, house flies, rabbits	**EXAMPLES** Oak trees, bluebirds, polar bears

55.8 Two Life History Strategies Species whose life histories are geared to achieve the maximum possible rate of population increase are referred to as *r*-strategists; those whose population dynamics are bounded by carrying capacity are *K*-strategists. The life histories of most species combine elements of both types.

yeasts and other microbes found on just about any kind of rotten fruit, reaches substantially higher population densities than do other fruit fly species that feed on the microbes found on specific fruits.

- *Species with small body sizes generally reach higher population densities than species with large body sizes.* In general, population density decreases as body size increases because, on a per capita basis, small individuals require less energy to survive than large individuals.

- *Complex social organization may facilitate high population densities.* Highly social species, including ants, termites, and humans, can achieve remarkably high population densities.

- *Some newly introduced species reach high population densities.* Species that are introduced into a new region, where their normal predators and pathogens are absent, sometimes reach population densities much higher than those in their native ranges. Sometimes these high population densities are only temporary; these densities decline if and when new mortality factors exert an influence. However, in many cases, in the absence of such factors, populations in the newly colonized habitat remain so dense that the introduced species becomes a major problem for native species.

The population of zebra mussels (*Dreissena polymorpha*) in North America demonstrates the speed with which newly introduced populations can grow. Zebra mussels first appeared in Lake St. Clair, between Lake Erie and Lake Huron, in 1988. They were probably transported there from their native Europe in the ballast water of transoceanic cargo ships. They spread rapidly and today occupy most of the Great Lakes and the Mississippi River drainage (**Figure 55.9**), reaching densities as high as 400,000 individuals per square meter in some places. Because they attach to any stable un-

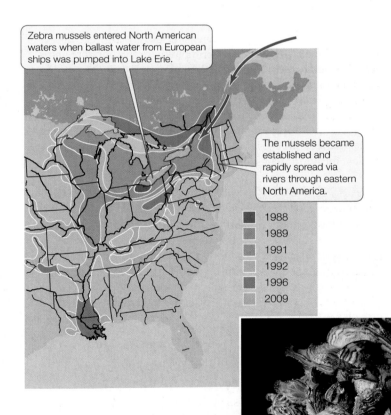

Zebra mussels entered North American waters when ballast water from European ships was pumped into Lake Erie.

The mussels became established and rapidly spread via rivers through eastern North America.

- 1988
- 1989
- 1991
- 1992
- 1996
- 2009

55.9 Newly Introduced Populations Can Grow Rapidly Between 1989 and 2005, the range of zebra mussels in eastern North America increased exponentially. Humans can unwittingly transport zebra mussel larvae from one body of water to another in their fishing boats and other watercraft, and in recent years the invasive pest has begun to appear in lakes and streams of the American West.

Dreissena polymorpha

derwater substratum, zebra mussels can cover the bottoms of boats and docks and clog municipal water supply intakes and power plant pipelines. They even settle on other aquatic organisms, causing problems for native mussels. Such high densities are never found in their native Europe, where over 100 species of predators and parasites keep their population densities under control.

Evolutionary history may explain species abundances

The four key factors that explain variation in population densities, as important as they are, cannot explain all differences in species abundances. For example, although Douglas firs and giant sequoias are both large trees that use the same sources of energy (sunlight) and nutrients, Douglas firs are widespread and abundant in western North America, whereas giant sequoias are restricted to a few groves in southern California. Similarly, each of several species of desert pupfish is restricted to a single spring in Death Valley, California, whereas smallmouth bass live in most of the rivers and lakes in eastern North America. To explain these differences, it is important to know not just the contemporary ecology of these organisms, but also their long-term evolutionary history.

As Section 23.2 describes, a new species can originate in several ways. Species that arise by polyploidy or by founder events inevitably begin with a very small, local population. Desert pupfish species appear to have evolved in isolation as increasing aridity in Death Valley over the past 50,000 years cut once continuous populations off from one another. Conversely, when a species is declining toward extinction (as may be happening to the giant sequoia), its range shrinks until it vanishes when the last individual dies.

55.3 RECAP

Population sizes are limited by the carrying capacity of the environment, which is determined by the availability of resources as well as by biotic interactions. Species associated with unpredictable habitats tend to be *r*-strategists, and species associated with predictable habitats tend to be *K*-strategists.

- Why can populations grow exponentially only for short periods? See p. 1175 and Figures 55.6 and 55.7

- What is the difference between density-dependent and density-independent factors? See p. 1176

- Describe the characteristics of *r*-strategists and *K*-strategists. See p. 1176 and Figure 55.8

All species, no matter how abundant, are found only in those habitats in which they can survive and reproduce well enough to persist over time. Yet a species is rarely found in all of the habitats that seem suitable for it. The evolutionary histories of species supply one explanation for this observation. The next section explores another explanation: spatial variation in habitat suitability.

55.4 How Does Habitat Variation Affect Population Dynamics?

Most natural history field guides display maps that show the geographic range over which a species is found. But no species, not even the most abundant, is found everywhere within its mapped range. Every species has particular habitat requirements that determine where within its range it will occur.

Many populations live in separated habitat patches

Most organisms live in distinct **habitat patches**: areas of a particular habitat type that are surrounded by other, less suitable habitats. Some populations living in separated habitat patches are effectively divided into separate, discrete *subpopulations*, linked together by regular movement of individuals between patches. The larger population to which such subpopulations belong is known as a **metapopulation**.

Each subpopulation has a probability of "birth" (colonization of its habitat patch) and "death" (extinction in that patch). Each subpopulation grows in the ways we have described, but because the subpopulations are much smaller than the metapopulation, local disturbances and random fluctuations in numbers of individuals are more likely to cause the extinction of a subpopulation than of the entire metapopulation. However, if individuals move frequently between subpopulations, immigration may prevent declining subpopulations from becoming extinct, a process called the **rescue effect**.

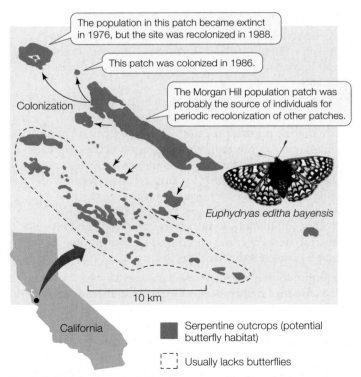

The population in this patch became extinct in 1976, but the site was recolonized in 1988.

This patch was colonized in 1986.

The Morgan Hill population patch was probably the source of individuals for periodic recolonization of other patches.

Colonization

Euphydryas editha bayensis

10 km

California

■ Serpentine outcrops (potential butterfly habitat)

⌐⌐ Usually lacks butterflies

55.10 A Checkerboard of Checkerspots The Bay checkerspot butterfly metapopulation is divided into a number of subpopulations confined to patches of habitat (serpentine rock outcrops) that contain its food plants.

The Bay checkerspot butterfly (*Euphydryas editha bayensis*) provides a dramatic illustration of the dynamics of metapopulations. The caterpillars of this butterfly feed on only two species of annual plants, California plantain and purple owl's clover, which are restricted to outcrops of serpentine rock on hills south of San Francisco, California. In 1960, Paul Ehrlich and his colleagues at Stanford University began studying a population of this butterfly in the Jasper Ridge Biological Preserve near the Stanford campus. They determined that the Jasper Ridge population was actually a subpopulation within a large, very fragmented metapopulation. They found that, over the years, three subpopulations within this metapopulation varied enormously and asynchronously in size. Larval survival depends on climatic factors, particularly temperature, the timing of rainfall, and host plant survival. During drought years, most host plants die early in spring, before the caterpillars have developed enough to enter their summer resting stage.

At least three butterfly subpopulations went extinct during a severe drought in 1975–1977. One of the empty patches was repopulated a few years later, most likely by individuals from the largest single subpopulation, near Morgan Hill, which as late as 1989 contained several hundred thousand butterflies (**Figure 55.10**). In 1998, however, the Morgan Hill subpopulation went extinct. Ehrlich and colleagues examined 70 years of climate data for the region and concluded that increasing climatic variation accounted for the extinction. Without a stable source subpopulation to provide emigrants for recolonization, it is likely that none of the other subpopulations will persist without human intervention.

Corridors allow subpopulations to persist

In any metapopulation, connections between patches, known as **corridors**, play a critical role in facilitating dispersal to maintain subpopulations. Studying corridors is experimentally daunting because long distances may separate patches; moreover, movements of animals, particularly those that fly, can be difficult to monitor. Therefore, to test the importance of corridors, ecologists conducted a small-scale, manipulative experiment using mosses growing on rocks, which provide habitat for a number of small arthropod species, including springtails (minute wingless hexapods) and mites.

yourBioPortal.com
GO TO Animated Tutorial 55.3 • Habitat Fragmentation

In one experiment, the investigators created patches of habitat by clearing away the mosses surrounding the patches (**Figure 55.11**). In small, completely isolated patches, the number of small arthropod species present declined about 40 percent within a year. The investigators also created patches that were connected by narrow corridors of moss. In some cases the corridors were left intact; in others, "pseudocorridors" were disrupted by a barrier 10 mm wide. A 10-mm barrier may seem small, but it presents a daunting obstacle to arthropods only 2 mm wide. Six months later, patches connected by unbroken corridors contained more small arthropod species than did patches connected by the disrupted pseudocorridors.

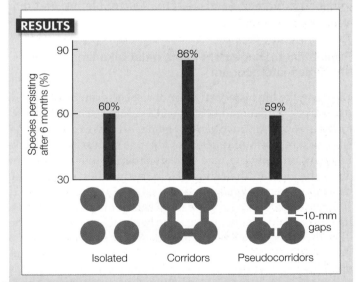

INVESTIGATING LIFE

55.11 Corridors Can Affect Metapopulations
The manipulative experiments summarized here suggest that corridors between patches of fragmented habitat increase the chances of recolonization and thus subpopulation persistence.

HYPOTHESIS Subpopulations of a fragmented metapopulation are more likely to persist if there is no barrier to recolonization.

METHOD
1. Determine the number of species of small organisms (mostly arthropods) living in an established habitat of mosses growing on a large rock.
2. Fragment the habitat by trimming the mosses into patches of various configurations, leaving bare rock surrounding the patches.
3. After 6 months, determine the number of species present in the various patch configurations.

RESULTS

CONCLUSION Even small barriers to recolonization reduce the number of subpopulations persisting in a fragmented habitat.

FURTHER INVESTIGATION: These studies involved the dispersal of very small organisms over very short distances over a very short span of time. How might the longer term effects of imposing barriers to dispersal be investigated?

Go to **yourBioPortal.com** for original citations, discussions, and relevant links for all INVESTIGATING LIFE figures.

For many centuries, people have tried to reduce populations of species they consider undesirable and maintain or increase populations of desirable or useful species. Such efforts to manage populations are more likely to be successful if they are based on knowledge of how those populations grow and what determines their densities.

55.5 How Can Populations Be Managed Scientifically?

If we wish to increase or decrease populations of other species, we need to understand the life histories and the population dynamics of the organisms we wish to manage. The principles of population dynamics can also help us understand the effects our own population and its activities are having on other species.

Population management plans must take life history strategies into account

Knowing the life history strategy of a species can be helpful in managing populations of commercial value. The black rockfish (*Sebastes melanops*), an important game fish that lives off the Pacific coast of North America, provides one such example. Rockfish have an indeterminate growth pattern—they continue to grow throughout their lives. As in many other animals, the number of eggs a female produces is proportional to her size, so larger females can produce more eggs than smaller females. In addition, older, larger females are better able to provision the eggs they produce with oil droplets, which provide energy to the newly hatched larvae, giving them a head start in life (**Figure 55.12**). Larvae from eggs with larger oil droplets, produced by larger females, grow faster and survive better than do larvae from eggs with smaller oil droplets.

These life history traits have important implications for the management of rockfish populations. Because fishermen prefer to catch big fish, intensive fishing off the Oregon coast from

1996 to 1999 reduced the average age of female rockfish from 9.5 to 6.5 years. Thus the females reproducing in 1999 were, on average, smaller than the females reproducing in 1996. This change decreased the average number of eggs produced by females and reduced the average growth rate of larvae by about 50 percent. Maintaining productive populations of rockfish may require setting aside no-fishing zones where some females can be protected from fishing and allowed to grow to large sizes.

Population management plans must be guided by the principles of population dynamics

If we look at a logistic growth curve (see Figure 55.7), we can see that the number of births tends to be highest when a population is well below its carrying capacity. Therefore, if we wish to maximize the number of individuals that can be harvested from a population, we should manage the population so that it is far enough below the carrying capacity to have a high birth rate. Hunting and fishing regulations are established with this objective in mind.

Populations that have high reproductive capacities can persist even if harvest rates are high. In such populations (which include many fish species), each female may produce thousands or millions of eggs. In many of these fast-reproducing populations, the growth rates of individuals are density-dependent. Therefore, if prereproductive individuals are harvested at a high rate, the remaining individuals may grow faster. Some fish populations can be harvested heavily on a sustained basis because a relatively small number of females can produce sufficient numbers of eggs to maintain the population.

(A) *Sebastes melanops*

(B)

Oil droplet

55.12 Energy Stocks Give Rockfish a Head Start (A) Among black rockfish, older, larger females are more reproductively successful, producing both more eggs and eggs with larger nutritive oil droplets. (B) The oil droplet attached to this rockfish larva provides it with nutrition to fuel its growth until it can feed on its own.

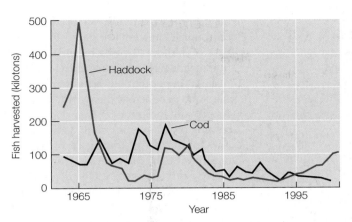

55.13 Overharvesting Can Reduce Fish Populations Populations of cod and haddock on Georges Bank have crashed due to overfishing.

Fish can, of course, be overharvested, as illustrated by the story of the black rockfish. Many fish populations have been greatly reduced because so many individuals were harvested that the few surviving reproductive adults could not maintain the population. Georges Bank, off the coast of New England—a source of cod, haddock, and other prime food fishes—was exploited so heavily during the twentieth century that many fish stocks have been reduced to levels insufficient to support a commercial fishery (**Figure 55.13**). The haddock population has rebounded enough to support a fishery because commercial fishing of that species ceased and was restarted only after the population had recovered. In contrast, managers reduced fishing pressure on cod only slowly, and the cod population has failed to increase.

Many rapidly reproducing species can recover if overharvesting is stopped, but recovery is more difficult for slowly reproducing species. Twentieth-century whalers hunted the blue whale (*Balaenoptera musculus*), Earth's largest animal, nearly to extinction. These whales reproduce very slowly: they live up to 10 years before becoming reproductively mature, produce only one offspring at a time, and have long intervals between births. Not surprisingly, the population has failed to recover.

Whether we want to manage the sizes of populations of desirable species for sustainable harvesting or of undesirable species for control purposes, the same principles apply. If the dynamics of a pest population are influenced primarily by density-dependent factors, then killing part of that population will only reduce it to a density at which it reproduces at a higher rate. A more effective approach to reducing such a population is to remove its resources, thereby lowering the carrying capacity of its environment. For example, we can rid our cities of rats more easily by making garbage unavailable (reducing the carrying capacity of the rats' environment) than by poisoning rats (which only increases their reproductive rate).

Biological control is the use of the natural enemies (predators, parasites, or pathogens) to reduce the population density of an economically damaging species (see the opening of Chapter 30). In many cases, the target species is a pest only because it has been introduced to a new area. Natural enemies used for

biological control are often obtained from the native region of the pest species. Biological control became popular in the nineteenth century after an outbreak of cottony-cushion scale, an Australian insect that attacks citrus, appeared in the citrus groves in California. A predaceous ladybeetle and a parasitic fly were then introduced from Australia and, within a year of their release, brought the scales under control.

On occasion, introduced predators and parasites not only fail to have any effect on the pest they were imported to control but, freed of their own enemies, they themselves become pests. This fact underlies the horror story of the cane toad (*Bufo marinus*) in Australia. These Central American toads (**Figure 55.14**) were introduced to control cane beetles attacking Australian sugarcane fields. But Australian cane beetles stay high on the upper stalks of the plants; the toads could not reach that high, and thus had no effect on the beetle population. Unfortunately, they had massive effects on other species.

All stages of the *B. marinus* life cycle are poisonous, and Australian reptiles (including snakes and lizards) and mammals that eat them usually die. With no enemies to limit their population growth, cane toads grow fast and outcompete native amphibian species for resources. The toads have spread from northern Australia down the east coast, where they threaten native frog species. The Australian government is forced to spend millions of dollars in attempting to reduce their numbers.

Human population increase has been exponential

In the nineteenth century, the historian Thomas Carlyle declared economics to be "the dismal science." He was referring in part to a famous essay written a century earlier by Thomas Malthus, in which Malthus pointed out that the human population was growing exponentially but its food supply was not, and at some point famine and death would be the ultimate fate of the

Bufo marinus

55.14 Biological Control Gone Awry The Central American cane toad not only failed to control destructive beetles in Australia's sugarcane fields, but increased dramatically in abundance and now threatens many native Australian species.

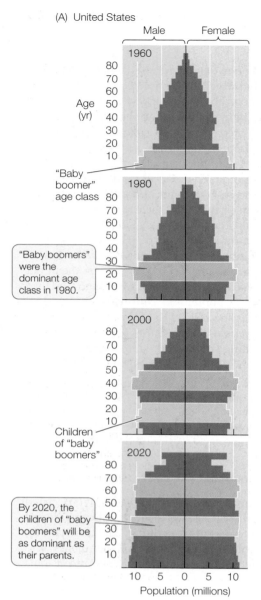

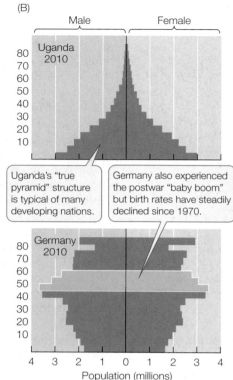

55.15 Population Pyramids
(A) Observed and predicted age distributions for the human population of the United States from 1960 to 2020 show how the birth rate during the "baby boom" has influenced the age structure of this country's population over many decades. (B) In Uganda, the largest proportion of the population is found in the youngest age groups, which means a greatly increased birth rate as these individuals achieve reproductive age. Conversely, a small but increasing number of highly developed nations have the population structure seen here for Germany, which presages a population decline.

the current worldwide rate of increase to be about 1.1 percent per year—but with a base of 7 billion, even a minimal growth rate means millions more individuals.

Human populations are not growing at the same pace across the world, however. As we saw in Figure 55.2, in populations of long-lived species such as elephants and humans, the timing of births and deaths affects the age distribution for many years. Between 1946 and 1964, the United States experienced a "baby boom." During those years, almost 75 million babies were born, and the average number of children per family grew from 2.5 to 3.8. Birth rates declined during the 1960s, but in the 1970s and 1980s, the baby boomers became parents, generating another demographic bulge—a "baby boom echo" (**Figure 55.15A**). Today, the echo babies are entering college in record numbers (and reading about themselves in introductory biology textbooks), and are on the threshold of becoming the dominant age class.

The age structure of the U.S. population is typical of many industrialized nations, but a few highly developed nations (particularly in Europe) are experiencing population declines. In the developing world, however, many countries are experiencing exponential growth rates and have populations highly skewed toward younger age classes, portending high rates of population growth in the future (**Figure 55.15B**).

human race. Neither Malthus nor Carlyle could have anticipated the technological innovations that greatly enhanced the capacity of humans to produce food. Today, however, the size of the human population is once again a serious concern as we confront the effects of our contributions to pollution, widespread habitat destruction, and the extinction of other species.

For thousands of years, Earth's carrying capacity for humans was low due to the relative inefficiency with which humans could obtain food and water. The development of social systems and communication, the domestication of plants and animals, ever-increasing crop and livestock yields through ongoing technological advances, and increasing proficiency at managing diseases have all contributed to unprecedented growth of the human population. It took more than 10,000 years for the population to reach 1 billion, which happened at some point in the late nineteenth century. Today, a mere 125 years later, the planet is home to approximately 7 billion human beings. Population growth has slowed somewhat from its post-World War II highs—the World Resources Institute estimates

55.5 RECAP

Efforts to manage populations are more likely to be successful if they are based on an understanding of life histories and population dynamics.

- Describe an effective strategy for reducing a pest population and explain why it is effective. See p. 1181
- How have humans changed the carrying capacity of Earth for our own population? See p. 1182

Earth's carrying capacity for humans is set in part by the biosphere's ability to absorb the by-products—especially carbon dioxide—of our enormous consumption of fossil fuel energy; by water availability; and by whether we are willing to tolerate the extinction of millions of other species in order to accommodate our prodigious use of Earth's resources. We enumerate some of the effects of human resource use on other species in Chapters 58 and 59, but first Chapters 56 and 57 discuss the nature of species interactions and the formation of ecological communities.

CHAPTER SUMMARY

55.1 How Do Ecologists Study Populations?

- A **population** consists of the individuals of a species that interact with one another within a particular area at a particular time.

- The **density** of a population is the number of individuals per unit of area or volume.

- The **age structure** and **dispersion** pattern of a population influence its dynamics. **Review Figures 55.2 and 55.3**

- **Life tables** provide summaries of demographic events in a population. A cohort life table tracks a **cohort** of individuals born at the same time and records the **survivorship** or **fecundity** of those individuals over time. **Review Tables 55.1 and 55.2**

- Life table data can be used to construct a **survivorship curve**. **Review Figure 55.4**

55.2 How Do Environmental Conditions Affect Life Histories?

- The **life history strategy** of an organism describes how it partitions its time and energy among growth, maintenance, and reproduction.

- A population's **intrinsic rate of increase** (*r*) is the difference between the per capita birth rate (*b*) and the per capita death rate (*d*).

- Life history traits within a species may vary with habitat and interactions with other species can influence the evolution of life history traits.

55.3 What Factors Limit Population Densities?

- Populations can exhibit **exponential growth** for short periods, but eventually their resources become depleted, causing birth rates to drop and death rates to rise. **Review Figure 55.6, ANIMATED TUTORIAL 55.1**

- **Logistic growth** is the pattern seen when the growth of a population slows as its density approaches the environmental **carrying capacity** (*K*). **Review Figure 55.7, ANIMATED TUTORIAL 55.2 and WEB ACTIVITY 55.1**

- Species that are *r*-strategists have life histories that allow for high intrinsic rates of increase. *K*-strategists persist at or near the carrying capacity (*K*) of their environment. Many species' life history strategies fall along a continuum between these two extremes. **Review Figure 55.8**

- Population densities are determined by both **density-dependent** and **density-independent** factors. Several factors—including resource abundance, body size, social organization, and the length of time a species has occupied an area—influence population densities.

55.4 How Does Habitat Variation Affect Population Dynamics?

- No species is found everywhere within its mapped range. Members of most species live in distinct **habitat patches**.

- A **metapopulation** consists of separate subpopulations among which some individuals move on a regular basis. **Review Figure 55.10**

- Extinction of a subpopulation may be prevented by immigration of individuals from another subpopulation, a process known as the **rescue effect**. **Corridors** between patches may facilitate such movement. **Review Figure 55.11, ANIMATED TUTORIAL 55.3**

55.5 How Can Populations Be Managed Scientifically?

- To manage populations, it is important to understand their life histories and population dynamics. To maximize the number of individuals that can be harvested from a population, the population should be kept well below carrying capacity.

- Reducing the carrying capacity of the environment for a pest species is a more effective way to reduce its population than killing its members.

- Earth's carrying capacity for humans depends on our use of resources and the effects of our activities on the environment. Human populations grow at different rates in different parts of the world. **Review Figure 55.15**

SELF-QUIZ

1. The way in which the individuals in a population are spread over the environment describes that population's
   a. dynamics.
   b. regulation.
   c. dispersion.
   d. strategy.
   e. survivorship.

2. The dispersion pattern in which the presence of one individual increases the likelihood of another individual being nearby is
   a. random.
   b. regular.
   c. clumped.
   d. uniform.
   e. irregular.

3. Which of the following is *not* a demographic event?
   a. Growth
   b. Birth

   c. Death
   d. Immigration
   e. Emigration

4. A group of individuals born at the same time is known as a
   a. deme.
   b. subpopulation.
   c. Mendelian population.
   d. cohort.
   e. taxon.

5. A population whose size levels off at its carrying capacity is exhibiting
   a. exponential growth.
   b. geometric growth.
   c. logistic growth.
   d. J-shaped growth.
   e. negative growth.

6. The process by which immigrants prevent a subpopulation from becoming extinct is called the
   a. colonization effect.
   b. rescue effect.
   c. metapopulation effect.
   d. genetic drift effect.
   e. salvage effect.

7. Density-dependent factors are population-regulating factors that
   a. affect the same proportion of individuals regardless of population size.
   b. are usually weather-related.
   c. affect only sessile organisms.
   d. influence only fecundity.
   e. have effects that vary depending on population size.

8. A metapopulation is
   a. an unusually large population.
   b. a population that is spread out over a very large area.
   c. a group of subpopulations among which some individuals move occasionally.
   d. a group of subpopulations that are isolated from one another.
   e. a group of subpopulations among which individuals move on a regular basis.

9. The best way to reduce the population of an undesirable species in the long term is to
   a. reduce the carrying capacity of the environment for the species.
   b. selectively kill reproducing adults.
   c. selectively kill prereproductive individuals.
   d. attempt to kill individuals of all ages.
   e. sterilize individuals.

10. Populations that are most readily overharvested are characterized by having
    a. very long-lived adults.
    b. short prereproductive periods and many offspring.
    c. short prereproductive periods and few offspring.
    d. long prereproductive periods and few offspring.
    e. long prereproductive periods and many offspring.

FOR DISCUSSION

1. Most organisms that humans manage for higher densities are long-lived and have low reproductive rates, whereas most organisms that humans want to reduce are short-lived but have high reproductive rates. What is the significance of these differences for management strategies and the effectiveness of management practices?

2. In the mid-nineteenth century, the human population of Ireland was largely dependent on a single food crop, the potato. When a disease caused the potato crop to fail, the Irish population declined drastically for three reasons: (1) a large percentage of the population emigrated to the United States and other countries; (2) the average age of a woman at marriage increased from about 20 to about 30 years; and (3) many families starved to death rather than accept food from Britain. None of these social changes was planned at the national level, yet all contributed to adjusting the population size to the new carrying capacity. Discuss the ecological principles involved, using examples from other species. What would you have done had you been in charge of the national population policy for Ireland at that time?

3. One method of controlling introduced pest species is to introduce a natural enemy (a predator, parasite, or pathogen) from the pest's native habitat to reduce its population density. However, some species introduced to control a pest have become pests themselves. Some scientists argue that biological controls should not be used under any circumstances for pest management. Others argue that, provided they are properly studied and thoroughly vetted, we should continue to use biological control organisms as part of our set of tools for managing pests. Which view do you support, and why?

ADDITIONAL INVESTIGATION

The arthropods whose metapopulation dynamics were studied in the experiment described in Figure 55.11 were all tiny animals with limited dispersal abilities. How could experiments be designed to test the influence of corridors on the metapopulation dynamics of species such as birds, lizards, and mammals, which readily disperse across large areas? Would you expect the results of such experiments to be similar to those using small arthropods on rocks? Why or why not?

WORKING WITH DATA (GO TO yourBioPortal.com)

Dead Sheep Tell a Tale Vertical life tables are constructed by sampling a population at a single point in time, across all age classes and life stages (thus their designation as "vertical"). One way to construct such a table is to record information from a *death assemblage*—the remains or fossils of the organisms in a given population. In this exercise, you will use data from a 1944 study of the skulls of more than 600 Dall sheep (*Ovis dalli dalli*) in Denali National Park, Alaska to create a vertical life table and survivorship curve for this population.

Habitat Fragmentation Figure 55.11 focuses on one of a series of experiments conducted by Andrew Gonzalez and Enrique Chaneton on population density and species diversity among moss-inhabiting mites. This exercise provides further data from these ingenious experiments and asks you to assess their implications for species diversity in fragmented environments.

Anty biotics

Familiar sights in Neotropical forest communities, "leaf-cutter" or "parasol" ants owe their name to their habit of clipping bits of leaf material from trees and carting them off to their nests. The leaves they collect, however, are not for their own consumption. Leaf-cutter ants are fungus farmers: they use the cut leaves as a substrate for growing the fungi on which they feed.

When a new queen ant is ready to leave the nest where she grew up, she packs her cheek pockets with a pellet containing pieces of the fungal mass that dominates her home. After mating, she lands, sheds her wings, and digs into the soil to form a tunnel with an enlarged end. There she ejects the fungal pellet and defecates in it. As the fungus begins to grow, the female lays eggs in it. With time, her eggs are engulfed by fungus, on which the hatching ant grubs feed. These first hatchlings develop into pint-sized workers that leave the nest to collect leaf material to "feed" the growing fungus.

As the colony increases in size, more nest chambers are constructed, and the fungus garden expands; by its third year, the total number of chambers in a colony can exceed several hundred. Colony size is one factor accounting for the reputation of leaf-cutters as pests; it takes more than 2 kg of plant material per day to maintain the enormous fungal garden that nourishes the colony, so these ants can easily strip an area of vegetation.

The ants diligently care for the fungi. In addition to supplying leaves to the garden, they add fertilizer (in the form of their own feces) and regulate the humidity of the chambers. They are even able to distinguish and exclude leaves containing chemicals with fungus-killing properties. The fungal gardens, however, are vulnerable to invasion by undesirable microbes, particularly the virulent green mold *Escovopsis*. To guard against mold invasion, the ants carry around a supply of bacteria in special organs on their exoskeletons, called crypts. These bacteria manufacture powerful antibiotics that suppress the unwelcome mold, but do no harm to the cultivated fungus in the nest.

In the first written description of leaf-cutter ants, the

Feeding a Partner Central American leaf-cutter ants (*Atta cephalotes*) transport leaf fragments to their nest. The leaves will serve as food for the fungus they grow.

A Fungal Garden Cutaway view of a South American leaf-cutter ant nest chamber filled with fungus. Several winged ants (*Atta colombica*) can be seen in the crevices of the fungal mass.

Spanish priest Bartolome de la Casas decried their assiduous leaf gathering and massive nests, "white as snow," as impediments to the cultivation of citrus and cassava in Hispaniola in 1559. The notion that these ants actually cultivate and eat the white fungus in their nests did not occur to anyone until the naturalist Thomas Belt introduced the idea in 1874. The existence and roles of the intrusive green mold and friendly bacteria were not documented until 1999, after another 125 years. It is not beyond the realm of possibility that even more interactions between species are involved in this system and remain to be revealed.

IN THIS CHAPTER we will examine how antagonistic interactions benefit one species at the expense of another, how mutualistic interactions benefit two or more interacting species, and how competition for the same resources affects the species involved. Throughout these discussions, we will see not only how these interactions influence the lives of these organisms, but how they shape species over the long range of evolution.

56.1 What Types of Interactions Do Ecologists Study?

One of life's certainties is that, at some point between birth and death, every individual organism will encounter and interact with individuals of other species. These interactions have consequences that can affect the individual's fitness. Thus they can influence the densities of populations, the distributions of species, and, over the long term, lead to evolutionary change in one or both of the interacting species.

Interactions among species can be grouped into several categories

Although the actual number of interactions that take place among living things on Earth is essentially limitless, ecologists group interactions between species into a few basic categories. These categories reflect whether the outcome of the interactions is positive (+), negative (−), or neutral (0) for each of the species involved (**Figure 56.1**). The five broad categories of species interactions that we introduce in this chapter are *antagonistic interactions*, *mutualism*, *competition*, *commensalism*, and *amensalism*.

Antagonistic interactions are those in which one species benefits and the other is harmed. Antagonistic interactions include *predation*, in which an individual of one species kills and consumes multiple individuals of other species during its lifetime; *herbivory*, in which an individual of another species consumes part or, more rarely, all of a plant; and *parasitism*, in which one species consumes only certain tissues in one or a few individuals of another species without necessarily killing those individuals—known as *hosts*. Some parasites are *pathogens*, which cause symptoms of disease in their hosts.

Mutualism is a type of interaction between species that benefits both species. The interaction between leaf-cutter ants and fungi described at the opening of this chapter is an example of mutualism: the ants feed and cultivate the fungi, and the fungi, in turn, serve as food for the ants. Mutualisms exist between widely varied pairs of partners, including not only animals and fungi, but also fungi and plants, animals and plants, animals and animals, and microbes and all other kinds of organisms.

Competition between species refers to interactions in which two or more species use the same resource. The outcomes of these interactions depend on resource availability. Competition can occur along with almost any other kind of interaction: between predators that depend on the same prey species, between herbivores that depend on the same host plant, or between pathogenic microbes attacking the same host. The limiting resource need not be food; species may compete for water, for space, or even, in the case of plants, for sunlight.

(A)

Major types of species interactions

TYPE OF INTERACTION	EFFECT ON SPECIES 1	EFFECT ON SPECIES 2
Predation (predator–prey)	+	–
Herbivory (plant–herbivore)	–	+
Parasitism (parasite/ pathogen–host)	+	–
Mutualism	+	+
Competition	–	–
Commensalism (commensal–host)	+	0
Amensalism	0	–

Antagonistic interactions { Predation, Herbivory, Parasitism }

(B)

Parasitism, Predation, Mutualism
The buffalo's hide is infested with parasitic ticks. Oxpecker birds eat the ticks, to the mutual benefit of the birds and the buffalo.

Herbivory
The African buffalo feeds on the grasses of the savanna.

Amensalism, Commensalism
The large mammal unwittingly destroys insects and their nests. The white cattle egrets feed on insects disturbed by the buffalo's passage.

Predation
Carnivores such as timberwolves hunt and kill herbivorous mammals.

Competition
The grizzly bear is attempting to take over the wolves' kill.

56.1 Types of Interactions (A) Interactions among species can be grouped into categories based on whether their influence on each of the interacting species is positive (+), negative (–), or neutral (0). These interactions can be broadly grouped as antagonistic or mutualistic. (B) Even a small scene can encompass many different species interactions.

yourBioPortal.com
GO TO **Web Activity 56.1 • Ecological Interactions**

Predation, mutualism, and competition all affect the fitness of both participants, but there are two other types of interactions that affect only one participant. **Commensalism** is a type of interaction in which one participant benefits, but the other is apparently unaffected. Most examples of commensalism (from the Latin, "eating together") involve one species feeding in, on, or around another species. For example, one species may associate with another species that, by virtue of its own feeding behavior, makes food more accessible. The brown-headed cowbird owes its name to its habit of following herds of grazing cattle, foraging on insects flushed from the vegetation by their hooves and teeth (in years past, it was called the buffalo bird because it followed the bison that were once abundant across the North American continent).

Another form of commensalism involves association for the purpose of transport, often to reach food resources that are rare and short-lived. Piles of mammal dung, for example, are a valuable resource for some detritivores, but they can be hard to find and never last long. Many kinds of detritivores that cannot fly—mites, nematodes, and even fungi—attach themselves to the bodies of dung beetles, which not only can fly but are also very good at locating fresh dung (as described in the opening story of Chapter 54). These hitchhikers have no effect on the dung beetle's fitness, nor do cowbirds have any on the cattle that flush insects for them.

Amensalism is a type of interaction in which one participant is unaffected while the other is harmed. A herd of elephants moving through a forest crushes insects and plants with each step, but the elephants are unaffected by this carnage. Amensal interactions tend to be more random, and thus less predictable, than other types of interactions.

Although ecologists find it useful to group interactions between species into a few basic categories, the boundaries between categories are not always clear. For example, sea anemones in the Pacific Ocean sting and eat small fish, but a select few fish species (mostly in the genus *Amphiprion*) live inside sea anemones and are unaffected by their stings. These anemonefish move freely among the stinging tentacles to scavenge the cnidarians' leavings and even steal their prey (**Figure 56.2**).

Anemonefish must acclimate to the anemone's venom, and the anemone, in turn, must acclimate to the fish. The acclimation process appears to involve a change in the mucus coat of the fish; wiping off the mucus of an acclimated fish results in immediate stinging, whereas anemones will not attack fish with intact mucus. Although the benefit to the anemonefish is clear—it escapes its own predators by hiding behind the anemone's

Amphiprion ocellaris

56.2 Interactions between Species Are Not Always Clear-Cut
Ecologists long believed that the relationship between sea anemones and anemonefish was a commensalism: that the fish, by living among the anemone's stinging tentacles, gained protection from its predators. But could it also be considered a mutualism—if the fish's feces provide the anemone with beneficial nutrients—or competition—if the fish occasionally steals the anemone's prey?

nematocysts, and it has no need to forage widely for food—the anemones may benefit from the association as well. By defecating while in residence, the anemonefish may provide nitrogen-rich nutrients to the anemones.

The interaction types described in this section are in reality part of a continuum, and their outcomes depend on both ecological and evolutionary circumstances, including the presence and influence of other interacting parties.

Some types of interactions result in coevolution

All types of interactions have the potential to influence the population densities of the interacting species. By contributing to the differential survival or reproduction of genotypes, they can also alter gene frequencies within the populations over time. Thus these interactions have ecological consequences, as when they affect the distribution and abundance of individuals within a species, as well as evolutionary consequences, as when they lead to adaptations. In some cases, an adaptation in one species may lead to the evolution of an adaptation in a species it interacts with, a process known as **reciprocal adaptation** or **coevolution**.

Darwin saw the fitness of an organism as a measure of whether it gains or loses from interactions with other species. He noted, too, that evolutionary change occurs not only in response to physical conditions, as we saw in Section 54.2, but also in response to interactions among organisms. In his introduction to *The Origin of Species*, Darwin pointed out that woodpeckers have feet, tails, beaks, and tongues "admirably adapted to catch insects under the bark of trees" as a result of their long-standing interactions with their insect prey. Organisms can thus influence the evolution of the organisms with which they interact.

While abiotic factors also act as agents of selection, they differ in a fundamental way from biotic agents of selection in that they do not themselves undergo change as a result of the interaction. Snow and ice cannot increase their killing power as a result of encountering cold-resistant organisms, but predators can, over evolutionary time, become swifter, more powerful, or more efficient at capturing their prey. In response, prey species can, over evolutionary time, become swifter, tougher, less conspicuous, or more poisonous to decrease the likelihood of being consumed. The insect prey of Darwin's woodpeckers, for example, might evolve features that make them more difficult for the woodpeckers to find and capture.

A series of reciprocal adaptations can lead to what has been dubbed a coevolutionary **arms race**. The arms race analogy, first used in the context of interactions between herbivores and plants, can be applied to most antagonistic interactions. The evolution of traits that increase the fitness of a predator or parasite species exerts selection pressure on its prey or host species to counter the consumer's adaptation. The prey or host adaptation, in turn, exerts selection pressure on the consumer to improve its fitness, resulting in an escalating arms race. The types of interactions most likely to lead to coevolution, then, are those that occur predictably with high frequency over time and that have a strong effect on the interacting species; thus most amensal and commensal interactions are less likely to coevolve than are many plant–herbivore, predator–prey, and mutualistic interactions.

56.1 RECAP

Species interactions can be grouped into five categories based on whether they benefit or harm the species involved. Some species interactions can lead to reciprocal adaptations, or coevolution.

- Describe the five categories of interactions between species. See pp. 1186–1187 and Figure 56.1

- Explain the concept of a coevolutionary arms race. See p. 1188

Now let's take a closer look at antagonistic interactions. In Section 31.3 we looked at a number of feeding strategies from the consumer's point of view. In this section we will see how the interactions of consumers with their resource species influence both species.

56.2 How Do Antagonistic Interactions Evolve?

Every species serves as a food resource, in one way or another, for at least one other species. Consumers can increase their fitness by acquiring food, whereas resource species can increase their fitness by avoiding being consumed. Thus the interests of consumer and resource species set up an antagonistic relationship that can lead to a coevolutionary arms race. These consumptive relationships need not, however, be fatal; organisms make meals of one another in many different ways.

(A) *Panthera tigris*

(B) *Cicindela campestris*

(C) *Platythelphusa* sp.

56.3 Predators Use Many Weapons (A) Tigers embody most people's image of a predator—a large animal that uses speed, strength, teeth, and claws to capture prey. (B) The 1.3-cm green tiger beetle is also formidable to its prey, including caterpillars. Its huge jaws account for most of its body length, and it is among the swiftest runners among the insects. (C) *Platythelphusa* crabs of Lake Tanganyika have tremendously strong "toothed" claws with which they crush their snail prey. In an evolutionary arms race, thick shells have evolved in the snails.

Predator–prey interactions result in a range of adaptations

Predator–prey interactions are probably the most familiar, and the most dramatic, type of antagonistic interaction. Predators invariably kill the individuals they consume—referred to as their *prey*—and over its lifetime, a predator kills and consumes many prey individuals. Predators tend to be less specialized than other types of consumers.

The fitness of predators depends on balancing the cost of pursuing, subduing, and handling prey against the energetic return from consuming it. Thus many predators are larger than their prey, and many of them use strength or swiftness to capture prey. This is true of predators of all sizes: tigers pursuing deer and tiger beetles pursuing smaller insects are both swift, powerful, and equipped with strong jaws (**Figure 56.3A,B**). The few predators that are smaller than their prey rely on other strategies that increase their efficiency. Spiders, for example, capture their prey in webs. The short-tailed shrew, among the smallest mammalian predators, produces venomous saliva that paralyzes not only earthworms and snails, but also prey much larger than itself, including mice and small birds.

Prey species have many different kinds of defenses against predators. Many animal species can escape from predators simply by running away. Others have morphological defenses. Tough skin, shells, spines, or hair can foil even a determined predator. A coevolutionary arms race between predators and prey can explain why the lumpy-clawed crabs with the great-

est claw strength are found in the vicinity of snail prey with the thickest, most crush-resistant shells (**Figure 56.3C**).

CHEMICAL DEFENSES Many animals use chemical defenses to escape or repel their predators. Chemical defenses are generally the province of animal prey that are small, weak, sessile, or otherwise unprotected. Among the mollusks, for example, the weaker a species' shell, the more likely it is to use chemical defenses (see Figure 56.4B). Some vertebrates also rely on chemicals to repel their predators.

Many insects produce sprays, oozes, or froths when attacked. Bombardier beetles, for example, possess a pair of glands near the anal opening. Each gland has two compartments lined with a protective cuticle. The inner compartment contains a mix of relatively nontoxic chemicals, along with hydrogen peroxide. The outer compartment contains enzymes. When the beetle is disturbed, it discharges the contents of the inner compartment into the outer compartment, which leads to an instant, energy-releasing chemical reaction. Oxygen is one of the end products generated by this reaction, and the resulting pressure discharges the mixture with an audible "pop." Due to the energy of the reaction, the temperature of the spray is approximately 100°C. The reaction of predators—including humans—to this hot, explosive secretion is predictable, and bombardier beetles have very few enemies.

But predators may evolve to overcome their prey's chemical defenses, as we saw in the case of the rough-skinned newt and the garter snakes that have become insensitive to its protective toxin (see Figure 21.22). Some predators are not only undeterred by their prey's defensive chemicals, but ingest them and sequester them in their bodies as defenses against their own predators. Some sea slug species, for example, acquire their defensive chemicals from the toxic sponges they can eat with impunity. Many prey species that defend themselves with toxicity advertise that fact, a phenomenon called *aposematism*.

APOSEMATISM Some prey species exploit the fact that predators can learn to avoid certain warning signals. Many toxic prey sport bright colors or striking patterns. Such warning coloration, or **aposematism**, increases the probability that a predator will

(A)

Danaus plexippus larva

Tetraopes tetraophthalamus

Aphis nerii

56.4 Some Prey Come with Warning Labels (A) Milkweed plants are toxic, and many of the insects that feed on them—including monarch butterfly larvae, milkweed beetles, and aphids—incorporate the plant's toxic chemicals into their systems. (B) Nudibranchs (sea slugs) are mollusks without protective shells, however, they may possess stinging nematocysts (acquired from their hydrozoan prey), and their bright coloration warns predators away. (C) Dart poison frogs of Central and South American sequester highly toxic chemicals in their brightly colored skin.

(B) *Chromodoris* sp.

(C) *Dendrobates reticulatus*

learn to recognize and avoid a toxic species (**Figure 56.4**). Certain visually hunting predators, particularly among the vertebrates, can learn quickly to associate certain color patterns with an unpleasant dining experience. Thus aposematic species are characteristically tough enough to survive a brief encounter with a predator. Any encounter that results in the death of the aposematic individual is unlikely to result in selection for that aposematic pattern. Aposematic butterflies in the field often sport beak marks, an indication of having survived a taste by an uneducated avian predator. And, as it turns out, even non-toxic species can benefit by mimicking warning coloration.

MIMICRY SYSTEMS We have seen that some prey species avoid consumption by mimicking inedible objects. Others do so by mimicking aposematic species. This strategy has led to the evolution of mimicry systems of two types. In **Batesian mimicry**, a nontoxic species (the *mimic*) resembles a toxic species (the *model*) and benefits from the avoidance behavior learned by the predator in response to the toxic model species. For example, venomous bees and wasps sport distinctive yellow and black stripes that are models for harmless hover flies (**Figure 56.5A**). Mimicry may extend beyond physical appearance; many mimics also simulate distinctive behaviors of their models. Some hover flies, for example, vibrate their wings to buzz in beelike fashion.

In **Müllerian mimicry**, a number of aposematic species converge on a common color pattern; all benefit from providing a stronger recognition signal to predators. Many of the Neotropical zebra butterflies, which feed on toxic passionflower plants

and incorporate their toxins into their bodies, are Müllerian mimics (**Figure 56.5B**).

AVOIDING DETECTION Many prey species escape predators by hiding. One form of hiding is camouflage, or background matching, also called **crypsis** (**Figure 56.6A**). Some animals can even change their coloration to match the substrate they find themselves on. The camouflage of some species not only hides them but allows them to resemble objects their predators consider inedible, a strategy called **homotypy** (**Figure 56.6B**). The dead-leaf butterfly looks very much like a dead leaf, even down to the likeness of a spot of fungal decay, and swallowtail caterpillars look like bird droppings.

Because the vision of many types of predators is adapted to spot moving prey, many prey species simply stop moving if they are being pursued. "Playing possum," a term that is sometimes applied to this strategy, refers to the ability of the opossum (*Didelphus virginianus*) to "play dead."

Herbivory is a widespread interaction

The most ubiquitous interaction on Earth is that between plants and the herbivores that eat them. Herbivores have a relatively easy time acquiring food: plants are sessile and cannot bite, scratch, or run away. A conservative estimate puts the number of herbivore species (the vast majority of which are insects) at

(A) Batesian mimics

Episyrphus balteatus

Vespula vulgaris

(B) Müllerian mimics

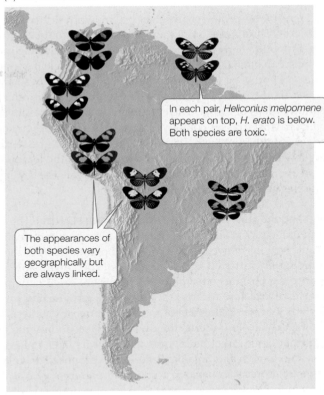

56.5 What You See Is What You Get—Sometimes
(A) Batesian mimics are harmless species that gain protection from mimicking dangerous ones. (B) The shared aposematic coloration of Müllerian mimics is an honest advertisement of their toxicity. All the zebra butterflies (genus *Heliconius*) of South America feed on toxic passionflower plants and incorporate the toxins into their bodies. *Heliconius* species living in a particular area have similar warning colors.

400,000 and the number of plant species they consume at over 300,000. Every major class of vertebrates includes at least a few herbivores. In marine systems, organisms that feed on plants and algae include mollusks, crustaceans, echinoderms, and annelids.

In terms of numbers of individuals as well as numbers of species, most of the world's herbivores are insects. Over 90 percent of these insects are *oligophagous*: specialists that dine on just one or a few, often taxonomically related, plant species. *Polyphagous* species, in contrast, feed on as many as dozens of unrelated plant species. Vertebrate herbivores are generally polyphagous; a cow grazing in a pasture, for example, can consume many different plant species in a single afternoon. There are exceptions to this pattern, however; Australian koalas famously feed exclusively on the foliage of eucalyptus trees, and the diet of giant pandas is made up almost entirely of bamboo.

Herbivores, particularly insects, generally consume only parts of their food plants, and usually do not kill them. In most natural ecosystems, insects rarely remove more than 20 percent of plant biomass. For that reason, questions occasionally arise as to the ability of insects to act as selective agents on plant traits. Mortality is not, however, the only form of selection that leads to evolutionary change; herbivores can reduce plant fitness if the plants they attack produce fewer offspring.

PLANT DEFENSES AGAINST HERBIVORES   The defenses of plants against their diverse consumers are necessarily highly diverse. For most plant species, chemistry is the principal defense mechanism. As we saw at the opening of this chapter, the leaves of

some trees contain chemicals that prevent them from being consumed by fungi—and thus, incidentally, from being harvested by leaf-cutter ants. The amazing variety of secondary metabolites produced by plants to defend themselves against herbivores is the topic of Section 39.2. Many plants, however, have additional defenses.

(A) *Hyla versicolor*

(B) *Mimetica* sp.

56.6 Avoiding Consumption by Avoiding Detection   (A) The gray tree frog can change its coloration to blend in with its substrate, an example of crypsis. (B) Homotypy—resemblance to an inedible object—can be an effective defense against visually hunting predators. Birds searching for insect prey are likely to pass by a katydid that looks like a leaf.

Some plants protect themselves by being physically difficult to ingest. Thorns and spines are effective deterrents to browsing vertebrate herbivores. Smaller herbivores, including many insects, can be deterred by small hooked hairs on leaf surfaces. The soft bodies of leafhoppers can be pierced by these hairs, which fix the insect in place until it eventually dies from starvation or loss of blood. The cuticle may also act as a physical barrier. Most grasses contain silica, which wears down sharp edges of herbivore teeth. Insects that feed only on grasses tend to have chisel-like mandibles to slice through the leaf tissue; moreover, their heads are larger to accommodate the larger jaw muscles needed to process their food.

RECIPROCAL ADAPTATIONS IN HERBIVORES AND PLANTS The concept of reciprocal adaptation was first described in the context of interactions between herbivores and plants. In 1959, the entomologist Gottfried Fraenkel reached the conclusion after many years of study that all green plants are essentially nutritionally equivalent for insects. Why, then, are so many insects such picky eaters? Fraenkel proposed the novel hypothesis that ecological factors underlie the diversity of secondary metabolites that deter insect herbivores. A few years later, entomologist Paul Ehrlich and botanist Peter Raven proposed the following evolutionary scenario to account for patterns of host plant use among herbivorous insects (specifically, in their case, butterfly families):

- Certain plants, by mutation or recombination, evolve a novel secondary metabolite.
- If the chemical reduces the plant's appeal to herbivores, then plant genotypes producing the chemical are favored.
- Freed from mortality associated with herbivory, plants possessing the novel chemical undergo an adaptive radiation.
- Certain herbivores, by mutation or recombination, evolve resistance to the chemical, and these resistant herbivores undergo their own adaptive radiation.
- With sufficient selection pressure, a resistant herbivore can evolve to use the chemical as a defense against its own predators.

This stepwise coevolutionary process explains not only the biochemical diversity of flowering plants, but also the tremendous diversity of herbivorous insects. The ecological scenario outlined by Ehrlich and Raven is an example of the type of coevolutionary arms race described earlier in the chapter.

A tremendous diversity of adaptations to plant defenses have evolved in herbivores. Many herbivores circumvent plant defenses by behavioral means. For example, the secondary metabolites produced by a plant called Saint-John's-wort (*Hypericum perforatum*) require exposure to sunlight for optimal toxicity, so insects that feed on this plant roll its leaves into a light-impervious cylinder and feed in comfort in the dark. The laticifer-cutting beetles described in Section 39.2 have a different method of detoxifying their food plant. Many large polyphagous herbivores, such as deer, horses and the like, graze on a wide variety of plant species, minimizing their exposure to any particular defensive chemical. Long-lived and with rel-

atively good memories, they can learn to avoid plants with an unpleasant taste.

Unlike large mammalian herbivores, caterpillars and many other insect herbivores may spend their entire lives feeding on a single individual plant. Such oligophagous diets are associated with highly specialized detoxification systems. The diamondback moth caterpillar eats plants in the cabbage family that are rich in toxic mustard oil glycosides. In its gut is an enzyme that breaks down the glycosides into harmless by-products, allowing it to eat these plants with impunity.

Some herbivores take resistance a step further by storing, or sequestering, plant toxins in specialized organs or tissues that are insensitive to those toxins. In this way, they can accumulate large quantities of toxins in their bodies with no ill effects. This strategy also makes the expropriated chemicals available for defense against the herbivores' own enemies. Caterpillars of the monarch butterfly, for example, are insensitive to the neurotoxic glycosides in their milkweed host plants, but most of their enemies, including insect-eating birds, cannot tolerate these compounds.

Yet the plants continue their side of the coevolutionary arms race. As we have seen, the principal consumers of passionflower plants are zebra butterflies. These oligophagous butterflies lay eggs only on passionflower plants, and their larvae sequester host plant toxins in their bodies as they feed on the leaves. Some passionflower species, however, have modified leaf structures that resemble the eggs of butterflies. Female butterflies will not lay eggs on plants already containing eggs, so the egg mimics reduce the plant's probability of being consumed (**Figure 56.7**).

Microparasite–host interactions may be pathogenic

Microparasites, such as viruses, bacteria, and protists, are many orders of magnitude smaller than their hosts and generally live and reproduce inside their hosts. Multiple generations of mi-

Spots on the leaves of *Passiflora* mimic butterfly eggs.

56.7 A Plant Uses Mimicry to Avoid Herbivory Passionflower leaves are modified to resemble the eggs of their principal herbivores, zebra butterflies (*Heliconius* spp.), which will not lay eggs on a plant already containing eggs. The egg mimics deter the female from laying eggs, thus protecting the plant from being eaten by hatchling caterpillars.

croparasites may reside within a single host individual, and a host may harbor thousands or millions of them. Many microparasites, in the process of acquiring nutrients at the expense of their host, cause symptoms of disease—that is, they are *pathogens*. Section 39.1 describes the array of secondary metabolites that plants produce to defend themselves against pathogens, and Chapter 42 describes the defenses of animals.

Infection by pathogens may in some cases result in the death of the host, but death is by no means an inevitable outcome of these interactions. If a pathogen strain is to persist in a host population, the pathogens must continually infect new host individuals. A less deadly strain that kills a smaller proportion of host individuals may succeed in infecting a larger number of new hosts. Thus pathogen and host may reach a state of coexistence as increased host *resistance* (ability to withstand the effects of a pathogen) and decreased pathogen *virulence* (ability to cause disease) evolve. Yet new, virulent strains may also arise, reminding us that the arms race goes on.

Hosts of pathogens fall into three classes: *susceptible* (capable of being infected), *infected*, or *recovered* (and thus, in many cases, *immune*; see Chapter 42). A pathogen can readily invade a host population dominated by susceptible individuals, but as the infection spreads, fewer susceptible individuals remain to be infected. Eventually a point is reached at which most infected individuals no longer transmit the infection to susceptible individuals. Thus rates of infection typically rise, then fall, and do not rise again until a sufficiently large population of susceptible host individuals has reappeared.

Most ectoparasites have adaptations for holding onto their hosts

While microparasites generally live and reproduce inside their hosts, larger parasites, called *macroparasites*, are associated with their hosts in a slightly less intimate way. Although macroparasites rarely cause the same kinds of disease symptoms that pathogenic microparasites cause, they may nevertheless affect host survival and reproduction and can thereby act as selective agents on their hosts.

Some ectoparasites (external parasites)—leeches, mosquitoes, and the like—are only casually associated with their hosts, interacting with them just long enough to eat their fill and then moving on. Others spend their entire lives in or on their hosts. These ectoparasites have a number of attributes designed to keep them attached to their hosts. Crab lice, ectoparasites that are generally found in the pubic region of their human hosts, have claws on the tips of their legs that clamp around pubic hairs with great precision (**Figure 56.8**). Pulling off a crab louse will often leave the leg behind, still firmly attached to the hair. Ectoparasites have other adaptations designed to reduce the ability of irritated hosts to remove them, such as flattened bodies and a thick, tough cuticle. Many ectoparasites also have adaptations to a sedentary existence, such as loss of sensory organs and wings. Most ectoparasitic insects are highly specialized, sometimes feeding on only a single host species.

Most hosts actively work to rid themselves of their ectoparasites. The Japanese macaque, for example, is prone to infesta-

Phthirus pubis

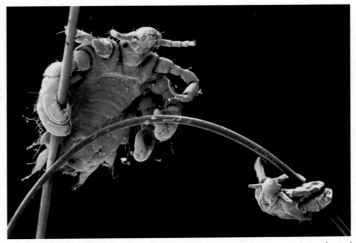

0.5 µm

56.8 Ectoparasites Can Make a Host Crabby Ectoparasites such as the human pubic (or "crab") lice in this scanning electron micrograph tend to be tiny, wingless, flattened, and equipped with strong claws for gripping their host.

tion by two species of lice, which tend to lay their eggs on the outer surfaces of the host's back, arms, and legs. An individual macaque may harbor more than 500 louse eggs, and would find it difficult to reach and eliminate all of them. To keep louse populations in check, macaques must form and maintain social bonds in order to ensure the consistent presence of grooming partners. Grooming behavior—an important component of the social interactions of many primates—may have evolved in response to ectoparasites. Some biologists believe that human hairlessness and bipedal posture (which freed the hands for manipulating small objects), as well as the opposable thumb, were evolutionary responses to ectoparasites.

56.2 RECAP

Predator–prey, herbivore–plant, and parasite–host interactions are all antagonistic. Consumers have adaptations for finding and utilizing their resource species with greater efficiency. The resource species in turn have adaptations that reduce their probability of being discovered, captured, and eaten.

- What are some of the adaptations that help prey species to avoid consumption by predators?
  See pp. 1189–1190

- How are aposematism and mimicry related?
  See pp. 1189–1190 and Figures 56.4 and 56.5

- Explain the scenario for coevolution between insect herbivores and their host plants proposed by Ehrlich and Raven. See p. 1192

Like antagonistic interactions, mutually beneficial interactions between species can result in reciprocal adaptation. A mutually beneficial exchange of goods or services can ensure the predictability and frequency of such interactions over evolutionary time; thus many mutualistic interactions are tightly coevolved.

56.3 How Do Mutualistic Interactions Evolve?

There are few, if any, taxonomic limits on the formation of mutualisms; animals, for example, can form mutualistic associations with other animals, with plants, and with a wide range of microorganisms. Mutualistic interactions often arise in environments where resources are in short supply. Consequently, many mutualisms involve an exchange of food for housing or defense. In another type of mutualism, many sessile organisms, particularly flowering plants, rely on more mobile species for mating or dispersal.

Many mutualisms are *asymmetrical*—in other words, one party benefits more than the other. One or both partners may evolve adaptations that ensure that the exchange benefits both of them. Reciprocal adaptations are most likely to arise in mutualistic interactions if an increase in dependency on a partner provides an increase in the benefits realized from the interaction. If increased dependence provides no selective advantage, mutualists (particularly those in asymmetrical mutualisms) may evolve into parasites, give up their partners for an independent existence, or even go extinct.

Plants and pollinators exchange food for pollen transport

For about three-quarters of the planet's 250,000 flowering plant species, reproduction requires the transport of pollen by an animal partner. A mutualistic pollination system requires several features:

- An *attractant* or reward that entices a pollinator to visit the plant
- *Behavior* that ensures that a pollinator visits more than one individual of a plant species
- *Anatomical features* that allow a pollinator to transport the plant's pollen

Flowers entice pollinators in many ways. The most direct reward for pollinators is the pollen itself, which sometimes serves as food. Pollen was probably the original attractant in the evolutionary history of plant–pollinator interactions. Plant reproduction would not be served, however, if pollinators were to eat *all* of a plant's pollen; thus plants have evolved various adaptations to ensure that they benefit from the exchange. For example, some plants have two types of anthers: feeding anthers to produce pollen for pollinators, and fertilization anthers to produce pollen for reproduction. These two types of anthers

56.9 Plants Sometimes Take Advantage of Their Pollinators This orchid (*Ophrys insectifera*) has flowers that look and smell like a female wasp. Male wasps (*Argogorytes mystaceus*) will pollinate the flower while trying to mate with it.

are shaped and positioned differently, so that as the pollinator dines on pollen from the feeding anthers, the fertilization anthers deposit pollen on a part of its body that will transfer it to the stigma of another flower of the same species.

Plants, like pollinators, may take advantage of their partners. Some species have evolved flowers that bear a striking resemblance to females of a particular wasp species (in some cases even producing the same chemical substance that the female wasp uses as a sexual attractant pheromone). Male wasps, in a futile effort to copulate with the flower, get pollen on their bodies, which they deliver to the next flower they visit in their effort to locate a genuine mate (**Figure 56.9**).

Compared with pollen, *nectar*, a sugar-rich solution produced by some angiosperms, is a relatively new evolutionary development. Of all floral rewards, nectar has the greatest appeal and is consumed by the widest range of animal pollinators, including not only insects but also birds (such as hummingbirds) and mammals (such as bats). While nectar serves to attract potential pollinators, it is prone to removal by *thieves*: flower visitors such as ants that can reach and consume the nectar without

TABLE 56.1
Pollination Syndromes Resulting from Diffuse Coevolution

PREFERRED POLLINATOR	SUITE OF ANATOMICAL TRAITS[a]			
	FLOWER SHAPE	FLOWER COLOR	REWARD	ODOR
Bees	Irregular	Many	Nectar, pollen	Sweet
Flesh flies	Irregular	Purplish	None	Carrion
Beetles	Bowl	White or pale	Pollen	Faint
Butterflies	Tubular platforms	Many	Nectar	Faint
Moths	Often pendant	White or pale	Nectar	Heavy
Hummingbirds	Tubular	Red	Nectar	Imperceptible
Bats	Cuplike	White or pale	Copious nectar	Musty

[a] In many plant groups, diffuse coevolution has led to suites of traits (as opposed to a single trait) that are characteristic of the interactions between the plants and their preferred pollinators.

Taraxacum officinale

56.10 See Like a Bee To normal human vision, a dandelion appears solid yellow. Ultraviolet photography reveals the patterns that attract bees to the region of the flower where pollen and nectar will be found. These patterns are invisible to humans and birds.

transporting pollen. Nectar thieves lower plant fitness by depleting nectar that would otherwise attract actual pollinators.

Plants not only need to attract pollinators, but must also ensure that those pollinators carry their pollen to other members of the same species. Repeat visits by a pollinator to different individuals of a particular plant species increase the likelihood that the pollen will end up on the appropriate stigma; thus some plants have adaptations to limit the diversity of their animal visitors. The depth and width of a flower can restrict the size and shape of the pollinator mouthparts that can gain access to their nectar. Timing of flowering can also restrict the number of potential pollinators and encourage pollinator fidelity. Floral characteristics influence the type of pollinator attracted initially. Ultraviolet wavelengths, for example, are highly attractive to bees (**Figure 56.10**) but invisible to most birds.

Most flowers can be successfully pollinated by a number of animal species. The evolution of broad suites of floral characteristics that attract certain groups of pollinators is an example of **diffuse coevolution**: the evolution of similar traits in suites of species experiencing similar selection pressures (**Table 56.1**). Scarlet gilia (*Ipomopsis aggregata*), a common wildflower in the Rocky Mountains, has successfully combined two strategies. Early in its growing season, it produces red flowers that attract hummingbirds; later in the season, the gilia shifts to producing white flowers because by then the most abundant pollinators are hawkmoths, which cannot see red but are attracted to white.

A few plant–pollinator relationships are much more exclusive; these lead to highly specific, rather than diffuse, coevolution. Yucca plants are pollinated only by a group of moths collectively known as yucca moths, whose larvae feed exclusively on yucca seeds. The stigma of the yucca flower is located fairly deep within the pistil, and fertilization will not occur unless pollen is physically placed there. The specialized mouthparts of female yucca moths have distinctive long tentacles, which the moths use to pack masses of pollen from one yucca flower into transportable balls that they then carry to another flower. The moth pushes the pollen ball deep into the recess in which the flower's stigma is tucked, then turns around and deposits her eggs inside the flower's ovule (**Figure 56.11**). When the eggs hatch, the caterpillars will consume some—but not all—of the

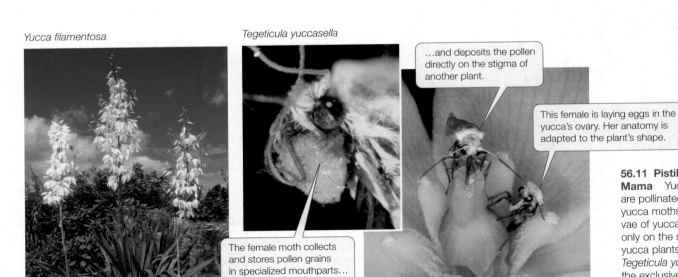

56.11 Pistil-Packing Mama Yucca flowers are pollinated only by yucca moths, and the larvae of yucca moths feed only on the seeds of yucca plants. The moth *Tegeticula yuccasella* is the exclusive pollinator of the *Yucca filamentosa*.

flower's developing seeds. Neither of these species can reproduce in the absence of the other.

Plants and frugivores exchange food for seed transport

Many animals that eat fruits provide a valuable service to the plants that produce them by dispersing seeds. Seed dispersal by animals offers plants the advantage of delivery to potential germination sites and comes with the bonus of organic fertilizer for the seeds. Interactions between plants and frugivores, however, are not always reciprocal; in many cases, one party benefits more than the other. Whereas the frugivore is paid "in advance" for transportation services, the seeds may never reach an appropriate destination for germination (your windshield, for example, will not do). From the plant's perspective, a partnership with frugivores requires a delicate balance between discouraging frugivores from eating fruits before the seeds are capable of germinating and attracting them when the seeds are ready. In addition, the plant must protect the seeds from destruction in the frugivore's digestive tract and defend them against inappropriate consumers that damage the seeds or fail to disperse them at all.

The chemical process of fruit ripening ensures that fruits are most attractive to frugivores when the seeds are mature and ready for dispersal. In many fruits, ripening is accompanied by a decrease in organic acids, which make many unripe fruits sour. Color changes, which result from loss of chlorophyll and the accumulation of other pigments (the conversion of peppers from green to red during ripening is an example), have enormous signal value to many frugivores. Green, unripe fruits are generally difficult for vertebrate frugivores to see against green foliage; red and bicolored red and black fruits contrast with foliage. Fruit softens as it ripens to allow for gentle processing by the frugi-

Dicaeum hirundinaceum

56.12 A Frugivore Plants and Fertilizes A Seed at the Same Time
After a mistletoe bird eats the fruit of the parasitic mistletoe plant, the seeds inside the fruit pass through the bird's digestive tract intact. As the seeds are voided, their sticky outer coat makes them stick to the bird's feathers. As the bird wipes itself clean on a branch, the seed sticks to the branch, where it germinates.

vore and rapid passage through its gut. Another conspicuous change in ripening fruits is an increase in sugar content—the "reward" most sought by frugivores. Seed coats, fruit pulp, and epidermis may all contain secondary chemicals designed to discourage inappropriate frugivores from consuming the fruit.

Because of the often asymmetrical nature of the mutualism between frugivores and plants, relatively few highly specialized frugivores exist. One apparently reciprocal interaction is between mistletoes—parasitic plants that grow on trees—and the mistletoe birds that serve as the plants' primary dispersal agents in Asia and Australia (**Figure 56.12**). These birds dine largely on the fleshy berries of mistletoe. The seeds, covered with a gluelike outer coat, experience little enzymatic or mechanical damage as they pass through the thin-walled guts of the birds that swallow them. When the seeds are voided with the bird's droppings, the sticky outer coat causes the seeds to adhere to the bird's vent feathers, prompting it to wipe its bottom across the tree branch on which it is perched. Once the seed is wiped on the branch, the gluey coat keeps it there—in an ideal location for a mistletoe seed to germinate.

Some mutualistic partners exchange food for care or transport

Some organisms, such as the leaf-cutter ants described at the opening of this chapter, get their food by "farming" fungus. Fungus farming has been documented in a wide variety of species, including bark beetles, termites, and even a snail. In most cases, the farmers provide housing, nutrition, and care for the fungal partner; the fungal species, in turn, provides food for the host. The fungus produces enzymes that can degrade plant proteins and cellulose, converting plant materials that the farmers cannot digest for themselves into an edible form.

Over the past 50 years, one fungus farmer, the southern pine bark beetle (*Dendroctonus frontalis*), has destroyed over a billion dollars' worth of pine forests in the southeastern United States (**Figure 56.13**). The beetle owes much of its efficiency to its mutualistic partners. Masses of adult beetles attack a pine tree at once, overwhelming the tree's ability to defend itself (the tree's defense is to release large quantities of resin under pressure to force out the beetles). The beetles then excavate a series of galleries through the vascular tissue underneath the bark, in which females lay their eggs. Female beetles also carry spores of their partner fungus into the galleries. The fungus grows on and breaks down the gallery walls; the beetles feed directly on the fungus and the partially digested wood. The beetles also transport a bacterium that produces an antibiotic to keep harmful bacteria from attacking the fungus. This insect–fungus–bacteria partnership can overcome the trees' antiherbivore defenses, to their mutual benefit.

Some mutualistic partners exchange food or housing for defense

Some plants are not only resources, they are also mutualistic partners for insects. The best-known of these interactions is that between ants and acacia trees in Central America. In 1874, in Nicaragua, the naturalist Thomas Belt observed a peculiar inter-

(A) *Dendroctonus frontalis*

Galleries in the bark vascular tissue

(B)

(C)

56.13 A Mutualistic Interaction Brings Death to Pine Trees (A) The southern pine bark beetle has a mutualistic relationship with a fungus, which it "farms" within the vascular tissue of pine trees. (B) The bark beetle excavates galleries inside the trees' vascular tissue. Here they lay eggs and farm fungus; the fungus overcomes the trees' defenses against the beetle and provides nutrition for the larvae. (C) Masses of bark beetles overwhelm pine forests, resulting in widespread death of pine trees.

action between bull's horn acacia (*Acacia cornigera*) and *Pseudomyrmex* ants, known as acacia ants because they are found only in association with acacias. Bull's horn acacias have enlarged hollow thorns in which the ants build nests. The trees produce nectar in specialized extrafloral structures and modified leaflet tips that are rich in oil and protein, on which the ants feed. These structures have no apparent purpose other than providing food for ants.

Belt suggested that the notoriously aggressive acacia ants defend the plants against their herbivores and competitors in exchange for food and shelter. But his idea was not tested until Daniel Janzen conducted an experiment in 1966. By removing ants from some acacias with insecticide, Janzen demonstrated that trees without ants suffered a reduction in growth and an increase in mortality (**Figure 56.14**). Ants also clipped weeds from around the base of the plants, presumably reducing competition for nutrients.

yourBioPortal.com
GO TO **Animated Tutorial 56.1** • Mutualism

INVESTIGATING LIFE

56.14 Are Ants and Acacias Mutualists?
Bull's horn acacia trees (*Acacia cornigera*) grow numerous structures that provide food and shelter for ants of the genus *Pseudomyrmex* (acacia ants). Daniel Janzen's experiments demonstrated that the trees benefit greatly from their association with these ants, and that the energy expended in growing ant-attractive structures is repaid with increased growth and survival.

HYPOTHESIS *Acacia cornigera* trees deprived of their *Pseudomyrmex* ant populations will not thrive in comparison with trees populated by ant colonies.

The "bull's horns" of these *Acacia* trees are enlarged, hollow thorns in which the ants build nests.

METHOD
1. Define a population of *A. cornigera* trees; designate some of them as untreated controls and the rest as experiment subjects.
2. Fumigate the experimental *A. cornigera* trees with insecticide to eliminate all *Pseudomyrmex* ants.
3. Apply Tanglefoot® (a sticky material) to the base of the fumigated experimental trees to prevent the ants from recolonizing them.
4. Record the survival and growth rates of the trees in both groups over a 10-month period.

RESULTS After 10 months, control trees (with ants) had considerably higher survival rates and greater growth than did trees without ant populations.

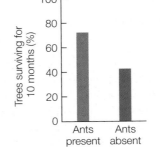
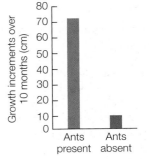

CONCLUSION *Pseudomyrmex* ants provide substantial survival benefits to *Acacia cornigera* trees.

Go to **yourBioPortal.com** for original citations, discussions, and relevant links for all INVESTIGATING LIFE figures.

From our discussion of interactions between species that benefit both, we move to a type of interaction that benefits neither: competition. This type of interaction is widespread because it can arise wherever two or more species require the same resources.

56.4 What Are the Outcomes of Competition?

The antagonistic interactions can be quite attention-getting; the scene of a lion stalking a gazelle is almost emblematic of the African savanna. But at the same time a predator is interacting with its prey, it may also be interacting with other predators that hunt the same prey species. Lions are not the only predators of gazelles; cheetahs, hyenas, and even crocodiles hunt and kill gazelles, potentially reducing the supply available for lions.

Whenever any resource is not sufficiently abundant to meet the needs of all the organisms with an interest in that resource, organisms must compete with one another to gain enough of that resource to survive. Competition not only influences the evolution of species, but also plays an important role in determining the structure and composition of communities, as we will see in the next chapter.

Competition is widespread because all species share resources

Virtually no species enjoys exclusive access to any given resource; the vast majority must share at least some resources with other species. As we saw in Section 55.3, populations do not grow indefinitely, largely because resources are limited. In the absence of sufficient resources, individuals in the population compete for those resources. Such **intraspecific competition**—competition among individuals of the same species—may result in reduced growth and reproductive rates for some individuals, may exclude some individuals from better habitats, and may cause the deaths of others. **Interspecific competition**—competition among individuals of different species—affects individuals in much the same way. In addition, it can influence the persistence and evolution of species.

The *principle of competitive exclusion* holds that no two species can long coexist sharing the same limiting resource. If one species can prevent all members of another species from utilizing a resource, the inferior competitor may go locally extinct, a result called **competitive exclusion**. In other cases, selection pressures resulting from interspecific competition cause changes in the ways in which the competing species use the limiting resource. If those changes allow them to coexist, the result is called **resource partitioning**.

Whether it is interspecific or intraspecific, competition occurs by two major mechanisms. **Interference competition** occurs when a competitor interferes with another competitor's access to a limiting resource. **Exploitation competition** occurs when a limiting resource is available to all competitors and the outcome of competition depends on the relative efficiency with which the competitors use up the resource.

Interference competition may restrict habitat use

Interference competition can take many forms. A graphic example involves the desert ant *Conomyrma bicolor* and the honeypot ant *Myrmecocystus mexicanus*. These two ant species occupy the same type of habitat—arid areas containing little vegetation—and feed on similar foods—the sugary excretions of aphids and other sap-feeding insects and occasional arthropods, none of which is in great supply. When *C. bicolor* workers find the entrance of a honeypot ant nest, they pick up small stones in their mandibles, carry them to the rim of the nest opening, and drop them down the hole—up to 200 stones in a 5-minute interval. This activity is enough to stop the honeypot ants from going out foraging. Some honeypot ant colonies, under constant stone-dropping attack for several weeks, may be almost entirely deprived of food.

Even microorganisms interfere with one another's use of resources. In the highly structured environment of the rhizosphere, or "root-world," of the soil, competitive interactions can be locally intense. Many soil bacteria produce substances that subdue their microbial competitors. Actinomycetes, for example, produce chemicals that interfere with essentially every life process in a bacterium. Many of the chemicals these remarkably well-defended microbes produce to defeat their competitors are used as antibiotics by mutualistic partners, such as the bark beetles described in Section 56.3, as well as in human pharmacology.

Exploitation competition may lead to coexistence

Exploitation competition may lead to coexistence, provided that the species relying on the same resource evolve ways to divide up, or partition, the resource. Resource partitioning can lead to the formation of **guilds**: groups of species that exploit the same resource, but in slightly different ways, making the resource less likely to be preempted. For example, in many Rocky Mountain communities, at least three species of bees consume the nectar of the shindagger agave (*Agave schottii*). The three bee species differ in where and when they collect shindagger nectar. Honey bees tend to forage in places with the greatest numbers of shindagger flowers, bumble bees in places with intermediate numbers of flowers, and carpenter bees where flowers are few and far between. Honey bees also tend to be most active when nectar output is greatest. With their larger nests and greater numbers of offspring to support, honey bees require greater efficiency and greater energy intake. Foraging sites that are not worth their while are left to the other bees.

Sometimes organisms respond to competition by leaving their competitors behind and finding new resources. Robert Denno examined a guild of seven sap-sucking insects called planthoppers living on salt marsh grass in marshes in New Jersey. Six of the seven species partition the resource by feeding on different parts of the plant or by feeding at different times during the season. All but one of the species are wingless and cannot fly. Whenever conditions become crowded in a particular patch of salt marsh grass, the seventh species—the only one with functional wings—flies away to find a new patch in which to feed.

Sometimes individuals within a species display different behaviors or morphologies depending on whether they are competing for resources with another species. Darwin remarked in *The Origin of Species* that "Natural Selection leads to divergence of character; for more living beings can be supported on the same area the more they diverge in structure, habits, and constitutions." This "divergence of character," today called **character displacement**, is most dramatic in cases in which the morphological attributes of a species vary depending on the presence or absence of a competitor. In some of the islands of the Galápagos archipelago, certain cactus species are pollinated exclusively by finches (**Figure 56.15**). On other islands, a carpenter bee (*Xylocopa darwinii*) competes with the finches for cactus nectar; the birds consequently feed more heavily on seeds and insects. On the islands where bees are absent, finches feed on nectar more often, and have significantly smaller wingspans, than finches on

islands where they share cacti with bees. On the bee-free islands, a smaller wingspan means a smaller body size, which allows the birds to reach the nectar of the cactus flowers more easily.

Species may compete indirectly for a resource

Species may compete indirectly for a resource even when they are not present in the same habitat at the same time. Sometimes a species so alters the quality of a resource that it is rendered less usable by other species that may encounter it afterward. For example, feeding by sap-sucking leafhoppers on potato plants early in the growing season can cause leaf curling and chlorosis (loss of chlorophyll); potato beetles that consume these damaged leaves later in the growing season suffer reduced growth and survival rates. Even though these two herbivores do not feed at the same time, one species influences the use of the shared food resource by its competitor.

Consumers may influence the outcome of competition

Indirect competition can also result when two species share a common predator. For example, the parasitoid wasp *Venturia canescens* is a consumer of two different species of caterpillars that infest stored food products such as flour: the Indianmeal moth caterpillar (*Plodia interpunctella*) and the Mediterranean flour moth caterpillar (*Ephestia kuehniella*). The two species can coexist in a flour bin, but when the wasp is present it preferentially attacks and kills the flour moth caterpillars. Thus, in the presence of the wasp, the competitive balance between the two caterpillar species is altered in the meal moth's favor. This type of competition is indirect because the outcome of competition depends not on how the two competitors utilize the shared resource, but on how the two competitors interact with a shared predator.

Competition may determine a species' niche

Competition is important in determining where a species can be found. A species' **niche** is the set of physical and biological conditions it requires to survive, grow, and reproduce. Thus a

56.15 Competition with Carpenter Bees Influences Finch Morphology On islands in the Galápagos archipelago where finches are the sole pollinators of cactus flowers, a small wingspan increases their ability to negotiate the flowers. On islands where carpenter bees compete with finches for cactus nectar, the birds have a larger wingspan and can also feed on other foods.

Geospiza fuliginosa

Nectar Use and Wingspan of *G. fuliginosa*		
ISLAND	TIME SPENT FEEDING ON FLOWER NECTAR (%)	MEAN WINGSPAN (mm)
BEES ABSENT		
Pinta	10	59.8
Marchena	28	58.2
BEES PRESENT		
Fernandina	1	64.8
Santa Cruz	14	64.0
San Salvador	0	63.8
Española	0	64.7
Isabela	7	64.5

Xylocopa darwinii on *Opuntia* flower

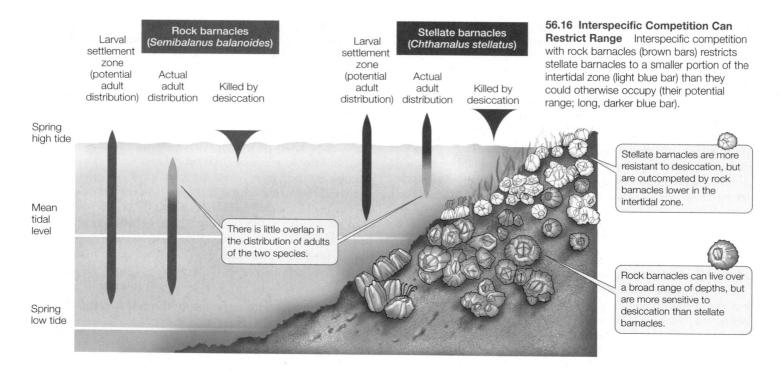

56.16 Interspecific Competition Can Restrict Range Interspecific competition with rock barnacles (brown bars) restricts stellate barnacles to a smaller portion of the intertidal zone (light blue bar) than they could otherwise occupy (their potential range; long, darker blue bar).

Stellate barnacles are more resistant to desiccation, but are outcompeted by rock barnacles lower in the intertidal zone.

Rock barnacles can live over a broad range of depths, but are more sensitive to desiccation than stellate barnacles.

There is little overlap in the distribution of adults of the two species.

species' niche is partly defined by the resources available in the environment. Although a species might be physiologically able to live under a wide range of resource conditions, competitors may restrict its use of resources in a particular location. Thus every species has a **fundamental niche**, defined by its physiological capabilities, and a **realized niche**, defined by interactions with other species.

Two species of barnacles, the rock barnacle (*Semibalanus balanoides*) and Poll's stellate barnacle (*Chthamalus stellatus*), compete for space on the rocky shorelines of the North Atlantic Ocean (**Figure 56.16**). The planktonic larvae of both species settle in the intertidal zone and metamorphose into sessile adults. The smaller stellate barnacles generally live at higher levels in the intertidal zone, where they face longer periods of exposure and desiccation (drying out) than do rock barnacles. There is little overlap between the areas occupied by adults of the two species. What explains their distinct distributions in the intertidal zone?

In a famous study conducted 50 years ago, Joseph Connell experimentally removed one or the other species from its characteristic zone and observed the response of the other species. Stellate barnacle larvae normally settle in large numbers throughout much of the intertidal zone, including the lower levels where rock barnacles are found, but they thrived at those lower levels only when rock barnacles were not present. The rock barnacles grow so fast that they smother, crush, or undercut the stellate barnacle larvae. In contrast, removing stellate barnacles from their spots higher in the intertidal zone did not lead to their replacement by rock barnacles; the rock barnacles are less tolerant of desiccation and fail to thrive there even when stellate barnacles are not around. The result of the competitive interaction between the two species is a pattern of *intertidal zonation*, with stellate barnacles restricted in their distribution by

competition and rock barnacles restricted in their distribution by their physiological limitations.

56.4 RECAP

Competition occurs when two or more species require a resource that is in limited supply. No two species can long coexist sharing a limiting resource. The outcome of competition may be competitive exclusion, in the form of local extinction, or coexistence, in the form of resource partitioning or character displacement.

- How does exploitation competition differ from interference competition? See pp. 1198–1199

- How can competition lead to character displacement? See p. 1199 and Figure 56.15

- Explain the difference between an organism's fundamental niche and its realized niche. See pp. 1199–1200 and Figure 56.16

The study of interactions among the species in a community is a large part of community ecology—the topic of the next chapter. Every kind of interaction we have studied in this chapter influences the nature and structure of communities. Competition helps determine which species persist and which go extinct, as well as dictating how many different species can be supported by a particular resource. Similarly, antagonistic interactions have important effects on the distribution and abundance of consumer and resource species, and the presence of mutualistic partners may dictate whether a particular species can exist in a particular community.

Dead reckoning

In March 1996 the body of an apparent suicide victim was discovered under the bushes near the railroad tracks in Cologne, Germany. The corpse was badly decomposed, and most of the abdominal organs had disintegrated completely. When an autopsy was performed, masses of maggots—fly larvae—about 8 mm long were found in what remained of the body cavity, and the leathery patches of dried skin were almost totally covered with a layer of pale yellow eggs. When he examined the body, Mark Benecke, a forensic entomologist, was able to recover a single adult fly, which he identified as *Piophila casei*, also known as the cheese skipper.

The community of species that colonizes a human corpse is diverse, and its composition varies predictably over time as decomposition progresses. Among the many insect species that infest dead bodies, cheese skippers are latecomers—female cheese skippers do not find a corpse attractive until the proteins in the body begin to break down, typically after 1 to 3 months, depending on season and locality. The abundance of *P. casei* eggs suggested to Benecke that the cheese skippers had undergone at least two generations in the decomposing corpse. Knowing that their development time from egg to adult ranged from 11 to 19 days under local weather conditions, he calculated that the first adult cheese skippers probably arrived and laid their eggs about 90 days after death, and that an additional 22 to 38 days would have been required for two generations to complete their development. Thus he calculated that death must have occurred between 112 and 128 days earlier. As it turned out, a 38-year-old woman had been reported missing about 4 months earlier; the estimated postmortem interval helped investigators to identify her as the suicide victim found by the railroad tracks.

Insects are found in decomposing corpses for many reasons. Some species, such as cheese skippers, consume the dead flesh; others, such as hide beetles, eat the hair and nails. Others prey on the insects consuming the corpse or eat their excrement and shed exoskeletons. Thus a dead body can support an entire community of organisms—one with complex connections among its inhabitants. Where and how an individual died can influence the composition of that community; bodies that are submerged, for example, are colonized by a different suite of insects than those that are buried in the ground.

Dead Reckoning   Attracted by the odor of decaying flesh, bluebottle flies feed on the fur of a dead mole. The insect species on a decomposing corpse change in such a predictable fashion over time that forensic entomologists can estimate the time of death by identifying the species that are present.

Prairie Potholes These wetland communities occur in depressions in the prairie landscapes of midwestern North America. The potholes, formed when the glaciers of the last ice age receded, support communities characterized by many aquatic plant species, including sedges, bulrushes, and pondweeds.

What species occur where and when is the concern of community ecologists. The composition of communities can change in a predictable way over time and over space. The spatial scales of these changes can vary from a dead body in a patch of shrubbery to a patch of Amazonian rainforest, and the time scales can range from days to millennia. But the ecological processes affecting a community are similar whatever its scale. The study of the seemingly esoteric patterns of change in the composition of carrion communities has not only allowed ecologists to help find evidence that can identify a missing person or convict a murderer, but has added greatly to our understanding of ecological communities.

IN THIS CHAPTER we will investigate how energy flows through communities and how that flow influences their structure, complexity, and composition. We will investigate how communities are assembled over time and how they are reassembled after a disturbance. Finally, we will consider the relationships between species diversity, productivity, and stability in communities.

57.1 What Are Ecological Communities?

An ecological **community** is a group of species that coexist and interact within a defined area. Although each species has unique interactions with the other species in its community, the community as a whole can be studied on the basis of the distribution of energy and biomass within it.

Communities vary greatly in size and scope. Whereas one ecologist might study the community of species inhabiting a dead body, another might study the community of species that inhabit Lake Superior, the largest of the Great Lakes. The boundaries defining a community are not always easy to recognize. The community of organisms within a pond, for example, is for the most part bounded by the borders of the pond. Pond species interact with one another to a far greater extent than with species outside the pond. The boundaries of other communities are not so clear-cut. For example, while prairie pothole communities (above left) may appear to be self-contained, many of their components originate far away. Seeds arrive at these relatively isolated wetlands via duck excrement; mallards and other ducks may consume, but not digest, seeds in one location and, in the 5–11 hours it takes for the seeds to move from one end of the digestive tract to the other, fly up to 1,400 km before depositing the seeds when they alight to feed again.

Communities may differ conspicuously in **species richness**—that is, in the number of species they contain. Even the same type of community may contain different numbers of species in different places. The number of plant species in prairie pothole communities, for example, may vary more than twofold across the upper Midwest. Although communities vary in size and complexity by orders of magnitude, ecologists have devised methods for quantifying basic properties of community structure and organization irrespective of their scale. These methods have revealed patterns that reflect underlying community assembly rules and general principles.

What determines how many species constitute a community in any particular place? One important factor is the amount of energy available to sustain organisms.

Energy enters communities through primary producers

Sunlight is the ultimate source of energy for most of Earth's communities. Sunlight makes photosynthesis possible, and photosynthesis, in the vast majority of communities, makes energy

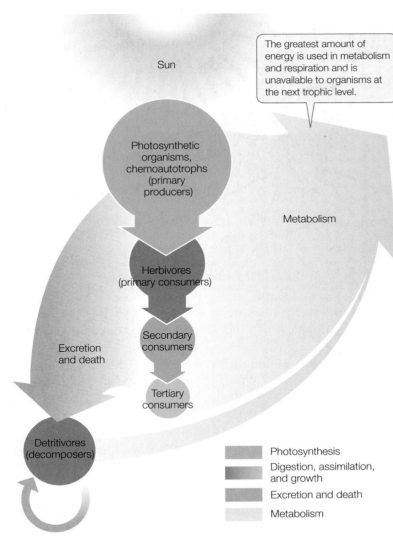

The greatest amount of energy is used in metabolism and respiration and is unavailable to organisms at the next trophic level.

Sun

Photosynthetic organisms, chemoautotrophs (primary producers)

Metabolism

Herbivores (primary consumers)

Secondary consumers

Excretion and death

Tertiary consumers

Detritivores (decomposers)

- Photosynthesis
- Digestion, assimilation, and growth
- Excretion and death
- Metabolism

57.1 Energy Flow through Trophic Levels Much of the energy of the biomass at each trophic level is lost (often as heat) to metabolism and respiration by the organisms at that level. In this diagram, the width of each arrow is roughly proportional to the amount of energy flowing through that channel. Arrows indicate directions of energy flow.

— **yourBioPortal.com** —

GO TO **Web Activity 57.1 • Energy Flow through an Ecological Community**

Gross primary productivity (**GPP**) is the rate at which all the primary producers in a particular community turn solar energy into stored chemical energy via photosynthesis. The energy that is accumulated is called **gross primary production**. (The terms "productivity" and "production" are often used interchangeably; *productivity* measures the rate of energy accumulation; *production* measures energy accumulation as a *product*.)

Not all GPP becomes available to heterotrophs, because primary producers use some of that energy for respiration and other metabolic processes. **Net primary productivity** (**NPP**) is the rate at which energy is incorporated into the primary producers' bodies through growth and reproduction. Thus **net primary production** is the amount of primary producer **biomass** (weight of organic matter) available for consumption by heterotrophs. This relationship is described mathematically as

$$NPP = GPP - R$$

where R is the energy lost through respiration.

Consumers use diverse sources of energy

An organism's **trophic level** indicates where in that sequence it obtains its energy (**Table 57.1**). Primary producers start the chain of trophic levels. At the next level are **primary consumers** — the herbivores that dine on the primary producers. Organisms that eat herbivores, called **secondary consumers**, are the next trophic level. Those that eat secondary consumers are *tertiary consumers*, and so on. The waste products and dead bodies of organisms provide another source of energy, as we saw at the

available to other organisms in an edible form. All nonphotosynthetic organisms (*heterotrophs*) consume, either directly or indirectly, the energy-rich organic molecules produced by the plants and other photosynthetic organisms that get their energy directly from sunlight (*autotrophs*). Photosynthetic autotrophs, along with a handful of *chemoautotrophs* (organisms that obtain chemical energy from inorganic molecules in their environment), are known as **primary producers** (**Figure 57.1**).

TABLE 57.1
The Major Trophic Levels

TROPHIC LEVEL	SOURCE OF ENERGY	EXAMPLES
Photosynthesizers (primary producers)	Solar energy	Green plants, photosynthetic bacteria and protists
Herbivores (primary consumers)	Tissues of primary producers	Termites, grasshoppers, gypsy moth larvae, anchovies, deer, geese, white-footed mice
Primary carnivores (secondary consumers)	Herbivores	Spiders, warblers, wolves, copepods
Secondary carnivores (tertiary consumers)	Primary carnivores	Tuna, falcons, killer whales
Omnivores	Several trophic levels	Humans, opossums, crabs, robins
Detritivores (decomposers)	Dead bodies and waste products of other organisms	Fungi, many bacteria, vultures, earthworms

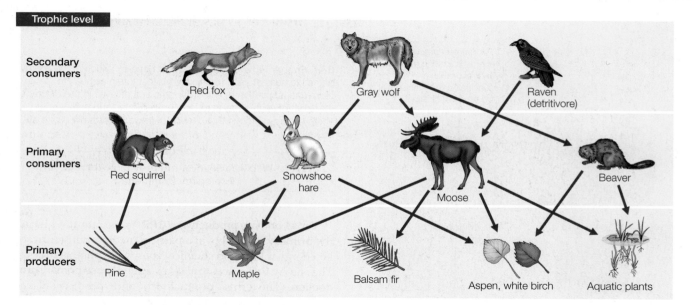

Trophic level

Secondary consumers — Red fox — Gray wolf — Raven (detritivore)

Primary consumers — Red squirrel — Snowshoe hare — Moose — Beaver

Primary producers — Pine — Maple — Balsam fir — Aspen, white birch — Aquatic plants

57.2 Food Webs Show Trophic Interactions in a Community This food web for Isle Royale National Park, located on a large island in Lake Superior, includes only large vertebrates and the plants on which they depend. Even with these restrictions, the web is complex. The arrows show who eats whom.

opening of this chapter. Organisms that consume such materials are called *detritivores* or *decomposers* (see Figure 57.1).

Some organisms, called *omnivores*, feed on multiple trophic levels. Opossums, for example, are famously omnivorous. Investigators in Portland, Oregon, dissected the stomachs of road-killed opossums and found remains of mammals, birds, insects, earthworms, snails, fruits, bulbs, seeds, leaves, grass, pet food, and garbage, along with some items they couldn't even identify.

Although few organisms are as omnivorous as opossums, most species in a community eat and are eaten by more than one other species. A **food chain** depicts the linear sequence of who eats whom in a given community; food chains are interwoven in a **food web** (**Figure 57.2**). Most communities contain so many species interacting in so many different ways that it is impossible to enumerate (or even identify) all of the links. Nevertheless, simplified diagrams of food webs are useful in envisioning how energy flows through a community.

── **yourBioPortal.com** ──

GO TO **Web Activity 57.2** • The Major Trophic Levels

Fewer individuals and less biomass can be supported at higher trophic levels

One important way to characterize a community is by the distribution of energy and biomass within it. The flow of energy through a food web is governed by the physical laws that regulate energy transformations, foremost among which are the first and second laws of thermodynamics. Recall from Section 8.1 that the amount of energy in the universe is constant, and that when energy is converted from one form to another, part of it becomes unavailable to do work. As Figure 57.1 illustrates, at each trophic level energy is lost to metabolism and respiration. On average, only about 10 percent of the energy

of one trophic level is transferred the next, for a number of reasons:

- *Heat loss.* Organisms incorporate much of the energy they accumulate into biomass, but much more is used for respiration and other metabolic processes. That energy is dissipated as heat and is lost to the community.

- *Biomass availability.* Not all biomass can be ingested. Grazers routinely miss blades of grass; effective plant defenses prevent herbivory; prey can escape predators or leave the community.

- *Indigestibility.* Not all biomass ingested can be assimilated by consumers. Tree bark, for example, contains lignin and cellulose, which cannot be digested by most herbivores.

The overall transfer of energy from one trophic level to the next (which can be expressed as the ratio of consumer production to producer production) is called **ecological efficiency**.

Pyramid diagrams such as **Figure 57.3A** can be used to illustrate the proportions of energy transferred from each successive trophic levels and to compare those proportions among different communities. Pyramid diagrams can also be used to illustrate the amount of biomass or numbers of individuals found at each level (**Figure 57.3B**). Progressive energy loss through the inefficiencies of energy transfer puts limits on the number of trophic levels in a food chain or food web; largely for this reason, most communities support only three to five trophic levels. Furthermore, each successive trophic level tends to have fewer species, and the species at each level tend to have lower reproductive rates, smaller populations, and larger body sizes.

Productivity and species richness are linked

Just as the diversity and complexity of trophic levels tend to be positively correlated with the amount of energy available to them, the species richness of a community tends to be positively correlated with its productivity, up to a point. A number of factors that influence productivity, and thus species richness, vary among communities.

The most obvious variation that influences productivity is energy input. The amount of solar energy reaching Earth's surface

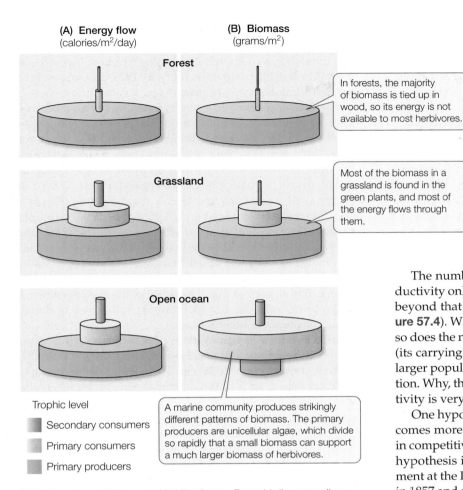

(A) Energy flow
(calories/m²/day)

(B) Biomass
(grams/m²)

Forest

In forests, the majority of biomass is tied up in wood, so its energy is not available to most herbivores.

Grassland

Most of the biomass in a grassland is found in the green plants, and most of the energy flows through them.

Open ocean

A marine community produces strikingly different patterns of biomass. The primary producers are unicellular algae, which divide so rapidly that a small biomass can support a much larger biomass of herbivores.

Trophic level

■ Secondary consumers

■ Primary consumers

■ Primary producers

57.3 Energy and Biomass Distributions Pyramid diagrams allow ecologists to compare (A) patterns of energy flow through trophic levels in different communities and (B) the amount of biomass present at the different trophic levels.

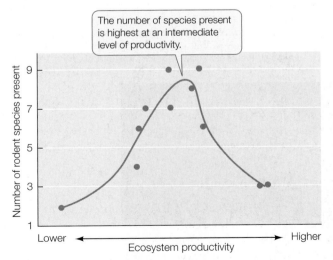

The number of species present is highest at an intermediate level of productivity.

Number of rodent species present

Lower ← Ecosystem productivity → Higher

57.4 Species Richness Peaks at Intermediate Productivity The number of rodent species living in ecosystems of varying productivity in the Gobi Desert exemplifies a pattern in which species richness increases only up to a certain point. Beyond that point, it can actually decline with productivity.

varies by latitude, as we saw in Figure 54.1. The ability of plants to photosynthesize, however, depends not only on the supply of energy from the sun, but also on the supply of water and nutrients. The species richness of trees across different regions of North America, for example, can be best predicted not by measuring incoming solar radiation, but by measuring an ecological attribute of those regions called *annual evapotranspiration*: the amount of water released from the land surface by evaporation from streams, lakes, and soil and by transpiration from plants. The annual evapotranspiration of a region is measure of the amount of water available to the organisms living there.

The number of species in a community increases with productivity only up to a point, however; if productivity increases beyond that point, species richness may actually decline (**Figure 57.4**). Why should that be? As local productivity increases, so does the number of individuals the local habitat can support (its carrying capacity). Thus populations can grow larger, and larger population sizes should reduce the risk of species extinction. Why, then, should species richness decrease when productivity is very high?

One hypothesis postulates that interspecific competition becomes more intense when productivity is very high, resulting in competitive exclusion of some species (see Section 56.4). This hypothesis is supported by the results of a long-term experiment at the Rothamsted Experiment Station in England, begun in 1857 and still ongoing. Fertilizer has been added regularly to selected plots of land to increase their productivity, and fertilized and unfertilized plots have been monitored continuously. Over 150 years, the number of plant species in the unfertilized plots has remained roughly constant, whereas that number has declined in the fertilized plots, supporting the premise that species richness can decline when productivity rises.

57.1 RECAP

An ecological community is a group of species that coexist and interact within a defined area. Net primary production—the amount of biomass that primary producers make available to heterotrophs through photosynthesis—is an important factor determining the species richness and number of trophic levels in a community.

● Explain the difference between gross primary productivity and net primary productivity. See p. 1205

● Describe three trophic levels that might appear in a food web. See pp. 1205–1206, Table 57.1, and Figure Figure 57.2

● What is a typical distribution of energy among the trophic levels of a community? See pp. 1205–1206 and Figures 57.1 and 57.3

Although energy and biomass distributions reveal much about the structure and dynamics of ecological communities, they reflect primarily the abiotic mechanisms that influence the community. In the next section we will see how the species interactions described in Chapter 56 shape community structure.

57.2 How Do Interactions among Species Influence Community Structure?

All of the interaction types discussed in Chapter 56, and especially the antagonistic interactions, have a strong influence on an ecological community. Species are not identical bags of biomass through which energy flows. Species in one part of a food web can affect many other species without necessarily eating them.

Species interactions can cause trophic cascades

The interactions of a single consumer with other species in its community can cause a progression of indirect effects across successive trophic levels, a pattern called a **trophic cascade**. The reintroduction of wolves into Lamar Valley, in Yellowstone National Park, initiated just such a pattern.

The food web in Yellowstone National Park is complex. Wolves in the park feed on moose, pronghorn, mule deer, elk, bison, and bighorn sheep. Although they share these prey with coyotes, mountain lions, and grizzly and black bears, wolves exert particularly strong effects on the park community's structure and dynamics, as demonstrated by the effects of their absence during most of the twentieth century. By 1926, unrestricted hunting had eliminated wolves from the park community.

Annual censuses of elk were initiated by park managers in 1920. To prevent elk from exceeding the park's carrying capacity, the park service culled elk herds (that is, they selectively killed some members of the herds) until 1968, when, in response to public pressure, the culls were stopped. When the culling ended, the elk population rapidly increased (**Figure 57.5A**). The elk browsed aspen trees so intensely that no young trees were added to the population after 1920 (**Figure 57.5B**). The elk also severely browsed streamside willows, with the result that beavers, which depend on willows for food, were nearly exterminated from Lamar Valley. In regions of the park where elk were absent, however, aspen and willow trees flourished. This observation suggested that the decline of the trees elsewhere in the park was indeed due to elk browsing, rather than to climatic conditions or some other factor.

In 1995, after a 70-year absence, park managers reintroduced wolves to Yellowstone, and their population grew rapidly. The wolves preyed primarily on elk. The elk population of Lamar Valley dropped, and elk avoided the aspen groves, where they were especially vulnerable to wolf predation. Young aspen began to grow, willows regrew along streams, and the number of beaver colonies increased from one in 1996 to seven in 2003. Thus the presence or absence of a single predator influenced not only populations of its prey, but also the structure of the vegetation and populations of other species that depend on that vegetation.

57.5 Wolves Initiated a Trophic Cascade (A) Number of elk and relative size of the wolf population in Wyoming's Yellowstone National Park. Once the wolves were gone, the elk population increased to the point where the young aspen trees were seriously overgrazed. (B) Aspen establishment in the presence and in the absence of wolves. With overgrazing by elk, young trees were not recruited; the photo shows a resulting stand of only mature trees.

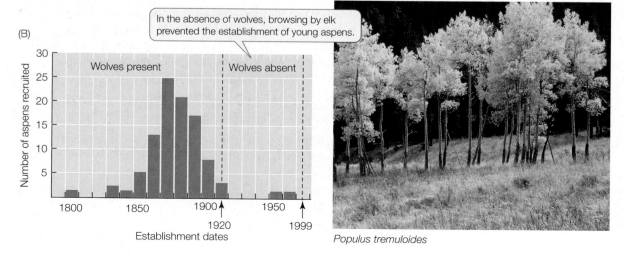

Populus tremuloides

Herbivores, too, can have indirect effects on other trophic levels, to produce an interaction cascade. The savannas of central Kenya are dominated by large grazing mammals such as zebras, eland, elephants, Grant's gazelles, giraffes, buffaloes, and hartebeests. A team of investigators used *exclosures* (areas protected by a barrier, such as a fence, designed to keep organisms out) to investigate the influence of these grazers on the savanna community. They created six exclosures and compared community structure within the exclosures with that in paired sites where large mammals could graze freely. Over 19 months of monitoring, they found that trees, beetles, and an insectivorous lizard species were all more abundant within the exclosures. The elimination of grazing increased the abundance of trees, which in turn increased the number of beetles by providing more food and habitat, which in turn increased the number of lizards, which feed preferentially on beetles.

Interaction cascades can be initiated by the behaviors of species as well as by their diets. Beavers preferentially cut down some species of trees to build their dams; by so doing, they alter the composition of the vegetation. In addition, the beaver dams create wetlands, meadows, and ponds that provide habitat for species that would otherwise not be able to live in the area. Organisms that build structures that alter existing habitats or create new habitats are called **ecosystem engineers**.

Keystone species have wide-ranging effects

A species that exerts an influence on a community disproportionate to its abundance is called a **keystone species**. Keystone species influence both the species richness and the number of trophic levels in a community. The ochre sea star (*Pisaster ochraceus*), which lives in rocky intertidal zones on the Pacific coast of North America, is a good example. Its preferred prey is the mussel *Mytilus californianus*. In the absence of sea stars, these mussels crowd out other organisms in a broad belt of the intertidal zone. By consuming mussels, *P. ochraceus* creates bare spaces on the rocky substratum that are taken over by a variety of other species (**Figure 57.6**).

In a classic experiment, Robert Paine of the University of Washington demonstrated the disproportionate influence of ochre sea stars on species richness by removing them from selected sites repeatedly over a 5-year period. Two major changes occurred in the areas where sea stars were absent. First, the lower edge of the mussel bed extended farther down into the intertidal zone, showing that sea stars are able to eliminate mussels completely in areas that are submerged most of the time. Second, and more dramatically, 28 species of animals and algae disappeared from the sea star removal sites. Eventually only *M. californianus*—the dominant competitor for space in the community—occupied the entire substratum. Through its effect on competitive relationships, predation by the sea star determines how many species can thrive in its rocky intertidal community.

Species other than consumers can be keystone species. A plant species that serves as food for many different animals can also be a keystone species. Fig trees in tropical forests produce fruits several times every year, so their fruits are abundant at times when few, if any, other trees are fruiting. Dozens of frugi-

Mussels

Pisaster ochraceus

57.6 Some Sea Stars Are Keystone Species Ochre sea stars (*Pisaster ochraceus*) have harvested all the mussels from the lower parts of these rocks on the Olympic Peninsula of Washington. By consuming mussels, the sea star creates bare spaces on the rocks that are occupied by a variety of other intertidal species.

vores depend on figs when no other fruits are present. Fig-eating animals include fruit bats, parrots, toucans, pigeons, flycatchers, trogons, orioles, rodents, howler monkeys, and even fish, which eat figs that fall into nearby streams. All of these animals provide prey for a diverse community of predators. Moreover, the trunks of fig trees provide habitat for several thousand species of insects, reptiles, rodents, and birds. Without fig trees, rainforest communities around the world would be profoundly diminished in species richness.

57.2 RECAP

Some interactions between consumers and other species result in a trophic cascade of indirect effects on species at successive trophic levels. Keystone species are species with especially strong effects on the species richness and number of trophic levels in communities.

- Describe an example of a consumer whose interactions with other species cause indirect effects across trophic levels.

- What are some of the ways in which keystone species can affect other species in their communities?
  See p. 1209

We have seen how certain keystone species influence the over-all species richness of their communities. But species richness is only one measure of community diversity. The next section takes a closer look at other ways in which ecologists measure species diversity.

57.3 What Patterns of Species Diversity Have Ecologists Observed?

Communities clearly vary in their diversity, both geographically on scales ranging from local to global, and over time on scales ranging from a day to centuries. Comparing the diversity of two or more communities can be challenging because diversity has many different components depending on the scale at which it is measured:

- **Alpha diversity** is diversity *within* a single community or habitat.
- **Beta diversity** is *between-habitat* diversity, a measure of the change in species composition from one community or habitat to another.
- **Gamma diversity** is the *regional* diversity found over a range of communities or habitats in a geographic region.

> Although the two communities have the same species richness, community A has a more even distribution of species, and is thus considered more diverse than community B.

Community A

Community B

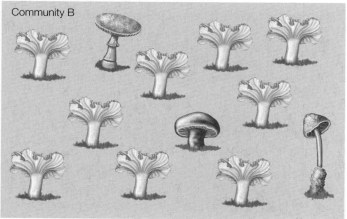

A community's diversity can be measured with a diversity index

The most straightforward way to quantify the alpha diversity of a community is simply to count the number of species present in a sample (that is, the species richness). The larger the area that is sampled, the greater the likelihood that rare species in the community will be found and that the resulting assessment of its species richness will be accurate.

To say that two communities of the same size have the same species richness tells only part of the story, however. Imagine that we take samples of 12 individuals in each of two communities. Our sample from community A contains 3 individuals of each of 4 species (an even distribution of individuals). Our sample from community B, however, contains 9 individuals of one species and only 1 individual each of the other 3 species (an uneven distribution). Even though the species richness of the two communities is the same (4), community B is considered less diverse because the less abundant species are encountered infrequently compared with the single most abundant species (**Figure 57.7**). Thus estimates of diversity should account for both species richness and species evenness.

Ecologists have devised mathematical formulas to quantify diversity that take both species richness and species evenness into account. One widely used measure is the **Shannon diversity index**, based on a mathematical expression of the certainty with which the next item sampled in a series can be predicted. If, for example, we picked individuals from community B in Figure 57.7 at random, we would be fairly certain which species would be picked first, second, third, and so on: most of the time, it would be the most abundant species. The Shannon diversity index (H) describes diversity as a measure of that certainty. The higher H is, the lower would be our certainty about the species composition of our sample and, thus, the greater the diversity of the community. In other words, when we know that the abundance of a particular species is high, we can be reasonably certain that we are likely to collect that species in a random sample.

We can calculate H in three steps:

- Determine the proportion of the total number of individuals in the community that are in each species (p_i) and multiply that proportion by its logarithm
- Sum the values obtained for all species in the community
- For mathematical reasons, multiply the result by –1

In algebraic notation, this calculation is written as

$$H = -[(p_1 \ln p_1) + (p_2 \ln p_2) + (p_3 \ln p_3) \dots (p_n \ln p_n)]$$

Estimating the alpha diversity of a single community with the Shannon diversity index requires only that we know the total number of individuals collected in a sample and the number of species in the sample. But describing the species diversity of a

Figure 57.7 Species Richness and Species Evenness Contribute to Diversity These two hypothetical mushroom communities are the same size (12) and have the same species richness (4) but differ in species evenness. The more even distribution results in greater diversity.

geographic region that comprises multiple communities requires a different approach. Recall that measures of beta diversity estimate the change in species composition from one community to another. **Sorenson's index** (C_S) is one such measure, calculated

$$C_S = \frac{2j}{a+b}$$

where $2j$ is twice the number of species found in both communities, a is the number of species in the first community, and b is the number of species in the second community.

Gamma diversity—the overall diversity across a geographic region containing many different communities—can be used to make comparisons of diversity among regions. Gamma diversity is equivalent to the alpha diversity of each community combined with the beta diversity between the communities.

Partitioning diversity within a community, between communities, and across an entire region can provide insights into the ecological characteristics of the species making up those communities as well as the processes by which those communities were assembled. High beta diversity, for example, shows that few species are shared between the communities being compared, suggesting that those communities consist mostly of specialists, species with unique habitat requirements, or species with poor dispersal abilities.

A study of diversity patterns in an agricultural area of southern England demonstrates how these three measures of diversity estimates are applied at different scales. Investigators sampled plants and macroinvertebrates in three types of freshwater communities—rivers, ponds, and ditches—in an area of English countryside. Rivers had high species richness (i.e., high alpha diversity), but most rivers in the area contained the same assortment of species, so beta diversity, from river to river, was low. In contrast, the ponds displayed a wide range of variation in species richness (i.e., the ponds, with high beta diversity, were more different from one another in species composition than the rivers were). Thus the ponds contributed more to gamma diversity—the overall diversity of the area—than the rivers did because they contained more unique or rare species. Surprisingly, the ditches, many of which contained water for only a short time, had the lowest alpha diversity (the fewest species overall) of the three community types, but many of the species found in ditches were found nowhere else in the area, including insects that live only in temporary bodies of water (including a very rare water beetle). Ponds and ditches, despite being relatively species-poor, contributed disproportionately to regional diversity (i.e., gamma diversity) because the few species they supported were found nowhere else in the region.

Latitudinal gradients in diversity are observed in both hemispheres

It has long been known that diversity varies with latitude. About 200 years ago, the German explorer and naturalist Alexander von Humboldt spent 5 years traveling around Latin America. He remarked in the account of his voyages that "the nearer we approach the tropics, the greater the increase in the variety of structure, grace of form, and mixture of colors, as also in perpetual youth and vigour of organic life." Humboldt would not have been surprised to learn that, if he had sailed toward the poles, the diversity he observed would have decreased. These latitudinal gradients in diversity have been observed in a wide variety of taxa, including birds, mammals, and insects (**Figure 57.8**).

Although most ecologists agree that latitudinal gradients in diversity exist, there is less consensus as to why they exist. At least four hypotheses have been advanced to account for latitudinal gradients in diversity:

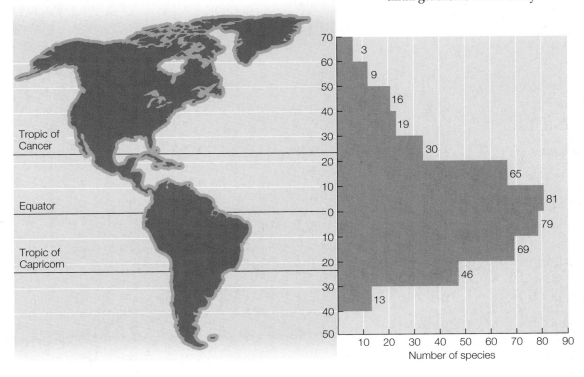

57.8 Latitudinal Gradients in Diversity Among swallowtail butterflies (Papilionidae), species diversity decreases with degree of latitude both north and south of the equator. Similar latitudinal gradients of diversity have been observed in many other taxa.

- The *time hypothesis* argues that organisms in tropical regions have had more time to diversify under relatively stable climatic conditions than have more temperate regions.

- The *spatial heterogeneity hypothesis* suggests that tropical regions have high spatial heterogeneity—more different types of microclimates, vegetation, soils, and so forth—and thus contain more different habitats and many more species.

- The *specialization hypothesis* attributes latitudinal gradients in diversity to greater biological competition in the tropics, which lead to narrower niches and more species.

- The *predation hypothesis* proposes that predation intensity is greater in the tropics (and thus conflicts with the specialization hypothesis). Where predation is high, it argues, prey populations are held to levels so low that interspecific competition never comes into play, and rare species can persist.

Why can't ecologists agree on the mechanism underlying latitudinal gradients in species diversity? Corroborative evidence can be found for each of these hypotheses, varying with taxon, locality, and scale. It may be that multiple factors are responsible for this widespread ecological pattern.

The theory of island biogeography suggests that species richness reaches an equilibrium

While latitudinal gradients prevail on a global scale, other factors influence species diversity within any latitudinal band. In the West Indies, for example, small islands tend to have fewer species of butterflies than large islands, irrespective of latitude. Furthermore, oceanic islands generally have fewer species, irrespective of taxon, than continental areas of comparable size. This **species–area relationship** is a well-established mathematical relationship between the size of an area of habitat and the number of species that area contains. The biologist Edward O. Wilson was struck by this relationship, which he encountered through his exhaustive collection of ant species from all over the world. With Robert MacArthur, Wilson developed the theory of **island biogeography**. They based their theory on just two processes: the immigration of new species to an island and the extinction of species already present on that island (**Figure 57.9A**).

The premise of island biogeography is that the number of species on an island (or in another geographically defined and isolated area) represents a balance, or equilibrium, between the rate at which species immigrate to the island and the rate at which resident species go locally extinct. The rate of immigration is determined in part by the number of species in the area providing the immigrants, known as the **species pool**. In the case of oceanic islands, the species pool comprises all the species on the nearest continent. Not all species that reach the island will persist, however. The more species there are on an island, the higher the likelihood that any one of those species will go extinct.

Rates of immigration and extinction are influenced by two other factors:

- *Distance of the island from the species pool.* The farther the island from the source of immigrants, the fewer species will successfully make the trip and the lower the immigration rate.

- *The size (area) of the island.* The smaller the island, the fewer resources it provides, and the higher the extinction rate. Larger islands provide greater habitat diversity and can sustain larger populations (which in general have lower extinction rates than small populations).

yourBioPortal.com

GO TO Animated Tutorial 57.1 • Biogeography Simulation

57.9 MacArthur and Wilson's Theory of Island Biogeography
(A) The rate of arrival of new species and the rate of extinction of species already present determine the equilibrium number of species on an island. (B) These rates are affected by the size of the island and its distance from the mainland.

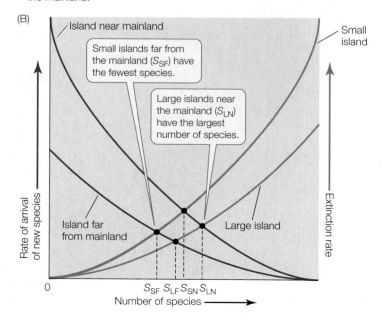

(A)

The number of species reaches an equilibrium (S) when the number of new species arriving equals the number becoming extinct.

Rate of arrival of new species

Extinction rate

Rate

Rate

Equilibrium (S)

0   S

Number of species →

(B)

Island near mainland

Small island

Small islands far from the mainland (S_{SF}) have the fewest species.

Large islands near the mainland (S_{LN}) have the largest number of species.

Island far from mainland

Large island

Rate of arrival of new species

Extinction rate

0   S_{SF} S_{LF} S_{SN} S_{LN}

Number of species →

INVESTIGATING LIFE

57.10 Testing the Theory of Island Biogeography

By experimentally removing all the arthropods on four small mangrove islands, Simberloff and Wilson were able to observe the rates at which arthropods recolonized the islands and compare these data with the predictions of island biogeography theory.

HYPOTHESIS Defaunated islands will be rapidly recolonized and will achieve about the same number of species that were present prior to defaunation.

METHOD
1. Census the terrestrial arthropod species on 4 small (11–25 m²) mangrove islets.
2. Erect scaffolding and tent the islets. Fumigate with methyl bromide, a chemical that kills arthropods but does not harm plants.

3. Remove tenting. Monitor recolonization for the following 2 years, periodically censusing terrestrial arthropod species.

RESULTS Recolonization was fastest on the closer islands, slowest on the one farthest from the mainland. Two years after defaunation, each island had about the same number of species it had before the experiment.

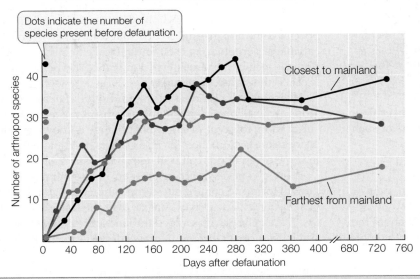

Dots indicate the number of species present before defaunation.

Closest to mainland

Farthest from mainland

Number of arthropod species

Days after defaunation

CONCLUSION The data support the premise that an island supports a certain equilibrium number of species.

FURTHER INVESTIGATION: This experiment assessed only arthropod species, and its duration (2 years) was brief in terms of evolutionary time. What experiments might you conduct to assess some of the broader predications of the theory?

Go to **yourBioPortal.com** for original citations, discussions, and relevant links for all INVESTIGATING LIFE figures.

At some point, the number of species arriving from the source and the number of resident species going extinct should balance, and the number should remain stable at this balance point—the *equilibrium number of species* (**Figure 57.9B**).

Wilson and his student Daniel Simberloff devised an ingenious experiment to test the theory of island biogeography. They used four small, isolated clumps of red mangrove (*Rhizophora mangel*), 11–18 m in diameter, in the Florida Keys. These mangrove islands were small enough that Simberloff and Wilson could count the arthropod species on each one, then enclose it in a tent and gas it with methyl bromide to kill all the arthropods. After defaunation, Simberloff and Wilson monitored and tracked recolonization of the islands by arthropods (**Figure 57.10**). Recolonization was generally rapid, but was slowest on the island farthest from the mainland, as predicted by the theory of island biogeography. Furthermore, all of the islands eventually supported roughly the same number of species that they had before defaunation—again, as predicted.

The theory of island biogeography has been widely accepted since Simberloff and Wilson's landmark study. In addition, it has been retrospectively tested by studies of recolonization after major disturbances that serve as "natural experiments." For example, in August 1883, the island of Krakatau was devastated by a series of volcanic eruptions that destroyed all life on its surface. After the lava cooled, plants and animals from Sumatra to the west and Java to the east recolonized Krakatau. By 1933, the island was again covered with a tropical evergreen forest, and 271 species of plants and 27 species of resident land birds were found there (**Table 57.2**). Today the numbers of plant and bird species are not increasing as fast as they did between 1900 and 1930, but colonizations and extinctions continue, as predicted by the theory of island biogeography.

The theory of island biogeography is so robust that it can be applied equally well to *habitat islands*—isolated patches of suitable habitat surrounded by extensive areas of unsuitable habitat. Thus an oasis in the desert or a prairie pothole surrounded by housing subdivisions may acquire an equilibrium number of species in much the same way an oceanic island does.

TABLE 57.2

Number of Land Plant and Resident Land Bird Species on Krakatau

TIME[a]	NUMBER OF PLANT SPECIES	NUMBER OF BIRD SPECIES
1883 ($t = 0$)	0	0
1886 ($t = 3$)	26	0
1897 ($t = 14$)	64	8
1908 ($t = 25$)	115	13
1920 ($t = 37$)	184	27
1928 ($t = 45$)	214	27
1934 ($t = 51$)	271	29

[a] Time is given as the year, followed by the number of years since the volcanic eruption that destroyed all life on the island.

57.3 RECAP

Species diversity encompasses both species richness and species evenness. It is highest in the tropics, decreasing in higher latitudes. Island biogeography theory states that species diversity on an island or other isolated habitat fragment represents a balance between immigration and extinction rates.

- Explain the concepts of alpha, beta, and gamma diversity. See p. 1210

- What is the difference between species richness and species evenness? See p. 1210 and Figure 57.7

- What factors influence the equilibrium number of species on an island, according to the theory of island biogeography? See p. 1212 and Figure 57.9

Droughts, fires, and volcanic eruptions are examples of ecological disturbances. The next section looks at the effects of such disturbances on ecological communities.

57.4 How Do Disturbances Affect Ecological Communities?

An ecological disturbance is any abiotic event that changes the probability of persistence of one or more species in an ecological community. Disturbances may remove some species from a community, but may open up space and resources for other species. The magnitude of the effect of a disturbance varies enormously. Some disturbances are limited to small areas—for example, a log carried by waves may crush algae and animals attached to rocks in an intertidal community. In contrast, hurricanes, forest fires, and volcanic eruptions can affect communities over hundreds or thousands of hectares. Although small-scale disturbances are far more frequent than large ones, a few large events may be responsible for most of the changes in a community. A single hurricane, for example, may fell more trees than years of "normal" storms.

A community's history of disturbance may explain patterns of species diversity that would otherwise be puzzling. The sub-Antarctic Province Islands, located in the South Indian Ocean, provide an example. The climate of these islands is cool, with monthly average temperatures above freezing for only 6 months of the year. Precipitation is high, and gale force winds are not uncommon; the vegetation is primarily tundra. South African entomologist S. L. Chown, an expert on life in the Antarctic, compared the insect faunas on the four largest of these islands and found that the two largest, Marion and Kerguelen, housed fewer arthropod species (16 and 22, respectively) than the significantly smaller Cochons and Possessions (26 and 38).

Why do the species diversity patterns on these islands fail to conform to the theory of island biogeography? One possible explanation is the different geological histories of the islands. The two smaller islands have escaped extensive glaciation, and accordingly have been accumulating species over a longer time than have the larger islands. Thus, the theory of island biogeography notwithstanding, island area may not always be the best predictor of species diversity. Geological history—particularly the history of disturbances—is often a better predictor.

Succession is the predictable pattern of change in a community after a disturbance

How does a community reassemble itself after a disturbance, particularly one as massive as a glacier? The pattern of change in community composition following a disturbance is known as **succession**. The most common type of succession is *directional* succession, which is characterized by an orderly (or at least predictable) progression of community assemblages. Species come and go until a particular community—one that is capable of perpetuating itself under local climatic and soil conditions—persists for a relatively long time. This persistent stage is called the **climax community**.

Directional succession is easiest to observe after a disturbance strips away all preexisting living organisms and exposes a bare substratum; this type of directional change is known as **primary succession**. Disturbances such as glaciers, volcanic activity, and in some cases, floods can initiate primary succession.

Primary succession can be seen in the change in plant growth form and community composition following the retreat of a glacier in Glacier Bay, Alaska, over the last 200 years (**Figure 57.11**). The glacier scraped the landscape down to bare rock and left a series of *moraines*—gravel deposits formed where the glacial front was stationary for a number of years. No human observer was present to measure changes over the entire 200-year period, but ecologists have inferred the temporal pattern of succession by studying the vegetation on moraines of different ages. The youngest moraines, closest to the current glacial front, are populated with bacteria, fungi, and photosynthetic microorganisms that can support themselves on bare rock. Slightly older moraines farther from the glacial front have lichens, mosses, and a few species of shallow-rooted herbs, such as mountain avens (*Dryas octopetala*). Still farther from the glacial front, successively older moraines have shrubby willows, alders, and spruces.

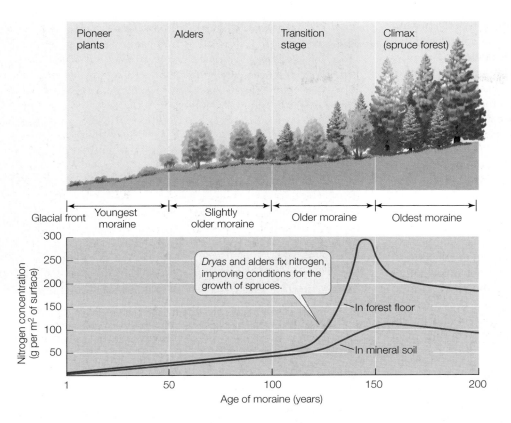

57.11 Primary Succession As the community occupying a glacial moraine at Glacier Bay, Alaska, changes from an assemblage of pioneer plants such as *Dryas* to a spruce forest, nitrogen accumulates in the soil.

── yourBioPortal.com ──

GO TO **Animated Tutorial 57.2 • Primary Succession on a Glacial Moraine**

tend to occur more frequently and progress more rapidly.

A typical sequence of secondary succession in eastern North America begins when forested land that had been cleared for agriculture is abandoned (**Figure 57.12**) The first plants to appear in old agricultural fields are fast-growing annuals such as pigweed, ragweed, and lamb's-quarter. These pioneer species are quickly replaced by more competitive biennials and perennials, such as milkweed, goldenrod, and thistles. Eventually, shrubby plants such as dogwood, eastern red cedar, and sumac become established, followed by tree species such as cottonwood, cherry, and red maple. Ultimately, shade-tolerant trees, including beech and sugar maple, dominate the landscape. The beech–maple forest is the climax community for much of the region.

Nitrogen is virtually absent from glacial moraines, so the plants that grow best on recently formed moraines at Glacier Bay are *Dryas* and alders (*Alnus*), both of which have nitrogen-fixing bacteria in nodules on their roots (see Figure 36.9). Nitrogen fixation by these plants improves the soil so that spruces can grow. Spruces then outcompete and displace the early colonists. If the local climate does not change dramatically, a climax community dominated by spruce trees is likely to persist for many centuries on old moraines at Glacier Bay.

Directional succession following a disturbance that some organisms, particularly those in the soil, survive is called **secondary succession**. Secondary successions are often initiated by human activities as well as by natural disasters. Generally easier to monitor than primary succession, secondary successions also

Directional succession, irrespective of where it takes place, is characterized by certain trends. In general, the colonizing species of early successional stages tend to be good dispersers with high intrinsic rates of reproduction (*r*-strategists). Early stages of succession are characterized by high productivity and simple food webs; most nutrients are present as detritus or in abiotic forms. As succession proceeds, nutrients accumulate in biomass, food webs become more complex, and abiotic sources of nutrients become less important. Species typical of late successional stages tend to be good competitors with relatively low intrinsic rates of reproduction (*K*-strategists).

Both facilitation and inhibition influence succession

To some extent, the progress of succession depends on the activity of successive colonists, each of which modifies the environment in such a way as to facilitate colonization by other species. Predators are unlikely to colonize a habitat with no prey species, nor can primary consumers exist before

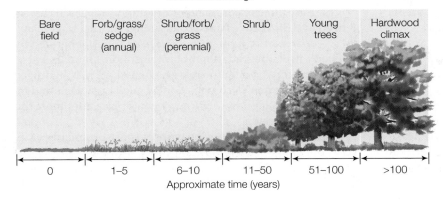

57.12 Secondary Succession Secondary succession is the process by which land that once supported an agricultural field can ultimately support a long-lasting climax community characterized by shade-tolerant trees.

plants are established. The fixation of nitrogen by *Dryas* and alders that allows spruces to become established in Glacier Bay (see Figure 57.11) is an example of such **facilitation**.

Although secondary succession is often described in terms of changes in plant species composition, colonization by plants is actually facilitated by heterotrophs. Often the first organisms to arrive on bare soil after a disturbance are detritivores, which process dead organic matter and release nutrients (especially nitrogen) and thus facilitate the establishment of plants. In a study of intensively burned 20-year-old pine plantations in northern Germany, the first organisms to colonize the burned forests were slime molds, liverworts, mosses, and mushrooms. These detritivores were followed by algae-feeding flies, fungus-feeding beetles, and moss-feeding springtails. Flowering plants such as fireweed moved in, at which point leaf-feeding insects appeared, soon followed by predaceous ground beetles and wolf spiders.

In other cases, the effect of early colonists is **inhibition**, rather than facilitation, of colonization by other species. Old-field species such as goldenrod and thistle produce root exudates that inhibit the germination and growth of potential competitors. Eventually, when these plants grow old and die, other plant species can become established.

Cyclical succession requires adaptation to periodic disturbances

Some patterns of succession are *cyclical*, rather than directional. In such cases, the climax community depends on periodic disturbance in order to persist. The lodgepole pine (*Pinus contorta*) forests that grow in the pumice plateau area of southern Oregon are maintained by periodic forest fires (**Figure 57.13**) In this fire-adapted community, fires return nutrients to the soil in the form of burned organic matter and provide proper conditions for seed germination. The cones of lodgepole pines are sealed shut by resins; only when they are subjected to high temperatures that melt the resins do they open and release their seeds. The lodgepole pine is attacked by *Dendroctonus ponderosae*, the mountain pine beetle, and is also prone to infection by a fungus, *Phaeolus schweinitzii*, which causes the roots and heartwood of the pines to rot. Trees that have lived long enough to experience and be scarred by a fire are much more likely to become infected by the fungus than are trees that have not been scarred. Fungus-infected, weakened trees are preferentially attacked by beetles. After a beetle outbreak, in which many fire-scarred mature trees are killed, the dead trees serve as potential fuel for a fire. The next fire frees up their nutrients for use by the remaining trees as well as new seedlings.

Heterotrophic succession generates distinctive communities

Plants play a vital role in most patterns of succession because, as autotrophs, they are the source of energy for the other organisms in the community. Successional changes, however, can take place without the participation of plants. Detritus-based communities—found in dung, carrion, or dead plants—undergo a

57.13 Some Communities are Adapted to Disturbance   Forest communities dominated by lodgepole pine (*Pinus contorta*) are adapted to periodic fires. Fire removes mature trees weakened by pest infestation, revitalizes the soil, and provides essential conditions for seed germination and new growth.

series of changes known as **heterotrophic succession**. These changes differ from those in other types of directional succession. First, succession begins heterotrophically: energy resources are greatest when the habitat first becomes available to colonists and are depleted as succession takes place. There is no mechanism, such as photosynthesis, for generating more energy. Second, total biomass and species diversity decrease over time as the resource base declines. Third, these temporary habitats are not really self-contained, so predators can outnumber primary consumers (detritivores, in these communities) in apparent violation of the laws of thermodynamics.

Because the habitats in which heterotrophic succession occurs are so conveniently small, many patterns of heterotrophic succession have been thoroughly investigated. One such pattern is faunal succession in human corpses, as we saw at the opening of this chapter. M. G. Motter's 1898 study of 150 disinterments, followed by more recent studies on pig decomposition (after disinterments of human bodies for scientific purposes were discouraged), allowed for precise reconstruction of the temporal sequence of species in these communities.

There are three major stages of decomposition, each characterized by a distinctive faunal community. The first stage, *autolysis*, is characterized by fermentative changes generated by the body itself: degradation of proteins, lipids, and carbohydrates, accompanied by the release of gases such as hydrogen sulfide. *Putrefaction*, decomposition by microorganisms and invasive saprobes, follows. *Dry decay* takes place after most fluids have evaporated.

The faunas of decomposing corpses consist primarily of insects, but their exact composition varies with those factors that influence the rate and nature of decomposition: climate, season, and the condition of the body, including whether it is immersed or buried, wrapped or exposed. Typically, immediately after death, as autolysis begins, blow flies, bluebottles, and house flies arrive to lay eggs. As a detectable odor develops, other flies, including greenbottles and flesh flies, arrive. Fat breakdown, with its accompanying release of volatile fatty acids, attracts a range of carrion-feeding beetles. As proteins decompose, cheese skippers move in. Other species have less interest in the corpse than in the corpse-eaters: rove beetles prey on the maggots that develop from the flies' eggs. During dry decay, skin beetles, hide beetles, and clothes moths (which can feed on the keratin in mammalian hair) dominate. In the final stages of decomposition, spider beetles and other scavengers arrive to feed on the excrement and shed exoskeletons of the insects consuming the corpse. This succession varies tremendously with climate and geography, but within any particular region it is sufficiently predictable that it is admissible in court as evidence of the time since death.

57.4 RECAP

Disturbances are abiotic events that change the probability of persistence of one or more species in an ecological community. Ecological succession—a predictable pattern of change in community composition—typically follows a disturbance.

- How do primary succession and secondary succession differ? See pp. 1214–1215 and Figures 57.11 and 57.12

- Describe some ways in which early colonists facilitate or inhibit colonization by the species that follow them in a pattern of succession. See pp. 1215–1216

Now that we have seen how communities change in composition over time we will look at how certain communities (such as climax communities) can persist over time and withstand disturbance with little change. In this context, we return to productivity as an important arbiter of community stability.

57.5 How Does Species Richness Influence Community Stability?

Up to a point, higher productivity favors higher species richness, as we saw in Section 57.1. Does species richness in turn influence productivity? And do both of those properties influence community stability? Species richness might enhance productivity because no two species in a community have the same relationship with the environment, so a mixture of more species might result in a more complete use of the available resources. Moreover, if environmental conditions should change, a species-rich community is more likely to contain some species that can persist under the new conditions. Thus a species-rich community should be more *stable*—that is, it should change less over time in either productivity or species composition—than a species-poor community.

Species-rich communities use resources more efficiently

To test the hypothesis that species-rich communities are more stable than species-poor communities, David Tilman and his colleagues at the University of Minnesota cleared several outdoor plots, in which they planted grasses in mixtures ranging from a few to 25 species. At the end of each growing season, they measured total plant cover (a measure of grass biomass, and thus of net primary production) and the population densities of all the grasses in each plot. Over a period of 11 years, which included a serious drought, the plots with more species were more productive, and their productivity was less variable from year to year. These findings were consistent with the hypothesis that species richness promotes productivity and stability (**Figure 57.14A**). Moreover, in the plots with greater species richness, soil nitrogen was used more efficiently (**Figure 57.14B**). However, the population densities of *individual* species in the plots were not stable over the years (regardless of a plot's species richness) because different species performed better during drought years and wet years.

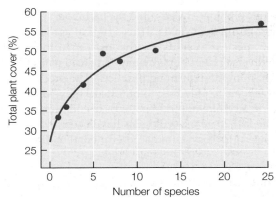

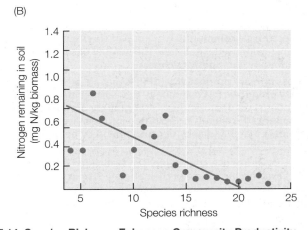

57.14 Species Richness Enhances Community Productivity A total of 120 grassland plots, containing from 2 to 22 grass species, were cultivated for 11 years. (A) The plots with greater species richness showed greater stability (with total plant cover being the measure of stability). (B) The plots with greater species richness utilized soil nitrogen more efficiently, as shown by the smaller amount of nitrogen remaining in the soil.

Researchers continue to debate whether species diversity is responsible for maintaining stability or is simply correlated with stability. This question is important because many of the alterations humans have made in the structure of natural communities have reduced their species richness, and many of these human-altered communities—notably, agricultural communities—are notoriously unstable.

Diversity, productivity, and stability differ between natural and managed communities

Although ecologists have been debating the relationships between community diversity, productivity, and stability for only a few decades, humans have been experimenting with those relationships, albeit inadvertently, for millennia—since plants were domesticated and agriculture was invented. Since the dawn of agriculture, crops have been susceptible to insect **outbreaks**: massive (often sudden) increases in the population sizes of species that destroy or damage crops.

The practice of growing crops as **monocultures**—plantings of only a single species—is one reason why modern agricultural ecosystems are particularly unstable. Most farmers have little tolerance for the presence of any potential competitors for their crops and actively eliminate weeds (and the herbivore species that live with them) from their fields. Thus a typical agroecosystem has very low species diversity. So the answer to the question of whether diversity causes or is merely correlated with stability may be sought in modern farming practices. The predisposition for agroecosystems to play host to outbreaks may well result from human-induced alterations in patterns of community structure.

For the last 20 years, ecologists have been using traditional subsistence agroecosystems as experimental models for testing the relationships between diversity and stability. Throughout the world, many farmers with small land holdings grow multiple crops on the same plot. In Costa Rica, for example, corn is often grown together with sweet potato. Corn–sweet potato dicultures contain fewer sweet potato pests and many more parasitoid wasps than corn monocultures do. Corn pollen provides nutrition for the wasps, and the tall corn plants act as a structural barrier, shade plant, and source of disruptive chemical signals that interfere with the ability of the sweet potato pests to find their host plants.

In recent years, such applications of community ecology have been paying dividends. Although monoculture is overwhelmingly the dominant agricultural practice, polycultures are under development for agricultural production systems as varied as carp and shrimp farming, vermicomposting (raising worms for compost), and biofuel feedstock production.

57.5 RECAP

Communities with higher species diversity tend to be more productive and more stable than less diverse communities because they use resources more efficiently. Modern agricultural monocultures are notoriously unstable.

- What relationships have ecologists observed between species diversity, community productivity, and community stability? See p. 1217 and Figure 57.14

- Describe some agricultural practices that might result in more stable ecological communities. See p. 1218

CHAPTER SUMMARY

57.1 What Are Ecological Communities?

- An ecological **community** is a group of species that coexist and interact within a defined area. The number of species living in a community is its **species richness**.

- **Gross primary productivity** (GPP) is the rate at which the **primary producers** in a community turn solar energy into chemical energy via photosynthesis. **Net primary production** (NPP) represents the energy incorporated into primary producer **biomass** that is available for consumption by heterotrophs. Review Figure 57.1, **WEB ACTIVITY 57.1**

- The organisms in a community can be divided into **trophic levels** based on the source of their energy. **Primary producers** get their energy from sunlight; **primary consumers** get their energy by eating primary producers; **secondary consumers** get their energy by eating primary consumers; and so on. Review Table 57.1, **WEB ACTIVITY 57.2**

- Organisms that consume the dead bodies of other organisms or their waste products are called detritivores or decomposers. Omnivores are organisms that feed on multiple trophic levels.

- A **food chain** diagrams who eats whom. A **food web** shows how the food chains in a community are interconnected. Review Figure 57.2

- **Ecological efficiency** is the overall transfer of energy from one trophic level to the next. Pyramid diagrams illustrate the proportions of energy or biomass that flow between trophic levels. Review Figure 57.3

- Species richness tends to increase with productivity up to a point; however; if productivity increases beyond that point, species richness may decline. Review Figure 57.4

57.2 How Do Interactions among Species Influence Community Structure?

- The interactions of a consumer with other species can result in a a **trophic cascade**: a series of indirect effects across successive trophic levels. Review Figure 57.5

- Organisms that build structures that create habitat for other species are known as **ecosystem engineers**.

- **Keystone species** have an influence on both the species richness and the number of trophic levels in communities out of proportion to their abundance.

57.3 What Patterns of Species Diversity Have Ecologists Observed?

- **Alpha diversity** is a measure of diversity within a single community or habitat, whereas **beta diversity** is a measure of the change in species composition from one community or habitat to another. **Gamma diversity** is the diversity found over a range of communities in a geographic region.

- Species diversity encompasses species evenness as well as species richness. **Review Figure 57.7**

- Diversity indexes are mathematical formulas used to quantify diversity. The **Shannon diversity index** is a measure of alpha diversity that takes species evenness into account. **Sorensen's index** is a measure of beta diversity.

- Latitudinal gradients in diversity, with the greatest diversity at low latitudes, have been observed in many taxa. **Review Figure 57.8**

- According to the theory of **island biogeography**, the equilibrium number of species on an island represents a balance between the rate at which species immigrate to the island from the mainland **species pool** and the rate at which resident species go extinct. **Review Figures 57.9 and 57.10, ANIMATED TUTORIAL 57.1**

57.4 How Do Disturbances Affect Ecological Communities?

- A **disturbance** is an abiotic event that changes the probability of persistence of one or more species in a community.

- **Succession** is a predictable pattern of change in community composition following a disturbance. In directional succession, species come and go in a predictable sequence until a **climax community** persists for an extended time.

- **Primary succession** begins on sites that lack living organisms. **Secondary succession** begins on sites where some organisms have survived a disturbance. **Review Figures 57.11 and 57.12, ANIMATED TUTORIAL 57.2**

- In any pattern of succession, species that have already become established may **facilitate** or **inhibit** colonization by other species.

- In **cyclical succession**, the climax community is maintained by periodic disturbances.

- **Heterotrophic succession** in detritus-based communities does not rely on photosynthesis and therefore differs in a number of ways from other types of succession.

57.5 How Does Species Richness Influence Community Stability?

- Species-rich communities use resources more efficiently, and thus tend to vary less in productivity, than do less diverse communities. **Review Figure 57.14**

- **Monocultures** are subject to pest outbreaks, whereas agroecosystems containing greater species diversity tend to be more stable.

SELF-QUIZ

1. An ecological community is
   a. a group of species that coexist and interact within a defined area.
   b. a group of species that coexist and interact in an area together with the abiotic environment.
   c. all the species in an area that belong to a particular trophic level.
   d. all the species that are members of a local food web.
   e. all of the above

2. A trophic level consists of the organisms
   a. whose energy has passed through the same number of steps to reach them.
   b. that use similar foraging methods to obtain food.
   c. that are eaten by a similar set of predators.
   d. that eat both plants and other animals.
   e. that compete with one another for food.

3. Net primary production is
   a. the total amount of photosynthesis in a community.
   b. the total amount of primary producer biomass available for consumption by heterotrophs.
   c. the total amount of biomass produced by all autotrophs and heterotrophs in a community.
   d. the total amount of biomass consumed by heterotrophs.
   e. gross primary productivity minus secondary productivity.

4. Pyramid diagrams of energy and biomass distribution for forests and grasslands differ because
   a. forests are more productive than grasslands.
   b. forests are less productive than grasslands.
   c. large mammals avoid living in forests.
   d. trees store much of their energy in difficult-to-digest wood, whereas grassland plants produce few difficult-to-digest tissues.
   e. grasses grow faster than trees.

5. Keystone species
   a. influence the communities in which they live more than expected on the basis of their abundance.
   b. may influence the species richness of communities.
   c. may influence the number of trophic levels in communities.
   d. are not necessarily predators.
   e. all of the above

6. Species diversity
   a. increases with latitude from the equator to the temperate zone.
   b. decreases with latitude from the equator to the temperate zone.
   c. shows no consistent pattern from the equator to the temperate zone.
   d. increases with latitude from the equator to the temperate zone for terrestrial organisms and decreases with latitude from the equator to the temperate zone for aquatic organisms.
   e. increases with latitude from the equator to the temperate zone for animals and decreases with latitude from the equator to the temperate zone for plants.

7. The theory of island biogeography
   a. predicts that the equilibrium number of species on an island is a balance between the rate of immigration of new species and the rate of extinction of resident species.
   b. predicts that the rate of immigration of new species will decline with island distance from the mainland species pool.
   c. predicts that the rate of extinction of resident species will decrease as island size increases.
   d. applies to isolated habitat patches as well as to oceanic islands
   e. all of the above

8. Ecological succession is
   a. the change in species over time.
   b. the change in community composition after a disturbance.
   c. the change in a forest as the trees grow larger.

d. the process by which a species becomes abundant.
   e. the buildup of soil nutrients.

9. Primary succession begins
   a. soon after a disturbance ends.
   b. at varying times after a disturbance ends.
   c. at sites where some organisms have survived a disturbance.
   d. at sites where no organisms have survived a disturbance.
   e. at sites where only primary producers have survived a disturbance.

10. Early stages of succession are characterized by
    a. species that are good dispersers.
    b. species with high intrinsic rates of reproduction.
    c. simple food webs.
    d. nutrients that are available primarily in abiotic sources.
    e. all of the above

FOR DISCUSSION

1. Marek Sammul, Lauri Oksanen, and M. Magi investigated the effect of productivity on species richness by conducting an experiment in 16 different plant tcommunities in western Estonia and northern Norway. They removed one perennial species, the goldenrod *Solidago virgaurea*, from these communities and found that its competitors, particularly the grass species *Anthoxanthum odoratum*, increased in biomass, most noticeably in communities with high productivity (where living plant biomass was greater than 200 g/m^2). In less productive communities, such increases could not be detected. How might interspecific competition lead to a decrease in species richness at high levels of productivity? What other hypotheses might explain this puzzling relationship? How would you test them?

2. The increased productivity and stability of species-rich communities could be explained by ecological differences among the species that allow the community as a whole to use resources more efficiently, or by the fact that the more species in a community, the greater the chance that it will contain an unusually productive species. How could you distinguish between these competing hypotheses?

3. If species-rich communities are more productive than species-poor communities, how can modern agriculture, which is based almost entirely on monocultures, be so productive?

4. Figures 57.11 and 57.12 illustrated succession with two examples from forests. How might ecological succession proceed in grasslands? In deserts? In the rocky intertidal zone?

ADDITIONAL INVESTIGATION

The species diversity of aphids (small sucking insects), parasitoid wasps, and kelps decreases, rather than increases, from the temperate zone to the equator. Can you propose any explanations to account for these unusual patterns?

WORKING WITH DATA (GO TO yourBioPortal.com)

Experimental Island Defaunation   This exercise presents the data from Simberloff and Wilson's experiments in the Florida Keys (see Figure 57.10), which were performed in the late 1960s and presented in support of island biogeography theory. You will be asked to analyze different aspects of the experiment and to draw conclusions as to how (and perhaps whether) the data offer significant support for this theory.

Ecologists swing in the canopy to measure carbon's fate

Photosynthesis by terrestrial plants transforms (*fixes*) the carbon atoms in atmospheric carbon dioxide into the carbohydrate sugars that fuel the fungal and animal worlds. Concentrations of CO_2 in the atmosphere have been rising steadily over the last century, largely as a result of human activities, and some scientists have hypothesized that we can expect rates of photosynthesis—and thus CO_2 fixation and carbon storage in ecosystems—to increase. But ecologists are discovering that this "silver lining" scenario is probably not going to be the case.

Forest studies are important for determining the responses of vegetation to atmospheric carbon enrichment. Most photosynthesis in any forest takes place high up in tree canopies where sunlight is most intense. Until recently, collecting data from forest canopies required climbing or using balloons. Then, in 1992, Alan Smith at the Smithsonian Tropical Research Institute in Panama realized that he could use a construction crane to gain access to the canopy. He recognized that by swinging the crane's boom in a circle, shuttling the gondola along its length, and lowering the cage to different heights, researchers could reach and measure biological processes over a large volume of the forest canopy. Smith's idea caught on quickly, and today ecologists on four continents use cranes to study forest canopies.

When they monitored growth, reproduction, and carbon storage in large trees, Swiss ecologists found that, as expected, canopy trees increased their growth rates in a carbon-enriched atmosphere; but unlike young, fast-growing saplings, they did not store large amounts of carbon in their trunks and branches. Instead, they transported more carbon first to the soil, and then to the atmosphere (via their roots). The leaf litter they produced was richer in starch and sugar, but poorer in lignin; the litter decomposed faster as soil microbes rapidly respired the carbon back to the atmosphere. These and other experiments suggest that young forests are likely to respond to greater atmospheric CO_2 concentrations by growing faster. However, mature forests, along with other types of vegetation such as grasslands, shrublands, and tundra, are not likely to store significantly more carbon in response to CO_2 enrichment.

Studying the Canopy A crane 42 meters high allows scientists to study the forest canopy trees of Panama's Smithsonian Tropical Research Institute. One aspect of these studies involves taking measurements that reveal the amount of carbon stored in tropical forest trees.

Enriching the Canopy Researchers from a team of Swiss ecologists install a mechanism of fine tubes that administer controlled amounts of CO_2 to enrich the trees' carbon supply. These experiments simulate what will happen as atmospheric levels of CO_2 increase.

As Chapter 54 described, the global ecosystem is composed of many different types of vegetation (biomes), many of which have few woody plants. To understand how ecosystems function at large scales, scientists must make observations and conduct experiments on many vegetation types. Moreover, because the responses of natural ecosystems to carbon enrichment will depend on many other factors—such as temperature, water, wind, and soil fertility—ecologists must study the interactions among these factors to understand how they influence ecosystem processes.

IN THIS CHAPTER we introduce the compartments of the physical environment through which energy flows and materials cycle. We will trace these events by following the movements of several chemical elements that are crucial to life. In so doing, we highlight many ways in which human activities have altered these cycles. Finally we will consider the many goods and services that ecosystems provide to human society.

58.1 What Are the Compartments of the Global Ecosystem?

An **ecosystem** is generally defined as any ecological system with defined boundaries. It includes all of the organisms within those boundaries as well as the physical and chemical factors that influence those organisms. In other words, all ecosystems have both biotic and abiotic components. Ecosystems can occupy a wide range of spatial scales, from the global ecosystem to a watershed, a specific forest, a lake or pond, or even a patch of lichen on a rock. In this chapter we focus on the global ecosystem.

Energy flows and materials cycle through ecosystems

Earth is essentially a closed system with respect to atomic matter, but it is an open system with respect to energy. The sun delivers an essentially constant amount of energy to Earth every day and has done so for billions of years. Energy from the sun, combined with energy from the radioactive decay in Earth's interior, drives the processes that move chemical elements around the planet. The rate at which energy moves through a system is called its *flux*. Elements may accumulate, or *pool*, in some component of an ecosystem; the term *flux* also applies to the movement of elements between pools. Some pools are *sinks*, in which an element is taken out of circulation and locked up for long periods.

Many elements move around the planet in a cyclic fashion (**Figure 58.1**). Nearly all the rocks that make up the continents, for example, have been transformed at least once through a complex cycle of plate tectonic processes (see Figure 25.2). Deep within the Earth, heat, pressure, and melting change the mineral composition of rocks. As the movements of the lithospheric plates push up mountains, the processes of weathering are already wearing them down by breaking rocks into soil and sediment particles (see Section 36.3). Streams carry sediments to the oceans, where they form sedimentary rocks on the seafloor. Eventually, those rocks may be returned to land in another episode of tectonic uplift.

Water is another material that moves cyclically. The water in the oceans today has evaporated, condensed in the atmosphere, precipitated as rain, and returned to the oceans via stream and groundwater flow millions of times. The carbon, nitrogen, oxygen, and sulfur in atmospheric gases have cycled repeatedly through living organisms, which use them to build the proteins that make up their bodies.

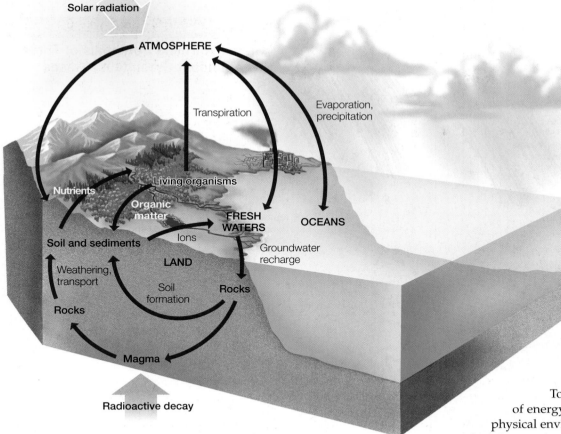

58.1 Compartments of the Global Ecosystem The four compartments—atmosphere, oceans, land, and fresh waters—are connected by the flux of materials (arrows). The atmosphere exchanges some components rapidly; exchange of materials within the other compartments (between rocks and soils, for example) is generally much slower.

Earth is a planet with many unusual features: lithospheric plates that move continuously, a moderate surface temperature, a large quantity of liquid water on its surface, and a tremendous diversity of living organisms, to name just a few. The plate tectonic processes that shape Earth's surface also release gases (including carbon dioxide) into the atmosphere through volcanic eruptions. Those atmospheric gases help to moderate planetary temperatures, and those moderate temperatures, in turn, keep water in a liquid state. And living organisms, by virtue of their activities, have an enormous influence on the chemical composition of the atmosphere, the water, and the land.

All organisms depend on inputs of energy (in the form of sunlight or high-energy molecules), liquid water, and nutrients for their metabolism and growth. As described in Section 57.1, energy flows through ecological communities from producers to consumers. At each trophic level, much of this energy is used to power metabolism; that energy is dissipated as heat and lost from the ecosystem (see Figure 57.1). Chemical elements, on the other hand, are not altered when they are transferred among organisms. The availability of the chemical elements of which living organisms are composed is strongly influenced by how organisms get them, how long they retain them, and what they do with them while they have them.

To understand the movements of energy and materials on Earth, the physical environment can be divided into four interacting compartments: the atmosphere, the oceans, fresh waters, and land. These four compartments, and the types of organisms living in them, are very different. Thus the quantities of different elements in each compartment, how long they remain in each compartment (i.e. their *residence time*), and the rates at which they enter and leave the compartments differ strikingly. After describing the four compartments, we will discuss how energy flows through them and elements cycle among them.

The atmosphere regulates temperatures close to Earth's surface

The outermost compartment of the global ecosystem, the *atmosphere*, is a thin layer of gases surrounding Earth. The atmosphere is 78.08% nitrogen gas (N_2), 20.95% oxygen gas (O_2), 1% water vapor, 0.93% argon, and 0.03% carbon dioxide (CO_2). It also contains traces of hydrogen gas, neon, helium, krypton, xenon, ozone, and methane. The atmosphere contains Earth's biggest pool of N_2 as well as a large proportion of its O_2. Although CO_2 constitutes a very small fraction of the atmosphere, atmospheric CO_2 is the source of the carbon used by terrestrial photosynthetic organisms, and atmospheric CO_2 that dissolves in ocean water is the source of the carbon used by marine primary producers.

About 80% of the mass of the atmosphere lies in its lowest layer, the **troposphere**. This layer extends upward from Earth's surface about 17 km in the tropics and subtropics, but only about 10 km at high latitudes. Most global air circulation takes place within the troposphere, and virtually all of the atmospheric water vapor is found there (**Figure 58.2**).

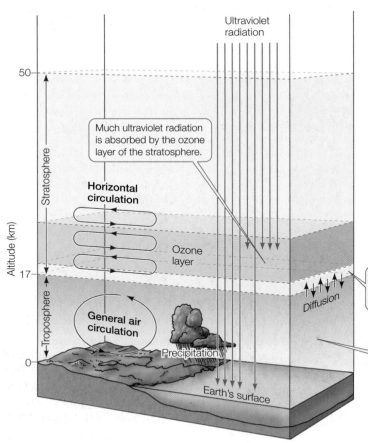

58.2 The Two Layers of Earth's Atmosphere The troposphere and the stratosphere differ in their circulation patterns, the amount of water vapor they contain, and the amount of ultraviolet radiation they receive.

TABLE 58.1
Global Warming Potentials of Three Abundant Greenhouse Gases

	ATMOSPHERIC LIFETIME[a]	GWP[b]
Carbon dioxide (CO_2)	50–200	1
Methane (CH_4)	12	21
Nitrous oxide (N_2O)	114	289

[a] *Atmospheric lifetime* is the number of years these molecules typically remain in the atmosphere before breaking up.

[b] *GWP* (global warming potential) is the ability of a given amount of gas relative to CO_2 (GWP = 1) to trap heat and warm the atmosphere over a specific time period (in this case, 100 years).

The **stratosphere**, which extends from the top of the troposphere up to about 50 kilometers above Earth's surface, contains very little water vapor. Most materials enter the stratosphere from the region of the troposphere that encircles the equator, where air heated by the sun rises to high altitudes (see Section 54.2).

The stratosphere contains a layer of ozone (O_3) that absorbs most of the biologically damaging ultraviolet radiation that enters the atmosphere. Over the last several decades, this **ozone layer** has been seriously damaged by human activities. The widespread release of chlorinated fluorocarbons (CFCs), such as the refrigerant freon, caused chemical reactions in the stratosphere that destroyed ozone molecules. This depletion of stratospheric ozone is a source of great concern, as increases in ultraviolet radiation at Earth's surface are associated with increased rates of skin cancer, cataract formation, and crop damage. In 1989, a group of nations signed a global treaty in which they agreed to phase out production of CFCs, so there is hope that the stratospheric ozone layer will one day be restored.

The atmosphere regulates temperatures at and close to Earth's surface by trapping heat. If Earth had no atmosphere, its average surface temperature would be about –18°C, rather than its actual +17°C. Carbon dioxide, methane, water vapor,

and certain other gases in the atmosphere are known as **greenhouse gases** because they are transparent to sunlight, but trap heat radiating from Earth's surface back toward space (**Figure 58.3**) The three major greenhouse gases and their potential effects on atmospheric warming are given in **Table 58.1**. Ozone, when it is present in the troposphere rather than the stratosphere, also acts as a greenhouse gas. As we'll see later in this chapter, human activities are increasing the concentrations of greenhouse gases in the atmosphere, and those increases are making the planet warmer.

The oceans receive materials from the other compartments

Over time scales of hundreds to thousands of years, most materials that cycle through the four compartments of the global ecosystem end up in the oceans. The oceans are enormous, but they exchange materials with the atmosphere only at their surface, so they respond very slowly to inputs from that compartment. They receive materials from land primarily in runoff from rivers.

Except on continental shelves—the shallow ocean waters surrounding large land masses—ocean waters mix slowly. Most materials that enter the oceans from other compartments gradually sink to the seafloor, where they may remain for millions of years until the seafloor sediments are uplifted above sea level by plate tectonic processes. Therefore, concentrations of mineral nutrients are very low in most ocean waters. Some nutrients are brought back to the surface near the coasts of continents, where offshore winds pushing surface waters away from shore cause cold bottom water to rise to the surface. The water in these

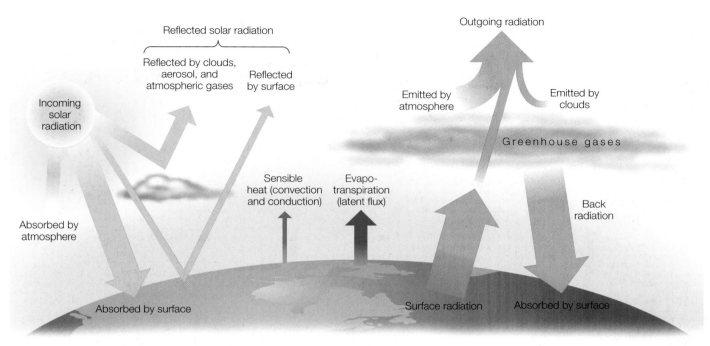

58.3 Radiant Energy Warms the Planet Solar energy (yellow arrows) drives the global climate system described in Chapter 54. Much of this energy (in the form of heat) along with the energy generated by Earth's biogeochemical processes (orange arrows) is prevented from radiating back into space by the atmospheric gases. The width of the arrows is roughly proportional to the size of energy flux.

upwelling zones is nutrient-rich because when organisms die their bodies sink to the seafloor and return the materials they contain to the bottom sediments. Upwelling zones support high rates of production by phytoplankton, which in turn support dense consumer populations (**Figure 58.4**). Most of the world's great fisheries are concentrated in upwelling zones. The up-

welling zone off the coast of Peru is, for example, the source of much of the world's supply of anchovies. In turn, this rich fish population supports vast seabird communities, which in turn produce enormous quantities of guano, or droppings, that has provided Peru with an important raw material to support a major fertilizer industry.

Water moves rapidly through lakes and streams

The freshwater compartment of the global ecosystem consists of streams, lakes, and **groundwater** (water occupying pore spaces in rock, sand, and soil). Only a small fraction of Earth's water resides in lakes and streams at any given time, but water moves rapidly through this compartment, so most of Earth's water spends some time there. Some mineral nutrients enter fresh waters in rainfall, but most are released by the weathering of rocks. They are carried into streams by surface runoff or by movement of groundwater.

After entering streams, mineral nutrients are usually carried rapidly to lakes or to the oceans. In lakes, they are taken up by organisms and incorporated into their bodies. Those organisms eventually die and sink to the lake bottom, taking their nutrients with them. Microbial decomposition of their tissues consumes most or all of the oxygen in the bottom water. The surface waters of lakes thus quickly become depleted of nutrients, while deeper waters become depleted of oxygen. This process is countered, however, by verti-

Primary production (mg of carbon per m² per day)

▢ <150   ▨ 150–250   ▨ >250

58.4 Upwelling Zones Are Highly Productive Biological productivity is greatest near continents where surface waters, driven by prevailing winds, move offshore and are replaced by cool, nutrient-rich water upwelling from below.

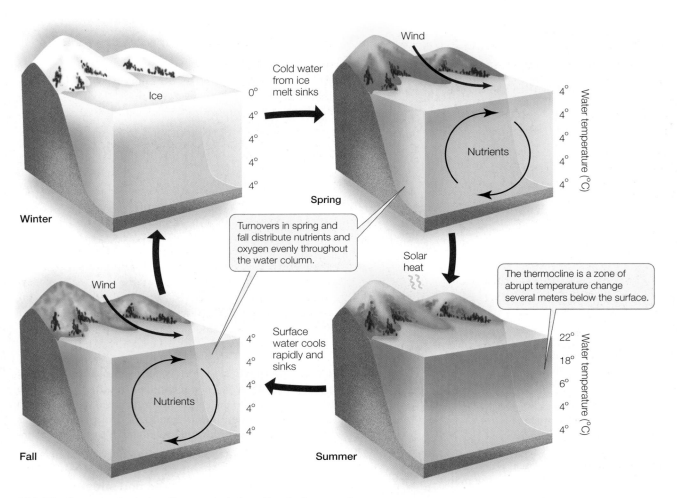

58.5 The Turnover Cycle in a Temperate Lake Regular turnovers in spring and fall allow nutrients and oxygen to become evenly distributed in the water column of a lake. The vertical temperature profiles shown here are typical of temperate-zone lakes that freeze in winter.

cal movements of water called **turnover**, which bring nutrients to the surface and oxygen to deeper water.

Wind is an important agent of turnover in shallow lakes, but in deeper lakes it mixes only surface waters. Deep lakes in temperate climates have an annual turnover cycle that depends on certain unique physical properties of water (**Figure 58.5**). Water is densest at 4°C, a few degrees above its freezing point; above and below that temperature, it expands. Thus, in winter, the coldest liquid water in a lake floats at its surface, often covered by a layer of ice. The waters below remain at 4°C. In spring, the sun melts the ice and warms the surface water to 4°C. At this point, water density is uniform throughout the lake, and even modest winds readily mix the entire water column.

As spring and summer progress, the surface water becomes warmer, and the depth of the warm layer gradually increases. However, there is still a well-defined region, called the *thermocline*, where the temperature changes abruptly. Only if the lake is shallow enough to warm right to the bottom does the temperature of the deepest water rise above 4°C. Another turnover occurs in fall as the process reverses itself. The surface of the lake cools until it is denser than the warmer water below it, at which point it sinks and is replaced by warmer water from below, with wind again mixing the entire water column.

Land covers about a quarter of Earth's surface

About one-fourth of Earth's surface, most of it in the Northern Hemisphere, is currently land above sea level. This *terrestrial* compartment of the physical environment is connected to the atmospheric compartment by organisms that take chemical elements from the air and ultimately release them back into the air. Chemical elements in rocks are released by weathering and carried in solution into groundwater, which transports them ultimately into the oceans, where they are inaccessible to most terrestrial organisms until an episode of tectonic uplift raises marine sediments above sea level and a new cycle of weathering begins.

Even though the global supply of chemical elements in the terrestrial compartment is constant, regional and local variations in the supply of particular elements influence ecosystem processes on land because elements move slowly there and usually over only short distances. The type of soil that exists in an area and the nutrients it contains depend on the underlying rock from which it forms, as well as on climate, topography, the local biota, and the length of time that soil-forming processes have been acting (see Section 36.3).

58.1 RECAP

Earth is a closed system with respect to atomic matter, but an open system with respect to energy. Earth's physical environment can be divided into four compartments—atmosphere, oceans, fresh waters, and land—through which energy flows and elements cycle.

- How does the atmosphere keep temperatures at and close to Earth's surface warmer than they would be in its absence? See pp. 1223–1224 and Figure 58.3
- Describe the process of turnover in a temperate-zone lake in fall. See p. 1226 and Figure 58.5

Elements cycle through the global ecosystem, but energy flows through it unidirectionally. In the next section we take a brief look at energy flow through ecosystems.

58.2 How Does Energy Flow through the Global Ecosystem?

Solar energy drives nearly all ecosystem processes on Earth. In this section we will examine how the geographic distribution of incoming solar radiation influences the amount of energy assimilated by primary producers, and how human activities are modifying energy flow through the global ecosystem.

The geographic distribution of energy flow is uneven

Nearly all energy utilized by organisms comes (or once came) from the sun. The only exceptions are found in those few ecosystems in which solar energy is not the main energy source (such as some caves and deep-sea hydrothermal vent ecosystems). Even the fossil fuels—coal, oil, and natural gas—on which the economy of modern human civilization is based are reserves of captured solar energy locked in the remains of organisms that lived millions of years ago.

As described in Section 57.1, solar energy enters ecosystems by way of plants and other photosynthetic organisms. These primary producers use some of the energy they assimilate for their own metabolism; the rest—net primary production—is stored in their bodies or used for their growth and reproduction (see Figure 57.1). Because only the energy of net primary production is potentially available to other organisms that consume the producers, we use net primary production as a rough measure of energy flow through ecosystems (**Figure 58.6**).

The geographic distribution of net primary production reflects geographic variation in incoming solar radiation, described in Section 53.2, and the climate patterns that result from it. The distribution of land masses, temperatures, and moisture on Earth makes some ecosystems more productive than others

58.6 Primary Production by Ecosystem Type The contributions of the different ecosystem types to primary productivity can be measured by (A) their geographic extent; (B) their net primary production; and (C) their percentage of Earth's total primary production.

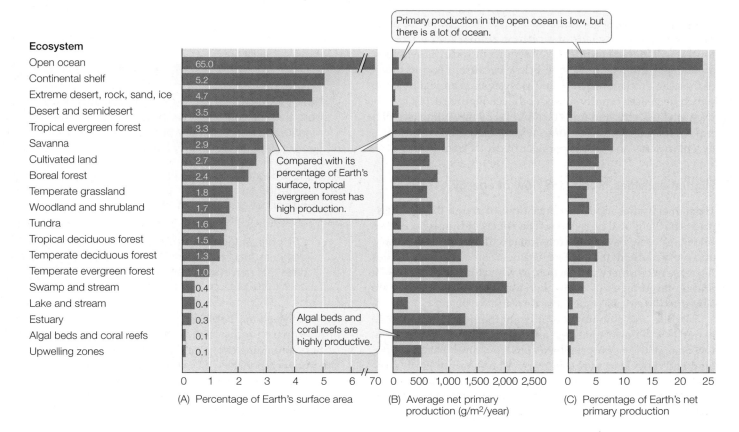

Ecosystem

Ecosystem	(A)
Open ocean	65.0
Continental shelf	5.2
Extreme desert, rock, sand, ice	4.7
Desert and semidesert	3.5
Tropical evergreen forest	3.3
Savanna	2.9
Cultivated land	2.7
Boreal forest	2.4
Temperate grassland	1.8
Woodland and shrubland	1.7
Tundra	1.6
Tropical deciduous forest	1.5
Temperate deciduous forest	1.3
Temperate evergreen forest	1.0
Swamp and stream	0.4
Lake and stream	0.4
Estuary	0.3
Algal beds and coral reefs	0.1
Upwelling zones	0.1

Primary production in the open ocean is low, but there is a lot of ocean.

Compared with its percentage of Earth's surface, tropical evergreen forest has high production.

Algal beds and coral reefs are highly productive.

(A) Percentage of Earth's surface area

(B) Average net primary production (g/m²/year)

(C) Percentage of Earth's net primary production

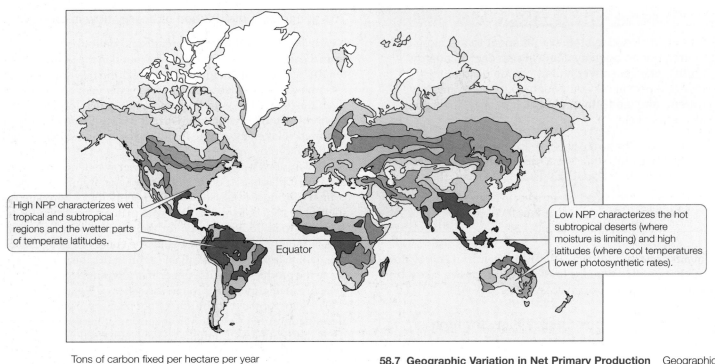

High NPP characterizes wet tropical and subtropical regions and the wetter parts of temperate latitudes.

Low NPP characterizes the hot subtropical deserts (where moisture is limiting) and high latitudes (where cool temperatures lower photosynthetic rates).

Equator

Tons of carbon fixed per hectare per year

☐ Ice caps 0   ☐ 0–2.5   ☐ 2.5–6.0   ■ 6.0–8.0
■ 8.0–10.0   ☐ 10.0–30.0   ■ >30.0

58.7 Geographic Variation in Net Primary Production   Geographic variation in solar energy input results in variation in temperature and moisture over Earth's land surface, and this variation, in turn, influences net primary production in terrestrial ecosystems.

(Figure 58.7). Close to the equator at sea level, temperatures are high throughout the year and the water supply is adequate for plant growth much of the time. In these climates, productive forests thrive. In lower- and mid-latitude deserts, where plant growth is limited by lack of moisture, primary production is low. At still higher latitudes, even though moisture is generally available, primary production is low because it is cold much of the year. Production in aquatic ecosystems is limited by light, which decreases rapidly with depth (see Figure 54.14); by nutrients, which sink and must be replaced by upwelling (see Figure 58.4); and by temperature, which decreases with depth, except in areas on the seafloor near hydrothermal vents.

Human activities modify the flow of energy

Human activities modify energy flows through the compartments of the global ecosystem. The effects of human activities on energy flow have accelerated markedly in the last century, and particularly in the last five decades. Some human activities decrease net primary production, as when forests are cut down and replaced by cities; some increase it, as when prairies are converted to extensive fields of corn and soybeans.

Satellite and climate data indicate that humans consume about 24% of Earth's average annual net primary production. Over half of this appropriation results from direct harvest (cropland and rangeland now occupy over one-third of Earth's ice-free surface); another 40% results from productivity changes brought about by alterations in land use; and 7% results from

fires caused by humans. The percentage varies strikingly among regions; urban areas consume as much as 300 times the net primary production they generate, but people in sparsely inhabited parts of the Amazon Basin appropriate vanishingly small amounts of the net primary production there.

58.2 RECAP

Nearly all energy utilized by living organisms comes from the sun. Energy flow through ecosystems, as measured by net primary production, varies with geographic location and climate.

● How does the distribution of land masses, temperature, and moisture influence the geographic distribution of net primary production? See pp. 1227–1228 and Figure 58.7

● What percentage of Earth's average annual net primary production is appropriated by humans, and how does that percentage vary regionally? See p. 1228

As we learned in Chapters 3 and 10, most of the biological energy primary producers transform from sunlight is stored in carbon-containing compounds. In the next section we will consider how carbon, as well as other chemical elements required by living organisms, cycle through the four compartments of Earth's physical environment.

58.3 How Do Materials Cycle through the Global Ecosystem?

The chemical elements that organisms need in large quantities—carbon, hydrogen, oxygen, nitrogen, phosphorus, and sulfur—cycle through the abiotic and biotic components of ecosystems. The pattern of movement of a chemical element through living organisms and the four compartments of the physical environment is called its **biogeochemical cycle**. Each element has a distinctive biogeochemical cycle whose properties depend on the physical and chemical nature of the element and the ways in which it is used by organisms.

Before we describe the biogeochemical cycles of individual elements, however, we must discuss the influences of water and fire on the rates at which elements cycle through ecosystems.

Water transfers materials from one compartment to another

The cycling of water through the four compartments of the physical environment is known as the **hydrologic cycle**. Water is an extraordinary molecule by Earth's standard in that it can be found in all three states of matter—gas, liquid, and solid—and can change state fairly readily. The sun powers the hydrologic cycle by causing evaporation from the vast surfaces of the oceans. The hydrologic cycle operates because more water is evaporated from the ocean surface than is returned to them as precipitation. Water also evaporates from soils, from lakes and rivers, and from the leaves of plants (by transpiration), but the total amount evaporated from those surfaces is less than the amount that falls on them as precipitation. The excess terrestrial precipitation eventually returns to the oceans via streams, coastal runoff, and groundwater flows (**Figure 58.8**). Earth's 16 largest rivers account for more than one-third of total runoff from the land. More than half of this total comes from the four largest rivers—the Amazon in South America, the Nile in Africa, the Mississippi in North America, and the Yangtze in Asia.

Despite their relatively small volume, rivers play a disproportionate role in the hydrologic cycle because the average residence time of a water molecule in rivers is only a few years. By comparison, the average residence time of a water molecule in lakes ranges from a few years to centuries. The larger the lake, the longer the residence time; the residence time for Lake Michigan, for example, is about 600 years, and the residence time of the top portion of Lake Superior is 1,500 to 2,000 years (and the brackish water at the bottom never cycles). In the ocean, the average residence time of a water molecule is about 3,000 years. Other accumulations of water include glaciers (with residence times of 20–100 years), seasonal snow cover (a few months), and soil moisture (a month or two). The average residence time of water in organisms is particularly brief—on average, just under a week.

Although large amounts of groundwater are present in underground pools called **aquifers**, this water has a long residence time underground and plays only a small role in the hydrologic

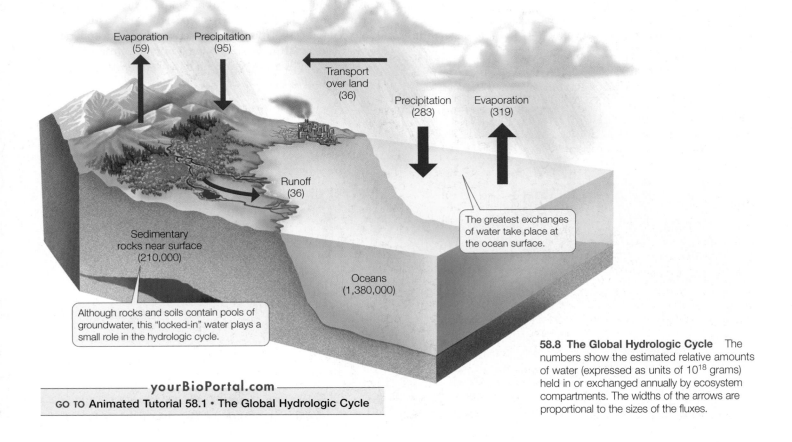

Evaporation (59) Precipitation (95)

Transport over land (36)

Precipitation (283) Evaporation (319)

Runoff (36)

The greatest exchanges of water take place at the ocean surface.

Sedimentary rocks near surface (210,000)

Oceans (1,380,000)

Although rocks and soils contain pools of groundwater, this "locked-in" water plays a small role in the hydrologic cycle.

58.8 The Global Hydrologic Cycle The numbers show the estimated relative amounts of water (expressed as units of 10^{18} grams) held in or exchanged annually by ecosystem compartments. The widths of the arrows are proportional to the sizes of the fluxes.

— **yourBioPortal.com** —
GO TO Animated Tutorial 58.1 • The Global Hydrologic Cycle

cycle. In some places, however, groundwater is being seriously depleted because humans are using it more rapidly than it can be replaced, primarily by pumping it for irrigation. Much of the groundwater being used today in the Northern Hemisphere was deposited during the most recent ice age, when regional precipitation was much greater than it is now. Using this groundwater for irrigation and other purposes has increased flows of water to the oceans and has contributed to the sea level rise of the past century.

The effects of groundwater depletion are already being felt. On the North China Plain, for example, depletion of shallow aquifers is forcing people to sink wells more than 1,000 meters deep to reach water. Worldwide, more than 1 billion people have no access to safe drinking water. If current water consumption patterns continue, by 2025 at least 3.5 billion people (48% of the current world population) will live in areas with inadequate water supplies. However, per capita water consumption in the United States and Europe is dropping as a result of increasing use of water-efficient home appliances and the increasing cost of water, as well as new regulations that restrict water use. If such programs are implemented vigorously, global water use in 2025 could be lower than it is today, despite continued population growth.

Fire is a major mover of elements

Every year, 200–400 million hectares of savannas, 5–15 million hectares of boreal forests, and smaller expanses of other biomes burn. Lightning ignites some fires, but humans start most fires to manage vegetation (as when they cut down and burn forests to clear land for growing crops). Fires consume the energy stored in, and release chemical elements from, the vegetation

they burn. Some nutrients, such as nitrogen, are easily vaporized by fire. They are discharged to the atmosphere in smoke or carried into groundwater by rain falling on burned ground.

Fires also release large amounts of carbon into the atmosphere. The global annual flux of carbon to the atmosphere from savanna and forest fires is estimated at 1.7 to 4.1 petagrams (1 pg = 10^{15} g or 10^{12} kg). Biomass burning (which includes combustion of wood and alcohol, wildfires, and land clearing, but not the burning of fossil fuels) is responsible for about 40% of Earth's annual flux of CO_2 into the atmosphere and contributes to the production of other greenhouse gases. Large-scale wildfires in the western and southeastern U.S. can release as much CO_2 into the atmosphere as motor vehicles in those regions release over an entire year.

Although biogeochemical cycles interact with one another, it is convenient to discuss the major cycles individually. Our discussion will focus on those elements whose cycles are best characterized and have the greatest effect on living organisms.

The carbon cycle has been altered by human activities

As described in Part One of this book, all the important macromolecules that compose living organisms contain carbon, and much of the energy that organisms need to fuel their metabolic activities comes from carbon-containing (organic) compounds. Carbon in the atmosphere, in the form of CO_2, is incorporated into organic molecules by photosynthesis in the cells of autotrophs. All heterotrophic organisms get their carbon by consuming autotrophs or other heterotrophs, their remains, or their waste products. Most of the carbon stored in the terrestrial compartment over the short term is stored as organic matter in soils. Biological processes move carbon between the atmospheric and terrestrial compartments, removing it from the atmosphere during photosynthesis and returning it to the atmosphere through metabolism (**Figure 58.9**).

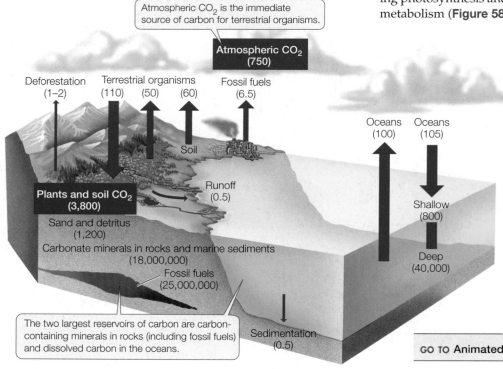

Atmospheric CO_2 is the immediate source of carbon for terrestrial organisms.

Atmospheric CO₂ (750)

Deforestation (1–2)
Terrestrial organisms (110) (50) (60)
Fossil fuels (6.5)

Soil

Plants and soil CO₂ (3,800)

Runoff (0.5)

Sand and detritus (1,200)

Carbonate minerals in rocks and marine sediments (18,000,000)

Fossil fuels (25,000,000)

The two largest reservoirs of carbon are carbon-containing minerals in rocks (including fossil fuels) and dissolved carbon in the oceans.

Sedimentation (0.5)

Oceans (100)   Oceans (105)

Shallow (800)

Deep (40,000)

58.9 The Global Carbon Cycle The numbers show the quantities of carbon (expressed as petagrams) held in or exchanged annually by ecosystem compartments. The widths of the arrows are proportional to the sizes of the fluxes.

yourBioPortal.com
GO TO Animated Tutorial 58.2 • The Global Carbon Cycle

The amount of carbon in the atmosphere in the form of CO_2, and the amount stored in the biosphere in living and dead organisms, are dwarfed by the amount of carbon stored in terrestrial soils and rocks, in marine sediments, and in dissolved form in ocean waters. Carbon dioxide moves into ocean waters from the atmosphere primarily by simple diffusion at the ocean surface; it is stored in seawater in the form of carbonate (CO_3^{-2}) or bicarbonate (HCO_3^-).

— yourBioPortal.com —

GO TO **Animated Tutorial 58.3 • The Ocean Conveyor Belt**

Today's oceans absorb 20–25 million tons of CO_2 every day—more than at any time during the past 20 million years. As a result, water near the ocean surface is becoming more acidic. An increase in the acidity of ocean surface waters could have negative effects on many marine organisms, particularly corals. The combination of decreasing pH and increasing water temperature kills their symbiotic algae, "bleaching" the corals and killing them as well (see Figure 27.8A). Because so many other reef species depend on corals and the structure they provide, the entire reef community collapses.

At times in the remote past, great quantities of carbon were removed from the global carbon cycle when organisms died in large numbers and were buried in sediments lacking oxygen. In such anaerobic environments, detritivores do not reduce organic carbon to CO_2. Instead, organic molecules accumulate and are eventually transformed into deposits of oil, natural gas, coal, or peat. Humans have discovered and used these deposits, known as **fossil fuels**, at ever-increasing rates during the past 150 years. As a result, CO_2, one of the final products of the burning of fossil fuels, is being released into the atmosphere faster today than it is dissolving in the oceans or being incorporated into terrestrial biomass. Based on a variety of calculations, atmospheric scientists believe that before the Industrial Revolution, the concentration of atmospheric CO_2 was probably about 265 parts per million. Today, it is 387 parts per million (**Figure 58.10**), representing a rate of increase more than 10 times faster than at any other time for millions of years.

WHERE HAS ALL THE CARBON GONE? Less than half of the vast amount of CO_2 released into the atmosphere by human activities remains in the atmosphere. Where is the rest? Much of it is dissolved in the oceans. Over decades to centuries, the oceans, which contain 50 times more dissolved inorganic carbon than the atmosphere, determine atmospheric CO_2 concentrations. The rate at which CO_2 moves from the atmosphere to the oceans depends, in part, on photosynthesis by phytoplankton in the surface waters. These organisms remove carbon from the water, thereby increasing its absorption from the atmosphere. In addition, many marine organisms, including clams, oysters, corals, and some algae, combine bicarbonate ions with calcium (Ca^{2+}) to form calcium carbonate, which they use to build shells and other structures. When these organisms die, they eventually sink to the ocean floor.

Photosynthesis by terrestrial vegetation, principally in forests and savannas, typically absorbs roughly the same amount of carbon that is released by metabolism—about half of it by plants and half of it by microbes in the soil that break down organic matter produced by plants. Currently, the photosynthetic consumption of CO_2 exceeds the metabolic production of CO_2, so Earth's terrestrial vegetation is storing carbon that would otherwise be increasing atmospheric CO_2 concentrations. Yet, as we saw at the opening of this chapter, we cannot count on terrestrial vegetation to store the extra CO_2 that human activities are producing. Furthermore, climate warming—another result of increasing CO_2 concentrations, as we'll see next—increases plant metabolism and is thus likely to increase the flux of CO_2 from vegetation to the atmosphere.

ATMOSPHERIC CO_2 AND GLOBAL WARMING What evidence do we have that the buildup of atmospheric CO_2 stemming from the burning of fossil fuels is raising Earth's temperature? Measurements of gases in air trapped in the Antarctic and Greenland ice caps show that atmospheric CO_2 levels have been higher when Earth has been warmer and lower when it has been cooler (**Figure 58.11**). For example, atmospheric CO_2 was very low during the most recent glaciation, between 30,000 and 15,000 years ago, when temperatures were presumably much colder than they are today. In contrast, during a warm interval 5,000 years ago, atmospheric CO_2 may have been slightly higher than it is today.

How global climates and ecosystems will change in response to this rapid warming is a subject of intense investigation. Complex computer models of the global ecosystem indicate that a doubling of the atmospheric CO_2 concentration from today's level would increase

Each year CO_2 concentrations rise during the Northern Hemisphere winter, when respiration exceeds photosynthesis.

CO_2 concentrations then fall during the summer, when photosynthesis exceeds respiration.

58.10 Atmospheric CO_2 Concentrations Are Increasing These carbon dioxide concentrations, expressed as parts per million by volume of dry air, were recorded on top of Mauna Loa, Hawaii, far from most sources of human-generated CO_2 emissions.

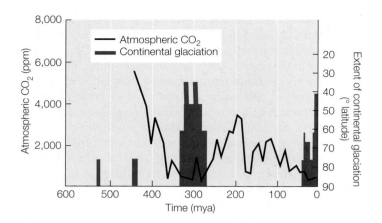

58.11 Higher Atmospheric CO₂ Concentrations Correlate with Warmer Temperatures The geological record of variation in atmospheric CO₂ concentrations over the last 500 million years shows that CO₂ levels have been lowest during times of extensive continental glaciation (red bars), and thus presumably lower temperatures.

mean annual temperatures worldwide and would probably cause droughts in the central regions of continents while increasing precipitation in coastal areas. Global warming has already resulted in the shrinking of Arctic sea ice (**Figure 58.12**). If the polar ice caps continue to melt, sea level will rise (due to both thermal expansion of ocean waters and the addition of glacial meltwater), flooding coastal cities and agricultural lands. Because the rate at which water evaporates from the ocean surface increases with temperature, tropical storms are likely to become more intense. Indeed, during the past two decades, the number of hurricanes generated per year worldwide has not increased, but the number of severe (category 4 and 5) hurricanes has.

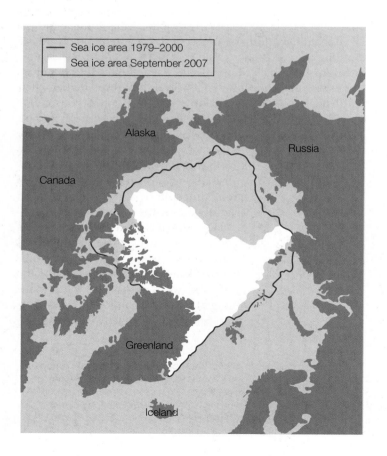

Global warming is having a profound effect on the distribution and abundance of interacting species. One clear example is an increase in the severity of insect infestations in certain temperate forest communities. Pine trees in these forests are attacked by pine bark beetles, which carry with them a symbiotic fungus that infects the trees and helps the beetles to overcome the trees' defenses (see Figure 56.13). In Colorado, cold winters have historically limited the ability of these beetles to kill many trees. In 2008, however, Colorado experienced an outbreak of mountain pine beetles larger than any previously recorded; more than a million acres of trees were killed. The lack of an extended period of below-zero temperatures in the previous winter had allowed large numbers of overwintering beetles to survive.

Another result of global warming may be an increase in the frequency of outbreaks of human diseases. Winter cold typically kills many pathogens, sometimes eliminating as much as 99% of a pathogen population. If climate warming reduces this population bottleneck, some diseases will spread to more temperate regions. For example, dengue fever and its mosquito vector are now spreading to higher latitudes where they were formerly absent.

The 2007 Assessment Report of the Intergovernmental Panel on Climate Change (IPCC) forecast these and many other negative effects of global warming on humans, warning, "Human beings are exposed to climate change through changing weather patterns (for example, more intense and frequent extreme events) and indirectly through changes in water, air, food quality and quantity, ecosystems, agriculture, and economy. At this early stage the effects are small but are projected to progressively increase in all countries and regions."

Recent perturbations of the nitrogen cycle have had adverse effects on ecosystems

Nitrogen gas makes up 78% of Earth's atmosphere, but most organisms cannot use nitrogen in its gaseous form. Only a few species of microorganisms can convert atmospheric N_2 into forms that are usable by plants, a process called *nitrogen fixation* (see Section 36.4). Nitrogen-fixing bacteria are often found as symbiotic associates in the root nodules of plants such as legumes. Other microorganisms carry out *denitrification*, the principal process that removes nitrogen from the biosphere and returns it to the atmosphere as N_2 (see Figure 36.11). These denitrifying microbes are found in places where oxygen levels are

58.12 The Ice Caps Are Melting Global warming is causing major changes in the sizes of the polar ice caps. In 2008, for the first time in recorded history, enough of the Arctic Ocean was ice-free that the North Pole could have been circumnavigated by ship.

58.13 The Global Nitrogen Cycle The numbers in parentheses show the quantities of nitrogen (expressed as teragrams) held in or exchanged annually by ecosystem compartments. Arrow widths are proportional to the sizes of the fluxes.

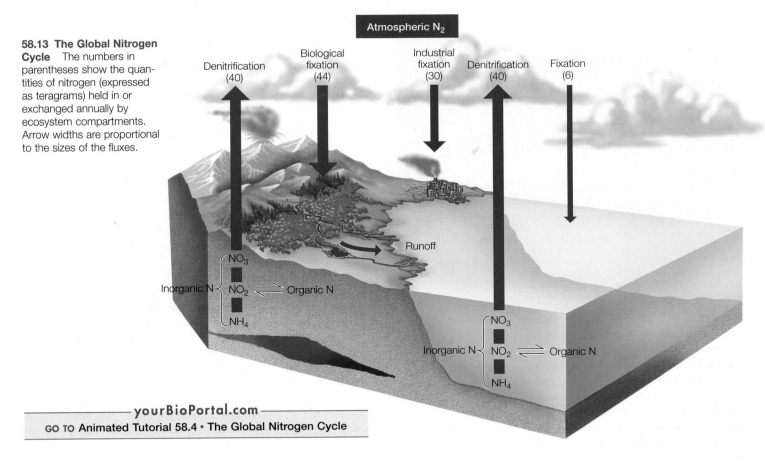

yourBioPortal.com

GO TO **Animated Tutorial 58.4 • The Global Nitrogen Cycle**

low, such as bogs, swamps, and marine sediments. Collectively, this microbial processing of nitrogen accounts for about 95 percent of all natural nitrogen flux on Earth (**Figure 58.13**).

All living organisms require nitrogen, and the inability of the vast majority of them to utilize N_2 means that usable nitrogen is often in short supply in ecosystems. Populations of nitrogen-fixing organisms rarely increase in abundance to the extent that nitrogen is no longer limiting because the end products of nitrogen fixation are rapidly lost from ecosystems: ammonia by

vaporization and denitrification, and the highly water-soluble nitrates by leaching. Furthermore, fixing nitrogen requires large amounts of energy. Thus nitrogen-fixing organisms often lose out in competition with non-fixers when nitrogen becomes more readily available.

The industrial production of inexpensive artificial fertilizers has had some unanticipated effects on the nitrogen cycle. As a result of extensive use of these fertilizers on agricultural crops, coupled with the burning of fossil fuels (which generates nitric oxide), total industrial nitrogen fixation by humans today is about equal to global natural nitrogen fixation. This human-generated nitrogen flux has been increasing over the past half-century and is expected to continue to increase during the coming decades (**Figure 58.14**).

These large perturbations of the nitrogen cycle have had numerous adverse effects. When more nitrogen is applied to croplands than can be taken up by the crops, the excess nitrogen moves out of the system in surface runoff, or downward into groundwater, and ultimately ends up in rivers, lakes, and oceans. The addition of nutrients to these bodies of water, known as **eutrophication** ("rich food"), can have dramatic effects on aquatic ecosystems. The "dead zone" that has formed

—— Total anthropogenic fixation
—— Nitrogen fixation in agroecosystems
—— Fossil fuel combustion
—— Fertilizer and industrial uses

58.14 Human Activities Have Increased Nitrogen Fixation Most of the industrially fixed nitrogen (produced by humans by combining, at high pressure and temperatures, atmospheric nitrogen with hydrogen derived from fossil fuels) is manufactured for use in agricultural fertilizers. Some nitrogen fixation is a by-product of fossil fuel combustion.

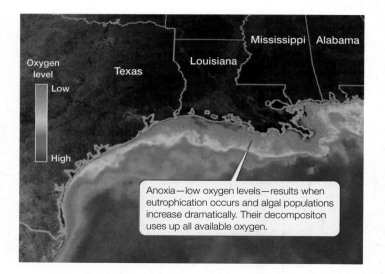

Oxygen
level

Low

High

Anoxia—low oxygen levels—results when eutrophication occurs and algal populations increase dramatically. Their decompositon uses up all available oxygen.

58.15 A Dead Zone   High concentrations of nitrogen and phosphorus in the runoff from agricultural lands in the midwestern United States are carried by the Mississippi River to the Gulf of Mexico. This nutrient enrichment causes excessive algal growth, depleting the oxygen in the water and creating a deoxygenated "dead zone" in which few aquatic organisms can survive.

near the mouth of the Mississippi River in the Gulf of Mexico is a result of flows of water from agricultural fields in the Upper Midwest carrying high concentrations of nitrogen from fertilizer (**Figure 58.15**). The high nutrient levels that result promote luxuriant algal growth. When the algae die, their decomposition by microbes uses up the oxygen in the water, and the resulting anaerobic conditions kill most other marine organisms. Given that the Gulf of Mexico is a key resource for the seafood industry, providing almost three-fourths of the shrimp and two-thirds of the oysters harvested in the United States, an expansion of the dead zone could have a devastating effect on the economies of the Gulf region, over and above its effects on biodiversity and ecosystem function.

The increase in nitrogen fixation has also increased atmospheric concentrations of the greenhouse gas nitrous oxide (N_2O; see Table 58.1), resulting in the production of tropospheric O_3 —also a greenhouse gas—and smog. Some of the nitrogen that enters the atmosphere falls back to land in precipitation or as dry particles. This deposition of nitrogen affects the composition of terrestrial vegetation by favoring those species that are best adapted to take advantage of high nutrient levels, which then crowd out other species. Atmospheric deposition of nitrogen has increased dramatically during recent decades due to human activity. Spatial variation in nitrogen deposition rates has allowed ecologists to determine that plant species richness in grasslands declines as the rate of nitrogen deposition increases. Rates of nitrogen deposition are high enough over much of Europe and eastern North America to cause substantial reductions in species richness in grasslands on both continents.

The burning of fossil fuels affects the sulfur cycle

Most of Earth's sulfur supply is locked up in rocks on land and as sulfate salts in deep-sea sediments. Emissions of the gases sulfur dioxide (SO_2) and hydrogen sulfide (H_2S) from volcanoes account for, on average, 10–20% of the total natural nonbiological flux of sulfur to the atmosphere, but they occur only intermittently. Although volcanic eruptions spew great quantities of sulfur over

broad areas, they are rare events. Sulfur in soil is taken up by plants and incorporated into proteins; ultimately, this sulfur is returned to the atmosphere via microbial decomposition. In marine systems, too, microbial decomposition is important in returning sulfur to the atmosphere. In particular, many marine phytoplankton and seaweeds manufacture large quantities of a sulfur-containing compound (DMSP, or dimethylsulfoniopropionate) to maintain their salt and water balance; when broken down, this compound releases dimethyl sulfide (CH_3SCH_3), the principal odorant of rotting seaweed stench. Because the quantities of phytoplankton in the vast oceans are enormous, dimethyl sulfide production accounts for about half of the biotic component of the global sulfur cycle.

Atmospheric sulfur plays an important role in global climate. Even if air is moist, clouds do not form readily unless there are small particles in the atmosphere around which water can condense. Dimethyl sulfide is the major component of such particles. Therefore, increases in atmospheric sulfur concentrations can increase cloud cover and reduce the amount of incoming solar radiation that ultimately reaches Earth's surface.

The burning of fossil fuels alters the sulfur cycle as well as the carbon and nitrogen cycles. The combustion of fossil fuels releases sulfur in the form of SO_2 and nitrogen in the form of nitrogen dioxide (NO_2) into the atmosphere, where they react with water molecules to form sulfuric acid (H_2SO_4) and nitric acid (HNO_3). These acids can travel hundreds of kilometers in the atmosphere before they settle to Earth as dry particles or in precipitation. Rain or snow containing sufficient levels of nitric and sulfuric acids to lower its pH is called **acid precipitation**. Acid precipitation now falls in all major industrialized countries and is particularly widespread in eastern North America and Europe. The normal pH of unpolluted precipitation is about 5.6, but precipitation in New England now averages about pH 4.5 and there have been occasional rainfalls and snowfalls with a pH as low as 3.0. Precipitation with a pH of about 3.5 or lower damages the leaves of plants and reduces rates of photosynthesis. Acidification of lakes in the Adirondack region of New York State has reduced fish species richness by causing the extinction of acid-sensitive species (**Figure 58.16**). Many invertebrates that are primary consumers in aquatic communities are sensitive to pH; in particular, multiple species of mayflies and caddisflies have experienced local population reductions in acidified streams around the world. Even when its effects are not lethal, acidification can have subtle effects on behavior that reduce the viability of aquatic organisms. Diving beetles, for example, lose their ability to regulate their underwater oxygen supply when pH levels drop significantly.

New regulations instituted by the 1990 Clean Air Act have raised the pH of precipitation in much of the eastern United

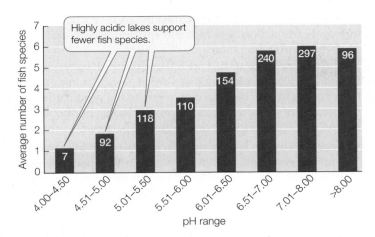

58.16 Acidification of Lakes Exterminates Fish Species The number of fish species found in lakes in the Adirondack region of New York State is directly correlated with pH. The numbers in the bars indicate the number of lakes in each pH range.

States, primarily by lowering sulfur emissions. There are indications that, once emissions have been reduced, acidified aquatic systems can recover quickly. David Schindler at the University of Alberta studied the effects of acid precipitation by adding enough H_2SO_4 to two small Canadian lakes to reduce their pH from about 6.6 to 5.2. In both lakes, nitrifying bacteria failed to survive under these moderately acidic conditions. As a result, the nitrogen cycle was blocked. When Schindler stopped adding acid to one of the lakes, its pH returned to its original value in about a year, and nitrification resumed. These experiments show that lake ecosystems are very sensitive to acidification but can recover rapidly, presumably because water in lakes is exchanged rapidly.

That said, the larger organisms in such systems may be slower to bounce back. Fourteen rivers in Wales were studied for a quarter-century, to determine the impacts of acid rain reductions. Investigators were disappointed that, even after 25 years, only four insect species—two mayfly species and two caddisfly species—had recolonized the area out of the 29 species expected to come back.

The global phosphorus cycle lacks a significant atmospheric component

Phosphorus accounts for only about 0.1% of Earth's crust, but it is an essential nutrient for all life forms. It is a key component of DNA, RNA, and ATP. Unlike the elemental cycles discussed thus far, the phosphorus cycle lacks a significant gaseous component. Some phosphorus is transported on dust particles, but very little of the phosphorus cycle takes place

in the atmospheric compartment. Most of Earth's phosphorus is present in the form of phosphate salts in rocks and deep-sea sediments. Phosphorus cycling takes millions of years because the processes of sedimentary rock formation, uplift, and weathering all take a long time (**Figure 58.17**). In contrast, phosphorus often cycles rapidly among organisms, and it is often a limiting factor for their growth, particularly for plants. In addition to nitrogen, fertilizers routinely include phosphorus.

Human activity has radically accelerated some parts of the phosphorus cycle. One consequence of the massive use of artificial fertilizers is that between 10.5 and 15.5 teragrams (1 tg = 10^{12} g or 10^9 kg) of phosphorus accumulate each year, primarily in agricultural soils. When the concentration of phosphorus in soils exceeds the capacity of plants to take it up, the excess moves into streams and lakes. Soil erosion due to the clearing of land for purposes such as agriculture and logging also increases the amount of phosphorus and other nutrients in runoff. Phosphorus is a limiting nutrient in many lakes, so when it enters those lakes through runoff, algae previously limited by relatively low levels of phosphorus respond with phenomenal growth. These algal blooms turn the water green and, depending on the metabolic capabilities of the algae, toxic. When the algae die, their decomposition by microbes consumes all of the oxygen in the lake, and anaerobic organisms dominate the bottom sediments. These anaerobic organisms do not break down carbon compounds all the way to CO_2, and their metabolic end products, many of which have unpleasant odors, build up.

The capacity of phosphorus-charged runoff to cause extensive eutrophication was graphically illustrated almost 50 years ago. A technological innovation in the formulation of laundry detergents in the early 1960s was the use of sodium tripolyphosphate (STPP) ($Na_5P_3O_{10}$) as a water softener and dirt-breaking agent to enhance cleaning efficiency. Widespread adoption of these new detergents led to massive phosphorus enrichment of lakes and streams and concomitant eutrophication and algal blooms. Water quality declined so much across the United States that phos-

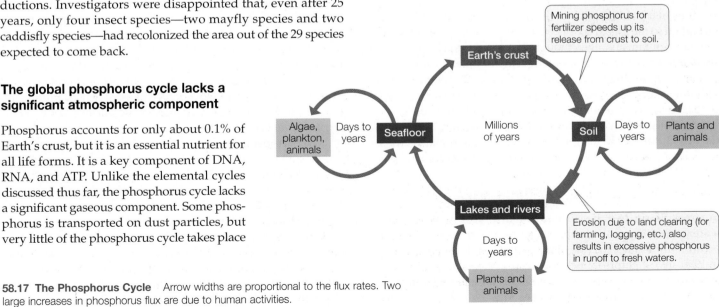

58.17 The Phosphorus Cycle Arrow widths are proportional to the flux rates. Two large increases in phosphorus flux are due to human activities.

phate-based detergents were banned in many states, as well as in Europe. In today's detergents, phosphonates—forms of phosphorus that do not appear to promote algal growth—and aluminum silicates perform the functions STPP used to serve.

Detergents, however, are only one agent contributing to eutrophication. Human waste is rich in phosphorus, as are manure from domesticated animals and industrial waste of various types. Two hundred years ago, Lake Erie, one of the Great Lakes on the border between the United States and Canada, had only moderate rates of photosynthesis and clear, oxygenated water. With increasing industrialization in the early part of the twentieth century, nutrient concentrations in the lake increased greatly, and algae proliferated. At the water filtration plant in Cleveland, Ohio, algae increased from 81 per milliliter in 1929 to 2,423 per milliliter in 1962. Populations of bacteria also increased: the numbers of *Escherichia coli* increased enough that many of the lake's beaches were closed and declared health hazards. Since 1972, the United States and Canada have invested more than U.S. $9 billion to improve municipal waste treatment facilities and reduce discharges of pollutants into Lake Erie. As a result, the amount of phosphorus added to Lake Erie has decreased more than 80% from its highest level, and phosphorus concentrations in the lake have declined substantially. The deeper waters of Lake Erie are still oxygen-poor during the summer months, but the rate of oxygen depletion is declining.

We could greatly reduce phosphorus pollution by recovering and recycling phosphorus. The phosphorus contained in sewage and animal wastes could supply much of the needs of the detergent and fertilizer industries. More careful application of fertilizers on agricultural lands could reduce the rate of phosphorus accumulation in soils without reducing crop yields. However, reduction of phosphorus concentrations in soils will take many decades after remedial actions are initiated, and eutrophication of lakes and streams may persist even after these actions are taken.

Other biogeochemical cycles are also important

Other elements, in addition to those we have just discussed, are important to the global ecosystem because they are essential nutrients for organisms, even though they are needed only in very small amounts. One such element is *iron* (Fe), an essential micronutrient for almost all organisms. It is a key component of the enzymes involved in chlorophyll synthesis, as well as an essential component of many animal enzymes. Iron confers oxygen-binding ability on hemoglobin in vertebrate blood and the cytochrome P450 monooxygenases, which in most aerobic organisms play a central role in detoxifying environmental poisons, rely upon iron for their catalytic activity.

Iron is readily available on land in rocks and minerals. It moves into coastal waters in streams and into the open ocean in atmospheric dust. Because iron is insoluble in oxygenated water, it rapidly sinks to the ocean floor. Therefore, in most ocean communities, the rate of photosynthesis is limited by iron. In 1996, investigators launched an ecosystem-scale experiment in which surface waters of the equatorial Pacific Ocean were seeded with dissolved iron. The response was a tremendous phytoplankton bloom, accompanied by massive uptake of nitrate and carbon dioxide, which had apparently been underutilized due to iron limitation.

Three micronutrients that are relatively rare globally—iodine (I), cobalt (Co), and selenium (Se)—are of particular importance to living organisms because they regulate the production and functioning of many metabolic agents, including hormones, vitamins, and enzymes that contain zinc and copper.

- *Iodine* is found on land in mineral deposits and in seawater as an inorganic salt; fresh water is a relatively poor source. Iodine, which is supplied to land primarily by marine algae that release it to the air, is often deficient in continental interiors. It is an essential component of the hormone thyroxine, which governs many metabolic processes, and a deficiency of iodine in humans leads to the medical condition called goiter (see Figure 41.10).

- *Cobalt* is a rare element found in soil and rocks. The bacteria that fix atmospheric nitrogen require coenzymes derived from cobalt. As an essential component of vitamin B_{12}, it is a dietary requirement for almost all animals and is important for the synthesis of proteins.

- *Selenium* is also found in soil and rocks. It is a component of several important mammalian enzymes, such as the antioxidant glutathione peroxidase (which protects tissues from oxidative damage) and several enzymes involved in synthesizing thyroid hormones.

Land plants require all three of these micronutrients in minimal concentrations, but animals, particularly endothermic vertebrates, require them in concentrations that exceed the supply in many environments. Herbivores sometimes obtain these nutrients by drinking water at mineralized springs or by eating enriched earth (see Figure 53.14). Deficiencies of these elements are most common in areas with high rainfall and soils that leach readily.

Biogeochemical cycles interact

The biogeochemical cycles of different elements interact with one another in complex ways, and perturbations to one cycle can have profound effects on other cycles. In recent years, human-induced perturbations have made these interactions glaringly apparent. For example, nitrate released by human activities can have profound effects on the biogeochemical cycles of iron and arsenic. Nitrate is a powerful oxidant, and nitrate pollutants can increase the movement of arsenic in lakes by causing its oxidation. Some urban lakes have arsenic levels in excess of 2,000 parts per million locked in sediments; oxidation releases arsenic into the water column in a form that is carcinogenic and has negative effects on development. Every year, scientists are discovering interactions of which they were previously unaware, and experiments exploring biogeochemical interactions are increasing in number.

Changes in atmospheric CO_2 concentrations have been a particular focus of investigation in recent years because of their potential for interacting with other biogeochemical cycles through photosynthesis. A case in point is a study of whether elevated atmospheric CO_2 concentrations affect rates of nitro-

INVESTIGATING LIFE

58.18 Effects of Atmospheric CO_2 Concentrations on Nitrogen Fixation

Although the original expectation was that CO_2 enrichment would enhance nitrogen fixation, the reverse—decreased fixation—was in fact being observed. In an attempt to explain this result, investigators studied the changes in concentrations of iron and molybdenum, two micronutrient components of enzymes that are essential to nitrogen fixation.

HYPOTHESIS Higher CO_2 levels should enhance nitrogen fixation.

METHOD

1. Grow nitrogen-fixing vines (*Galactia elliottii*) in plots under control conditions (normal levels of CO_2) and experimental conditions of artificially heightened levels of atmospheric CO_2.
2. Measure levels of nitrogen fixation over a period of 7 years.
3. Determine the concentrations of iron and molybdenum in the leaves of *G. elliottii* at year 7.

RESULTS

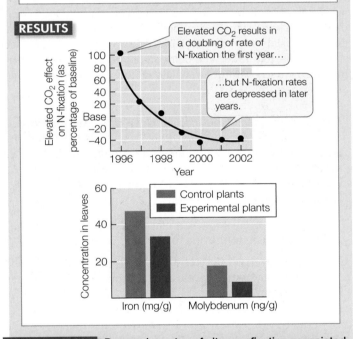

Elevated CO_2 results in a doubling of rate of N-fixation the first year...

...but N-fixation rates are depressed in later years.

CONCLUSION Decreasing rates of nitrogen fixation associated with increased concentrations of atmospheric CO_2 correlate with lower concentrations of essential iron and molybdenum in plant tissues; thus the initial increase in the rate of fixation can result in its eventual depression.

Go to **yourBioPortal.com** for original citations, discussions, and relevant links for all INVESTIGATING LIFE figures.

gen fixation by microorganisms associated with plant roots. A group of ecologists grew a nitrogen-fixing vine called Elliott's milkpea (*Galactia elliottii*) under artificially increased CO_2 concentrations. Higher CO_2 concentrations led to an enhancement

of nitrogen fixation during the first year of the experiment, but surprisingly, the positive effect disappeared by the third year, and elevated CO_2 concentrations actually *reduced* nitrogen fixation below normal levels during the fifth, sixth, and seventh years of the experiment (**Figure 58.18**). The experimenters suspected that depletion of micronutrients such as iron and molybdenum caused the reduction in nitrogen fixation, so they measured concentrations of those elements in the leaves of *G. elliottii*. They found that concentrations of molybdenum were particularly low. Elevated CO_2 can increase the acidity of water in soil by enhancing the rate of carbonic acid formation and, by enhancing photosynthesis, increase the accumulation of soil organic matter; both types of change increase the tendency of molybdenum to adsorb onto soil particles, reducing its availability to nitrogen-fixing bacteria and causing a decrease in nitrogen fixation rates.

58.3 RECAP

The pattern of movement of a chemical element through organisms and the physical environment is its biogeochemical cycle. Human activities have affected many biogeochemical cycles, especially those of water, carbon, nitrogen, sulfur, and phosphorus. Increasing concentrations of carbon dioxide and other greenhouse gases in the atmosphere are implicated in global warming.

- Describe the global hydrologic cycle and explain what drives it. See pp. 1229–1230 and Figure 58.8
- How do biological processes move carbon from the atmospheric compartment to the terrestrial compartment and then return it to the atmosphere? See pp. 1230–1231 and Figure 58.9
- What are some results of human-induced alterations of the sulfur and nitrogen cycles? See pp. 1232–1234
- Name two elements that can cause eutrophication in aquatic ecosystems and describe their effects on those systems. See pp. 1235–1236

The biogeochemical cycles of chemical elements are intimately involved in ecosystem function. Just as human alterations of those cycles are having many effects on ecosystems worldwide, the resulting changes in those ecosystems are having profound effects on human lives.

58.4 What Services Do Ecosystems Provide?

Although it seems obvious today that humans depend on natural ecosystems for survival, explicit recognition of the value of those ecosystems is rather recent. Environmental writers introduced the idea of "natural capital" in the 1940s; it was in 1970 that ecosystems were first said to provide people with a variety of "goods and services." The goods include food, clean water, clean air, fiber, building materials, and fuel; the services include flood control, soil stabilization, pollination, and climate regulation. Most of these benefits either are irreplaceable, or the technology necessary to replace them is prohibitively expensive. For example, potable fresh water can be provided by desalinating seawater, but only at great cost. The aesthetic, psychological, spiritual, and recreational benefits of ecosystems are less tangible, but no less important, and no more easily replaced.

Humans have increasingly altered Earth's ecosystems to increase their capacity to provide us with goods such as food, fresh water, timber, fiber, and fuel. The benefits have not, however, been equally distributed, and some human populations have actually been harmed by our manipulations of natural ecosystems. Moreover, short-term increases in some ecosystem goods and services have come at the cost of the long-term degradation of others.

The most important driver of change in ecosystems has been changes in land use as natural ecosystems have been converted to other, more intensive uses. Although humans have been altering ecosystems for millennia, the pace and scope of the alterations have increased considerably in recent years. More land was converted to cropland between 1950 and 1980 than in the 150 years between 1700 and 1850. Ecosystem conversions have been particularly rapid in tropical and subtropical biomes. Aquatic ecosystems have suffered losses at an increasing pace as well. Between 1970 and 2000, about 20% of the world's coral reefs were lost, and an additional 20% were degraded. In freshwater ecosystems, so much water is now impounded behind dams that artificial reservoirs hold about six times as much water today as do natural rivers. These freshwater systems are being rapidly depleted: the amount of water withdrawn from rivers—most of it for agriculture—has doubled since 1960. In terms of nutrient cycles, more than half of all the artificial nitrogen fertilizer ever used on Earth has been used since 1985.

Human alteration of ecosystems has had many positive effects on human health and prosperity, but it necessarily involves trade-offs. Agriculture, for example, feeds and employs huge numbers of people. But the spread of agriculture into marginal lands may degrade soils and compromise ecosystem delivery of clean water, as when overuse of artificial fertilizers results in eutrophication. Extensive use of pesticides controls insect pests, but also reduces populations of pollinators and the services they provide to both crops and native plants.

Similarly, the loss of wetlands and other natural buffers has reduced the ability of ecosystems to regulate flooding and other natural hazards. The damage from the tsunami that hit Indonesia and other Southeast Asian countries in December 2004 was greater in many places than it would have been had the mangrove forests that protect the coast not been cut down and converted to cropland. Hurricane Katrina, which struck the U.S. Gulf Coast less than a year later, would not have caused as much flooding in New Orleans had the wetlands surrounding the city been intact. Katrina's devastating effects were due in part to a situation that had been developing for decades.

New Orleans is located on the Mississippi River delta. Much of the city lies below sea level, buffered by dams and levees constructed by the Army Corps of Engineers. The upstream dams that protect New Orleans from flooding also prevent the river

from depositing the sediments that have sustained the surrounding delta wetlands for centuries. Oil and natural gas producers have cut thousands of small canals through those wetlands in order to lay pipelines and drilling rigs, and the extraction of oil and gas from beneath the land has caused it to sink. Increased dredging of shipping lanes and rising sea levels due to global warming have contributed to a rise in salinity, killing off many of the great cypress tree swamps. These extensive alterations resulted in the loss of over 80% (more than 1.2 million acres) of the delta wetlands between 1930 and 2005. By the time Katrina made landfall, those wetlands could no longer protect New Orleans from flooding. Storm surges raced along the paths carved by canals and shipping lanes to breach the levees, inundating much of the city.

As it flooded New Orleans, Hurricane Katrina raised awareness and appreciation of an ecosystem hitherto taken for granted by most people. Crucial not only for their flood control services, the delta wetlands provide winter habitat for some 70% of the migrating birds in the huge Mississippi Valley. They are also the spawning grounds for marine organisms, some of which are commercially valuable. The delta's famous shrimping industry contributes about 30% by weight of the total commercial fish harvest in the continental United States. The importance of coastal wetlands—and, indeed, of a wide range of other ecosystems—to human well-being mandates careful ecosystem management to guarantee a sustained flow of ecosystem goods and services.

58.4 RECAP

Ecosystems provide human society with indispensable goods and services. Altering ecosystems can compromise their ability to provide these goods and services.

- What are some of the essential goods and services that ecosystems provide to humans? **See p. 1238**

- How do human efforts to increase the provision of some ecosystem services cause degradation of other ecosystem services?

How can we meet the challenge of obtaining goods and services from ecosystems without compromising their ability to provide those goods and services over the long term? Fortunately, options exist for sustainable management of ecosystems.

58.5 How Can Ecosystems Be Sustainably Managed?

Practices that allow us to conserve or enhance ecosystems so as to benefit from specific ecosystem goods and services over the long term without compromising others are referred to as **sustainable**. In many cases, the total economic value of a sustainably managed ecosystem is higher than that of an intensively exploited ecosystem (**Figure 58.19**). The long-term economic benefits of preventing overexploitation of marine fish popula-

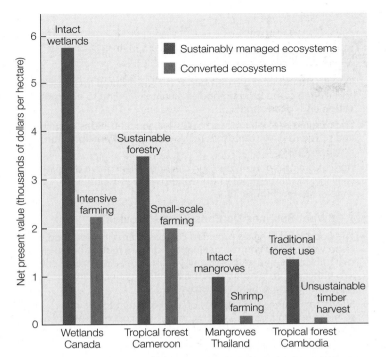

58.19 Economic Values of Sustainably Managed and Converted Ecosystems   Many types of ecosystems can provide more goods and services when they are sustainably managed than when they are converted and intensively exploited.

tions, for example, are enormous. The collapse of the cod fishery on Georges Bank due to overfishing (see Figure 55.13) resulted in the loss of tens of thousands of jobs and cost at least $2 billion in income support and retraining, and it can take a long time for such fisheries to recover.

Impeding the establishment of policies that encourage sustainable management is the impression that ecosystem services are "public goods" with no market value. People who do not stand to profit from the goods and services provided by natural ecosystems have no incentive to pay for them, whereas individuals who own converted ecosystems can reap great eco-nomic benefits from them. As with other "public goods," government action may be needed to create incentives encouraging sustainable ecosystem management. Examples of such action might include:

- Elimination of subsidies that promote damaging exploitation of ecosystems. For example, the billions of dollars paid by governments in developed nations to subsidize domestic agriculture (with the hope of insulating farmers from economic risk) have led to greater food production than the global market warrants, promoted excessive use of fertilizers, and reduced the profitability of agriculture in developing countries.

- More sustainable use of fresh water could be achieved by charging users the full cost of providing the water, by developing methods to use water more efficiently, and by altering the allocation of water rights so that the incentives favor conservation.

- More sustainable use of marine fisheries could be achieved by establishing more protected marine reserves and "no-take" zones where fish can grow to reproductive age. Recent discussions of marine fisheries management center on what is colloquially known as the BOFFF hypothesis—for "Big Old Fat Female Fish"—that the largest and oldest females in a population are the most important to protect because they outproduce younger fish by an enormous margin.

Raising public awareness of endangered ecosystems is essential to the successful implementation of sustainable management programs. Most people do not understand the long-term value of ecosystem goods and services or realize how human activities affect the functioning of ecosystems, so public education programs are needed.

Maintaining and enhancing ecosystem goods and services that have no established market value is especially difficult. Probably the most difficult service to maintain in the face of increasingly intensive human use of the global ecosystem will be biological diversity. The final chapter of this book is devoted to this important topic.

CHAPTER SUMMARY

58.1 What Are the Compartments of the Global Ecosystem?

- Earth is a closed system with respect to atomic matter, but an open system with respect to energy.

- Earth's physical environment can be divided into four compartments: atmosphere, oceans, fresh waters, and land. **Review Figure 58.1**

- Most global air circulation takes place in the lowest layer of the atmosphere, the **troposphere**. A layer of ozone in the **stratosphere** absorbs ultraviolet radiation. **Review Figure 58.2**

- Water vapor, carbon dioxide, and other **greenhouse gases** in the atmosphere are transparent to sunlight but trap heat, thus warming Earth's surface. **Review Figure 58.3**

- Most materials that cycle through the four compartments end up in the oceans, where they eventually sink to the bottom. Cold, nutrient-rich water rises and returns nutrients to the ocean surface in coastal **upwelling zones**. **Review Figure 58.4**

- Only a small fraction of Earth's water resides in the freshwater compartment, but water cycles rapidly through it. Nutrients reach lakes and streams by surface flow or by the movement of **groundwater**.

- **Turnover** brings nutrients and dissolved carbon dioxide to the surface and oxygen to the deeper waters of freshwater lakes. Turnover occurs regularly in temperate-zone lakes in spring in fall. **Review Figure 58.5**

- Land covers only one-fourth of Earth's surface but is connected to the atmospheric compartment by living organisms, which remove chemical elements from the atmosphere and release them back.

58.2 How Does Energy Flow through the Global Ecosystem?

- Net primary production varies across the globe, reflecting differences in temperature, moisture, and the distribution of land masses. **Review Figures 58.6 and 58.7**

- Humans appropriate approximately 24% of Earth's average annual net primary production, although the percentage varies regionally.

58.3 How Do Materials Cycle through the Global Ecosystem?

- The pattern of movement of a chemical element through organisms and the compartments of the global ecosystem is its **biogeochemical cycle**.

- The **hydrologic cycle** is driven by the sun, which evaporates water from ocean surface than it returns by precipitation. The excess precipitation that falls on land eventually returns to the oceans, primarily in rivers. **Review Figure 58.8, ANIMATED TUTORIAL 58.1**

- Groundwater plays a minor role in the hydrologic cycle, but is being seriously depleted by human activities.

- Carbon is removed from the atmosphere by photosynthesis and returned to the atmosphere by metabolism. The rate at which carbon moves from the atmosphere to the oceans depends in part on photosynthesis by phytoplankton. **Review Figure 58.9, ANIMATED TUTORIALS 58.2 and 58.3**

- The concentration of CO_2 in the atmosphere has greatly increased in the last 150 years, largely due to the burning of **fossil fuels**. This buildup of CO_2 is warming the global climate. **Review Figures 58.10–58.12**

- As a result of agricultural use of fertilizers and the burning of fossil fuels, total nitrogen fixation by humans is about equal to natural nitrogen fixation. **Review Figures 58.13 and 58.14, ANIMATED TUTORIAL 58.4**

- Human alteration of the nitrogen cycle has resulted in excesses of nitrogen compounds in bodies of water, leading to **eutrophication** and "dead zones."

- The burning of fossil fuels increases the concentrations of sulfur and nitrogen in the atmosphere, leading to **acid precipitation**. **Review Figure 58.16**

- Agricultural use of fertilizers and clearing of land have dramatically increased the input of phosphorus into soils and fresh waters. **Review Figure 58.17**

58.4 What Services Do Ecosystems Provide?

- The goods and services provided by ecosystems include food, clean water, flood control, pollination, climate regulation, spiritual fulfillment, and aesthetic enjoyment. Most ecosystem services either are irreplaceable or the technology necessary to replace them is prohibitively expensive.

- Efforts to enhance the capacity of an ecosystem to provide some goods and services often come at the cost of its ability to provide others.

- The most important driver of change in ecosystems has been changes in land use, as natural ecosystems have been converted to other, more intensive uses.

58.5 How Can Ecosystems Be Sustainably Managed?

- In many cases, the total economic value of a sustainably managed ecosystem is higher than that of a converted or intensively exploited ecosystem. **Review Figure 58.19**

- Recognition of the value of ecosystem goods and services that are now perceived as "public goods" may induce government action to protect them. Public education is needed to make people aware of how much they benefit from ecosystem goods and services and how human activities affect the ecosystems that provide them.

SEE WEB ACTIVITY 58.1 for a concept review of this chapter.

SELF-QUIZ

1. Unusual features of Earth as a planet that affect its ecosystem dynamics include
   a. lithospheric plates that move continuously.
   b. atmospheric gases that moderate surface temperatures.
   c. large amounts of water in liquid form.
   d. a diversity of living organisms.
   e. All of the above

2. Marine upwelling zones are important because
   a. they help scientists measure the chemistry of deep ocean water.
   b. they bring to the surface organisms that are difficult to observe elsewhere.
   c. ships can sail faster in these zones.
   d. they increase marine productivity by bringing nutrients back to surface ocean waters.
   e. they bring oxygenated water to the surface.

3. Which of the following is *not* true of the troposphere?
   a. It contains nearly all of the atmospheric water vapor.
   b. It contains a layer of ozone, which filters out ultraviolet light.
   c. It is about 17 kilometers thick in the tropics and subtropics.
   d. Most global air circulation takes place there.
   e. It contains about 80 percent of the mass of the atmosphere.

4. The hydrologic cycle is driven by
   a. the flow of water into the oceans via rivers.
   b. evaporation (transpiration) of water from the leaves of plants.
   c. evaporation of water from the surface of the oceans.
   d. precipitation falling on land.
   e. the fact that more water falls on the ocean as precipitation than evaporates from its surface.

5. Carbon dioxide is called a greenhouse gas because
   a. it is used in greenhouses to increase plant growth.
   b. it is transparent to heat, but traps sunlight.
   c. it is transparent to sunlight, but traps heat.
   d. it is transparent to both sunlight and heat.
   e. it traps both sunlight and heat.

6. Iron is an essential micronutrient that
   a. is insoluble in oxygenated water.
   b. is abundant in rocks and minerals.
   c. moves into the ocean via atmospheric dust.
   d. is a limiting factor for photosynthesis in oceans.
   e. All of the above

7. The biogeochemical cycle of phosphorus differs from the cycles of carbon and nitrogen in that
   a. phosphorus lacks an atmospheric component.
   b. phosphorus lacks a liquid phase.
   c. only phosphorus is cycled through marine organisms.
   d. living organisms do not need phosphorus.
   e. The phosphorus cycle does not differ importantly from the carbon and nitrogen cycles.

8. The sulfur cycle influences the global climate because
   a. sulfur compounds are important greenhouse gases.
   b. sulfur compounds help transfer carbon from the atmosphere to the oceans.
   c. sulfur compounds in the atmosphere are components of particles around which water condenses to form clouds.
   d. sulfur compounds contribute to acid precipitation.
   e. Both c and d

9. Acid precipitation results from human modifications of
   a. the carbon and nitrogen cycles.
   b. the carbon and sulfur cycles.
   c. the carbon and phosphorus cycles.
   d. the nitrogen and sulfur cycles.
   e. the nitrogen and phosphorus cycles.

10. Maintaining the capacity of ecosystems to provide goods and services is important because
    a. most ecosystem services cannot be replicated by any other means.
    b. replacing them with technological substitutes is possible only in developed nations
    c. technological substitutes take up valuable land.
    d. governments cannot function without taxing ecosystem services.
    e. It is not important. Humans could survive quite well even if ecosystem services declined greatly.

FOR DISCUSSION

1. A string of powerful hurricanes struck the east coast of the United States over the course of a single year's hurricane season. Some people claim that this disaster was due to warming of the oceans caused by greenhouse gases in the atmosphere. Others assert that global warming is not responsible because hurricanes have occurred for many centuries. How would you evaluate these conflicting claims?

2. The waters of Lake Washington, adjacent to the city of Seattle, rapidly returned to their preindustrial condition when sewage was diverted from the lake to Puget Sound, an arm of the Pacific Ocean. Would all lakes being polluted with sewage clean themselves up as quickly as Lake Washington if sewage inputs were stopped? What characteristics of a lake are most important to its rate of recovery following reduction of nutrient inputs?

3. Tropical forests are being cut down and burned at a rapid rate to clear land for agriculture. Does this necessarily mean that deforestation is a major source of CO_2 inputs to the atmosphere? Why or why not?

4. What types of experiments would you conduct to assess the likely consequences of fertilization of the oceans with iron to increase rates of photosynthesis? At what spatial and temporal scales should they be conducted?

5. A government official authorizes the construction of a large coal-burning power plant in a former wilderness area. Its smokestacks discharge great quantities of combustion wastes. List and describe all the likely effects on ecosystems at local, regional, and global levels. If the wastes were thoroughly scrubbed from the stack gases, which of the effects you have just outlined would still happen?

6. Many nations have signed the Kyoto Accord, which commits them to reducing their emissions of CO_2 to the atmosphere. This agreement, however, is set to expire in 2012. In 2009, the United Nations drafted a new treaty to control greenhouse gases, proposing a goal of reducing global average annual greenhouse gas emissions to 2 metric tons per person (in 2006, average annual emissions per U.S. citizen were 19.78 tons). One mechanism for achieving this goal is called "cap and trade," whereby a government sets a "cap," or limit, on emissions by polluters, but allows facilities that emit less than their allowance to sell their excess credits to polluters who would otherwise exceed their allowance. What are the relative benefits and drawbacks you can see to such an approach to reducing emissions?

ADDITIONAL INVESTIGATION

The experiment described in Figure 58.18 showed that interactions between elevated atmospheric CO_2 concentrations and nitrogen fixation can reduce the availability of other essential nutrients. The investigators suggested an expansion in the suite of elements that should be studied to determine whether elevated concentrations of CO_2 are likely to result in a large carbon sink in terrestrial ecosystems. What other elements should be considered, and how might their influence be investigated?

WORKING WITH DATA (GO TO yourBioPortal.com)

Atmospheric CO_2 and Nitrogen Fixation This exercise offers further details and data from the experiments by Bruce Hungate and colleagues illustrated in Figure 58.18. You will be asked to analyze the data and come to your own conclusions about the effects of increasing atmospheric carbon dioxide on global fixation of nitrogen by green plants.

59 Conservation Biology

No laughing matter

By the time the U.S. Congress passed the Endangered Species Act in 1973, Blackburn's sphinx moth (*Manduca blackburni*) was already believed to be extinct. The largest native insect in the Hawaiian Islands, with a wingspan of over 120 mm, this species was once widespread throughout the islands, occurring in dry forests from sea level up to 760 m elevation. As a caterpillar, it fed on native shrubs in the tomato family. Nobody knew exactly what led to its disappearance, but one suspect was a parasitoid wasp imported to the islands for the biological control of the tobacco hornworm (*Manduca sexta*), a close relative of Blackburn's sphinx moth that itself had been accidentally introduced to the islands and had become a pest in commercial tomato fields.

Miraculously, a population of Blackburn's sphinx moths was discovered on Maui in 1984, in an area used by the Hawaii National Guard for military training. Since that time, several other populations have been found on the island of Hawaii and on Kaho'olawe, an island adjacent to Maui. In February 2000, Blackburn's sphinx moth became the first Hawaiian insect species to be protected under the U.S. Endangered Species Act.

To date, no Blackburn's sphinx moths have been rediscovered on the islands of Kauai or Niihau. These islands, however, are home to another endangered species, the Hawaiian alula plant (*Brighamia insignis*). This odd-shaped shrub, one of several species called by the Hawaiian name of "haha," is found in dry forests ranging from sea level to 480 m on sheer cliffs and rocky outcrops. Its bulbous stem, typically 1–2 m long, produces a whorl of leaves at the top, along with clusters of yellow flowers with petals that are fused into a trumpet-shaped tube up to 14 cm in length. Today the species persists only with the assistance of dedicated botanists and conservation biologists who hand-pollinate the flowers, often rappelling down steep cliff faces to do so. Its most important (and possibly only) native pollinator likely went extinct.

The long, tubular flowers are typical of those pollinated by long-tongued moths in the hawk moth family. Short of traveling back in time, there is no sure way to identify the

Extinction and Back The large Blackburn's sphinx moth (*Manduca blackburni*) was believed to be extinct, possibly the victim of mortality due to a parasitoid wasp imported to Hawaii to control the tobacco hornworm (*M. sexta*)— another imported insect. A population of *M. blackburni* was rediscovered in 1984.

One Extinction May Lead to Another The endangered Hawaiian shrub alula (*Brighamia insignis*), also known as "haha," persists only because humans hand-pollinate the flowers; its natural pollinators may have gone extinct.

native pollinator of *B. insignis*, but evidence points to Blackburn's sphinx moth—whose existence today hangs by a thread on islands many kilometers away, where no alula plants remain. It is no coincidence that both species are endangered.

This contemporary example of the unanticipated effects of species loss is far from unique. Given what we've learned about species interactions, it's not surprising that the loss of one species from a community can result in the endangerment of another. While the effects of biodiversity losses can be difficult to predict, anticipating those effects is a high priority in view of today's escalating rates of species extinctions.

IN THIS CHAPTER we will describe the rapidly expanding field of conservation biology, which is devoted to protecting biodiversity. We will show how conservation biologists predict changes in biodiversity and see how some human activities are causing those changes. Finally, we will describe the strategies conservation biologists use to reduce extinction rates, protect species from threats to their persistence, and help populations of endangered species recover.

59.1 What Is Conservation Biology?

Virtually all natural ecosystems on Earth have been altered by human activities. Many habitats have disappeared completely, and many others have been greatly modified. Even Earth's climate and its great biogeochemical cycles have been altered. One consequence of these changes has been a rapid increase in the rate at which species go extinct. **Conservation biology** is an applied scientific discipline devoted to protecting and managing Earth's biodiversity. The discipline draws heavily upon principles of ecology, ethology, and evolutionary biology, particularly with respect to elucidating the factors that determine whether a given population will persist.

Early conservation efforts were characterized by tensions between those whose principal goal was to conserve natural resources for their economic benefits and those who believed that nature has intrinsic value independent of human economic interests. Today conservation biologists study the full array of goods and services that humans derive from species and ecosystems, including aesthetic and psychological benefits. Understanding the global ecosystem and the effects of human activities on that system is now understood to be essential to the long-term well-being of *Homo sapiens*.

Conservation biology is an *applied* scientific discipline, which is to say that it involves the practical application of knowledge to solve problems. Workers in this field are guided by three basic principles:

- *Evolution is the process that unites all of biology.* To be effective in protecting and managing biodiversity, it is essential to understand the evolutionary processes that generate and maintain it.

- *The ecological world is dynamic.* Because populations and communities change continuously over time, there is no static "balance of nature" that can serve as a goal of conservation activities.

- *Humans are a part of ecosystems.* Human interests and activities must be incorporated into conservation goals and practices.

Conservation biology aims to protect and manage biodiversity

The term "biodiversity" has multiple meanings. We may speak of biodiversity as the degree of genetic variation within a species. Genetic variation can be measured as the number of alleles at a locus, the number of polymorphic loci in a genome, or the number of individuals in a population that are polymorphic at given loci. As we have seen throughout this book, genetic variation allows organisms to adapt to environmental change. Biodiversity can also be defined in terms of species richness in a particular community. At a larger scale, biodiversity also embraces ecosystem diversity—the complex interactions within and between ecosystems. While we may study these components of biodiversity separately, in life they are intimately interconnected.

The most conspicuous manifestation of biodiversity loss is species extinction. Extinction is a constant theme in the history of life; most of the species that have lived on Earth over the ages are extinct today. Consider, for example, the anaerobic organisms that were lost as early photosynthetic prokaryotes and eukaryotes added oxygen to Earth's atmosphere, as described in Section 25.2. Extinctions have occurred throughout Earth's history at what is referred to as a "background" rate, as changes in environmental conditions favor some species and negatively affect others. But the rate of extinctions taking place today rivals those of the five great *mass extinctions* of life's history (see Table 25.1 and Figure 25.12). Those historical extinction episodes were the result of cataclysmic natural disturbances, whereas the majority of modern extinctions can be attributed to effects of the human population.

Humans have a tremendous capacity to alter ecosystems and, accordingly, to cause extinctions. When humans first arrived in North America from Siberia about 14,000 years ago, they encountered a diverse and spectacular fauna of large mammals, including saber-toothed cats, dire wolves, mammoths,

mastodons, giant ground sloths, and giant beavers (**Figure 59.1**). Most of this megafauna went extinct within a few thousand years after humans arrived. Although several hypotheses have been advanced to account for the geologically rapid, simultaneous disappearance of so many large animals, overhunting by humans is widely embraced as the principal cause. Losses of megafauna coinciding with the arrival of humans have also been documented in Australia and Hawaii.

As Chapter 25 described, several mass extinctions have occurred on Earth, each providing ecological opportunities for other groups, which subsequently underwent adaptive radiations. These extinction events occurred relatively far apart over the course of the history of life (see Figure 25.12). Over the past 400 years, increasing industrialization and urbanization have accelerated the rate of species extinctions astronomically. The renowned evolutionary biologist Edward O. Wilson estimates that Earth is losing on the order of 30,000 species per year, putting us in the midst of a sixth mass extinction. Protecting Earth's biodiversity requires maintaining the processes that generate new species as well as providing conditions that bring extinction rates closer to background levels.

Biodiversity has great value to human society

Conservation biologists are concerned about the escalating loss of Earth's biodiversity for many reasons:

- *Losing species can threaten the functioning of ecosystems.* The preceding chapters have all described the complex interactions among species. When species are lost, entire communities and ecosystems may change or be lost completely. When that happens, humans may lose the goods and services those ecosystems provide (see Section 58.4).

- *Humans depend on thousands of other species for food, fiber, and medicine.* Over 2,000 species of plants are used for fiber worldwide, and countless more have been domesticated as a source of food. In India alone, over 7,000 species of plants are utilized in traditional medicine, and in the United States more than a quarter of all medical prescriptions contain or are based on plant products. Hundreds of animal species also supply us with food, clothing, and medicine. In addition to goods provided, many organisms provide important services; a wide variety of microbes, for example, provide the fermentation services that render many foods more nutritious and palatable (see Section 51.2).

- *Humans derive enormous psychological benefits, including aesthetic pleasure, from interacting with other organisms.* Aesthetic benefits actually confer economic value on biodiversity. Trees growing on a residential lot, for example, can increase the lot's property value to a point where this added value may be greater than the lumber from the trees could provide.

59.1 Extinct North American Megafauna   The extinction of some Pleistocene megafauna in North America may have been partially driven by the arrival of *Homo sapiens*. This reconstruction of California's La Brea Tar Pits shows a dire wolfe (*Canis dirus*, left) and a sabertooth tiger (*Smilodon* sp).

- *Extinctions deprive the scientific community of opportunities to study and understand ecological relationships among organisms.* The more species that are lost, the more difficult it will be to understand the structure and functioning of ecological communities and ecosystems.

- *Living in ways that cause the extinction of other species raises ethical issues.* Losses of biodiversity increasingly concern philosophers, ethicists, and religious leaders because in such circles species are judged to have intrinsic value.

All of these concerns, to varying degrees, may be integrated by conservation biologists into strategies for protecting biodiversity.

59.1 RECAP

Conservation biology is an applied scientific discipline aimed at protecting and managing biodiversity, which is rapidly decreasing due to extinctions caused by human activities.

- Explain the distinctions among the multiple meanings of the term "biodiversity." See pp. 1243–1244

- What are some of the ways in which biodiversity is valuable to humans? See pp. 1244–1245

To protect biodiversity, conservation biologists must understand biodiversity as it exists today as well as the ways in which it is changing. Thus one important enterprise for conservation biologists is to predict rates of species extinctions.

59.2 How Do Biologists Predict Changes in Biodiversity?

How many, and which, species will go extinct will depend both on human activities and on natural events. Conservation biologists attempt to track the extinctions that are occurring and to predict the number that will occur during the coming century.

Our knowledge of biodiversity is incomplete

Predicting extinctions is difficult for a number of reasons. First, we do not know how many species are living on Earth today. Many of the species that may go extinct in the near future have not even been named and described. Insects provide a case in point: although over 900,000 species have been described, estimates of the number of species yet to be discovered range from about 2 million to more than 50 million (see Section 32.4). Even in the case of larger organisms, our understanding of biodiversity is far from complete. For example, in an 18-month period in 2005–2006, over 50 species of animals and plants previously unknown to science were discovered in rainforests in Borneo.

Second, we do not know where species live. The ranges of most described species, particularly those that are small, reclusive, and rare to start with, are poorly known. One tiny North American true bug, *Corixidea major* (so rare that it has no common name), had been found in only one location near Clarksville, Tennessee, until entomologists collecting insects attracted to lights at night discovered it in Virginia and Florida, extending its known range by over 1,000 km.

Third, it is difficult to determine whether a species is truly extinct. Rarely is the death of the last surviving member of a species recorded with certainty, as it was in the case of the last passenger pigeon (*Ectopistes migratorius*), a female named Martha who died in the Cincinnati Zoo on September 1, 1914. The status of rare, reclusive species with poorly known life histories is much more difficult to determine, as has been the case with the ivory-billed woodpecker (*Campephilus principalis*) in the United States (**Figure 57.2A**). Pygmy tarsiers (*Tarsius pumilus*, tiny primates weighing less than 60 grams) were thought to have gone extinct from their native cloud forests on the island of Sulewesi in Indonesia. In 2008—85 years after the last live specimen had been sighted—a research team from Texas A&M University discovered individuals of this species living in one of the island's national parks (**Figure 57.2B**).

Fourth, we rarely know all of the connections among species. At the beginning of this chapter, we saw that the extinction of a hawk moth species put at risk the survival of a plant species dependent upon the hawk moth as a pollinator. How many other species have been placed at risk by the extinction of the hawk moth is unknown because the ecological associations of this species were not thoroughly documented before its extinction.

(A) *Campephilus principalis*

(B) *Tarsius pumilus*

59.2 Is It Really Extinct? (A) The ivory-billed woodpecker (shown here in a nineteenth-century John James Audubon print) was presumed to be extinct until reports of sightings in 2004. Clear proof of the living bird has so far eluded watchers. (B) Among vertebrates, pygmy tarsiers were recently rediscovered in Indonesia after having been declared extinct.

We can predict the effects of human activities on biodiversity

Despite these gaps in our understanding of biodiversity, methods exist for estimating probable rates of extinction resulting from human activities. To make such predictions, conservation biologists have applied the principles of the species–area relationship and the theory of island biogeography, which we discussed in Section 57.3. When they have measured the rate at which species richness decreases with decreasing habitat patch size, they have found that, on average, a 90 percent loss of habitat area results in the loss of half of the species that live in and depend on that habitat.

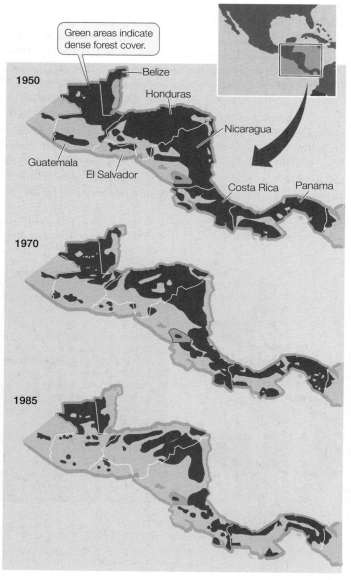

59.3 Deforestation Rates in Tropical Forests   Less than half the tropical forest that existed in Central America in 1950 remained as of 1985, most of it in small patches. Although the rate of deforestation slowed somewhat after 1985, by 2005 an additional 19 percent had disappeared.

Similar calculations can be made for the total global area of a habitat type. The current rate of loss of tropical evergreen forest—Earth's most species-rich biome—is about 2 percent of the remaining forest each year, due to the increasing demand for forest resources by a rapidly expanding human population (**Figure 59.3**). If this rate of loss continues, at least 1 million species that live in tropical evergreen forests could become extinct this century.

To estimate the risk that a particular population will become extinct, conservation biologists develop statistical models that incorporate information about a population's size, its genetic variation, and the morphology, physiology, and behavior of its members. Species in imminent danger of extinction in all or most of their range are labeled *endangered*. *Threatened* species are those that are likely to become endangered in the near future.

Species with small populations face particular risks. Rarity itself is not always a cause for concern, because some species that live in unusual habitats have probably always been rare and are well adapted to being rare. However, species whose populations suddenly shrink rapidly—the "newly rare"—are usually at high risk, as such reductions in population size can lead to genetic drift and loss of genetic variation. Species with specialized habitat or dietary requirements are more likely to become extinct than species with more generalized requirements.

In addition, populations reduced to a small size or confined to a small range can easily be eliminated by local disturbances. For example, numbers of the Cozumel thrasher (*Toxostoma guttatum*), a member of the mockingbird family known only on the island of Cozumel off the coast of Mexico, had been declining for years due to a combination of factors (including the introduction of bird-eating boa constrictors to the island in 1971). A series of hurricanes, beginning with Hurricane Gilbert in 1988, had a catastrophic effect on the remaining populations. Surveys in 2006 failed to document any surviving individuals, and the species may now be extinct.

59.2 RECAP

Predicting changes in biodiversity is difficult because our knowledge of biodiversity is incomplete. The species–area relationship can be used to predict rates of extinction in areas that are subject to habitat loss.

- What are some of the gaps in our current knowledge of biodiversity? See p. 1245

- Why are species with rapidly shrinking populations especially vulnerable to extinction? See p. 1246

Now that we've examined some of the factors that place species at risk of extinction, let's see how human activities contribute to those risks. We begin with habitat loss, the primary threat to species persistence today.

59.3 What Factors Threaten Species Persistence?

Human activities that threaten the persistence of species include habitat alteration, introduction of exotic species, overexploitation, and climate alteration. Conservation biologists determine how these activities affect species and use that information to devise strategies to protect species that are endangered or threatened.

Species are endangered by the degradation, destruction, and fragmentation of their habitats

Loss of suitable habitat by degradation, fragmentation, or destruction is the most important cause of species endangerment in the United States (**Figure 59.4**) and around the world. Many habitats—particularly freshwater habitats—are degraded by pollution. Toxic substances released by human activities have negative effects on the behavior, reproduction, and development of species, reducing both survivorship and competitive

ability. Habitat loss can also occur through outright removal. Physical destruction of particular habitat types (such as tropical rainforest) eliminates species that cannot survive anywhere else. Habitat loss also affects nearby habitats that are not destroyed. As portions of a habitat are progressively lost to human activities, the remaining habitat becomes increasingly fragmented: habitat patches become smaller and more isolated.

Small habitat patches are qualitatively different from larger patches of the same habitat in ways that affect species persistence. Small patches cannot maintain populations of species that require large areas, and they support only small populations of those species that can survive in small patches. In addition, the fraction of a patch influenced by external factors increases rapidly as patch size decreases (**Figure 59.5**). Close to the edges of a forest patch, for example, winds are stronger, temperatures are higher, humidity is lower, and light levels are higher than they are farther inside the forest. Species from surrounding habitats often colonize edges to compete with or prey upon the species living there. The effects of such factors are known as **edge effects**.

One effect of forest fragmentation in midwestern North America has been an increase in the abundance of the brown-headed cowbird (*Molothrus ater*). This bird is a brood parasite—that is, it lays its eggs in the nests of other bird species. When a cowbird egg hatches, the hatchling is raised by the host parents, often to the detriment of their own young. Historically, cowbirds followed bison and other grazing mammals, feeding on insects kicked up by the herds; thus their eggs were laid primarily in nests of grassland host species. Forest fragmentation, however, opened up new opportunities for the cowbirds, which can now lay their eggs in the nests of forest birds in forest edges. Fragmented forests, with relatively more edge than intact forests, thus favor the proliferation of cowbirds at the expense of forest species.

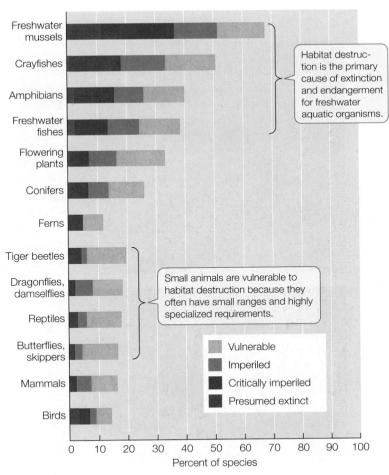

Habitat destruction is the primary cause of extinction and endangerment for freshwater aquatic organisms.

Small animals are vulnerable to habitat destruction because they often have small ranges and highly specialized requirements.

Legend:
- Vulnerable
- Imperiled
- Critically imperiled
- Presumed extinct

59.4 Proportions of U.S. Species Extinct or Threatened The species most threatened with extinction all live in freshwater and wetlands habitats, which are highly vulnerable and have been extensively degraded, polluted, and even destroyed for human development.

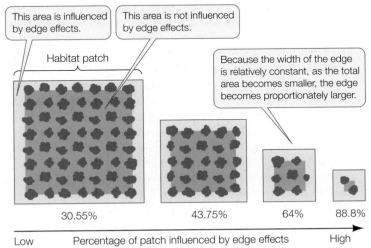

This area is influenced by edge effects.

This area is not influenced by edge effects.

Habitat patch

Because the width of the edge is relatively constant, as the total area becomes smaller, the edge becomes proportionately larger.

30.55%   43.75%   64%   88.8%

Low   Percentage of patch influenced by edge effects   High

59.5 Edge Effects The smaller a patch of habitat, the greater the proportion of that patch that is influenced by conditions in the surrounding environment.

yourBioPortal.com
GO TO **Animated Tutorial 59.1 • Edge Effects**

> Isolated patches lost species much more quickly...

> ...than patches connected to unfragmented forest.

> Even large patches lost some species of animals.

59.6 Species Losses in Brazilian Forest Fragments Biologists studied patches of tropical evergreen forest near Manaus, Brazil, before and after they were isolated by forest clearing. The results of the study demonstrated that small, isolated habitat fragments support fewer species than larger patches of the same habitat.

Because so many habitats have already undergone fragmentation by the time they are first investigated by ecologists, determining the effects of fragmentation on the original communities can be difficult. Timely surveys can provide some of this information. For example, a major research project was launched in 1979 in tropical evergreen forest near Manaus, Brazil, that was slated for conversion to pasture. The landowners agreed to preserve forest patches of certain sizes and configurations laid out by biologists (**Figure 59.6**). The biologists counted the species in the future "patches" while they were still part of the continuous forest, then monitored the patches after the surrounding forest was cut. Species soon began to disappear from isolated patches. The first species to be eliminated were monkeys that travel over large areas. Army ants and the birds that follow army ant swarms also disappeared quickly.

Species that are lost from small habitat fragments are unlikely to become reestablished there because dispersing individuals are unlikely to find the isolated fragments. As Section 55.4 points out, however, a species may persist in a small patch if it is connected to other patches by habitat corridors through which individuals can disperse. Among the experimental forest fragments in Brazil, those that were completely isolated lost species more rapidly than did those that were connected to unfragmented forest by corridors. Since the experiment began, some of the pastures that surrounded the experimental fragments have been abandoned, and young forests now grow in them. Within 7–9 years of abandonment, army ants and some of the birds that follow them recolonized forest fragments connected to larger forest patches by young forests. Other birds that forage in the forest canopy also reestablished themselves. Young forest is not a suitable permanent habitat for most of these species, but they can disperse through it to find more suitable habitat.

Insight into the importance of corridors has led to new conservation initiatives, among the most notable of which is the Yellowstone to Yukon initiative. A joint Canada–United States not-for-profit organization has as its goal sustainable management of the mountain ecosystem extending from Yellowstone National Park to the Yukon Territory. This stretch of land, the largest intact ecosystem of its kind on the planet, contains high-quality habitat for many of North America's most imperiled animals, including grizzly bears, gray wolves, lynx, and native fish. Managing the entire region not only provides corridors to connect habitats, it also provides room for populations to shift in response to global climate change.

Overexploitation has driven many species to extinction

Although habitat loss presents a greater threat to more species today than does overexploitation, many species are still threatened by overexploitation today. Elephants and rhinoceroses are at risk in much of Africa and Asia because poachers kill them for their tusks and horns, which are used for ornaments and knife handles; in addition, some men believe that powdered rhinoceros horn boosts their sexual potency.

The principal threat to the continued survival of tigers, whose numbers have declined by over 90 percent over the past

(A)

(B)

59.7 Endangered by Exploitation (A) The principal threat to the continued survival of tigers is the use of their body parts in Asian traditional medicine. (B) The endangered Banggai cardinalfish of Indonesia's coral reefs has been overexploited because of its value to the private aquarium trade.

century, is the use of their body parts in traditional medicine—bones to treat rheumatism, eyes to cure epilepsy, and penises to enhance virility (**Figure 59.7A**). In 2009, a bowl of tiger penis soup could be obtained for $300 in Taiwan. There is some hope that the development of drugs for treating erectile dysfunction will reduce the incidence of poaching of these endangered species, but as yet the drugs are not widely available in Asia, where demand for animal aphrodisiacs is greatest.

Massive international trade in exotic pets and aquarium fishes, ornamental plants, and tropical forest hardwoods has also decimated many species. The Banggai cardinalfish *Pterapogon kauderni* (**Figure 59.7B**) is on the brink of extinction entirely due to the pet trade; almost a million of these critically endangered fish are hauled out of the waters near Sulawesi, Indonesia, to satisfy the demand of saltwater aquarium enthusiasts.

Invasive predators, competitors, and pathogens threaten many species

As people travel, they deliberately or inadvertently move species to regions outside their original range. Some of these exotic (nonnative) species become **invasive**—that is, they reproduce rapidly, spread widely, and have negative effects on the native species of the region. As we saw in Section 55.3, species that are introduced into a region where their natural enemies are absent may reach very high population densities. Moreover, the native species in an invader's new range may not have evolved specific defenses against these new antagonists and competitors.

Invasive species are spread in a number of ways. Marine organisms have been spread throughout the oceans by ballast water from ships, taken on at the port of departure and discharged at the destination port along with its content of surviving animals and plants. The notorious zebra mussel (*Dreissena polymorpha*) is thought to have arrived in North America in this way (see Figure 55.9). The brown tree snake (*Boiga irregularis*; **Figure 59.8**) arrived on Guam in air cargo shortly after World War II. Until then, the only snake on Guam was a tiny, insect-eating species. Although brown tree snakes remained rare for 20 years, in the 1960s they began to multiply and today can be found at densities as high as 5,000 individuals per km^2. The snake has exterminated 15 species of land birds, including 3 found only on Guam.

Over the past 400 years, Europeans colonizing new continents deliberately introduced plants and animals to their new homes in an effort to reconstruct their familiar surroundings. Many of these introductions have had disastrous effects on native flora and fauna. In Australia, the introduction of European rabbits and foxes for sport hunting and of dogs and cats as pets led to the extermination of nearly half the small to medium-sized native marsupials over the last 100 years, by a combination of competition with rabbits and predation by foxes, dogs, and cats. Some species deliberately introduced to control other invasive species have themselves become invasive and caused further problems. One example was the case of *Manduca sexta* and its wasp parasite in Hawaii, mentioned at the opening of this chapter; another such case is that of the cane toad (*Bufo marinus*), introduced in Australia to control sugarcane pests (see Figure 55.14).

59.8 An Agent of Extinction Since it was accidentally introduced onto the tiny Pacific island of Guam, the brown tree snake (*Boiga irregularis*) has eaten 15 species of land birds to extinction.

Invasive plants may also have negative effects on ecosystems. While native plants must devote considerable energy and resources to defending themselves against native herbivores, invasive plants are less prone to attack, in part because their natural enemies have been left behind in their home range. Therefore, invasive plants can devote their resources to growth and reproduction rather than to producing defensive secondary compounds.

Introduced pathogens have also wreaked havoc among native species, as exemplified by the effects of avian malaria in the Hawaiian Islands. Before the arrival of Europeans, no mosquitoes existed anywhere in the islands. The first mosquito species was found there in 1827, and over the next century several others followed. At the start of the twentieth century, the microbial pathogen that causes avian malaria arrived, most likely carried by imported caged birds. Having had no exposure to malaria over the course of their evolutionary history, Hawaii's many endemic bird species were exceptionally vulnerable to infection. Today, nearly all species living below 1,500 m elevation—the current upper limit of the range of the mosquito vectors—have been eliminated by avian malaria. Species living at higher elevations have fared better, but the range of the mosquitoes appears to be expanding upward as the climate warms, placing the surviving endemic species at risk.

Rapid climate change can cause species extinctions

Human emissions of greenhouse gases are causing global climate warming, as we saw in Section 58.3. Average temperatures in North America, for example, are expected to increase 2°C–5°C by the end of the twenty-first century. If the climate warms to that extent, then the average temperature found at any given location in North America today will be found 500–800 km to the north, and species adapted to particular temperatures will have to shift their ranges by that distance, or evolve new temperature adaptations within a single century. Some habitats, such as alpine tundra, could disappear entirely as temperate forests expand up mountain slopes.

Conservation biologists cannot alter rates of global warming, but they can predict how the resulting climate changes may affect organisms and look for ways to mitigate those effects. Their research activities include analyses of past climate changes and studies of sites currently undergoing rapid climate change. It would be helpful to know, for example, how rapidly species' ranges shifted after the end of the most recent ice age, about 10,000 years ago. Which species did and did not keep pace with climate change? How much, and in what ways, did past ecological communities differ from those of today as a result of differences in the rates at which species ranges shifted?

Species that can disperse easily, such as most birds, may be able to shift their ranges as rapidly as the climate changes, provided that appropriate habitats can be found. However, the ranges of other species are likely to shift more slowly. After the glaciers retreated in North America about 8,000 years ago, for example, the ranges of some coniferous trees such as pine trees, with lightweight seeds that can be carried great distances by wind, expanded northward, so that today they grow as far north as the current climate permits (**Figure 59.9**). Earthworms, on the other hand, fared less well: the glaciers eliminated all earthworm species in Canada and the northern United States, and they have not been replaced by other North American species, which have moved their ranges northward only very slowly. (The earthworms found in many parts of North America today are actually exotic species accidentally introduced from Europe.)

If Earth's surface warms as predicted, entirely new climates will develop, especially at low elevations in the tropics, where a warming of even 2°C would result in conditions warmer than those found anywhere in the humid tropics today. Adaptation to those climates may prove difficult even for many tropical organisms. Since the mid-1980s, the average minimum nightly temperature at the La Selva Biological Station, in the Caribbean lowlands of Costa Rica, has increased from about 20°C to 22°C. On warmer nights, trees use more of their energy reserves just to maintain themselves. As a result, even this small rise in temperature has reduced the average growth rates of six different tree species by about 20 percent.

59.3 RECAP

A number of human activities threaten the persistence of species, including habitat degradation, fragmentation, and destruction; overexploitation; introductions of invasive species; and activities that cause rapid climate change.

- Describe three ways in which habitat loss is occurring today. See pp. 1247–1248

- Why are extinction rates high in small habitat patches? See pp. 1247–1248 and Figures 59.5 and 59.6

- How can dispersal ability and climate change interact to affect the probability of extinction? See pp. 1249–1250

Demonstrating that species are endangered is an empty exercise if we cannot implement a plan of action to save them. In the next section we consider some of the positive steps that can be taken to protect biodiversity.

59.9 Some Species Have Expanded Their Range   Stands of lodgepole pines expanded their range northward as the continental glaciers that covered North America during the Pleistocene retreated.

The numbers indicate the time (thousands of years ago) when lodgepole pine entered the area.

0.4
1.1
2.5
5.6
5.0
8.0
8.0
10.7
11.2
12.2

Never glaciated

Never glaciated

Canada

U.S.A.

Maximum extent of glacier

Current range of inland lodgepole pine

● Fossil sample collection site

59.4 What Strategies Do Biologists Use to Protect Biodiversity?

Conservation biologists use scientific theory, empirical data, and tools from a variety of disciplines to help protect endangered and threatened species and ecosystems. They identify the factors that present risks to species and use that information to devise action plans. Implementing those plans, however, often requires the cooperation of many different groups of people, so conservation biologists also work with landowners, politicians, lawyers, environmental activists, and the general public. Let's look at some of the actions that conservation biologists take to protect biodiversity.

Protected areas preserve habitat and prevent overexploitation

The establishment of *protected areas* in which habitat alteration and exploitation are restricted or prohibited is an important component of efforts to conserve biological diversity. Protected areas preserve habitat and may serve as nurseries from which individuals can disperse into exploited areas, replenishing populations that might otherwise become extinct.

Deciding which areas to protect is a challenging enterprise. Two robust criteria are the total number of species living in an area (its *species richness*; see Chapter 57) and the number of *endemic species* in an area (a measure of its uniqueness). Using these criteria, biologists have identified a number of biodiversity "hotspots" of unusual richness and endemism (**Figure 59.10**). Hotspots occupy only 15.7 percent of Earth's land surface but are home to 77 percent of its terrestrial vertebrate species. Most hotspots are also areas of high human population density, so habitat loss is ongoing. Developing a conservation strategy for any of these regions requires not only a detailed analysis of the distributions of species and the locations of special resources (such as caves, freshwater springs, or migratory

(A) Tropical rainforest hotspots

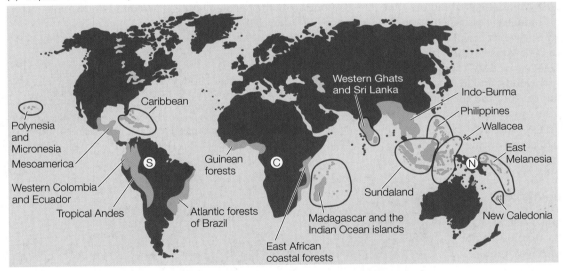

(B) Hotspots in other biomes

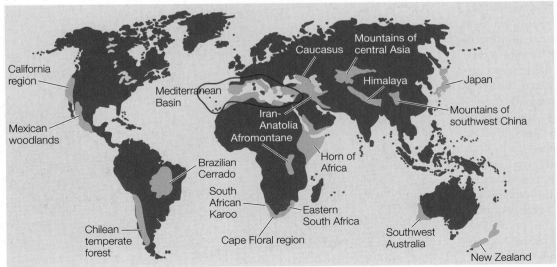

59.10 Hotspots of Biodiversity
(A) Almost half of the world's highest priority "hotspots" for biodiversity protection are regions of tropical rainforest habitat. Areas circled in red encompass island groups. The circled letters represent the only three remaining areas of extensive unbroken rainforest: S, South American Amazonia; C, the Congo Basin of Africa; and N, the island of New Guinea. (B) Eighteen hotspots representing other (non-rainforest) ecosystems.

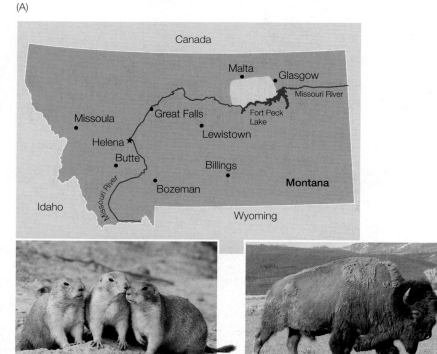

"Centers of imminent extinction" are concentrated in tropical forests, islands, and mountainous regions.

59.11 Centers of Imminent Extinction   The areas shown in yellow include 595 "centers of imminent extinction." These regions are home to 794 species that are at serious risk of extinction.

stopover areas for birds), but also an analysis of factors that threaten and factors that support biodiversity in the region.

In an effort to further pinpoint sites with threatened species that are found nowhere else, in 2005 conservation biologists identified 595 "centers of imminent extinction." These sites are concentrated in tropical forests, on islands, and in mountainous regions (**Figure 59.11**). Only one-third of the sites are legally protected, and most of them are surrounded by land that is undergoing rapid development. Unless protective actions are taken soon, species extinctions are inevitable.

Degraded ecosystems can be restored

When a species is endangered as a consequence of ecosystem degradation rather than outright habitat loss, protecting the species may require restoring the habitat to a more natural state. Practitioners of **restoration ecology** are developing methods aimed at just such reconstitution, because many degraded ecosystems recover only slowly, if at all, without human assistance.

Because the soil that supports them is so rich, grasslands all over the world have been converted to agriculture. By the middle of the twentieth century, for example, most North American prairies had been converted to cropland or were heavily grazed by domestic livestock. The herds of large mammals that roamed the prairies when European settlers arrived have been reduced to tiny remnants confined to small areas. Most of these remaining populations are too small to maintain their genetic diversity or play their original ecological roles. The species *have* survived, however, so opportunities exist to reintroduce them if their habitat can be restored.

A major prairie restoration project is under way in northeastern Montana. When Lewis and Clark mapped this region 200 years ago, they saw large herds of bison, elk, deer, and pronghorn, as well as abundant populations of their predators. The goal of the restoration project, run by the World Wildlife Fund and the American Prairie Foundation in cooperation with public land managers, is to restore the native prairie and its fauna in a 15,000-km^2 area near the Missouri River (**Figure 59.12**).

This ambitious project is feasible for three reasons. First, the private land in the area is owned by a small number of ranchers, each of whom owns extensive grazing leases on public lands administered by either U.S. federal agencies or the State of Montana. Second, most of the land has never been plowed, so native vegetation can recover rapidly when grazing pressures are reduced. Third, the area's human population is decreasing. The ranchers are aging, and their children are abandoning the hard work and uncertain profits of ranching for careers in urban settings. Once a free-ranging herd of several thousand bison and large numbers of elk—along with their predators (wolves)—has been established, nature-minded tourists are expected to flock to the area to view the wildlife spectacle. Over the long term, the restored ecosystem should deliver major economic benefits to the region.

59.12 Restoring a North American Prairie   (A) A major prairie restoration project is underway north of the Missouri River in the state of Montana (yellow area). (B) Native prairie dogs maintain the vegetation by digging extensive burrows and clipping plants. (C) The first bison were reintroduced to the area in 2005.

(A)

Canada

Malta
Glasgow
Missouri River
Great Falls
Fort Peck
Lake
Missoula
Lewistown
Helena
Butte
Billings
Bozeman
Montana
Idaho
Wyoming

(B) *Cynomys ludovicianus*

(C) *Bison bison*

59.13 A Wetlands Laboratory Tijuana Estuary, near San Diego, is a shallow-water wetlands habitat. Experiments at this natural research reserve shed light on ways to restore this valuable ecosystem.

In the United States, the notion that humans are capable of creating functioning replacement ecosystems underlies policies that allow developers to destroy habitats. Destruction of wetlands in particular is often permitted because of developers' assertions that they can be replaced. However, creating new wet-lands requires detailed ecological knowledge that almost invariably far surpasses that which is currently available.

In southern California, where 90 percent of the coastal wetlands have been destroyed, wetland restoration is a high priority. Species have been lost from degraded coastal wetlands, so restoration requires species introductions. Early attempts, in which one or two common wetland species were introduced, did not succeed—other wetland-associated species failed to re-colonize the "rehabilitated" wetlands. To understand why, conservation biologists established a large field experiment at the Tijuana Estuary (**Figure 59.13**). Here they found that experimental plots planted with species-rich mixtures developed a more complex vegetation structure (which is important to insects and birds) and also accumulated nitrogen (required for plant growth) faster than did species-poor plots (**Figure 59.14**).

Disturbance patterns sometimes need to be restored

Many species depend on particular patterns of disturbance—such as fires, windstorms, or grazing—to maintain their populations (see Section 57.4). Recognition of the need for periodic disturbance to maintain healthy ecosystems is a rather new dimension of conservation biology. For example, although many

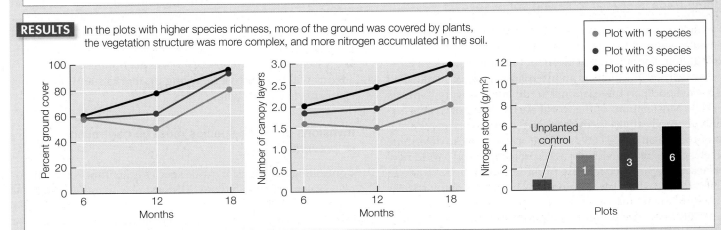

INVESTIGATING LIFE

59.14 Species Richness Enhances Wetland Restoration
When biologists attempted to restore degraded areas of the Tijuana Estuary, they found that several measures of wetland ecosystem health improved more rapidly in species-rich than in species-poor plantings.

HYPOTHESIS When attempting to restore degraded wetlands, planting a mixture of species will approach the ecosystem's original condition more rapidly than planting a single species.

METHOD
1. Delimit experimental plots of degraded wetland habitat, leaving control plots unplanted (i.e., no human intervention).
2. Plant some experimental plots with a single plant species typical of wetlands in the region. Plant other plots with randomly chosen assemblages of 3 and of 6 typical species. Plant the same density of seedlings in all plots.
3. Over the subsequent 18 months, replant and weed the experimental tracts as necessary to compensate for early mortality of seedlings.

RESULTS In the plots with higher species richness, more of the ground was covered by plants, the vegetation structure was more complex, and more nitrogen accumulated in the soil.

- Plot with 1 species
- Plot with 3 species
- Plot with 6 species

CONCLUSION Planting a rich mixture of species should enhance wetland restoration attempts.

Go to **yourBioPortal.com** for original citations, discussions, and relevant links for all INVESTIGATING LIFE figures.

59.15 Mimicking Natural Disturbance Patterns As revealed by scars (arrows) in the growth rings of this Ponderosa pine, low-intensity non-lethal ground fires were frequent in the pine forests of the southwestern United States prior to fire suppression.

plant species require periodic fires for successful establishment and survival, for many years the official policy of the U.S. Forest Service, symbolized by the iconic mascot Smokey Bear, was to suppress all forest fires. Today, however, controlled burning is common, particularly in western North America. In order to use fire as an ecosystem management tool, it is important to know the historical pattern of fires in an area, which can be determined in part by studies of annual growth rings of trees (**Figure 59.15**). A schedule of controlled burning that recreates the historical pattern can reduce forest floor litter, avoiding a buildup of fuel that can lead to intense, tree-killing canopy fires.

Ending trade is crucial to saving some species

Most endangered species cannot survive any further reductions in their breeding populations, so it is important to prevent their exploitation. The legal mechanism for prohibiting trade in these species or their products is an international agreement called the Convention on International Trade in Endangered Species (CITES). Most nations of the world are members of CITES. National representatives meet every two years to review the status of species on the CITES protected species lists, to determine which species may no longer need protection, and to add new species. CITES rules currently prohibit international trade in items such as whale meat, rhinoceros horn, and many species of parrots and orchids.

CITES instituted a ban on international trade in elephant ivory in 1989, but demand for ivory remains strong, especially in Japan and China. As a result, poaching of elephants continues in the forests of central Africa and in East Africa, where the animals are threatened. Southern African countries

(Botswana, Namibia, South Africa, Zambia, and Zimbabwe), however, have so many elephants that government officials must kill many of them to control populations in the limited areas where they are allowed to roam to prevent them from dispersing to populated areas and damaging crops. These countries would like to sell the ivory from the culled elephants to fund their conservation efforts. Other countries are worried that if trade restrictions on ivory are relaxed, poaching will escalate over the entire continent.

Control and regulation of ivory trade might be possible if scientists could determine where the ivory comes from. Samuel Wasser and his colleagues at the University of Washington have identified 16 DNA markers that can be extracted from elephant feces. Park rangers in Malawi and Zambia were able to sample the elephant populations in their countries by collecting fresh scat while they were on routine patrols. The source of an elephant tusk can then be determined by matching DNA extracted from the ivory with the geographically based frequencies of the 16 DNA markers found in the dung samples. In June 2002, 6.5 tons of illegal ivory were seized in Singapore. DNA analyses showed that the elephants were killed in Zambia.

Such safeguards were partially responsible for the controversial decision to sanction legal sales of ivory from Namibia, Botswana, Zimbabwe, and South Africa beginning in 2008, the first such sale in close to a decade. Over 100 tons of elephant tusks, the equivalent of over 20,000 dead elephants, were auctioned off over a 2-week period to authorized buyers from China and Japan, for use primarily in folk medicine. The sale of this ivory, all of which was obtained legally from animals that had died of natural causes, generated over $15 million for elephant conservation efforts. Although the sales were monitored by CITES, and although the last sanctioned sale, in 1999, did not appear to increase poaching activities, other African nations and many wildlife conservation groups are concerned that in the absence of oversight and enforcement, the flood of legal ivory will be intermingled with poached ivory in international markets and encourage more illegal elephant slaughter.

One promising development in curbing illegal sales was the decision by eBay, the international online marketplace, to ban sales of ivory on its platform effective January 2009. An investigation by the International Fund for Animal Welfare revealed that two-thirds of online sales of protected wildlife products take place on eBay, so conservationists are optimistic that eBay's actions will be effective in drying up markets and protecting elephants from poaching.

Invasions of exotic species must be controlled or prevented

Controlling invasions of exotic species is an important component of conservation biology. The best way to reduce the damage caused by invasive species, of course, is to prevent their introduction in the first place. Given the tremendous volume of global trade, it might seem impossible to curtail their spread. Some promising strategies, however, do exist. For example, transoceanic transport of invasive species in ballast water could be largely eliminated by the simple procedure of deoxygenating ballast water before it is pumped out. This practice not only kills most organisms in the water, but also extends the life of ballast tanks, providing an economic benefit to shippers. Although the U.S. House of Representatives passed (by a vote of 395 to 7) a Coast Guard funding bill with language stipulating that freighters operating in the Great Lakes must disinfect their ballast water, the Senate vote has been impeded by legal wrangling. In the absence of federal regulation, some states have already implemented their own ballast water treatment regulations.

Regulating the importation and sale of exotic species can reduce deliberate introductions. In 2002, some members of the American horticulture industry crafted a voluntary code of conduct for their profession, stating that the invasive potential of a plant should be assessed prior to its introduction and marketing. Horticulturists will work with conservation biologists to determine which species are currently invasive, or likely to become so, and to identify suitable alternative species. Stocks of invasive species will be phased out, and gardeners will be encouraged to use noninvasive plants.

Conservation biologists have developed a decision tree, based on the traits that characterize plant species that have become invasive, to help horticulturists and regulators determine whether an exotic plant species should be allowed into North America (**Figure 59.16**). Although the protocols stipulated by this decision tree cannot eliminate all introductions of invasive species, if they are followed conscientiously, they can greatly reduce the risk of such introductions.

Biodiversity can have market value

Much of the value of ecosystems to human society depends on their biodiversity, but it has been difficult to assess the value of biodiversity in monetary terms. When biodiversity is perceived to have economic value, industries and government agencies have a greater incentive to protect it. The interdisciplinary field of *ecological economics* has provided tools for assessing the economic value of biodiversity.

Many studies have demonstrated the market value of protecting biodiversity. Such cases show that the argument for biodiversity conservation is compelling not only from an ecological or ethical perspective, but also from an economic perspective. Let's take a closer look at two of them.

WILD DOGS AND ECOTOURISM *Ecotourism*—environmentally responsible travel to natural areas that contributes to conserving their biodiversity and to the economic well-being of the local communities—is a major source

of income for many developing nations. For example, tourists visiting Africa often express interest in seeing wild dogs (*Lycaon pictus*; **Figure 59.17**). However, diseases such as rabies and canine distemper, along with habitat loss, road traffic, deliberate extermination due to a perceived threat to livestock, and many other factors, have decimated wild dog populations throughout the continent, making them the second most endangered carnivore in Africa. Once found in 39 countries, they are today found in only 14 countries. Their endangered status has piqued interest among tourists in catching glimpses of these charismatic animals. South Africa is home to about 400 of Africa's remaining 5,000 dogs, most of which live in Kruger National Park. A survey of visitors to South Africa revealed that nearly three-fourths of them would be willing to pay an extra U.S. $12 for the opportunity to see wild dogs. Conservation biologists are

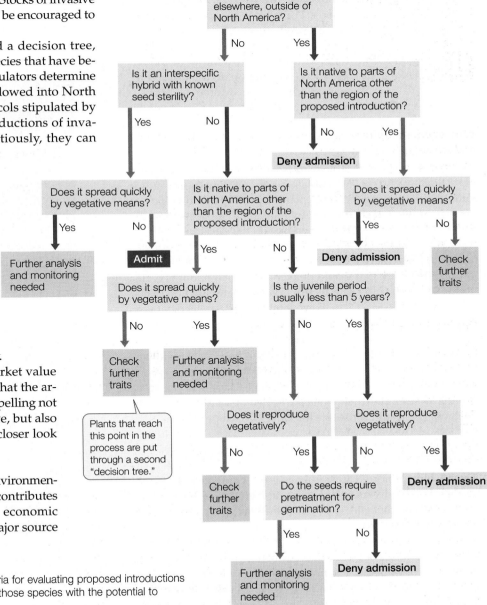

59.16 A "Decision Tree" This diagram sets criteria for evaluating proposed introductions of exotic plant species and helps regulators identify those species with the potential to become invasive.

working with lodge owners and ranchers elsewhere in South Africa and in Kenya to encourage them to reestablish wild dogs in areas from which they have disappeared.

THE FYNBOS Studies by a group of economists, ecologists, and land managers have attempted to calculate the value of the economic benefits provided by the spectacularly species-rich shrubland community called *fynbos* in the Western Cape Province of South Africa (**Figure 59.18A**). Of the 8,500 plant species in this region, over two-thirds are endemics. Fynbos plants thrive despite the summer droughts, nutrient-poor soils, and periodic fires of the region.

The highlands where fynbos vegetation thrives provide about two-thirds of the Western Cape's water supply. In addition, some of the endemic plants, including proteas, are harvested for cut and dried flowers. An international market has developed for rooibos, a fynbos shrub used for herbal tea. Income also comes from hundreds of thousands of ecotourists who visit the region every year to see the fynbos. The fynbos also provides recreational opportunities for local residents in urban areas nearby.

Recently, several trees and shrubs from other continents have invaded the fynbos. Taller and faster-growing than the endemics, they displace the native vegetation, in the process increasing the intensity and severity of fires. Moreover, because the invaders transpire more water than the endemics, they decrease stream flows to less than half those from areas covered with native plants, reducing the water supply for people in the region (**Figure 59.18B**).

We can get an idea of the market value of fynbos biodiversity by estimating the cost of maintaining or replacing the services it provides. Removing the exotic plants by felling and digging out invasive trees and shrubs and by managing fire costs between $140 and $830 per hectare, depending on the densities of the invaders. Alternatively, the services provided by fynbos vegetation could be replaced, but only at a much higher cost. A sewage purification plant that would deliver the same amount of water to the Western Cape as a well-managed watershed of 10,000 hectares would cost $135 million to build and $2.6 million per year to operate. Desalination of seawater would cost four times as much. Thus the available alternatives would deliver water at a cost between 1.8 and 6.7 times higher than the cost of maintaining natural vegetation in the watershed. Maintaining the fynbos vegetation would be less expensive and more labor-intensive (generating more employment) than the technologically sophisticated methods that could substitute for the services it provides.

Simple changes can help protect biodiversity

Establishing protected areas is an essential component of efforts to maintain biodiversity, but this action alone is insufficient to stem global biodiversity loss. The extensive landscapes in which people live and extract resources must also play important roles in biodiversity conservation. The good news is that, carefully used, these lands can contribute much more to conservation than they currently do. The practice of using ecosystems for residences, resources, or recreation in ways that sustain their biodiversity is known as **reconciliation ecology**.

Lycaon pictus

59.17 A Sight for Travelers' Eyes Tourists visiting Africa to experience its wildlife often express a desire to see the rare African wild dogs. Protecting such species can be in the economic interest of the region.

(A)

(B)

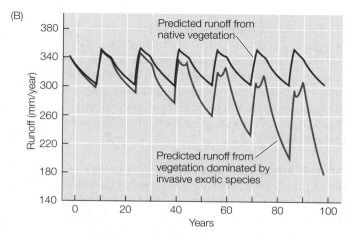

59.18 Biodiversity Maintains Ecosystem Functioning (A) The unique fynbos ecosystem of the Western Cape Province of South Africa provides much of the area's water. (B) A computer simulation of stream flows from fynbos watersheds that have and have not been invaded by exotic trees and shrubs.

Reconciliation ecology is based on the principle that most ecosystem services are provided locally, and that people are more motivated to work to protect their local interests than they are to work on national or global issues. The National Wildlife Federation has established a very successful program in which people petition to have their backyards certified as wildlife-friendly. Criteria for certification include planting shrubs that provide food for birds and refraining from applying pesticides on lawns. The city of Tucson, Arizona, has launched a major project to make the city a suitable habitat for many species of birds, not just the typical urban birds that live in most cities in North America.

Even some industrial sites can support biodiversity. The Turkey Point power plant in southern Florida uses large amounts of water to cool its generating units. To cool the heated water before discharging it, the Florida Power & Light Company dug a system of 38 canals that covers 6,000 acres. These cooling canals are separated by low-lying berms that support a variety of native and exotic plants. Red mangroves grow along the edges of the canals. Today, they support a thriving population of American crocodiles, a highly endangered species. Crocodiles living in the canals yield about 10 percent of all young crocodiles in the United States. Having discovered the biodiversity value of its cooling system, the company employs biologists to monitor the crocodiles and works actively to ensure their continued reproductive success.

Captive breeding programs can maintain a few species

A few of the world's many endangered species can be maintained in captivity while the external threats to their existence are reduced or removed. However, captive propagation is only a temporary measure that buys time to deal with those threats. Zoos, aquariums, and botanical gardens do not have enough space to maintain adequate populations of more than a small fraction of Earth's rare and endangered species. Nonetheless, captive propagation can play an important role by maintaining species during critical periods, providing a source of individuals for reintroduction into the wild, and raising public awareness of threatened and endangered species.

The California condor, North America's largest bird, survives today only because of captive propagation (**Figure 59.19**). Two centuries ago, condors ranged from British Columbia to northern Mexico, but by 1978, the wild population was plunging toward extinction. Many of the birds, which are scavengers, died from ingesting animal carcasses containing lead shot. To save the condor from certain extinction, biologists captured all known remaining condors, which numbered only 22 individuals, and initiated a captive breeding program in 1983.

The first captive-bred birds were released in the mountains north of Los Angeles in 1992. Since that time, captive-bred birds have also been released in northern Arizona and Baja California. These birds are provided with lead-free food in remote areas, and they are trained to avoid power lines prior to release, thereby reducing a major source of mortality. Today, the captive-bred birds use the same roosting sites, bathing pools, and mountain ridges that their wild-born predecessors did. In December 2008, there were 327 known condors in the world, half of which were living in the wild. In 2003, a wild-born chick fledged in the wild for the first time in over two decades.

Most of the major threats to condor survival, including power lines, pesticides, and museum collectors, have been mitigated. Lead poisoning is still a problem, but as of July 1, 2008, under the Ridley–Tree Condor Preservation Act, California hunters are required to use non-lead bullets when hunting within the condor's range. Passage of this legislation marks a change in public attitudes from the days when cattle ranchers, in the mistaken belief that the birds killed livestock, vociferously opposed their reintroduction into the wild.

The legacy of Samuel Plimsoll

During the nineteenth century, many British merchant ships sailed the world's oceans. Once these ships left harbor, they

(A)

(B) *Gymnogyps californianus*

59.19 A California Condor Soars (A) California condors raised in captivity are fed by humans wearing hand puppets so they will not imprint on their captors and will be able to survive in the wild. (B) Numbered wing tags allow conservation biologists to identify and track released adult condors. The survival of North America's largest bird species depends on this captive propagation project.

were out of contact with the rest of the world; in the case of a shipwreck, signaling for rescue was impossible. But ship owners could maximize their profits by overloading their ships, even though this practice caused some of them to become unseaworthy and sink.

Samuel Plimsoll, a member of Parliament, became concerned about the rate of loss of British vessels and sailors. He convinced Parliament to pass legislation requiring that a "load line" be painted on the hull of every large oceangoing vessel. The position of the line was calculated using factors such as the structural strength of the vessel and the shape of its hull. If the load line was under water, the ship was not permitted to leave the harbor. The "Plimsoll line," as it came to be known, dramatically reduced the rate of loss of British ships and sailors. In 1930, the international community agreed on universal adoption of load line regulations.

The increasing loss of Earth's species suggests that the load of human activities has pushed the planet below its Plimsoll line. But where and how should society draw that line? Our decision should be based on scientific information, but, just as in Samuel Plimsoll's time, science cannot determine an "acceptable rate of loss." Ethical considerations will figure prominently in the decisions we make about changing the way we use ecosystems so that other species can survive on Earth with us.

59.4 RECAP

To conserve biodiversity, it is necessary to set aside protected areas, restore ecosystems and natural disturbance patterns, restrict trade in endangered species and transport of invasive species, increase populations of endangered species, and otherwise recognize the benefits of maintaining functioning ecosystems.

- What are some of the priorities conservation biologists consider when establishing protected areas? **See pp. 1251–1252 and Figures 59.11–59.13**

- Explain the difference between restoration ecology and reconciliation ecology. **See pp. 1252 and 1257**

- Do you think the analogy of the Plimsoll line is a good one to apply to protecting the world's biodiversity? Why or why not?

CHAPTER SUMMARY

59.1 What Is Conservation Biology?

- **Conservation biology** is an applied scientific discipline devoted to protecting and managing biodiversity.
- Conservation biologists recognize that an understanding of the evolutionary processes that generate biodiversity is essential to protecting it. They also understand that ecosystems are dynamic, and that humans are part of those ecosystems.
- Species extinctions have always occurred, but today they are occurring at an alarming rate.
- There are many compelling reasons for protecting biodiversity, including the maintenance of the ecosystems whose many functions provide humans with goods and services, including air and water purification.

59.2 How Do Biologists Predict Changes in Biodiversity?

- Our understanding of biodiversity is incomplete. We do not know how many species there are, where they live, or which of the species known to science have gone extinct.
- Biologists can use the species–area relationship and the theory of island biogeography to estimate rates of extinction likely to be caused by habitat loss.
- To estimate a species' risk of extinction, statistical models take into account data on population sizes, genetic variation, physiology, morphology, and behavior.
- Rarity is not always a cause for concern, but species whose populations are shrinking rapidly are usually at risk of extinction.

59.3 What Factors Threaten Species Persistence?

- Habitat loss is the most important cause of species endangerment worldwide. As habitats become increasingly fragmented, more species are lost from those habitats. Small habitat patches can support only small populations and are adversely influ-

enced by **edge effects**. Review Figures 59.5 and 59.6, **ANIMATED TUTORIAL 59.1**
- Overexploitation has historically been the most important cause of species extinctions, and it is still a major threat to biodiversity today.
- Some species introduced to regions outside their original range become **invasive**, causing extinctions of native species by competing with them, eating them, or transmitting diseases to them.
- Climate is likely to become an increasingly important cause of extinctions for those species that cannot shift their ranges as rapidly as the climate warms. **Review Figure 59.9**

59.4 What Strategies Do Biologists Use to Protect Biodiversity?

- Establishing protected areas is crucial to conserving biodiversity. Protected areas are selected by taking into account species richness, endemism, imminence of threats, and the need to protect representative ecosystems.
- **Restoration ecology** is an important conservation strategy because many degraded ecosystems will not recover, or will do so only very slowly, without human assistance. **Review Figure 59.15**
- International trade in endangered species is controlled by regulations that most countries endorse.
- Conservation biologists work to determine which species are likely to become invasive and prevent their introduction to new areas.
- Even within landscapes where people live and extract resources, steps may be taken to protect biodiversity. This approach is known as **reconciliation ecology**.
- Captive breeding programs can maintain some endangered species for the short term while threats to their persistence are reduced or removed.

SEE WEB ACTIVITY 59.1 for a concept review of this chapter.

SELF-QUIZ

1. Which of the following is *not* currently a major cause of species extinctions?
   a. Habitat destruction
   b. Rising sea levels
   c. Overexploitation
   d. Introduction of exotic predators
   e. Introduction of exotic pathogens

2. The most important cause of endangerment of species in the United States currently is
   a. climate change.
   b. invasive species.
   c. overexploitation.
   d. habitat loss.
   e. loss of mutualists.

3. Species extinctions matter to human society because
   a. more than a quarter of the medical prescriptions written in the United States contain a plant product.
   b. people derive aesthetic pleasure from interacting with other organisms.
   c. causing species extinctions raises serious ethical issues.
   d. biodiversity helps maintain valuable ecosystem services.
   e. All of the above

4. As a habitat patch gets smaller, it
   a. cannot support populations of species that require large areas.
   b. supports only small populations of many species.
   c. is influenced to an increasing degree by edge effects.
   d. is invaded by species from surrounding habitats.
   e. All of the above

5. A plant species is most likely to become invasive when introduced to a new area if it
   a. grows tall.
   b. has become invasive in other places where it has been introduced.
   c. is closely related to species living in the area where it has been introduced.
   d. has specialized disseminators of its seeds.
   e. has a long life span.

6. Global warming is a concern because
   a. the rate of change in climate is projected to be faster than the rate at which many species can shift their ranges.
   b. it is already too hot in the tropics.
   c. climates have been so stable for thousands of years that many species lack the ability to tolerate variable temperatures.
   d. climate change will be especially harmful to rare species.
   e. None of the above

7. Scientists can determine the historical frequency of fires in an area by
   a. examining charcoal in sites of ancient villages.
   b. measuring carbon in soils.
   c. radioactively dating fallen tree trunks.
   d. examining fire scars in growth rings of living trees.
   e. determining the age structure of forests.

8. Captive propagation is a useful conservation tool, when
   a. there is space in zoos, aquariums, and botanical gardens for breeding a few individuals.
   b. the areas of origin of the captive individuals are known.
   c. the threats that endangered the species are being alleviated so that captive-reared individuals can later be released back into the wild.
   d. there are sufficient caretakers.
   e. None of the above; captive propagation should never be used because it directs attention away from the need to protect the species in their natural habitats.

9. Restoration ecology is an important field because
   a. many areas have been highly degraded.
   b. many areas are vulnerable to global climate change.
   c. many species suffer from demographic stochasticity.
   d. many species are genetically impoverished.
   e. fire is a threat to many areas.

10. The field of reconciliation ecology has developed because
    a. all other methods of preserving biodiversity have failed.
    b. protected areas should be able to maintain biodiversity.
    c. protected areas alone are insufficient to maintain biodiversity.
    d. scientists are unable to control diseases today.
    e. we are not reconciled with other species.

FOR DISCUSSION

1. Most species driven to extinction by humans in the past were large vertebrates. Do you expect this pattern to persist into the future? If not, why not?

2. Conservation biologists have debated extensively which is better: many small protected areas (which may contain more species) or a few large protected areas (which may be the only ones that can support populations of species that require large areas). What ecological processes should be evaluated in making judgments about the sizes and locations of protected areas?

3. During World War I, doctors adopted a "triage" system for dealing with wounded soldiers. The wounded were divided into three categories: those almost certain to die no matter what was done to help them, those likely to recover even if not assisted, and those whose probability of survival was greatly increased if they were given immediate medical attention. Limited medical resources were directed primarily at the third category. What are some implications of adopting a similar attitude toward species preservation?

4. Utilitarian arguments dominate discussions about the importance of preserving biological richness. In your opinion, what role should ethical and moral arguments play?

5. The desert bighorn sheep of the southwestern United States is endangered. Its major predator, the puma, is also threatened in the region. Under what conditions, if any, would it be appropriate to suppress the population of one rare species to assist another rare species?

ADDITIONAL INVESTIGATION

Pollination by forest-dwelling bees was found to increase seed production in coffee plants growing within a kilometer of a forest patch in Costa Rica. What forest organisms are likely to provide other benefits (such as control of insect pests) to nearby agricultural crops? Over what distances are those effects likely to be felt? What experiments could be designed to test the importance of those effects and how they vary with distance from forest patches?

Appendix A: The Tree of Life

Phylogeny is the organizing principle of modern biological taxonomy. A guiding principle of modern phylogeny is *monophyly*. A monophyletic group is considered to be one that contains an ancestral lineage and *all* of its descendants. Any such group can be extracted from a phylogenetic tree with a single cut.

The tree shown here provides a guide to the relationships among the major groups of extant (living) organisms in the tree of life as we have presented them throughout this book. The position of the branching "splits" indicates the relative branching order of the lineages of life, but the time scale is not meant to be uniform. In addition, the groups appearing at the branch tips do not necessarily carry equal phylogenetic "weight." For example, the ginkgo [63] is indeed at the apex of its lineage; this gymnosperm group consists of a single living species. In contrast, a phylogeny of the eudicots [55] could continue on from this point to fill many more trees the size of this one.

The glossary entries that follow are informal descriptions of some major features of the organisms described in Part Seven of this book. Each entry gives the group's common name, followed by the formal scientific name of the group (in parentheses). Numbers in square brackets reference the location of the respective groups on the tree.

It is sometimes convenient to use an informal name to refer to a collection of organisms that are not monophyletic but nonetheless all share (or all lack) some common attribute. We call these "convenience terms"; such groups are indicated in these entries by quotation marks, and we do not give them formal scientific names. Examples include "prokaryotes," "protists," and "algae." Note that these groups cannot be removed with a single cut; they represent a collection of distantly related groups that appear in different parts of the tree. We also use quotation marks here to designate two groups of fungi that are not believed to be monophyletic.

An interactive version of this tree, with links to much greater detail (such as photos, distribution maps, species lists, and identification keys), can be found at yourBioPortal.com.

– A –

acorn worms (*Enteropneusta*) Benthic marine hemichordates [120] with an acorn-shaped proboscis, a short collar (neck), and a long trunk.

"algae" A convenience term encompassing various distantly related groups of aquatic, photosynthetic chromalveolates [5] and certain members of the Plantae [8].

alveolates (*Alveolata*) [7] Unicellular eukaryotes with a layer of flattened vesicles (alveoli) supporting the plasma membrane. Major alveolate groups include the dinoflagellates [52], apicomplexans [53], and ciliates [54].

amborella (*Amborella*) [60] An understory shrub or small tree found in New Caledonia. Thought to be the sister-group of the remaining living angiosperms [14].

ambulacrarians (*Ambulacraria*) [30] The echinoderms [119] and hemichordates [120].

amniotes (*Amniota*) [37] Mammals, reptiles, and their extinct close relatives. Characterized by many adaptations to terrestrial life, including an amniotic egg (with a unique set of membranes—the amnion, chorion, and allantois), a water-repellant epidermis (with epidermal scales, hair, or feathers), and, in males, a penis that allows internal fertilization.

amoebozoans (*Amoebozoa*) [85] A group of eukaryotes [4] that use lobe-shaped pseudopods for locomotion and to engulf food. Major amoebozoan groups include the loboseans, plasmodial slime molds, and cellular slime molds.

amphibians (*Amphibia*) [129] Tetrapods [36] with glandular skin that lacks epidermal scales, feathers, or hair. Many amphibian species undergo a complete metamorphosis from an aquatic larval form to a terrestrial adult form, although direct development is also common. Major amphibian groups include frogs and toads (anurans), salamanders, and caecilians.

amphipods (*Amphipoda*) Small crustaceans [117] that are abundant in many marine and freshwater habitats. They are important herbivores, scavengers, and micropredators, and are an important food source for many aquatic organisms.

angiosperms (*Anthophyta* or *Magnoliophyta*) [14] The flowering plants. Major angiosperm groups include the monocots, eudicots, and magnoliids.

animals (*Animalia* or *Metazoa*) [21] Multicellular heterotrophic eukaryotes. The majority of animals are bilaterians [24]. Other groups of animals include the cnidarians [98], ctenophores [97], placozoans [96], and sponges [22]. The closest living relatives of the animals are the choanoflagellates [92].

annelids (*Annelida*) [105] Segmented worms, including earthworms, leeches, and polychaetes. One of the major groups of lophotrochozoans [26].

anthozoans (*Anthozoa*) One of the major groups of cnidarians [98]. Includes the sea anemones, sea pens, and corals.

anurans (*Anura*) Comprising the frogs and toads, this is the largest group of living amphibians [129]. They are tail-less, with a shortened vertebral column and elongate hind legs modified for jumping. Many species have an aquatic larval form known as a tadpole.

apicomplexans (*Apicomplexa*) [53] Parasitic alveolates [7] characterized by the possession of an apical complex at some stage in the life cycle.

arachnids (*Arachnida*) Chelicerates [115] with a body divided into two parts: a cephalothorax that bears six pairs of appendages (four pairs of which are usually used as legs) and an abdomen that bears the genital opening. Familiar arachnids include spiders, scorpions, mites and ticks, and harvestmen.

arbuscular mycorrhizal fungi (*Glomeromycota*) [88] A group of fungi [19] that associate with plant roots in a close symbiotic relationship.

archaeans (*Archaea*) [3] Unicellular organisms lacking a nucleus and lacking peptidoglycan in the cell wall. Once grouped with the bacteria, archaeans possess distinctive membrane lipids.

archosaurs (*Archosauria*) [39] A group of reptiles [38] that includes dinosaurs and crocodilians [134]. Most dinosaur groups became extinct at the end of the Cretaceous; birds [133] are the only surviving dinosaurs.

arrow worms (*Chaetognatha*) [107] Small planktonic or benthic predatory marine worms with fins and a pair of hooked, prey-grasping spines on each side of the head.

arthropods (*Arthropoda*) The largest group of ecdysozoans [27]. Arthropods are characterized by a stiff exoskeleton, segmented bodies, and jointed appendages. Includes the chelicerates [115], myriapods [116], crustaceans [117], and hexapods (insects and their relatives) [118].

ascidians (*Ascidiacea*) "Sea squirts"; the largest group of urochordates [122]. Also known as tunicates, they are sessile (as adults), marine, saclike filter feeders.

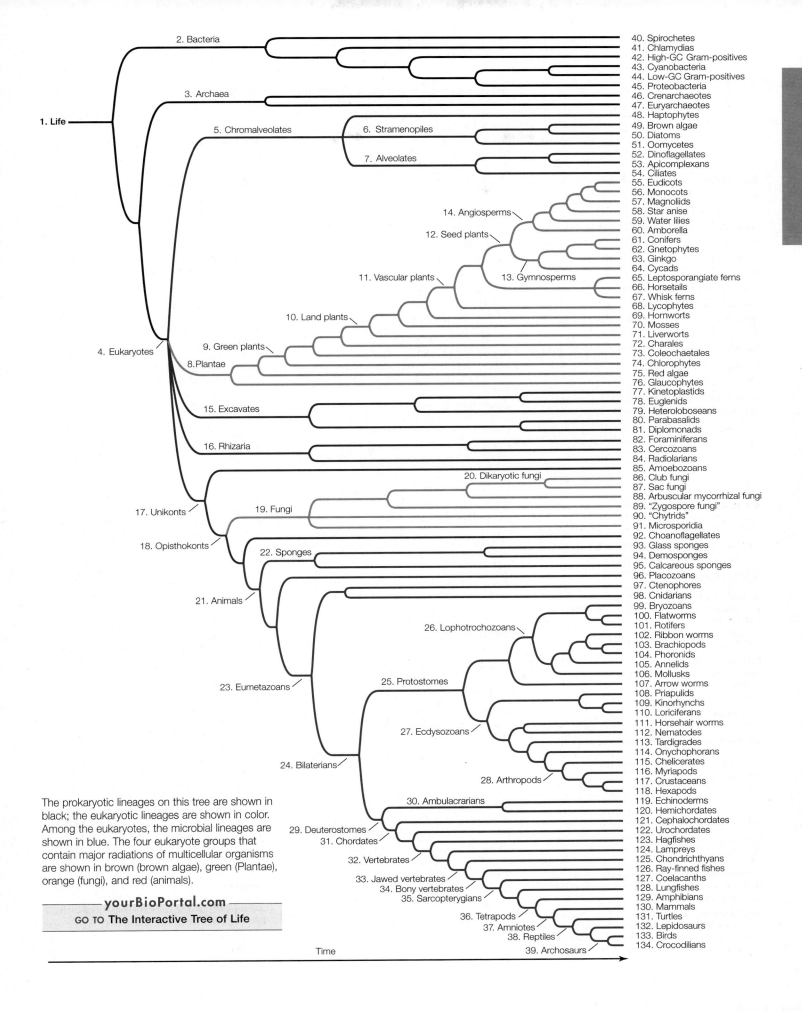

The prokaryotic lineages on this tree are shown in black; the eukaryotic lineages are shown in color. Among the eukaryotes, the microbial lineages are shown in blue. The four eukaryote groups that contain major radiations of multicellular organisms are shown in brown (brown algae), green (Plantae), orange (fungi), and red (animals).

yourBioPortal.com
GO TO The Interactive Tree of Life

Time

1. Life
2. Bacteria
3. Archaea
4. Eukaryotes
5. Chromalveolates
6. Stramenopiles
7. Alveolates
8. Plantae
9. Green plants
10. Land plants
11. Vascular plants
12. Seed plants
13. Gymnosperms
14. Angiosperms
15. Excavates
16. Rhizaria
17. Unikonts
18. Opisthokonts
19. Fungi
20. Dikaryotic fungi
21. Animals
22. Sponges
23. Eumetazoans
24. Bilaterians
25. Protostomes
26. Lophotrochozoans
27. Ecdysozoans
28. Arthropods
29. Deuterostomes
30. Ambulacrarians
31. Chordates
32. Vertebrates
33. Jawed vertebrates
34. Bony vertebrates
35. Sarcopterygians
36. Tetrapods
37. Amniotes
38. Reptiles
39. Archosaurs

40. Spirochetes
41. Chlamydias
42. High-GC Gram-positives
43. Cyanobacteria
44. Low-GC Gram-positives
45. Proteobacteria
46. Crenarchaeotes
47. Euryarchaeotes
48. Haptophytes
49. Brown algae
50. Diatoms
51. Oomycetes
52. Dinoflagellates
53. Apicomplexans
54. Ciliates
55. Eudicots
56. Monocots
57. Magnoliids
58. Star anise
59. Water lilies
60. Amborella
61. Conifers
62. Gnetophytes
63. Ginkgo
64. Cycads
65. Leptosporangiate ferns
66. Horsetails
67. Whisk ferns
68. Lycophytes
69. Hornworts
70. Mosses
71. Liverworts
72. Charales
73. Coleochaetales
74. Chlorophytes
75. Red algae
76. Glaucophytes
77. Kinetoplastids
78. Euglenids
79. Heteroloboseans
80. Parabasalids
81. Diplomonads
82. Foraminiferans
83. Cercozoans
84. Radiolarians
85. Amoebozoans
86. Club fungi
87. Sac fungi
88. Arbuscular mycorrhizal fungi
89. "Zygospore fungi"
90. "Chytrids"
91. Microsporidia
92. Choanoflagellates
93. Glass sponges
94. Demosponges
95. Calcareous sponges
96. Placozoans
97. Ctenophores
98. Cnidarians
99. Bryozoans
100. Flatworms
101. Rotifers
102. Ribbon worms
103. Brachiopods
104. Phoronids
105. Annelids
106. Mollusks
107. Arrow worms
108. Priapulids
109. Kinorhynchs
110. Loriciferans
111. Horsehair worms
112. Nematodes
113. Tardigrades
114. Onychophorans
115. Chelicerates
116. Myriapods
117. Crustaceans
118. Hexapods
119. Echinoderms
120. Hemichordates
121. Cephalochordates
122. Urochordates
123. Hagfishes
124. Lampreys
125. Chondrichthyans
126. Ray-finned fishes
127. Coelacanths
128. Lungfishes
129. Amphibians
130. Mammals
131. Turtles
132. Lepidosaurs
133. Birds
134. Crocodilians

– B –

bacteria (*Eubacteria*) [2]   Unicellular organisms lacking a nucleus, possessing distinctive ribosomes and initiator tRNA, and generally containing peptidoglycan in the cell wall. Different bacterial groups are distinguished primarily on nucleotide sequence data.

barnacles (*Cirripedia*)   Crustaceans [117] that undergo two metamorphoses—first from a feeding planktonic larva to a nonfeeding swimming larva, and then to a sessile adult that forms a "shell" composed of four to eight plates cemented to a hard substrate.

bilaterians (*Bilateria*) [24]   Those animal groups characterized by bilateral symmetry and three distinct tissue types (endoderm, ectoderm, and mesoderm). Includes the protostomes [25] and deuterostomes [29].

birds (*Aves*) [133]   Feathered, flying (or secondarily flightless) tetrapods [36].

bivalves (*Bivalvia*)   Major mollusk [106] group; clams and mussels. Bivalves typically have two similar hinged shells that are each asymmetrical across the midline.

bony vertebrates (*Osteichthyes*) [34]   Vertebrates [32] in which the skeleton is usually ossified to form bone. Includes the ray-finned fishes [126], coelacanths [127], lungfishes [128], and tetrapods [36].

brachiopods (*Brachiopoda*) [103]   Lophotrochozoans [26] with two similar hinged shells that are each symmetrical across the midline. Superficially resemble bivalve mollusks, except for the shell symmetry.

brittle stars (*Ophiuroidea*)   Echinoderms [119] with five long, whip-like arms radiating from a distinct central disk that contains the reproductive and digestive organs.

brown algae (*Phaeophyta*) [49]   Multicellular, almost exclusively marine stramenopiles [6] generally containing the pigment fucoxanthin as well as chlorophylls *a* and *c* in their chloroplasts.

bryozoans (*Ectoprocta* or *Bryozoa*) [99]   A group of marine and freshwater lophotrochozoans [26] that live in colonies attached to substrata; also known as ectoprocts or moss animals.

– C –

caecilians (*Gymnophiona*)   A group of burrowing or aquatic amphibians [129]. They are elongate, legless, with a short tail (or none at all), reduced eyes covered with skin or bone, and a pair of sensory tentacles on the head.

calcareous sponges (*Calcarea*) [95]   Filter-feeding marine sponges with spicules composed of calcium carbonate.

cellular slime molds (*Dictyostelida*)   Amoebozoans [85] in which individual amoebas aggregate under stress to form a multicellular pseudoplasmodium.

cephalochordates (*Cephalochordata*) [121]   A group of weakly swimming, eel-like benthic marine chordates [31]; also called lancelets.

cephalopods (*Cephalopoda*)   Active, predatory mollusks [106] in which the molluscan foot has been modified into muscular hydrostatic arms or tentacles. Includes octopuses, squids, and nautiluses.

cercozoans (*Cercozoa*) [83]   Unicellular eukaryotes [4] that feed by means of threadlike pseudopods. Group together with foraminiferans [82] and radiolarians [84] to comprise the rhizaria [16].

Charales [72]   Multicellular green algae with branching, apical growth and plasmodesmata between adjacent cells. The closest living relatives of the land plants [10], they retain the egg in the parent organism.

chelicerates (*Chelicerata*) [115]   A major group of arthropods [28] with pointed appendages (chelicerae) used to grasp food (as opposed to the chewing mandibles of most other arthropods). Includes the arachnids, horseshoe crabs, pycnogonids, and extinct sea scorpions.

chimaeras (*Holocephali*)   A group of bottom-dwelling, marine, scaleless chondrichthyan fishes [125] with large, permanent, grinding tooth plates (rather than the replaceable teeth found in other chondrichthyans).

chitons (*Polyplacophora*)   Flattened, slow-moving mollusks [106] with a dorsal protective calcareous covering made up of eight articulating plates.

chlamydias (*Chlamydiae*) [41]   A group of very small Gram-negative bacteria; they live as intracellular parasites of other organisms.

chlorophytes (*Chlorophyta*) [74]   The most abundant and diverse group of green algae, including freshwater, marine, and terrestrial forms; some are unicellular, others colonial, and still others multicellular. Chlorophytes use chlorophylls *a* and *c* in their photosynthesis.

choanoflagellates (*Choanozoa*) [92]   Unicellular eukaryotes [4] with a single flagellum surrounded by a collar. Most are sessile, some are colonial. The closest living relatives of the animals [21].

chondrichthyans (*Chondrichthyes*) [125]   One of the two main groups of jawed vertebrates [33]; includes sharks, rays, and chimaeras. They have cartilaginous skeletons and paired fins.

chordates (*Chordata*) [31]   One of the two major groups of deuterostomes [29], characterized by the presence (at some point in development) of a notochord, a hollow dorsal nerve cord, and a post-anal tail. Includes the cephalochordates [121], urochordates [122], and vertebrates [32].

chromalveolates (*Chromalveolata*) [5]   A contested group, said to have arisen from a common ancestor with chloroplasts derived from a red alga and supported by some molecular evidence. Major chromalveolate groups include the alveolates [7], stramenopiles [6], and haptophytes [48].

"chytrids" [90]   A convenience term used for a paraphyletic group of mostly aquatic, microscopic fungi [19] with flagellated gametes. Some exhibit alternation of generations.

ciliates (*Ciliophora*) [54]   Alveolates [7] with numerous cilia and two types of nuclei (micronuclei and macronuclei).

clitellates (*Clitellata*)   Annelids [105] with gonads contained in a swelling (called a clitellum) toward the head of the animal. Includes earthworms (oligochaetes) and leeches.

club fungi (*Basidiomycota*) [86]   Fungi [19] that, if multicellular, bear the products of meiosis on club-shaped basidia and possess a long-lasting dikaryotic stage. Some are unicellular.

club mosses (*Lycopodiophyta*) [68]   Vascular plants [11] characterized by microphylls. See lycophytes.

cnidarians (*Cnidaria*) [98]   Aquatic, mostly marine eumetazoans [23] with specialized stinging organelles (nematocysts) used for prey capture and defense, and a blind gastrovascular cavity. The closest living relatives of the ctenophores [97].

coelacanths (*Actinista*) [127]   A group of marine sarcopterygians [35] that was diverse from the Middle Devonian to the Cretaceous, but is now known from just two living species. The pectoral and anal fins are on fleshy stalks supported by skeletal elements, so they are also called lobe-finned fishes.

Coleochaetales [73]   Multicellular green algae characterized by flattened growth form composed of thin-walled cells. Thought to be the sister-group to the Charales [72] plus land plants [10].

conifers (*Pinophyta* or *Coniferophyta*) [61]   Cone-bearing, woody seed plants [12].

copepods (*Copepoda*)   Small, abundant crustaceans [117] found in marine, freshwater, or wet terrestrial habitats. They have a single eye, long antennae, and a body shaped like a teardrop.

craniates (*Craniata*)   Some biologist exclude the hagfishes [123] from the vertebrates [32], and use the term craniates to refer to the two groups combined.

crenarchaeotes (*Crenarchaeota*) [46]   A major and diverse group of archaeans [3], defined on the basis of rRNA base sequences. Many are extremophiles (inhabit extreme environments), but the group may also be the most abundant archaeans in the marine environment.

crinoids (*Crinoidea*)   Echinoderms [119] with a mouth surrounded by feeding arms, and a U-shaped gut with the mouth next to the anus. They attach to the substratum by a stalk or are free-swimming. Crinoids were abundant in the middle and late Paleozoic, but only a few hundred species have survived to the present. Includes the sea lilies and feather stars.

crocodilians (*Crocodylia*) [134]   A group of large, predatory, aquatic archosaurs [39]. The closest living relatives of birds [133]. Includes alligators, caimans, crocodiles, and gharials.

crustaceans (*Crustacea*) [117]   Major group of marine, freshwater, and terrestrial arthropods [28] with a head, thorax, and abdomen (although the head and thorax may be fused), covered with a thick exoskeleton, and with two-part appendages. Crustaceans undergo metamorphosis from a nauplius larva. Includes decapods, isopods, krill, barnacles, amphipods, copepods, and ostracods.

ctenophores (*Ctenophora*) [97]   Radially symmetrical, diploblastic marine animals [21], with a complete gut and eight rows of fused plates of cilia (called ctenes).

cyanobacteria (*Cyanobacteria*) [43]   A group of unicellular, colonial, or filamentous bacteria that conduct photosynthesis using chlorophyll *a*.

cycads (*Cycadophyta*) [64] Palmlike gymnosperms with large, compound leaves.

cyclostomes (*Cyclostomata*) This term refers to the possibly monophyletic group of lampreys [124] and hagfishes [123]. Molecular data support this group, but morphological data suggest that lampreys are more closely related to jawed vertebrates [33] than to hagfishes.

– D –

decapods (*Decapoda*) A group of marine, freshwater, and semiterrestrial crustaceans [117] in which five of the eight pairs of thoracic appendages function as legs (the other three pairs, called maxillipeds, function as mouthparts). Includes crabs, lobsters, crayfishes, and shrimps.

demosponges (*Demospongiae*) [94] The largest of the three groups of sponges [22], accounting for 90 percent of all sponge species. Demosponges have spicules made of silica, spongin fiber (a protein), or both.

deuterostomes (*Deuterostomia*) [29] One of the two major groups of bilaterians [24], in which the mouth forms at the opposite end of the embryo from the blastopore in early development (contrast with protostomes). Includes the ambulacrarians [30] and chordates [31].

diatoms (*Bacillariophyta*) [50] Unicellular, photosynthetic stramenopiles [6] with glassy cell walls in two parts.

dikaryotic fungi (*Dikarya*) [20] A group of fungi [19] in which two genetically different haploid nuclei coexist and divide within the same hypha; includes club fungi [86] and sac fungi [87].

dinoflagellates (*Dinoflagellata*) [52] A group of alveolates [7] usually possessing two flagella, one in an equatorial groove and the other in a longitudinal groove; many are photosynthetic.

diplomonads (*Diplomonadida*) [81] A group of eukaryotes [4] lacking mitochondria; most have two nuclei, each with four associated flagella.

– E –

ecdysozoans (*Ecdysozoa*) [27] One of the two major groups of protostomes [25], characterized by periodic molting of their exoskeletons. Nematodes [112] and arthropods [28] are the largest ecdysozoan groups.

echinoderms (*Echinodermata*) [119] A major group of marine deuterostomes [29] with fivefold radial symmetry (at some stage of life) and an endoskeleton made of calcified plates and spines. Includes sea stars, crinoids, sea urchins, sea cucumbers, and brittle stars.

elasmobranchs (*Elasmobranchii*) The largest group of chondrichthyan fishes [125]. Includes sharks, skates, and rays. In contrast to the other group of living chondrichthyans (the chimaeras), they have replaceable teeth.

embryophytes *See* land plants [10].

eudicots (*Eudicotyledones*)[55] A group of angiosperms [14] with pollen grains possessing three openings. Typically with two cotyledons, net-veined leaves, taproots, and floral organs typically in multiples of four or five.

euglenids (*Euglenida*) [78] Flagellate excavates characterized by a pellicle composed of spiraling strips of protein under the plasma membrane; the mitochondria have disk-shaped cristae. Some are photosynthetic.

eukaryotes (*Eukarya*) [4] Organisms made up of one or more complex cells in which the genetic material is contained in nuclei. Contrast with archaeans [3] and bacteria [2].

eumetazoans (*Eumetazoa*) [23] Those animals [21] characterized by body symmetry, a gut, a nervous system, specialized types of cell junctions, and well-organized tissues in distinct cell layers (although there have been secondary losses of some of these characteristics in some eumetazoans).

euphyllophytes (*Euphyllophyta*) The group of vascular plants [11] that is sister to the lycophytes [68] and which includes all plants with megaphylls.

euryarchaeotes (*Euryarchaeota*) [47] A major group of archaeans [3], diagnosed on the basis of rRNA sequences. Includes many methanogens, extreme halophiles, and thermophiles.

eutherians (*Eutheria*) A group of viviparous mammals [130], eutherians are well developed at birth (contrast to prototherians and marsupials, the other two groups of mammals). Most familiar mammals outside the Australian and South American regions are eutherians (see Table 33.1).

excavates (*Excavata*) [15] Diverse group of unicellular, flagellate eukaryotes, many of which possess a feeding groove; some lack mitochondria.

– F –

"ferns" Vascular plants [11] usually possessing large, frondlike leaves that unfold from a "fiddlehead." Not a monophyletic group, although most fern species are encompassed in a monophyletic clade, the leptosporangiate ferns [65].

flatworms (*Platyhelminthes*) [100] A group of dorsoventrally flattened and generally elongate soft-bodied lophotrochozoans [26]. May be free-living or parasitic, found in marine, freshwater, or damp terrestrial environments. Major flatworm groups include the tapeworms, flukes, monogeneans, and turbellarians.

flowering plants *See* angiosperms [14].

flukes (*Trematoda*) A group of wormlike parasitic flatworms [100] with complex life cycles that involve several different host species. May be paraphyletic with respect to tapeworms.

foraminiferans (*Foraminifera*) [82] Amoeboid organisms with fine, branched pseudopods that form a food-trapping net. Most produce external shells of calcium carbonate.

fungi (*Fungi*) [19] Eukaryotic heterotrophs with absorptive nutrition based on extracellular digestion; cell walls contain chitin. Major fungal groups include the microsporidia [91], "chytrids" [90], "zygospore fungi" [89], arbuscular mycorrhizal fungi [88], sac fungi [87], and club fungi [86].

– G –

gastropods (*Gastropoda*) The largest group of mollusks [106]. Gastropods possess a well-defined head with two or four sensory tentacles (often terminating in eyes) and a ventral foot. Most species have a single coiled or spiraled shell. Common in marine, freshwater, and terrestrial environments.

ginkgo (*Ginkgophyta*) [63] A gymnosperm [13] group with only one living species. The ginkgo seed is surrounded by a fleshy tissue not derived from an ovary wall and hence not a fruit.

glass sponges (*Hexactinellida*) [93] Sponges [22] with a skeleton composed of four- and/or six-pointed spicules made of silica.

glaucophytes (*Glaucophyta*) [76] Unicellular freshwater algae with chloroplasts containing traces of peptidoglycan, the characteristic cell wall material of bacteria.

gnathostomes (*Gnathostomata*) *See* jawed vertebrates [33].

gnetophytes (*Gnetophyta*) [62] A gymnosperm [13] group with three very different lineages; all have wood with vessels, unlike other gymnosperms.

green plants (*Viridiplantae*) [9] Organisms with chlorophylls *a* and *b*, cellulose-containing cell walls, starch as a carbohydrate storage product, and chloroplasts surrounded by two membranes.

gymnosperms (*Gymnospermae*) [13] Seed plants [12] with seeds "naked" (i.e., not enclosed in carpels). Probably monophyletic, but status still in doubt. Includes the conifers [61], gnetophytes [62], ginkgo [63], and cycads [64].

– H –

hagfishes (*Myxini*) [123] Elongate, slimy-skinned vertebrates [32] with three small accessory hearts, a partial cranium, and no stomach or paired fins. *See also* craniata; cyclostomes.

haptophytes (*Haptophyta*) [48] Unicellular, photosynthetic chromalveolates [5] with two slightly unequal, smooth flagella. Abundant as phytoplankton, some form marine algal blooms.

hemichordates (*Hemichordata*) [120] One of the two primary groups of ambulacrarians [30]; marine wormlike organisms with a three-part body plan.

heteroloboseans (*Heterolobosea*) [79] Colorless excavates [15] that can transform among amoeboid, flagellate, and encysted stages.

hexapods (*Hexapoda*) [118] Major group of arthropods [28] characterized by a reduction (from the ancestral arthropod condition) to six walking appendages, and the consolidation of three body segments to form a thorax. Includes insects and their relatives (see Table 32.2).

high-GC Gram-positives (*Actinobacteria*) [42] Gram-positive bacteria with a relatively high (G+C)/(A+T) ratio of their DNA, with a filamentous growth habit.

hornworts (*Anthocerophyta*) [69] Nonvascular plants with sporophytes that grow from the base. Cells contain a single large, platelike chloroplast.

horsehair worms (*Nematomorpha*) [111] A group of very thin, elongate, wormlike freshwater ecdysozoans [27]. Largely nonfeeding as adults, they are parasites of insects and crayfish as larvae.

horseshoe crabs (*Xiphosura*) Marine chelicerates [115] with a large outer shell in three parts: a carapace, an abdomen, and a tail-like telson. There are only five living species, but many additional species are known from fossils.

horsetails (*Sphenophyta* or *Equisetophyta*) [66] Vascular plants [11] with reduced megaphylls in whorls.

hydrozoans (*Hydrozoa*) A group of cnidarians [98]. Most species go through both polyp and mesuda stages, although one stage or the other is eliminated in some species.

– I –

insects (*Insecta*) The largest group within the hexapods [118]. Insects are characterized by exposed mouthparts and one pair of antennae containing a sensory receptor called a Johnston's organ. Most have two pairs of wings as adults. There are more described species of insects than all other groups of life [1] combined, and many species remain to be discovered. The major insect groups are described in Table 32.2.

"invertebrates" A convenience term that encompasses any animal [21] that is not a vertebrate [32].

isopods (*Isopoda*) Crustaceans [117] characterized by a compact head, unstalked compound eyes, and mouthparts consisting of four pairs of appendages. Isopods are abundant and widespread in salt, fresh, and brackish water, although some species (the sow bugs) are terrestrial.

– J –

jawed vertebrates (*Gnathostomata*) [33] A major group of vertebrates [32] with jawed mouths. Includes chondrichthyans [125], ray-finned fishes [126], and sarcopterygians [35].

– K –

kinetoplastids (*Kinetoplastida*) [77] Unicellular, flagellate organisms characterized by the presence in their single mitochondrion of a kinetoplast (a structure containing multiple, circular DNA molecules).

kinorhynchs (*Kinorhyncha*) [109] Small (< 1 mm) marine ecdysozoans [27] with bodies in 13 segments and a retractable proboscis.

korarchaeotes (*Korarchaeota*) A group of archaeans [3] known only by evidence from nucleic acids derived from hot springs. Its phylogenetic relationships within the Archaea are unknown.

krill (*Euphausiacea*) A group of shrimplike marine crustaceans [117] that are important components of the zooplankton.

– L –

lampreys (*Petromyzontiformes*) [124] Elongate, eel-like vertebrates [32] that often have rasping and sucking disks for mouths.

lancelets (*Cephalochordata*) *See* cephalochordates [121].

land plants (*Embryophyta*) [10] Plants with embryos that develop within protective structures; also called embryophytes. Sporophytes and gametophytes are multicellular. Land plants possess a cuticle. Major groups are the liverworts [71], mosses [70], hornworts [69], and vascular plants [11].

larvaceans (*Larvacea*) Solitary, planktonic urochordates [122] that retain both notochords and nerve cords throughout their lives.

lepidosaurs (*Lepidosauria*) [132] Reptiles [38] with overlapping scales. Includes tuataras and squamates (lizards, snakes, and amphisbaenians).

leptosporangiate ferns (*Pteridopsida* or *Polypodiopsida*) [65] Vascular plants [11] usually possessing large, frondlike leaves that unfold from a "fiddlehead," and possessing thin-walled sporangia.

life (*Life*) [1] The monophyletic group that includes all known living organisms. Characterized by a nucleic-acid based genetic system (DNA or RNA), metabolism, and cellular structure. Some parasitic forms, such as viruses, have secondarily lost some of these features and rely on the cellular environment of their host.

liverworts (*Hepatophyta*) [71] Nonvascular plants lacking stomata; stalk of sporophyte elongates along its entire length.

loboseans (*Lobosea*) A group of unicellular amoebozoans [85]; includes the most familiar amoebas (e.g., *Amoeba proteus*).

"lophophorates" Not a monophyletic group. A convenience term used to describe several groups of lophotrochozoans [26] that have a feeding structure called a lophophore (a circular or U-shaped ridge around the mouth that bears one or two rows of ciliated, hollow tentacles).

lophotrochozoans (*Lophotrochozoa*) [26] One of the two main groups of protostomes [25]. This group is morphologically diverse, and is supported primarily on information from gene sequences. Includes bryozoans [99], flatworms [100], rotifers [101], ribbon worms [102], brachiopods [103], phoronids [104], annelids [105], and mollusks [106].

loriciferans (*Loricifera*) [110] Small (< 1 mm) ecdysozoans [27] with bodies in four parts, covered with six plates.

low-GC Gram-positives (*Firmicutes*) [44] A diverse group of bacteria [2] with a relatively low (G+C)/(A+T) ratio of their DNA, often but not always Gram-positive, some producing endospores.

lungfishes (*Dipnoi*) [128] A group of aquatic sarcopterygians [35] that are the closest living relatives of the tetrapods [36]. They have a modified swim bladder used to absorb oxygen from air, so some species can survive the temporary drying of their habitat.

lycophytes (*Lycopodiophyta*) [68] Vascular plants [11] characterized by microphylls; includes club mosses, spike mosses, and quillworts.

– M –

magnoliids (*Magnoliidae*) [57] A major group of angiosperms [14] possessing two cotyledons and pollen grains with a single opening. The group is defined primarily by nucleotide sequence data; it is more closely related to the eudicots and monocots than to three other small angiosperm groups.

mammals (*Mammalia*) [130] A group of tetrapods [36] with hair covering all or part of their skin; females produce milk to feed their developing young. Includes the prototherians, marsupials, and eutherians.

marsupials (*Marsupialia*) Mammals [130] in which the female typically has a marsupium (a pouch for rearing young, which are born at an extremely early stage in development). Includes such familiar mammals as opossums, koalas, and kangaroos.

metazoans (*Metazoa*) *See* animals [21].

microbial eukaryotes *See* "protists."

microsporidia (*Microsporidia*) [91] A group of parasitic unicellular fungi [19] that lack mitochondria and have walls that contain chitin.

mollusks (*Mollusca*) [106] One of the major groups of lophotrochozoans [26], mollusks have bodies composed of a foot, a mantle (which often secretes a hard, calcareous shell), and a visceral mass. Includes monoplacophorans, chitons, bivalves, gastropods, and cephalopods.

monilophytes (*Monilophyta*) A group of vascular plants [11], sister to the seed plants [12], characterized by overtopping and possession of megaphylls; includes the ferns [65], horsetails [66], and whisk ferns [67].

monocots (*Monocotyledones*) [56] Angiosperms [14] characterized by possession of a single cotyledon, usually parallel leaf veins, a fibrous root system, pollen grains with a single opening, and floral organs usually in multiples of three.

monogeneans (*Monogenea*) A group of ectoparasitic flatworms [100].

monoplacophorans (*Monoplacophora*) Mollusks [106] with segmented body parts and a single, thin, flat, rounded, bilateral shell.

mosses (*Bryophyta*) [70] Nonvascular plants with true stomata and erect, "leafy" gametophytes; sporophytes elongate by apical cell division.

moss animals *See* bryozoans [99].

myriapods (*Myriapoda*) [116] Arthropods [28] characterized by an elongate, segmented trunk with many legs. Includes centipedes and millipedes.

– N –

nanoarchaeotes (*Nanoarchaeota*) A group of extremely small, thermophilic archaeans [3] with a much-reduced genome. The only described example can survive only when attached to a host organism.

nematodes (*Nematoda*) [112] A very large group of elongate, unsegmented ecdysozoans [27] with thick, multilayer cuticles. They are among the most abundant and diverse animals, although most species have not yet been described. Include free-living predators and scavengers, as well as parasites of most species of land plants [10] and animals [21].

neognaths (*Neognathae*) The main group of birds [133], including all living species except the ratites (ostrich, emu, rheas, kiwis, cassowaries) and tinamous. *See* palaeognaths.

– O –

oligochaetes (*Oligochaeta*) An annelid [105] group whose members lack parapodia, eyes, and anterior tentacles, and have few setae. Earthworms are the most familiar oligochaetes.

onychophorans (*Onychophora*) [114] Elongate, segmented ecdysozoans [27] with many pairs of soft, unjointed, claw-bearing legs. Also known as velvet worms.

oomycetes (*Oomycota*) [51] Water molds and relatives; absorptive heterotrophs with nutrient-absorbing, filamentous hyphae.

opisthokonts (*Opisthokonta*) [18] A group of unikonts [17] in which the flagellum on motile cells, if present, is posterior. The opisthokonts include the fungi [19], animals [21], and choanoflagellates [92].

ostracods (*Ostracoda*) Marine and freshwater crustaceans [117] that are laterally compressed and protected by two clamlike calcareous or chitinous shells.

– P –

palaeognaths (*Palaeognathae*) A group of secondarily flightless or weakly flying birds [133]. Includes the flightless ratites (ostrich, emu, rheas, kiwis, cassowaries) and the weakly flying tinamous.

parabasalids (*Parabasalia*) [80] A group of unicellular eukaryotes [4] that lack mitochondria; they possess flagella in clusters near the anterior of the cell.

phoronids (*Phoronida*) [104] A small group of sessile, wormlike marine lophotrochozoans [26] that secrete chitinous tubes and feed using a lophophore.

placoderms (*Placodermi*) An extinct group of jawed vertebrates [33] that lacked teeth. Placoderms were the dominant predators in Devonian oceans.

placozoans (*Placozoa*) [96] A poorly known group of structurally simple, asymmetrical, flattened, transparent animals found in coastal marine tropical and subtropical seas. Most evidence suggests that placozoans are the sister-group of eumetazoans [23].

Plantae [8] The most broadly defined plant group. In most parts of this book, we use the word "plant" as synonymous with "land plant" [10], a more restrictive definition.

plasmodial slime molds (*Myxogastrida*) Amoebozoans [85] that in their feeding stage consist of a coenocyte called a plasmodium.

pogonophorans (*Pogonophora*) Deep-sea annelids [105] that lack a mouth or digestive tract; they feed by taking up dissolved organic matter, facilitated by endosymbiotic bacteria in a specialized organ (the trophosome).

polychaetes (*Polychaeta*) A group of mostly marine annelids [105] with one or more pairs of eyes and one or more pairs of feeding tentacles; parapodia and setae extend from most body segments. May be paraphyletic with respect to the clitellates.

priapulids (*Priapulida*) [108] A small group of cylindrical, unsegmented, wormlike marine ecdysozoans [27] that takes its name from its phallic appearance.

"progymnosperms" Paraphyletic group of extinct vascular plants [11] that flourished from the Devonian through the Mississippian periods. The first truly woody plants, and the first with vascular cambium that produced both secondary xylem and secondary phloem, they reproduced by spores rather than by seeds.

"prokaryotes" Not a monophyletic group; as commonly used, includes the bacteria [2] and archaeans [3]. A term of convenience encompassing all cellular organisms that are not eukaryotes.

proteobacteria (*Proteobacteria*) [45] A large and extremely diverse group of Gram-negative bacteria that includes many pathogens, nitrogen fixers, and photosynthesizers. Includes the alpha, beta, gamma, delta, and epsilon proteobacteria.

"protists" This term of convenience does not describe a monophyletic group but is used to encompass a large number of distinct and distantly related groups of eukaryotes, many but far from all of which are microbial and unicellular. Essentially a "catch-all" term for any eukaryote group not contained within the land plants [10], fungi [19], or animals [21].

protostomes (*Protostomia*) [25] One of the two major groups of bilaterians [24]. In protostomes, the mouth typically forms from the blastopore (if present) in early development (contrast with deuterostomes). The major protostome groups are the lophotrochozoans [26] and ecdysozoans [27].

prototherians (*Prototheria*) A mostly extinct group of mammals [130], common during the Cretaceous and early Cenozoic. The five living species—the echidnas and the duck-billed platypus—are the only extant egg-laying mammals.

pterobranchs (*Pterobranchia*) A small group of sedentary marine hemichordates [120] that live in tubes secreted by the proboscis. They have one to nine pairs of arms, each bearing long tentacles that capture prey and function in gas exchange.

pycnogonids (*Pycnogonida*) Treated in this book as a group of chelicerates [115], but sometimes considered an independent group of arthropods [28]. Pycnogonids have reduced bodies and very long, slender legs. Also called sea spiders.

– R –

radiolarians (*Radiolaria*) [84] Amoeboid organisms with needlelike pseudopods supported by microtubules. Most have glassy internal skeletons.

ray-finned fishes (*Actinopterygii*) [126] A highly diverse group of freshwater and marine bony vertebrates [34]. They have reduced swim bladders that often function as hydrostatic organs and fins supported by soft rays (lepidotrichia). Includes most familiar fishes.

red algae (*Rhodophyta*) [75] Mostly multicellular, marine and freshwater algae characterized by the presence of phycoerythrin in their chloroplasts.

reptiles (*Reptilia*) [38] One of the two major groups of extant amniotes [37], supported on the basis of similar skull structure and gene sequences. The term "reptiles" traditionally excluded the birds [133], but the resulting group is then clearly paraphyletic. As used in this book, the reptiles include turtles [131], lepidosaurs [132], birds [133], and crocodilians [134].

rhizaria (*Rhizaria*) [16] Mostly amoeboid unicellular eukaryotes with pseudopods, many with external or internal shells. Includes the foraminiferans [82], cercozoans [83], and radiolarians [84].

rhyniophytes (*Rhyniophyta*) A group of early vascular plants [11] that appeared in the Silurian and became extinct in the Devonian. Possessed dichotomously branching stems with terminal sporangia but no true leaves or roots.

ribbon worms (*Nemertea*) [102] A group of unsegmented lophotrochozoans [26] with an eversible proboscis used to capture prey. Mostly marine, but some species live in fresh water or on land.

rotifers (*Rotifera*) [101] Tiny (< 0.5 mm) lophotrochozoans [26] with a pseudocoelomic body cavity that functions as a hydrostatic organ and a ciliated feeding organ called the corona that surrounds the head. They live in freshwater and wet terrestrial habitats.

roundworms (*Nematoda*) [112] *See* nematodes.

– S –

sac fungi (*Ascomycota*) [87] Fungi that bear the products of meiosis within sacs (asci) if the organism is multicellular. Some are unicellular.

salamanders (*Caudata*) A group of amphibians [129] with distinct tails in both larvae and adults and limbs set at right angles to the body.

salps *See* thaliaceans.

sarcopterygians (*Sarcopterygii*) [35] One of the two major groups of bony vertebrates [34], characterized by jointed appendages (paired fins or limbs).

scyphozoans (*Scyphozoa*) [98] Marine cnidarians [98] in which the medusa stage dominates the life cycle. Commonly known as jellyfish.

sea cucumbers (*Holothuroidea*) Echinoderms [119] with an elongate, cucumber-shaped body and leathery skin. They are scavengers on the ocean floor.

sea spiders *See* pycnogonids.

sea squirts *See* ascidians.

sea stars (*Asteroidea*) Echinoderms [119] with five (or more) fleshy "arms" radiating from an indistinct central disk. Also called starfishes.

sea urchins (*Echinoidea*) Echinoderms [119] with a test (shell) that is covered in spines. Most are globular in shape, although some groups (such as the sand dollars) are flattened.

"seed ferns" A paraphyletic group of loosely related, extinct seed plants that flourished in the Devonian and Carboniferous. Characterized by large, frondlike leaves that bore seeds.

seed plants (*Spermatophyta*) [12] Heterosporous vascular plants [11] that produce seeds; most produce wood; branching is axillary (not dichotomous). The major seed plant groups are gymnosperms [13] and angiosperms [14].

sow bugs *See* isopods.

spirochetes (*Spirochaetes*) [40] Motile, Gram-negative bacteria with a helically coiled structure and characterized by axial filaments.

sponges (*Porifera*) [22] A group of relatively asymmetric, filter-feeding animals that lack a gut or nervous system and generally lack differentiated tissues. Includes glass sponges [93], demosponges [94], and calcareous sponges [95].

springtails (*Collembola*) Wingless hexapods [118] with springing structures on the third and fourth segments of their bodies. Springtails are extremely abundant in some environments (especially in soil, leaf litter, and vegetation).

squamates (*Squamata*) The major group of lepidosaurs [132], characterized by the possession of movable quadrate bones (which allow the upper jaw to move independently of the rest of the skull) and hemipenes (a paired set of eversible penises, or penes) in males. Includes the lizards (a paraphyletic group), snakes, and amphisbaenians.

star anise (*Austrobaileyales*) [58]   A group of woody angiosperms [14] thought to be the sister-group of the clade of flowering plants that includes eudicots [55], monocots [56], and magnoliids [57].

starfish (*Asteroidea*)   *See* sea stars.

stramenopiles (*Heterokonta* or *Stramenopila*) [6] Organisms having, at some stage in their life cycle, two unequal flagella, the longer possessing rows of tubular hairs. Chloroplasts, when present, surrounded by four membranes. Major stramenopile groups include the brown algae [49], diatoms [50], and oomycetes [51].

– T –

tapeworms (*Cestoda*)   Parasitic flatworms [100] that live in the digestive tracts of vertebrates as adults, and usually in various other species of animals as juveniles.

tardigrades (*Tardigrada*) [113]   Small (< 0.5 mm) ecdysozoans [27] with fleshy, unjointed legs and no circulatory or gas exchange organs. They live in marine sands, in temporary freshwater pools, and on the water films of plants. Also called water bears.

tetrapods (*Tetrapoda*) [36]   The major group of sarcopterygians [35]; includes the amphibians [129] and the amniotes [37]. Named for the presence of four jointed limbs (although limbs have been secondarily reduced or lost completely in several tetrapod groups).

thaliaceans (*Thaliacea*)   A group of solitary or colonial planktonic marine urochordates [122]. Also called salps.

therians (*Theria*)   Mammals [130] characterized by viviparity (live birth). Includes eutherians and marsupials.

theropods (*Theropoda*)   Archosaurs [39] with bipedal stance, hollow bones, a furcula ("wishbone"), elongated metatarsals with three-fingered feet, and a pelvis that points backwards. Includes many well-known extinct dinosaurs (such as *Tyrannosaurus rex*), as well as the living birds [133].

tracheophytes   *See* vascular plants [11].

trilobites (*Trilobita*)   An extinct group of arthropods [28] related to the chelicerates [115]. Trilobites flourished from the Cambrian through the Permian.

tuataras (*Rhyncocephalia*)   A group of lepidosaurs [132] known mostly from fossils; there are just two living tuatara species. The quadrate bone of the upper jaw is fixed firmly to the skull. Sister group of the squamates.

tunicates   *See* ascidians.

turbellarians (*Turbellaria*)   A group of free-living, generally carnivorous flatworms [100]. Their monophyly is questionable.

turtles (*Testudines*) [131]   A group of reptiles [38] with a bony carapace (upper shell) and plastron (lower shell) that encase the body.

– U –

unikonts (*Unikonta*) [17]   A group of eukaryotes [4] whose motile cells possess a single flagellum. Major unikont groups include the amoebozoans [85], fungi [19], and animals [21].

urochordates (*Urochordata*) [122]   A group of chordates [31] that are mostly saclike filter feeders as adults, with motile larvae stages that resemble a tadpole.

– V –

vascular plants (*Tracheophyta*) [11]   Plants with xylem and phloem. Major groups include the lycophytes [68] and euphyllophytes.

vertebrates (*Vertebrata*) [32]   The largest group of chordates [31], characterized by a rigid endoskeleton supported by the vertebral column and an anterior skull encasing a brain. Includes hagfishes [123], lampreys [124], and the jawed vertebrates [33], although some biologists exclude the hagfishes from this group. *See also* craniates.

– W –

water bears   *See* tardigrades.

water lilies (*Nymphaeaceae*) [59]   A group of aquatic, freshwater angiosperms [14] that are rooted in soil in shallow water, with round floating leaves and flowers that extend above the water's surface. They are the sister-group to most of the remaining flowering plants, with the exception of the genus *Amborella* [60].

whisk ferns (*Psilotophyta*) [67]   Vascular plants [11] lacking leaves and roots.

– Y –

"yeasts"   A convenience term for several distantly related groups of unicellular fungi [19].

– Z –

"zygospore fungi" (*Zygomycota*, if monophyletic) [89]   A convenience term for a probably paraphyletic group of fungi [19] in which hyphae of differing mating types conjugate to form a zygosporangium.

Appendix B: Some Measurements Used in Biology

MEASURES OF	UNIT	EQUIVALENTS	METRIC → ENGLISH CONVERSION
Length	meter (m)	base unit	1 m = 39.37 inches = 3.28 feet
	kilometer (km)	1 km = 1000 (10^3) m	1 km = 0.62 miles
	centimeter (cm)	1 cm = 0.01 (10^{-2}) m	1 cm = 0.39 inches
	millimeter (mm)	1 mm = 0.1 cm = 10^{-3} m	1 mm = 0.039 inches
	micrometer (µm)	1 µm = 0.001 mm = 10^{-6} m	
	nanometer (nm)	1 nm = 0.001 µm = 10^{-9} m	
Area	square meter (m^2)	base unit	1 m^2 = 1.196 square yards
	hectare (ha)	1 ha = 10,000 m^2	1 ha = 2.47 acres
Volume	liter (L)	base unit	1 L = 1.06 quarts
	milliliter (mL)	1 mL = 0.001 L = 10^{-3} L	1 mL = 0.034 fluid ounces
	microliter (µL)	1 µL = 0.001 mL = 10^{-6} L	
Mass	gram (g)	base unit	1 g = 0.035 ounces
	kilogram (kg)	1 kg = 1000 g	1 kg = 2.20 pounds
	metric ton (mt)	1 mt = 1000 kg	1 mt = 2,200 pounds = 1.10 ton
	milligram (mg)	1 mg = 0.001 g = 10^{-3} g	
	microgram (µg)	1 µg = 0.001 mg = 10^{-6} g	
Temperature	degree Celsius (°C)	base unit	°C = (°F − 32)/1.8
			0°C = 32°F (water freezes)
			100°C = 212°F (water boils)
			20°C = 68°F ("room temperature")
			37°C = 98.6°F (human internal body temperature)
	Kelvin (K)*	K = °C + 273	0 K = −460°F
Energy	joule (J)		1 J ≈ 0.24 calorie = 0.00024 kilocalorie[†]

*0 K (−273°C) is "absolute zero," a temperature at which molecular oscillations approach 0—that is, the point at which motion all but stops.

[†]A *calorie* is the amount of heat necessary to raise the temperature of 1 gram of water 1°C. The *kilocalorie*, or nutritionist's calorie, is what we commonly think of as a calorie in terms of food.

Answers to Self-Quizzes

Chapter 2
1. b 6. a
2. d 7. d
3. c 8. a
4. c 9. c
5. d 10. b

Chapter 3
1. e 6. a
2. e 7. c
3. c 8. e
4. d 9. b
5. b 10. d

Chapter 4
1. c 6. c
2. c 7. e
3. c 8. b
4. d 9. c
5. e 10. b

Chapter 5
1. b 6. e
2. d 7. a
3. c 8. d
4. e 9. b
5. a 10. d

Chapter 6
1. e 6. c
2. c 7. c
3. a 8. b
4. d 9. e
5. c 10. c

Chapter 7
1. d 6. a
2. d 7. e
3. c 8. d
4. c 9. c
5. d 10. a

Chapter 8
1. c 6. c
2. e 7. d
3. b 8. b
4. c 9. d
5. c 10. e

Chapter 9
1. d 6. d
2. d 7. e
3. e 8. d
4. e 9. a
5. c 10. e

Chapter 10
1. e 6. d
2. b 7. d
3. d 8. d
4. b 9. d
5. e 10. b

Chapter 11
1. d 6. d
2. c 7. e
3. b 8. d
4. d 9. d
5. c 10. c

Chapter 12*
1. e 6. d
2. a 7. b
3. d 8. b
4. d 9. b
5. d 10. b

Chapter 13
1. c 6. b
2. a 7. d
3. c 8. d
4. b 9. c
5. e 10. c

Chapter 14
1. c 6. d
2. d 7. b
3. e 8. d
4. b 9. d
5. a 10. c

Chapter 15
1. a 6. b
2. c 7. d
3. b 8. d
4. b 9. d
5. d 10. b

Chapter 16
1. b 6. c
2. e 7. c
3. a 8. d
4. e 9. b
5. b 10. a

Chapter 17
1. c 6. e
2. c 7. b
3. a 8. b
4. e 9. b
5. e 10. a

Chapter 18
1. b 6. b
2. b 7. c
3. a 8. a
4. c 9. e
5. e 10. d

Chapter 19
1. c 6. c
2. b 7. e
3. a 8. b
4. b 9. a
5. d 10. b

Chapter 20
1. c 6. b
2. a 7. a
3. c 8. e
4. b 9. d
5. c 10. b

Chapter 21
1. d 6. d
2. e 7. a
3. d 8. e
4. c 9. b
5. d 10. c

Chapter 22
1. b 6. b
2. e 7. e
3. a 8. a
4. b 9. e
5. e 10. d

Chapter 23
1. c 6. c
2. e 7. a
3. d 8. e
4. c 9. c
5. a 10. e

Chapter 24
1. a 6. e
2. a 7. e
3. d 8. e
4. a 9. b
5. b 10. e

Chapter 25
1. d 6. a
2. b 7. c
3. e 8. b
4. c 9. c
5. a 10. d

Chapter 26
1. e 6. b
2. e 7. d
3. a 8. d
4. c 9. b
5. e 10. d

*Answers to Chapter 12 Genetics Problems

1. $BB \times bb$; $bb \times bb$; $Bb \times bb$; $Bb \times Bb$

2. 1/32

3a. Autosomal dominant

3b. 1/4

4a. Males (XY) contain only one allele and will show only one color, black (X^BY) or yellow (X^bY). Females can be heterozygous (X^BX^b).

4b. X^bY, yellow

5. The body color (G/g) and wing size (A/a) genes are linked; eye color (R/r) is unlinked to the other two genes. The map distance is 18.5 units.

6. Yellow, blue, and white in a 1:2:1 ratio

7. F_1 all wild-type, $PpSwsw$; F_2 will have phenotypes in the ratio 9:3:3:1. See Figure 12.7 (p. 244) for analogous genotypes.

8a. Ratio of phenotypes in F_2 is 3:1 (double dominant to double recessive).

8b. The F_1 are $PpByby$; they produce just two kinds of gametes (Pby and pBy). Combine them carefully and see the 1:2:1 phenotypic ratio fall out in the F_2.

8c. Pink-blistery

8d. See Figures 11.17 and 11.19 (pp. 224–226). Crossing over took place in the F_1 generation.

9. $Rraa$ and $RrAa$

10a. $w^+ > w^e > w$

10b. Parents are w^ew and w^+Y. Progeny are w^+w^e, w^+w, w^eY, and wY.

11a. BX^a, BY, bX^a, bY

11b. Mother bbX^AX^a, father BbX^aY, son BbX^aY, daughter bbX^aX^a

12. 75 percent

13. Because the gene is carried on mitochondrial DNA, it is passed through the mother only. Thus if the woman does not have the disease but her husband does, their child will not be affected. On the other hand, if the woman has the disease but her husband does not, their child will have the disease.

Chapter 27
1. a   6. b
2. e   7. d
3. c   8. b
4. d   9. a
5. c   10. d

Chapter 28
1. d   6. e
2. c   7. c
3. e   8. b
4. b   9. b
5. b   10. d

Chapter 29
1. d   6. c
2. c   7. a
3. d   8. e
4. a   9. c
5. d   10. a

Chapter 30
1. b   6. a
2. d   7. e
3. e   8. a
4. c   9. e
5. d   10. c

Chapter 31
1. c   6. b
2. d   7. c
3. c   8. d
4. d   9. e
5. e   10. d

Chapter 32
1. b   6. e
2. e   7. e
3. c   8. d
4. d   9. e
5. a   10. e

Chapter 33
1. d   6. e
2. a   7. a
3. c   8. e
4. d   9. c
5. d   10. b

Chapter 34
1. c   6. b
2. b   7. b
3. e   8. c
4. e   9. a
5. a   10. d

Chapter 35
1. c   6. d
2. d   7. d
3. b   8. e
4. b   9. e
5. b   10. a

Chapter 36
1. d   6. c
2. d   7. e
3. c   8. d
4. a   9. d
5. c   10. e

Chapter 37
1. a   6. c
2. e   7. b
3. c   8. c
4. d   9. a
5. b   10. b

Chapter 38
1. d   6. e
2. b   7. a
3. e   8. c
4. b   9. c
5. d   10. d

Chapter 39
1. e   6. a
2. b   7. b
3. c   8. c
4. c   9. e
5. d   10. a

Chapter 40
1. c   6. b
2. c   7. e
3. a   8. a
4. d   9. e
5. b   10. b

Chapter 41
1. b   6. b
2. a   7. d
3. b   8. e
4. e   9. c
5. e   10. c

Chapter 42
1. a   6. a
2. b   7. d
3. e   8. d
4. e   9. a
5. c   10. d

Chapter 43
1. c   6. d
2. e   7. d
3. a   8. d
4. d   9. d
5. d   10. a

Chapter 44
1. a   6. c
2. c   7. b
3. e   8. b
4. a   9. b
5. d   10. c

Chapter 45
1. d   6. e
2. d   7. c
3. c   8. c
4. c   9. d
5. e   10. d

Chapter 46
1. d   6. e
2. d   7. b
3. a   8. c
4. e   9. c
5. e   10. d

Chapter 47
1. c   6. c
2. a   7. a
3. e   8. c
4. d   9. a
5. d   10. c

Chapter 48
1. e   6. d
2. a   7. e
3. b   8. a
4. c   9. a
5. b   10. e

Chapter 49
1. e   6. b
2. d   7. c
3. a   8. c
4. b   9. a
5. c   10. d

Chapter 50
1. d   6. d
2. a   7. b
3. c   8. d
4. d   9. c
5. c   10. e

Chapter 51
1. b   6. d
2. e   7. a
3. c   8. b
4. a   9. d
5. b   10. d

Chapter 52
1. d   6. b
2. a   7. e
3. d   8. a
4. a   9. c
5. d   10. e

Chapter 53
1. b   6. d
2. e   7. e
3. c   8. e
4. a   9. a
5. d   10. e

Chapter 54
1. d   6. c
2. a   7. c
3. a   8. c
4. d   9. d
5. a   10. b

Chapter 55
1. c   6. b
2. c   7. e
3. a   8. c
4. d   9. e
5. c   10. d

Chapter 56
1. a   6. a
2. c   7. b
3. b   8. d
4. c   9. e
5. d   10. b

Chapter 57
1. a   6. b
2. a   7. e
3. b   8. b
4. d   9. d
5. e   10. e

Chapter 58
1. e   6. e
2. d   7. a
3. b   8. e
4. c   9. d
5. c   10. b

Chapter 59
1. b   6. a
2. d   7. d
3. e   8. c
4. e   9. a
5. b   10. c

Glossary

- A -

abiotic (a′ bye ah tick) [Gk. *a*: not + *bios*: life] Nonliving. (Contrast with biotic.)

abscisic acid (ABA) (ab sighs′ ik) A plant growth substance with growth-inhibiting action. Causes stomata to close; involved in a plant's response to salt and drought stress.

abscission (ab sizh′ un) [L. *abscissio*: break off] The process by which leaves, petals, and fruits separate from a plant.

absorption (1) Of light: complete retention, without reflection or transmission. (2) Of water or other molecules: soaking up (taking in through pores or by diffusion).

absorption spectrum A graph of light absorption versus wavelength of light; shows how much light is absorbed at each wavelength.

absorptive heterotroph An organism (usually a fungus) that obtains its food by secreting digestive enzymes into the environment to break down large food molecules, then absorbing the breakdown products.

abyssal plain (uh biss′ ul) [Gk. *abyssos*: bottomless] The deep ocean floor.

accessory pigments Pigments that absorb light and transfer energy to chlorophylls for photosynthesis.

acetylcholine (ACh) A neurotransmitter that carries information across vertebrate neuromuscular junctions and some other synapses. It is then broken down by the enzyme acetylcholinesterase (AChE).

acetyl coenzyme A (acetyl CoA) A compound that reacts with oxaloacetate to produce citrate at the beginning of the citric acid cycle; a key metabolic intermediate in the formation of many compounds.

acid [L. *acidus*: sharp, sour] A substance that can release a proton in solution. (Contrast with base.)

acid growth hypothesis The hypothesis that auxin increases proton pumping, thereby lowering the pH of the cell wall and activating enzymes that loosen polysaccharides. Proposed to explain auxin-induced cell expansion in plants.

acid precipitation Precipitation that has a lower pH than normal as a result of acid-forming precursor molecules introduced into the atmosphere by human activities.

acidic Having a pH of less than 7.0 (a hydrogen ion concentration greater than 10^{-7} molar). (Contrast with basic.)

acoelomate An animal that does not have a coelom.

acrosome (a′ krow soam) [Gk. *akros*: highest + *soma*: body] The structure at the forward tip of an animal sperm which is the first to fuse with the egg membrane and enter the egg cell.

ACTH *See* corticotropin.

actin [Gk. *aktis*: ray] A protein that makes up the cytoskeletal microfilaments in eukaryotic cells and is one of the two contractile proteins in muscle.

action potential An impulse in a neuron taking the form of a wave of depolarization or hyperpolarization.

action spectrum A graph of a biological process versus light wavelength; shows which wavelengths are involved in the process.

activation energy (E_a) The energy barrier that blocks the tendency for a chemical reaction to occur.

active site The region on the surface of an enzyme or ribozyme where the substrate binds, and where catalysis occurs.

active transport The energy-dependent transport of a substance across a biological membrane against a concentration gradient—that is, from a region of low concentration (of that substance) to one of high concentration. (*See also* primary active transport, secondary active transport; contrast with facilitated diffusion, passive transport.)

adaptation (a dap tay′ shun) (1) In evolutionary biology, a particular structure, physiological process, or behavior that makes an organism better able to survive and reproduce. Also, the evolutionary process that leads to the development or persistence of such a trait. (2) In sensory neurophysiology, a sensory cell's loss of sensitivity as a result of repeated stimulation.

adaptive radiation An evolutionary radiation that results in an array of related species that live in a variety of environments and differ in the characteristics they use to exploit those environments.

adenine (A) (a′ den een) A nitrogen-containing base found in nucleic acids, ATP, NAD, and other compounds.

adenosine triphosphate *See* ATP.

adrenal gland (a dree′ nal) [L. *ad*: toward + *renes*: kidneys] An endocrine gland located near the kidneys of vertebrates, consisting of two glandular parts, the cortex and medulla.

adrenaline *See* epinephrine.

adrenocorticotropic hormone *See* corticotropin.

adsorption Binding of a gas or a solute to the surface of a solid.

adventitious roots (ad ven ti′ shus) [L. *adventitius*: arriving from outside] Roots originating from the stem at ground level or below; typical of the fibrous root system of monocots.

aerenchyma In plants, parenchymal tissue containing air spaces.

aerobic (air oh′ bic) [Gk. *aer*: air + *bios*: life] In the presence of oxygen; requiring oxygen. (Contrast with anaerobic.)

afferent (af′ ur unt) [L. *ad*: toward + *ferre*: to carry] Carrying to, as in a neuron that carries impulses to the central nervous system (afferent neuron), or a blood vessel that carries blood to a structure. (Contrast with efferent.)

age structure The distribution of the individuals in a population across all age groups.

AIDS Acquired immune deficiency syndrome, a condition caused by human immunodeficiency virus (HIV) in which the body's T-helper cells are reduced, leaving the victim subject to opportunistic diseases.

air sacs Structures in the respiratory system of birds that receive inhaled air; they keep fresh air flowing unidirectionally through the lungs, but are not themselves gas exchange surfaces.

aldosterone (al dohs′ ter own) A steroid hormone produced in the adrenal cortex of mammals. Promotes secretion of potassium and reabsorption of sodium in the kidney.

aleurone layer In some seeds, a tissue that lies beneath the seed coat and surrounds the endosperm. Secretes digestive enzymes that break down macromolecules stored in the endosperm.

allantoic membrane In animal development, an outgrowth of extraembryonic endoderm plus adjacent mesoderm that forms the allantois, a saclike structure that stores metabolic wastes produced by the embryo.

allantois (al lun twah′) [Gk. *allant*: sausage] An extraembryonic membrane enclosing a sausage-shaped sac that stores the embryo's nitrogenous wastes.

allele (a leel′) [Gk. *allos*: other] The alternate form of a genetic character found at a given locus on a chromosome.

allele frequency The relative proportion of a particular allele in a specific population.

allergic reaction [Ger. *allergie*: altered] An overreaction of the immune system to amounts of an antigen that do not affect most people; often involves IgE antibodies.

allometric growth A pattern of growth in which some parts of the body of an organism grow faster than others, resulting in a change in body proportions as the organism grows.

allopatric speciation (al′ lo pat′ rick) [Gk. *allos*: other + *patria*: homeland] The formation of two species from one when reproductive isolation occurs because of the interposition of (or crossing of) a physical geographic barrier such as a river. Also called geographic speciation. (Contrast with sympatric speciation.)

allopolyploidy The possession of more than two chromosome sets that are derived from more than one species.

allosteric regulation (al lo steer′ ik) [Gk. *allos*: other + *stereos*: structure] Regulation of the activity of a protein (usually an enzyme) by the binding of an effector molecule to a site other than the active site.

alpha diversity Species diversity within a single community or habitat. (Contrast with beta diversity, gamma diversity.)

α (alpha) helix A prevalent type of secondary protein structure; a right-handed spiral.

alternation of generations The succession of multicellular haploid and diploid phases in some sexually reproducing organisms, notably plants.

alternative splicing A process for generating different mature mRNAs from a single gene by splicing together different sets of exons during RNA processing.

altruism   Pertaining to behavior that benefits other individuals at a cost to the individual who performs it.

alveolus (al ve′ o lus) (plural: alveoli) [L. *alveus:* cavity]   A small, baglike cavity, especially the blind sacs of the lung.

amensalism (a men′ sul ism)   Interaction in which one animal is harmed and the other is unaffected. (Contrast with commensalism, mutualism.)

amine   An organic compound containing an amino group (NH_2).

amino acid   An organic compound containing both NH_2 and $COOH$ groups. Proteins are polymers of amino acids.

amino acid replacement   A change in the nucleotide sequence that results in one amino acid being replaced by another.

ammonotelic (am moan′ o teel′ ic) [Gk. *telos:* end]   Pertaining to an organism in which the final product of breakdown of nitrogen-containing compounds (primarily proteins) is *ammonia.* (Contrast with ureotelic, uricotelic.)

amnion (am′ nee on)   The fluid-filled sac within which the embryos of reptiles (including birds) and mammals develop.

amniote egg   A shelled egg surrounding four extraembryonic membranes and embryo-nourishing yolk. This evolutionary adaptation permitted mammals and reptiles to live and reproduce in drier environments than can most amphibians.

amphipathic (am′ fi path′ ic) [Gk. *amphi:* both + *pathos:* emotion]   Of a molecule, having both hydrophilic and hydrophobic regions.

amplitude   The magnitude of change over the course of a regular cycle.

amygdala   A component of the limbic system that is involved in fear and fear memory.

amylase (am′ ill ase)   An enzyme that catalyzes the hydrolysis of starch, usually to maltose or glucose.

anabolic reaction (an uh bah′ lik) [Gk. *ana:* upward + *ballein:* to throw]   A synthetic reaction in which simple molecules are linked to form more complex ones; requires an input of energy and captures it in the chemical bonds that are formed. (Contrast with catabolic reaction.)

anaerobic (an ur row′ bic) [Gk. *an:* not + *aer:* air + *bios:* life]   Occurring without the use of molecular oxygen, O_2. (Contrast with aerobic.)

analogy (a nal′ o jee) [Gk. *analogia:* resembling]   A resemblance between two features that is due to convergent evolution rather than to common ancestry. The structures are said to be *analogous,* and each is an *analog* of the others. (Contrast with homology.)

anaphase (an′ a phase) [Gk. *ana:* upward]   The stage in cell nuclear division at which the first separation of sister chromatids (or, in the first meiotic division, of paired homologs) occurs.

ancestral trait   The trait originally present in the ancestor of a given group; may be retained or changed in the descendants of that ancestor.

androgen (an′ dro jen)   Any of the several male sex steroids (most notably testosterone).

aneuploidy (an′ you ploy dee)   A condition in which one or more chromosomes or pieces of chromosomes are either lacking or present in excess.

angiotensin (an′ jee oh ten′ sin)   A peptide hormone that raises blood pressure by causing peripheral vessels to constrict. Also maintains glomerular filtration by constricting efferent vessels and stimulates thirst and the release of aldosterone.

angular gyrus   A part of the human brain believed to be essential for integrating spoken and written language.

animal hemisphere   The metabolically active upper portion of some animal eggs, zygotes, and embryos; does not contain the dense nutrient yolk. (Contrast with vegetal hemisphere.)

anion (an′ eye on) [Gk. *ana:* upward progress]   A negatively charged ion. (Contrast with cation.)

anisogamous (an eye sog′ a muss) [Gk. *aniso:* unequal + *gamos:* marriage]   Having morphologically dissimilar male and female gametes. (Contrast with isogamous.)

annual   A plant whose life cycle is completed in one growing season. (Contrast with biennial, perennial.)

antagonistic interactions   Interactions between two species in which one species benefits and the other is harmed. Includes predation, herbivory, and parasitism.

antenna system   In photosynthesis, a group of different molecules that cooperate to absorb light energy and transfer it to a reaction center.

anterior   Toward or pertaining to the tip or headward region of the body axis. (Contrast with posterior.)

anterior pituitary   The portion of the vertebrate pituitary gland that derives from gut epithelium and produces tropic hormones.

anther (an′ thur) [Gk. *anthos:* flower]   A pollen-bearing portion of the stamen of a flower.

antheridium (an′ thur id′ ee um) [Gk. *antheros:* blooming]   The multicellular structure that produces the sperm in nonvascular land plants and ferns.

antibody   One of the myriad proteins produced by the immune system that specifically binds to a foreign substance in blood or other tissue fluids and initiates its removal from the body.

anticodon   The three nucleotides in transfer RNA that pair with a complementary triplet (a codon) in messenger RNA.

antidiuretic hormone (ADH)   A hormone that promotes water reabsorption by the kidney. ADH is produced by neurons in the hypothalamus and released from nerve terminals in the posterior pituitary. Also called vasopressin.

antigen (an′ ti jun)   Any substance that stimulates the production of an antibody or antibodies in the body of a vertebrate.

antigenic determinant   The specific region of an antigen that is recognized and bound by a specific antibody. Also called an epitope.

antiparallel   Pertaining to molecular orientation in which a molecule or parts of a molecule have opposing directions.

antiporter   A membrane transport protein that moves one substance in one direction and another in the opposite direction. (Contrast with symporter, uniporter.)

antisense RNA   A single-stranded RNA molecule complementary to, and thus targeted against, an mRNA of interest to block its translation.

anus (a′ nus)   An opening through which solid digestive wastes are expelled, located at the posterior end of a tubular gut.

aorta (a or′ tah) [Gk. *aorte:* aorta]   The main trunk of the arteries leading to the systemic (as opposed to the pulmonary) circulation.

aortic body   A chemosensor in the aorta that senses a decrease in blood supply or a dramatic decrease in partial pressure of oxygen in the blood.

aortic valve   A one-way valve between the left ventricle of the heart and the aorta that prevents backflow of blood into the ventricle when it relaxes.

apex (a′ pecks)   The tip or highest point of a structure, as of a growing stem or root.

aphasia   a deficit in the ability to use or understand words.

apical (a′ pi kul)   Pertaining to the *apex,* or tip, usually in reference to plants.

apical dominance   In plants, inhibition by the apical bud of the growth of axillary buds.

apical hook   A form taken by the stems of many eudicot seedlings that protects the delicate shoot apex while the stem grows through the soil.

apical meristem   The meristem at the tip of a shoot or root; responsible for a plant's primary growth.

apomixis (ap oh mix′ is) [Gk. *apo:* away from + *mixis:* sexual intercourse]   The asexual production of seeds.

apoplast (ap′ oh plast)   In plants, the continuous meshwork of cell walls and extracellular spaces through which material can pass without crossing a plasma membrane. (Contrast with symplast.)

apoptosis (ap uh toh′ sis)   A series of genetically programmed events leading to cell death.

aposematism   Warning coloration; bright colors or striking patterns of toxic or mimetic prey species that act as a warning to predators.

appendix   In the human digestive system, the vestigial equivalent of the cecum, which serves no digestive function.

aquaporin   A transport protein in plant and animal cell membranes through which water passes in osmosis.

aquatic (a kwa′ tic) [L. *aqua:* water]   Pertaining to or living in water. (Contrast with marine, terrestrial.)

aqueous (a′ kwee us)   Pertaining to water or a watery solution.

aquifer   A large pool of groundwater.

archegonium (ar′ ke go′ nee um)   The multicellular structure that produces eggs in nonvascular land plants, ferns, and gymnosperms.

archenteron (ark en′ ter on) [Gk. *archos:* first + *enteron:* bowel]   The earliest primordial animal digestive tract.

area phylogenies   Phylogenies in which the names of the taxa are replaced with the names of the places where those taxa live or lived.

arms race   A series of reciprocal adaptations between species involved in antagonistic interactions, in which adaptations that increase the fitness of a consumer species exert selection pressure on its resource species to counter the consumer's adaptation, and vice versa.

arteriole   A small blood vessel arising from an artery that feeds blood into a capillary bed.

artery   A muscular blood vessel carrying oxygenated blood away from the heart to other parts of the body. (Contrast with vein.)

artificial selection   The selection by plant and animal breeders of individuals with certain desirable traits.

ascus (ass' cus) (plural: asci) [Gk. *askos*: bladder] In sac fungi, the club-shaped sporangium within which spores (ascospores) are produced by meiosis.

asexual reproduction   Reproduction without sex.

assisted reproductive technologies (ARTs)   Any of several procedures that remove unfertilized eggs from the ovary, combine them with sperm outside the body, and then place fertilized eggs or egg–sperm mixtures in the appropriate location in a female's reproductive tract for development.

association cortex   In the vertebrate brain, the portion of the cortex involved in higher-order information processing, so named because it integrates, or associates, information from different sensory modalities and from memory.

associative learning   A form of learning in which two unrelated stimuli become linked to the same response.

astrocyte [Gk. *astron*: star]   A type of glial cell that contributes to the blood–brain barrier by surrounding the smallest, most permeable blood vessels in the brain.

atherosclerosis (ath' er oh sklair oh' sis) [Gk. *athero*: gruel, porridge + *skleros*: hard]   A disease of the lining of the arteries characterized by fatty, cholesterol-rich deposits in the walls of the arteries. When fibroblasts infiltrate these deposits and calcium precipitates in them, the disease become arteriosclerosis, or "hardening of the arteries."

atom [Gk. *atomos*: indivisible]   The smallest unit of a chemical element. Consists of a nucleus and one or more electrons.

atomic mass   *See* atomic weight.

atomic number   The number of protons in the nucleus of an atom; also equals the number of electrons around the neutral atom. Determines the chemical properties of the atom.

atomic weight   The average of the mass numbers of a representative sample of atoms of an element, with all the isotopes in their normally occurring proportions. Also called atomic mass.

ATP (adenosine triphosphate)   An energy-storage compound containing adenine, ribose, and three phosphate groups. When it is formed from ADP, useful energy is stored; when it is broken down (to ADP or AMP), energy is released to drive endergonic reactions.

ATP synthase   An integral membrane protein that couples the transport of protons with the formation of ATP.

atrial natriuretic peptide   A hormone released by the atrial muscle fibers of the heart when they are overly stretched, which decreases reabsorption of sodium by the kidney and thus blood volume.

atrioventricular node   A modified node of cardiac muscle that organizes the action potentials that control contraction of the ventricles.

atrium (a' tree um) [L. *atrium*: central hall]   An internal chamber. In the hearts of vertebrates, the thin-walled chamber(s) entered by blood on its way to the ventricle(s). Also, the outer ear.

auditory system   A sensory system that uses mechanoreceptors to convert pressure waves into receptor potentials; includes structures that gather sound waves, direct them to a sensory or-gan, and amplify their effect on the mechanoreceptors.

autocatalysis [Gk. *autos*: self + *kata*: to break down]   A positive feedback process in which an activated enzyme acts on other inactive molecules of the same enzyme to activate them.

autocrine   A chemical signal that binds to and affects the cell that makes it. (Contrast with paracrine.)

autoimmunity   An immune response by an organism to its own molecules or cells.

autonomic nervous system (ANS)   The portion of the peripheral nervous system that controls such involuntary functions as those of guts and glands. Also called the involuntary nervous system.

autopolyploidy   The possession of more than two entire chromosomes sets that are derived from a single species.

autosome   Any chromosome (in a eukaryote) other than a sex chromosome.

autotroph (au' tow trowf') [Gk. *autos*: self + *trophe*: food]   An organism that is capable of living exclusively on inorganic materials, water, and some energy source such as sunlight (photoautotrophs) or chemically reduced matter (see chemolithotrophs). (Contrast with heterotroph.)

auxin (awk' sin) [Gk. *auxein*: to grow]   In plants, a substance (the most common being indoleacetic acid) that regulates growth and various aspects of development.

avirulence (*Avr*) genes   Genes in a pathogen that may trigger defenses in plants. *See* gene-for-gene resistance.

Avogadro's number   The number of atoms or molecules in a mole (weighed out in grams) of a substance, calculated to be 6.022×10^{23}.

axillary bud   A bud that forms in the angle (axil) where a leaf meets a stem.

axon [Gk. axle]   The part of a neuron that conducts action potentials away from the cell body.

axon hillock   The junction between an axon and its cell body, where action potentials are generated.

axon terminals   The endings of an axon; they form synapses and release neurotransmitter.

- B -

B cell   A type of lymphocyte involved in the humoral immune response of vertebrates. Upon recognizing an antigenic determinant, a B cell develops into a plasma cell, which secretes an antibody. (Contrast with T cell.)

bacillus (bah sil' us) [L: little rod]   Any of various rod-shaped bacteria.

bacterial artificial chromosome (BAC)   A DNA cloning vector used in bacteria that can carry up to 150,000 base pairs of foreign DNA.

bacteriophage (bak teer' ee o fayj) [Gk. *bakterion*: little rod + *phagein*: to eat]   Any of a group of viruses that infect bacteria. Also called phage.

bacteroids   Nitrogen-fixing organelles that develop from endosymbiotic bacteria.

bark   All tissues external to the vascular cambium of a plant.

baroreceptor [Gk. *baros*: weight]   A pressure-sensing cell or organ. Sometimes called a stress receptor.

basal body   A centriole found at the base of a eukaryotic flagellum or cilium.

basal metabolic rate (BMR)   The minimum rate of energy turnover in an awake (but resting) bird or mammal that is not expending energy for thermoregulation.

base   (1) A substance that can accept a hydrogen ion in solution. (Contrast with acid.) (2) In nucleic acids, the purine or pyrimidine that is attached to each sugar in the sugar–phosphate backbone.

base pair (bp)   In double-stranded DNA, a pair of nucleotides formed by the *complementary base pairing* of a purine on one strand and a pyrimidine on the other.

basic   Having a pH greater than 7.0 (i.e., having a hydrogen ion concentration lower than 10^{-7} molar). (Contrast with acidic.)

basidiocarp   A fruiting structure produced by club fungi.

basidium (bass id' ee yum)   In club fungi, the characteristic sporangium in which four spores are formed by meiosis and then borne externally before being shed.

basilar membrane   A membrane in the human inner ear whose flexion in response to sound waves activates hair cells; flexes at different locations in response to different pitches of sound.

basophil   A type of phagocytic white blood cell that releases histamine and may promote T cell development.

Batesian mimicry   The convergence in appearance of an edible species (mimic) with an unpalatable species (model).

benefit   An improvement in survival and reproductive success resulting from performing a behavior or having a trait.

benthic zone [Gk. *benthos*: bottom]   The bottom of the ocean.

beta diversity   Between-habitat diversity; a measure of the change in species composition from one community or habitat to another. (Contrast with alpha diveristy, gamma diversity.)

β (beta) pleated sheet   A type of protein secondary structure; results from hydrogen bonding between polypeptide regions running antiparallel to each other.

biennial   A plant whose life cycle includes vegetative growth in the first year and flowering and senescence in the second year. (Contrast with annual, perennial.)

bilateral symmetry   The condition in which only the right and left sides of an organism, divided by a single plane through the midline, are mirror images of each other.

bilayer   A structure that is two layers in thickness. In biology, most often refers to the phospholipid bilayer of membranes. (*See* phospholipid bilayer.)

bile   A secretion of the liver made up of bile salts synthesized from cholesterol, various phospholipids, and bilirubin (the breakdown product of hemoglobin). Emulsifies fats in the small intestine.

binary fission   Reproduction of a prokaryote by division of a cell into two comparable progeny cells.

binocular vision   Overlapping visual fields of an animal's two eyes; allows the animal to see in three dimensions.

binomial nomenclature   A taxonomic naming system in which each species is given two names (a genus name followed by a species name).

biofilm A community of microorganisms embedded in a polysaccharide matrix, forming a highly resistant coating on almost any moist surface.

biogeochemical cycle Movement of inorganic elements such as nitrogen, phosphorus, and carbon through living organisms and the physical environment.

biogeographic region One of several defined, continental-scale regions of Earth, each of which has a biota distinct from that of the others. (Contrast with biome.)

biogeography The scientific study of the patterns of distribution of populations, species, and ecological communities across Earth.

bioinformatics The use of computers and/or mathematics to analyze complex biological information, such as DNA sequences.

biological species concept The definition of a species as a group of actually or potentially interbreeding natural populations that are reproductively isolated from other such groups. (Contrast with lineage species concept; morphological species concept.)

biology [Gk. *bios*: life + *logos*: study] The scientific study of living things.

bioluminescence The production of light by biochemical processes in an organism.

biomass The total weight of all the organisms, or some designated group of organisms, in a given area.

biome (bye' ome) A major division of the ecological communities of Earth, characterized primarily by distinctive vegetation. A given biogeographic region contains many different biomes.

bioremediation The use by humans of other organisms to remove contaminants from the environment.

biosphere (bye' oh sphere) All regions of Earth (terrestrial and aquatic) and Earth's atmosphere in which organisms can live.

biota (bye oh' tah) All of the organisms—animals, plants, fungi, and microorganisms—found in a given area. (Contrast with flora, fauna.)

biotechnology The use of cells or living organisms to produce materials useful to humans.

biotic (bye ah' tick) [Gk. *bios*: life] Alive. (Contrast with abiotic.)

biotic interchange The dispersal of species from two different biotas into the region they had not previously inhabited, as when two formerly separated land masses fuse.

blade The thin, flat portion of a leaf.

blastocoel (blass' toe seal) [Gk. *blastos*: sprout + *koilos*: hollow] The central, hollow cavity of a blastula.

blastocyst (blass' toe cist) An early embryo formed by the first divisions of the fertilized egg (zygote). In mammals, a hollow ball of cells.

blastodisc (blass' toe disk) An embryo that forms as a disk of cells on the surface of a large yolk mass; comparable to a blastula, but occurring in animals such as birds and reptiles, in which the massive yolk restricts complete cleavage.

blastomere Any of the cells produced by the early divisions of a fertilized animal egg.

blastula (blass' chu luh) An early stage of the animal embryo; in many species, a hollow sphere of cells surrounding a central cavity, the blastocoel. (Contrast with blastodisc.)

block to polyspermy Any of several responses to entry of a sperm into an egg that prevent more than one sperm from entering the egg.

blood A fluid tissue that is pumped around the body; a component of the circulatory system.

blood–brain barrier A property of blood vessels in the brain that prevents most chemicals from diffusing from the blood into the brain.

blue-light receptors Pigments in plants that absorb blue light (400–500 nm). These pigments mediate many plant responses including phototropism, stomatal movements, and expression of some genes.

body plan The general structure of an animal, the arrangement of its organ systems, and the integrated functioning of its parts.

bond See chemical bond.

bottleneck See population bottleneck.

bone A rigid component of vertebrate skeletal systems that contains an extracellular matrix of insoluble calcium phosphate crystals as well as collagen fibers.

Bowman's capsule An elaboration of the renal tubule, composed of podocytes, that surrounds and collects the filtrate from the glomerulus.

brain The centralized integrative center of a nervous system.

brainstem The portion of the vertebrate brain between the spinal cord and the forebrain, made up of the medulla, pons, and midbrain.

brassinosteroids Plant steroid hormones that mediate light effects promoting the elongation of stems and pollen tubes.

Broca's area A portion of the human brain essential for speech. Located in the frontal lobe just in front of the primary motor cortex.

bronchioles The smallest airways in a vertebrate lung, branching off the bronchi.

bronchus (plural: bronchi) The major airway(s) branching off the trachea into the vertebrate lung.

brown fat In mammals, fat tissue that is specialized to produce heat. It has many mitochondria and capillaries, and a protein that uncouples oxidative phosphorylation.

budding Asexual reproduction in which a more or less complete new organism grows from the body of the parent organism, eventually detaching itself.

buffer A substance that can transiently accept or release hydrogen ions and thereby resist changes in pH.

bulbourethral glands Secretory structures of the human male reproductive system that produce a small volume of an alkaline, mucoid secretion that helps neutralize acidity in the urethra and lubricate it to facilitate the passage of semen.

bulk flow The movement of a solution from a region of higher pressure potential to a region of lower pressure potential.

bundle of His Fibers of modified cardiac muscle that conduct action potentials from the atria to the ventricular muscle mass.

bundle sheath cell Part of a tissue that surrounds the veins of plants; contains chloroplasts in C_4 plants.

- C -

C3 plants Plants that produce 3PG as the first stable product of carbon fixation in photosynthesis and use ribulose bisphosphate as a CO_2 receptor.

C4 plants Plants that produce oxaloacetate as the first stable product of carbon fixation in photosynthesis and use phosphoenolpyruvate as CO_2 acceptor. C_4 plants also perform the reactions of C_3 photosynthesis.

calcitonin Hormone produced by the thyroid gland; lowers blood calcium and promotes bone formation. (Contrast with parathyroid hormone.)

calorie [L. *calor*: heat] The amount of heat required to raise the temperature of 1 gram of water by 1°C. Physiologists commonly use the kilocalorie (kcal) as a unit of measure (1 kcal = 1,000 calories). Nutritionists also use the kilocalorie, but refer to it as the *Calorie* (capital C).

Calvin cycle The stage of photosynthesis in which CO_2 reacts with RuBP to form 3PG, 3PG is reduced to a sugar, and RuBP is regenerated, while other products are released to the rest of the plant. Also known as the Calvin–Benson cycle.

calyx (kay' licks) [Gk. *kalyx*: cup] All of the sepals of a flower, collectively.

CAM See crassulacean acid metabolism.

Cambrian explosion The rapid diversification of multicellular life that took place during the Cambrian period.

cAMP (cyclic AMP) A compound formed from ATP that acts as a second messenger.

cancellous bone A type of bone with numerous internal cavities that make it appear spongy, although it is rigid. (Contrast with compact bone.)

canopy The leaf-bearing part of a tree. Collectively, the aggregate of the leaves and branches of the larger woody plants of an ecological community.

capillaries [L. *capillaris*: hair] Very small tubes, especially the smallest blood-carrying vessels of animals between the termination of the arteries and the beginnings of the veins. Capillaries are the site of exchange of materials between the blood and the interstitial fluid.

capsid The outer shell of a virus that encloses its nucleic acid.

carbohydrates Organic compounds containing carbon, hydrogen, and oxygen in the ratio 1:2:1 (i.e., with the general formula $C_nH_{2n}O_n$). Common examples are sugars, starch, and cellulose.

carbon skeleton The chains or rings of carbon atoms that form the structural basis of organic molecules. Other atoms or functional groups are attached to the carbon atoms.

carboxylase An enzyme that catalyzes the addition of carboxyl groups to a substrate.

cardiac (kar' dee ak) [Gk. *kardia*: heart] Pertaining to the heart and its functions.

cardiac cycle Contraction of the two atria of the heart, followed by contraction of the two ventricles and then relaxation.

cardiac muscle A type of muscle tissue that makes up, and is responsible for the beating of, the heart. Characterized by branching cells with single nuclei and a striated (striped) appearance. (Contrast with smooth muscle, skeletal muscle.)

cardiovascular system [Gk. *kardia*: heart + L. *vasculum*: small vessel] The heart, blood, and vessels of a circulatory system.

carnivore [L. *carn*: flesh + *vorare*: to devour] An organism that eats animal tissues. (Contrast with detritivore, herbivore, omnivore.)

carotenoid (ka rah' tuh noid) A yellow, orange, or red lipid pigment commonly found as an accessory pigment in photosynthesis; also found in fungi.

carotid body   A chemosensor in the carotid artery that senses a decrease in blood supply or a dramatic decrease in partial pressure of oxygen in the blood.

carpel (kar' pel) [Gk. *karpos*: fruit]   The organ of the flower that contains one or more ovules.

carrier   (1) In facilitated diffusion, a membrane protein that binds a specific molecule and transports it through the membrane. (2) In respiratory and photosynthetic electron transport, a participating substance such as NAD that exists in both oxidized and reduced forms. (3) In genetics, a person heterozygous for a recessive trait.

carrying capacity (K)   The number of individuals in a population that the resources of its environment can support.

cartilage   In vertebrates, a tough connective tissue found in joints, the outer ear, and elsewhere. Forms the entire skeleton in some animal groups.

cartilage bone   A type of bone that begins its development as a cartilaginous structure resembling the future mature bone, then gradually hardens into mature bone. (Contrast with membranous bone.)

Casparian strip   A band of cell wall containing suberin and lignin, found in the endodermis. Restricts the movement of water across the endodermis.

caspase   One of a group of proteases that catalyze cleavage of target proteins and are active in apoptosis.

catabolic reaction (kat uh bah' lik) [Gk. *kata*: to break down + *ballein*: to throw]   A synthetic reaction in which complex molecules are broken down into simpler ones and energy is released. (Contrast with anabolic reaction.)

catabolite repression   In the presence of abundant glucose, the diminished synthesis of catabolic enzymes for other energy sources.

catalyst (kat' a list) [Gk. *kata*: to break down]   A chemical substance that accelerates a reaction without itself being consumed in the overall course of the reaction. Catalysts lower the activation energy of a reaction. Enzymes are biological catalysts.

cation (cat' eye on)   An ion with one or more positive charges. (Contrast with anion.)

caudal [L. *cauda*: tail]   Pertaining to the tail, or to the posterior part of the body.

cDNA   *See* complementary DNA.

cDNA library   A collection of complementary DNAs derived from mRNAs of a particular tissue at a particular time in the life cycle of an organism.

cecum (see' cum) [L. blind]   A blind branch off the large intestine. In many nonruminant mammals, the cecum contains a colony of microorganisms that contribute to the digestion of food.

cell   The simplest structural unit of a living organism. In multicellular organisms, the building blocks of tissues and organs.

cell adhesion molecules   Molecules on animal cell surfaces that affect the selective association of cells into tissues during development of the embryo.

cell cycle   The stages through which a cell passes between one division and the next. Includes all stages of interphase and mitosis.

cell division   The reproduction of a cell to produce two new cells. In eukaryotes, this process involves nuclear division (mitosis) and cytoplasmic division (cytokinesis).

cell fate   The type of cell that an undifferentiated cell in an embryo will become in the adult.

cell junctions   Specialized structures associated with the plasma membranes of epithelial cells. Some contribute to cell adhesion, others to intercellular communication.

cell recognition   Binding of cells to one another mediated by membrane proteins or carbohydrates.

cell theory   States that cells are the basic structural and physiological units of all living organisms, and that all cells come from preexisting cells.

cell wall   A relatively rigid structure that encloses cells of plants, fungi, many protists, and most prokaryotes, and which gives these cells their shape and limits their expansion in hypotonic media.

cellular immune response   Immune system response mediated by T cells and directed against parasites, fungi, intracellular viruses, and foreign tissues (grafts). (Contrast with humoral immune response.)

cellular respiration   The catabolic pathways by which electrons are removed from various molecules and passed through intermediate electron carriers to O_2, generating H_2O and releasing energy.

cellulose (sell' you lowss)   A straight-chain polymer of glucose molecules, used by plants as a structural supporting material.

central dogma   The premise that information flows from DNA to RNA to polypeptide.

central nervous system (CNS)   That portion of the nervous system that is the site of most information processing, storage, and retrieval; in vertebrates, the brain and spinal cord. (Contrast with peripheral nervous system.)

central vacuole   In plant cells, a large organelle that stores the waste products of metabolism and maintains turgor.

centrifuge [L. *centrum*: center + *fugere*: to flee]   A laboratory device in which a sample is spun around a central axis at high speed. Used to separate suspended materials of different densities.

centriole (sen' tree ole)   A paired organelle that helps organize the microtubules in animal and protist cells during nuclear division.

centromere (sen' tro meer) [Gk. *centron*: center + *meros*: part]   The region where sister chromatids join.

centrosome (sen' tro soam)   The major microtubule organizing center of an animal cell.

cephalization (sef ah luh zay' shun) [Gk. *kephale*: head]   The evolutionary trend toward increasing concentration of brain and sensory organs at the anterior end of the animal.

cerebellum (sair uh bell' um) [L. diminutive of *cerebrum*, brain]   The brain region that controls muscular coordination; located at the anterior end of the hindbrain.

cerebral cortex   The thin layer of gray matter (neuronal cell bodies) that overlies the cerebrum.

cerebrum (su ree' brum) [L. brain]   The dorsal anterior portion of the forebrain, making up the largest part of the brain of mammals; the chief coordination center of the nervous system; consists of two *cerebral hemispheres*.

cervix (sir' vix) [L. neck]   The opening of the uterus into the vagina.

cGMP (cyclic guanosine monophosphate)   An intracellular messenger that is part of signal transmission pathways involving G proteins. (*See* G protein.)

channel protein   An integral membrane protein that forms an aqueous passageway across the membrane in which it is inserted and through which specific solutes may pass.

chaperone   A protein that guards other proteins by counteracting molecular interactions that threaten their three-dimensional structure.

character   In genetics, an observable feature, such as eye color. (Contrast with trait.)

character displacement   An evolutionary phenomenon in which species that compete for the same resources within the same territory tend to diverge in morphology and/or behavior.

chemical bond   An attractive force stably linking two atoms.

chemical evolution   The theory that life originated through the chemical transformation of inanimate substances.

chemical reaction   The change in the composition or distribution of atoms of a substance with consequent alterations in properties.

chemical synapse   Neural junction at which neurotransmitter molecules released from a presynaptic cell induce changes in a postsynaptic cell. (Contrast with electrical synapse.)

chemically gated channel   A type of gated channel that opens or closes depending on the presence or absence of a specific molecule, which binds to the channel protein or to a separate receptor that in turn alters the three-dimensional shape of channel protein.

chemiosmosis   Formation of ATP in mitochondria and chloroplasts, resulting from a pumping of protons across a membrane (against a gradient of electrical charge and of pH), followed by the return of the protons through a protein channel with ATP synthase activity.

chemoautotroph   *See* chemolithotroph.

chemoheterotroph   An organism that must obtain both carbon and energy from organic substances. (Contrast with chemolithotroph, photoautotroph, photoheterotroph.)

chemolithotroph [Gk. *lithos*: stone, rock]   An organism that uses carbon dioxide as a carbon source and obtains energy by oxidizing inorganic substances from its environment; also called chemoautotroph. (Contrast with chemoheterotroph, photoautotroph, photoheterotroph.)

chemoreceptor   A sensory receptor cell that senses specific molecules (such as odorant molecules or pheromones) in the environment.

chiasma (kie az' muh) (plural: chiasmata) [Gk. cross]   An X-shaped connection between paired homologous chromosomes in prophase I of meiosis. A chiasma is the visible manifestation of crossing over between homologous chromosomes.

chitin (kye' tin) [Gk. *kiton*: tunic]   The characteristic tough but flexible organic component of the exoskeleton of arthropods, consisting of a complex, nitrogen-containing polysaccharide. Also found in cell walls of fungi.

chlorophyll (klor' o fill) [Gk. *kloros*: green + *phyllon*: leaf]   Any of several green pigments associated with chloroplasts or with certain bacterial membranes; responsible for trapping light energy for photosynthesis.

chloroplast [Gk. *kloros*: green + *plast*: a particle] An organelle bounded by a double membrane containing the enzymes and pigments that perform photosynthesis. Chloroplasts occur only in eukaryotes.

choanocyte (ko' an uh site)   The collared, flagellated feeding cells of sponges.

cholecystokinin (ko' luh sis tuh kai' nin)   A hormone produced and released by the lining of the duodenum when it is stimulated by undigested fats and proteins. It stimulates the gallbladder to release bile and slows stomach activity.

chorion (kor' ee on) [Gk. *khorion*: afterbirth]   The outermost of the membranes protecting mammal, bird, and reptile embryos; in mammals it forms part of the placenta.

chromatid (kro' ma tid)   A newly replicated chromosome, from the time molecular duplication occurs until the time the centromeres separate (during anaphase of mitosis or of meiosis II).

chromatin   The nucleic acid–protein complex that makes up eukaryotic chromosomes.

chromosomal mutation   Loss of or changes in position/direction of a DNA segment on a chromosome.

chromosome (krome' o sowm) [Gk. *kroma*: color + *soma*: body]   In bacteria and viruses, the DNA molecule that contains most or all of the genetic information of the cell or virus. In eukaryotes, a structure composed of DNA and proteins that bears part of the genetic information of the cell.

chylomicron (ky low my' cron)   Particles of lipid coated with protein, produced in the gut from dietary fats and secreted into the extracellular fluids.

chyme (kime) [Gk. *kymus*: juice]   Created in the stomach; a mixture of ingested food with the digestive juices secreted by the salivary glands and the stomach lining.

cilium (sil' ee um) (plural: cilia) [L. eyelash]   Hairlike organelle used for locomotion by many unicellular organisms and for moving water and mucus by many multicellular organisms. Generally shorter than a flagellum.

circadian rhythm (sir kade' ee an) [L. *circa*: approximately + *dies*: day]   A rhythm of growth or activity that recurs about every 24 hours.

circannual rhythm [L. *circa*: + *annus*: year]   A rhythm of growth or activity that recurs on a yearly basis.

circulatory system   A system consisting of a muscular pump (heart), a fluid (blood or hemolymph), and a series of conduits (blood vessels) that transports materials around the body.

citric acid cycle   In cellular respiration, a set of chemical reactions whereby acetyl CoA is oxidized to carbon dioxide and hydrogen atoms are stored as NADH and FADH$_2$. Also called the Krebs cycle.

clade [Gk. *klados*: branch]   A monophyletic group made up of an ancestor and all of its descendants.

class I MHC molecules   Cell surface proteins that participate in the cellular immune response directed against virus-infected cells.

class II MHC molecules   Cell surface proteins that participate in the cell–cell interactions (of T-helper cells, macrophages, and B cells) of the humoral immune response.

cleavage   The first few cell divisions of an animal zygote. *See also* complete cleavage, incomplete cleavage.

climate   The long-term average atmospheric conditions (temperature, precipitation, humidity, wind direction and velocity) found in a region.

climax community   The final stage of succession; a community that is capable of perpetuating itself under local climatic and soil conditions and persists for a relatively long time.

clinal variation [Gk. *klinein*: to lean]   Gradual change in the phenotype of a species over a geographic gradient.

clonal deletion   Inactivation or destruction of lymphocyte clones that would produce immune reactions against the animal's own body.

clonal selection   Mechanism by which exposure to antigen results in the activation of selected T- or B-cell clones, resulting in an immune response.

clone [Gk. *klon*: twig, shoot]   (1) Genetically identical cells or organisms produced from a common ancestor by asexual means. (2) To produce many identical copies of a DNA sequence by its introduction into, and subsequent asexual reproduction of, a cell or organism.

closed circulatory system   Circulatory system in which the circulating fluid is contained within a continuous system of vessels. (Contrast with open circulatory system.)

coastal zone   The marine life zone that extends from the shoreline to the edge of the continental shelf. Characterized by relatively shallow, well-oxygenated water and relatively stable temperatures and salinities.

coccus (kock' us) (plural: cocci) [Gk. *kokkos*: berry, pit]   Any of various spherical or spheroidal bacteria.

cochlea (kock' lee uh) [Gk. *kokhlos*: snail]   A spiral tube in the inner ear of vertebrates; it contains the sensory cells involved in hearing.

codominance   A condition in which two alleles at a locus produce different phenotypic effects and both effects appear in heterozygotes.

codon   Three nucleotides in messenger RNA that direct the placement of a particular amino acid into a polypeptide chain. (Contrast with anticodon.)

coelom (see' loam) [Gk. *koiloma*: cavity]   An animal body cavity, enclosed by muscular mesoderm and lined with a mesodermal layer called peritoneum that also surrounds the internal organs.

coenocytic (seen' a sit ik) [Gk. *koinos*: common + *kytos*: container]   Referring to the condition, found in some fungal hyphae, of "cells" containing many nuclei but enclosed by a single plasma membrane. Results from nuclear division without cytokinesis.

coenzyme   A nonprotein organic molecule that plays a role in catalysis by an enzyme.

coevolution   Evolutionary processes in which an adaptation in one species leads to the evolution of an adaptation in a species with which it interacts; also known as reciprocal adaptation.

cofactor   An inorganic ion that is weakly bound to an enzyme and required for its activity.

cohesin   A protein involved in binding chromatids together.

cohesion   The tendency of molecules (or any substances) to stick together.

cohort (co' hort) [L. *cohors*: company of soldiers]   A group of similar-aged organisms.

coleoptile   A sheath that surrounds and protects the shoot apical meristem and young primary leaves of a grass seedling as they move through the soil.

collagen [Gk. *kolla*: glue]   A fibrous protein found extensively in bone and connective tissue.

collecting duct   In vertebrates, a tubule that receives urine produced in the nephrons of the kidney and delivers that fluid to the ureter for excretion.

collenchyma (cull eng' kyma) [Gk. *kolla*: glue + *enchyma*: infusion]   A type of plant cell, living at functional maturity, which lends flexible support by virtue of primary cell walls thickened at the corners. (Contrast with parenchyma, sclerenchyma.)

colon [Gk. *kolon*]   The large intestine.

commensalism [L. *com*: together + *mensa*: table]   A type of interaction between species in which one participant benefits while the other is unaffected.

communication   A signal from one organism (or cell) that alters the functioning or behavior of another organism (or cell).

community   Any ecologically integrated group of species of microorganisms, plants, and animals inhabiting a given area.

compact bone   A type of bone with a solid, hard structure. (Contrast with cancellous bone.)

companion cell   In angiosperms, a specialized cell found adjacent to a sieve tube element.

comparative experiment   Experimental design in which data from various unmanipulated samples or populations are compared, but in which variables are not controlled or even necessarily identified. (Contrast with controlled experiment.)

comparative genomics   Computer-aided comparison of DNA sequences between different organisms to reveal genes with related functions.

competition   In ecology, use of the same resource by two or more species when the resource is present in insufficient supply for the combined needs of the species.

competitive exclusion   A result of competition between species for a limiting resource in which one species completely eliminates the other.

competitive inhibitor   A nonsubstrate that binds to the active site of an enzyme and thereby inhibits binding of its substrate. (Contrast with noncompetitive inhibitor.)

complement system   A group of eleven proteins that play a role in some reactions of the immune system. The complement proteins are not immunoglobulins.

complementary base pairing   The AT (or AU), TA (or UA), CG, and GC pairing of bases in double-stranded DNA, in transcription, and between tRNA and mRNA.

complementary DNA (cDNA)   DNA formed by reverse transcriptase acting with an RNA template; essential intermediate in the reproduction of retroviruses; used as a tool in recombinant DNA technology; lacks introns.

complete cleavage   Pattern of cleavage that occurs in eggs that have little yolk. Early cleavage furrows divide the egg completely and the blastomeres are of similar size. (Contrast with incomplete cleavage.)

complete metamorphosis   A change of state during the life cycle of an organism in which the body is almost completely rebuilt to produce an individual with a very different body form. Characteristic of insects such as butterflies, moths, beetles, ants, wasps, and flies.

compound (1) A substance made up of atoms of more than one element. (2) Made up of many units, as in the *compound eyes* of arthropods.

concerted evolution The common evolution of a family of repeated genes, such that changes in one copy of the gene family are replicated in other copies of the gene family.

condensation reaction A chemical reaction in which two molecules become connected by a covalent bond and a molecule of water is released (AH + BOH → AB + H_2O.) (Contrast with hydrolysis reaction.)

conditional mutation A mutation that results in a characteristic phenotype only under certain environmental conditions.

conduction The transfer of heat from one object to another through direct contact.

cone (1) In conifers, a reproductive structure consisting of spore-bearing scales extending from a central axis. (Contrast with strobilus.) (2) In the vertebrate retina, a type of photoreceptor cell responsible for color vision.

conidium (ko nid' ee um) (plural: conidia) [Gk. *konis*: dust] A type of haploid fungal spore borne at the tips of hyphae, not enclosed in sporangia.

conjugation (kon ju gay' shun) [L. *conjugare*: yoke together] (1) A process by which DNA is passed from one cell to another through a *conjugation tube*, as in bacteria. (2) A nonreproductive sexual process by which *Paramecium* and other ciliates exchange genetic material.

connective tissue A type of tissue that connects or surrounds other tissues; its cells are embedded in a collagen-containing matrix. One of the four major tissue types in multicellular animals.

connexon In a gap junction, a protein channel linking adjacent animal cells.

consensus sequences Short stretches of DNA that appear, with little variation, in many different genes.

conservation biology An applied science that carries out investigations with the aim of maintaining the diversity of life on Earth.

conserved Pertaining to a gene or trait that has evolved very slowly and is similar or even identical in individuals of highly divergent groups.

conspecifics Individuals of the same species.

constant region The portion of an immunoglobulin molecule whose amino acid composition determines its class and does not vary among immunoglobulins in that class. (Contrast with variable region.)

constitutive Always present; produced continually at a constant rate. (Contrast with inducible.)

consumer An organism that eats the tissues of some other organism.

continental drift The gradual movements of the world's continents that have occurred over billions of years.

contractile vacuole (kon trak' tul) A specialized vacuole that collects excess water taken in by osmosis, then contracts to expel the water from the cell.

controlled experiment An experiment in which a sample is divided into groups whereby experimental groups are exposed to manipulations of an independent variable while one group serves as an untreated control. The data from the various groups are then compared to see if there are changes in a dependent variable as a result of the

experimental manipulation. (Contrast with comparative experiment.)

convection The transfer of heat to or from a surface via a moving stream of air or fluid.

convergent evolution Independent evolution of similar features from different ancestral traits.

copulation Reproductive behavior that results in a male depositing sperm in the reproductive tract of a female.

cork cambium [L. *cambiare*: to exchange] In plants, a lateral meristem that produces secondary growth, mainly in the form of waxy-walled protective cells, including some of the cells that become bark.

cornea The clear, transparent tissue that covers the eye and allows light to pass through to the retina.

corolla (ko role' lah) [L. *corolla*: a small crown] All of the petals of a flower, collectively.

coronary artery (kor' oh nair ee) An artery that supplies blood to the heart muscle.

coronary thrombosis A fibrous clot that blocks a coronary artery.

corpus luteum (kor' pus loo' tee um) (plural: corpora lutea) [L. yellow body] A structure formed from a follicle after ovulation; produces hormones important to the maintenance of pregnancy.

corridor A connection between habitat patches through which organisms can disperse; plays a critical role in maintaining subpopulations.

cortex [L. *cortex*: covering, rind] (1) In plants, the tissue between the epidermis and the vascular tissue of a stem or root. (2) In animals, the outer tissue of certain organs, such as the adrenal gland (adrenal cortex) and the brain (cerebral cortex).

corticosteroids Steroid hormones produced and released by the cortex of the adrenal gland.

corticotropin A tropic hormone produced by the anterior pituitary hormone that stimulates cortisol release from the adrenal cortex. Also called adrenocorticotropic hormone (ACTH).

corticotropin-releasing hormone A releasing hormone produced by the hypothalamus that controls the release of cortisol from the anterior pituitary.

cortisol A corticosteroid that mediates stress responses.

cost–benefit analysis An approach to evolutionary studies that assumes an animal has a limited amount of time and energy to devote to each of its activities, and that each activity has fitness costs as well as benefits. (*See also* trade-off.)

cotyledon (kot' ul lee' dun) [Gk. *kotyledon*: hollow space] A "seed leaf." An embryonic organ that stores and digests reserve materials; may expand when seed germinates.

countercurrent flow An arrangement that promotes the maximum exchange of heat, or of a diffusible substance, between two fluids by having the fluids flow in opposite directions through parallel vessels close together.

countercurrent multiplier The mechanism that increases the concentration of the interstitial fluid in the mammalian kidney through countercurrent flow in the loops of Henle and selective permeability and active transport of ions by segments of the loops of Henle.

covalent bond Chemical bond based on the sharing of electrons between two atoms.

CpG islands DNA regions rich in C residues adjacent to G residues. Especially abundant in promoters, these regions are where methylation of cytosine usually occurs.

crassulacean acid metabolism (CAM) A metabolic pathway enabling the plants that possess it to store carbon dioxide at night and then perform photosynthesis during the day with stomata closed.

critical night length In the photoperiodic flowering response of short-day plants, the length of night above which flowering occurs and below which the plant remains vegetative. (The reverse applies in the case of long-day plants.)

critical period *See* sensitive period.

cross section A section taken perpendicular to the longest axis of a structure. Also called a transverse section.

crossing over The mechanism by which linked genes undergo recombination. In general, the term refers to the reciprocal exchange of corresponding segments between two homologous chromatids.

crypsis [Gk. *kryptos*: hidden] The resemblance of an organism to some part of its environment, which helps it to escape detection by enemies.

cryptochromes [Gk. *kryptos*: hidden + *kroma*: color] Photoreceptors mediating some blue-light effects in plants and animals.

ctene (teen) [Gk. *cteis*: comb] In ctenophores, a comblike row of cilia-bearing plates. Ctenophores move by beating the cilia on their eight ctenes.

culture (1) A laboratory association of organisms under controlled conditions. (2) The collection of knowledge, tools, values, and rules that characterize a human society.

cuticle (1) In plants, a waxy layer on the outer body surface that retards water loss. (2) In ecdysozoans, an outer body covering that provides protection and support and is periodically molted.

cyclic AMP *See* cAMP.

cyclic electron transport In photosynthetic light reactions, the flow of electrons that produces ATP but no NADPH or O_2.

cyclin A protein that activates a cyclin-dependent kinase, bringing about transitions in the cell cycle.

cyclin-dependent kinase (Cdk) A protein kinase whose target proteins are involved in transitions in the cell cycle and which is active only when complexed with additional protein subunits, called cyclins.

cytokine A regulatory protein made by immune system cells that affects other target cells in the immune system.

cytokinesis (sy' toe kine ee' sis) [Gk. *kytos*: container + *kinein*: to move] The division of the cytoplasm of a dividing cell. (Contrast with mitosis.)

cytokinin (sy' toe kine' in) A member of a class of plant growth substances that plays roles in senescence, cell division, and other phenomena.

cytoplasm The contents of the cell, excluding the nucleus.

cytoplasmic determinants In animal development, gene products whose spatial distribution may determine such things as embryonic axes.

cytoplasmic segregation The asymmetrical distribution of cytoplasmic determinants in a developing animal embryo.

cytosine (C) (site' oh seen) A nitrogen-containing base found in DNA and RNA.

cytoskeleton The network of microtubules and microfilaments that gives a eukaryotic cell its shape and its capacity to arrange its organelles and to move.

cytosol The fluid portion of the cytoplasm, excluding organelles and other solids.

cytotoxic T cells (T$_C$) Cells of the cellular immune system that recognize and directly eliminate virus-infected cells. (Contrast with T-helper cells.)

- D -

DAG *See* diacylglycerol.

daughter chromosomes During mitosis, the separated chromatids from the beginning of anaphase onward.

dead space The lung volume that fails to be ventilated with fresh air (because the lungs are never completely emptied during exhalation).

deciduous [L. *deciduus*: falling off] Pertaining to a woody plant that sheds its leaves but does not die.

declarative memory Memory of people, places, events, and things that can be consciously recalled and described. (Contrast with procedural memory.)

decomposer An organism that metabolizes organic compounds in debris and dead organisms, releasing inorganic material; found among the bacteria, protists, and fungi. *See also* detritivore, saprobe.

defensin A type of protein made by phagocytes that kills bacteria and enveloped viruses by insertion into their plasma membranes.

degeneracy The situation in which a single amino acid may be represented by any of two or more different codons in messenger RNA. Most of the amino acids can be represented by more than one codon.

deletion A mutation resulting from the loss of a continuous segment of a gene or chromosome. Such mutations almost never revert to wild type. (Contrast with duplication, point mutation.)

demethylase An enzyme that catalyzes the removal of the methyl group from cytosine, reversing DNA methylation.

demography The study of population structure and of the processes by which it changes.

denaturation Loss of activity of an enzyme or nucleic acid molecule as a result of structural changes induced by heat or other means.

dendrite [Gk. *dendron*: tree] A fiber of a neuron which often cannot carry action potentials. Usually much branched and relatively short compared with the axon, and commonly carries information to the cell body of the neuron.

denitrification Metabolic activity by which nitrate and nitrite ions are reduced to form nitrogen gas; carried out by certain soil bacteria.

denitrifiers Bacteria that release nitrogen to the atmosphere as nitrogen gas (N_2).

density-dependent Pertaining to a factor with an effect on population size that increases in proportion to population density.

density-independent Pertaining to a factor with an effect on population size that acts independently of population density.

deoxyribonucleic acid *See* DNA.

deoxyribose A five-carbon sugar found in nucleotides and DNA.

depolarization A change in the resting potential across a membrane so that the inside of the cell becomes less negative, or even positive, compared with the outside of the cell. (Contrast with hyperpolarization.)

derived trait A trait that differs from the ancestral trait. (Contrast with synapomorphy.)

dermal tissue system The outer covering of a plant, consisting of epidermis in the young plant and periderm in a plant with extensive secondary growth. (Contrast with ground tissue system and vascular tissue system.)

desmosome (dez' mo sowm) [Gk. *desmos*: bond + *soma*: body] An adhering junction between animal cells.

desmotubule A membrane extension connecting the endoplasmic retituclum of two plant cells that traverses the plasmodesma.

determination In development, the process whereby the fate of an embryonic cell or group of cells (e.g., to become epidermal cells or neurons) is set.

determinate growth A growth pattern in which the growth of an organism or organ ceases when an adult state is reached; characteristic of most animals and some plant organs. (Contrast with indeterminate growth.)

detritivore (di try' ti vore) [L. *detritus*: worn away + *vorare*: to devour] An organism that obtains its energy from the dead bodies or waste products of other organisms.

developmental module A functional entity in the embryo encompassing genes and signaling pathways that determine a physical structure independently of other such modules.

developmental plasticity The capacity of an organism to alter its pattern of development in response to environmental conditions.

diacylglycerol (DAG) In hormone action, the second messenger produced by hydrolytic removal of the head group of certain phospholipids.

diapause A period of developmental or reproductive arrest, entered in response to day length, that enables an organism to better survive.

diaphragm (dye' uh fram) [Gk. *diaphrassein*: barricade] (1) A sheet of muscle that separates the thoracic and abdominal cavities in mammals; responsible for breathing. (2) A method of birth control in which a sheet of rubber is fitted over the woman's cervix, blocking the entry of sperm.

diastole (dye ass' toll ee) [Gk. dilation] The portion of the cardiac cycle when the heart muscle relaxes. (Contrast with systole.)

diencephalon The portion of the vertebrate forebrain that develops into the thalamus and hypothalamus.

differential gene expression The hypothesis that, given that all cells contain all genes, what makes one cell type different from another is the difference in transcription and translation of those genes.

differentiation The process whereby originally similar cells follow different developmental pathways; the actual expression of determination.

diffuse coevolution The evolution of similar traits in suites of species experiencing similar selection pressures imposed by other suites of species with which they interact.

diffusion Random movement of molecules or other particles, resulting in even distribution of the particles when no barriers are present.

dihybrid cross A mating in which the parents differ with respect to the alleles of two loci of interest.

dikaryon (di care' ee ahn) [Gk. *di*: two + *karyon*: kernel] A cell or organism carrying two genetically distinguishable nuclei. Common in fungi.

dioecious (die eesh' us) [Gk. *di*: two + *oikos*: house] Pertaining to organisms in which the two sexes are "housed" in two different individuals, so that eggs and sperm are not produced in the same individuals. Examples: humans, fruit flies, date palms. (Contrast with monoecious.)

diploblastic Having two cell layers. (Contrast with triploblastic.)

diploid (dip' loid) [Gk. *diplos*: double] Having a chromosome complement consisting of two copies (homologs) of each chromosome. Designated $2n$. (Contrast with haploid.)

diplontic A type of life cycle in which gametes are the only haploid cells and mitosis occurs only in diploid cells. (Contrast with haplontic.)

direct transduction A cell signaling mechanism in which the receptor acts as the effector in the cellular response. (Contrast with indirect transduction.)

directional selection Selection in which phenotypes at one extreme of the population distribution are favored. (Contrast with disruptive selection, stabilizing selection.)

disaccharide A carbohydrate made up of two monosaccharides (simple sugars).

dispersal Movement of organisms away from a parent organism or from an existing population.

dispersion The distribution of individuals in space within a population.

disruptive selection Selection in which phenotypes at both extremes of the population distribution are favored. (Contrast with directional selection; stabilizing selection.)

distal Away from the point of attachment or other reference point. (Contrast with proximal.)

distal convoluted tubule The portion of a renal tubule from where it reaches the renal cortex, just past the loop of Henle to where it joins a collecting duct. (Compare with proximal convoluted tubule.)

disturbance A short-term event that disrupts populations, communities, or ecosystems by changing the environment.

disulfide bridge The covalent bond between two sulfur atoms (–S—S–) linking two molecules or remote parts of the same molecule.

DNA (deoxyribonucleic acid) The fundamental hereditary material of all living organisms. In eukaryotes, stored primarily in the cell nucleus. A nucleic acid using deoxyribose rather than ribose.

DNA fingerprint An individual's unique pattern of allele sequences, commonly short tandem repeats and single nucleotide polymorphisms.

DNA helicase An enzyme that functions to unwind the double helix.

DNA ligase Enzyme that unites broken DNA strands during replication and recombination.

DNA methylation The addition of methyl groups to to bases in DNA, usually cytosine or guanine.

DNA methyltransferase   An enzyme that catalyzes the methylation of DNA.

DNA microarray   A small glass or plastic square onto which thousands of single-stranded DNA sequences are fixed so that hybridization of cell-derived RNA or DNA to the target sequences can be performed.

DNA polymerase   Any of a group of enzymes that catalyze the formation of DNA strands from a DNA template.

DNA topoisomerase   An enzyme that unwinds and winds coils of DNA that form during replication and transcription.

docking protein   A receptor protein that binds (docks) a ribosome to the membrane of the endoplasmic reticulum by binding the signal sequence attached to a new protein being made at the ribosome.

domain   (1) An independent structural element within a protein. Encoded by recognizable nucleotide sequences, a domain often folds separately from the rest of the protein. Similar domains can appear in a variety of different proteins across phylogenetic groups (e.g., "homeobox domain"; "calcium-binding domain"). (2) In phylogenetics, the three monophyletic branches of life (Bacteria, Archaea, and Eukarya).

dominance   In genetics, the ability of one allelic form of a gene to determine the phenotype of a heterozygous individual in which the homologous chromosomes carry both it and a different (recessive) allele. (Contrast with recessive.)

dormancy   A condition in which normal activity is suspended, as in some spores, seeds, and buds.

dorsal   [L. *dorsum*: back]   Toward or pertaining to the back or upper surface. (Contrast with ventral.)

dorsal lip   In amphibian embryos, the dorsal segment of the blastopore. Also called the "organizer," this region directs the development of nearby embryonic regions.

double fertilization   In angiosperms, a process in which the nuclei of two sperm fertilize one egg. One sperm's nucleus combines with the egg nucleus to produce a zygote, while the other combines with the same egg's two polar nuclei to produce the first cell of the triploid endosperm (the tissue that will nourish the growing plant embryo).

double helix   Refers to DNA and the (usually right-handed) coil configuration of two complementary, antiparallel strands.

downregulation   A negative feedback process in which continuous high concentrations of a hormone can decrease the number of its receptors. (Contrast with upregulation.)

duodenum   (do' uh dee' num)   The beginning portion of the vertebrate small intestine. (Contrast with ileum, jejunum.)

duplication   A mutation in which a segment of a chromosome is duplicated, often by the attachment of a segment lost from its homolog. (Contrast with deletion.)

- E -

ecdysone   (eck die' sone) [Gk. *ek*: out of + *dyo*: to clothe]   In insects, a hormone that induces molting.

ecological efficiency   The overall transfer of energy from one trophic level to the next, expressed as the ratio of consumer production to producer production.

ecology   [Gk. *oikos*: house]   The scientific study of the interaction of organisms with their living (biotic) and nonliving (abiotic) environments.

ecosystem   (eek' oh sis tum)   The organisms of a particular habitat, such as a pond or forest, together with the physical environment in which they live.

ecosystem engineer   An organism that builds structures that alter existing habitats or create new habitats.

ecosystem services   Processes by which ecosystems maintain resources that benefit human society.

ectoderm   [Gk. *ektos*: outside + *derma*: skin]   The outermost of the three embryonic germ layers first delineated during gastrulation. Gives rise to the skin, sense organs, and nervous system.

ectotherm   [Gk. *ektos*: outside + *thermos*: heat]   An animal that is dependent on external heat sources for regulating its body temperature (Contrast with endotherm.)

edema   (i dee' mah) [Gk. *oidema*: swelling]   Tissue swelling caused by the accumulation of fluid.

edge effect   The changes in ecological processes in a community caused by physical and biological factors originating in an adjacent community.

effector protein   In cell signaling, a protein responsible for the cellular reponse to a signal transduction pathway.

efferent   (ef' ur unt) [L. *ex*: out + *ferre*: to bear]   Carrying outward or away from, as in a neuron that carries impulses outward from the central nervous system (efferent neuron), or a blood vessel that carries blood away from a structure. (Contrast with afferent.)

egg   In all sexually reproducing organisms, the female gamete; in birds, reptiles, and some other vertebrates, a structure within which early embryonic development occurs. *See also* amniote egg, ovum.

electrical synapse   A type of synapse at which action potentials spread directly from presynaptic cell to postsynaptic cell. (Contrast with chemical synapse.)

electrocardiogram (ECG or EKG)   A graphic recording of electrical potentials from the heart.

electrochemical gradient   The concentration gradient of an ion across a membrane plus the voltage difference across that membrane.

electroencephalogram (EEG)   A graphic recording of electrical potentials from the brain.

electromagnetic radiation   A self-propagating wave that travels though space and has both electrical and magnetic properties.

electron   A subatomic particle outside the nucleus carrying a negative charge and very little mass.

electron shell   The region surrounding the atomic nucleus at a fixed energy level in which electrons orbit.

electron transport   The passage of electrons through a series of proteins with a release of energy which may be captured in a concentration gradient or chemical form such as NADH or ATP.

electronegativity   The tendency of an atom to attract electrons when it occurs as part of a compound.

electrophoresis   *See* gel electrophoresis.

element   A substance that cannot be converted to simpler substances by ordinary chemical means.

elongation   (1) In molecular biology, the addition of monomers to make a longer RNA or protein during transcription or translation. (2) Growth of a plant axis or cell primarily in the longitudinal direction.

embolus   (em' buh lus) [Gk. *embolos*: stopper]   A circulating blood clot. Blockage of a blood vessel by an embolus or by a bubble of gas is referred to as an *embolism*. (Contrast with thrombus.)

embryo   [Gk. *en*: within + *bryein*: to grow]   A young animal, or young plant sporophyte, while it is still contained within a protective structure such as a seed, egg, or uterus.

embryonic stem cell (ESC)   A pluripotent cell in the blastocyst.

embryo sac   In angiosperms, the female gametophyte. Found within the ovule, it consists of eight or fewer cells, membrane bounded, but without cellulose walls between them.

emergent property   A property of a complex system that is not exhibited by its individual component parts.

emigration   The deliberate and usually oriented departure of an organism from the habitat in which it has been living.

3′ end   (3 prime)   The end of a DNA or RNA strand that has a free hydroxyl group at the 3′ carbon of the sugar (deoxyribose or ribose).

5′ end   (5 prime)   The end of a DNA or RNA strand that has a free phosphate group at the 5′ carbon of the sugar (deoxyribose or ribose).

endemic   (en dem' ik) [Gk. *endemos*: native]   Confined to a particular region, thus often having a comparatively restricted distribution.

endergonic   A chemical reaction in which the products have higher free energy than the reactants, thereby requiring free energy input to occur. (Contrast with exergonic.)

endocrine gland   (en' doh krin) [Gk. *endo*: within + *krinein*: to separate]   An aggregation of secretory cells that secretes hormones into the blood. The *endocrine system* consists of all *endocrine cells* and endocrine glands in the body that produce and release hormones. (Contrast with exocrine gland.)

endocytosis   A process by which liquids or solid particles are taken up by a cell through invagination of the plasma membrane. (Contrast with exocytosis.)

endoderm   [Gk. *endo*: within + *derma*: skin]   The innermost of the three embryonic germ layers delineated during gastrulation. Gives rise to the digestive and respiratory tracts and structures associated with them.

endodermis   In plants, a specialized cell layer marking the inside of the cortex in roots and some stems. Frequently a barrier to free diffusion of solutes.

endomembrane system   A system of intracellular membranes that exchange material with one another, consisting of the Golgi apparatus, endoplasmic reticulum, and lysosomes when present.

endometrium   The epithelial lining of the uterus.

endoplasmic reticulum (ER)   [Gk. *endo*: within + L. *reticulum*: net]   A system of membranous tubes and flattened sacs found in the cytoplasm of eukaryotes. Exists in two forms: rough ER, studded with ribosomes; and smooth ER, lacking ribosomes.

endorphins   Molecules in the mammalian brain that act as neurotransmitters in pathways that control pain.

endoskeleton [Gk. *endo*: within + *skleros*: hard] An internal skeleton covered by other, soft body tissues. (Contrast with exoskeleton.)

endosperm [Gk. *endo*: within + *sperma*: seed]   A specialized triploid seed tissue found only in angiosperms; contains stored nutrients for the developing embryo.

endospore   [Gk. *endo*: within + *spora*: to sow] In some bacteria, a resting structure that can survive harsh environmental conditions.

endosymbiosis theory [Gk. *endo*: within + *sym*: together + *bios*: life]   The theory that the eukaryotic cell evolved via the engulfing of one prokaryotic cell by another.

endothelium   The single layer of epithelial cells lining the interior of a blood vessel.

endotherm [Gk. *endo*: within + *thermos*: heat] An animal that can control its body temperature by the expenditure of its own metabolic energy. (Contrast with ectotherm.)

endotoxin   A lipopolysaccharide that forms part of the outer membrane of certain Gram-negative bacteria that is released when the bacteria grow or lyse. (Contrast with exotoxin.)

energetic cost   The difference between the energy an animal expends in performing a behavior and the energy it would have expended had it rested.

energy   The capacity to do work or move matter against an opposing force. The capacity to accomplish change in physical and chemical systems.

energy budget   A quantitative description of all paths of energy exchange between an animal and its environment.

enkephalins   Molecules in the mammalian brain that act as neurotransmitters in pathways that control pain.

enthalpy (*H*)   The total energy of a system.

entropy (*S*) (en' tro pee) [Gk. *tropein*: to change] A measure of the degree of disorder in any system. Spontaneous reactions in a closed system are always accompanied by an increase in entropy.

enveloped virus   A virus enclosed within a phospholipid membrane derived from its host cell.

environment   Whatever surrounds and interacts with or otherwise affects a population, organism, or cell. May be external or internal.

environmentalism   The use of ecological knowledge, along with economics, ethics, and many other considerations, to inform both personal decisions and public policy relating to stewardship of natural resources and ecosystems.

enzyme (en' zime) [Gk. *zyme*: to leaven (as in yeast bread)]   A catalytic protein that speeds up a biochemical reaction.

epi- [Gk. upon, over]   A prefix used to designate a structure located on top of another; for example, epidermis, epiphyte.

epiblast   The upper or overlying portion of the avian blastula which is joined to the hypoblast at the margins of the blastodisc.

epiboly   The movement of cells over the surface of the blastula toward the forming blastopore.

epitope   *See* antigenic determinant.

epidermis [Gk. *epi*: over + *derma*: skin]   In plants and animals, the outermost cell layers. (Only one cell layer thick in plants.)

epididymis (epuh did' uh mus) [Gk. *epi*: over + *didymos*: testicle]   Coiled tubules in the testes that store sperm and conduct sperm from the seminiferous tubules to the vas deferens.

epigenetics   The scientific study of changes in the expression of a gene or set of genes that occur without change in the DNA sequence.

epinephrine (ep i nef' rin) [Gk. *epi*: over + *nephros*: kidney]   The "fight or flight" hormone produced by the medulla of the adrenal gland; it also functions as a neurotransmitter. (Also known as adrenaline.)

epistasis   Interaction between genes in which the presence of a particular allele of one gene determines whether another gene will be expressed.

epithelium   A type of animal tissue made up of sheets of cells that lines or covers organs, makes up tubules, and covers the surface of the body; one of the four major tissue types in multicellular animals.

equilibrium   Any state of balanced opposing forces and no net change.

ER   *See* endoplasmic reticulum.

error signal   In regulatory systems, any difference between the set point of the system and its current condition.

erythrocyte (ur rith' row site) [Gk. *erythros*: red + *kytos*: container]   A red blood cell.

erythropoietin   A hormone produced by the kidney in response to lack of oxygen that stimulates the production of red blood cells.

esophagus (i soff' i gus) [Gk. *oisophagos*: gullet] That part of the gut between the pharynx and the stomach.

essential acids   Amino acids or fatty acids that an animal cannot synthesize for itself and must obtain from its food.

essential element   A mineral nutrient required for normal growth and reproduction in plants and animals.

ester linkage   A condensation (water-releasing) reaction in which the carboxyl group of a fatty acid reacts with the hydroxyl group of an alcohol. Lipids are formed in this way.

estivation (ess tuh vay' shun) [L. *aestivalis*: summer]   A state of dormancy and hypometabolism that occurs during the summer; usually a means of surviving drought and/or intense heat. (Contrast with hibernation.)

estrogen   Any of several steroid sex hormones; produced chiefly by the ovaries in mammals.

estrus (es' trus) [L. *oestrus*: frenzy]   The period of heat, or maximum sexual receptivity, in some female mammals. Ordinarily, the estrus is also the time of release of eggs in the female.

ethology [Gk. *ethos*: character + *logos*: study]   An approach to the study of animal behavior that focuses on studying many species in natural environments and addresses questions about the evolution of behavior.

ethylene   One of the plant growth hormones, the gas $H_2C=CH_2$. Involved in fruit ripening and other growth and developmental responses.

eukaryotes (yew car' ree oats) [Gk. *eu*: true + *karyon*: kernel or nucleus]   Organisms whose cells contain their genetic material inside a nucleus. Includes all life other than the viruses, archaea, and bacteria.

eusocial   Pertaining to a social group that includes nonreproductive individuals, as in honey bees.

eutrophication (yoo trofe' ik ay' shun) [Gk. *eu*: truly + *trephein*: to flourish]   The addition of nutrient materials to a body of water, resulting in changes in ecological processes and species composition therein.

evaporation   The transition of water from the liquid to the gaseous phase.

evolution   Any gradual change. Most often refers to organic or Darwinian evolution, which is the genetic and resulting phenotypic change in populations of organisms from generation to generation. (*See* macroevolution, microevolution; contrast with speciation.)

evolutionary radiation   The proliferation of many species within a single evolutionary lineage.

evolutionary reversal   The reappearance of an ancestral trait in a group that had previously acquired a derived trait.

excision repair   A mechanism that removes damaged DNA and replaces it with the appropriate nucleotide.

excited state   The state of an atom or molecule when, after absorbing energy, it has more energy than in its normal, ground state.

excretion   Release of metabolic wastes by an organism.

exergonic   A chemical reaction in which the products of the reaction have lower free energy than the reactants, resulting in a release of free energy. (Contrast with endergonic.)

exocrine gland (eks' oh krin) [Gk. *exo*: outside + *krinein*: to separate]   Any gland, such as a salivary gland, that secretes to the outside of the body or into the gut. (Contrast with endocrine gland.)

exocytosis   A process by which a vesicle within a cell fuses with the plasma membrane and releases its contents to the outside. (Contrast with endocytosis.)

exon   A portion of a DNA molecule, in eukaryotes, that codes for part of a polypeptide. (Contrast with intron.)

exoskeleton (eks' oh skel' e ton) [Gk. *exos*: outside + *skleros*: hard]   A hard covering on the outside of the body to which muscles are attached. (Contrast with endoskeleton.)

exotoxin   A highly toxic, usually soluble protein released by living, multiplying bacteria. (Contrast with endotoxin.)

expanding triplet repeat   A three-base-pair sequence in a human gene that is unstable and can be repeated a few to hundreds of times. Often, the more the repeats, the less the activity of the gene involved. Expanding triplet repeats occur in some human diseases such as Huntington's disease and fragile-X syndrome.

experiment   A testing process to support or disprove hypotheses and to answer questions. The basis of the scientific method. *See* comparative experiment, controlled experiment.

expiratory reserve volume   The amount of air that can be forcefully exhaled beyond the normal tidal expiration. (Contrast with inspiratory reserve volume, tidal volume, vital capacity.)

exploitation competition Competition in which individuals reduce the quantities of their shared resources. (Contrast with interference competition.)

exponential growth Growth, especially in the number of organisms in a population, which is a geometric function of the size of the growing entity: the larger the entity, the faster it grows. (Contrast with logistic growth.)

expression vector A DNA vector, such as a plasmid, that carries a DNA sequence for the expression of an inserted gene into mRNA and protein in a host cell.

expressivity The degree to which a genotype is expressed in the phenotype; may be affected by the environment.

extensor A muscle that extends an appendage.

external fertilization The release of gametes into the environment; typical of aquatic animals. Also called spawning. (Contrast with internal fertilization.)

external gills Highly branched and folded extensions of the body surface that provide a large surface area for gas exchange with water; typical of larval amphibians and many larval insects.

extinction The termination of a lineage of organisms.

extracellular matrix A material of heterogeneous composition surrounding cells and performing many functions including adhesion of cells.

extraembryonic membranes Four membranes that support but are not part of the developing embryos of reptiles, birds, and mammals, defining these groups phylogenetically as amniotes. (See amnion, allantois, chorion, and yolk sac.)

- F -

F_1 The first filial generation; the immediate progeny of a parental (P) mating.

F_2 The second filial generation; the immediate progeny of a mating between members of the F_1 generation.

facilitated diffusion Passive movement through a membrane involving a specific carrier protein; does not proceed against a concentration gradient. (Contrast with active transport, diffusion.)

facilitation In succession, modification of the environment by a colonizing species in a way that allows colonization by other species. (Contrast with inhibition.)

facultative anaerobe A prokaryote that can shift its metabolism between anaerobic and aerobic modes depending on the presence or absence of O_2. (Alternatively, facultative aerobe.)

fast-twitch fibers Skeletal muscle fibers that can generate high tension rapidly, but fatigue rapidly ("sprinter" fibers). Characterized by an abundance of enzymes of glycolysis.

fat A triglyceride that is solid at room temperature. (Contrast with oil.)

fate map A diagram of the blastula showing which cells (blastomeres) are "fated " to contribute to specific tissues and organs in the mature body.

fatty acid A molecule made up of a long nonpolar hydrocarbon chain and a polar carboxyl group. Found in many lipids.

fauna (faw' nah) All the animals found in a given area. (Contrast with flora.)

feces [L. *faeces*: dregs] Waste excreted from the digestive system.

fecundity (m_x) The average number of offspring produced by each female.

feedback information In regulatory systems, information about the relationship between the set point of the system and its current state.

feedforward information In regulatory systems, information that changes the set point of the system.

fermentation (fur men tay' shun) [L. *fermentum*: yeast] The anaerobic degradation of a substance such as glucose to smaller molecules such as lactic acid or alcohol with the extraction of energy.

fertilization Union of gametes. Also known as syngamy.

fetus Medical and legal term for the stages of a developing human embryo from about the eighth week of pregnancy (the point at which all major organ systems have formed) to the moment of birth.

fiber In angiosperms, an elongated, tapering sclerenchyma cell, usually with a thick cell wall, that serves as a support function in xylem. (See also muscle fiber.)

fibrin A protein that polymerizes to form long threads that provide structure to a blood clot.

fibrinogen A circulating protein that can be stimulated to fall out of solution and provide the structure for a blood clot.

fibrous root system A root system typical of monocots composed of numerous thin adventitious roots that are all roughly equal in diameter. (Contrast with taproot system.)

Fick's law of diffusion An equation that describes the factors that determine the rate of diffusion of a molecule from an area of higher concentration to an area of lower concentration.

fight-or-flight response A rapid physiological response to a sudden threat mediated by the hormone epinephrine.

filter feeder An organism that feeds on organisms much smaller than itself that are suspended in water or air by means of a straining device.

first law of thermodynamics The principle that energy can be neither created nor destroyed.

fission See binary fission.

fitness The contribution of a genotype or phenotype to the genetic composition of subsequent generations, relative to the contribution of other genotypes or phenotypes. (See also inclusive fitness.)

flagellum (fla jell' um) (plural: flagella) [L. *flagellum*: whip] Long, whiplike appendage that propels cells. Prokaryotic flagella differ sharply from those found in eukaryotes.

fixed action pattern In ethology, a genetically determined behavior that is performed without learning, stereotypic (performed the same way each time), and not modifiable by learning.

flexor A muscle that flexes an appendage.

flora (flore' ah) All of the plants found in a given area. (Contrast with fauna.)

floral meristem In angiosperms, a meristem that forms the floral organs (sepals, petals, stamens, and carpels).

floral organ identity genes In angiosperms, genes that determine the fates of floral meristem cells; their expression is triggered by the products of meristem identity genes.

florigen A plant hormone involved in the conversion of a vegetative shoot apex to a flower.

flower The sexual structure of an angiosperm.

fluid feeder An animal that feeds on fluids it extracts from the bodies of other organisms; examples include nectar-feeding birds and blood-sucking insects.

fluid mosaic model A molecular model for the structure of biological membranes consisting of a fluid phospholipid bilayer in which suspended proteins are free to move in the plane of the bilayer.

follicle [L. *folliculus*: little bag] In female mammals, an immature egg surrounded by nutritive cells.

follicle-stimulating hormone (FSH) A gonadotropin produced by the anterior pituitary.

food chain A portion of a food web, most commonly a simple sequence of prey species and the predators that consume them.

food vacuole Membrane-enclosed structure formed by phagocytosis in which engulfed food particles are digested by the action of lysosomal enzymes.

food web The complete set of food links between species in a community; a diagram indicating which ones are the eaters and which are eaten.

forebrain The region of the vertebrate brain that comprises the cerebrum, thalamus, and hypothalamus.

fossil Any recognizable structure originating from an organism, or any impression from such a structure, that has been preserved over geological time.

fossil fuels Fuels, including oil, natural gas, coal, and peat, formed over geologic time from organic material buried in anaerobic sediments.

founder effect Random changes in allele frequencies resulting from establishment of a population by a very small number of individuals.

fovea [L. *fovea*: a small pit] In the vertebrate retina, the area of most distinct vision.

frame-shift mutation The addition or deletion of a single or two adjacent nucleotides in a gene's sequence. Results in the misreading of mRNA during translation and the production of a nonfunctional protein. (Contrast with missense mutation, nonsense mutation, silent mutation.)

Frank–Starling law The stroke volume of the heart increases with increased return of blood to the heart.

free energy (G) Energy that is available for doing useful work, after allowance has been made for the increase or decrease of disorder.

freeze-fracturing Method of tissue preparation for transmission and scanning electron microscopy in which a tissue is frozen and a knife is then used to crack open the tissue; the fracture often occurs in the path of least resistance, within a membrane.

frequency-dependent selection Selection that changes in intensity with the proportion of individuals in a population having the trait.

fruit In angiosperms, a ripened and mature ovary (or group of ovaries) containing the seeds. Sometimes applied to reproductive structures of other groups of plants.

functional genomics The assignment of functional roles to the proteins encoded by genes identified by sequencing entire genomes.

functional group A characteristic combination of atoms that contribute specific properties when attached to larger molecules.

fundamental niche   A species' niche as defined by its physiological capabilities. (Contrast with realized niche.)

- G -

G cap   A chemically modified GTP added to the 5′ end of mRNA; facilitates binding of mRNA to ribosome and prevents mRNA breakdown.

G1   In the cell cycle, the gap between the end of mitosis and the onset of the S phase.

G2   In the cell cycle, the gap between the S (synthesis) phase and the onset of mitosis.

G protein   A membrane protein involved in signal transduction; characterized by binding GDP or GTP.

gain of function mutation   A mutation that results in a protein with a new function. (Contrast with loss of function mutation.)

gallbladder   In the human digestive system, an organ in which bile is stored.

gametangium   (gam uh tan′ gee um) (plural: gametangia) [Gk. *gamos*: marriage + *angeion*: vessel] Any plant or fungal structure within which a gamete is formed.

gamete   (gam′ eet) [Gk. *gamete/gametes*: wife, husband]   The mature sexual reproductive cell: the egg or the sperm.

gametogenesis   (ga meet′ oh jen′ e sis)   The specialized series of cellular divisions that leads to the production of gametes. (*See also* oogenesis, spermatogenesis.)

gametophyte   (ga meet′ oh fyte)   In plants and photosynthetic protists with alternation of generations, the multicellular haploid phase that produces the gametes. (Contrast with sporophyte.)

gamma diversity   The regional diversity found over a range of communities or habitats in a geographic region. (Contrast with alpha diveristy, beta diversity.)

ganglion   (gang′ glee un) (plural: ganglia) [Gk. tumor]   A cluster of neurons that have similar characteristics or function.

ganglion cells   Cells at the front of the human retina that transmit information from the bipolar cells to the brain.

gap junction   A 2.7-nanometer gap between plasma membranes of two animal cells, spanned by protein channels. Gap junctions allow chemical substances or electrical signals to pass from cell to cell.

gastric pits   Deep infoldings in the walls of the stomach lined with secretory cells.

gastrin   A hormone secreted by cells in the lower region of the stomach that stimulates the secretion of digestive juices as well as movements of the stomach.

gastrovascular cavity   Serving for both digestion (gastro) and circulation (vascular); in particular, the central cavity of the body of jellyfish and other cnidarians.

gastrulation   Development of a blastula into a gastrula. In embryonic development, the process by which a blastula is transformed by massive movements of cells into a *gastrula*, an embryo with three germ layers and distinct body axes.

gated channel   A membrane protein that changes its three-dimensional shape, and therefore its ion conductance, in response to a stimulus. When open, it allows specific ions to move across the membrane.

gel electrophoresis   (e lek′ tro fo ree′ sis) [L. *electrum*: amber + Gk. *phorein*: to bear]   A technique for separating molecules (such as DNA fragments) from one another on the basis of their electric charges and molecular weights by applying an electric field to a gel.

gene   [Gk. *genes*: to produce]   A unit of heredity. Used here as the unit of genetic function which carries the information for a single polypeptide or RNA.

gene family   A set of similar genes derived from a single parent gene; need not be on the same chromosomes. The vertebrate globin genes constitute a classic example of a gene family.

gene flow   Exchange of genes between populations through migration of individuals or movements of gametes.

gene-for-gene resistance   In plants, a mechanism of resistance to pathogens in which resistance is triggered by the specific interaction of the products of a pathogen's *Avr* genes and a plant's *R* genes.

gene pool   All of the different alleles of all of the genes existing in all individuals of a population.

gene therapy   Treatment of a genetic disease by providing patients with cells containing functioning alleles of the genes that are nonfunctional in their bodies.

gene tree   A graphic representation of the evolutionary relationships of a single gene in different species or of the members of a gene family.

genetic code   The set of instructions, in the form of nucleotide triplets, that translate a linear sequence of nucleotides in mRNA into a linear sequence of amino acids in a protein.

genetic drift   Changes in gene frequencies from generation to generation as a result of random (chance) processes.

genetic map   The positions of genes along a chromosome as revealed by recombination frequencies.

genetic marker   (1) In gene cloning, a gene of identifiable phenotype that indicates the presence of another gene, DNA segment, or chromosome fragment. (2) In general, a DNA sequence such as a single nucleotide polymorphism whose presence is correlated with the presence of other linked genes on that chromosome.

genetic structure   The frequencies of different alleles at each locus and the frequencies of different genotypes in a Mendelian population.

genetic switches   Mechanisms that control how the genetic toolkit is used, such as promoters and the transcription factors that bind them. The signal cascades that converge on and operate these switches determine when and where genes will be turned on and off.

genetic toolkit   In evolutionary developmental biology, DNA sequences controlling developmental mechanisms that have been conserved over evolutionary time.

genetics   The scientific study of the structure, functioning, and inheritance of genes, the units of hereditary information.

genome   (jee′ nome)   The complete DNA sequence for a particular organism or individual.

genomic equivalence   The principle that no information is lost from the nuclei of cells as they pass through the early stages of embryonic development.

genomic imprinting   The form of a gene's expression is determined by parental source (i.e., whether the gene is inherited from the male or female parent).

genomic library   All of the cloned DNA fragments generated by the action of a restriction endonuclease on a genome.

genomics   The scientific study of entire sets of genes and their interactions.

genotype   (jean′ oh type) [Gk. *gen*: to produce + *typos*: impression]   An exact description of the genetic constitution of an individual, either with respect to a single trait or with respect to a larger set of traits. (Contrast with phenotype.)

genus   (jean′ us) (plural: genera) [Gk. *genos*: stock, kind]   A group of related, similar species recognized by taxonomists with a distinct name used in binomial nomenclature.

germ cell   [L. *germen*: to beget]   A reproductive cell or gamete of a multicellular organism. (Contrast with somatic cell.)

germ layers   The three embryonic layers formed during gastrulation (ectoderm, mesoderm, and endoderm). Also called cell layers or tissue layers.

germ line mutation   Mutation in a cell that produces gametes (i.e., a germ line cell). (Contrast with somatic mutation.)

germination   Sprouting of a seed or spore.

gestation   (jes tay′ shun) [L. *gestare*: to bear]   The period during which the embryo of a mammal develops within the uterus. Also known as pregnancy.

ghrelin   A hormone produced and secreted by cells in the stomach that stimulates appetite.

gibberellin   (jib er el′ lin)   A class of plant growth hormones playing roles in stem elongation, seed germination, flowering of certain plants, etc.

gill   An organ specialized for gas exchange with water.

gizzard   (giz′ erd) [L. *gigeria*: cooked chicken parts]   A muscular port of the stomach of birds that grinds up food, sometimes with the aid of fragments of stone.

glia   (glee′ uh) [Gk. *glia*: glue]   Cells of the nervous system that do not conduct action potentials.

glomerular filtration rate (GFR)   The rate at which the blood is filtered in the glomeruli of the kidney.

glomerulus   (glo mare′ yew lus) [L. *glomus*: ball]   Sites in the kidney where blood filtration takes place. Each glomerulus consists of a knot of capillaries served by afferent and efferent arterioles.

glucagon   Hormone produced by alpha cells of the pancreatic islets of Langerhans. Glucagon stimulates the liver to break down glycogen and release glucose into the circulation.

gluconeogenesis   The biochemical synthesis of glucose from other substances, such as amino acids, lactate, and glycerol.

glucose   [Gk. *gleukos*: sugar, sweet]   The most common monosaccharide; the monomer of the polysaccharides starch, glycogen, and cellulose.

glycerol   (gliss′ er ole)   A three-carbon alcohol with three hydroxyl groups; a component of phospholipids and triglycerides.

glycogen   (gly′ ko jen)   An energy storage polysaccharide found in animals and fungi; a branched-chain polymer of glucose, similar to starch.

glycolipid   A lipid to which sugars are attached.

glycolysis (gly kol′ li sis) [Gk. *gleukos*: sugar + *lysis*: break apart]   The enzymatic breakdown of glucose to pyruvic acid.

glycoprotein   A protein to which sugars are attached.

glycosidic linkage   Bond between carbohydrate (sugar) molecules through an intervening oxygen atom (–O–).

glycosylation   The addition of carbohydrates to another type of molecule, such as a protein.

glyoxysome (gly ox′ ee soam)   An organelle found in plants, in which stored lipids are converted to carbohydrates.

Golgi apparatus (goal′ jee)   A system of concentrically folded membranes found in the cytoplasm of eukaryotic cells; functions in secretion from the cell by exocytosis.

gonad (go′ nad) [Gk. *gone*: seed]   An organ that produces gametes in animals: either an ovary (female gonad) or testis (male gonad).

gonadotropin   A type of trophic hormone that stimulates the gonads.

gonadotropin-releasing hormone (GnRH)   Hormone produced by the hypothalamus that stimulates the anterior pituitary to secrete ("release") gonadotropins.

Gondwana   The large southern land mass that existed from the Cambrian (540 mya) to the Jurassic (138 mya). Present-day remnants are South America, Africa, India, Australia, and Antarctica.

grafting   Artificial transplantation of tissue from one organism to another. In horticulture, the transfer of a bud or stem segment from one plant onto the root of another as a form of asexual reproduction.

Gram stain   A differential purple stain useful in characterizing bacteria. The peptidoglycan-rich cell walls of Gram-positive bacteria stain purple; cell walls of Gram-negative bacteria generally stain orange.

gravitropism [Gk. *tropos*: to turn]   A directed plant growth response to gravity.

gray matter   In the nervous system, tissue that is rich in neuronal cell bodies. (Contrast with white matter.)

greenhouse gases   Gases in the atmosphere, such as carbon dioxide and methane, that are transparent to sunlight, but trap heat radiating from Earth's surface, causing heat to build up at Earth's surface.

gross primary production   The amount of energy captured by the primary producers in a community.

gross primary productivity (GPP)   The rate at which the primary producers in a community turn solar energy into stored chemical energy via photosynthesis.

ground meristem   That part of an apical meristem that gives rise to the ground tissue system of the primary plant body.

ground tissue system   Those parts of the plant body not included in the dermal or vascular tissue systems. Ground tissues function in storage, photosynthesis, and support.

growth   An increase in the size of the body and its organs by cell division and cell expansion.

growth factor   A chemical signal that stimulates cells to divide.

growth hormone   A peptide hormone released by the anterior pituitary that stimulates many anabolic processes.

guanine (G) (gwan′ een)   A nitrogen-containing base found in DNA, RNA, and GTP.

guard cells   In plants, specialized, paired epidermal cells that surround and control the opening of a stoma (pore). *See* stoma.

guild   In ecology, a group of species that exploit the same resource, but in slightly different ways.

gustation   The sense of taste.

gut   An animal's digestive tract.

- H -

habitat   The particular environment in which an organism lives. A *habitat patch* is an area of a particular habitat surrounded by other habitat types that may be less suitable for that organism.

hair cell   A type of mechanoreceptor in animals. Detects sound waves and other forms of motion in air or water.

half-life   The time required for half of a sample of a radioactive isotope to decay to its stable, nonradioactive form, or for a drug or other substance to reach half its initial dosage.

halophyte (hal′ oh fyte) [Gk. *halos*: salt + *phyton*: plant]   A plant that grows in a saline (salty) environment.

Hamilton's rule   The principle that, for an apparent altruistic behavior to be adaptive, the fitness benefit of that act to the recipient times the degree of relatedness of the performer and the recipient must be greater than the cost to the performer.

haplodiploidy   A sex determination mechanism in which diploid individuals (which develop from fertilized eggs) are female and haploid individuals (which develop from unfertilized eggs) are male; typical of hymenopterans.

haploid (hap′ loid) [Gk. *haploeides*: single]   Having a chromosome complement consisting of just one copy of each chromosome; designated $1n$ or n. (Contrast with diploid.)

haplontic   A type of life cycle in which the zygote is the only diploid cell and mitosis occurs only in haploid cells. (Contrast with diplontic.)

haplotype   Linked nucleotide sequences that are usually inherited as a unit (as a "sentence" rather than as individual "words").

Hardy–Weinberg equililbrium   In a sexually reproducing population, the allele frequency at a given locus that is not being acted on by agents of evolution; the conditions that would result in no evolution in a population.

haustorium (haw stor′ ee um) (plural: haustoria)[L. *haustus*: draw up]   A specialized hypha or other structure by which fungi and some parasitic plants draw nutrients from a host plant.

Haversian systems   Units of organization in compact bone that reflect the action of intercommunicating osteoblasts.

heart   In circulatory systems, a muscular pump that moves extracellular fluid around the body.

heat of vaporization   The energy that must be supplied to convert a molecule from a liquid to a gas at its boiling point.

heat shock proteins   Chaperone proteins expressed in cells exposed to high or low temperatures or other forms of environmental stress.

helical   Shaped like a screw or spring; this shape occurs in DNA and proteins.

helper T cells   *See* T-helper cells.

hemiparasite   A parasitic plant that can photosynthesize, but derives water and mineral nutrients from the living body of another plant. (Contrast with holoparasite.)

hemizygous (hem′ ee zie′ gus) [Gk. *hemi*: half + *zygotos*: joined]   In a diploid organism, having only one allele for a given trait, typically the case for X-linked genes in male mammals and Z-linked genes in female birds. (Contrast with homozygous, heterozygous.)

hemoglobin (hee′ mo glow bin) [Gk. *heaema*: blood + L. *globus*: globe]   Oxygen-transporting protein found in the red blood cells of vertebrates (and found in some invertebrates).

Hensen's node   In avian embryos, a structure at the anterior end of the primitive groove; determines the fates of cells passing over it during gastrulation.

hepatic (heh pat′ ik) [Gk. *hepar*: liver]   Pertaining to the liver.

herbivore (ur′ bi vore) [L. *herba*: plant + *vorare*: to devour]   An animal that eats plant tissues. (Contrast with carnivore, detritivore, omnivore.)

heritable trait   A trait that is at least partly determined by genes.

hermaphroditism (her maf′ row dite ism)   The coexistence of both female and male sex organs in the same organism.

hetero- [Gk.: *heteros*: other, different]   A prefix indicating two or more different conditions, structures, or processes. (Contrast with homo-.)

heterochrony   Alteration in the timing of developmental events, leading to different results in the adult organism.

heterocyst   A large, thick-walled cell type in the filaments of certain cyanobacteria that performs nitrogen fixation.

heteromorphic (het′ er oh more′ fik) [Gk. *heteros*: different + *morphe*: form]   Having a different form or appearance, as two heteromorphic life stages of a plant. (Contrast with isomorphic.)

heterosporous (het′ er os′ por us)   Producing two types of spores, one of which gives rise to a female megaspore and the other to a male microspore. (Contrast with homosporous.)

heterosis   The superior fitness of heterozygous offspring as compared with that of their dissimilar homozygous parents. Also called hybrid vigor.

heterotherm   An animal that regulates its body temperature at a constant level at some times but not others, such as a hibernator.

heterotroph (het′ er oh trof) [Gk. *heteros*: different + *trophe*: feed]   An organism that requires preformed organic molecules as food. (Contrast with autotroph.)

heterotrophic succession   Succession in detritus-based communities, which differs from other types of succession in taking place without the participation of plants.

heterotypic   Pertaining to adhesion of cells of different types. (Contrast with homotypic.)

heterozygous (het′ er oh zie′ gus) [Gk. *heteros*: different + *zygotos*: joined]   In diploid organisms, having different alleles of a given gene on the pair of homologs carrying that gene. (Contrast with homozygous.)

hexose [Gk. *hex*: six]   A sugar containing six carbon atoms.

hibernation [L. *hibernum*: winter]   The state of inactivity of some animals during winter; marked by a drop in body temperature and metabolic rate.

hierarchical sequencing   An approach to DNA sequencing in which genetic markers are mapped and DNA sequences are aligned by matching overlapping sites of known sequence. (Contrast with shotgun sequencing.)

high-density lipoproteins (HDLs)   Lipoproteins that remove cholesterol from tissues and carry it to the liver; HDLs are the "good" lipoproteins associated with good cardiovascular health.

high-throughput sequencing   Rapid DNA sequencing on a micro scale in which many fragments of DNA are sequenced in parallel.

highly repetitive sequences   Short (less than 100 bp), nontranscribed DNA sequences, repeated thousands of times in tandem arrangements.

hindbrain   The region of the developing vertebrate brain that gives rise to the medulla, pons, and cerebellum.

hippocampus [Gk. sea horse]   A part of the forebrain that takes part in long-term memory formation.

histamine (hiss′ tah meen)   A substance released by damaged tissue, or by mast cells in response to allergens. Histamine increases vascular permeability, leading to edema (swelling).

histone   Any one of a group of proteins forming the core of a nucleosome, the structural unit of a eukaryotic chromosome.

HIV   Human immunodeficiency virus, the retrovirus that causes acquired immune deficiency syndrome (AIDS).

holoparasite   A fully parasitic plant (i.e., one that does not perform photosynthesis).

homeobox   180-base-pair segment of DNA found in certain homeotic genes; regulates the expression of other genes and thus controls large-scale developmental processes.

homeostasis (home′ ee o sta′ sis) [Gk. *homos*: same + *stasis*: position]   The maintenance of a steady state, such as a constant temperature or a stable social structure, by means of physiological or behavioral feedback responses.

homeotic genes   Genes that act during development to determine the formation of an organ from a region of the embryo.

homeotic mutation   Mutation in a homeotic gene that results in the formation of a different organ than that normally made by a region of the embryo.

homo- [Gk. *homos*: same]   A prefix indicating two or more similar conditions, structures, or processes. (Contrast with hetero-.)

homolog   (1) In cytogenetics, one of a pair (or larger set) of chromosomes having the same overall genetic composition and sequence. In diploid organisms, each chromosome inherited from one parent is matched by an identical (except for mutational changes) chromosome—its homolog—from the other parent. (2) In evolutionary biology, one of two or more features in different species that are similar by reason of descent from a common ancestor.

homology (ho mol′ o jee) [Gk. *homologia*: of one mind; agreement]   A similarity between two or more features that is due to inheritance from a common ancestor. The structures are said to be *homologous*, and each is a *homolog* of the others. (Contrast with analogy.)

homoplasy (home′ uh play zee) [Gk. *homos*: same + *plastikos*: shape, mold]   The presence in multiple groups of a trait that is not inherited from the common ancestor of those groups. Can result from convergent evolution, evolutionary reversal, or parallel evolution.

homosporous   Producing a single type of spore that gives rise to a single type of gametophyte, bearing both female and male reproductive organs. (Contrast with heterosporous.)

homotypic   Pertaining to adhesion of cells of the same type. (Contrast with heterotypic.)

homozygous (home′ oh zie′ gus) [Gk. *homos*: same + *zygotos*: joined]   In diploid organisms, having identical alleles of a given gene on both homologous chromosomes. An individual may be a homozygote with respect to one gene and a heterozygote with respect to another. (Contrast with heterozygous.)

horizons   The horizontal layers of a soil profile, including the topsoil (A horizon), subsoil (B horizon) and parent rock or bedrock (C horizon).

hormone (hore′ mone) [Gk. *hormon*: to excite, stimulate]   A chemical signal produced in minute amounts at one site in a multicellular organism and transported to another site where it acts on target cells.

host   An organism that harbors a parasite or symbiont and provides it with nourishment.

Hox genes   Conserved homeotic genes found in vertebrates, *Drosophila*, and other animal groups. Hox genes contain the homeobox domain and specify pattern and axis formation in these animals.

human chorionic gonadotropin (hCG)   A hormone secreted by the placenta which sustains the corpus luteum and helps maintain pregnancy.

humoral immune response   The response of the immune system mediated by B cells that produces circulating antibodies active against extracellular bacterial and viral infections. (Contrast with cellular immune response.)

humus (hew′ mus)   The partly decomposed remains of plants and animals on the surface of a soil.

hybrid (high′ brid) [L. *hybrida*: mongrel]   (1) The offspring of genetically dissimilar parents. (2) In molecular biology, a double helix formed of nucleic acids from different sources.

hybridize   (1) In genetics, to combine the genetic material of two distinct species or of two distinguishable populations within a species. (2) In molecular biology, to form a double-stranded nucleic acid in which the two strands originate from different sources.

hybrid vigor   *See* heterosis.

hybridoma   A cell produced by the fusion of an antibody-producing cell with a myeloma (tumor) cell; it produces monoclonal antibodies.

hydrocarbon   A compound containing only carbon and hydrogen atoms.

hydrogen bond   A weak electrostatic bond which arises from the attraction between the slight positive charge on a hydrogen atom and a slight negative charge on a nearby oxygen or nitrogen atom.

hydrologic cycle   The movement of water from the oceans to the atmosphere, to the soil, and back to the oceans.

hydrolysis reaction (high drol′ uh sis) [Gk. *hydro*: water + *lysis*: break apart]   A chemical reaction that breaks a bond by inserting the components of water ($AB + H_2O \rightarrow AH + BOH$). (Contrast with condensation reaction.)

hydrophilic (high dro fill′ ik) [Gk. *hydro*: water + *philia*: love]   Having an affinity for water. (Contrast with hydrophobic.)

hydrophobic (high dro foe′ bik) [Gk. *hydro*: water + *phobia*: fear]   Having no affinity for water. Uncharged and nonpolar groups of atoms are hydrophobic. (Contrast with hydrophilic.)

hydroponic   Pertaining to a method of growing plants with their roots suspended in nutrient solutions instead of soil.

hydrostatic pressure   Pressure generated by compression of liquid in a confined space. Generated in plants, fungi, and some protists with cell walls by the osmotic uptake of water. Generated in animals with closed circulatory systems by the beating of a heart.

hydrostatic skeleton   A fluid-filled body cavity that transfers forces from one part of the body to another when acted on by surrounding muscles.

hydroxyl group   The —OH group found on alcohols and sugars.

hyper- [Gk. *hyper*: above, over]   Prefix indicating above, higher, more. (Contrast with hypo-.)

hyperpolarization   A change in the resting potential across a membrane so that the inside of a cell becomes more negative compared with the outside of the cell. (Contrast with depolarization.)

hypersensitive response   A defensive response of plants to microbial infection in which phytoalexins and pathogenesis-related proteins are produced and the infected tissue undergoes apoptosis to isolate the pathogen from the rest of the plant.

hypertonic   Having a greater solute concentration. Said of one solution compared with another. (Contrast with hypotonic, isotonic.)

hypha (high′ fuh) (plural: hyphae) [Gk. *hyphe*: web]   In the fungi and oomycetes, any single filament.

hypo- [Gk. *hypo*: beneath, under]   Prefix indicating underneath, below, less. (Contrast with hyper-.)

hypoblast   The lower tissue portion of the avian blastula which is joined to the epiblast at the margins of the blastodisc.

hypothalamus   The part of the brain lying below the thalamus; it coordinates water balance, reproduction, temperature regulation, and metabolism.

hypothermia   Below-normal body temperature.

hypothesis   A tentative answer to a question, from which testable predictions can be generated. (Contrast with theory.)

hypotonic   Having a lesser solute concentration. Said of one solution in comparing it to another. (Contrast with hypertonic, isotonic.)

hypoxia   A deficiency of oxygen.

- I -

ileum   The final segment of the small intestine.

imbibition   Water uptake by a seed; first step in germination.

immediate hypersensitivity   A rapid, extensive overreaction of the immune system against an allergen, resuting in the release of large amounts of histamine. (Contrast with delayed hypersensitivity.)

immediate memory   A form of memory for events happening in the present that is almost perfectly photographic, but lasts only seconds. (Contrast with short-term memory, long-term memory.)

immune system [L. *immunis*: exempt from] A system in an animal that recognizes and attempts to eliminate or neutralize foreign substances such as bacteria, viruses, and pollutants.

immunoassay The use of antibodies to measure the concentration of an antigen in a sample.

immunoglobulins A class of proteins containing a tetramer consisting of four polypeptide chains—two identical light chains and two identical heavy chains—held together by disulfide bonds; active as receptors and effectors in the immune system.

immunological memory The capacity to more rapidly and massively respond to a second exposure to an antigen than occurred on first exposure.

imperfect flower A flower lacking either functional stamens or functional carpels. (Contrast with perfect flower.)

implantation The process by which the early mammalian embryo becomes attached to and embedded in the lining of the uterus.

imprinting In animal behavior, a rapid form of learning in which an animal learns, during a brief critical period, to make a particular response, which is maintained for life, to some object or other organism. *See also* genomic imprinting.

in vitro [L. in glass] A biological process occurring outside of the organism, in the laboratory. (Contrast with in vivo.)

in vivo [L. alive] A biological process occurring within a living organism or cell. (Contrast with in vitro.)

inbreeding Breeding among close relatives.

inclusive fitness The sum of an individual's genetic contribution to subsequent generations both via production of its own offspring and via its influence on the survival of relatives who are not direct descendants.

incomplete cleavage A pattern of cleavage that occurs in many eggs that have a lot of yolk, in which the cleavage furrows do not penetrate all of it. (*See also* discoidal cleavage, superficial cleavage; contrast with complete cleavage.)

incomplete dominance Condition in which the heterozygous phenotype is intermediate between the two homozygous phenotypes.

incomplete metamorphosis Insect development in which changes between instars are gradual.

independent assortment During meiosis, the random separation of genes carried on nonhomologous chromosomes into gametes so that inheritance of these genes is random. This principle was articulated by Mendel as his second law.

indeterminate growth A open-ended growth pattern in which an organism or organ continues to grow as long as it lives; characteristic of some animals and of plant shoots and roots. (Contrast with determinate growth.)

indirect transduction Cell signaling mechanism in which a second messenger mediates the interaction between receptor binding and cellular response. (Contrast with direct transduction.)

individual fitness That component of inclusive fitness resulting from an organism producing its own offspring. (Contrast with kin selection.)

induced fit A change in the shape of an enzyme caused by binding to its substrate that exposes the active site of the enzyme.

induced mutation A mutation resulting from exposure to a mutagen from outside the cell. (Contrast with spontaneous mutation.)

induced pluripotent stem cells (iPS cells) Multipotent or pluripotent animal stem cells produced from differentiated cells in vitro by the addition of several genes that are expressed.

inducer (1) A compound that stimulates the synthesis of a protein. (2) In embryonic development, a substance that causes a group of target cells to differentiate in a particular way.

inducible Produced only in the presence of a particular compound or under particular circumstances. (Contrast with constitutive.)

induction In embryonic development, the process by which a factor produced and secreted by certain cells determines the fates other cells.

inflammation A nonspecific defense against pathogens; characterized by redness, swelling, pain, and increased temperature.

inflorescence A structure composed of several to many flowers.

inflorescence meristem A meristem that produces floral meristems as well as other small leafy structures (bracts).

ingroup In a phylogenetic study, the group of organisms of primary interest. (Contrast with outgroup.)

inhibitor A substance that blocks a biological process.

initials Cells that perpetuate plant meristems, comparable to animal stem cells. When an initial divides, one daughter cell develops into another initial, while the other differentiates into a more specialized cell.

initiation In molecular biology, the beginning of transcription or translation.

initiation complex In protein translation, a combination of a small ribosomal subunit, an mRNA molecule, and the tRNA charged with the first amino acid coded for by the mRNA; formed at the onset of translation.

initiation site The part of a promoter where transcription begins.

inner cell mass Derived from the mammalian blastula (bastocyst), the inner cell mass will give rise to the yolk sac (via hypoblast) and embryo (via epiblast).

inositol trisphosphate (IP$_3$) An intracellular second messenger derived from membrane phospholipids.

inspiratory reserve volume The amount of air that can be inhaled above the normal tidal inspiration. (Contrast with expiratory reserve volume, tidal volume, vital capacity.)

instar (in' star) An immature stage of an insect between molts.

insulin (in' su lin) [L. *insula*: island] A hormone synthesized in islet cells of the pancreas that promotes the conversion of glucose into the storage material, glycogen.

integrin In animals, a transmembrane protein that mediates the attachment of epithelial cells to the extracellular matrix.

integument [L. *integumentum*: covering] A protective surface structure. In gymnosperms and angiosperms, a layer of tissue around the ovule which will become the seed coat.

intercostal muscles Muscles between the ribs that can augment breathing movements by elevating and suppressing the rib cage.

interference competition Competition in which individuals actively interfere with one another's access to resources. (Contrast with exploitation competition.)

interference RNA (RNAi) *See* RNA interference.

interferon A glycoprotein produced by virus-infected animal cells; increases the resistance of neighboring cells to the virus.

intermediate filaments Components of the cytoskeleton whose diameters fall between those of the larger microtubules and those of the smaller microfilaments.

internal environment In multicelluar organisms, the extracellular fluid surrounding the cells.

internal fertilization The release of sperm into the female reproductive tract; typical of most terrestrial animals. (Contrast with external fertilization.)

internal gills Gills enclosed in protective body cavities; typical of mollusks, arthropods, and fishes.

interneuron A neuron that communicates information between two other neurons.

internode The region between two nodes of a plant stem.

interphase In the cell cycle, the period between successive nuclear divisions during which the chromosomes are diffuse and the nuclear envelope is intact. During interphase the cell is most active in transcribing and translating genetic information.

interspecific competition Competition between members of two or more species. (Contrast with intraspecific competition.)

interstitial fluid Extracellular fluid that is not contained in the vessels of a circulatory system.

intestine The portion of the gut following the stomach, in which most digestion and absorption occurs.

intraspecific competition Competition among members of the same species. (Contrast with interspecific competition.)

intrinsic rate of increase (r) The rate at which a population can grow when its density is low and environmental conditions are highly favorable.

intron Portion of a of a gene within the coding region that is transcribed into pre-mRNA but is spliced out prior to translation. (Contrast with exon.)

invasive species An exotic species that reproduces rapidly, spreads widely, and has negative effects on the native species of the region to which it has been introduced.

invasiveness The ability of a pathogen to multiply in a host's body. (Contrast with toxigenicity).

inversion A rare 180° reversal of the order of genes within a segment of a chromosome.

ion (eye' on) [Gk. *ion*: wanderer] An electrically charged particle that forms when an atom gains or loses one or more electrons.

ion channel An integral membrane protein that allows ions to diffuse across the membrane in which it is embedded.

ionic bond An electrostatic attraction between positively and negatively charged ions.

ionotropic receptors A receptor that directly alters membrane permeability to a type of ion when it combines with its ligand.

iris (eye' ris) [Gk. *iris*: rainbow] The round, pigmented membrane that surrounds the pupil of the eye and adjusts its aperture to regulate the amount of light entering the eye.

island biogeography A theory proposing that the number of species on an island (or in another geographically defined and isolated area) represents a balance, or equilibrium, between the rate at which species immigrate to the island and the rate at which resident species go extinct.

islets of Langerhans Clusters of hormone-producing cells in the pancreas.

iso- [Gk. *iso*: equal] Prefix used for two separate entities that share some element of identity.

isogamous Having male and female gametes that are morphologically identical. (Contrast with anisogamous.)

isomers Molecules consisting of the same numbers and kinds of atoms, but differing in the bonding patterns by which the atoms are held together.

isomorphic (eye so more' fik) [Gk. *isos*: equal + *morphe*: form] Having the same form or appearance, as when the haploid and diploid life stages of an organism appear identical. (Contrast with heteromorphic.)

isotonic Having the same solute concentration; said of two solutions. (Contrast with hypertonic, hypotonic.)

isotope (eye' so tope) [Gk. *isos*: equal + *topos*: place] Isotopes of a given chemical element have the same number of protons in their nuclei (and thus are in the same position on the periodic table), but differ in the number of neutrons.

isozymes Enzymes of an organism that have somewhat different amino acid sequences but catalyze the same reaction.

iteroparous [L. *itero*, to repeat + *pario*, to beget] Reproducing multiple times in a lifetime. (Contrast with semelparous.)

- J -

jejunum (jih jew' num) The middle division of the small intestine, where most absorption of nutrients occurs. (*See* duodenum, ileum.)

jelly coat The outer protective layer of a sea urchin egg, which triggers an acrosomal reaction in sperm.

joint In skeletal systems, a junction between two or more bones.

juvenile hormone In insects, a hormone maintaining larval growth and preventing maturation or pupation.

- K -

K-strategist A species whose life history strategy allows it to persist at or near the carrying capacity (*K*) of its environment. (Contrast with *r*-strategist.)

karyogamy The fusion of nuclei of two cells. (Contrast with plasmogamy.)

karyotype The number, forms, and types of chromosomes in a cell.

keystone species Species that have a dominant influence on the composition of a community.

kidneys A pair of excretory organs in vertebrates.

kin selection That component of inclusive fitness resulting from helping the survival of rela-tives containing the same alleles by descent from a common ancestor. (Contrast with individual fitness.)

kinase *See* protein kinase.

kinetic energy (kuh-net' ik) [Gk. *kinetos*: moving] The energy associated with movement. (Contrast with potential energy.)

kinetochore (kuh net' oh core) Specialized structure on a centromere to which microtubules attach.

knockout A molecular genetic method in which a single gene of an organism is permanently inactivated.

Koch's postulates A set of rules for establishing that a particular microorganism causes a particular disease.

Krebs cycle *See* citric acid cycle.

- L -

lagging strand In DNA replication, the daughter strand that is synthesized in discontinuous stretches. (*See* Okazaki fragments.)

larva (plural: larvae) [L. *lares*: guiding spirits] An immature stage of any animal that differs dramatically in appearance from the adult.

lateral [L. *latus*: side] Pertaining to the side.

lateral gene transfer The transfer of genes from one species to another, common among bacteria and archaea.

lateral line A sensory system in fishes consisting of a canal filled with water and hair cells running down each side under the surface of the skin, which senses disturbances in the surrounding water.

lateral meristem Either of the two meristems, the vascular cambium and the cork cambium, that give rise to a plant's secondary growth.

lateral root A root extending outward from the taproot in a taproot system; typical of eudicots.

laticifers (luh tiss' uh furs) In some plants, elongated cells containing secondary plant products such as latex.

Laurasia The northernmost of the two large continents produced by the breakup of Pangaea.

laws of thermodynamics [Gk. *thermos*: heat + *dynamis*: power] Laws derived from studies of the physical properties of energy and the ways energy interacts with matter. (*See also* first law of thermodynamics, second law of thermodynamics.)

leaching In soils, a process by which mineral nutrients in upper soil horizons are dissolved in water and carried to deeper horizons, where they are unavailable to plant roots.

leading strand In DNA replication, the daughter strand that is synthesized continuously. (Contrast with lagging strand.)

leaf (plural: leaves) In plants, the chief organ of photosynthesis.

leaf primordium (plural: primordia) An outgrowth on the side of the shoot apical meristem that will eventually develop into a leaf.

leghemoglobin In nitrogen-fixing plants, an oxygen-carrying protein in the cytoplasm of nodule cells that transports enough oxygen to the nitrogen-fixing bacteria to support their respiration, while keeping free oxygen concentrations low enough to protect nitrogenase.

lek A display ground within which male animals compete for and defend small display areas as a means of demonstrating their territorial prowess and winning opportunities to mate.

lens In the vertebrate eye, a crystalline protein structure that makes fine adjustments in the focus of images falling on the retina.

lenticel (len' ti sill) In plants, a spongy region in the periderm that allows gas exchange.

leptin A hormone produced by fat cells that is believed to provide feedback information to the brain about the status of the body's fat reserves.

leukocyte *See* white blood cell.

lichen (lie' kun) An organism resulting from the symbiotic association of a fungus and either a cyanobacterium or a unicellular alga.

life cycle The entire span of the life of an organism from the moment of fertilization (or asexual generation) to the time it reproduces in turn.

life history strategy The way in which an organism partitions its time and energy among growth, maintenance, and reproduction.

ligament A band of connective tissue linking two bones in a joint.

ligand (lig' and) Any molecule that binds to a receptor site of another (usually larger) molecule.

light reactions The initial phase of photosynthesis, in which light energy is converted into chemical energy.

light-independent reactions The phase of photosynthesis in which chemical energy captured in the light reactions is used to drive the reduction of CO_2 to form carbohydrates.

lignin A complex, hydrophobic polyphenolic polymer in plant cell walls that crosslinks other wall polymers, strengthening the walls, especially in wood.

limbic system A group of evolutionarily primitive structures in the vertebrate telencephalon that are involved in emotions, drives, instinctive behaviors, learning, and memory.

liming Application of compounds such as calcium carbonate, calcium hydroxide, or magnesium carbonate—commonly known as *lime*— to soil to reverse its acidification and increase the availability of calcium to plants.

limiting resource The required resource whose supply most strongly influences the size of a population.

lineage species concept The definition of a species as a branch on the tree of life, which has a history that starts at a speciation event and ends either at extinction or at another speciation event. (Contrast with biological species concept; morphological species concept.)

linkage Association between genes on the same chromosome such that they do not show random assortment and seldom recombine; the closer the genes, the lower the frequency of recombination.

lipase (lip' ase; lye' pase) An enzyme that digests fats.

lipid (lip' id) [Gk. *lipos*: fat] Nonpolar, hydrophobic molecules that include fats, oils, waxes, steroids, and the phospholipids that make up biological membranes.

lipid bilayer *See* phospholipid bilayer.

liver A large digestive gland. In vertebrates, it secretes bile and is involved in the formation of blood.

loam A type of soil consisting of a mixture of sand, silt, clay, and organic matter. One of the best soil types for agriculture.

locus (low' kus) (plural: loci, low' sigh) In genetics, a specific location on a chromosome. May be considered synonymous with *gene*.

logistic growth   Growth, especially in the size of an organism or in the number of organisms in a population, that slows steadily as the entity approaches its maximum size. (Contrast with exponential growth.)

long-day plant (LDP)   A plant that requires long days (actually, short nights) in order to flower.

long-term depression (LTD)   A long-lasting decrease in the responsiveness resulting from continuous, repetitive, low-level stimulation. (Contrast with long-term potentiation.)

long-term potentiation (LTP)   A long-lasting increase in the responsiveness of a neuron resulting from a period of intense stimulation. (Contrast with long-term depression.)

loop of Henle (hen' lee)   Long, hairpin loop of the mammalian renal tubule that runs from the cortex down into the medulla and back to the cortex; creates a concentration gradient in the interstitial fluids in the medulla.

lophophore   A U-shaped fold of the body wall with hollow, ciliated tentacles that encircles the mouth of animals in several different groups. Used for filtering prey from the surrounding water.

loss of function mutation   A mutation that results in the loss of a functional protein. (Contrast with gain of function mutation.)

low-density lipoproteins (LDLs)   Lipoproteins that transport cholesterol around the body for use in biosynthesis and for storage; LDLs are the "bad" lipoproteins associated with a high risk of cardiovascular disease.

lumen (loo' men) [L. *lumen*: light]   The open cavity inside any tubular organ or structure, such as the gut or a renal tubule.

lung   An internal organ specialized for respiratory gas exchange with air.

luteinizing hormone (LH)   A gonadotropin produced by the anterior pituitary that stimulates the gonads to produce sex hormones.

lymph [L. *lympha*: liquid]   A fluid derived from blood and other tissues that accumulates in intercellular spaces throughout the body and is returned to the blood by the lymphatic system.

lymph node   A specialized structure in the vessels of the lymphatic system. Lymph nodes contain lymphocytes, which encounter and respond to foreign cells and molecules in the lymph as it passes through the vessels.

lymphatic system   A system of vessels that returns interstitial fluid to the blood.

lymphocyte   One of the two major classes of white blood cells; includes T cells, B cells, and other cell types important in the immune system.

lymphoid tissues   Tissues of the immune system that are dispersed throughout the body, consisting of the thymus, spleen, bone marrow, and lymph nodes.

lysis (lie' sis) [Gk. *lysis*: break apart]   Bursting of a cell.

lysogeny   A form of viral replication in which the virus becomes incorporated into the host chromosome and remains inactive. Also called a lysogenic cycle. (Contrast with lytic cycle.)

lysosome (lie' so soam) [Gk. *lysis*: break away + *soma*: body]   A membrane-enclosed organelle originating from the Golgi apparatus and containing hydrolytic enzymes. (Contrast with secondary lysosome.)

lysozyme (lie' so zyme)   An enzyme in saliva, tears, and nasal secretions that hydrolyzes bacterial cell walls.

lytic cycle   A viral reproductive cycle in which the virus takes over a host cell's synthetic machinery to replicate itself, then bursts (lyses) the host cell, releasing the new viruses. (Contrast with lysogeny.)

- M -

M phase   The portion of the cell cycle in which mitosis takes place.

macroevolution [Gk. *makros*: large]   Evolutionary changes occurring over long time spans and usually involving changes in many traits. (Contrast with microevolution.)

macromolecule   A giant (molecular weight > 1,000) polymeric molecule. The macromolecules are the proteins, polysaccharides, and nucleic acids.

macronutrient   In plants, a mineral element required in concentrations of at least 1 milligram per gram of plant dry matter; in animals, a mineral element required in large amounts. (Contrast with micronutrient.)

macrophage (mac' roh faj)   Phagocyte that engulfs pathogens by endocytosis.

MADS box   DNA-binding domain in many plant transcription factors that is active in development.

maintenance methylase   An enzyme that catalyzes the methylation of the new DNA strand when DNA is replicated.

major histocompatibility complex (MHC)   A complex of linked genes, with multiple alleles, that control a number of cell surface antigens that identify self and can lead to graft rejection.

malignant   Pertaining to a tumor that can grow indefinitely and/or spread from the original site of growth to other locations in the body. (Contrast with benign.)

Malpighian tubule (mal pee' gy un)   A type of protonephridium found in insects.

map unit   The distance between two genes as calculated from genetic crosses; a recombination frequency.

marine [L. *mare*: sea, ocean]   Pertaining to or living in the ocean. (Contrast with aquatic, terrestrial.)

mark–recapture method   A method of estimating population sizes of mobile organisms by capturing, marking, and releasing a sample of individuals, then capturing another sample at a later time.

mass extinction   A period of evolutionary history during which rates of extinction are much higher than during intervening times.

mass number   The sum of the number of protons and neutrons in an atom's nucleus.

mast cells   Cells, typically found in connective tissue, that release histamine in response to tissue damage.

maternal effect genes   Genes coding for morphogens that determine the polarity of the egg and larva in fruit flies.

mating type   A particular strain of a species that is incapable of sexual reproduction with another member of the same strain but capable of sexual reproduction with members of other strains of the same species.

maximum likelihood   A statistical method of determining which of two or more hypotheses (such as phylogenetic trees) best fit the observed data, given an explicit model of how the data were generated.

mechanically gated channel   A molecular channel that opens or closes in response to mechanical force applied to the plasma membrane in which it is inserted.

mechanoreceptor   A cell that is sensitive to physical movement and generates action potentials in response.

medulla (meh dull' luh)   (1) The inner, core region of an organ, as in the adrenal medulla (adrenal gland) or the renal medulla (kidneys). (2) The portion of the brainstem that connects to the spinal cord.

medusa (plural: medusae)   In cnidarians, a freeswimming, sexual life cycle stage shaped like a bell or an umbrella.

megaphyll   The generally large leaf of a fern, horsetail, or seed plant, with several to many veins. (Contrast with microphyll.)

megaspore [Gk. *megas*: large + *spora*: to sow]   In plants, a haploid spore that produces a female gametophyte.

megastrobilus   In conifers, the female (seedbearing) cone. (Contrast with microstrobilus.)

meiosis (my oh' sis) [Gk. *meiosis*: diminution]   Division of a diploid nucleus to produce four haploid daughter cells. The process consists of two successive nuclear divisions with only one cycle of chromosome replication. In *meiosis I*, homologous chromosomes separate but retain their chromatids. The second division *meiosis II*, is similar to mitosis, in which chromatids separate.

melatonin   A hormone released by the pineal gland. Involved in photoperiodicity and circadian rhythms.

membrane potential   The difference in electrical charge between the inside and the outside of a cell, caused by a difference in the distribution of ions.

membranous bone   A type of bone that develops by forming on a scaffold of connective tissue. (Contrast with cartilage bone.)

memory cells   Long-lived lymphocytes produced after exposure to antigen. They persist in the body and are able to mount a rapid response to subsequent exposures to the antigen.

Mendel's laws   *See* independent assortment; segregation.

meristem [Gk. *meristos*: divided]   Plant tissue made up of undifferentiated actively dividing cells.

meristem identity genes   In angiosperms, a group of genes whose expression initiates flower formation, probably by switching meristem cells from a vegetative to a reproductive fate.

mesenchyme (mez' en kyme) [Gk. *mesos*: middle + *enchyma*: infusion]   Embryonic or unspecialized cells derived from the mesoderm.

mesoderm [Gk. *mesos*: middle + *derma*: skin]   The middle of the three embryonic germ layers first delineated during gastrulation. Gives rise to the skeleton, circulatory system, muscles, excretory system, and most of the reproductive system.

mesoglea (mez' uh glee uh) [Gk. *mesos*: middle + *gloia*, glue]   A thick, gelatinous noncellular layer that separates the two cellular tissue layers of ctenophores, cnidarians, and scyphozoans.

mesophyll (mez' uh fill) [Gk. *mesos*: middle + *phyllon*: leaf] Chloroplast-containing, photosynthetic cells in the interior of leaves.

messenger RNA (mRNA) Transcript of a region of one of the strands of DNA; carries information (as a sequence of codons) for the synthesis of one or more proteins.

meta- [Gk.: between, along with, beyond] Prefix denoting a change or a shift to a new form or level; for example, as used in metamorphosis.

metabolic pathway A series of enzyme-catalyzed reactions so arranged that the product of one reaction is the substrate of the next.

metabolism (meh tab' a lizm) [Gk. *metabole*: change] The sum total of the chemical reactions that occur in an organism, or some subset of that total (as in respiratory metabolism).

metabolome The quantitative description of all the small molecules in a cell or organism.

metabotropic receptor A receptor that that indirectly alters membrane permeability to a type of ion when it combines with its ligand.

metagenomics The practice of analyzing DNA from environmental samples without isolating intact organisms.

metamorphosis (met' a mor' fo sis) [Gk. *meta*: between + *morphe*: form, shape] A change occurring between one developmental stage and another, as for example from a tadpole to a frog. (*See* complete metamorphosis, incomplete metamorphosis.)

metanephridia The paired excretory organs of annelids.

metaphase (met' a phase) The stage in nuclear division at which the centromeres of the highly supercoiled chromosomes are all lying on a plane (the metaphase plane or plate) perpendicular to a line connecting the division poles.

metapopulation A population divided into subpopulations, among which there are occasional exchanges of individuals.

metastasis (meh tass' tuh sis) The spread of cancer cells from their original site to other parts of the body.

methylation The addition of a methyl group (—CH_3) to a molecule.

MHC *See* major histocompatibility complex.

micelle A particle of lipid covered with bile salts that is produced in the duodenum and facilitates digestion and absorption of lipids.

microclimate A subset of climatic conditions in a small specific area, which generally differ from those in the environment at large, as in an animal's underground burrow.

microevolution Evolutionary changes below the species level, affecting allele frequencies. (Contrast with macroevolution.)

microfibril Crosslinked cellulose polymers, forming strong aggregates in the plant cell wall.

microfilament In eukaryotic cells, a fibrous structure made up of actin monomers. Microfilaments play roles in the cytoskeleton, in cell movement, and in muscle contraction.

microglia Glial cells that act as macrophages and mediators of inflammatory responses in the central nervous system.

micronutrient In plants, a mineral element required in concentrations of less than 100 micrograms per gram of plant dry matter; in animals, a mineral element required in concentrations of less than 100 micrograms per day. (Contrast with macronutrient.)

microphyll A small leaf with a single vein, found in club mosses and their relatives. (Contrast with megaphyll.)

micropyle (mike' roh pile) [Gk. *mikros*: small + *pylon*: gate] Opening in the integument(s) of a seed plant ovule through which pollen grows to reach the female gametophyte within.

microRNA A small, noncoding RNA molecule, typically about 21 bases long, that binds to mRNA to inhibit its translation.

microspore [Gk. *mikros*: small + *spora*: to sow] In plants, a haploid spore that produces a male gametophyte.

microstrobilus In conifers, male pollen-bearing cone. (Contrast with megastrobilus.)

microtubules Tubular structures found in centrioles, spindle apparatus, cilia, flagella, and cytoskeleton of eukaryotic cells. These tubules play roles in the motion and maintenance of shape of eukaryotic cells.

microvilli (sing.: microvillus) Projections of epithelial cells, such as the cells lining the small intestine, that increase their surface area.

midbrain One of the three regions of the vertebrate brain. Part of the brainstem, it serves as a relay station for sensory signals sent to the cerebral hemispheres.

middle lamella (la mell' ah) [L. *lamina*: thin sheet] A layer of polysaccharides that separates plant cells; a shared middle lamella lies outside the primary walls of the two cells.

mineral nutrients Inorganic ions required by organisms for normal growth and reproduction.

mismatch repair A mechanism that scans DNA after it has been replicated and corrects any base-pairing mismatches.

missense mutation A change in a gene's sequence that results in a change in the sequence of the amino acid specified by the corresponding codon. (Contrast with frame-shift mutation, nonsense mutation, silent mutation.)

mitochondrial matrix The fluid interior of the mitochondrion, enclosed by the inner mitochondrial membrane.

mitochondrion (my' toe kon' dree un) (plural: mitochondria) [Gk. *mitos*: thread + *chondros*: grain] An organelle in eukaryotic cells that contains the enzymes of the citric acid cycle, the respiratory chain, and oxidative phosphorylation.

mitosis (my toe' sis) [Gk. *mitos*: thread] Nuclear division in eukaryotes leading to the formation of two daughter nuclei, each with a chromosome complement identical to that of the original nucleus.

model systems Also known as model organisms, these include the small group of species that are the subject of extensive research. They are organisms that adapt well to laboratory situations and findings from experiments on them can apply across a broad range of species. Classic examples include white rats and the fruit fly *Drosophila*.

moderately repetitive sequences DNA sequences repeated 10–1,000 times in the eukaryotic genome. They include the genes that code for rRNAs and tRNAs, as well as the DNA in telomeres.

Modern Synthesis An understanding of evolutionary biology that emerged in the early twentieth century as the principles of evolution were integrated with the principles of modern genetics.

modularity In evolutionary developmental biology, the principle that the molecular pathways that determine different developmental processes operate independently from one another. *See also* developmental module.

mole A quantity of a compound whose weight in grams is numerically equal to its molecular weight expressed in atomic mass units. Avogadro's number of molecules: 6.023×10^{23} molecules.

molecular clock The approximately constant rate of divergence of macromolecules from one another over evolutionary time; used to date past events in evolutionary history.

molecular evolution The scientific study of the mechanisms and consequences of the evolution of macromolecules.

molecular tool kit A set of developmental genes and proteins that is common to most animals and is hypothesized to be responsible for the evolution of their differing developmental pathways.

molecular weight The sum of the atomic weights of the atoms in a molecule.

molecule A chemical substance made up of two or more atoms joined by covalent bonds or ionic attractions.

molting The process of shedding part or all of an outer covering, such as the shedding of feathers by birds or of the entire exoskeleton by arthropods.

monoclonal antibody Antibody produced in the laboratory from a clone of hybridoma cells, each of which produces the same specific antibody.

monoculture In agriculture, a large-scale planting of a single species of domesticated crop plant.

monoecious (mo nee' shus) [Gk. *mono*: one + *oikos*: house] Pertaining to organisms in which both sexes are "housed" in a single individual that produces both eggs and sperm. (In some plants, these are found in different flowers within the same plant.) Examples include corn, peas, earthworms, hydras. (Contrast with dioecious.)

monohybrid cross A mating in which the parents differ with respect to the alleles of only one locus of interest.

monomer [Gk. *mono*: one + *meros*: unit] A small molecule, two or more of which can be combined to form oligomers (consisting of a few monomers) or polymers (consisting of many monomers).

monophyletic (mon' oh fih leht' ik) [Gk. *mono*: one + *phylon*: tribe] Pertaining to a group that consists of an ancestor and all of its descendants. (Contrast with paraphyletic, polyphyletic.)

monosaccharide A simple sugar. Oligosaccharides and polysaccharides are made up of monosaccharides.

monosomic Pertaining to an organism with one less than the normal diploid number of chromosomes.

monosynaptic reflex A neural reflex that begins in a sensory neuron and makes a single synapse before activating a motor neuron.

morphogen A diffusible substance whose concentration gradient determines a developmental pattern in animals and plants.

morphogenesis (more' fo jen' e sis) [Gk. *morphe*: form + *genesis*: origin]   The development of form; the overall consequence of determination, differentiation, and growth.

morphological species concept   The definition of a species as a group of individuals that look alike. (Contrast with biological species concept; lineage species concept.)

morphology (more fol' o jee) [Gk. *morphe*: form + *logos*: study, discourse]   The scientific study of organic form, including both its development and function.

mosaic development   Pattern of animal embryonic development in which each blastomere contributes a specific part of the adult body. (Contrast with regulative development.)

motif   *See* structural motif.

motile (mo' tul)   Able to move from one place to another. (Contrast with sessile.)

motor cortex   The region of the cerebral cortex that contains motor neurons that directly stimulate specific muscle fibers to contract.

motor neuron   A neuron carrying information from the central nervous system to a cell that produces movement.

motor proteins   Specialized proteins that use energy to change shape and move cells or structures within cells.

motor unit   A motor neuron and the muscle fibers it controls.

mouth   An opening through which food is taken in, located at the anterior end of a tubular gut.

mRNA   *See* messenger RNA.

mucosal epithelium   An epithelial cell layer containing cells that secrete mucus; found in the digestive and respiratory tracts. Also called mucosa.

Müllerian mimicry   Convergence in appearance of two or more unpalatable species.

multipotent   Having the ability to differentiate into a limited number of cell types. (Contrast with pluripotent, totipotent.)

muscle fiber   A single muscle cell. In the case of skeletal muscle, a syncitial, multinucleate cell.

muscle tissue   Excitable tissue that can contract through the interactions of actin and myosin; one of the four major tissue types in multicellular animals. There are three types of muscle tissue: skeletal, smooth, and cardiac.

mutagen (mute' ah jen) [L. *mutare*: change + Gk. *genesis*: source]   Any agent (e.g., a chemical, radiation) that increases the mutation rate.

mutation   A change in the genetic material not caused by recombination.

mutualism   A type of interaction between species that benefits both species.

mycelium (my seel' ee yum) [Gk. *mykes*: fungus]   In the fungi, a mass of hyphae.

mycorrhiza (my' ko rye' za) (plural: mycorrhizae) [Gk. *mykes*: fungus + *rhiza*: root]   An association of the root of a plant with the mycelium of a fungus.

myelin (my' a lin)   Concentric layers of plasma membrane that form a sheath around some axons; myelin provides the axon with electrical insulation and increases the rate of transmission of action potentials.

myocardial infarction   Blockage of an artery that carries blood to the heart muscle.

myofibril (my' oh fy' bril) [Gk. *mys*: muscle + L. *fibrilla*: small fiber]   A polymeric unit of actin or myosin in a muscle.

myoglobin (my' oh globe' in) [Gk. *mys*: muscle + L. *globus*: sphere]   An oxygen-binding molecule found in muscle. Consists of a heme unit and a single globin chain; carries less oxygen than hemoglobin.

myosin   One of the two contractile proteins of muscle.

- N -

natural killer cell   A type of lymphocyte that attacks virus-infected cells and some tumor cells as well as antibody-labeled target cells.

natural selection   The differential contribution of offspring to the next generation by various genetic types belonging to the same population. The mechanism of evolution proposed by Charles Darwin.

nauplius (naw' plee us) [Gk. *nauplios*: shellfish]   A bilaterally symmetrical larval form typical of crustaceans.

necrosis (nec roh' sis) [Gk. *nekros*: death]   Premature cell death caused by external agents such as toxins.

negative feedback   In regulatory systems, information that decreases a regulatory response, returning the system to the set point. (Contrast with positive feedback.)

negative regulation   A type of gene regulation in which a gene is normally transcribed, and the binding of a repressor protein to the promoter prevents transcription. (Contrast with positive regulation.)

nematocyst (ne mat' o sist) [Gk. *nema*: thread + *kystis*: cell]   An elaborate, threadlike structure produced by cells of jellyfishes and other cnidarians, used chiefly to paralyze and capture prey.

nephron (nef' ron) [Gk. *nephros*: kidney]   The functional unit of the kidney, consisting of a structure for receiving a filtrate of blood and a tubule that reabsorbs selected parts of the filtrate.

Nernst equation   A mathematical statement that calculates the potential across a membrane permeable to a single type of ion that differs in concentration on the two sides of the membrane.

nerve   A structure consisting of many neuronal axons and connective tissue.

nervous tissue   Tissue specialized for processing and communicating information; one of the four major tissue types in multicellular animals.

net primary productivity (NPP)   The rate at which energy captured by photosynthesis is incorporated into the bodies of primary producers through growth and reproduction.

net primary production   The amount of primary producer biomass made available for consumption by heterotrophs.

neural network   An organized group of neurons that contains three functional categories of neurons—afferent neurons, interneurons, and efferent neurons—and is capable of processing information.

neural tube   An early stage in the development of the vertebrate nervous system consisting of a hollow tube created by two opposing folds of the dorsal ectoderm along the anterior–posterior body axis.

neurohormone   A chemical signal produced and released by neurons that subsequently acts as a hormone.

neuromuscular junction   Synapse (point of contact) where a motor neuron axon stimulates a muscle fiber cell.

neuron (noor' on) [Gk. *neuron*: nerve]   A nervous system cell that can generate and conduct action potentials along an axon to a synapse with another cell.

neurotransmitter   A substance produced in and released by a neuron (the presynaptic cell) that diffuses across a synapse and excites or inhibits another cell (the postsynaptic cell).

neurulation   Stage in vertebrate development during which the nervous system begins to form.

neutral allele   An allele that does not alter the functioning of the proteins for which it codes.

neutral theory   A view of molecular evolution that postulates that most mutations do not affect the amino acid being coded for, and that such mutations accumulate in a population at rates driven by genetic drift and mutation rates.

neutron (new' tron)   One of the three fundamental particles of matter (along with protons and electrons), with mass approximately 1 amu and no electrical charge.

niche (nitch) [L. *nidus*: nest]   The set of physical and biological conditions a species requires to survive, grow, and reproduce. (*See also* fundamental niche, realized niche.)

nitrate reduction   The process by which nitrate (NO_3^-) is reduced to ammonia (NH_3).

nitric oxide (NO)   An unstable molecule (a gas) that serves as a second messenger causing smooth muscle to relax. In the nervous system it operates as a neurotransmitter.

nitrifiers   Chemolithotrophic bacteria that oxidize ammonia to nitrate in soil and in seawater.

nitrogen fixation   Conversion of atmospheric nitrogen gas (N_2) into a more reactive and biologically useful form (ammonia), which makes nitrogen available to living things. Carried out by nitrogen-fixing bacteria, some of them free-living and others living within plant roots.

nitrogenase   An enzyme complex found in nitrogen-fixing bacteria that mediates the stepwise reduction of atmospheric N_2 to ammonia and which is strongly inhibited by oxygen.

node [L. *nodus*: knob, knot]   In plants, a (sometimes enlarged) point on a stem where a leaf is or was attached.

node of Ranvier   A gap in the myelin sheath covering an axon; the point where the axonal membrane can fire action potentials.

nodule   A specialized structure in the roots of nitrogen-fixing plants that houses nitrogen-fixing bacteria, in which oxygen is maintained at a low level by leghemoglobin.

noncompetitive inhibitor   A nonsubstrate that inhibits the activity of an enzyme by binding to a site other than its active site. (Contrast with competitive inhibitor.)

noncyclic electron transport   In photosynthesis, the flow of electrons that forms ATP, NADPH, and O_2.

nondisjunction   Failure of sister chromatids to separate in meiosis II or mitosis, or failure of homologous chromosomes to separate in meiosis I. Results in aneuploidy.

nonpolar   Having electric charges that are evenly balanced from one end to the other. (Contrast with polar.)

nonrandom mating   Selection of mates on the basis of a particular trait or group of traits.

nonsense mutation   Change in a gene's sequence that prematurely terminates translation by changing one of its codons to a stop codon.

nonsynonymous substitution   A change in a gene from one nucleotide to another that changes the amino acid specified by the corresponding codon (i.e., AGC →AGA, or serine → arginine). (Contrast with synonymous substitution.)

norepinephrine   A neurotransmitter found in the central nervous system and also at the post-ganglionic nerve endings of the sympathetic nervous system. Also called noradrenaline.

normal flora   Microorganisms that normally live and reproduce on or in the body without causing disease, and which form a nonspecific defense against pathogens by competing with them for space and nutrients.

notochord (no' tow kord) [Gk. *notos*: back + *chorde*: string]   A flexible rod of gelatinous material serving as a support in the embryos of all chordates and in the adults of tunicates and lancelets.

nucleic acid (new klay' ik)   A polymer made up of nucleotides, specialized for the storage, transmission, and expression of genetic information. DNA and RNA are nucleic acids.

nucleic acid hybridization   A technique in which a single-stranded nucleic acid probe is made that is complementary to, and binds to, a target sequence, either DNA or RNA. The resulting double-stranded molecule is a hybrid.

nucleoid (new' klee oid)   The region that harbors the chromosomes of a prokaryotic cell. Unlike the eukaryotic nucleus, it is not bounded by a membrane.

nucleolus (new klee' oh lus)   A small, generally spherical body found within the nucleus of eukaryotic cells. The site of synthesis of ribosomal RNA.

nucleoside   A nucleotide without the phosphate group; a nitrogenous base attached to a sugar.

nucleosome   A portion of a eukaryotic chromosome, consisting of part of the DNA molecule wrapped around a group of histone molecules, and held together by another type of histone molecule. The chromosome is made up of many nucleosomes.

nucleotide   The basic chemical unit in nucleic acids, consisting of a pentose sugar, a phosphate group, and a nitrogen-containing base.

nucleotide substitution   A change of one base pair to another in a DNA sequence.

nucleus (new' klee us) [L. *nux*: kernel or nut]   (1) In cells, the centrally located compartment of eukaryotic cells that is bounded by a double membrane and contains the chromosomes. (2) In the brain, an identifiable group of neurons that share common characteristics or functions.

nutrient   A food substance; or, in the case of mineral nutrients, an inorganic element required for completion of the life cycle of an organism.

- O -

obligate anaerobe   An anaerobic prokaryote that cannot survive exposure to O_2.

odorant   A molecule that can bind to an olfactory receptor.

oil   A triglyceride that is liquid at room temperature. (Contrast with fat.)

Okazaki fragments   Newly formed DNA making up the lagging strand in DNA replication. DNA ligase links Okazaki fragments together to give a continuous strand.

olfactory [L. *olfacere*: to smell]   Pertaining to the sense of smell (*olfaction*).

oligodendrocyte   A type of glial cell that myelinates axons in the central nervous system.

oligosaccharide   A polymer containing a small number of monosaccharides.

ommatidia [Gk. *omma*: eye]   The units that make up the compound eye of some arthropods.

omnivore [L. *omnis*: everything + *vorare*: to devour]   An organism that eats both animal and plant material. (Contrast with carnivore, detritivore, herbivore.)

oncogene [Gk. *onkos*: mass, tumor + *genes*: born]   A gene that codes for a protein product that stimulates cell proliferation. Mutations in oncogenes that result in excessive cell proliferation can give rise to cancer.

oocyte   *See* primary oocyte, secondary oocyte.

oogenesis (oh' eh jen e sis) [Gk. *oon*: egg + *genesis*: source]   Gametogenesis leading to production of an ovum.

oogonium (oh' eh go' nee um) (plural: oogonia)   (1) In some algae and fungi, a cell in which an egg is produced. (2) In animals, the diploid progeny of a germ cell in females.

operator   The region of an operon that acts as the binding site for the repressor.

open circulatory system   Circulatory system in which extracellular fluid leaves the vessels of the circulatory system, percolates between cells and through tissues, and then flows back into the circulatory system to be pumped out again. (Contrast with closed circulatory system.)

operon   A genetic unit of transcription, typically consisting of several structural genes that are transcribed together; the operon contains at least two control regions: the promoter and the operator.

opportunity cost   The sum of the benefits an animal forfeits by not being able to perform some other behavior during the time when it is performing a given behavior.

opsin (op' sin) [Gk. *opsis*: sight]   The protein portion of the visual pigment rhodopsin. (*See* rhodopsin.)

optic chiasm [Gk. *chiasma*: cross]   Structure on the lower surface of the vertebrate brain where the two optic nerves come together.

optical isomers   Two isomers that are mirror images of each other.

optimal foraging theory   The application of a cost–benefit approach to feeding behavior to identify the fitness value of feeding choices.

orbital   A region in space surrounding the atomic nucleus in which an electron is most likely to be found.

organ [Gk. *organon*: tool]   A body part, such as the heart, liver, brain, root, or leaf. Organs are composed of different tissues integrated to perform a distinct function. Organs, in turn, are integrated into organ systems.

organ identity genes   In angiosperms, genes that specify the different organs of the flower. (Compare with homeotic genes.)

organ of Corti   Structure in the inner ear that transforms mechanical forces produced from pressure waves ("sound waves") into action potentials that are sensed as sound.

organ system   An interrelated and integrated group of tissues and organs that work together in a physiological function.

organelle (or gan el')   Any of the membrane-enclosed structures within a eukaryotic cell. Examples include the nucleus, endoplasmic reticulum, and mitochondria.

organic   (1) Pertaining to any chemical compound that contains carbon. (2) Pertaining to any aspect of living matter, e.g., to its evolution, structure, or chemistry.

organism   Any living entity.

organizer   Region of the early amphibian embryo that directs early embryonic development. Also known as the primary embryonic organizer.

organogenesis   The formation of organs and organ systems during development.

origin of replication (*ori*)   DNA sequence at which helicase unwinds the DNA double helix and DNA polymerase binds to initiate DNA replication.

orthology (or thol' o jee)   Type of homology in which the divergence of homologous genes can be traced to speciation events. (Contrast with paralogy.)

osmoconformer   An aquatic animal that equilibrates the osmolarity of its extracellular fluid to be the same as that of the external environment.

osmolarity   The concentration of osmotically active particles in a solution.

osmoregulation   Regulation of the chemical composition of the body fluids of an organism.

osmosis (oz mo' sis) [Gk. *osmos*: to push]   Movement of water across a differentially permeable membrane, from one region to another region where the water potential is more negative.

ossicle (oss' ick ul) [L. *os*: bone]   The calcified construction unit of echinoderm skeletons.

osteoblast (oss' tee oh blast) [Gk. *osteon*: bone + *blastos*: sprout]   A cell that lays down the protein matrix of bone.

osteoclast (oss' tee oh clast) [Gk. *osteon*: bone + *klastos*: broken]   A cell that dissolves bone.

osteocyte   An osteoblast that has become enclosed in lacunae within the bone it has built.

outgroup   In phylogenetics, a group of organisms used as a point of reference for comparison with the groups of primary interest (the ingroup).

oval window   The flexible membrane that, when moved by the bones of the middle ear, produces pressure waves in the inner ear.

ovarian cycle   In human females, the monthly cycle of events by which eggs and hormones are produced. (Contrast with uterine cycle).

ovary (oh' var ee) [L. *ovum*: egg]   Any female organ, in plants or animals, that produces an egg.

overtopping   Plant growth pattern in which one branch differentiates from and grows beyond the others.

oviduct   In mammals, the tube serving to transport eggs to the uterus or to the outside of the body.

oviparity   Reproduction in which eggs are released by the female and development is external to the mother's body. (Contrast with viviparity.)

ovoviviparity   Pertaining to reproduction in which fertilized eggs develop and hatch within the mother's body but are not attached to the mother by means of a placenta.

ovulation   Release of an egg from an ovary.

ovule (oh′ vule)   In plants, a structure comprising the megasporangium and the integument, which develops into a seed after fertilization.

ovum (oh′ vum) (plural: ova) [L. egg]   The female gamete.

oxidation (ox i day′ shun)   Relative loss of electrons in a chemical reaction; either outright removal to form an ion, or the sharing of electrons with substances having a greater affinity for them, such as oxygen. Most oxidations, including biological ones, are associated with the liberation of energy. (Contrast with reduction.)

oxidative phosphorylation   ATP formation in the mitochondrion, associated with flow of electrons through the respiratory chain.

oxygenase   An enzyme that catalyzes the addition of oxygen to a substrate from O_2.

oxytocin   A hormone released by the posterior pituitary that promotes social bonding.

- P -

pancreas (pan′ cree us)   A gland located near the stomach of vertebrates that secretes digestive enzymes into the small intestine and releases insulin into the bloodstream.

Pangaea (pan jee′ uh) [Gk. *pan*: all, every]   The single land mass formed when all the continents came together in the Permian period.

para- [Gk. *para*: akin to, beside]   Prefix indicating association in being along side or accessory to.

parabronchi   Passages in the lungs of birds through which air flows.

paracrine [Gk. *para*: near]   Pertaining to a chemical signal, such as a hormone, that acts locally, near the site of its secretion. (Contrast with autocrine.)

paralogy (par al′ o jee)   Type of homology in which the divergence of homologous genes can be traced to gene duplication events. (Contrast with orthology.)

paraphyletic (par′ a fih leht′ ik) [Gk. *para*: beside + *phylon*: tribe]   Pertaining to a group that consists of an ancestor and some, but not all, of its descendants. (Contrast with monophyletic, polyphyletic.)

parasite   An organism that consumes parts of an organism much larger than itself (known as its host). Parasites sometimes, but not always, kill their host.

parasympathetic nervous system   The division of the autonomic nervous system that works in opposition to the sympathetic nervous system. (Contrast with sympathetic nervous system.)

parathyroid glands   Four glands on the posterior surface of the thyroid gland that produce and release parathyroid hormone.

parathyroid hormone (PTH)   A hormone secreted by the parathyroid glands that stimulates osteoclast activity and raises blood calcium levels. Also called parathormone.

parenchyma (pair eng′ kyma)   A plant tissue composed of relatively unspecialized cells without secondary walls.

parent rock   The soil horizon consisting of the rock that is breaking down to form the soil. Also called bedrock, or the C horizon.

parental (P) generation   The individuals that mate in a genetic cross. Their offspring are the first filial (F_1) generation.

parsimony   Preferring the simplest among a set of plausible explanations of any phenomenon.

parthenocarpy   Formation of fruit from a flower without fertilization.

parthenogenesis [Gk. *parthenos*: virgin]   Production of an organism from an unfertilized egg.

particulate theory   In genetics, the theory that genes are physical entities that retain their identities after fertilization.

passive transport   Diffusion across a membrane; may or may not require a channel or carrier protein. (Contrast with active transport.)

patch clamping   A technique for isolating a tiny patch of membrane to allow the study of ion movement through a particular channel.

pathogen (path′ o jen) [Gk. *pathos*: suffering + *genesis*: source]   An organism that causes disease.

pattern formation   In animal embryonic development, the organization of differentiated tissues into specific structures such as wings.

pedigree   The pattern of transmission of a genetic trait within a family.

pelagic zone [Gk. *pelagos*: sea]   The open ocean.

penetrance   The proportion of individuals with a particular genotype that show the expected phenotype.

penis   An accessory sex organ of male animals that enables the male to deposit sperm in the female's reproductive tract.

pentose [Gk. *penta*: five]   A sugar containing five carbon atoms.

PEP carboxylase   The enzyme that combines carbon dioxide with PEP to form a 4-carbon dicarboxylic acid at the start of C_4 photosynthesis or of crassulacean acid metabolism (CAM).

pepsin [Gk. *pepsis*: digestion]   An enzyme in gastric juice that digests protein.

pepsinogen   Inactive secretory product that is converted into pepsin by low pH or by enzymatic action.

peptide linkage   The bond between amino acids in a protein; formed between a carboxyl group and amino group (—CO—NH—) with the loss of water molecules.

peptidoglycan   The cell wall material of many bacteria, consisting of a single enormous molecule that surrounds the entire cell.

perennial (per ren′ ee al) [L. *per*: throughout + *annus*: year]   A plant that survives from year to year. (Contrast with annual, biennial.)

perfect flower   A flower with both stamens and carpels; a hermaphroditic flower. (Contrast with imperfect flower.)

pericycle [Gk. *peri*: around + *kyklos*: ring or circle]   In plant roots, tissue just within the endodermis, but outside of the root vascular tissue. Meristematic activity of pericycle cells produces lateral root primordia.

periderm   The outer tissue of the secondary plant body, consisting primarily of cork.

period   (1) A category in the geological time scale. (2) The duration of a single cycle in a cyclical event, such as a circadian rhythm.

peripheral nervous system (PNS)   The portion of the nervous system that transmits information to and from the central nervous system, consisting of neurons that extend or reside outside the brain or spinal cord and their supporting cells. (Contrast with central nervous system.)

peristalsis (pair′ i stall′ sis)   Wavelike muscular contractions proceeding along a tubular organ, propelling the contents along the tube.

peritoneum   The mesodermal lining of the body cavity in coelomate animals.

peroxisome   An organelle that houses reactions in which toxic peroxides are formed and then converted to water.

petal [Gk. *petalon*: spread out]   In an angiosperm flower, a sterile modified leaf, nonphotosynthetic, frequently brightly colored, and often serving to attract pollinating insects.

petiole (pet′ ee ole) [L. *petiolus*: small foot]   The stalk of a leaf.

pH   The negative logarithm of the hydrogen ion concentration; a measure of the acidity of a solution. A solution with pH = 7 is said to be neutral; pH values higher than 7 characterize basic solutions, while acidic solutions have pH values less than 7.

phage (fayj)   *See* bacteriophage.

phagocyte [Gk. *phagein*: to eat + *kystos*: sac]   One of two major classes of white blood cells; one of the nonspecific defenses of animals; ingests invading microorganisms by phagocytosis.

phagocytosis   Endocytosis by a cell of another cell or large particle.

pharming   The use of genetically modified animals to produce medically useful products in their milk.

pharynx [Gk. throat]   The part of the gut between the mouth and the esophagus.

phenotype (fee′ no type) [Gk. *phanein*: to show]   The observable properties of an individual resulting from both genetic and environmental factors. (Contrast with genotype.)

phenotypic plasticity   *See* developmental plasticity.

pheromone (feer′ o mone) [Gk. *pheros*: carry + *hormon*: excite, arouse]   A chemical substance used in communication between organisms of the same species.

phloem (flo′ um) [Gk. *phloos*: bark]   In vascular plants, the vascular tissue that transports sugars and other solutes from sources to sinks.

phosphate group   The functional group —OPO_3H_2.

phosphodiester linkage   The connection in a nucleic acid strand, formed by linking two nucleotides.

phospholipid   A lipid containing a phosphate group; an important constituent of cellular membranes. (*See* lipid.)

phospholipid bilayer   The basic structural unit of biological membranes; a sheet of phospholipids two molecules thick in which the phospholipids are lined up with their hydrophobic "tails" packed tightly together and their hydrophilic, phosphate-containing "heads" facing outward. Also called lipid bilayer.

phosphorylation   Addition of a phosphate group.

photoautotroph   An organism that obtains energy from light and carbon from carbon dioxide. (Contrast with chemolithotroph, chemoheterotroph, photoheterotroph.)

photoheterotroph   An organism that obtains energy from light but must obtain its carbon from organic compounds. (Contrast with chemolithotroph, chemoheterotroph, photoautotroph.)

photon (foe′ ton) [Gk. *photos*: light]   A quantum of visible radiation; a "packet" of light energy.

photoperiodicity   Control of an organism's physiological or behavioral responses by the length of the day or night.

photoreceptor   (1) In plants, a pigment that triggers a physiological response when it absorbs a photon. (2) In animals, a sensory receptor cell that senses and responds to light energy.

photorespiration   Light-driven uptake of oxygen and release of carbon dioxide, the carbon being derived from the early reactions of photosynthesis.

photosynthesis (foe tow sin' the sis) [literally, "synthesis from light"]   Metabolic processes, carried out by green plants and cyanobacteria, by which visible light is trapped and the energy used to synthesize compounds such as ATP and glucose.

photosystem [Gk. *phos*: light + *systema*: assembly]   A light-harvesting complex in the chloroplast thylakoid composed of pigments and proteins.

photosystem I   In photosynthesis, the complex that absorbs light at 700 nm, passing electrons to ferrodoxin and thence to NADPH.

photosystem II   In photosynthesis, the complex that absorbs light at 680 nm, passing electrons to the electron transport chain in the chloroplast.

phototropins   A class of blue light receptors that mediate phototropism and other plant responses.

phototropism [Gk. *photos*: light + *trope*: turning]   A directed plant growth response to light.

phycobilin   Photosynthetic pigment that absorbs red, yellow, orange, and green light and is found in cyanobacteria and some red algae.

phylogenetic tree   A graphic representation of lines of descent among organisms or their genes.

phylogeny (fy loj' e nee) [Gk. *phylon*: tribe, race + *genesis*: source]   The evolutionary history of a particular group of organisms or their genes.

physiology (fiz' ee ol' o jee) [Gk. *physis*: natural form]   The scientific study of the functions of living organisms and the individual organs, tissues, and cells of which they are composed.

phytoalexins   Substances toxic to pathogens, produced by plants in response to fungal or bacterial infection.

phytochrome (fy' tow krome) [Gk. *phyton*: plant + *chroma*: color]   A plant pigment regulating a large number of developmental and other phenomena in plants.

phytomers   In plants, the repeating modules that compose a shoot, each consisting of one or more leaves, attached to the stem at a node; an internode; and one or more axillary buds.

phytoplankton   Photosynthetic plankton.

phytoremediation   A form of bioremediation that uses plants to clean up environmental pollution.

pigment   A substance that absorbs visible light.

pineal gland   Gland located between the cerebral hemispheres that secretes melatonin.

pinocytosis   Endocytosis by a cell of liquid containing dissolved substances.

pistil [L. *pistillum*: pestle]   The structure of an angiosperm flower within which the ovules are borne. May consist of a single carpel, or of several carpels fused into a single structure. Usually differentiated into ovary, style, and stigma.

pith   In plants, relatively unspecialized tissue found within a cylinder of vascular tissue.

pituitary gland   A small gland attached to the base of the brain in vertebrates. Its hormones control the activities of other glands. Also known as the hypophysis.

placenta (pla sen' ta)   The organ in female mammals that provides for the nourishment of the fetus and elimination of the fetal waste products.

plankton   Free-floating small aquatic organisms. Photosynthetic members of the plankton are referred to as phytoplankton.

planula (plan' yew la) [L. *planum*: flat]   A free-swimming, ciliated larval form typical of the cnidarians.

plaque (plack) [Fr.: a metal plate or coin]   (1) A circular clearing in a layer (lawn) of bacteria growing on the surface of a nutrient agar gel. (2) An accumulation of prokaryotic organisms on tooth enamel. Acids produced by these microorganisms cause tooth decay. (3) A region of arterial wall invaded by fibroblasts and fatty deposits. (*See* atherosclerosis.)

plasma (plaz' muh)   The liquid portion of blood, in which blood cells and other particulates are suspended.

plasma cell   An antibody-secreting cell that develops from a B cell; the effector cell of the humoral immune system.

plasma membrane   The membrane that surrounds the cell, regulating the entry and exit of molecules and ions. Every cell has a plasma membrane.

plasmid   A DNA molecule distinct from the chromosome(s); that is, an extrachromosomal element; found in many bacteria. May replicate independently of the chromosome.

plasmodesma (plural: plasmodesmata) [Gk. *plassein*: to mold + *desmos*: band]   A cytoplasmic strand connecting two adjacent plant cells.

plasmogamy   The fusion of the cytoplasm of two cells. (Contrast with karyogamy.)

plastid   Any of the plant cell organelles that house biochemical pathways for photosynthesis.

plate tectonics [Gk. *tekton*: builder]   The scientific study of the structure and movements of Earth's lithospheric plates, which are the cause of continental drift.

platelet   A membrane-bounded body without a nucleus, arising as a fragment of a cell in the bone marrow of mammals. Important to blood-clotting action.

pleiotropy (plee' a tro pee) [Gk. *pleion*: more]   The determination of more than one character by a single gene.

pleural membrane [Gk. *pleuras*: rib, side]   The membrane lining the outside of the lungs and the walls of the thoracic cavity. Inflammation of these membranes is a condition known as pleurisy.

pluripotent [L. *pluri*: many + *potens*: powerful]   Having the ability to form all of the cells in the body. (Contrast with multipotent, totipotent.)

podocytes   Cells of Bowman's capsule of the nephron that cover the capillaries of the glomerulus, forming filtration slits.

point mutation   A mutation that results from the gain, loss, or substitution of a single nucleotide.

polar   Having separate and opposite electric charges at two ends, or poles. (Contrast with nonpolar.)

polar body   A nonfunctional nucleus produced by meiosis during oogenesis.

polar nuclei   In angiosperms, the two nuclei in the central cell of the megagametophyte; following fertilization they give rise to the endosperm.

polarity   (1) In chemistry, the property of unequal electron sharing in a covalent bond that defines a polar molecule. (2) In development, the difference between one end of an organism or structure and the other.

pollen [L. *pollin*: fine flour]   In seed plants, microscopic grains that contain the male gametophyte (microgametophyte) and gamete (microspore).

pollen tube   A structure that develops from a pollen grain through which sperm are released into the megagametophyte.

pollination   The process of transferring pollen from an anther to the stigma of a pistil in an angiosperm or from a strobilus to an ovule in a gymnosperm.

poly- [Gk. *poly*: many]   A prefix denoting multiple entities.

poly A tail   A long sequence of adenine nucleotides (50–250) added after transcription to the 3' end of most eukaryotic mRNAs.

polyandry   Mating system in which one female mates with multiple males.

polygyny   Mating system in which one male mates with multiple females.

polymer [Gk. *poly*: many + *meros*: unit]   A large molecule made up of similar or identical subunits called monomers. (Contrast with monomer.)

polymerase chain reaction (PCR)   An enzymatic technique for the rapid production of millions of copies of a particular stretch of DNA where only a small amount of the parent molecule is available.

polymorphic (pol' lee mor' fik) [Gk. *poly*: many + *morphe*: form, shape]   Coexistence in a population of two or more distinct traits.

polyp (pah' lip) [Gk. *poly*: many + *pous*: foot]   In cnidarians, a sessile, asexual life cycle stage.

polypeptide   A large molecule made up of many amino acids joined by peptide linkages. Large polypeptides are called proteins.

polyphyletic (pol' lee fih leht' ik) [Gk. *poly*: many + *phylon*: tribe]   Pertaining to a group that consists of multiple distantly related organisms, and does not include the common ancestor of the group. (Contrast with monophyletic, paraphyletic.)

polyploidy (pol' lee ploid ee)   The possession of more than two entire sets of chromosomes.

polyribosome (polysome)   A complex consisting of a threadlike molecule of messenger RNA and several (or many) ribosomes. The ribosomes move along the mRNA, synthesizing polypeptide chains as they proceed.

polysaccharide   A macromolecule composed of many monosaccharides (simple sugars). Common examples are cellulose and starch.

pons [L. *pons*: bridge]   Region of the brainstem anterior to the medulla.

population   Any group of organisms coexisting at the same time and in the same place and capable of interbreeding with one another.

population bottleneck   A period during which only a few individuals of a normally large population survive.

population density   The number of individuals in a population per unit of area or volume.

population dynamics The patterns and processes of change in populations.

population genetics The study of genetic variation and its causes within populations.

portal blood vessels Blood vessels that begin and end in capillary beds.

positive cooperativity Occurs when a molecule can bind several ligands and each one that binds alters the conformation of the molecule so that it can bind the next ligand more easily. The binding of four molecules of O_2 by hemoglobin is an example of positive cooperativity.

positive feedback In regulatory systems, information that amplifies a regulatory response, increasing the deviation of the system from the set point. (Contrast with negative feedback.)

positive regulation A form of gene regulation in which a regulatory macromolecule is needed to turn on the transcription of a structural gene; in its absence, transcription will not occur. (Contrast with negative regulation.)

post- [L. *postere*: behind, following after] Prefix denoting something that comes after.

postabsorptive state State in which no food remains in the gut and thus no nutrients are being absorbed. (Contrast with absorptive state.)

posterior Toward or pertaining to the rear. (Contrast with anterior.)

postsynaptic cell The cell that receives information from a neuron at a synapse. (Contrast with presynaptic neuron.)

postzygotic reproductive barriers Barriers to the reproductive process that occur after the union of the nuclei of two gametes. (Contrast with prezygotic reproductive barriers.)

potential energy Energy not doing work, such as the energy stored in chemical bonds. (Contrast with kinetic energy.)

precapillary sphincter A cuff of smooth muscle that can shut off the blood flow to a capillary bed.

pre-mRNA (precursor mRNA) Initial gene transcript before it is modified to produce functional mRNA. Also known as the primary transcript.

predator An organism that kills and eats other organisms.

prereplication complex In eukaryotes, a complex of proteins that binds to DNA at the initiation of DNA replication.

pressure flow model An effective model for phloem transport in angiosperms. It holds that sieve element transport is driven by an osmotically generated pressure gradient between source and sink.

pressure potential The hydrostatic pressure of an enclosed solution in excess of the surrounding atmospheric pressure. (Contrast with solute potential, water potential.)

presynaptic neuron The neuron that transmits information to another cell at a synapse. (Contrast with postsynaptic cell.)

prey [L. *praeda*: booty] An organism consumed by a predator as an energy source.

prezygotic reproductive barriers Barriers to the reproductive process that occur before the union of the nuclei of two gametes (Contrast with postzygotic reproductive barriers.)

primary active transport Active transport in which ATP is hydrolyzed, yielding the energy required to transport an ion or molecule against its concentration gradient. (Contrast with secondary active transport.)

primary cell wall In plant cells, a structure that forms at the middle lamella after cytokinesis, made up of cellulose microfibrils, hemicelluloses, and pectins. (Contrast with secondary cell wall.)

primary consumer An organism (herbivore) that eats plant tissues.

primary growth In plants, growth that is characterized by the lengthening of roots and shoots and by the proliferation of new roots and shoots through branching. (Contrast with secondary growth.)

primary immune response The first response of the immune system to an antigen, involving recognition by lymphocytes and the production of effector cells and memory cells. (Contrast with secondary immune response.)

primary lysosome *See* lysosome.

primary meristem Meristem that produces the tissues of the primary plant body.

primary oocyte (oh' eh site) [Gk. *oon*: egg + *kytos*: container] The diploid progeny of an oogonium. In many species, a primary oocyte enters prophase of the first meiotic division, then remains in developmental arrest for a long time before resuming meiosis to form a secondary oocyte and a polar body.

primary plant body That part of a plant produced by primary growth. Consists of all the *nonwoody* parts of a plant; many herbaceous plants consist entirely of a primary plant body. (Contrast with secondary plant body.)

primary producer A photosynthetic or chemosynthetic organism that synthesizes complex organic molecules from simple inorganic ones.

primary sex determination Genetic determination of gametic sex, male or female. (Contrast with secondary sex determination.)

primary spermatocyte The diploid progeny of a spermatogonium; undergoes the first meiotic division to form secondary spermatocytes.

primary succession Succession that begins in an area initially devoid of life, such as on recently exposed glacial till or lava flows. (Contrast with secondary succession.)

primary structure The specific sequence of amino acids in a protein.

primase An enzyme that catalyzes the synthesis of a primer for DNA replication.

primer Strand of nucleic acid, usually RNA, that is the necessary starting material for the synthesis of a new DNA strand, which is synthesized from the 3′ end of the primer.

primordium (plural: primordia) [L. origin] The most rudimentary stage of an organ or other part.

prion An infectious protein that can proliferate by converting the inactive form of a particular protein into an active protein.

pro- [L.: first, before, favoring] A prefix often used in biology to denote a developmental stage that comes first or an evolutionary form that appeared earlier than another. For example, prokaryote, prophase.

probe A segment of single-stranded nucleic acid used to identify DNA molecules containing the complementary sequence.

procambium Primary meristem that produces the vascular tissue.

procedural memory Memory of motor tasks. Cannot be consciously recalled and described. (Contrast with declarative memory.)

processive Pertaining to an enzyme that catalyzes many reactions each time it binds to a substrate, as DNA polymerase does during DNA replication.

progesterone [L. *pro*: favoring + *gestare*: to bear] A female sex hormone that maintains pregnancy.

prolactin A hormone released by the anterior pituitary, one of whose functions is the stimulation of milk production in female mammals.

proliferating cell nuclear antigen (PCNA) A protein complex that ensures processivity of DNA replication in eukaryotes.

prometaphase The phase of nuclear division that begins with the disintegration of the nuclear envelope.

promoter A DNA sequence to which RNA polymerase binds to initiate transcription.

prophage (pro' fayj) The noninfectious units that are linked with the chromosomes of the host bacteria and multiply with them but do not cause dissolution of the cell. Prophage can later enter into the lytic phase to complete the virus life cycle.

prophase (pro' phase) The first stage of nuclear division, during which chromosomes condense from diffuse, threadlike material to discrete, compact bodies.

prostaglandin Any one of a group of specialized lipids with hormone-like functions. It is not clear that they act at any considerable distance from the site of their production.

prostate gland In male humans, surrounds the urethra at its junction with the vas deferens; supplies an acid-neutralizing fluid to the semen.

prosthetic group Any nonprotein portion of an enzyme.

proteasome In the eukaryotic cytoplasm, a huge protein structure that binds to and digests cellular proteins that have been tagged by ubiquitin.

protein (pro' teen) [Gk. *protos*: first] Long-chain polymer of amino acids with twenty different common side chains. Occurs with its polymer chain extended in fibrous proteins, or coiled into a compact macromolecule in enzymes and other globular proteins.

protein kinase (kye' nase) An enzyme that catalyzes the addition of a phosphate group from ATP to a target protein.

protein kinase cascade A series of reactions in response to a molecular signal, in which a series of protein kinases activate one another in sequence, amplifying the signal at each step.

proteoglycan A glycoprotein containing a protein core with attached long, linear carbohydrate chains.

proteolysis [protein + Gk. *lysis*: break apart] An enzymatic digestion of a protein or polypeptide.

proteome The set of proteins that can be made by an organism. Because of alternative splicing of pre-mRNA, the number of proteins that can be made is usually much larger than the number of protein-coding genes present in the organism's genome.

protoderm Primary meristem that gives rise to the plant epidermis.

proton (pro' ton) [Gk. *protos*: first, before] (1) A subatomic particle with a single positive charge.

The number of protons in the nucleus of an atom determine its element. (2) A hydrogen ion, H^+.

proton pump An active transport system that uses ATP energy to move hydrogen ions across a membrane, generating an electric potential.

proton-motive force Force generated across a membrane having two components: a chemical potential (difference in proton concentration) plus an electrical potential due to the electrostatic charge on the proton.

protonephridium The excretory organ of flatworms, made up of a tubule and a flame cell.

protoplast The living contents of a plant cell; the plasma membrane and everything contained within it.

provirus Double-stranded DNA made by a virus that is integrated into the host's chromosome and contains promoters that are recognized by the host cell's transcription apparatus.

proximal Near the point of attachment or other reference point. (Contrast with distal.)

proximal convoluted tubule The initial segment of a renal tubule, closest to the glomerulus. (Compare with distal convoluted tubule.)

proximate cause The immediate genetic, physiological, neurological, and developmental mechanisms responsible for a behavior or morphology. (Contrast with ultimate cause.)

pseudocoelomate (soo' do see' low mate) [Gk. *pseudes*: false + *koiloma*: cavity] Having a body cavity, called a pseudocoel, consisting of a fluid-filled space in which many of the internal organs are suspended, but which is enclosed by mesoderm only on its outside.

pseudogene [Gk. *pseudes*: false] A DNA segment that is homologous to a functional gene but is not expressed because of changes to its sequence or changes to its location in the genome.

pseudopod (soo' do pod) [Gk. *pseudes*: false + *podos*: foot] A temporary, soft extension of the cell body that is used in location, attachment to surfaces, or engulfing particles.

pulmonary [L. *pulmo*: lung] Pertaining to the lungs.

pulmonary circuit The portion of the circulatory system by which blood is pumped from the heart to the lungs or gills for oxygenation and back to the heart for distribution. (Contrast with systemic circuit.)

pulmonary valve A one-way valve between the right ventricle of the heart and the pulmonary artery that prevents backflow of blood into the ventricle when it relaxes.

Punnett square Method of predicting the results of a genetic cross by arranging the gametes of each parent at the edges of a square.

pupa (pew' pa) [L. *pupa*: doll, puppet] In certain insects (the Holometabola), the encased developmental stage between the larva and the adult.

pupil The opening in the vertebrate eye through which light passes.

purine (pure' een) One of the two types of nitrogenous bases in nucleic acids. Each of the purines—adenine and guanine—pairs with a specific pyrimidine.

Purkinje fibers Specialized heart muscle cells that conduct excitation throughout the ventricular muscle.

pyrimidine (per im' a deen) One of the two types of nitrogenous bases in nucleic acids. Each

of the pyrimidines—cytosine, thymine, and uracil—pairs with a specific purine.

pyrogen Molecule that produces a rise in body temperature (fever); may be produced by an invading pathogen or by cells of the immune system in response to infection.

pyruvate The ionized form of pyruvic acid, a three-carbon acid; the end product of glycolysis and the raw material for the citric acid cycle.

pyruvate oxidation Conversion of pyruvate to acetyl CoA and CO_2 that occurs in the mitochondrial matrix in the presence of O_2.

- Q -

Q_{10} A value that compares the rate of a biochemical process or reaction over $10^{\circ}C$ temperature ranges. A process that is not temperature-sensitive has a Q_{10} of 1; values of 2 or 3 mean the reaction speeds up as temperature increases.

quantitative trait loci A set of genes that determines a complex character that exhibits quantitative variation.

quaternary structure The specific three-dimensional arrangement of protein subunits.

- R -

R group The distinguishing group of atoms of a particular amino acid; also known as a side chain.

r-strategist A species whose life history strategy allows for a high intrinsic rate of population increase (r). (Contrast with K-strategist.)

radial symmetry The condition in which any two halves of a body are mirror images of each other, providing the cut passes through the center; a cylinder cut lengthwise down its center displays this form of symmetry.

radiation The transfer of heat from warmer objects to cooler ones via the exchange of infrared radiation. *See also* electromagnetic radiation; evolutionary radiation.

radicle An embryonic root.

radioisotope A radioactive isotope of an element. Examples are carbon-14 (^{14}C) and hydrogen-3, or tritium (^{3}H).

reactant A chemical substance that enters into a chemical reaction with another substance.

reaction center A group of electron transfer proteins that receive energy from light-absorbing pigments and convert it to chemical energy by redox reactions.

realized niche A species' niche as defined by its interactions with other species. (Contrast with fundamental niche.)

receptive field The area of visual space that activates a particular cell in the visual system.

receptor *See* receptor protein, sensory receptor cell.

receptor-mediated endocytosis Endocytosis initiated by macromolecular binding to a specific membrane receptor.

receptor potential The change in the resting potential of a sensory cell when it is stimulated.

receptor protein A protein that can bind to a specific molecule, or detect a specific stimulus, within the cell or in the cell's external environment.

recessive In genetics, an allele that does not determine phenotype in the presence of a dominant allele. (Contrast with dominance.)

reciprocal adaptation *See* coevolution.

reciprocal crosses A pair of matings in one of which a female of genotype A mates with a male of genotype B and in the other of which a female of genotype B mates with a male of genotype A.

recognition sequence *See* restriction site.

recombinant Pertaining to an individual, meiotic product, or chromosome in which genetic materials originally present in two individuals end up in the same haploid complement of genes.

recombinant DNA A DNA molecule made in the laboratory that is derived from two or more genetic sources.

recombinant frequency The proportion of offspring of a genetic cross that have phenotypes different from the parental phenotypes due to crossing over between linked genes during gamete formation.

reconciliation ecology The practice of making exploited lands more biodiversity-friendly.

rectum The terminal portion of the gut, ending at the anus.

redox reaction A chemical reaction in which one reactant becomes oxidized and the other becomes reduced.

reduction Gain of electrons by a chemical reactant; any reduction is accompanied by an oxidation. (Contrast with oxidation.)

refractory period The time interval after an action potential during which another action potential cannot be elicited from an excitable membrane.

regeneration The development of a complete individual from a fragment of an organism.

regulative development A pattern of animal embryonic development in which the fates of the first blastomeres are not absolutely fixed. (Contrast with mosaic development.)

regulatory sequence A DNA sequence to which the protein product of a regulatory gene binds.

regulatory gene A gene that codes for a protein (or RNA) that in turn controls the expression of another gene.

regulatory system A system that uses feedback information to maintain a physiological function or parameter at an optimal level. (Contrast with controlled system.)

regulatory T cells (T_{reg}) The class of T cells that mediates tolerance to self antigens.

reinforcement The evolution of enhanced reproductive isolation between populations due to natural selection for greater isolation.

releaser Sensory stimulus that triggers performance of a stereotyped behavior pattern.

REM (rapid-eye-movement) sleep A sleep state characterized by vivid dreams, skeletal muscle relaxation, and rapid eye movements. (Contrast with slow-wave sleep.)

renal [L. *renes*: kidneys] Relating to the kidneys.

renal tubule A structural unit of the kidney that collects filtrate from the blood, reabsorbs specific ions, nutrients, and water and returns them to the blood, and concentrates excess ions and waste products such as urea for excretion from the body.

replication The duplication of genetic material.

replication complex The close association of several proteins operating in the replication of DNA.

replication fork   A point at which a DNA molecule is replicating. The fork forms by the unwinding of the parent molecule.

replicon   A region of DNA replicated from a single origin of replication.

reporter gene   A genetic marker included in recombinant DNA to indicate the presence of the recombinant DNA in a host cell.

repressor   A protein encoded by a regulatory gene that can bind to a specific operator and prevent transcription of the operon. (Contrast with activator.)

reproductive isolation   Condition in which two divergent populations are no longer exchanging genes. Can lead to speciation.

rescue effect   The process by which individuals moving between subpopulations of a metapopulation may prevent declining subpopulations from becoming extinct.

resistance (R) genes   Plant genes that confer resistance to specific strains of pathogens.

resource   Something in the environment required by an organism for its maintenance and growth that is consumed in the process of being used.

resource partitioning   A situation in which selection pressures resulting from interspecific competition cause changes in the ways in which the competing species use the limiting resource, thereby allowing them to coexist.

respiration (res pi ra' shun) [L. *spirare*: to breathe]   (1) Cellular respiration. (2) Breathing.

respiratory chain   The terminal reactions of cellular respiration, in which electrons are passed from NAD or FAD, through a series of intermediate carriers, to molecular oxygen, with the concomitant production of ATP.

respiratory gases   Oxygen (O_2) and carbon dioxide (CO_2); the gases that an animal must exchange between its internal body fluids and the outside medium (air or water).

resting potential   The membrane potential of a living cell at rest. In cells at rest, the interior is negative to the exterior. (Contrast with action potential.)

restoration ecology   The science and practice of restoring damaged or degraded ecosystems.

restriction enzyme   Any of a type of enzyme that cleaves double-stranded DNA at specific sites; extensively used in recombinant DNA technology. Also called a restriction endonuclease.

restriction fragment length polymorphism   See RFLP.

restriction point (R)   The specific time during G1 of the cell cycle at which the cell becomes committed to undergo the rest of the cell cycle.

restriction site   A specific DNA base sequence that is recognized and acted on by a restriction endonuclease.

reticular system   A central region of the vertebrate brainstem that includes complex fiber tracts conveying neural signals between the forebrain and the spinal cord, with collateral fibers to a variety of nuclei that are involved in autonomic functions, including arousal from sleep.

retina (rett' in uh) [L. *rete*: net]   The light-sensitive layer of cells in the vertebrate or cephalopod eye.

retinoblastoma protein   A protein that inhibits an animal cell from passing through the restriction point; inactivation of this protein is necessary for the cell cycle to proceed.

retrovirus   An RNA virus that contains reverse transcriptase. Its RNA serves as a template for cDNA production, and the cDNA is integrated into a chromosome of the host cell.

reverse genetics   Method of genetic analysis in which a phenotype is first related to a DNA variation, then the protein involved is identified.

reverse transcriptase   An enzyme that catalyzes the production of DNA (cDNA), using RNA as a template; essential to the reproduction of retroviruses.

RFLP   Restriction fragment length polymorphism, the coexistence of two or more patterns of restriction fragments resulting from underlying differences in DNA sequence.

rhizoids (rye' zoids) [Gk. root]   Hairlike extensions of cells in mosses, liverworts, and a few vascular plants that serve the same function as roots and root hairs in vascular plants. The term is also applied to branched, rootlike extensions of some fungi and algae.

rhizome (rye' zome)   An underground stem (as opposed to a root) that runs horizontally beneath the ground.

rhodopsin   A photopigment used in the visual process of transducing photons of light into changes in the membrane potential of photoreceptor cells.

ribonucleic acid   See RNA.

ribose   A five-carbon sugar in nucleotides and RNA.

ribosomal RNA (rRNA)   Several species of RNA that are incorporated into the ribosome. Involved in peptide bond formation.

ribosome   A small particle in the cell that is the site of protein synthesis.

ribozyme   An RNA molecule with catalytic activity.

risk cost   The increased chance of being injured or killed as a result of performing a behavior, compared to resting.

RNA (ribonucleic acid)   An often single stranded nucleic acid whose nucleotides use ribose rather than deoxyribose and in which the base uracil replaces thymine found in DNA. Serves as genome from some viruses. (See ribosomal RNA, transfer RNA, mRNA, and ribozyme.)

RNA interference (RNAi)   A mechanism for reducing mRNA translation whereby a double-stranded RNA, made by the cell or synthetically, is processed into a small, single-stranded RNA, whose binding to a target mRNA results in the latter's breakdown.

RNA polymerase   An enzyme that catalyzes the formation of RNA from a DNA template.

RNA splicing   The last stage of RNA processing in eukaryotes, in which the transcripts of introns are excised through the action of small nuclear ribonucleoprotein particles (snRNP).

rod cells   Light-sensitive cell in the vertebrate retina; these sensory receptor cells are sensitive in extremely dim light and are responsible for dim light, black and white vision.

root   The organ responsible for anchoring the plant in the soil, absorbing water and minerals, and producing certain hormones. Some roots are storage organs.

root cap   A thimble-shaped mass of cells, produced by the root apical meristem, that protects the meristem; the organ that perceives the gravitational stimulus in root gravitropism.

root hair   A long, thin process from a root epidermal cell that absorbs water and minerals from the soil solution.

root system   The organ system that anchors a plant in place, absorbs water and dissolved minerals, and may store products of photosynthesis from the shoot system.

rough endoplasmic reticulum (RER)   The portion of the endoplasmic reticulum whose outer surface has attached ribosomes. (Contrast with smooth endoplasmic reticulum.)

rRNA   See ribosomal RNA.

rubisco   Contraction of ribulose bisphosphate carboxylase/oxygenase, the enzyme that combines carbon dioxide or oxygen with ribulose bisphosphate to catalyze the first step of photosynthetic carbon fixation or photorespiration, respectively.

ruminant   Herbivorous, cud-chewing mammals such as cows or sheep, characterized by a stomach that consists of four compartments: the rumen, reticulum, omasum, and abomasum.

- S -

S phase   In the cell cycle, the stage of interphase during which DNA is replicated. (Contrast with G1 phase, G2 phase, M phase.)

saltatory conduction   [L. *saltare*: to jump]   The rapid conduction of action potentials in myelinated axons; so called because action potentials appear to "jump" between nodes of Ranvier along the axon.

saprobe [Gk. *sapros*: rotten]   An organism (usually a bacterium or fungus) that obtains its carbon and energy by absorbing nutrients from dead organic matter.

sarcomere (sark' o meer) [Gk. *sark*: flesh + *meros*: unit]   The contractile unit of a skeletal muscle.

sarcoplasm   The cytoplasm of a muscle cell.

sarcoplasmic reticulum   The endoplasmic reticulum of a muscle cell.

saturated fatty acid   A fatty acid in which all the bonds between carbon atoms in the hydrocarbon chain are single bonds—that is, all the bonds are saturated with hydrogen atoms. (Contrast with unsaturated fatty acid.)

scientific method   A means of gaining knowledge about the natural world by making observations, posing hypotheses, and conducting experiments to test those hypotheses.

Schwann cell   A type of glial cell that myelinates axons in the peripheral nervous system.

scion   In horticulture, the bud or stem from one plant that is grafted to a root or root-bearing stem of another plant (the stock).

sclerenchyma (skler eng' kyma) [Gk. *skleros*: hard + *kymus*: juice]   A plant tissue composed of cells with heavily thickened cell walls. The cells are dead at functional maturity. The principal types of sclerenchyma cells are fibers and sclereids.

second law of thermodynamics   The principle that when energy is converted from one form to another, some of that energy becomes unavailable for doing work.

second messenger   A compound, such as cAMP, that is released within a target cell after a hormone (the first messenger) has bound to a surface receptor on a cell; the second messenger triggers further reactions within the cell.

secondary active transport   A form of active transport that does not use ATP as an energy source; rather, transport is coupled to ion diffusion down a concentration gradient established by primary active transport.

secondary cell wall   A thick, cellulosic structure internal to the primary cell wall formed in some plant cells after cell expansion stops (Contrast with primary cell wall.)

secondary consumer   An organism that eats primary consumers.

secondary growth   In plants, growth that contributes to an increase in girth. (Contrast with primary growth.)

secondary immune response   A rapid and intense response to a second or subsequent exposure to an antigen, initiated by memory cells. (Contrast with primary immune response.)

secondary lysosome   Membrane-enclosed organelle formed by the fusion of a primary lysosome with a phagosome, in which macromolecules taken up by phagocytosis are hydrolyzed into their monomers. (Contrast with lysosome.)

secondary metabolite   A compound synthesized by a plant that is not needed for basic cellular metabolism. Typically has an antiherbivore or antiparasite function.

secondary plant body   That part of a plant produced by secondary growth; consists of woody tissues. (Contrast with primary plant body.)

secondary sex determination   Formation of secondary sexual characteristics (i.e., those other than gonads), such as external sex organs and body hair. (Contrast with primary sex determination.)

secondary spermatocyte   One of the products of the first meiotic division of a primary spermatocyte.

secondary structure   Of a protein, localized regularities of structure, such as the α helix and the β pleated sheet.

secondary succession   Succession after a disturbance that did not eliminate all the organisms originally living on the site. (Contrast with primary succession.)

secretin (si kreet' in)   A peptide hormone secreted by the upper region of the small intestine when acidic chyme is present. Stimulates the pancreatic duct to secrete bicarbonate ions.

sedimentary rock   Rock formed by the accumulation of sediment grains on the bottom of a body of water.

seed   A fertilized, ripened ovule of a gymnosperm or angiosperm. Consists of the embryo, nutritive tissue, and a seed coat.

segmentation   Division of an animal body into segments.

segmentation genes   Genes that determine the number and polarity of body segments.

segregation   In genetics, the separation of alleles, or of homologous chromosomes, from each other during meiosis so that each of the haploid daughter nuclei produced contains one or the other member of the pair found in the diploid parent cell, but never both. This principle was articulated by Mendel as his first law.

selective permeability   Allowing certain substances to pass through while other substances are excluded; a characteristic of membranes.

self-incompatability   In plants, the possession of mechanisms that prevent self-fertilization.

semelparous   [L. *semel*: once + *pario*: to beget] Reproducing only once in a lifetime. (Contrast with iteroparous.)

semen (see' men) [L. *semin*: seed]   The thick, whitish liquid produced by the male reproductive system in mammals, containing the sperm.

semiconservative replication   The way in which DNA is synthesized. Each of the two partner strands in a double helix acts as a template for a new partner strand. Hence, after replication, each double helix consists of one old and one new strand.

seminiferous tubules   The tubules within the testes within which sperm production occurs.

senescence [L. *senescere*: to grow old]   Aging; deteriorative changes with aging; the increased probability of dying with increasing age.

sensitive period   The life stage during which some particular type of learning must take place, or during which it occurs much more easily than at other times. Typical of song learning among birds.

sensor   *See* sensory receptor cell.

sensory neuron   A specialized neuron that transduces a particular type of sensory stimulus into action potentials.

sensory receptor cell   Cell that is responsive to a particular type of physical or chemical stimulation.

sensory transduction   The transformation of environmental stimuli or information into neural signals.

sepal (see' pul) [L. *sepalum*: covering]   One of the outermost structures of the flower, usually protective in function and enclosing the rest of the flower in the bud stage.

septum (plural: septa) [L. wall]   (1) A partition or cross-wall appearing in the hyphae of some fungi. (2) The bony structure dividing the nasal passages.

sequence alignment   A method of identifying homologous positions in DNA or amino acid sequences by pinpointing the locations of deletions and insertions that have occurred since two (or more) organisms diverged from a common ancestor.

Sertoli cells   Cells in the seminiferous tubules that nurture the developing sperm.

sessile (sess' ul) [L. *sedere*: to sit]   Permanently attached; not able to move from one place to another. (Contrast with motile.)

set point   In a regulatory system, the threshold sensitivity to the feedback stimulus.

sex chromosome   In organisms with a chromosomal mechanism of sex determination, one of the chromosomes involved in sex determination.

sex linkage   The pattern of inheritance characteristic of genes located on the sex chromosomes of organisms having a chromosomal mechanism for sex determination.

sexual reproduction   Reproduction involving the union of gametes.

sexual selection   Selection by one sex of characteristics in individuals of the opposite sex. Also, the favoring of characteristics in one sex as a result of competition among individuals of that sex for mates.

Shannon diversity index   A formula for quantifying diversity that takes both species richness and species evenness into account; based on a mathematical expression of the certainty with which the next item sampled in a series can be predicted.

shared derived trait   *See* synapomorphy.

shoot system   In plants, the organ system consisting of the leaves, stem(s), and flowers.

short tandem repeat (STR)   A short (1–5 base pairs), moderately repetitive sequence of DNA. The number of copies of an STR at a particular location varies between individuals and is inherited.

shotgun sequencing   A relatively rapid method of DNA sequencing in which a DNA molecule is broken up into overlapping fragments, each fragment is sequenced, and high-speed computers analyze and realign the fragments. (Contrast with hierarchical sequencing.)

side chain   *See* R group.

sieve tube element   The characteristic cell of the phloem in angiosperms, which contains cytoplasm but relatively few organelles, and whose end walls (*sieve plates*) contain pores that form connections with neighboring cells.

signal sequence   The sequence within a protein that directs the protein to a particular organelle.

signal transduction pathway   The series of biochemical steps whereby a stimulus to a cell (such as a hormone or neurotransmitter binding to a receptor) is translated into a response of the cell.

silent mutation   A change in a gene's sequence that has no effect on the amino acid sequence of a protein because it occurs in noncoding DNA or because it does not change the amino acid specified by the corresponding codon. (Contrast with frame-shift mutation, missense mutation, nonsense mutation.)

similarity matrix   A matrix used to compare the degree of divergence among pairs of objects. For molecular sequences, constructed by summing the number or percentage of nucleotides or amino acids that are identical in each pair of sequences.

single nucleotide polymorphisms (SNPs)   Inherited variations in a single nucleotide base in DNA that differ between individuals.

single-strand binding protein   In DNA replication, a protein that binds to single strands of DNA after they have been separated from each other, keeping the two strands separate for replication.

sink   In plants, any organ that imports the products of photosynthesis, such as roots, developing fruits, and immature leaves. (Contrast with source.)

sinoatrial node (sigh' no ay' tree al) [L. *sinus*: curve + *atrium*: chamber]   The pacemaker of the mammalian heart.

siRNAs (small interfering RNAs)   Short, double-stranded RNA molecules used in RNA interference.

sister chromatid   Each of a pair of newly replicated chromatids.

sister groups   Two phylogenetic groups that are each other's closest relatives.

skeletal muscle   A type of muscle tissue characterized by multinucleated cells containing highly ordered arrangements of actin and myosin microfilaments. Also called striated muscle. (Contrast with cardiac muscle, smooth muscle.)

skeletal systems   Organ systems that provide rigid supports against which muscles can pull to create directed movements.

sliding DNA clamp   Protein complex that keeps DNA polymerase bound to DNA during replication.

sliding filament theory Mechanism of muscle contraction based on the formation and breaking of crossbridges between actin and myosin filaments, causing the filaments to slide together.

slow-wave sleep A state of deep, restorative sleep characterized by high-amplitude slow waves in the EEG. (Contrast with REM sleep.)

small intestine The portion of the gut between the stomach and the colon; consists of the duodenum, the jejunum, and the ileum.

small nuclear ribonucleoprotein particle (snRNP) A complex of an enzyme and a small nuclear RNA molecule, functioning in RNA splicing.

smooth endoplasmic reticulum (SER) Portion of the endoplasmic reticulum that lacks ribosomes and has a tubular appearance. (Contrast with rough endoplasmic reticulum.)

smooth muscle Muscle tissue consisting of sheets of mononucleated cells innervated by the autonomic nervous system. (Contrast with cardiac muscle, skeletal muscle.)

sodium–potassium (Na⁺–K⁺) pump Antiporter responsible for primary active transport; it pumps sodium ions out of the cell and potassium ions into the cell, both against their concentration gradients. Also called a sodium–potassium ATPase.

soil horizon *See* horizon.

solute A substance that is dissolved in a liquid (solvent) to form a solution.

solute potential A property of any solution, resulting from its solute contents; it may be zero or have a negative value. The more negative the solute potential, the greater the tendency of the solution to take up water through a differentially permeable membrane. (Contrast with pressure potential, water potential.)

solution A liquid (the solvent) and its dissolved solutes.

solvent Liquid in which a substance (solute) is dissolved to form a solution.

somatic cell [Gk. *soma*: body] All the cells of the body that are not specialized for reproduction. (Contrast with germ cell.)

somatic mutation Permanent genetic change in a somatic cell. These mutations affect the individual only; they are not passed on to offspring. (Contrast with germ line mutation.)

somatosensory cortex The region of the cerebral cortex that receives input from mechanosensors distributed throughout the body.

somatostatin Peptide hormone made in the hypothalamus that inhibits the release of other hormones from the pituitary and intestine.

somite (so' might) One of the segments into which an embryo becomes divided longitudinally, leading to the eventual segmentation of the animal as illustrated by the spinal column, ribs, and associated muscles.

Sorenson's index Mathematical formula that measures beta diversity.

source In plants, any organ that exports the products of photosynthesis in excess of its own needs, such as a mature leaf or storage organ. (Contrast with sink.)

spatial summation In the production or inhibition of action potentials in a postsynaptic cell, the interaction of depolarizations and hyperpolarizations produced at different sites on the postsynaptic cell. (Contrast with temporal summation.)

spawning *See* external fertilization.

speciation (spee' see ay' shun) The process of splitting one population into two populations that are reproductively isolated from one another.

species (spee' sees) [L. kind] The base unit of taxonomic classification, consisting of an ancestor–descendant group of populations of evolutionarily closely related, similar organisms. The more narrowly defined "biological species" consists of individuals capable of interbreeding with each other but not with members of other species.

species–area relationship The relationship between the size of an area and the numbers of species it supports.

species richness The total number of species living in a region.

specific defenses Defensive reactions of the vertebrate immune system that are based on the reaction of an antibody to a specific antigen. (Contrast with nonspecific defenses.)

specific heat The amount of energy that must be absorbed by a gram of a substance to raise its temperature by one degree centigrade. By convention, water is assigned a specific heat of one.

sperm [Gk. *sperma*: seed] The male gamete.

spermatid One of the products of the second meiotic division of a primary spermatocyte; four haploid spermatids, which remain connected by cytoplasmic bridges, are produced for each primary spermatocyte that enters meiosis.

spermatogenesis (spur mat' oh jen' e sis) [Gk. *sperma*: seed + *genesis*: source] Gametogenesis leading to the production of sperm.

spermatogonia In animals, the diploid progeny of a germ cell in males.

spherical symmetry The simplest form of symmetry, in which body parts radiate out from a central point such that an infinite number of planes passing through that central point can divide the organism into similar halves.

spicule [L. arrowhead] A hard, calcareous skeletal element typical of sponges.

spinal reflex The conversion of afferent to efferent information in the spinal cord without participation of the brain.

sphincter (sfink' ter) [Gk. *sphinkter*: something that binds tightly] A ring of muscle that can close an orifice, for example, at the anus.

spindle Array of microtubules emanating from both poles of a dividing cell during mitosis and playing a role in the movement of chromosomes at nuclear division. Named for its shape.

spleen Organ that serves as a reservoir for venous blood and eliminates old, damaged red blood cells from the circulation.

spliceosome RNA–protein complex that splices out introns from eukaryotic pre-mRNAs.

splicing *See* RNA splicing.

spontaneous mutation A genetic change caused by internal cellular mechanisms, such as an error in DNA replication. (Contrast with induced mutation.)

sporangiophore A stalked reproductive structure produced by zygospore fungi that extends from a hypha and bears one or many sporangia.

sporangium (spor an' gee um) (plural: sporangia) [Gk. *spora*: seed + *angeion*: vessel or reservoir] In plants and fungi, any specialized structure within which one or more spores are formed.

spore [Gk. *spora*: seed] (1) Any asexual reproductive cell capable of developing into an adult organism without gametic fusion. In plants, haploid spores develop into gametophytes, diploid spores into sporophytes. (2) In prokaryotes, a resistant cell capable of surviving unfavorable periods.

sporocyte Specialized cells of the diploid sporophyte that will divide by meiosis to produce four haploid spores. Germination of these spores produces the haploid gametophyte.

sporophyte (spor' o fyte) [Gk. *spora*: seed + *phyton*: plant] In plants and protists with alternation of generations, the diploid phase that produces the spores. (Contrast with gametophyte.)

stabilizing selection Selection against the extreme phenotypes in a population, so that the intermediate types are favored. (Contrast with disruptive selection.)

stamen (stay' men) [L. *stamen*: thread] A male (pollen-producing) unit of a flower, usually composed of an anther, which bears the pollen, and a filament, which is a stalk supporting the anther.

starch [O.E. *stearc*: stiff] A polymer of glucose; used by plants to store energy.

Starling's forces The two opposing forces responsible for water movement across capillary walls: blood pressure, which squeezes water and small solutes out of the capillaries, and osmotic pressure, which pulls water back into the capillaries.

start codon The mRNA triplet (AUG) that acts as a signal for the beginning of translation at the ribosome. (Contrast with stop codon.)

stele (steel) [Gk. *stylos*: pillar] The central cylinder of vascular tissue in a plant stem.

stem In plants, the organ that holds leaves and/or flowers and transports and distributes materials among the other organs of the plant.

stem cell In animals, an undifferentiated cell that is capable of continuous proliferation. A stem cell generates more stem cells and a large clone of differentiated progeny cells. (*See also* embryonic stem cell.)

steroid Any of a family of lipids whose multiple rings share carbons. The steroid cholesterol is an important constituent of membranes; other steroids function as hormones.

sticky ends On a piece of two-stranded DNA, short, complementary, one-stranded regions produced by the action of a restriction endonuclease. Sticky ends facilitate the joining of segments of DNA from different sources.

stigma [L. *stigma*: mark, brand] The part of the pistil at the apex of the style that is receptive to pollen, and on which pollen germinates.

stimulus [L. *stimulare*: to goad] Something causing a response; something in the environment detected by a receptor.

stock In horticulture, the root or root-bearing stem to which a bud or piece of stem from another plant (the scion) is grafted.

stoma (plural: stomata) [Gk. *stoma*: mouth, opening] Small opening in the plant epidermis that permits gas exchange; bounded by a pair of guard cells whose osmotic status regulates the size of the opening.

stomatal crypt In plants, a sunken cavity below the leaf surface in which a stoma is sheltered from the drying effects of air currents.

stop codon   Any of the three mRNA codons that signal the end of protein translation at the ribosome: UAG, UGA, UAA.

stratosphere   The upper part of Earth's atmosphere, above the troposphere; extends from approximately 18 kilometers upward to approximately 50 kilometers above Earth's surface.

stratum (plural strata) [L. *stratos*: layer]   A layer of sedimentary rock laid down at a particular time in the past.

stretch receptor   A modified muscle cell embedded in the connective tissue of a muscle that acts as a mechanoreceptor in response to stretching of that muscle.

striated muscle   *See* skeletal muscle.

strobilus (plural: strobili)   One of several cone-like structures in various groups of plants (including club mosses, horsetails, and conifers) associated with the production and dispersal of reproductive products.

stroma   The fluid contents of an organelle such as a chloroplast or mitochondrion.

structural gene   A gene that encodes the primary structure of a protein not involved in the regulation of gene expression.

structural isomers   Molecules made up of the same kinds and numbers of atoms, in which the atoms are bonded differently.

structural motif   A three-dimensional structural element that is part of a larger molecule. For example, there are four common motifs in DNA-binding proteins: helix-turn-helix, zinc finger, leucine zipper, and helix-loop-helix.

style [Gk. *stylos*: pillar or column]   In the angiosperm flower, a column of tissue extending from the tip of the ovary, and bearing the stigma or receptive surface for pollen at its apex.

sub- [L. under]   A prefix used to designate a structure that lies beneath another or is less than another. For example, subcutaneous (beneath the skin); subspecies.

suberin   A waxlike lipid that is a barrier to water and solute movement across the Casparian strip of the endodermis.

submucosa (sub mew koe' sah)   The tissue layer just under the epithelial lining of the lumen of the digestive tract.

subsoil   The soil horizon lying below the topsoil and above the parent rock (bedrock); the zone of infiltration and accumulation of materials leached from the topsoil. Also called the B horizon.

substrate (sub' strayte)   The molecule or molecules on which an enzyme exerts catalytic action.

substratum (plural: substrata)   The base material on which a sessile organism lives.

succession   The gradual, sequential series of changes in the species composition of a community following a disturbance.

succulence   In plants, possession of fleshy, water-storing leaves or stems; an adaptation to dry environments.

superficial cleavage   A variation of incomplete cleavage in which cycles of mitosis occur without cell division, producing a syncytium (a single cell with many nuclei).

suprachiasmatic nuclei (SCN)   In mammals, two clusters of neurons just above the optic chiasm that act as the master circadian clock.

surface area-to-volume ratio   For any cell, organism, or geometrical solid, the ratio of surface area to volume; this is an important factor in setting an upper limit on the size a cell or organism can attain.

surface tension   The attractive intermolecular forces at the surface of liquid; especially important in water.

surfactant   A substance that decreases the surface tension of a liquid. Lung surfactant, secreted by cells of the alveoli, is mostly phospholipid and decreases the amount of work necessary to inflate the lungs.

survivorship (l_x)   In life tables, the proportion of individuals in a cohort that are alive at age x. A graph of this data is a survivorship curve.

suspensor   In the embryos of seed plants, the stalk of cells that pushes the embryo into the endosperm and is a source of nutrient transport to the embryo.

sustainable   Pertaining to the use and management of ecosystems in such a way that humans benefit over the long term from specific ecosystem goods and services without compromising others.

symbiosis (sim' bee oh' sis) [Gk. *sym*: together + *bios*: living]   The living together of two or more species in a prolonged and intimate relationship.

symmetry   Pertaining to an attribute of an animal body in which at least one plane can divide the body into similar, mirror-image halves. (*See* bilateral symmetry, radial symmetry.)

sympathetic nervous system   The division of the autonomic nervous system that works in opposition to the parasympathetic nervous system. (Contrast with parasympathetic nervous system.)

sympatric speciation (sim pat' rik) [Gk. *sym*: same + *patria*: homeland]   Speciation due to reproductive isolation without any physical separation of the subpopulation. (Contrast with allopatric speciation.)

symplast   The continuous meshwork of the interiors of living cells in the plant body, resulting from the presence of plasmodesmata. (Contrast with apoplast.)

symporter   A membrane transport protein that carries two substances in the same direction. (Contrast with antiporter, uniporter.)

synapomorphy   A trait that arose in the ancestor of a phylogenetic group and is present (sometimes in modified form) in all of its members, thus helping to delimit and identify that group. Also called a shared derived trait.

synapse (sin' aps) [Gk. *syn*: together + *haptein*: to fasten]   A specialized type of junction where a neuron meets its target cell (which can be another neuron or some other type of cell) and information in the form of neurotransmitter molecules is exchanged across a synaptic cleft.

synapsis (sin ap' sis)   The highly specific parallel alignment (pairing) of homologous chromosomes during the first division of meiosis.

synergids [Gk. *syn*: together + *ergos*: work]   In angiosperms, the two cells accompanying the egg cell at one end of the megagametophyte.

syngamy   *See* fertilization.

synonymous (silent) substitution   A change of one nucleotide in a sequence to another when that change does not affect the amino acid specified (i.e., UUA → UUG, both specifying leucine). (Contrast with nonsynonymous substitution, missense mutation, nonsense mutation.)

systematics   The scientific study of the diversity and relationships among organisms.

systemic acquired resistance   A general resistance to many plant pathogens following infection by a single agent.

systemic circuit   Portion of the circulatory system by which oxygenated blood from the lungs or gills is distributed throughout the rest of the body and returned to the heart. (Contrast with pulmonary circuit.)

systems biology   The scientific study of an organism as an integrated and interacting system of genes, proteins, and biochemical reactions.

systole (sis' tuh lee) [Gk. *systole*: contraction]   Contraction of a chamber of the heart, driving blood forward in the circulatory system. (Contrast with diastole.)

- T -

T cell   A type of lymphocyte involved in the cellular immune response. The final stages of its development occur in the thymus gland. (Contrast with B cell; *see also* cytotoxic T cell, T-helper cell.)

T cell receptor   A protein on the surface of a T cell that recognizes the antigenic determinant for which the cell is specific.

T-helper cell (T_H)   Type of T cell that stimulates events in both the cellular and humoral immune responses by binding to the antigen on an antigen-presenting cell; target of the HIV-I virus, the agent of AIDS. (Contrast with cytotoxic T cells.)

T tubules   A system of tubules that runs throughout the cytoplasm of a muscle fiber, through which action potentials spread.

taproot system   A root system typical of eudicots consisting of a primary root (*taproot*) that extends downward by tip growth and outward by initiating lateral roots. (Contrast with fibrous root system.)

target cell   A cell with the appropriate receptors to bind and respond to a particular hormone or other chemical mediator.

taste bud   A structure in the epithelium of the tongue that includes a cluster of chemoreceptors innervated by sensory neurons.

TATA box   An eight-base-pair sequence, found about 25 base pairs before the starting point for transcription in many eukaryotic promoters, that binds a transcription factor and thus helps initiate transcription.

taxon (plural: taxa) [Gk. *taxis*: arrange, put in order]   A biological group (typically a species or a clade) that is given a name.

telencephalon   The outer, surrounding structure of the embryonic vertebrate forebrain, which develops into the cerebrum.

telomerase   An enzyme that catalyzes the addition of telomeric sequences lost from chromosomes during DNA replication.

telomeres (tee' lo merz) [Gk. *telos*: end + *meros*: units, segments]   Repeated DNA sequences at the ends of eukaryotic chromosomes.

telophase (tee' lo phase) [Gk. *telos*: end]   The final phase of mitosis or meiosis during which chromosomes became diffuse, nuclear envelopes re-form, and nucleoli begin to reappear in the daughter nuclei.

template   A molecule or surface on which another molecule is synthesized in complementary fashion, as in the replication of DNA.

template strand   In double-stranded DNA, the strand that is transcribed to create an RNA transcript that will be processed into a protein. Also refers to a strand of RNA that is used to create a complementary RNA.

temporal summation In the production or inhibition of action potentials in a postsynaptic cell, the interaction of depolarizations or hyperpolarizations produced by rapidly repeated stimulation of a single point on the postsynaptic cell. (Contrast with spatial summation.)

tendon A collagen-containing band of tissue that connects a muscle with a bone.

tepal A sterile, modified, nonphotosynthetic leaf of an angiosperm flower that cannot be distinguished as a petal or a sepal.

termination In molecular biology, the end of transcription or translation.

terminator A sequence at the 3' end of mRNA that causes the RNA strand to be released from the transcription complex.

terrestrial (ter res' tree al) [L. *terra*: earth] Pertaining to or living on land. (Contrast with aquatic, marine.)

tertiary structure In reference to a protein, the relative locations in three-dimensional space of all the atoms in the molecule. The overall shape of a protein. (Contrast with primary, secondary, and quaternary structures.)

test cross Mating of a dominant-phenotype individual (who may be either heterozygous or homozygous) with a homozygous-recessive individual.

testis (tes' tis) (plural: testes) [L. *testis*: witness] The male gonad; the organ that produces the male gametes.

tetanus [Gk. *tetanos*: stretched] (1) A state of sustained maximal muscular contraction caused by rapidly repeated stimulation. (2) In medicine, an often fatal disease ("lockjaw") caused by the bacterium *Clostridium tetani*.

tetrad [Gk. *tettares*: four] During prophase I of meiosis, the association of a pair of homologous chromosomes or four chromatids.

thalamus [Gk. *thalamos*: chamber] A region of the vertebrate forebrain; involved in integration of sensory input.

theory [Gk. *theoria*: analysis of facts] A far-reaching explanation of observed facts that is supported by such a wide body of evidence, with no significant contradictory evidence, that it is scientifically accepted as a factual framework. Examples are Newton's theory of gravity and Darwin's theory of evolution. (Contrast with hypothesis.)

thermoneutral zone [Gk. *thermos*: temperature] The range of temperatures over which an endotherm does not have to expend extra energy to thermoregulate.

thermophile (ther' muh fyle)[Gk. *thermos*: temperature + *philos*: loving] An organism that lives exclusively in hot environments.

thoracic cavity [Gk. *thorax*: breastplate] The portion of the mammalian body cavity bounded by the ribs, shoulders, and diaphragm. Contains the heart and the lungs.

thoracic duct The connection between the lymphatic system and the circulatory system.

threshold The level of depolarization that causes an electrically excitable membrane to fire an action potential.

thrombus (throm' bus) [Gk. *thrombos*: clot] A blood clot that forms within a blood vessel and remains attached to the wall of the vessel. (Contrast with embolus.)

thylakoid (thigh la koid) [Gk. *thylakos*: sack or pouch] A flattened sac within a chloroplast. Thylakoid membranes contain all of the chlorophyll in a plant, in addition to the electron carriers of photophosphorylation. Thylakoids stack to form grana.

thymine (T) Nitrogen-containing base found in DNA.

thymus [Gk. *thymos*: warty] A ductless, glandular lymphoid tissue, involved in development of the immune system of vertebrates. In humans, the thymus degenerates during puberty.

thyroid gland [Gk. *thyreos*: door-shaped] A two-lobed gland in vertebrates. Produces the hormone thyroxine.

thyrotropin Hormone produced by the anterior pituitary that stimulates the thyroid gland to produce and release thyroxine. Also called thyroid-stimulating hormone (TSH).

thyrotropin-releasing hormone (TRH) Hormone produced by the hypothalamus that stimulates the anterior pituitary to release thyrotropin.

thyroxine Hormone produced by the thyroid gland; controls many metabolic processes.

tidal volume The amount of air that is exchanged during each breath when a person is at rest.

tight junction A junction between epithelial cells in which there is no gap between adjacent cells.

tissue A group of similar cells organized into a functional unit; usually integrated with other tissues to form part of an organ.

tissue system In plants, any of three organized groups of tissues—dermal tissue, vascular tissue, and ground tissue—that are established during embryogenesis and have distinct functions.

titin A protein that holds bundles of myosin filaments in a centered position within the sarcomeres of muscle cells. The largest protein in the human body.

tonoplast The membrane of the plant central vacuole.

topsoil The uppermost soil horizon; contains most of the organic matter of soil, but may be depleted of most mineral nutrients by leaching. Also called the A horizon.

totipotent [L. *toto*: whole, entire + *potens*: powerful] Possessing all the genetic information and other capacities necessary to form an entire individual. (Contrast with multipotent, pluripotent.)

trachea (tray' kee ah) [Gk. *trakhoia*: tube] A tube that carries air to the bronchi of the lungs of vertebrates. When plural (*tracheae*), refers to the major airways of insects.

tracheary element Either of two types of xylem cells—tracheids and vessel elements—that undergo apoptosis before assuming their transport function.

tracheid (tray' kee id) A type of tracheary element found in the xylem of nearly all vascular plants, characterized by tapering ends and walls that are pitted but not perforated. (Contrast with vessel element.)

trade-off The relationship between the fitness benefits conferred by an adaptation and the fitness costs it imposes. For an adaptation to be favored by natural selection, the benefits must exceed the costs.

trait In genetics, a specific form of a character: eye color is a character; brown eyes and blue eyes are traits. (Contrast with character.)

transcription The synthesis of RNA using one strand of DNA as a template.

transcription factors Proteins that assemble on a eukaryotic chromosome, allowing RNA polymerase II to perform transcription.

transduction (1) Transfer of genes from one bacterium to another by a bacteriophage. (2) In sensory cells, the transformation of a stimulus (e.g., light energy, sound pressure waves, chemical or electrical stimulants) into action potentials.

transfection Insertion of recombinant DNA into animal cells.

transfer RNA (tRNA) A family of folded RNA molecules. Each tRNA carries a specific amino acid and anticodon that will pair with the complementary codon in mRNA during translation.

transformation (1) A mechanism for transfer of genetic information in bacteria in which pure DNA from a bacterium of one genotype is taken in through the cell surface of a bacterium of a different genotype and incorporated into the chromosome of the recipient cell. (2) Insertion of recombinant DNA into a host cell.

transgenic Containing recombinant DNA incorporated into the genetic material.

translation The synthesis of a protein (polypeptide). Takes place on ribosomes, using the information encoded in messenger RNA.

translocation (1) In genetics, a rare mutational event that moves a portion of a chromosome to a new location, generally on a nonhomologous chromosome. (2) In vascular plants, movement of solutes in the phloem.

transmembrane protein An integral membrane protein that spans the phospholipid bilayer.

transpiration [L. *spirare*: to breathe] The evaporation of water from plant leaves and stem, driven by heat from the sun, and providing the motive force to raise water (plus mineral nutrients) from the roots.

transpiration–cohesion–tension mechanism Theoretical basis for water movement in plants: evaporation of water from cells within leaves (transpiration) causes an increase in surface tension, pulling water up through the xylem. Cohesion of water occurs because of hydrogen bonding.

transposable element A segment of DNA that can move to, or give rise to copies at, another locus on the same or a different chromosome.

transposon Mobile DNA segment that can insert into a chromosome and cause genetic change.

triglyceride A simple lipid in which three fatty acids are combined with one molecule of glycerol.

triploblastic Having three cell layers.

trisomic Containing three rather than two members of a chromosome pair.

tRNA *See* transfer RNA.

trochophore (troke' o fore) [Gk. *trochos*: wheel + *phoreus*: bearer] A radially symmetrical larval form typical of annelids and mollusks, distinguished by a wheel-like band of cilia around the middle.

trophic cascade The progression over successively lower trophic levels of the indirect effects of a predator.

trophic level [Gk *trophes*: nourishment]   A group of organisms united by obtaining their energy from the same part of the food web of a biological community.

trophoblast [Gk *trophes*: nourishment + *blastos*: sprout]   At the 32-cell stage of mammalian development, the outer group of cells that will become part of the placenta and thus nourish the growing embryo. (Contrast with inner cell mass.)

tropic hormones   Hormones produced by the anterior pituitary that control the secretion of hormones by other endocrine glands.

tropomyosin [troe poe my' oh sin]   One of the three protein components of an actin filament; controls the interactions of actin and myosin necessary for muscle contraction.

troponin   One of the three components of an actin filament; binds to actin, tropomyosin, and Ca^{2+}.

troposphere   The lowest atmospheric zone, reaching upward from the Earth's surface approximately 10–17 km. Zone in which virtually all water vapor is located.

true-breeding   A genetic cross in which the same result occurs every time with respect to the trait(s) under consideration, due to homozygous parents.

trypsin   A protein-digesting enzyme. Secreted by the pancreas in its inactive form (trypsinogen), it becomes active in the duodenum of the small intestine.

tubulin   A protein that polymerizes to form microtubules.

tumor [L. *tumor*: a swollen mass]   A disorganized mass of cells. Malignant tumors spread to other parts of the body.

tumor necrosis factor   A family of cytokines (growth factors) that causes cell death and is involved in inflammation.

tumor suppressor   A gene that codes for a protein product that inhibits cell proliferation; inactive in cancer cells. (Contrast with oncogene.)

turgor pressure [L. *turgidus*: swollen]   *See* pressure potential.

turnover   In freshwater ecosystems, vertical movements of water that bring nutrients and dissolved CO_2 to the surface and O_2 to deeper water.

tympanic membrane [Gk. *tympanum*: drum]   The eardrum.

- U -

ubiquinone (yoo bic' kwi known) [L. *ubique*: everywhere]   A mobile electron carrier of the mitochondrial respiratory chain. Similar to plastoquinone found in chloroplasts.

ubiquitin   A small protein that is covalently linked to other cellular proteins identified for breakdown by the proteosome.

ultimate cause   In ethology, the evolutionary processes that produced an animal's capacity and tendency to behave in particular ways. (Contrast with proximate cause.)

uniporter [L. *unus*: one + *portal*: doorway]   A membrane transport protein that carries a single substance in one direction. (Contrast with antiporter, symporter.)

unsaturated fatty acid   A fatty acid whose hydrocarbon chain contains one or more double bonds. (Contrast with saturated fatty acid.)

upregulation   A process by which the abundance of receptors for a hormone increases when

hormone secretion is suppressed. (Contrast with downregulation.)

upwelling zones   Areas of the ocean where cool, nutrient-rich water from deeper layers rises to the surface.

uracil (U)   A pyrimidine base found in nucleotides of RNA.

urea   A compound that is the main form of nitrogen excreted by many animals, including mammals.

ureotelic   Pertaining to an organism in which the final product of the breakdown of nitrogen-containing compounds (primarily proteins) is urea. (Contrast with ammonotelic, uricotelic.)

ureter (your' uh tur)   Long duct leading from the vertebrate kidney to the urinary bladder or the cloaca.

urethra (you ree' thra)   In most mammals, the canal through which urine is discharged from the bladder and which serves as the genital duct in males.

uric acid   A compound that serves as the main excreted form of nitrogen in some animals, particularly those which must conserve water, such as birds, insects, and reptiles.

uricotelic   Pertaining to an organism in which the final product of the breakdown of nitrogen-containing compounds (primarily proteins) is uric acid. (Contrast with ammonotelic, ureotelic.)

urinary bladder   A structure in which urine is stored until it can be excreted to the outside of the body.

urine (you' rin)   In vertebrates, the fluid waste product containing the toxic nitrogenous by-products of protein and nucleic acid metabolism.

uterine cycle   In human females, the monthly cycle of events by which the endometrium is prepared for the arrival of a blastocyst. (Contrast with ovarian cycle).

uterus (yoo' ter us) [L. *utero*: womb]   A specialized portion of the female reproductive tract in mammals that receives the fertilized egg and nurtures the embryo in its early development. Also called the womb.

- V -

vaccination   Injection of virus or bacteria or their proteins into the body, to induce immunity. The injected material is usually attenuated (weakened) before injection.

vacuole (vac' yew ole)   Membrane-enclosed organelle in plant cells that can function for storage, water concentration for turgor, or hydrolysis of stored macromolecules.

vagina (vuh jine' uh) [L. *sheath*]   In female animals, the entry to the reproductive tract.

van der Waals forces   Weak attractions between atoms resulting from the interaction of the electrons of one atom with the nucleus of another. This type of attraction is about one-fourth as strong as a hydrogen bond.

variable region   The portion of an immunoglobulin molecule or T cell receptor that includes the antigen-binding site and is responsible for its specificity. (Contrast with constant region.)

vas deferens (plural: vasa deferentia)   Duct that transfers sperm from the epididymis to the urethra.

vasa recta   Blood vessels that parallel the loops of Henle and the collecting ducts in the renal medulla of the kidney.

vascular (vas' kew lar) [L. *vasculum*: a small vessel]   Pertaining to organs and tissues that conduct fluid, such as blood vessels in animals and xylem and phloem in plants.

vascular bundle   In vascular plants, a strand of vascular tissue, including xylem and phloem as well as thick-walled fibers.

vascular cambium (kam' bee um) [L. *cambiare*: to exchange]   In plants, a lateral meristem that gives rise to secondary xylem and phloem.

vascular tissue system   The transport system of a vascular plant, consisting primarily of xylem and phloem.

vasopressin   *See* antidiuretic hormone.

vector   (1) An agent, such as an insect, that carries a pathogen affecting another species. (2) A plasmid or virus that carries an inserted piece of DNA into a bacterium for cloning purposes in recombinant DNA technology.

vegetal hemisphere   The lower portion of some animal eggs, zygotes, and embryos, in which the dense nutrient yolk settles. The *vegetal pole* is to the very bottom of the egg or embryro. (Contrast with animal hemisphere.)

vegetative   Nonreproductive, nonflowering, or asexual.

vegetative meristem   An apical meristem that produces leaves.

vegetative reproduction   Asexual reproduction through the modification of stems, leaves, or roots.

vein [L. *vena*: channel]   A blood vessel that returns blood to the heart. (Contrast with artery.)

ventral [L. *venter*: belly, womb]   Toward or pertaining to the belly or lower side. (Contrast with dorsal.)

ventricle   A muscular heart chamber that pumps blood through the lungs or through the body.

venule   A small blood vessel draining a capillary bed that joins others of its kind to form a vein. (Contrast with arteriole.)

vernalization [L. *vernalis*: spring]   Events occurring during a required chilling period, leading eventually to flowering.

vertebral column [L. *vertere*: to turn]   The jointed, dorsal column that is the primary support structure of vertebrates.

very low-density lipoproteins (VLDLs)   Lipoproteins that consist mainly of triglyceride fats, which they transport to fat cells in adipose tissues throughout the body; associated with excessive fat deposition and high risk for cardiovascular disease.

vesicle   Within the cytoplasm, a membrane-enclosed compartment that is associated with other organelles; the Golgi complex is one example.

vessel element   A type of tracheary element with perforated end walls; found only in angiosperms. (Contrast with tracheid.)

vestibular system (ves tib' yew lar) [L. *vestibulum*: an enclosed passage]   Structures within the inner ear that sense changes in position or momentum of the head, affecting balance and motor skills.

vicariant event (vye care' ee unt) [L. *vicus*: change]   The splitting of a taxon's range by the imposition of some barrier to dispersal.

villus (vil' lus) (plural: villi) [L. *villus*: shaggy hair or beard]   A hairlike projection from a membrane; for example, from many gut walls.

virion (veer' e on)   The virus particle, the minimum unit capable of infecting a cell.

virulence [L. *virus*: poison, slimy liquid]   The ability of a pathogen to cause disease and death.

virus   Any of a group of ultramicroscopic particles constructed of nucleic acid and protein (and, sometimes, lipid) that require living cells in order to reproduce. Viruses evolved multiple times from different cellular species.

vital capacity   The maximum capacity for air exchange in one breath; the sum of the tidal volume and the inspiratory and expiratory reserve volumes.

vitamin [L. *vita*: life]   An organic compound that an organism cannot synthesize, but nevertheless requires in small quantities for normal growth and metabolism.

vitelline envelope   The inner, proteinaceous protective layer of a sea urchin egg.

viviparity (vye vi par' uh tee)   Reproduction in which fertilization of the egg and development of the embryo occur inside the mother's body. (Contrast with oviparity.)

vivipary   Premature germination in plants.

voltage-gated channel   A type of gated channel that opens or closes when a certain voltage exists across the membrane in which it is inserted.

vomeronasal organ (VNO)   Chemosensory structure embedded in the nasal epithelium of amphibians, reptiles, and many mammals. Often specialized for detecting pheromones.

- W -

water potential   In osmosis, the tendency for a system (a cell or solution) to take up water from pure water through a differentially permeable membrane. Water flows toward the system with a more negative water potential. (Contrast with solute potential, pressure potential.)

water vascular system   In echinoderms, a network of water-filled canals that functions in gas exchange, locomotion, and feeding.

wavelength   The distance between successive peaks of a wave train, such as electromagnetic radiation.

weathering   The mechanical and chemical processes by which rocks are broken down into soil particles.

Wernicke's area   A region in the temporal lobe of the human brain that is involved with the sensory aspects of language.

white blood cells   Cells in the blood plasma that play defensive roles in the immune system. Also called leukocytes.

white matter   In the central nervous system, tissue that is rich in axons. (Contrast with gray matter.)

wild-type   Geneticists' term for standard or reference type. Deviants from this standard, even if the deviants are found in the wild, are usually referred to as mutant. (Note that this terminology is not usually applied to human genes.)

wood   Secondary xylem tissue.

- X -

xerophyte (zee' row fyte) [Gk. *xerox*: dry + *phyton*: plant]   A plant adapted to an environment with limited water supply.

xylem (zy' lum) [Gk. *xylon*: wood]   In vascular plants, the tissue that conducts water and minerals; xylem consists, in various plants, of tracheids, vessel elements, fibers, and other highly specialized cells.

- Y -

yolk [M.E. *yolke*: yellow]   The stored food material in animal eggs, rich in protein and lipids.

yolk sac   In reptiles, birds, and mammals, the extraembryonic membrane that forms from the endoderm of the hypoblast; it encloses and digests the yolk.

- Z -

zeaxanthin   A blue-light receptor involved in the opening of plant stomata.

zona pellucida   A jellylike substance that surrounds the mammalian ovum when it is released from the ovary.

zoospore (zoe' o spore) [Gk. *zoon*: animal + *spora*: seed]   In algae and fungi, any swimming spore. May be diploid or haploid.

zygote (zye' gote) [Gk. *zygotos*: yoked]   The cell created by the union of two gametes, in which the gamete nuclei are also fused. The earliest stage of the diploid generation.

zymogen   The inactive precursor of a digestive enzyme; secreted into the lumen of the gut, where a protease cleaves it to form the active enzyme.

Illustration Credits

Cover: © Dr. Merlin D. Tuttle/Photo Researchers, Inc.
Frontispiece: © Art Wolfe/www.artwolfe.com.

Photographs appearing behind chapter numbers:
Part 1 *HMS Beagle*: Painting by Ronald Dean, reproduced by permission of the artist and Richard Johnson, Esquire.
Part 2 *Plant cells*: © Ed Reschke/Peter Arnold Inc.
Part 3 *Stoma*: © Andrew Syred/SPL/Photo Researchers, Inc.
Part 4 *Anaphase*: © Nasser Rusan.
Part 5 *Sheep*: © Roddy Field, the Roslin Institute.
Part 6 *Trilobite*: Courtesy of the Amherst College Museum of Natural History and the Trustees of Amherst College.
Part 7 *Diatoms*: © Scenics & Science/Alamy.
Part 8 *Flowers*: © Ed Reschke/Peter Arnold Inc.
Part 9 *Leopard*: Courtesy of Andrew D. Sinauer.
Part 10 *Hummingbird*: © Yufeng Zhou/istockphoto.com.

Table of Contents
p. xxi: © Corbis RF/AGE Fotostock.
p. xxiii: © Dennis Kunkel Microscopy, Inc.
p. xxiv: © G. Brad Lewis/AGE Fotostock.
p. xxvi: © Dr. Alexey Khodjakov/Photo Researchers, Inc.
p. xxvii: © louise murray/Alamy.
p. xxviii: From B. Palenik et al. 2003. *Nature* 424: 1037.
p. xxx: Courtesy of Andrew D. Sinauer.
p. xxxii: © Ervin Monn/Shutterstock.
p. xxxiv: © Muriel Lasure/Shutterstock.
p. xxxv: © Heather A. Craig/Shutterstock.
p. xxxvii: Courtesy of Jane Sinauer.
p. xxxviii: © Christian Anthony/istockphoto.com.
p. xxxix: © PaulTessier/istockphoto.com.
p. xli: © Marina Cano Trueba/Shutterstock.
p. xlii: © Shutterstock.
p. xliii: © Herbert Hopfensperger/AGE Fotostock.

Chapter 1 *Frogs*: © Pete Oxford/Minden Pictures. *T. Hayes*: © Pamela S. Turner. 1.1A: © Eye of Science/SPL/Photo Researchers, Inc. 1.1B: Science Photo Library/Photolibrary.com. 1.1C: © Steve Gschmeissner/Photo Researchers, Inc. 1.1D: © Frans Lanting. 1.1E: © Glen Threlfo/Auscape/Minden Pictures. 1.1F: © Piotr Naskrecki/Minden Pictures. 1.1G: © Tui De Roy/Minden Pictures. 1.2A: From R. Hooke, 1664. *Micrographia*. 1.2B: © John Durham/SPL/Photo Researchers, Inc. 1.2C: © Biophoto Associates/Photo Researchers, Inc. 1.3 *Maple*: © Simon Colmer & Abby Rex/Alamy. 1.3 *Spruce*: David McIntyre. 1.3 *Lily pad*: © Pete Oxford/Naturepl.com. 1.3 *Cucumber*: David McIntyre. 1.3 *Pitcher plant*: © Nick Garbutt/Naturepl.com. 1.5A: © Frans Lanting. 1.5B: © Stefan Huwiler/Rolfnp/Alamy. 1.6 *Organism*: © Nico Smit/istockphoto.com. 1.6 *Population*: © blickwinkel/Alamy. 1.6 *Community*: © Georgie Holland/AGE Fotostock. 1.6 *Biosphere*: Courtesy of NASA. 1.7: © P&R Fotos/AGE Fotostock. 1.9: © Michael Abbey/Visuals Unlimited, Inc. 1.11: Courtesy of Wayne Whippen. 1.13: From T. Hayes et al., 2003. *Environ. Health Perspect*. 111: 568.

Chapter 2 *Hair*: © Steve Gschmeissner/Photo Researchers, Inc. *Barber*: © Digital Vision/Alamy. 2.3: From N. D. Volkow et al., 2001. *Am. J. Psychiatry* 158: 377. 2.14: © Pablo H Caridad/Shutterstock. 2.15: © Denis Miraniuk/Shutterstock.

Chapter 3 *T. Rex*: © The Natural History Museum, London. *Chicken*: David McIntyre. 3.8: Data from PDB 1IVM. T. Obita, T. Ueda, & T. Imoto, 2003. *Cell. Mol. Life Sci*. 60: 176. 3.10A: Data from PDB 2HHB. G. Fermi et al., 1984. *J. Mol. Biol*. 175: 159. 3.16C *left*: © Biophoto Associates/Photo Researchers, Inc. 3.16C *middle*: © Ken Wagner/Visuals Unlimited. 3.16C *right*: © CNRI/SPL/Photo Researchers, Inc. 3.17 *Cartilage*: © Robert Brons/Biological Photo Service. 3.17 *Beetle*: © Scott Bauer/USDA.

Chapter 4 *Phoenix*: Courtesy of NASA/JPL/UA/Lockheed Martin. *Ice*: Courtesy of NASA/JPL-Caltech/University of Arizona/Texas A&M University. 4.4: Data from S. Arnott & D. W. Hukins, 1972. *Biochem. Biophys. Res. Commun*. 47(6): 1504. 4.8: Courtesy of the Argonne National Laboratory. 4.13B: Courtesy of Janet Iwasa, Szostak group, MGH/Harvard. 4.14: © Stanley M. Awramik/Biological Photo Service. 4.14 *inset*: © Dennis Kunkel Microscopy, Inc.

Chapter 5 *Heart cell*: © Roger J. Bick & Brian J. Poindexter/UT-Houston Medical School/Photo Researchers, Inc. *Surgery*: © The Stock Asylum, LLC/Alamy. 5.1: After N. Campbell, 1990. *Biology*, 2nd Ed., Benjamin Cummings. 5.1 *Protein*: Data from PDB 1IVM. T. Obita, T. Ueda, & T. Imoto, 2003. *Cell. Mol. Life Sci*. 60: 176. 5.1 *T4*: © Dept. of Microbiology, Biozentrum/SPL/Photo Researchers, Inc. 5.1 *Bacterium*: © Jim Biddle/Centers for Disease Control. 5.1 *Plant cells*: © Michael Eichelberger/Visuals Unlimited, Inc. 5.1 *Frog egg*: David McIntyre. 5.1 *Bird*: © Steve Byland/Shutterstock. 5.1 *Baby*: Courtesy of Sebastian Grey Miller. 5.3 *Light microscope*: © Radu Razvan/Shutterstock. 5.3 *Bright-field*: Courtesy of the IST Cell Bank, Genoa. 5.3 *Phase-contrast*: © Michael W. Davidson, Florida State U. 5.3 *Stained*: © Richard J. Green/SPL/Photo Researchers, Inc. 5.3 *Confocal*: © Dr. Gopal Murti/SPL/Photo Researchers, Inc. 5.3 *Electron microscope*: © Sinclair Stammers/Photo Researchers, Inc. 5.3 TEM: © Dr. Gopal Murti/Visuals Unlimited. 5.3 SEM: © K. R. Porter/SPL/Photo Researchers, Inc. 5.3 *Freeze-fracture*: © D. W. Fawcett/Photo Researchers, Inc. 5.4: © J. J. Cardamone Jr. & B. K. Pugashetti/Biological Photo Service. 5.5A: © Dennis Kunkel Microscopy, Inc. 5.5B: Courtesy of David DeRosier, Brandeis U. 5.7 *Mitochondrion*: © K. Porter, D. Fawcett/Visuals Unlimited. 5.7 *Cytoskeleton*: © Don Fawcett, John Heuser/Photo Researchers, Inc. 5.7 *Nucleolus*: © Richard Rodewald/Biological Photo Service. 5.7 *Peroxisome*: © E. H. Newcomb & S. E. Frederick/Biological Photo Service. 5.7 *Cell wall*: © Biophoto Associates/Photo Researchers, Inc. 5.7 *Ribosome*: From M. Boublik et al., 1990. *The Ribosome*, p. 177. Courtesy of American Society for Microbiology. 5.7 *Centrioles*: © Barry F. King/Biological Photo Service. 5.7 *Plasma membrane*: Courtesy of J. David Robertson, Duke U. Medical Center. 5.7 *Rough ER*: © Don Fawcett/Science Source/Photo Researchers, Inc. 5.7 *Smooth ER*: © Don Fawcett, D. Friend/Science Source/Photo Researchers, Inc. 5.7 *Chloroplast*: © W. P. Wergin, E. H. Newcomb/Biological Photo Service. 5.7 *Golgi apparatus*: Courtesy of L. Andrew Staehelin, U. Colorado. 5.8: © D. W. Fawcett/Photo Researchers, Inc. 5.9A: © Barry King, U. California, Davis/Biological Photo Service. 5.9B: © Biophoto Associates/Science Source/Photo Researchers, Inc. 5.10: © B. Bowers/Photo Researchers, Inc. 5.11: © Sanders/Biological Photo Service. 5.12: © K. Porter, D. Fawcett/Visuals Unlimited. 5.13 *left*: © W. P. Wergin, E. H. Newcomb/Biological Photo Service. 5.13 *right*: © W. P. Wergin/Biological Photo Service. 5.14A: © Michael Eichelberger/Visuals Unlimited, Inc. 5.14B: © Ed Reschke/Peter Arnold Inc. 5.14C: © Gerald & Buff Corsi/Visuals Unlimited. 5.15A: David McIntyre. 5.15A *inset*: © Richard Green/Photo Researchers, Inc. 5.15B: David McIntyre. 5.15B *inset*: Courtesy of R. R. Dute. 5.16: Courtesy of M. C. Ledbetter, Brookhaven National Laboratory. 5.17: Courtesy of Vic Small, Austrian Academy of Sciences, Salzburg, Austria. 5.19: Courtesy of N. Hirokawa. 5.20A *upper*: © SPL/Photo Researchers, Inc. 5.20A *lower*, 5.20B: © W. L. Dentler/Biological Photo Service. 5.22: From N. Pollock et al., 1999. *J. Cell Biol*. 147: 493. Courtesy of R. D. Vale. 5.23: © Michael Abbey/Visuals Unlimited. 5.24: © Biophoto Associates/Photo Researchers, Inc. 5.25 *left*: Courtesy of David Sadava. 5.25 *upper right*: From J. A. Buckwalter & L. Rosenberg, 1983. *Coll. Rel. Res*. 3: 489. Courtesy of L. Rosenberg. 5.25 *lower right*: © J. Gross, Biozentrum/SPL/Photo Researchers, Inc. 5.27: Courtesy of Noriko Okamoto and Isao Inouye.

Chapter 6 *Patient*: From Alzheimer, A. 1906. Über einen eigenartigen schweren Erkrankungsprozess der Hirnrinde. *Neurologisches Centralblatt* 23: 1129. *Plaques*: © G. W. Willis/Photolibrary.com. 6.2: After L. Stryer, 1981. *Biochemistry*, 2nd Ed., W. H. Freeman. 6.4: © D. W. Fawcett/Photo Researchers, Inc. 6.7A: Courtesy of D. S. Friend, U. California, San Francisco. 6.7B: Courtesy of Darcy E. Kelly, U. Washington. 6.7C: Courtesy of C. Peracchia. 6.10A *left*: © Stanley Flegler/Visuals Unlimited, Inc. 6.10A *center*: © David M. Phillips/Photo Researchers, Inc. 6.10B *left*: © Ed Reschke/Peter Arnold Inc. 6.13: From G. M. Preston et al., 1992. *Science* 256: 385. 6.19: From M. M. Perry, 1979. *J. Cell Sci*. 39: 26.

Chapter 7 *Voles*: Courtesy of Todd Ahern. *Oxytocin*: Data from PDB 1NPO. J. P. Rose et al., 1996. *Nat. Struct .Biol.* 3: 163. 7.3: © Biophoto Associates/Photo Researchers, Inc. 7.4: Data from PDB 3EML. V. P. Jaakola et al., 2008. *Science* 322: 1211. 7.16: © Stephen A. Stricker, courtesy of Molecular Probes, Inc.

Chapter 8 *Dairy*: © Bob Randall/ istockphoto.com. *Maasai*: ©blickwinkel/Alamy. 8.1: Courtesy of Violet Bedell-McIntyre. 8.5B: © Satoshi Kuribayashi/OSF/Photolibrary.com. 8.11A: Data from PDB 1AL6. B. Schwartz et al., 1997. 8.11B: Data from PDB 1BB6. V. B. Vollan et al., 1999. *Acta Crystallogr. D. Biol. Crystallogr.* 55: 60. 8.11C: Data from PDB 1AB9. N. H. Yennawar, H. P. Yennawar, & G. K. Farber, 1994. *Biochemistry* 33: 7326. 8.13: Data from PDB 1IG8 (P. R. Kuser et al., 2000. *J. Biol. Chem.* 275: 20814) and 1BDG (A. M. Mulichak et al., 1998 *Nat. Struct. Biol.* 5: 555).

Chapter 9 *Marathoners*: © Chuck Franklin/ Alamy. *Mouse*: © Royalty-Free/Corbis. 9.9: From Y. H. Ko et al., 2003. *J. Biol. Chem.* 278: 12305. Courtesy of P. Pedersen.

Chapter 10 *Rainforest*: © Jon Arnold Images/Photolibrary.com. *FACE*: Courtesy of David F. Karnosky. 10.1: © Andrew Syred/ SPL/Photo Researchers, Inc. 10.6: © Martin Shields/Alamy. 10.13: Courtesy of Lawrence Berkeley National Laboratory. 10.17A: © E. H. Newcomb & S. E. Frederick/Biological Photo Service. 10.21: © Aflo Foto Agency/Alamy.

Chapter 11 *Cells*: © Parviz M. Pour/Photo Researchers, Inc. *Model*: Data from PDB 1GUX. J. Lee et al., 1998. *Nature* 391: 859. 11.1A: © SPL/Photo Researchers, Inc. 11.1B: © Biodisc/ Visuals Unlimited/Alamy. 11.1C: © Robert Valentic/Naturepl.com. 11.2B: © John J. Cardamone Jr./Biological Photo Service. 11.8 *Chromosome*: Courtesy of G. F. Bahr. 11.8 *Nucleus*: © D. W. Fawcett/Photo Researchers, Inc. 11.9 inset: © Biophoto Associates/Science Source/Photo Researchers, Inc. 11.10B: © Conly L. Rieder/Biological Photo Service. 11.11: © Nasser Rusan. 11.13A: © T. E. Schroeder/Biological Photo Service. 11.13B: © B. A. Palevitz, E. H. Newcomb/Biological Photo Service. 11.14A: © Dr. John Cunningham/ Visuals Unlimited. 11.14B: © Steve Gschmeissner/Photo Researchers, Inc. 11.15 *left*: © Andrew Syred/SPL/Photo Researchers, Inc. 11.15 *center*: David McIntyre. 11.15 *right*: © Gerry Ellis, DigitalVision/PictureQuest. 11.16: Courtesy of Dr. Thomas Ried and Dr. Evelin Schröck, NIH. 11.17: © C. A. Hasenkampf/Biological Photo Service. 11.18: Courtesy of J. Kezer. 11.22A: © Gopal Murti/ Photo Researchers, Inc. 11.23: © Dennis Kunkel Microscopy, Inc.

Chapter 12 *Vavilov*: Courtesy of the Library of Congress/New York World-Telegram & Sun Collection. *Harvest*: © Paul Nevin/ Photolibrary.com. 12.1: © the Mendelianum. 12.11: Courtesy the American Netherland Dwarf Rabbit Club. 12.14: Courtesy of Madison, Hannah, and Walnut. 12.15: Courtesy of the Plant and Soil Sciences eLibrary (http:// plantandsoil.unl.edu); used with permission from the Institute of Agriculture and Natural Resources at the University of Nebraska. 12.16: © Grant Heilman Photography/Alamy. 12.17:

© Peter Morenus/U. of Connecticut. 12.26A: Courtesy of L. Caro and R. Curtiss.

Chapter 13 *Velociraptor*: © Joe Tucciarone/ Photo Researchers, Inc. *DNA art*: © Simon Fraser/Karen Rann/SPL/Photo Researchers, Inc. 13.3: © Biozentrum, U. Basel/SPL/Photo Researchers, Inc. 13.6 *X-ray crystallograph*: Courtesy of Prof. M. H. F. Wilkins, Dept. of Biophysics, King's College, U. London. 13.6B: © CSHL Archives/Peter Arnold, Inc. 13.8A: © A. Barrington Brown/Photo Researchers, Inc. 13.8B: Data from S. Arnott & D. W. Hukins, 1972. *Biochem. Biophys. Res. Commun.* 47(6): 1504. 13.14A: Data from PDB 1SKW. Y. Li et al., 2001. *Nat. Struct. Mol. Biol.* 11: 784. 13.20C: © Dr. Peter Lansdorp/Visuals Unlimited.

Chapter 14 *Mycobacterium*: © Dennis Kunkel Microscopy, Inc. *Ethiopia*: © Jenny Matthews/ Alamy. 14.3: Data from PDB 1I3Q. P. Cramer et al., 2001. *Science* 292: 1863. 14.9: From D. C. Tiemeier et al., 1978. *Cell* 14: 237. 14.12: Data from PDB 1EHZ. H. Shi & P. B. Moore, 2000. *RNA* 6: 1091. 14.14: Data from PDB 1GIX and 1G1Y. M. M. Yusupov et al., 2001. *Science* 292: 883. 14.18B: Courtesy of J. E. Edström and *EMBO J.*

Chapter 15 *Destruction*: © Suzanne Plunkett/ AP Images. *Baby 81*: © Gemunu Amarasinghe/ AP Images. 15.3: © Stanley Flegler/Visuals Unlimited, Inc. 15.8: © Philippe Plailly/Photo Researchers, Inc. 15.10: Bettmann/CORBIS. 15.11 *Butterfly*: © Bershadsky Yuri/ Shutterstock. 15.11 *Bacteria*: Courtesy of Janice Haney Carr/CDC. 15.11 *Fungus*: © Warwick Lister-Kaye/istockphoto.com. 15.17: From C. Harrison et al., 1983. *J. Med. Genet.* 20: 280. 15.19: © Simon Fraser/Photo Researchers, Inc.

Chapter 16 *Alcoholic*: © LJSphotography/ Alamy. *CREB*: Data from PDB 1T2K. D. Panne et al., 2004. *EMBO J.* 23: 4384. 16.2A: © Dennis Kunkel Microscopy, Inc. 16.2B: © Lee D. Simon/ Photo Researchers, Inc. 16.6: © NIBSC/Photo Researchers, Inc. 16.21A: Courtesy of the Centers for Disease Control.

Chapter 17 *Mastiff and chihuahua*: © moodboard RF/Photolibrary.com. *Whippets*: © Stuart Isett/Anzenberger/Eyevine. 17.3: © Sam Ogden/Photo Researchers, Inc. 17.5: Based on an illustration by Anthony R. Kerlavage, Institute for Genomic Research. *Science* 269: 449 (1995). 17.12: Courtesy of O. L. Miller, Jr. 17.17: From P. H. O'Farrell, 1975. *J. Biol. Chem.* 250: 4007. Courtesy of Patrick H. O'Farrell.

Chapter 18 *Kuwait*: © K. Bry—UNEP/Peter Arnold Inc. *A. Chakrabarty*: © Ted Spiegel/ Corbis. 18.5: © Martin Shields/Alamy. 18.15: Courtesy of the Golden Rice Humanitarian Board, www.goldenrice.org. 18.16: Courtesy of Eduardo Blumwald.

Chapter 19 *Horse*: © Benoit Photo. *Centrifuge*: Courtesy of Cytori Therapeutics. 19.4: © Roddy Field, the Roslin Institute. 19.11A: From J. E. Sulston & H. R. Horvitz, 1977. *Dev. Bio.* 56: 100. 19.14C: Courtesy of J. Bowman. 19.17: Courtesy of W. Driever and C. Nüsslein-Vollhard. 19.18B: Courtesy of C. Rushlow and M. Levine. 19.18C: Courtesy of T. Karr. 19.18D: Courtesy of S. Carroll and S. Paddock. 19.20: Courtesy of F. R. Turner, Indiana U.

Chapter 20 *Fly head*: © Science Photo Library RF/Photolibrary.com. *Mutant leg*: From G. Halder et al., 1995. *Science* 267: 1788. Courtesy of W. J. Gehring and G. Halder. 20.1 *Mouse*: © orionmystery@flickr/Shutterstock. 20.1 *Fly*: David McIntyre. 20.1 *Shark*: © Kristian Sekulic/ Shutterstock. 20.1 *Squid*: © Gergo Orban/ Shutterstock. 20.3: © David M. Phillips/Photo Researchers, Inc. 20.5: © Bone Clones, www.boneclones.com. 20.6: Courtesy of J. Hurle and E. Laufer. 20.7: Courtesy of J. Hurle. 20.8 *Cladogram*: After R. Galant & S. Carroll, 2002. *Nature* 415: 910. 20.8 *Insect*: © Stockbyte/ PictureQuest. 20.8 *Centipede*: © Burke/Triolo/ Brand X Pictures/PictureQuest. 20.9: From M. Kmita and D. Duboule, 2003. *Science* 301: 331. 20.10: © Neil Hardwick/Alamy. 20.11: © Rob Valentic/ANTPhoto.com. 20.12A: © Erick Greene. 20.12B: Courtesy of John Gruber. 20.13: © Nigel Cattlin, Holt Studios International/ Photo Researchers, Inc. 20.15: Courtesy of Mike Shapiro and David Kingsley.

Chapter 21 *Pandemic*: Courtesy of the Naval Historical Foundation. *Researcher*: Courtesy of James Gathany/Centers for Disease Control. 21.1A: Painting by Ronald Dean, reproduced by permission of the artist and Richard Johnson, Esquire. 21.2A: © Luis César Tejo/ Shutterstock. 21.2B: © Duncan Usher/Alamy. 21.2C: © PetStockBoys/Alamy. 21.2D: © Arco Images GmbH/Alamy. 21.9A: © Tom Ulrich/ OSF/Photolibrary.com. 21.9B: © Kim Karpeles/ Alamy. 21.11: David McIntyre. 21.14: Courtesy of David Hillis. 21.16: © Jason Gallier/Alamy. 21.17: David McIntyre. 21.21A: © Reinhard Dirscherl/WaterFrame - Underwater Images/ Photolibrary.com. 21.21B: © Marevision/AGE Fotostock. 21.22 *Snake*: © Joseph T. Collins/ Photo Researchers, Inc. 21.22 *Newt*: © Robert Clay/Visuals Unlimited.

Chapter 22 *HIV*: © James Cavallini/Photo Researchers, Inc. *Chimpanzees*: © John Cancalosi/Naturepl.com. 22.6A: Courtesy of William Jeffery. 22.6B: © Jurgen Freund/ Naturepl.com. 22.6C: © Michael & Patricia Fogden/Minden Pictures. 22.6D: © Mark Kostich/Shutterstock. 22.8: © Larry Jon Friesen. 22.10: © Alexandra Basolo. 22.13A: © Krieger, C./AGE Fotostock. 22.13B: © Krieger, C./ AGE Fotostock. 22.13C: © Ed Reschke/Peter Arnold Inc.

Chapter 23 *Cuatro Cienegas*: © George Grall/National Geographic. *Fly*: © Dr. David Phillips/Visuals Unlimited, Inc. 23.1A left: © R. L. Hambley/Shutterstock. 23.1A right: © Frank Leung/istockphoto.com. 23.1B: © Norman Bateman/istockphoto.com. 23.10: © OSF/ Photolibrary.com. 23.11 *G. olivacea*: © Charles Melton/Visuals Unlimited, Inc. 23.11 *G. carolinensis*: © Michael Redmer/Visuals Unlimited, Inc. 23.12A: © Yufeng Zhou/istockphoto.com. 23.12B: © P01017 Desmette Frede/AGE Fotostock. 23.12C: © J. S. Sira/ Photolibrary.com. 23.12D: © Daniel L. Geiger/SNAP/Alamy. 23.13: Courtesy of Donald A. Levin. 23.14: © Boris I. Timofeev/ Pensoft. 23.16A: © Tony Tilford/ Photolibrary.com. 23.16B: © W. Peckover/ VIREO. 23.17 *Madia*: © Peter K. Ziminsky/ Visuals Unlimited. 23.17 *Argyroxiphium*: © Ron Dahlquist/Pacific Stock/Photolibrary.com. 23.17 *Wilkesia*: © Gerald D. Carr. 23.17 *Dubautia*: © Noble Proctor/The National

Index

in regulation of arterial pressure, 1063
in signal transduction affecting blood glucose, 143
signal transduction and, *866*
source of, 865
Epiphyseal cartilage, *1021*
Epiphytes, 1156
Epistasis, 250, *251*
Episyrphus balteatus, *1191*
Epithelial tissue
cell junctions, 112–113
functions, 835
skin, 835–836
Epitopes, 880–881
Epstein–Barr virus, 896
Eptatretus stoutii, *700*
Equation for exponential growth, 1175
Equatorial Countercurrent, *1144*
Equatorial plate, *219*
meiosis I, *225*
meiosis II, *224*
Equilibrium
across membranes, 115
of solutions, 114
Equisetum, 602
E. arvense, *602*
E. pratense, *602*
Equus
E. africanus asinus, *1152*
E. quagga, *1154*
Eras, *520–521*, 521
Erectile dysfunction, 911
Erectile penis, *908*
Erection, 911
Eridophyllum, *529*
Erinaceomorpha, *712*
Erithizon dorsatum, *712*
Error signals, 834
Erwin, Terry, 687
Erwinia uredovora, 401
Erysiphe, *642*
Erythrocytes
function, *876*, 1056–1057
life cycle, 1057
See also Red blood cells
Erythropoietin, 215, 1057
Escherichia, 555
Escherichia coli, *4*
amount of DNA in, 211
with bacteriophage T, *344*
conjugation, 260
DNA replication experiments, 277–278
EnvZ-OmpR signal transduction pathway, 130–132
genome, 372, *376*, *508*
in Lake Erie, 1236
as a model organism, 292
noncoding DNA, *509*
number of DNA polymerases in, 280
proteobacteria, 551
rate of DNA replication, 281, 283
rate of reproduction in, 210
recombinant DNA and, 387
regulation of lactose metabolism, 348–351, *352*
strain O157:H7, 373
Eschirchtius robustus, *1129*
Escovopsis, 1185
Esophagitis, 632
Esophagus, 695, *1077*, *1078*, *1079*
Essay on the Principle of Population, An (Malthus), 443
Essential amino acids, 1070–1071
Essential elements, in plant nutrition, 757–759

Essential fatty acids, 1071
Essential minerals, animal foraging behavior and, 1126, *1127*
Ester linkages, 54, 552
Estivation, 1098, 1144
Estradiol, *866*, 867
Estrogens
in childbirth, 916
female reproductive cycles and, 914–915
as a ligand, 133
overview, *858*
during pregnancy, 916
source of, 867
temperature-controlled synthesis, 433
Estuaries, 1164
Ether linkages, 552
Ethnobotanists, 624
Ethology, 1115–1117
Ethyl alcohol (ethanol), 182
Ethylene, *775*, 786, *787*
Etiolation, 789
Euarchaea, 552–553
Eublepharis macularius, *707*
Eucalyptus, 1074
Eucalyptus regnans, 745
Euchromatin, 359
Eudicots, 620, 622
embryonic development, *800*
general characteristics, 721
germination and early shoot development, *773*
leaf anatomy, 732–733
root anatomy, *731*
root system, 721
secondary growth, 733–375
vascular bundle, *732*
vegetative organs and systems, *721*
Eudistylia, *676*
Euglena, *563*, 581–582
Euglenids, *562*, *563*, 566, 581–582
Eukarya (domain), 12, 81, *538*
Eukaryotes
cell organization, 81
cellular locations of glucose metabolism, *172*
commonalities with other living forms, 537
evolution from prokaryotes, 11
major clades, *563*
monophyly of, 561, 563
phylogenetic tree, *562*
prokaryotes compared to, 537–538
Eukaryotic cells
animal cell structure, *86*
compartmentalization in, 84
cytoskeleton, 95–99
dimensions, 84
endomembrane system, 89–92
endosymbiosis, 565–566
extracellular structures, 100–101
flagella, 97, 98
flexible cell surface and, 564
genetic transformation, 270–271, *272*
mitochondria, 92–93
nucleus, 85, 88–89
origin, 101–103, 564–566
other membrane-enclosed organelles, 94
overview, 81
plant cell structure, *87*
plastids, 93–94
ribosomes, 84, 85
Eukaryotic viruses, 346–347, *348*
Eulemur coronatus, *714*

Eumeces tasciatus, *1173*
Eumetazoans, *647*, 658
Eupholus magnificus, *688*
Euphorbia pulcherrima, 794–795
Euphydryas editha bayensis, *1178*, 1179
Euphyllophytes, *595*, 596
Euplectella aspergillum, *658*
Euplotes, *576*
European bee-eaters, *908*
European toads, 492–493
Euryarchaeota, 552
Eurycea waterlooensis, *705*
Eurylepta californica, *672*
Eusociality, 1136
Eustachian tube, *972*, 973
Eutherians
major groups, *712*
overview, 711–712
primates, 713, *714* (*see also* Primates)
Eutrophication, *549*, 1233–1234
phosphorus-related, 1235–1236
Evans, Martin, 394
Evans, Ron, 168, 169
Evaporation
defined, 32
in heat transfer, 841
in thermoregulation, 846
Evapotranspiration, 1207
Evergreen trees
boreal forests, 1148
tropical forests, 1156
Evo-devo (evolutionary developmental biology)
defined, 427
developmental modules, 429–431
developmental plasticity, 433–435
evolution of differences among species, 432–433
genetic constraints on evolution, 436–437
principles of, 427
similarity in developmental genes, 427–428
Evolution
adaptation, 444
atmospheric oxygen and, 523
constraints on, 459–460
defined, 5
DNA and, 64
emergence of multicellularity, 11
of eukaryotes from prokaryotes, 11
genetic constraints, 436–437
Hardy–Weinberg equilibrium and, 448
impact of photosynthesis on life, 10–11
mutations and, 499
natural selection and, 5 (*see also* Natural selection)
"opportunistic," 768
parallel, 436–437
population genetics and, 444–445
Evolutionary developmental biology. *See* Evo-devo
Evolutionary mechanisms
gene flow, 449
genetic drift, 449–450
mutation, 448–449
nonrandom mating, 451
overview, 448
Evolutionary radiations, 494–495
Evolutionary reversal, 467
Evolutionary theory
Darwinian, 5–6
Darwin's development of, 442–444
major propositions of, 443

meaning of the term "theory," 441
Modern Synthesis, 444
significance of, 441–442
Evolutionary tree of life, 11–12
Evolutionary trend, 452–453
Excavates, *562*, *563*, 581–582
Excision repair, 285, 286
Excitatory synapses, 958
Exclosures, 1209
Excretory systems
excretion of nitrogen, 1094–1095
invertebrate, 1095–1097
maintenance of homeostasis, 1092–1094
nephrons, 1097, 1098–1100
vampire bat, 1091–1092
See also Kidneys
Excurrent siphon, 677, 679
Exercise, effects on muscle, 1017
Exergonic reactions, 152, 155, 156
Exhalation, 1035–1036
Exocrine glands, 853
Exocytosis, *123*, 124
Exons, 300, 301, *302*, *380*
Exoskeleton, 669–670, 1019
Exotic species, controlling, 1254–1255
Exotoxins, *549*, 555
Expanding triplet repeats, 333, *334*
Expansins, 724, 783
Experiments, 14–15
Expiratory reserve volume (ERV), 1033
Exploitation competition, 1198–1199
Exponential growth, in populations, 1174–1175
Expression vectors, 397–398
Expressivity, 252
Extensin, 816
Extensors, 988, 1022
External fertilization, 906
External gills, 1029
External intercostal muscles, 1036
Extinctions
"centers of imminent extinction," 1252
climate change and, 1249–1250
humans and, 1244
in island biogeography theory, 1212
problems in predicting, 1245
through overexploitation, 1248–1249
See also Mass extinctions
Extra-nuclear genes, 259, *260*
Extracellular fluid (ECF)
closed circulatory systems, 1047
open circulatory systems, 1046–1047
overview, 833, *834*
Extracellular matrix
adherence of the plasma membrane to, 113
connective tissues and, 836
functional role in eukaryotes, 84
sponges, 659
structure and function, 100, *101*
Extraembryonic membranes
amniote egg, 706
chicken, 933, 937–938
hormones from, 915–916
mammals, 934, 938
Extreme halophiles, 552–553
Exxon Valdez (oil tanker), 386
Eye cups, 978
Eye infections, 548
eyeless gene, 426, *428*
Eyes